AF443495

# MECHANOBIOLOGY: CARTILAGE AND CHONDROCYTE

# Biomedical and Health Research

## Volume 68

*Recently published in this series:*



ISSN 0929-6743

# Mechanobiology:
# Cartilage and Chondrocyte

## Volume 4

Edited by

## J.-F. Stoltz

*Laboratoire de Mécanique et Ingénierie Cellulaire et Tissulaire,*
*UMR CNRS 7563, Faculté de Médecine, UHP, Vandoeuvre lès Nancy, France*

IOS
*Press*

Amsterdam • Berlin • Oxford • Tokyo • Washington, DC

ISBN 1-58603-648-3
Library of Congress Control Number: 2006930131

This is the book edition in the BHR book series of the journal *Biorheology*, Volume 43, Nos 3,4 (2006), ISSN 0006-355X

*Publisher*
IOS Press
Nieuwe Hemweg 6B
1013 BG  Amsterdam
The Netherlands
fax: +31 20 687 0019
e-mail: order@iospress.nl


*Distributor in the UK and Ireland*
Gazelle Books Services Ltd.
White Cross Mills
Hightown
Lancaster LA1 4XS
United Kingdom
fax: +44 1524 63232
e-mail: sales@gazellebooks.co.uk

*Distributor in the USA and Canada*
IOS Press, Inc.
4502 Rachael Manor Drive
Fairfax, VA 22032
USA
fax: +1 703 323 3668
e-mail: iosbooks@iospress.com

LEGAL NOTICE
The publisher is not responsible for the use which might be made of the following information.


PRINTED IN THE NETHERLANDS

# Contents

Biorheology 43 (2006) 169–170
IOS Press



# Inaugural address

Dear Colleagues, Dear Friends,

It is a great pleasure for me to welcome you to this "Fourth International Symposium on Mechanobiology of Cartilage and Chondrocyte". After Sainte-Maxime (1999), Paris (2001) and Brussels (2003), this fourth meeting provided an in depth opportunity to review the recent advances in cartilage mechanobiology and clinical applications.

Indeed, due to populations ageing in industrialized countries, an increase in osteoarticular pathologies has been observed. These pathologies are now a public health challenge to be taken into account by the physicians in their daily practice. The knowledge and comprehension of old and new mechanisms involved in the pathophysiology of osteoarthritic complications is the only weapon for efficient, early and successful prevention.

There has been increasing interest over recent years in the fundamental role played by local mechanical parameters in chondrocyte regulation and cartilage dysfunction as a first step in the development of osteoarthritis. This is how the idea of mechanobiology and the concept of mechanotransduction were born in the 90's. Indeed, a broad diversity of physiological phenomena is induced by mechanical stimuli (hearing, orientation to gravity, touch, tissue remodelling . . .) but the mechanism by which mechanical forces may regulate a physiological response is still unknown.

In other respects, the concept of regenerative medicine has recently developed in parallel to this. Regenerative medicine is an emerging multidisciplinary field involving medicine, biology, chemistry, mechanics and engineering that is likely to revolutionize the ways we improve the health and quality of life by restoring, maintaining or enhancing tissue and organ functions. Indeed, human tissues do not regenerate spontaneously and healing is only a stopgap that may be associated with contraction which in turn may prevent regeneration. Tissue engineering through the *in vitro* preparation of biotissues presents an interesting alternative. Today, the in vitro preparation of biological tissues such as cartilage, bone, tendon, vessel, heart muscle, skin, brings out major expectations for the next decade. However, to each type of tissue correspond a large number of potential methods related to the support, the cells used (tissular cells or stem cells) and to the conditions of the environment (culture medium and mechanical forces). For instance, the quality of cartilage tissue obtained depends on the intensity, the magnitude and the frequency of mechanical stimuli. Optimization of the mechanical forces is fundamental for biotissue remodelling or synthesis, in particular to prevent cellular phenotypic degeneration. The problem of tissue or in vitro cell culture therefore has several aspects: genetic in the choice of the initial cells (stem or differentiated cells or genetically modified cells), biochemical (choice of the polymeric support, composition of the medium, oxygen control) and mechanical (magnitude and frequency of the mechanical stimuli).

For cartilage, different therapeutic approaches have been considered. In the case of severe and deep osteochondral lesions, potential treatments use surgery, but other methods have been suggested among which can be mentioned abrasion, medullar stimulation and chondrocyte grafts.

However, these techniques quite often lead to the creation of a fibrocartilaginous repair tissue with poor stability in time. Chondrocyte grafts seem to produce encouraging results but remain marginal because of their high cost and of the surgical skills they require. The development of biotissues could therefore represent an interesting alternative especially as it would also make possible the transport of therapeutic agents likely to prevent cellular phenotypic degeneration (pharmacological approaches).

The aim of this 4th symposium is to provide a thoughtful and balanced dissertation on new crucial concepts with clinical implications.

During the symposium, I will have the pleasure to give for the second time the *Negma-Lerads Prize* to four colleagues. This prize is awarded for outstanding research that represents original contributions in the field of clinical and basic mechanobiology. The choice of the winners is entirely in the hands of a panel of international scientists from Europe and USA, independent of the NEGMA-LERADS group.

I would not finish without expressing my thanks to all participants for their enthusiastic response to my invitation and their outstanding contributions. I would like to thank my coworker Natalia De ISLA who acted as scientific secretary of the symposium and Doctor Martine BURGER for the extensive and always enthusiastic investment of time in the preparation of the program and organization of the symposium.

J.F. Stoltz
Budapest, 20 May 2005

Biorheology 43 (2006) 171–180
IOS Press



# Mechanobiology and cartilage engineering: The underlying pathophysiological phenomena

J.F. Stoltz [a,b,*], N. de Isla [a,*], C. Huselstein [a], D. Bensoussan [b], S. Muller [a] and V. Decot [b]
[a] *UMR CNRS 7563 LEMTA, Faculté de Médecine, Université Henri Poincaré, Nancy 1, CHU-54500 Vandoeuvre les Nancy, France*
[b] *Unité de Thérapie cellulaire tissus, CHU-54500 Vandoeuvre les Nancy, France*

## 1. Introduction

At the end of the 20th century, various data have been collected in biology through molecular approaches without being able however to explain the underlying pathophysiological processes. Today, recent advances in cell biology (confocal imaging, optical tweezers, NMR . . .) and molecular biology (proteomics, genomics . . .) can give the option of new therapeutic approaches taking into account the central role of mechanical forces on tissue differentiation, development and remodelling. A broad spectrum of physiological phenomena is induced by mechanical stimuli (hearing, orientation to gravity, touch, tissue remodelling . . .), however the accurate mechanisms underlying the way mechanical factors mediate the physiological responses remain globally unknown (membrane deformation secondary to mechanical forces, modifications of local osmolarity, conformational changes of transmembrane proteins, direct impact on nucleus . . .).

## 2. Mechanobiology of cell and tissue and applications to cartilage

The etymological definition of biomechanics is mechanics applied to biosciences. Traditionally, biomechanics explored the structural/mechanical performance of biological tissue. Mechanobiology corresponds more to a global vision and is involved not only in cells and tissues properties through their structures but also in cell function and in the pathophysiological consequences of exerting mechanical forces. This approach requires not only to develop physical models but also to master the most recent knowledge in cell biology, in molecular biology, in genomics and proteomics.

If there is a growing interest in this global approach, it must be noticed that, as soon as 1638, Galileo hypothesized that gravity and mechanical forces are limiting factors on the growth and architecture of living organisms. However, the first major works and hypotheses on the impact of mechanical forces on morphogenesis and on tissues adaptation have been made at the end of the 19th century (Roux, 1881; Wolff, 1892) and the development of the concept of mechanobiology and its associated research works will be only realized in the 1990s.

---

*Both authors contributed equally to this work.

## 2.1. What do we know?

For a quarter of a century, there has been a conceptual revolution in cell mechanics. Today nobody can deny the major role played by mechanical factors on the environment of cellular physiology. It is now accepted that the organization of cartilage, bone, blood vessels, heart, muscles . . . is under the control of regulatory genes which mediate the expression of downstream genes and that the mechanical forces on fetal tissues are fundamental on musculoskeletal system growth and differentiation. While the biological effects of mechanical forces on various cell types (endothelial cells, chondrocytes . . .) are now well reported, the molecular mechanisms being able to explain these phenomena are still poorly understood and it is often difficult to explain the transition between a mechanical stimulus and its physiological response. Concerning specifically the cartilage, we know now that the ratio proteoglycans/collagen is essential to understand its particular mechanical properties which enable the cartilage to resist to the great variations of compression forces. In addition, the presence of proteoglycans-associated negative charges results in elevated concentrations in mobile cations and in an acid hyperosmolar environment for the chondrocyte. More, the biological effects are dependent on the type of compression. Thus, a static compression can reduce GAG and collagen production while a cyclic compression can induce a reverse effect.

Our current models take usually into account the fact that a stimulation of the matrix synthesis is rather associated with fluid circulation through the induction of transport phenomena without truly understanding the involved cellular mechanisms.

As a matter of fact, mechanical forces exerted *in vivo* onto the joints during these movements (0 to 20 MPa at the hip site) are the result of a complex combination between tension mechanisms, stress and compression forces, this last phenomenon being the most important within the cartilage. The chondrocyte is particularly reactive to these forces which could ultimately alter its metabolism and therefore the mechanical properties of the extracellular matrix and of the cell–extracellular matrix interactions (focal adhesion mechanisms). Thus, a motionless cartilage can lose its mechanical resistance properties and an excessive strain on cartilage, without any resting phase, could accelerate degeneration. Fundamentally, it is today well accepted that the chondrocytes of the articular cartilage mediate their metabolic activities through mechanical signals. The application of a mechanical force, such as pressure, will specifically result in chondrocyte deformation and in stimulating cellular and sub-cellular events, thus interfering with the cell physiology. The magnitude of the characteristics (rate, time, and intensity) of this mechanical force on chondrocyte properties such as its differentiation, its ability to synthesize the extracellular matrix or such as the cell stimulation is now demonstrated.

## 2.2. What do we need to know?

For more than 30 years, the objectives of all our efforts have been turned towards giving a molecular explanation of the living. We must now return to a more global approach. The understanding of the mechanotransduction process is then fundamental as we will not be able to understand such a complex pathology as osteoarthritis (OA) without crossing the different metabolic pathways with their relations to environment. This knowledge will also be necessary to confirm the new approaches on cell/tissue therapy and to validate the production of surrogate biotissues.

In this context, the understanding of the influence of the mechanical forces on the different ways a chondrocyte reacts to inflammation will be fundamental (PGE2, COX, NO, phenotypic modifications, regulating factors [HSP, chemokines, cytokines such as IL-1$\beta$]) . . .

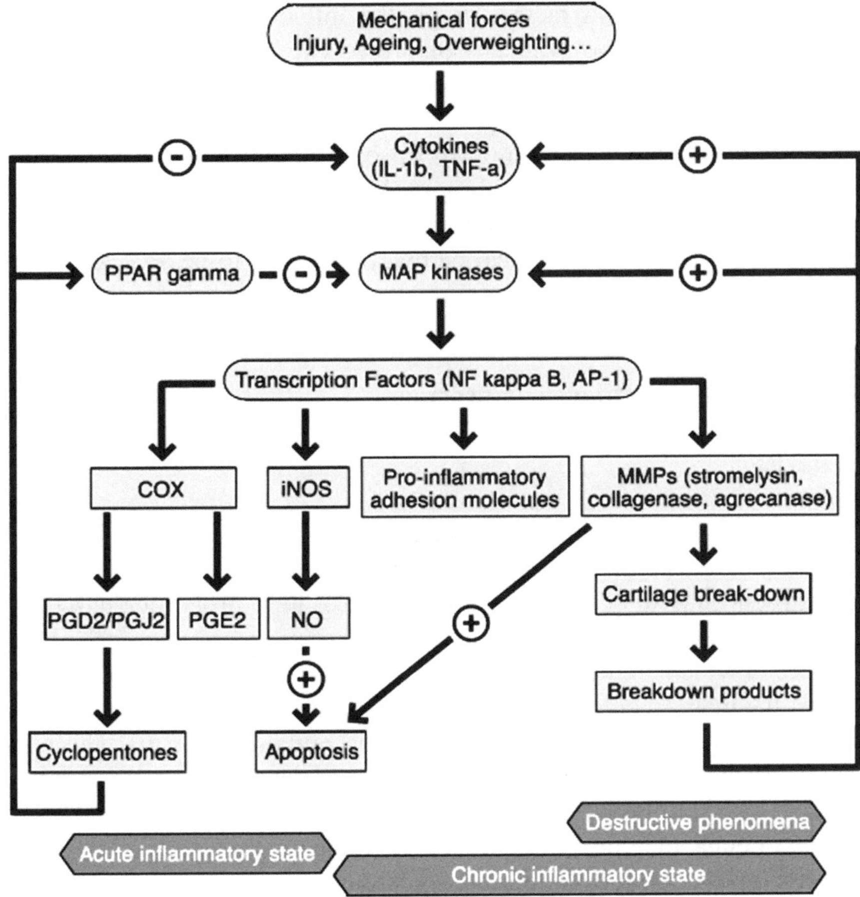


Fig. 1. Schematic representation of the potential pathways of injury involved in OA.

The metabolic activity of the chondrocyte is influenced not only by growth factors and cytokines (TNF-$\alpha$, IL-1$\beta$) but also by the activity of the environmental cellular tissue (synovial membrane, sub-chondral bone). The chondrocyte is the centre of a stable or unstable balance. In osteoarthritis (OA), a breakdown of the cartilage homeostasis has been observed and the role of mechanical factors is major.

A schematic representation (Fig. 1) would be of interest to explain the role of the mechanical forces on chondrocyte reaction and the mechanisms leading to the acute inflammatory phases of OA. It could help to shape our thinking on the metabolic pathways involved in OA.

One of the major issues is the relationship and the balance between the various metabolic factors and the mechanical forces. These issues are still poorly understood, so is the timing of these phenomena. Thus, after a mechanical strain, a stimulation of different cytokines such as TNF-$\alpha$ and IL-1$\beta$ can be observed, inducing the release of MAP kinases with a transcription of pro-inflammatory genes such as NF kappa B and AP-1 and a release of PGE-2 as soon as the first hours following the injury. PGE-2 disappears within hours while cyclopentones, a prostaglandin-class component (PGD2, PGJ2) is released and linked to PPAR gamma to inhibit the activation of NF kappa B and counteracting the influence of the pro-inflammatory factors such as TNF-$\alpha$, IL-1$\beta$, Il-6, MMP9 and their concomitant pro-apoptotic action.

At the same time, the transcription factors under the influence of TNF $\alpha$ and IL-1 activate the production of iNOS and stimulate the synthesis of proteolytic enzymes (collagenase, stromelysin 1, aggrecanase...) which will result in breaking down the cartilage matrix. These degradation products will act as pro-inflammatory mediators at the level of the cytokines and the MAP kinases, leading to a chronic inflammatory state. Thus, OA settled down in some sort of vicious cycle resulting after many years in cartilage injury, then subchondral bone lesions. We have to know more on these phenomena. We have to know more on each step of this destructive pathway. This means we have to find the ways linking the mechanical forces to the pathophysiological pathways in order to be able to propose a therapeutic solution at some stage of this cascade.

## 3. Regenerative medicine and cartilage engineering

### 3.1. Regenerative medicine

Most of human tissues do not regenerate spontaneously explaining why "cell therapy and tissue engineering" represent today promising alternative treatments. Their principle is simple. Cells are collected from a patient and introduced with or without modification of their properties into the injured tissue or on a porous 3D material where they are cultivated in a bioreactive environment the physicochemical and mechanical parameters of which are kept stable. After reaching their maturity, these tissues or cells can be implanted. Among the main middle-term therapeutic applications, cartilage and vessels engineering could be considered in cardiac insufficiency, atherosclerosis, and osteoarthritis. This concept of regenerative medicine is an emerging multidisciplinary field involving medicine, biology, chemistry, mechanics and engineering that is likely to revolutionize the ways "to improve the health and quality of life by restoring, maintaining or enhancing tissue and organs functions".

Currently, stem cells represent a full booming research field given the expectations we have as far as tissues repair is concerned, or in the therapeutic field in oncology. It is now sure that embryonic stem cells are potentially the most interesting cells as they are the only "omnipotent" cells able to differentiate in any adult cellular type. Conversely, their collection can only be realized at the very early stages of embryonic development. That means that we have today either to utilize extra-embryos produced via an *in vitro* fecundation or to create embryos according to the nuclear transfer technology. However, we must remind that using embryonic stem cells is controlled by law many countries (such as France).

Adult stem cells have only limited potential (they are "multipotent"), but their collection is in no way an ethical issue. Since the 60s, we know that bone marrow through its hematopoietic stem cells confers some regenerative capabilites to the blood cells. However bone marrow stem cells have equally others cells such as endothelial stem cells and mesenchymal stem cells. These cells can provide a support to the growth of the hematopoietic stem cells through the secretion of cytokines and through the creation of cellular interactions either directly (adhesion molecules) or indirectly (production of the extra-cellular matrix components). Mesenchymal stem cells are also able to differentiate into various cells. In addition, some stem cells could be more immature, without any specificity in any tissular specialization and their existence has been suspected in human beings. This research pathway is potentially very interesting as it could prevent from initiating any research work from an embryo.

At last, fetal stem cells are represented by umbilical cord cells. These cells could possess some precursor properties of the hematopoietic system, but their capability in forming other tissues has not been clearly shown.

## 3.2. Cartilage engineering

As far as cartilage is concerned, developing biocartilages through engineering is a full growing field. Various approaches have been proposed according to the type of cells (chondrocytes or mesenchymal stem cells) and to the matrix support. A good knowledge of cells–matrix interactions according to the applied mechanical forces will be required. Thus, it has been shown that implanting chondrocytes-including biomaterial improves cartilage injury in animals. It has been proposed in man as soon as 1994 to use adult chondrocytes obtained via culture from cartilage tissue. However, on the one hand using mature adult cells from a single layer culture exhibits a moderate potential of proliferation able to induce a dedifferentiation of the chondrocyte into a fibroblast and on the other hand requires collecting healthy cartilage in biopsy samples. Besides, the presence of fibrocartilage seems to be predominant (Roberts et al., 2003). Recently, various reports have shown that mesenchymal stem cells collected from bone marrow could differentiate *in vitro* in adding TGF $\beta$, d'IGF, BMP2 or FGF into chondroblasts, able to synthesize type II collagen. Thus, given their capabilities of differentiation, mesenchymal stem cells seem to be the most interesting way to collect cartilaginous cells.

Other components seem also to play a role in the differentiation of these cells into chondrocytes. Indeed, cartilage is a nonvascular connective tissue, able to react to the mechanical requirements of the locomotive apparatus, the restoration capacity of which is however poor. The permanent pressure forces impacting on this system are an essential regulatory factor of its functionality. Chondrocytes synthesize an extracellular matrix containing various types of collagen, initially a type II collagen then later on a type X collagen, proteoglycans (more specifically aggrecan and hyaluronates [HA]). Structural and functional integrity of cartilage is kept stable through the quantitative and qualitative maintaining of these components. In addition, of note, the fact that this extracellular matrix is in contact with the synovial fluid, the redox state of which, as well as the content in ionic charges and proteins, could influence chondrogenesis. Thus, osteoarthritis or any other inflammatory condition of the synovial fluid could result in pro-inflammatory cytokines release such as Il-1 the effect of which is negative on cartilaginous synthesis. Inversely, hypoxia seems to be beneficial.

Recently some authors have shown that mesenchymal stem cells differentiation into chondroblasts was even more efficient when mechanical forces were applied onto the cells. To do so, TGF$\beta$-cultivated mesenchymal stem cells were laid down on 3D containing-hyaluronate gel matrix and submitted or not to a daily compression for several weeks. After 3 weeks, results showed that the applied mechanical force could accelerate the differentiation of mesenchymal stem cells into chondroblasts validated by an increased synthesis in aggrecan and type II collagen.

Various synthetic, organic supports or hybrid biomaterials have been proposed in cartilage engineering. Among synthetic or biological components are polylactic acid and polyglycolic acid polymers, collagen-based material, fibrin or polysaccharide polymers such as hydrogels. Hydrogels are composed of reticulated polymers able to absorb a great amount of water. Mechanistically, hydrogels have the advantage to use water like cartilage. Under compression, water is released from hydrogel, allowing hydrogel to absorb a strain; once this strain over, water can come back to its initial place within the material and this latter returns to its initial volume. Biologically, hydrogels have a sufficient 3D porous environment to allow cellular proliferation as well as nutriments transportation. Among hydrogels, sodium alginate hydrogels are a standard in cellular morphology trials, in proteoglycans and collagen synthesis trials, as well as a standard as a natural component of extracellular matrix. It has a similar structure to cartilaginous glycosaminoglycans and has been proposed as an interesting material to sustain the chondrocyte phenotype. Recently, a MIT research group has developed an injectable polysaccharide-based gel integrating a photosensitive molecule able after injection to be photopolymerized by UV radiation.

In summary, as far as cartilage engineering is concerned, one of the major issues will be to have at our disposal a biomaterial with optimal mechanical characteristics and able to integrate the targeted cells. Many research works must be developed in order to characterize the optimal biocartilage:

- Choosing the cells (autologous chondrocytes or stem cells);
- Developing a porous material having properties able to interact with cells and to resorb;
- Evaluating the influence of mechanical stimuli observed in the joints (magnitude and rate);
- Exploring the inhibition of the potential inflammatory response after cells implantation in evaluating the numerous involved metabolic pathways with a possible pharmacological approach during the phase of biotissue preparation.

Clinically, the available data are encouraging, but their interpretation remains difficult owing to the lack of control groups. Therefore evolution of chondrocyte cell therapy technology will depend upon several aspects:

- Clarify matters about mechanisms of graft-stimulated restoration.
- Improvement of biomaterial.
- On the long term genetic engineering with the introduction of specific growth factors genes involved in the development of chondrocytes or precursor cells.

## 4. Conclusions

Different possibilities seem now to be available to manufacture a biocartilage. Most of the above concepts are not really new, but the recent advances in cell biology, polymers, genomics, synthesis and surface processing are encouraging. In addition, the characterization of biocartilage must go along with a better knowledge of the impact of mechanical forces upon the various pathways contributing to the cartilage synthesis and directly involved in the inflammation process and its inhibition.

The main requirements to fulfil cartilage manufacturing are the following: a 3D scaffold allowing consistent cell attachment, maintenance of cell phenotype and *in vitro* optimisation of cell culture in a biochemical environment. All our studies must be aimed at improving cartilage morphogenesis and *in vitro* tissue cultures in view to develop clinical and pharmacological applications.

## References

[1] D. Adams and S.A. Swanson, Direct measurement of local pressures in the cadaveric human hip joint during simulated level walking, *Ann. Rheum. Dis.* **44** (1985), 658–686.

[2] N.Y. Afoke, P.D. Byers and W.C. Hutton, Contact pressures in the human hip joint, *J. Bone Joint. Surg. Br.* **69** (1987), 536–541.

[3] K.N. An, Y.L. Sun and Z.P. Luo, Flexibility of type I collagen and mechanical property of connective tissue, *Biorheology* **41** (2004), 239–246.

[4] P. Angele, J.U. Yoo, C. Smith, J. Mansour, K.J. Jepsen et al., Cyclic hydrostatic pressure enhances the chondrogenic phenotype of human mesenchymal progenitor cells differentiated in vitro, *J. Orthop. Res.* **21** (2003), 451–457.

[5] P. Angele, D. Schumann, M. Angele, B. Kinner, C. Englert, R. Hente, B. Füchtmeier, M. Nerlich, C. Neumann and R. Kujat, Cyclic, mechanical compression enhances chondrogenesis of mesenchymal progenitor cells in tissue engineering scaffolds, *Biorheology* **41** (2004), 335–348.

[6] J.P. Arokoski, M.M. Hyttinen, T. Lapvetelainen, P. Takacs, B. Kosztaczk et al., Decreased birefringence of the superficial zone collagen network in the canine knee (stifle) articular cartilage after long distance running training, detected by quantitative polarised light microscopy, *Ann. Rheum. Dis.* **55** (1996), 253–264.

[7] G. Bentley, L.C. Biant, R.W.J. Carrington, M. Akmal, A. Goldbergand, A.M. Williams et al., A prospective, randomised comparison of autologous chondrocyte implantation *versus* mosaicplasty for osteochondral defects in the knee, *J. Bone Joint Surg. Br.* **85** (2003), 223–230.

[8] G. Bergmann, G. Deuretzbacher, M. Heller, F. Graichen, A. Rohlmann, J. Strauss et al., Hip contact forces and gait patterns from routine activities, *Journal Biomechanics* **34** (2001), 859–871.

[9] C. Bernardeau, P. Richette and P. Bizot, Greffes chondrales, greffes de chondrocytes, *Actual Rheumatol.* (2000), 453–459.

[10] F. Boschetti, G. Pennati, F. Gervaso, G.M. Peretti and G. Dubini, Biomechanical properties of human articular cartilage under compressive loads, *Biorheology* **41** (2004), 159–165.

[11] H.A. Breinan, T. Minas and H.P. Hsu, Effect of cultured antologous chondrocytes on repair of chondral defects in a canine model, *J. Bone Joint Surg. Am.* **79** (1997), 1439–1451.

[12] T.W.R. Briggs, S. Mahroof, L.A. David, J. Flannely, J. Pringle and M. Bayliss, Histological evaluation of chondral defects after autologous chondrocyte implantation of the knee, *J. Bone Joint Surg. Br.* **85**B(7) (2003), 1077–1083.

[13] P. Brittberg, L. Peterson, N.E. Sjogën-Jansson, T. Tallheden and A. Lindahl, Articular cartilage engineering with autologous chondrocyte transplantation. A review of recent developments, *J. Bone Joint Surg. Am.* **85**-A(Suppl. 3) (2003), 109–115.

[14] M. Brittberg, A. Lindahl, A. Nilsson, C. Ohlsson, O. Isaksson and L. Peterson, Treatment of deep cartilage defects in the knee with autologous chondrocyte transplantation, *N. Engl. J. Med.* **331** (1994), 889–895.

[15] J.A. Browning, K. Saunders, J.P.G. Urban and R.J. Wilkins, The influence and interactions of hydrostatic and osmotic pressures on the intracellular milieu of chondrocytes, *Biorheology* **41** (2004), 299–308.

[16] J.A. Buckwalter, Regenerating articular cartilage: why the sudden interest?, *Orthopedics Today* **16** (1996), 4–5.

[17] J.A. Buckwalter and H.J. Mankin, Articular cartilage: II degeneration and osteoarthritis, repair, regeneration and transplantation, *J. Bone Joint Surg. Am.* **79** (1997), 612–632.

[18] D.L. Butler, S.A. Goldstein and F. Guilak, Functionnal tissue engineering: the role of biomechanics, *Journal of Biomechanical Engineering* **122** (2000), 570–575.

[19] R. Cancedda, B. Dozin, P. Giannoni and R. Quarto, Tissue engineering and cell therapy of cartilage and bone, *Matrix Biol.* **22** (2003), 81–91.

[20] D.L. Cecil, K. Johnson, J. Rediske, M. Lotz, A.M. Schmidt and R. Terkeltaub, Inflammation-induced chondrocyte hypertrophy is driven by receptor for advanced glycation end products, *J. Immunol.* **175** (2005), 8296–8302.

[21] P. Cherubino, F.A. Grassi, P. Bulgheroni and M. Ronga, Autologous chondrocyte implantation using a bilayer collagen membrane: a preliminary report, *J. Orthop. Surg.* **11** (2003), 10–15.

[22] J.T. Connelly, E.J. Vanderploeg and M.E. Levenston, The influence of cyclic tension amplitude on chondrocyte matrix synthesis: experimental and finite element analyses, *Biorheology* **41** (2004), 377–388.

[23] P.R. Colville-Nash and D.A. Willoughby, COX-1, CO-2 and articular joint disease: a role of chondroprotective agents, *Biorheology* **39** (2002), 171–179.

[24] P.H. Corkhill, J.H. Fitton and B.J. Tighe, Towards a synthetic articular cartilage, *J. Biomater. Sci. Polym. Ed.* **4** (1993), 615–630.

[25] M.T. Corvol, La thérapie cellulaire dans ses applications cliniques : thérapie cellulaire du cartilage, présent et futur, *J. Soc. Biol.* **195** (2001), 79–82.

[26] P. D'andrea and F. Vittur, $Ca^{2+}$ oscillations and intercellular $Ca^{2+}$ waves in ATP-stimulated articular chondrocytes, *J. Bone Miner. Res.* **11** (1996), 946–954.

[27] S. Daouti, B. Latario, S. Nagulapalli, F. Buxton, S. Uziel-Fusi, G.W. Chirn, D. Bodian, C. Song, M. Labow, M. Lotz, J. Quintavalla and C. Kumar, Development of comprehensive functional genomic screens to identify novel mediators of osteoarthritis, *Osteoarthritis Cartilage* **13** (2005), 508–518.

[28] P. Dieppe and J. Kirwan, The localization of osteoarthritis, *Br. J. Rheumatol.* **33** (1994), 201–203.

[29] E.H. Frank, M. Jin, A.M. Loening, M.E. Levenston and A.J. Grodzinsky, A versatile shear and compression apparatus for mechanical stimulation of tissue culture explants, *J. Biomech.* **33** (2000), 1523–1527.

[30] E. Fragonas, M. Valente, M. Pozzi-Mucelli, R. Toffanin, R. Rizzo et al., Articular cartilage repair in rabbits by using suspensions of allogenic chondrocytes in alginate, *Biomaterials* **21** (2000), 795–801.

[31] T. Fujisawa, T. Hattori, K. Takahashi, T. Kuboki, A. Yamashita and M. Takigawa, Cyclic mechanical stress induces extracellular matrix degradation in cultured chondrocytes via gene expression of matrix metalloproteinases and interleukin-1, *J. Biochem.* **125** (1999), 966–975.

[32] L. Galois, A.M. Freyria, L. Grossin, P. Hubert, D. Mainard, D. Herbage, J.F. Stoltz, P. Netter, E. Dellacherie and E. Payan, Cartilage repair: Surgical techniques and tissue engineering using polysaccharide- and collagen-based biomaterials, *Biorheology* **41** (2004), 433–444.

[33] R. Gassner, M.J. Buckley, H. Georgescu, R. Studer, M. Stefanovich-Racic et al., Cyclic tensile stress exerts anti-inflammatory actions on chondrocytes by inhibiting inducible nitric oxide synthase, *J. Immunol.* **163** (1999), 2187–2192.

[34] C. Gigant-Huselstein, D. Dumas, P. Hubert, D. Baptiste, E. Dellacherie et al., Influence of mechanical stress on extracellular matrixes synthezed by chondrocytes seeded onto alginate and hyaluronate-based 3D bio systems, *JMMB* **3** (2003), 59–70.

[35] C. Gigant-Huselstein, P. Hubert, D. Dumas, E. Dellacherie, P. Netter, E. Payan and J.F. Stoltz, Expression of adhesion molecules and collagen on rat chondrocyte seeded into alginate and hyaluronate based 3D bio systems. Influence of mechanical stresses, *Biorheology* **41** (2004), 423–432.

[36] M.L. Gray, A.M. Pizzanelli, A.J. Grodzinsky and R.C. Lee, Mechanical and physiochemical determinants of the chondrocyte biosynthetic response, *J. Orthop. Res.* **6** (1988), 777–792.

[37] F. Guilak, Volume and surface area measurement of viable chondrocytes in situ using geometric modelling of serial confocal sections, *J. Microsc.* **173** (1994), 245–256.

[38] F. Guilak, Compression-induced changes in the shape and volume of the chondrocyte nucleus, *J. Biomech.* **28** (1995), 1529–1541.

[39] F. Guilak, R.A. Zell, G.R. Erickson, D.A. Grande, C.T. Rubin et al., Mechanically induced calcium waves in articular chondrocytes are inhibited by gadolinium and amiloride, *J. Orthop. Res.* **17** (1999), 421–429.

[40] F. Guilak, H.A. Awad, B. Fermor, H.A. Leddy and J.M. Gimble, Adipose-derived adult stem cells for cartilage tissue engineering, *Biorheology* **41** (2004), 389–400.

[41] H.J. Hauselmann, M.B. Aydelotte, B.L. Schumacher, K.E. Kuettner, S.H. Gitelis and E.J. Thonar, Synthesis and turnover of proteoglycans by human and bovine adult articular chondrocytes cultured in alginate beads, *Matrix* **12** (1992), 116–129.

[42] J.H. Heegaard, G.S. Beaupre and D.R. Carter, Mechanically modulated cartilage growth may regulate joint surface morphogenesis, *Journal of Orthopaedics Research* **17** (1999), 509–577.

[43] I.J.P. Henderson, B. Tuy, D. Connell, B. Oakes and W.H. Hettwer, Prospective clinical study of autologous chondrocyte implantation and correlation with MRI at three and 12 months, *J. Bone Joint Surg. Br.* **85B**(7) (2003), 1060–1066.

[44] K. Hjelle, E. Solheim, T. Strand, R. Muri and M. Brittberg, Articular cartilage defects in 1,000 knee arthroscopies, *Arthroscopy* **18** (2002), 730–734.

[45] W.A. Hodge, K.L. Carlson, R.S. Fijan, R.G. Burgess, P.O. Riley et al., Contact pressures from an instrumented hip endoprosthesis, *J. Bone Joint. Surg. Am.* **71** (1989), 1378–1386.

[46] U. Horas, D. Pelinkovic, G. Herr, T. Aigner and R. Schnettler, Autologous chondrocyte implantation and osteochondral cylinder transplantation in cartilage repair of the knee joint: a prospective, comparative trial, *J. Bone Joint Surg. Am.* **85-A**(2) (2003), 185–192.

[47] J.Y. Jouzeau, S. Pacquelet, C. Boileau, E. Nedelec, N. Presle, P. Netter and B. Terlain, Nitric oxide (NO) and cartilage metabolism: NO effects are modulated bys superoxide in response to IL-1, *Biorheology* **39** (2002), 201–214.

[48] M.J. Kaab, R.G. Richards, K. Ito, I. Gwynn and H.P. Notzli, Deformation of chondrocytes in articular cartilage under compressive load: a morphological study, *Cells Tissues Organs* **175** (2003), 133–139.

[49] P.J. King, T. Bryant and T. Minas, Autologous chondrocyte implantation for chondral defects of the knee: indications and technique, *J. Knee Surg.* **15** (2002), 177–184.

[50] I. Kiviranta, M. Tammi, J. Jurvelin, A.M. Saamanen and H.J. Helminen, Moderate running exercise augments glycosaminoglycans and thickness of articular cartilage in the knee joint of young beagle dogs, *J. Orthop. Res.* **6** (1988), 188–195.

[51] M. Lammi, M.A. Elo, R.S. Sironen, H.M. Karjalainen, K. Kaarniranta and H.J. Helminen, Hydrostatic pressure-induced changes in cellular protein synthesis, *Biorheology* **41** (2004), 309–314.

[52] M.J. Lammi, Current perspectives on cartilage and chondrocyte mechanobiology, *Biorheology* **41** (2004), 593–596.

[53] D.A. Lee, M.M. Knight, J.F. Bolton, B.D. Idowu, M.V. Kayser and D.L. Bader, Chondrocyte deformation within compressed agarose constructs at the cellular and sub-cellular levels, *Journal Biomechanics* **33** (2000), 81–95.

[54] D.A. Lee and D.L. Bader, Compressive strains at physiological frequencies influence the metabolism of chondrocytes seeded in agarose, *J. Orthop. Res.* **15** (1997), 181–188.

[55] L.P. Li and W. Herzog, The role of viscoelasticity of collagen fibers in articular cartilage: theory and numerical formulation, *Biorheology* **41** (2004), 181–194.

[56] E.G. Lima, R.L. Mauck, S.H. Han, S. Park, K.W. Ng, G.A. Ateshian and C.T. Hung, Functional tissue engineering of chondral and osteochondral constructs, *Biorheology* **41** (2004), 577–592.

[57] E. Lucchinetti, M.M. Bhargava and P.A. Torzilli, The effect of mechanical load on integrin subunits alpha5 and beta1 in chondrocytes from mature and immature cartilage explants, *Cell Tissue Res.* **315** (2004), 385–391.

[58] Z.P. Luo, Y.L. Sun, T. Fujii and K.N. An, Single molecule mechanical properties of type II collagen and hyaluronan measured by optical tzeezers, *Biorheology* **41** (2004), 247–254.

[59] J.A. Martin, T. Brown, A. Heiner and J.A. Buckwalter, Post traumatic osteoarthritis: The role of accelerated chondrocyte Senescence, *Biorheology* **41** (2004), 479–492.

[60] S.J. Millward-Sadler, M.O. Wright, H. Lee, K. Nishida, H. Caldwell et al., Integrin-regulated secretion of interleukin 4: A novel pathway of mechanotransduction in human articular chondrocytes, *J. Cell Biol.* **145** (1999), 183–189.

[61] S.J. Millward-Sadler and D.M. Salter, Integrin-dependent signal cascades in chondrocyte mechanotransduction, *Ann. Biomed. Eng.* **32** (2004), 435–446.

[62] S.J. Millward Sadler, M.O. Wright, P.W. Flatman and D.M. Salter, ATP in the mechanotransduction pathway of normal human chondrocytes, *Biorheology* **41** (2004), 567–576.

[63] G. Miralles, R. Baudoin, D. Dumas, D. Baptiste, P. Hubert, J.F. Stoltz and E. Dellacherie, Sodium alginate sponges with or without sodium hyaluronate: in vitro engineering of cartilage, *J. Biomed. Mater. Res.* **57** (2001), 268–278.

[64] R.W. Moskowitz, D.S. Howell, R.D. Altman, J.A. Buckwalter and V.M. Goldberg, *Ostearthritis: Diagnostic and Medical Surgical Management*, 3rd edn, Sauders, London, 2001, 674 pp.

[65] V.C. Mow, C.C.B. Wang and C.T. Hung, The extracellular matrix, interstitial fluid and ions as a mechanical signal transducer in articular cartilage, *Osteoarthritis and Cartilage* **7** (1999), 41–58.

[66] O'B.P. Hara, J.P. Urban and A. Maroudas, Influence of cyclic loading on the nutrition of articular cartilage, *Ann. Rheum. Dis.* **49** (1990), 536–539.

[67] M. Ochi, Y. Uchio, K. Kawasaki, S. Wakitani and J. Iwasa, Transplantation of cartilage-like tissue made by tissue engineering in the treatment of cartilage defects of the knee, *J. Bone Joint Surg. Am.* **84-B** (2003), 571–578.

[68] L. Peterson, M. Brittberg, I. Kiviranta, El. Akerlund and A. Lindahl, Autologous chondrocyte transplantation: biomechanics and long-term durability, *Am. J. Sport. Med.* **30** (2002), 2–12.

[69] L. Peterson, T. Minas, M. Brittberg and A. Lindahl, Treatment of osteochondritis dissecans of the knee with autologous chondrocyte transplantation: results at two to ten years, *J. Bone Joint Surg. Am.* **85-A**(Supll. 2) (2003), 17–24.

[70] P.M. Ragan, A.M. Badger, M. Cook, V.I. Chin, M. Gowen et al., Down-regulation of chondrocyte aggrecan and type-II collagen gene expression correlates with increases in static compression magnitude and duration, *J. Orthop. Res.* **17** (1999), 836–842.

[71] P.M. Ragan, V.I. Chin, H.H. Hung, K. Masuda, E.J. Thonar et al., Chondrocyte extracellular matrix synthesis and turnover are influenced by static compression in a new alginate disk culture system, *Arch. Biochem. Biophys.* **383** (2000), 256–264.

[72] H. Robert and J. Bahuaud, Autogreffe de chondrocytes. Revue des techniques et premiers résultats, *Rev. Rheum.* **66** (1999), 835–838.

[73] P.P. Rooij Der, M. Siebrecht, M. Tagil and P. Aspenberg, The fate of mechanically induced cartilage in an unloaded environment, *J. Biomech.* **34** (2001), 961–966.

[74] C.A. Roufosse, N.C. Direkze, W.R. Otto and N.A. Wright, Circulating mesenchymal stem cells, *Int. J. Biochem. Cell. Biol.* **36** (2004), 585–597.

[75] E. Ruoshlati and M.D. Pierschbacher, New perspectives in cell adhesion: RGD and integrins, *Sciences* **238** (1987), 491–497.

[76] A.M. Saamamen, I. Kiviranta, J. Jurvelin, H.J. Helminen and M. Tammi, Proteoglycan and collagen alterations in canine knee articular cartilage following 20 km daily running exercise for 15 weeks, *Connect. Tissue Res.* **30** (1994), 191–201.

[77] S. Saarakkala, R.K. Korkonen, M.S. Laasanen, J. Toyras, J. Rieppo and J.S. Herzog, Mechano-acoustic determination of Young's modulus of articular cartilage, *Biorheology* **41** (2004), 167–180.

[78] D.M. Salter, S.J. Millward-Sadler, G. Nuki and M.O. Wright, Differential responses of chondrocytes from normal and osteoarthritic human articular cartilage to mechanical stimulation, *Biorheology* **39** (2002), 97–108.

[79] D.M. Salter, M.O. Wright and S.J. Millward Sadler, NMDA receptor expression and roles in human articular chondrocyte mechanotransduction, *Biorheology* **41** (2004), 273–282.

[80] A.E. Sams and A.J. Nixon, Chondrocyte – laden collagen scaffolds for resurfacing extensive articular, cartilage defects, *Osteoarthritis Cartilage* **3** (1995), 47–59.

[81] B. Schmitt, J. Ringe, T. Haupl, M. Notter, R. Manz et al., BMP2 initiates chondrogenic lineage development of adult human mesenchymal stem cells in high-density culture, *Differentiation* **71** (2003), 567–577.

[82] J.O. Seidel, M. Pei, M.L. Gray, R. Langer, L.E. Freed and G. Vunjak Novakovic, Long-term culture of tissue engineered cartilage in a perfused chamber with mechanical stimulation, *Biorheology* **41** (2004), 445–458

[83] I. Sekiya, D.C. Colter and D.J. Prockop, BMP-6 enhances chondrogenesis in a subpopulation of human marrow stromal cells, *Biochem. Biophys. Res. Commun.* **284** (2001), 411–418.

[84] R.A. Sellards, S.J. Nho and B.J. Cole, Chondral injuries, *Curr. Opin. Rheumatol.* **14** (2002), 134–141.

[85] J.F. Stoltz, *Mechanobiology: Cartilage and Chondrocyte*, Vol. 2, Proceedings of the second symposium, Paris, 28/29 April 2002, in Biomedical and Health Research, IOS, Amsterdam, 2003, pp. 52–288.

[86] J.F. Stoltz, *Mechanobiology: Cartilage and Chondrocyte*, Vol. 3, Proceedings of the third symposium, Brussels, 16/17 May 2003, in Biomedical and Health Research, IOS, Amsterdam, 2005, pp. 61–445.

[87] T. Sugimoto, M. Yoshino, M. Nagao, S. Ishii and H. Yabu, Voltage-gated ionic channels in cultured rabbit articular chondrocytes, *Comp. Biochem. Physiol. C. Pharmacol. Toxicol. Endocrinol.* **115** (1996), 223–232.

[88] M.C.H. Van Der Meulern and R. Huiskes, Why Mechanobiology? A survey article, *J. Biomech.* **35** (2002), 401–414.

[89] K.Y. Volokh, Mathematical framework for modelling tissue growth, *Biorheology* **41** (2004), 263–272.

[90] C.C. Wang, X.E. Guo, D. Sun, V.C. Mow, G.A. Ateshian and C.T. Hung, The functional environment of chondrocytes within cartilage subjected to compressive loading: a theoretical and experimental approach, *Biorheology* **39** (2002), 11–25.

[91] R.J. Wilkins, J.A. Browning and J.P. Urban, Chondrocyte regulation by mechanical load, *Biorheology* **37** (2000), 67–74.

[92] M. Wong, M. Siegrist, X. Wang and E. Hunziker, Development of mechanically stable alginate/chondrocyte constructs: effects of guluronic acid content and matrix synthesis, *J. Orthop. Res.* **19** (2001), 493–499.

[93] M. Wright, P. Jobanputra, C. Bavington, D.M. Salter and G. Nuki, Effects of intermittent pressure-induced strain on the electrophysiology of cultured human chondrocytes: evidence for the presence of stretch-activated membrane ion channels, *Clin. Sci. (Lond.)* **90** (1996), 61–71.

[94] R.R. Wroble, Articular cartilage injury and autologous chondrocyte implantation. Which patients might benefit?, *Phys. Sportsmed* **28** (2000), 43–49.

[95] C.E. Yellowley, C.R. Jacobs and H.J. Donahue, Mechanisms contributing to fluid-flowinduced $Ca^{2+}$ mobilization in articular chondrocytes, *J. Cell. Physiol.* **180** (1999), 402–408.

[96] Z. Xu, M.J. Buckley, C.H. Evans and S. Agarwal, Cyclic tensile strain acts as an antagonist of IL-1 beta actions in chondrocytes, *J. Immunol.* **165** (2000), 453–460.

# Part I:
# Mechanobiology-Mechanotransduction

Biorheology 43 (2006) 183–190
IOS Press



# Force-mediated dissociation of proteoglycan aggregate in articular cartilage

Xuhui Liu [a], Jian Q. Sun [b], Michael H. Heggeness [a], Ming-Long Yeh [a] and Zong-Ping Luo [a,*]

[a] *Department of Orthopedic Surgery, Baylor College of Medicine, 6550 Fannin, Suite 451, Houston, TX 77030, USA*
[b] *Department of Mechanical Engineering, University of Delaware, Newark, DE 19716, USA*

**Abstract.** Proteoglycan aggregate is the primary component in articular cartilage responsible for resisting compressive loading. It consists of a core molecule of hyaluronan and a number of side chains of aggrecan bound to hyaluronan non-covalently. The loss of aggrecan from articular cartilage is considered to be a major factor in the development of osteoarthritis. Though enzymatic digestion of aggrecan is believed to be responsible for the release of aggrecan from osteoarthritic cartilage, other mechanisms, such as direct force-mediated detachment of aggrecan from hyaluronan may also be involved. In this study, the rupture force of the single bond between hyaluronan and aggrecan in articular cartilage was directly quantified using experimental measurement and Monte Carlo simulation. Low rupture force of this bond, as determined in this study suggested a possible direct force-mediated detachment of aggrecan from proteoglycan aggregate in osteoarthritic cartilage.

Keywords: Force, dissociation, proteoglycan aggregate, hyaluronan, aggrecan, articular cartilage, laser tweezers, Monte-Carlo simulation

## 1. Introduction

Proteoglycan (PG) aggregate is a major component of the extracellular matrix in articular cartilage, which is considered to be responsible for resistance to compression. It consists of a core molecule of hyaluronan (HA) and a number of branches of molecules of aggrecan, which bind to the HA non-covalently with the help of link protein, a small protein able to bind to both aggrecan and HA [7,14]. Like most HA-binding molecules, aggrecan and link protein contain proteoglycan tandem repeats as functional sites to interact with HA [3,16,17,26].

Loss of aggrecan from articular cartilage is considered to be a major factor in the development of osteoarthritis (OA), a progressive degenerative joint disease, which is characterized by the degeneration of articular cartilage in involved joints [8]. Enzymatic fragmentation of aggrecan is considered to be responsible for the release of aggrecan from the aggregate [20]. However, since the cartilage is always in a mechanical environment in joints during daily activities, the direct mechanical failure of PG aggregate structure may also contribute to the loss of aggrecan in OA. It is well known that covalent bonds are mechanically much stronger than non-covalent bonds. Thus, the only non-covalent bond within PG aggregate – the bond between HA, aggrecan and link protein – is the most vulnerable breaking point in

---

*Address for correspondence: Zong-Ping Luo, PhD, Sport Medicine Research Center, Department of Orthopedic Surgery, Baylor College of Medicine, 6550 Fannin, Suite 451, Houston, TX 77030, USA. Tel.: +1 713 441 5764; Fax: +1 713 790 2134; E-mail: luo@bcm.tmc.edu.

force-mediated destruction of PG aggregate in articular cartilage. The strength of this bond determines the integrity of PG aggregate in articular cartilage.

However, to our knowledge, the strength of the bond between HA, aggrecan and link protein has not yet been determined. In articular cartilage, the linkage of aggrecan to HA is intermediated by the binding of its hyaluronan-binding domain (G1 domain) to HA. Since the other components of aggrecan are not involved in its binding to HA, the linkage of hyaluronan-binding protein (HABP)-complex of G1 domain of aggrecan and link protein-to HA represents the linkage of aggrecan, link protein and HA. The goals of this study are to (1) quantify the bond strength between HA, aggrecan and link protein by directly measuring the rupture force of single bond between HA and HABP using a nanomechanical testing system and comparing those results to a theoretical prediction based on a smart version of Monte-Carlo simulation; and (2) evaluate the strength of this bond by comparing its rupture force to rupture forces of known covalent and non-covalent bonds.

## 2. Materials and methods

Mechanical testing was performed using a single molecule nanomechanical testing system, previously developed in our laboratory [11,12]. In this system, a laser tweezers workstation (Cell Robotics Inc., Albuquerque, NM), which consisted of a laser tweezers 980/1000 module, with a laser wavelength of 980 nm and a continuous wave power diode laser up to 1000 mW, was built in an inverted microscope (Axiovert 135, Carl Zeiss Inc., Thornwood, NY). A piezo-stage with a resolution of 1 nm (P-731.20, Polytec PI, Inc., Auburn, MA) and an interferometry system (Micro Development Inc., Zimmerman, MN) were included in order to improve the resolution of the stage movement and molecule displacement measurement to nanometer level. A two-bead system was employed to perform the mechanical test on the single molecule nanomechanical testing system. Descriptive details have been provided previously [5]. In brief, a large bead-molecule-small bead linkage was sandwiched between two coverglasses, and was then mounted on the stage. The large bead was immobilized between coverglasses. The small bead was free in the solution except for its linkage to the molecule.

In order to form the bead-HA-HABP-bead linkage, the molecules of HA and HABP were each separately coated onto two different sizes of polystyrene beads and then mixed together to allow the formation of the linkage prior to the mechanical testing. Both molecules were obtained from commercial resource. Biotinylated HABP was purchased from Seikagaku Corp., Tokyo, Japan. HA with a molecule weight $>10^6$ Dalton, was obtained from Acros Organics BVBA, Geel, Belgium. Biotinylated HABP was bound to streptavidin-coated polystyrene beads (5.6 $\mu$m in diameter, Bangs Laboratories Inc., Fishers, IN) through biotin/streptavidin binding. 5 $\mu$l of 0.25 mM biotinylated HABP was combined with 30 $\mu$l 0.5 g/ml streptavidin-coated beads in 0.1 M phosphate buffer saline (pH 7.4) with 1% bovine serum albumin for 1 hour at room temperature [15]. In order to reduce the density of HABP on the beads, excessive biotin (Sigma-Aldrich Corp. St. Louis, MO) was also added to impede the binding sites of the streptavidin on the beads surfaces. Afterward, the unbound biotinylated HABP and biotin were removed by centrifugation. In the control group, only biotin was used. HA was bound to another size of amino group-coated polystyrene bead (3.1 $\mu$m in diameter, Bangs Laboratories Inc., Fishers, IN) through amino group/carboxyl group binding. 1 $\mu$l of 50 nM HA was reacted with 1-ethyl-3-(3-dimethylaminopropyl)-carbodiimide hydrochloride and N-hydroxysuccimide (Pierce Biotechnology Inc., Rockford, IL) in phosphate sodium buffer (pH 5.5) with 0.5 M NaCl. After 15 minutes, 30 $\mu$l 1 g/ml amino group-functioned beads were added and the pH was adjusted to 8. The reaction was quenched by hydroxylamine (Sigma Aldrich Corp. St. Louis, MO) at a final concentration of 10 nM 10 minutes later [5]. Unbound HA was removed through centrifugation. The two molecule-linked beads were then mixed in an associated buffer of 0.15 M sodium chloride, 0.01 M 2-[N-Morpholino]-ethanesulfonic acid (MES), 0.005 M ethylenediaminetetraacetic acid (EDTA) at pH 7.0 and left overnight at 4°C to allow the formation of the binding of HA to HABP [25].

The bond between HA and HABP was ruptured by manipulating the beads during the mechanical test. After a large bead-molecule-small bead linkage was identified under microscopy, the laser was turned on to trap the small bead. The large bead was moved away from the small bead using the stage and the molecules were stretched between the beads until the bond between the molecules was broken. A computer program based on the LabVIEW® program (National Instruments Corp., Austin, TX) was created to control the piezo-stage movement and to measure the trapped bead displacement. The deformation of the molecules was measured as the displacement between the beads. The force applied on the molecules was equal to the bead trapping force, which can be calculated as the trapping stiffness times the departure distance of the small bead from the laser beam center [23,24]. The force–displacement curve was noted and the rupture force was recorded as the highest force the curve reached. A pulling velocity of 0.5 $\mu$m/s was adopted in the mechanical test.

Based on the concept that the rupture process of a bond can be modeled as a random event, a smart version of Monte-Carlo simulations was made to predict the rupture forces, following the steps previously described by Fritz et al. [4]. The wormlike chain (WLC) model presented by Bustamante et al. [2] was used to fit the rupture force and extension data of the HA/HABP bond. The model was given by

$$F = \frac{k_\mathrm{B}T}{p}\left(\frac{1}{4(1 - x/L_\mathrm{contour})^2} - \frac{1}{4} + \frac{x}{L_\mathrm{contour}}\right),$$

where $k_\mathrm{B}$ is the Boltzmann constant, $T$ is the temperature in Kelvin, $p$ is the persistence length, and $L_\mathrm{contour}$ is the contour length. In the Monte-Carlo simulation, the pulling speed $v_\mathrm{pull}$ is given to be 0.5 $\mu$m/s. At each small time interval ($\Delta t = 0.1$ ms), the extension is calculated by $x(t) = v_\mathrm{pull} \cdot t$, and the actual force is calculated from the above WLC model. The off-rate is determined from the following equation [4]

$$k_\mathrm{off}(F) = k_\mathrm{off}^0 \cdot \mathrm{e}^{F(t)\cdot S_\mathrm{pot}/(k_\mathrm{B}T)},$$

where $k_\mathrm{off}^0$ is the off-rate at zero external force, and $S_\mathrm{pot}$ is the mean width of the binding potential. The probability of a rupture of the binding at the force level $F$ during the current time interval $\Delta t$ is given by $P_\mathrm{rupture} = k_\mathrm{off}(F)\Delta t$. In the Monte-Carlo simulation, the occurrence of the bond rupture at this force level is determined by comparing $P_\mathrm{rupture}$ with a random number $P_\mathrm{U}$ within a uniform distribution in [0,1]. When $P_\mathrm{rupture} > P_\mathrm{U}$, the rupture occurs.

## 3. Results

The rupture force of single HA/HABP bond was measured as $40 \pm 11$ pN (mean $\pm$ standard deviation) at a pulling velocity of 0.5 $\mu$m/s. In total, 75 samples were tested at this pulling velocity. A typical force–displacement curve of the rupture of one single HA/HABP bond is shown in Fig. 1. The persistence length $p$ and the contour length $L_\mathrm{contour}$ of the molecule were determined as 3.7 nm and 3.1 $\mu$m, respectively, using the WLC model [5].

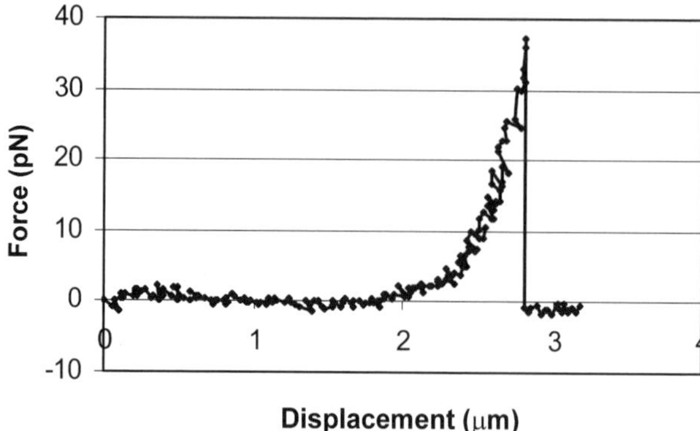


Fig. 1. A typical force–displacement curve of HA elongation. The maximal force represents the rupture force of the bond between HA and HABP. The pulling velocity was 0.5 $\mu$m/s.

A smart version of Monte-Carlo simulation was found to agree well with the experimental data. From such a fit, an off-rate $k_{\text{off}}^0 = 0.0044$ s$^{-1}$ at zero external force and a mean width of the HA/HABP bond potential $s_{\text{pot}} = 4.1$ nm were determined. Six hundred samples have been generated in the Monte-Carlo simulation. The probability distribution of the rupture force is shown and compared with experimental data in Fig. 2. The simulated rupture forces have a mean value 40 pN with a standard deviation 11 pN.

## 4. Discussion

In this study, we measured the rupture force of the single bond between HA and HABP – the critical intermolecular interaction that ensures the integrity in the proteoglycan aggregate in the extracellular matrix of articular cartilage – as $40 \pm 11$ pN at the pulling velocity of 0.5 $\mu$m/s using a nanomechanical testing system. The results showed a good fit to a theoretical prediction based on a smart version of the Monte-Carlo simulation.

Bond rupture force is the minimum external forced needed to apply in order to rupture a bond. It reflexes the mechanical strength of the bonds, which differs from its thermokinetic properties. Major thermokinetic paramenters of the bond between HA and aggrecan, such as the equilibrium dissociation constant ($K_{\text{D}}$), associate rate constant ($K_{\text{on}}$) and dissociation rate constant ($K_{\text{off}}$) had been studied previously. Watanabe et al. measured the $K_{\text{D}}$ of the binding between HA and aggrecan as $2.26 \times 10^{-7}$ M, and the $K_{\text{D}}$ of the HA/link protein bond as $0.89 \times 10^{-7}$ M [26]. These thermokinetic paramenters reflect thermokinetic properties of the bond between HA and aggrecan under no external force. However, previous studies have shown that for many non-covalent bonds, their properties under external force are quite different from those under extended force. Fritz et al. has shown that for the non-covalent bond between P-selectin and its ligand PSGL-1, dissociation rate constant ($K_{\text{off}}$) increased four orders of magnitude under extended pulling force compared to that under no extended force [4]. Therefore, the rupture force of the bond between HA, aggrecan and link protein measured in this study, is more suitable than $K_{\text{D}}$, $K_{\text{on}}$ and $K_{\text{off}}$ to reflect mechanical properties of this bond under pulling force. The rupture force of this bond could be neither derived from nor substituted by thermokinetic parameters, such as $K_{\text{D}}$, $K_{\text{on}}$, $K_{\text{off}}$ from traditional thermokinetic experiments.

Compared with rupture force of other covalent and non-covalent bonds, the bond between HA, aggrecan and link protein is weak. As discussed above, rupture force could be used as criteria in evaluating mechanical strength of bonds. Thus, strength of bonds could be ranked according to their rupture forces.

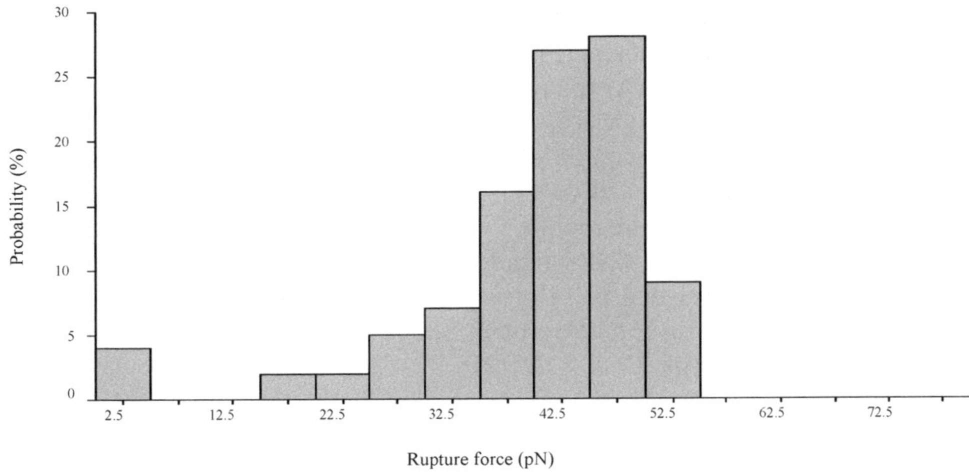

(A)

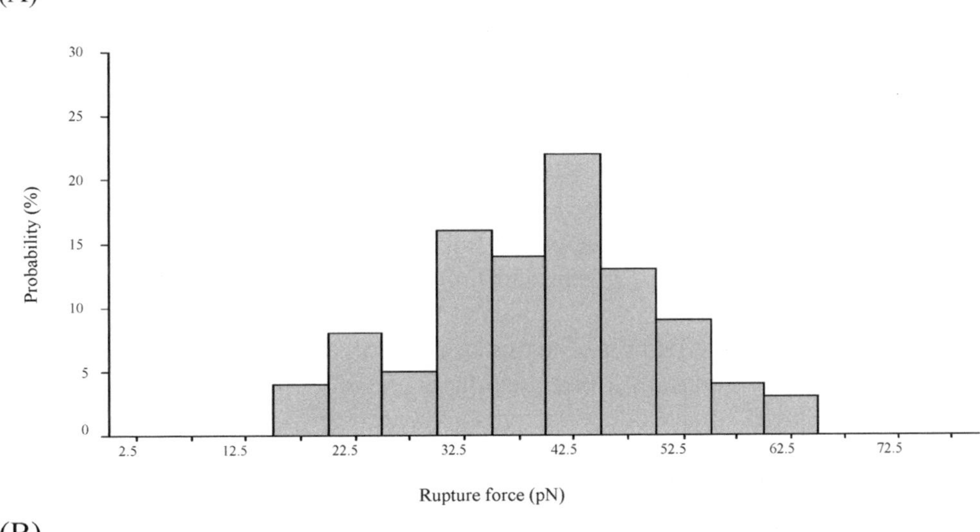

(B)

Fig. 2. Force-dependent bond rupture probabilities for the single bond between HA/HABP predicted by Monte-Carlo simulation (A) was compared to that measured in the experiment (B).

Table 1

Summary of rupture forces of some single covalent and non-covalent bonds

| | Bond | | | | | |
|---|---|---|---|---|---|---|
| | Covalent | | Non-covalent | | | |
| | Au–S | C–Si | Biotin/ streptavidin | P-selectin/ PSGL-1 | Titin unfolding | HA/HABP |
| Rupture force (pN) | 1400 | 2000 | 257 | ~125 | 190 | 40 |
| Pulling velocity | N/A | N/A | N/A | 500 nm/s | 500 nm/s | 500 nm/s |
| Reference | 18 | 18 | 19 | 16 | 20 | |

Grandbois et al. has determined that the rupture force for covalent bonds was at least 1400 pN [6]. Moy et al. has quantified the rupture force of single bond between biotin and streptavidin, one of the strongest non-covalent bonds, as 257 pN [13]. At the pulling velocity of 500 nm/s, the same velocity adopted in this study, Fritz et al. demonstrated that the rupture force of single non-covalent bond between P-selectin and its ligand – PSGL-1 was at ~125 pN [4]. Rief et al. determined the rupture force of another non-covalent bond – titin unfolding as 190 pN [19]. Compared with the rupture force of covalent and non-covalent bonds described above, the rupture force of the bond between HA, aggrecan and link protein, which was determined as 40 pN, was two orders of magnitude lower than those of covalent bonds and one order of magnitude lower that those of non-covalent bonds described above. This suggested that the bond between HA, aggrecan and link protein, which determines the integrity of PG aggregate in articular cartilage, was mechanically weak.

Weak mechanical strength of the bond between HA, aggrecan and link protein determined in this experiment suggested a possible direct force-mediated detachment of aggrecan from PG aggregate in articular cartilage. Findings in biochemical studies in osteoarthritic joints also support this hypothesis. It has been found that concentrations of G1 domain of aggrecan [21] and link protein [18] were increased in synovial fluid from osteoarthritic joints compared to normal joints. This suggested increased release of aggrecan and link protein from cartilage in OA. However, other experiments showed decreased concentration of HA in osteoarthritic joint synovial fluid [1,9], which suggested less HA is released from cartilage to synovial fluid in OA joints. Taken together, all these findings inferred one possibility – aggrecan and link protein were detached from HA in the cartilage before they were released into synovial fluid. As to the mechanism of detachments, since there was no clue suggesting enzymatic digestion was involved (HA binding domain of aggrecan and link protein found in synovial fluid were "intact"), force-mediated direct detachment of aggrecan and link protein from HA might be the most possible mechanism involved in this process.

Some technical details need to be addressed in this study. First, single HA/HABP bonds were achieved by adopting some specific measures described previously [5,10,22] in this experiment. These includes (1) the molecules were dissolved and diluted at low concentrations in order to maximize the presence of single molecules in solution; (2) biotin was added to impede the binding of biotinylated HABP to the streptavidin-coated beads to help reduce the density of the HABP on the beads; and (3) the persistence length of the molecule of each sample was measured using the WLC model in order to assure the HA between the two beads was a single molecule. The persistence length of the molecules between the beads was found to be 3.7 ± 1.4 nm, which was similar to that of single HA reported previously by Fujii et al. (4.5 ± 1.2 nm) [5]. These measures assured that HA/HABP bonds tested in this experiment were single ones. Second, it was critical to ensure that the bond broken during the experiment was the bond between HA and HABP. In order to achieve this, we linked the molecules of HA and HABP to the beads with stronger bonds. HA was attached on the bead with a covalent bond, whose rupture force was at least 1400 pN [6]. Biotin was linked to HABP also through a covalent bond according to the manufacturer's protocol. The linkage between HABP and the bead was the bond between biotin and streptavidin, one of the strongest known non-covalent bonds. The rupture force of single bond between biotin and streptavidin was reported as 257 pN [13], which was much higher than the rupture force we measured in this experiment. Therefore, it is reasonable to conclude that the bond ruptured during the experiment was the non-covalent bond between HA and HABP.

A smart version of Monte-Carlo simulation was used to predict the rupture force of HA/HABP at the pulling velocity of 0.5 $\mu$m/s and compared to experimental data. A mean width of the HA/HABP bond of 4.1 nm was predicted by Monte-Carlo simulation. This value is higher than published data of mean width of other bonds [4,19]. Since the HABP includes two parts – hyaluronan binding region of aggrecan and the link protein, both of which can bind to HA independently – the bond between HA/HABP may include two bonds: the bond of hyaluronan binding region of aggrecan to HA and the bond of link protein to HA. Therefore, the mean width of HA/HABP may include those of both bonds, and probably some distance between. The mean value and standard deviation of predicted data showed good agreements to the experimental data. However, there was some difference between the distributions of predicted rupture force and experimental data. There is no clear cut-off at $\sim$50 pN in the latter compared to predicted results. One explanation is that the experimental data included few multiple bonds. Although various measures were adopted to achieve single molecule binding, a small number of multiple bonds still may form. Those multiple bonds can be ruptured simultaneously, showing single peaks on the force–displacement curve with higher rupture force than that of single bonds [27].

In summary, we directly determined the strength of the bond between HA, aggrecan and link protein in articular cartilage by measuring the rupture force of the single bond between HA and HABP both experimentally and theoretically in this study. Low rupture force determined in this study compared with those of other covalent and non-covalent bonds reported in literature suggested that the bond between HA, aggrecan and link protein was mechanically weak. Combined with other evidence reported in literature, our results suggest a possible direct force-mediated detachment of aggrecan from PG aggregate in osteoarthritic cartilage.

## Acknowledgement

This study was supported by a grant from the Arthritis Foundation. We also would like to thank Carolyn Adams for her assistance in the preparation of this manuscript.

## References

[1] C. Belcher, R. Yaqub, F. Fawthrop, M. Bayliss and M. Doherty, *Ann. Rheum. Dis.* **56** (1997), 299–307.
[2] C. Bustamante, J.F. Marko, E.D. Siggia and S. Smith, *Science* **265** (1994), 1599–1600.
[3] K.J. Doege, M. Sasaki, T. Kimura and Y.J. Yamada, *Biol. Chem.* **266** (1991), 894–902.
[4] J. Fritz, A.G. Katopodis, F. Kolbinger and D. Anselmetti, *Proc. Natl. Acad. Sci. USA* **95** (1998), 12283–12288.
[5] T. Fujii, Y.L. Sun, K.N. An and Z.P. Luo, *J. Biomech.* **35** (2002), 527–531.
[6] M. Grandbois, M. Beyer, M. Rief, M. Clausen-Schaumann and H.E. Gaub, *Science* **283** (1999), 1727–1730.
[7] V.C. Hascall, *J. Supramol. Struc.* **7** (1977), 101–120.
[8] A.J. Hough, Jr., in: *Osteoarthritis*, 3rd edn, *Diagnosis and Medical/Surgical Management*, R.W. Moskowitz, D.S. Howell, R.D. Altman, J.A. Buckwalter and V.M. Goldberg, eds, W.B. Saunders Company, Philadelphia, PN, 2001, pp. 69–100.
[9] J.I. Ishimaru, N. Ogi, S. Mizuno and A.N. Goss, *Osteoarthritis Cartilage* **9** (2001), 365–370.
[10] M.S.Z. Kellermayer, S.B. Smith, H.L. Granzier and C. Bustamante, *Science* **276** (1997), 1112–1116.
[11] Z.P. Luo and K.N. An, *J. Biomech.* **31** (1998), 1075–1079.
[12] Z.P. Luo, M.E. Bolander and K.N. An, *Biochem. Biophys. Res. Commun.* **232** (1997), 251–254.
[13] V.T. Moy, E. Florin and H.E. Gaub, *Science* **266** (1994), 257–259.
[14] H. Muir, *Biochem. Soc. Trans.* **11** (1983), 613–622.
[15] J.J. Parkkinen, T.P. Häkkinen, S. Savolainen, C. Wang, R. Tammi, U.M. Ågren, M.J. Lammi, J. Arokoski, H.J. Helminen and M.I. Tammi, *Histochem. Cell Biol.* **105** (1996), 187–194.
[16] J.P. Perin, F. Bonnet, C. Thurieau and P. Jolles, *J. Biol. Chem.* **262** (1987), 13269–13272.
[17] S.J. Perkins, A.S. Nealis, J. Dudhia and T.E. Hardingham, *J. Mol. Biol.* **206** (1989), 737–753.
[18] A. Ratcliffe, P.J. Beauvais, F. Saed-Nejad, W. Shurety and B. Caterson, *Agents Actions Suppl.* **39** (1993), 63–67.
[19] M. Rief, M. Gautel, F. Oesterhelt, J.M. Fernandez and H.E. Gaud, *Science* **276** (1997), 1109–1116.
[20] L.J. Sandell and T.M. Hering, in: *Osteoarthritis*, 3rd edn, *Diagnosis and Medical/Surgical Management*, R.W. Moskowitz, D.S. Howell, R.D. Altman, J.A. Buckwalter and V.M. Goldberg, eds, W.B. Saunders Company, Philadelphia, PN, 2001, pp. 115–143.

[21] J.D. Sandy and C. Verscharen, *Biochem. J.* **258** (2001), 615–626.

[22] A.L. Stout, *Biophys. J.* **80** (2001), 2976–2986.

[23] K. Svoboda and S.M. Block, *Cell* **77** (1994), 773–784.

[24] K. Svoboda and S.M. Block, *Annu. Rev. Biophys. Biomol. Struct.* **23** (1994), 247–285.

[25] L.H. Tang, J.A. Buckwalter and L.C. Rosenberg, *J. Orthop. Res.* **14** (1996), 334–339.

[26] H. Watanabe, S.C. Cheung, N. Itano, K. Kimata and Y.J. Yamada, *Biol. Chem.* **272** (1997), 28057–28065.

[27] X. Zhang and V.T. Moy, *Biophys. Chem.* **104** (2003), 271–278.

Biorheology 43 (2006) 191–200
IOS Press



# Static and dynamic compression regulate cartilage metabolism of PRoteoGlycan 4 (PRG4)

G.E. Nugent [a], T.A. Schmidt [a], B.L. Schumacher [a], M.S. Voegtline [a], W.C. Bae [a], K.D. Jadin [a] and R.L. Sah [a,b,*]

[a] *Department of Bioengineering, University of California, San Diego, La Jolla, CA, USA*
[b] *Whitaker Institute of Biomedical Engineering, University of California, San Diego, La Jolla, CA, USA*

**Abstract.** The boundary lubrication function of articular cartilage is mediated in part by molecules at the articular surface and in synovial fluid, encoded by *Prg4*. The objective of this study was to determine whether static and dynamic compression regulate PRG4 biosynthesis by cartilage explants. Articular cartilage disks were harvested to include the articular surface from immature bovines. Some disks were subjected to 24 h (day 1) of loading, followed by 72 h (days 2–4) of free-swelling culture to assess chondrocyte responses following unloading. Loading consisted of 6 or 100 kPa of static compression, with or without superimposed dynamic compression (10 or 300 kPa peak amplitude, 0.01 Hz). Other disks were cultured free-swelling as controls. PRG4 secretion into culture medium was inhibited by all compression protocols during day 1. Following unloading, cartilage previously subjected to dynamic compression to 300 kPa exhibited a rebound effect, secreting more PRG4 than did controls, while cartilage previously subjected to 100 kPa static loading secreted less PRG4. Immunohistochemistry revealed that all compression protocols also affected the number of cells expressing PRG4. The paradigm that mechanical stimuli regulate biosynthesis in cartilage appears operative not only for load bearing matrix constituents, but also for PRG4 molecules mediating lubrication.

Keywords: Mechanobiology, chondrocyte, PRG4, lubrication

## 1. Introduction

Articular cartilage functions to provide a low-friction, load bearing surface which allows the bones of diarthrodial joints to slide smoothly against each other while transmitting load [31]. Cartilage tissue has classically been divided into three zones: superficial, middle, and deep, with distinct biochemical content and organization that impart specific functions to each zone. The middle and deep zones provide load bearing, while the superficial zone mediates low-friction sliding at the cartilage surface [3]. This lubrication function is mediated in part by proteoglycan molecules synthesized from the proteoglycan 4 (*Prg4*) gene by tissues surrounding the joint cavity [11–13,27,32,33].

The *Prg4* gene encodes multiple similar proteins, which have been identified in a number of tissues and are therefore known by several names: MSF, CACP, HAPO, Lubricin, SZP, and PRG4. Megakaryocyte stimulating factor (MSF) was originally purified from urine, and stimulates platelet-forming cells [21,36]. Mutations in the PRG4 gene cause camptodactyly-arthropathy-coxa vara-pericarditis (CACP)

*Address for correspondence: Dr. Robert L. Sah, Department of Bioengineering, Mail Code 0412, 9500 Gilman Dr., La Jolla, CA 92093-0412, USA. Tel.: +1 858 534 0821; Fax: +1 858 822 1614; E-mail: rsah@ucsd.edu.

syndrome in humans [20], which results in early onset non-inflammatory joint failure [1], demonstrating the functional importance of expression of this gene *in vivo*. Hemangiopoietin (HAPO), an alternatively spliced isoform, simulates both hematopoietic progenitor and endothelial cells [19]. Lubricin, originally purified from synovial fluid [33–35], is highly expressed by cells of the synovial lining, and functions to reduce friction in latex-on-glass [11–13] and cartilage-on-glass [32,33] friction assays. Superficial zone protein (SZP) was first isolated from conditioned medium of cultured superficial zone (but not deep zone) cartilage explants [28], and also reduces friction in cartilage-on-cartilage sliding [27]. In this paper these molecules will be collectively referred to as PRG4 [10].

Biomechanical regulation of metabolism of matrix molecules such as aggrecan and collagen, by chondrocytes is well documented for both *in vivo* and *in vitro* stimulation protocols (reviewed in [7,9]). In general, static compression inhibits matrix biosynthesis, while dynamic compression and dynamic shear at certain frequencies and amplitudes can stimulate matrix metabolism in cartilage explants [14,23]. While effects of mechanical stimuli on lubricant molecules are not as widely studied, recent evidence suggests that mechanical stimuli might also regulate PRG4 metabolism. During embryonic development of the mouse elbow joint, PRG4 mRNA expression begins at the onset of joint cavitation [22], suggesting that PRG4 expression might be turned on by the initiation of loading of the articular surfaces. A similar pattern is seen during post-natal growth, as fetal bovine cartilage exhibits inconsistent PRG4 staining at the articular surface and in cells near the surface, in contrast with adult tissue, which stains intensely for PRG4 in these locations [29]. In both cases (*in utero* and *in vivo*), increased expression of PRG4 coincides with increased joint loading. Regulatory effects of mechanical loading *in vitro* have also been reported, as dynamic tension [37] and dynamic surface motion [6] can upregulate PRG4 mRNA levels in chondrocyte-seeded constructs.

While these studies implicate a role for mechanical stimuli in regulating chondrocyte expression of PRG4, the effects of such stimulation on PRG4 expression by chondrocytes within their native extracellular matrix remain unknown. Thus the objective of this study was to determine whether graded levels of static and dynamic compression regulate PRG4 metabolism in cartilage explants, assessed by (1) PRG4 secretion into culture media, and (2) depth-associated variation in localization of PRG4 expressing cells.

## 2. Methods

### 2.1. Cartilage explant

Cartilage disks were obtained as described previously [18]. Briefly, knees from 4 immature (1–3 week old) bovines were obtained from an abattoir. Under sterile conditions and with irrigation using phosphate buffered saline (PBS) supplemented with antibiotics (100 units/ml penicillin, 100 $\mu$g/ml streptomycin, and 0.25 $\mu$g/ml amphotericin B; all from Gibco BRL, Grand Island, NY), 9 mm-diameter osteochondral cores were harvested from the patellofemoral groove using Osteochondral Autograft Transfer System (Arthrex, Naples, FL). These cores were cut parallel to the articular surface with a sledge microtome (Microm, Waldorf, Germany) to obtain slices (1 mm thick) including the intact articular surface. From these slices, smaller disks (3 mm diameter) were obtained using a stainless steel dermal punch (Miltex GmbH, Tuttlingen, Germany).

### 2.2. Culture and mechanical stimulation of cartilage explants

Cartilage disks were incubated in a humidified atmosphere of 5% $CO_2$, 95% air at 37°C with medium (low-glucose Dulbecco's modified Eagle's medium, 10 mM HEPES buffer, 0.1 mM non-essential amino

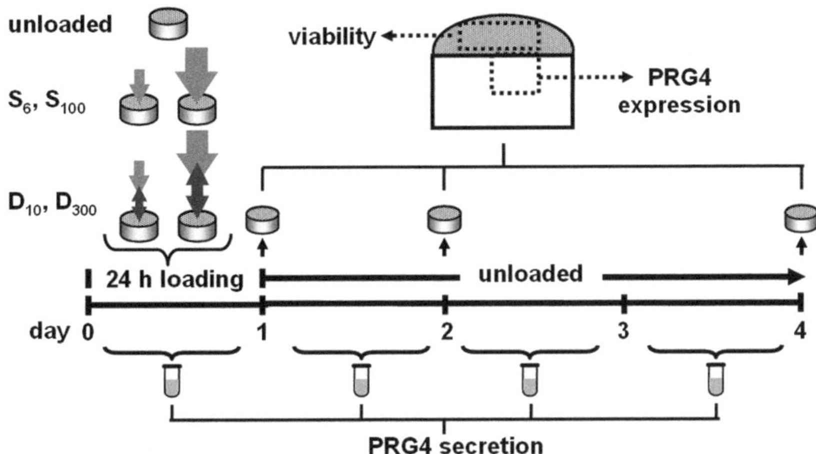


Fig. 1. Experimental design. Cartilage was loaded for 24 h (day 1), followed by 72 h (days 2, 3, 4) unloaded culture. PRG4 secretion (by indirect ELISA) into culture media was determined at the end of each 24 h period, while chondrocyte viability (by Live-Dead stain) and PRG4 expression (by immunohistochemistry) were assessed at the end of days 1, 2, and 4. Loading protocols included static compression to 6 kPa ($S_6$) or 100 kPa ($S_{100}$), alone or with superimposed dynamic (0.01 Hz) loading with peak amplitudes of 10 kPa ($D_{10}$) or 300 kPa ($D_{300}$).

acids, 0.4 mM L-proline, 2 mM L-glutamine, 100 units/ml penicillin, 100 $\mu$g/ml streptomycin, and 0.25 $\mu$g/ml amphotericin B; 1 ml per disk per day), supplemented with 10% FBS and 50 $\mu$g/ml ascorbate. Some disks were subjected to 24 h (day 1) of graded levels of continuous static (6 kPa ($S_6$) or 100 kPa ($S_{100}$)) or dynamic (3–10 kPa ($D_{10}$) or 3–300 kPa ($D_{300}$), 0.01 Hz) axial loading (Fig. 1) in load-controlled unconfined compression with fluid-impermeable polysulphone platens using custom-compression chambers described previously [18]. Static compression was applied as a step-load, and amplitudes were equal to the static offset of the dynamic waveforms. Other disks were cultured free-swelling (unloaded) as controls. All disks were then placed into free-swelling culture for a subsequent, 72 h recovery period (days 2, 3, 4) to assess chondrocyte responses following unloading (Fig. 1). Control studies, with staining for live and dead cells using calcein AM and ethidium homodimer (Live-Dead®, Molecular Probes, Eugene, OR) and threshold-based image analysis with Matlab 6.5 software (The Mathworks, Inc., Natick, MD), confirmed that the loading protocols did not affect chondrocyte viability (84–95%, $p = 0.62$). The automated image analysis software was validated by comparison with manual counts of live and dead cells.

## 2.3. Metabolic analysis

### 2.3.1. PRG4 secretion

Conditioned medium samples, collected and replaced after each 24 h period (Fig. 1), were quantitatively analyzed for PRG4 by indirect ELISA, as previously described [16] using mAb 3-A-4 (a generous gift from Dr. Bruce Caterson, University of Wales, Cardiff, UK) [29]. Briefly, samples were diluted serially, adsorbed, and then reacted with mAb 3-A-4, horseradish peroxidase-conjugated secondary antibody, and ABTS substrate, with 3 washes with PBS–0.1% Tween (Bio-Rad, Hercules, CA) between each step. A standard curve was generated from samples containing known amounts of PRG4, obtained from conditioned medium from explants from the superficial zone of bovine calf cartilage as previously described [28]. The protein-equivalent amount of PRG4 in each sample was calculated from the linear region of the standard curve (between 0.078 and 5 $\mu$g/ml of PRG4), as described in [16]. Sample PRG4 protein content is expressed normalized to cartilage surface area. Control studies indicated that cartilage disks contained PRG4 in amounts, $\sim 1$ $\mu$g/cm$^2$, that were small relative to that secreted into the medium, so that the secreted quantities were representative of biosynthesis levels. Cartilage from the middle zone (1 mm thick, from 1 to 2 mm below the articular surface) subjected to the same loading protocols did not secrete PRG4 at levels above the detection of our assay for any of the loading conditions, and therefore were not analyzed further.

### 2.3.2. PRG4 immunolocalization

The presence of PRG4 within chondrocytes was determined qualitatively in disks from cultures terminated at 24, 48, and 72 h (i.e., at the end of loading, and 24 or 36 h after unloading, Fig. 1). For the 4 h just prior to termination, these disks were incubated with medium supplemented with 1 $\mu$M monensin. Upon termination, the disks were frozen in Tissue Tek OCT (Sakura USA, Torrance, CA) and sectioned (5 $\mu$m slices) normal to the articular surface. Immunohistochemistry (IHC) was performed as described previously [16]. The sections were reacted with mAb 3-A-4, and detected with a peroxidase-based system. The stained samples were viewed to identify immunoreactive cells, indicating synthesis of PRG4. Sections probed with a non-specific mouse IgG antibody served as negative controls. Qualitative results were documented by photomicroscopy. The number of cells that were PRG4+ was determined by manual identification, and counted as a function of depth from the articular surface using custom Matlab code. Briefly, PRG4+ cells were counted in a representative 300 $\mu$m wide $\times$ 400 $\mu$m deep region of each section. Results are expressed as number of PRG4+ cells per area in each successive 50 $\mu$m bin below the articular surface, and as cumulative number of PRG4+ per area with depth from the articular surface. Since the thickness of free-swelling cartilage increased (approximately 20%, data not shown) during culture, and that of compressed tissues did not (neither during loading nor after unloading), bin sizes were adjusted for compressed tissues. For these tissues, bins near the articular surface were 35 $\mu$m thick, and increased with depth from the articular surface to 49 $\mu$m thick at 320 $\mu$m total depth (equivalent to the 400 $\mu$m depth on unloaded tissues that swelled 20%). This modified strain distribution was approximated for unconfined compression based on axial strain profiles determined previously [15] for axially compressed immature bovine cartilage explants including the articular surface.

### 2.4. Statistical analysis

Data are expressed as mean $\pm$ SEM. $N = 5$–12, from 2–4 animals. PRG4 secretion data were log-transformed for statistical analysis to improve the uniformity of variance, but results are presented untransformed (Fig. 2A) and normalized to unloaded secretion levels (Fig. 2B) for ease of interpretation. Effects of loading condition on PRG4 secretion and number of PRG4+ cells were assessed using ANOVA with repeated measures for day of culture (PRG4 secretion) and depth from articular surface (number of PRG4+ cells). Dunnett's post hoc tests were used to compare secretion and expression by control cartilage to that of each loading condition. Statistical analysis was implemented with Systat 10.2 (Systat Software, Inc., Point Richmond, CA).

## 3. Results

PRG4 secretion into culture media by cartilage disks was affected by both static and dynamic compression at both amplitudes tested (Fig. 2). Secretion was inhibited by all compression protocols (from $21.0 \pm 1.7$ $\mu$g/(cm$^2$ day) by free-swelling disks to an average $10.5 \pm 1.3$ $\mu$g/(cm$^2$ day) by loaded disks,

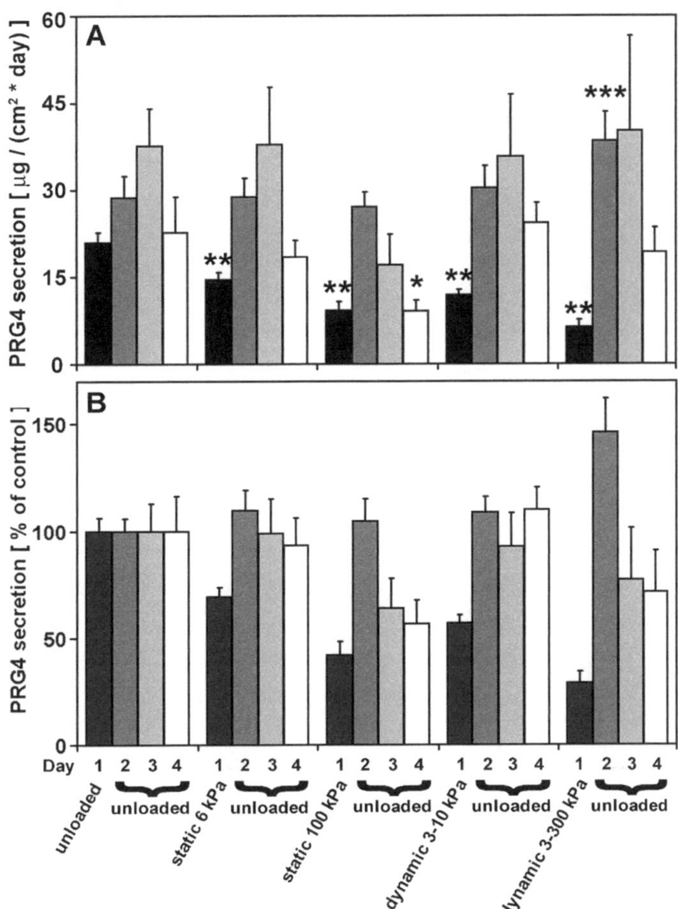

Fig. 2. Absolute (A) and normalized (B) PRG4 secretion by cartilage during 24 hours of continuous static or dynamic compression of various magnitudes (day 1), followed by 72 hours (days 2–4) of unloaded culture. Mean ± SEM, $n = 5$–12. Differences from unloaded values for a given day are represented by $^*p < 0.05$, $^{**}p < 0.01$, $^{***}p < 0.001$.

$p < 0.01$ each) during the 24 hours of continuous static and dynamic loading (day 1). Following unloading (day 2), secretion rates returned to control levels (28.7 ± 3.7 $\mu g/(cm^2$ day)) for all cartilage except that in group $D_{300}$, which on average secreted 46% more than controls ($p < 0.001$). During subsequent days following unloading (days 3 & 4) all cartilage continued to secrete similar levels of PRG4 regardless of day 1 loading condition, except $S_{100}$, which on average secreted only 45% of the control level on day 3 ($p = 0.057$), and 40% on day 4 ($p < 0.05$).

Depth-associated variation in chondrocyte expression of PRG4 was also modulated by static and dynamic compression. In free-swelling tissue, many cells were PRG4+ in the top 0 to 100–200 $\mu$m (Figs 3B,H,N and 4), with very few cells deeper than 200 $\mu$m expressing PRG4. This pattern was consistent throughout the duration of culture (Fig. 4). In contrast, during the 24 h loading period (day 1, Figs 3B–F, 4A,B), all compressed tissues had more PRG4+ cells deeper than 200 $\mu$m from the articular surface than did controls, a trend appearing opposite to the decreased secretion of PRG4 by compressed tissues during loading. During the 24 h following unloading (Figs 3H–L, 4C,D), the number of PRG4+ cells below 200 $\mu$m remained elevated over control levels for all previously loaded cartilage, except that previously statically loaded to 100 kPa, which had very few PRG4+ cells, even in the top 100 $\mu$m. By

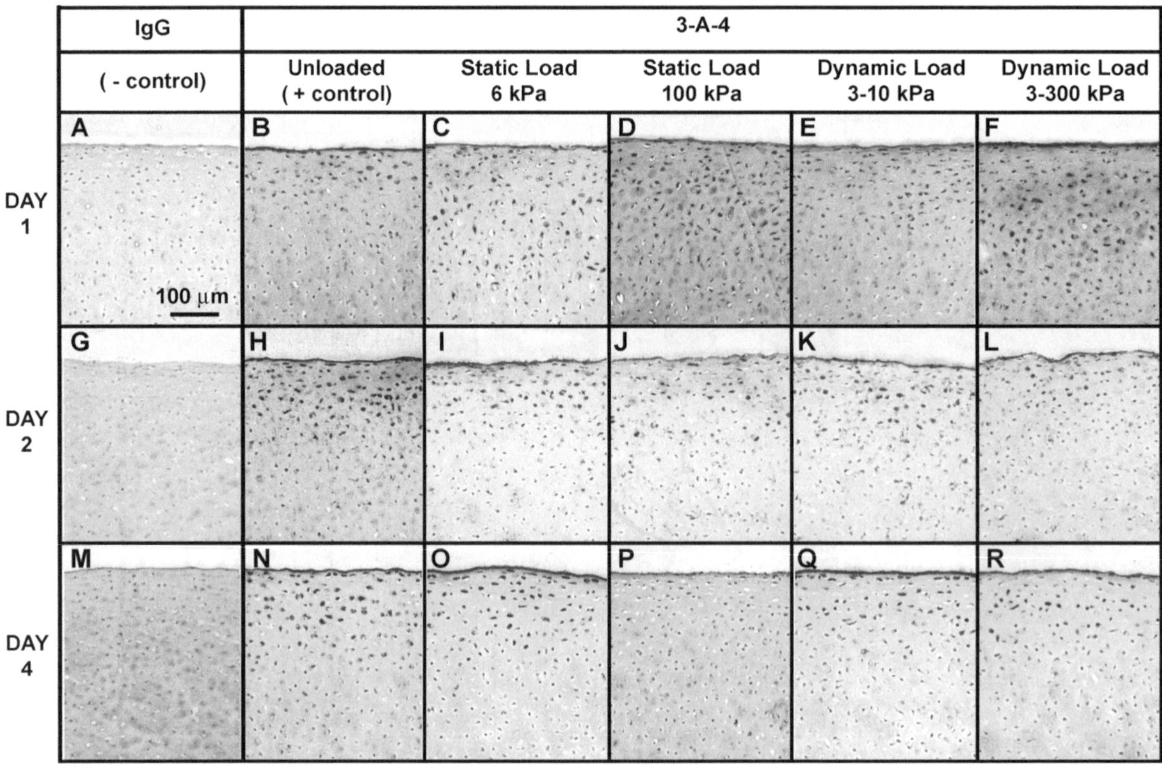

Fig. 3. PRG4 immunolocalization in tissue subjected to 24 hours of continuous static or dynamic loading of various magnitudes, immediately following loading (A–F), 1 day after removal of load (G–L), and 3 days after removal of load (M–R). IgG labeled controls (A,G,M) show PRG4 negative surfaces and cells, while unloaded controls labeled with 3-A-4 antibody (B,H,N) show PRG4 positive surfaces and PRG4 positive cells in the top 100 $\mu$m.

day 4, the profile of PRG4 expressing cells was again similar to that of controls (PRG4+ cells only in the top $\sim$ 100 $\mu$m) for all loaded groups (Figs 3N–R, 4E,F), except in $S_{100}$, where very few cells were PRG4+ anywhere in the tissue (Figs 3P, 4E,F), consistent with the return of PRG4 secretion to control levels all by groups except $S_{100}$, which secreted less PRG4 than controls. Chondrocytes below 400 $\mu$m of depth did not express PRG4 under any of the imposed conditions (data not shown). IgG controls were appropriately PRG4 negative (Fig. 2A,G,M) for all conditions.

The total number of PRG4+ chondrocytes (deepest bins, Fig. 4B) in the area counted was higher ($p < 0.01$) in cartilage subjected to $S_{100}$ and $D_{300}$ than in unloaded cartilage during day 1. Cartilage subjected to $S_6$ and $D_{10}$ showed similar, though not statistically significant trends. All loading groups exhibited total numbers of PRG4+ cells statistically similar to those of controls on day 2 (Fig. 4D) and day 4 (Fig. 4F), except cartilage subjected to $S_{100}$, which had fewer ($p < 0.01$) PRG4+ cells than controls.

## 4. Discussion

These results demonstrate marked magnitude- and time-dependent regulatory effects of mechanical stimuli on cartilage biosynthesis of PRG4 lubricant molecules. PRG4 secretion was inhibited by all compression protocols during the 24 h loading period (Fig. 2), but chondrocyte response following

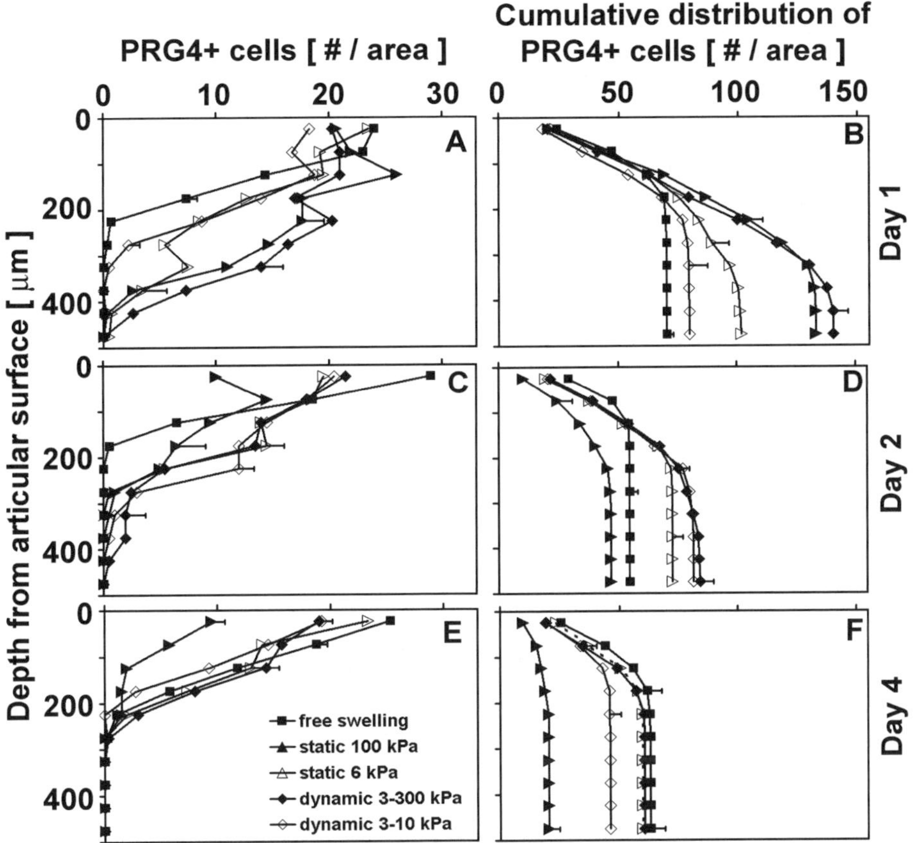


Fig. 4. Depth-associated variation in chondrocyte expression of PRG4 in unloaded and loaded cartilage, represented as number of cells per area (A,C,E), and as cumulative number of PRG4+ cells with depth from surface (B,D,F), on day 1 (A,B), 2 (C,D), and 4 (E,F). Mean ± SEM, with representative SEM shown for each loading condition, $n = 5$–12.

unloading depended on previous loading condition, with secretion returning to ($S_6$ and $D_{10}$), exceeding ($D_{300}$), or remaining below ($S_{100}$) control levels. All compression protocols increased the number of PRG4+ cells below 200 $\mu$m during loading, with expression patterns more similar to those of controls following unloading, except for cartilage previously subjected to $S_{100}$, in which expression remained low throughout the duration of culture (Figs 3, 4). Thus the paradigm that mechanical stimuli regulate the synthesis, assembly, and release of molecules in cartilage tissue appears operative not only for aggrecan and collagen constituents involved in load bearing, but also for PRG4 molecules that mediate lubrication.

These results support the hypothesis that chondrocyte metabolism depends on local microenvironment, which is determined by a complex combination of imposed chemical and mechanical stimuli. Previous work has shown that PRG4 synthesis and expression by chondrocytes in culture are markedly regulated by chemical factors. Inclusion of fetal bovine serum (FBS) and ascorbate in culture media upregulates PRG4 secretion levels [26]. Certain growth factors and cytokines that are present at high levels in injury and arthritis also affect PRG4 metabolism. For example, transforming growth factor-$\beta$1 (TGF-$\beta$1) increases PRG4 mRNA synthesis by chondrocytes cultured in monolayer [2], as well as protein secretion by chondrocytes in explants [25] and 3D constructs [5]. Conversely, interleukin-1 (IL-1$\alpha$) has inhibitory effects on PRG4 mRNA [5] and protein [25] synthesis. These results, combined with the present work, demonstrate that both chemical and mechanical stimuli can independently up- or down-regulate PRG4 metabolism by chondrocytes.

Compression protocols were selected based on previous work [18], in which $D_{300}$ and $S_{100}$ regulated biosynthesis of matrix molecules for immature bovine cartilage explants of this geometry. While 0.01 Hz is not a loading frequency experienced in common daily activities such as running or walking, it is within the range of frequencies that consistently upregulate matrix biosynthesis over a range of loading amplitudes [23] throughout 3 mm diameter cartilage disks, whereas compression at 0.1 Hz increases biosynthesis only in the outer radial periphery [14]. In addition, continuous passive motion treatments *in vivo* use frequencies (0.025 Hz) of this order of magnitude [24]. The load magnitudes used here are low to intermediate within the range used typically for stimulating cartilage explants [4,8,17,30], but since $D_{300}$ and $S_{100}$ had significant effects on matrix biosynthesis in previous studies [18], their effects on PRG4 metabolism in this explant loading configuration were deemed relevant. Lower level static and dynamic loading protocols ($S_6$, $D_{10}$) were also employed to determine whether the effects of compression were dose- (magnitude-) dependent. Compression with impermeable platens may have decreased the ability of PRG4 to exit the tissue into the culture medium. However, if this was the only effect resulting in decreased secretion during loading, presumably all of the PRG4 synthesized during this time would be free to exit the tissue upon unloading, resulting in day 2 secretion levels high enough to make up for the decrease seen in day 1, which did not occur (Fig. 2).

These results are consistent with the paradigm that in general, static compression to a high enough level inhibits cartilage biosynthesis, while dynamic compression can have stimulatory effects, and expand this paradigm to include regulation of PRG4 lubricant biosynthesis. The findings also expand upon the results from a recent study [6] in which chondrocyte-seeded polyurethane constructs were subjected to compressive stimulation to 10, 20, or 40% peak strain. PRG4 mRNA expression was decreased somewhat by loading to 10% peak strain whereas higher level loading did not affect expression. Together, this result and the current data suggest that compressive regulation of PRG4 secretion may occur at the translation level, while transcription remains relatively unchanged. In contrast, in that same study Grad et al did not detect PRG4 in conditioned medium from compressed or unloaded constructs, though this discrepancy may be due to differences in the sensitivity of the ELISA method used here, and the Western blot assay following purification steps used in that study.

As IHC results provide snapshots in time of the pattern of chondrocyte PRG4 expression, while PRG4 detected in medium samples represents an average of secretion over 1 day, it is not surprising that increased secretion concurred with increased expression by deeper chondrocytes (Fig. 4A,C,E), and total number of PRG4+ cells in the areas counted (Fig. 4B,D,F) in some cases, but not in others. A possible explanation results from the qualitative nature of IHC methods: even though compression clearly stimulates cells deeper in the tissue to express PRG4, the quantity of PRG4 produced may in fact be very small relative to that produced by more superficial chondrocytes, and thus would not affect the quantity measured in the media. Alternatively, PRG4 released from cells deeper in the tissue may get trapped in the tissue as matrix molecule concentrations increase with depth from the surface, and may prevent large molecules such as PRG4 from being easily transported out of the tissue.

Nevertheless, the marked increase in expression by deeper cells due to compression suggests that these cells are receiving some signal that unloaded cells at this depth do not receive. Since the deformation experienced by cartilage explants under compression is not homogeneously distributed throughout the depth of the tissue [15], with highest strain at the articular surface that decreases rapidly with depth to approximately 400 $\mu$m, perhaps the chondrocytes alter PRG4 synthesis in response to strain above a certain threshold. This would be consistent with the fact chondrocytes deeper than 400 $\mu$m did not express PRG4 under any conditions tested.

Understanding the role of mechanical stimuli in regulating PRG4 metabolism could have implications for tissue engineering strategies that aim to replicate lubrication function at the surface of cartilaginous constructs formed *in vitro*, and for understanding the role of mechanical stimuli in regulating the processes of wear and degeneration. If PRG4 accumulation in the synovial fluid determines its function as a boundary lubricant at the articular surface, it is important to understand the contribution by cartilage, and how this contribution can be regulated by mechanical factors. Loss of PRG4 molecules in the synovial fluid, due to decreased secretion by the surrounding tissues, could lead to tissue damage or even degeneration, so mechanical stimuli that increase PRG4 secretion could help maintain a healthy joint, while those that decrease secretion could result in damage to the articular surface.

## Acknowledgments

Arthritis Foundation, NASA, NIH, NSF, Whitaker Foundation (pre-doctoral fellowship to G.E.N.).

## References

[1] S.A. Bahabri, W.M. Suwairi, R.M. Laxer, A. Polinkovsky, A.A. Dalaan and M.L. Warman, The camptodactyly-arthropathy-coxa vara-pericarditis syndrome: clinical features and genetic mapping to human chromosome 1, *Arthritis Rheum.* **41** (1998), 730–735.

[2] P.D. Benya, B. Qiao and S.R. Padilla, Synthesis of superficial zone protein/lubricin is synergistically stimulated by tgf-beta and adenoviral expression of tak1a in rabbit articular chondrocytes, *Trans. Orthop. Res. Soc.* **28** (2003), 135.

[3] J. Buckwalter, E. Hunziker, L. Rosenberg, R. Coutts, M. Adams and D. Eyre, Articular cartilage: composition and structure, in: *Injury and Repair of the Musculoskeletal Soft Tissues*, S.L.-Y. Woo and J.A. Buckwalter, eds, American Academy of Orthopaedic Surgeons, Park Ridge, IL, 1988, pp. 405–425.

[4] N. Burton-Wurster, M. Vernier-Singer, T. Farquhar and G. Lust, Effect of compressive loading and unloading on the synthesis of total protein, proteoglycan, and fibronectin by canine cartilage explants, *J. Orthop. Res.* **11** (1993), 717–729.

[5] C.R. Flannery, C.E. Hughes, B.L. Schumacher, D. Tudor, M.B. Aydelotte, K.E. Kuettner and B. Caterson, Articular cartilage superficial zone protein (SZP) is homologous to megakaryocyte stimulating factor precursor and Is a multifunctional proteoglycan with potential growth-promoting, cytoprotective, and lubricating properties in cartilage metabolism, *Biochem. Biophys. Res. Commun.* **254** (1999), 535–541.

[6] S. Grad, C.R. Lee, K. Gorna, S. Gogolewski, M.A. Wimmer and M. Alini, Surface motion upregulates superficial zone protein and hyaluronan production in chondrocyte-seeded three-dimensional scaffolds, *Tissue Eng.* **11** (2005), 249–256.

[7] A.J. Grodzinsky, M.E. Levenston, M. Jin and E.H. Frank, Cartilage tissue remodeling in response to mechanical forces, *Annu. Rev. Biomed. Eng.* **2** (2000), 691–713.

[8] F. Guilak, B.C. Meyer, A. Ratcliffe and V.C. Mow, The effects of matrix compression on proteoglycan metabolism in articular cartilage explants, *Osteoarthritis Cartilage* **2** (1994), 91–101.

[9] F. Guilak, R.L. Sah and L.A. Setton, Physical regulation of cartilage metabolism, in: *Basic Orthopaedic Biomechanics*, V.C. Mow and W.C. Hayes, eds, Raven Press, New York, 1997, pp. 179–207.

[10] S. Ikegawa, M. Sano, Y. Koshizuka and Y. Nakamura, Isolation, characterization and mapping of the mouse and human PRG4 (proteoglycan 4) genes, *Cytogenet. Cell Genet.* **90** (2000), 291–297.

[11] G.D. Jay, Characterization of a bovine synovial fluid lubricating factor. I. Chemical, surface activity and lubricating properties, *Connect. Tissue Res.* **28** (1992), 71–88.

[12] G.D. Jay, K. Haberstroh and C.-J. Cha, Comparison of the boundary-lubricating ability of bovine synovial fluid, lubricin, and Healon, *J. Biomed. Mater. Res.* **40** (1998), 414–418.

[13] G.D. Jay and B.-S. Hong, Characterization of a bovine synovial fluid lubricating factor. II. Comparison with purified ocular and salivary mucin, *Connect. Tissue Res.* **28** (1992), 89–98.

[14] Y.J. Kim, R.L. Sah, A.J. Grodzinsky, A.H.K. Plaas and J.D. Sandy, Mechanical regulation of cartilage biosynthetic behavior: physical stimuli, *Arch. Biochem. Biophys.* **311** (1994), 1–12.

[15] T.J. Klein, M. Chaudhry, W.C. Bae and R.L. Sah, Depth-dependent compressive modulus in immature native and tissue engineered articular cartilage, *Trans. Orthop. Res. Soc.* **29** (2004), 170.

[16] T.J. Klein, B.L. Schumacher, T.A. Schmidt, K.W. Li, M.S. Voegtline, K. Masuda, E.J.-M.A. Thonar and R.L. Sah, Tissue engineering of stratified articular cartilage from chondrocyte subpopulations, *Osteoarthritis Cartilage* **11** (2003), 595–602.

[17] T. Larsson, R.M. Aspden and D. Heinegard, Effects of mechanical load on cartilage matrix biosynthesis *in vitro*, *Matrix* **11** (1991), 388–394.

[18] K.W. Li, A.K. Williamson, A.S. Wang and R.L. Sah, Growth responses of cartilage to static and dynamic compression, *Clin. Orthop.* **391S** (2001), 34–48.

[19] Y.J. Liu, S.H. Lu, B. Xu, R.C. Yang, Q. Ren, B. Liu, B. Li, M. Lu, F.Y. Yan, Z.B. Han and Z.C. Han, Hemangiopoietin, a novel human growth factor for the primitive cells of both hematopoietic and endothelial cell lineages, *Blood* **103** (2004), 4449–4456.

[20] J. Marcelino, J.D. Carpten, W.M. Suwairi, O.M. Gutierrez, S. Schwartz, C. Robbins, R. Sood, I. Makalowska, A. Baxevanis, B. Johnstone, R.M. Laxer, L. Zemel, C.A. Kim, J.K. Herd, J. Ihle, C. Williams, M. Johnson, V. Raman, L.G. Alonso, D. Brunoni, A. Gerstein, N. Papadopoulos, S.A. Bahabri, J.M. Trent and M.L. Warman, CACP, encoding a secreted proteoglycan, is mutated in camptodactyly-arthropathy-coxa vara-pericarditis syndrome, *Nat. Genet.* **23** (1999), 319–322.

[21] D.M. Merberg, L.J. Fitz, P. Temple, J.W. Giannotti, P. Murtha, M. Fitzgerald, J. Scaltreto, K. Kelleher, K. Preissner, R. Kriz, K. Jacobs and K.A. Turner, Comparison of vitronectin and megakaryocyte stimulating factor, in: *Biology of Vitronectins and Their Receptors*, K.T. Preissner, S. Rosenblatt, C. Kost, J. Wegerhoff and D.F. Mosher, eds, Elsevier Science Publishers, New York, 1993, pp. 45–52.

[22] D.K. Rhee, J. Marcelino, M. Baker, Y. Gong, P. Smits, V. Lefebvre, G.D. Jay, M. Stewart, H. Wang, M.L. Warman and J.D. Carpten, The secreted glycoprotein lubricin protects cartilage surfaces and inhibits synovial cell overgrowth, *J. Clin. Invest.* (2005).

[23] R.L. Sah, Y.J. Kim, J.H. Doong, A.J. Grodzinsky, A.H.K. Plaas and J.D. Sandy, Biosynthetic response of cartilage explants to dynamic compression, *J. Orthop. Res.* **7** (1989), 619–636.

[24] R.B. Salter, D.F. Simmonds, B.W. Malcolm, E.J. Rumble, D. MacMichael and N.D. Clements, The biological effect of continuous passive motion on the healing of full-thickness defects in articular cartilage, *J. Bone Joint Surg. Am.* **62-A** (1980), 1232–1251.

[25] T.A. Schmidt, B.L. Schumacher, E.H. Han, T.J. Klein, M.S. Voegtline and R.L. Sah, Chemo-mechanical coupling in articular cartilage: IL-1a and TGF-$\beta$1 regulate chondrocyte synthesis and secretion of lubricin/superficial zone protein, in: *Physical Regulation of Skeletal Repair*, R.K. Aaron and M.E. Bolander, eds, American Academy of Orthopaedic Surgeons, Chicago, 2004 (in press).

[26] T.A. Schmidt, B.L. Schumacher, T.J. Klein, M.S. Voegtline and R.L. Sah, Synthesis of proteoglycan 4 by chondrocyte subpopulations in cartilage explants, monolayer cultures, and resurfaced cartilage cultures, *Arthritis Rheum.* **50** (2004), 2849–2857.

[27] T.A. Schmidt, B.L. Schumacher, G.E. Nugent, N.S. Gastelum and R.L. Sah, PRG4 contributes to a "sacrificial layer" mechanism of boundary lubrication of articular cartilage, *Trans. Orthop. Res. Soc.* **30** (2005), 900.

[28] B.L. Schumacher, J.A. Block, T.M. Schmid, M.B. Aydelotte and K.E. Kuettner, A novel proteoglycan synthesized and secreted by chondrocytes of the superficial zone of articular cartilage, *Arch. Biochem. Biophys.* **311** (1994), 144–152.

[29] B.L. Schumacher, C.E. Hughes, K.E. Kuettner, B. Caterson and M.B. Aydelotte, Immunodetection and partial cDNA sequence of the proteoglycan, superficial zone protein, synthesized by cells lining synovial joints, *J. Orthop. Res.* **17** (1999), 110–120.

[30] J. Steinmeyer, S. Knue, R.X. Raiss and I. Pelzer, Effects of intermittently applied cyclic loading on proteoglycan metabolism and swelling behaviour of articular cartilage explants, *Osteoarthritis Cartilage* **7** (1999), 155–164.

[31] R.A. Stockwell, *Biology of Cartilage Cells*, Cambridge University Press, New York, 1979, 316 pp.

[32] D.A. Swann, R.B. Hendren, E.L. Radin, S.L. Sotman and E.A. Duda, The lubricating activity of synovial fluid glycoproteins, *Arthritis Rheum.* **24** (1981), 22–30.

[33] D.A. Swann, F.H. Silver, H.S. Slayter, W. Stafford and E. Shore, The molecular structure and lubricating activity of lubricin isolated from bovine and human synovial fluids, *Biochem. J.* **225** (1985), 195–201.

[34] D.A. Swann, H.S. Slayter and F.H. Silver, The molecular structure of lubricating glycoprotein-I, the boundary lubricant for articular cartilage, *J. Biol. Chem.* **256** (1981), 5921–5925.

[35] D.A. Swann, S. Sotman, M. Dixon and C. Brooks, The isolation and partial characterization of the major glycoprotein (LGP-I) from the articular lubricating fraction of synovial fluid, *Biochem. J.* **161** (1977), 473–485.

[36] K.J. Turner, L.J. Fitz, P. Temple, K. Jacobs, D. Larson, A.C. Leary, K. Kelleher, J. Giannotti, J. Calvetti, M. Fitzgerald, M.J. Kriz, C. Ferenz, J. Grobholz, H. Fraser, K. Bean, C.R. Norton, T. Gesner, S. Bhatia, R. Kriz, R. Hewick and S.C. Clark, Purification, biochemical characterization and cloning of a novel megakaryocyte stimulating factor that has megakaryocyte stimulating activity, *Blood* **78S1** (1991), 279.

[37] M. Wong, M. Siegrist and K. Goodwin, Cyclic tensile strain and cyclic hydrostatic pressure differentially regulate expression of hypertrophic markers in primary chondrocytes, *Bone* **33** (2003), 685–693.

Biorheology 43 (2006) 201–214
IOS Press



# Intracellular mechanics and mechanotransduction associated with chondrocyte deformation during pipette aspiration

T. Ohashi [a,b], M. Hagiwara [a,c], D.L. Bader [a] and M.M. Knight [a,*]

[a] *Medical Engineering Division, Department of Engineering, Queen Mary University of London, London, UK*
[b] *Graduate School of Engineering, Tohoku University, Japan*
[c] *Graduate School of Systems Life Sciences, Kyusyu University, Japan*

**Abstract.** The present study utilised pipette aspiration and simultaneous confocal microscopy to test the hypothesis that chondrocyte deformation is associated with distortion of intracellular organelles and activation of calcium signalling. Aspiration pressure was applied to isolated articular chondrocytes in increments of 2 cm of water every 60 seconds up to a maximum of 10 cm of water. At each pressure increment, confocal microscopy was used to visualise the mitochondria and nucleus labelled with JC-1 and Syto-16, respectively. To investigate intracellular calcium signalling, separate cells were labelled with Fluo 4, rapidly aspirated to 5 cm of water and then imaged for 5 minutes at a tare pressure of 0.1 cm of water. Partial cell aspiration was associated with distortion of the mitochondrial network, elongation of the nucleus and movement towards the pipette mouth. Treatment with cytochalasin D or nocodazole produced an increase in cell aspiration indicating that both the actin microfilaments and microtubules provide mechanical integrity to the cell. When the data was normalised to account for the increased cell deformation, both actin microfilaments and microtubules were shown to be necessary for strain transfer to the intracellular organelles. Mitochondria and nucleus deformation may both be involved in chondrocyte mechanotransduction as well as cellular and intracellular mechanics. In addition, pipette aspiration induced intracellular calcium signalling which may also form part of a mechanotransduction pathway. Alternatively calcium mobilisation may serve to modify actin polymerisation, thereby changing cell mechanics and membrane rigidity in order to facilitate localised cell deformation. These findings have important implications for our understanding of cell mechanics and mechanotransduction as well as interpretation and modelling of pipette aspiration data.

Keywords: Cartilage, cytoskeleton, calcium, mitochondria, signalling, micropipette

## 1. Introduction

Physiological mechanical loading is essential for the development, health and homeostasis of articular cartilage through the control of extracellular matrix synthesis and catabolism [19]. Consequently mechanical loading has been proposed as a mechanism for stimulating the growth of tissue engineered cartilage within a controlled bioreactor environment [35,41,47]. However, the biomechanical behaviour

---

*Address for correspondence: Dr. Martin Knight, Medical Engineering Division, Department of Engineering, Queen Mary University of London, Mile End Rd., London, E1 4NS, UK. Tel.: +44 207 882 5512; E-mail: m.m.knight@qmul.ac.uk.

of living chondrocytes and the underlying mechanotransduction pathways remain unclear. Cell deformation is believed to be one of the principle primary signalling events, although other factors such as fluid shear, hydrostatic pressure, electrical streaming potentials and osmotic pressure may also be involved (for review see [46]). Cell deformation may activate intracellular mechanotransduction events associated with distortion of organelles [10,20,34,43], integrin mediated signalling [11,36,38] and calcium ($Ca^{2+}$) mobilisation [1,13,22,40,49]. However, in order to elucidate the precise intracellular signalling pathways it will be necessary to further understand the inter-relationship between cellular and intracellular mechanics and physiology.

Pipette aspiration is a well-established technique for examining cell mechanics. The length of cell aspiration into the micropipette at different aspiration pressures can be incorporated into theoretical models to estimate the apparent cell modulus [21,23,26,42,44]. The present study utilises pipette aspiration to mechanically deform individual viable chondrocytes and examine the intracellular mechanical environment and the role of the cytoskeleton in both gross cell mechanics and intracellular strain transfer. This is achieved through the use of cytochalasin D and nocodazole which disrupt actin and tubulin respectively. In addition, the study investigates the influence of cell deformation on potential mechanotransduction signalling events involving mitochondria and nucleus deformation and intracellular $Ca^{2+}$ signalling. The aim of the study is therefore to test the hypothesis that chondrocyte deformation during pipette aspiration is associated with distortion of intracellular organelles and activation of calcium signalling.

## 2. Materials and methods

### 2.1. Chondrocyte isolation

Articular chondrocytes were isolated from bovine metacarpal–phalangeal joints by a process of sequential enzyme digestion as previously described [40]. To review briefly, full depth cartilage from the entire proximal surface of a joint was removed under sterile conditions and digested at $37°C$ for 1 hour in 10 ml pronase (Type E, 700 units/ml, BDH Industries Ltd, Poole, UK) followed by 16 hours in 40 ml or collagenase (Type XI, 100 units/ml, Sigma-Aldrich, Poole, UK). Enzymes were prepared in supplemented Dulbecco's Minimal Essential Medium (DMEM, Gibco Ltd, Paisley, UK) with 20% (v/v) foetal calf serum (FCS) and sterilised by passing through a 0.22 $\mu$m pore cellulose acetate filter. The digest was filtered through a 200 $\mu$m filter and the resulting cell suspension washed three times in fresh DMEM + 20% FCS. Cell suspensions from several joints were pooled and resuspended in fresh medium.

### 2.2. Micropipette aspiration system

A micropipette aspiration system, similar to that previously described [42,44], was used in conjunction with a confocal laser scanning microscope (Ultra View, Perkin Elmer, UK). The confocal system was associated with an inverted microscope (TE eclipse, Nikon, UK) with a $\times$60/1.4 NA objective lens. A schematic diagram of the experimental configuration is shown in Fig. 1a. A syringe pump was used to precisely control the aspiration pressure which was calibrated using a water filled manometer. All experiments used blunt ended, polished glass micropipettes with a diameter of approximately 4 $\mu$m (New Objective, MA, USA). The micropipettes were coated with silicone solution (Sigmacote, Sigma, UK) to prevent cell adhesion.

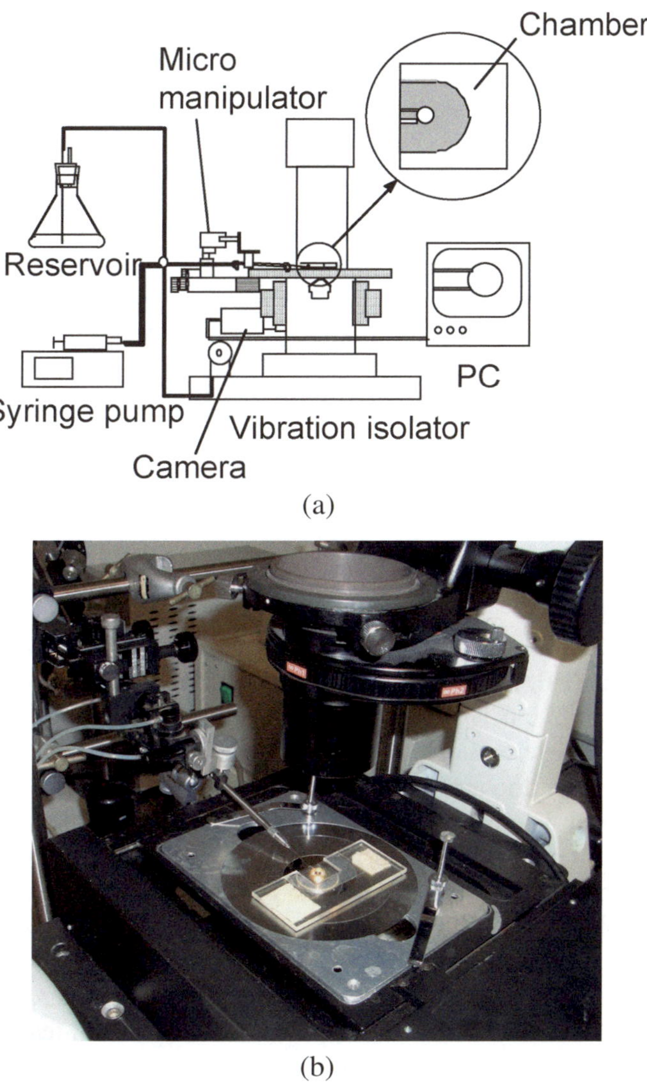


Fig. 1. Schematic diagram of pipette aspiration system (a) and photograph showing the cell chamber mounted on the stage of an inverted microscope (b).

Approximately 1 ml of cell suspension was placed in a custom-made chamber ($2 \times 20 \times 20$ mm), which was designed to allow entry of the micropipette from the side and simultaneous microscopy of the cells (Fig. 1b). The end of the micropipette was positioned adjacent to a selected cell using 3-way micromanipulators. A tare pressure of 0.1 cm of water was then applied so that the cell was maintained in a constant position in contact with the micropipette mouth.

## 2.3. Mitochondria and nucleus deformation

Prior to pipette aspiration, isolated chondrocytes were incubated at 37°C for 30 minutes in a solution of JC-1 (5 $\mu$M, Molecular Probes) to fluorescently label the mitochondria. This was followed by 15 minutes in Syto-16 (1 $\mu$M, Molecular Probes) to label the cell nucleus. Staining solutions were pre-

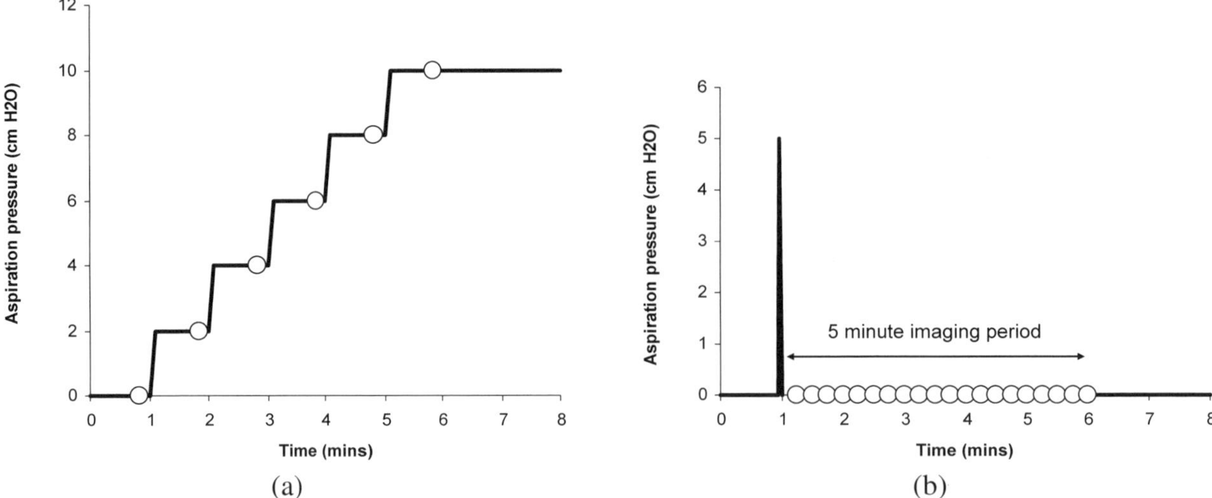

Fig. 2. Schematic diagrams indicating the pipette aspiration and imaging protocols adopted for the study of mitochondria and nucleus deformation (a) and intracellular calcium signalling (b). The open circles indicate the time of imaging.

pared in DMEM + 20% FCS. Following staining, cells were rinsed in DMEM + 20% FCS, resuspended in Phosphate Buffered Saline (PBS, Sigma, UK) and subjected to pipette aspiration. Aspiration pressure was applied in increments of 2 cm $H_2O$ every 60 seconds to a maximum of 10 cm $H_2O$ as shown schematically in Fig. 2a. Cells were allowed to equilibrate for 50 seconds at each pressure increment. Confocal images of the mitochondria and nucleus were then obtained at laser excitations of 488 nm and 564 nm respectively. Confocal z-series were made throughout the thickness of the cell with a nominal step size of 0.5 $\mu$m. These z-series were reconstructed to create a single image with an x–y resolution of 0.1 $\mu$m. In addition, brightfield images of the cell and micropipette were recorded at each pressure increment. Non-aspirated control cells were held at the tare pressure and visualised in an identical manner every 60 seconds over a 5 minute period.

At each pressure increment, the aspiration length of the cell into the micropipette was measured from the brightfield image. For each cell, measurements of the cell aspiration length were normalised to the internal diameter of the micropipette and plotted against aspiration pressure. A linear regression was fitted to the data and the gradient used to estimate the apparent Young's modulus of the cell based on the following well-established theoretical model [44]:

$$\text{Cell Young's modulus,} \quad E = \frac{3\phi(\eta)}{2\pi}\left(\frac{\Delta P}{L/a}\right) \quad \text{with } \eta = \frac{b-a}{a},$$

where $a$ and $b$ are the inner and outer radii of the micropipette, respectively. The parameter, $\phi(\eta)$ is termed the wall function with a value of 2.1 for the micropipettes used in this study.

To investigate the role of the cytoskeleton in cellular and intracellular mechanics, cells were pretreated with Cytochalasin D (2 $\mu$M, 3 hrs, Sigma, UK) and nocodazole (10 $\mu$g/ml, 5 mins, Sigma, UK) to disrupt the actin microfilaments and microtubules respectively. The procedure was repeated for samples of 10–20 cells for each treatment group.

Analysis of mitochondrial distortion was based on co-localisation in a similar technique to that used to examine cytoskeletal movement [24]. At each time interval for both aspirated and non-aspirated cells, confocal software (Ultra View, Perkin Elmer, UK) was used to calculate the percentage co-localisation relative to the initial image at $t = 0$ min. The end of the micropipette was used as a fixed reference point. Movement and/or distortion of the mitochondrial network was reflected by a decrease in co-localisation. Nucleus deformation was quantified by the percentage change in nucleus diameter parallel to the axis of the micropipette with a positive value indicating an elongation of the nucleus. Nucleus movement was given by the percentage change in distance from the micropipette, such that a positive value represented movement toward the pipette mouth.

## 2.4. Calcium signalling

Isolated chondrocytes in suspension were labelled for 1 hour at 37°C with the fluorescent intracellular $Ca^{2+}$ probe, Fluo-4 AM (5 $\mu$M, Molecular Probes). The cell suspension was transferred to the pipette aspiration chamber on the inverted microscope (Fig. 1). Individual isolated chondrocytes was subjected to a transient aspiration pressure of 5 cm of $H_2O$ applied over 2 seconds as shown schematically in Fig. 2b. The aspiration pressure was returned to the tare pressure and the cell imaged every 4 seconds for a 5 minute period. Laser excitation was set at 488 nm with fluorescent emission detected above 500 nm.

$Ca^{2+}$ signalling behaviour was analysed using UltraView software (Perkin Elmer, UK) associated with the confocal system. A circular region of interest (ROI) was created around the cell and the mean intensity within the ROI calculated at each time point. The intensity was plotted against time to reveal the temporal variation in intracellular $Ca^{2+}$ concentration. The procedure was repeated yielding a sample of 15 cells with a further 15 non-aspirated control cells imaged over an identical 5 minute period.

## 3. Results

### 3.1. Mitochondria and nucleus deformation

Representative brightfield and fluorescence confocal images of the nuclei and mitochondria are shown in Fig. 3 for both an aspirated cell and a non-aspirated control cell. Cell aspiration was associated with movement and/or distortion of the mitochondrial network, particularly close to the site of contact with the micropipette. This was reflected by a loss of co-localisation from the initial non-aspirated state with increasing aspiration pressure, as shown in Fig. 4a for a representative cell. The mean mitochondrial co-localisation was significantly lower in cells aspirated to cells 1 cm of $H_2O$ and above, compared to in non-aspirated control cells over the same time period (Fig. 4b).

Cell aspiration at pressures of 2 cm of $H_2O$ and above resulted in nucleus deformation. This was characterised by a statistically significant increase in nucleus diameter parallel to the axis of the micropipette compared to that in non-aspirated cells (Fig. 5a). Consequently at an aspiration pressure of 10 cm of $H_2O$, the nuclei exhibited a mean elongation of approximately 7%. Aspiration also caused movement of the nucleus towards the pipette mouth. For aspiration pressures of 2 cm of $H_2O$ and above, the percentage change in nucleus position was greater than that which occurred in non-aspirated cells at the same time period, the differences being statistically significant (Fig. 5b).

### 3.2. Cell mechanics and cytoskeletal disruption

For all cells tested, the aspiration length measured from the brightfield images, was linearly related to the aspiration pressure enabling the calculation of cell modulus. Untreated cells had a mean apparent

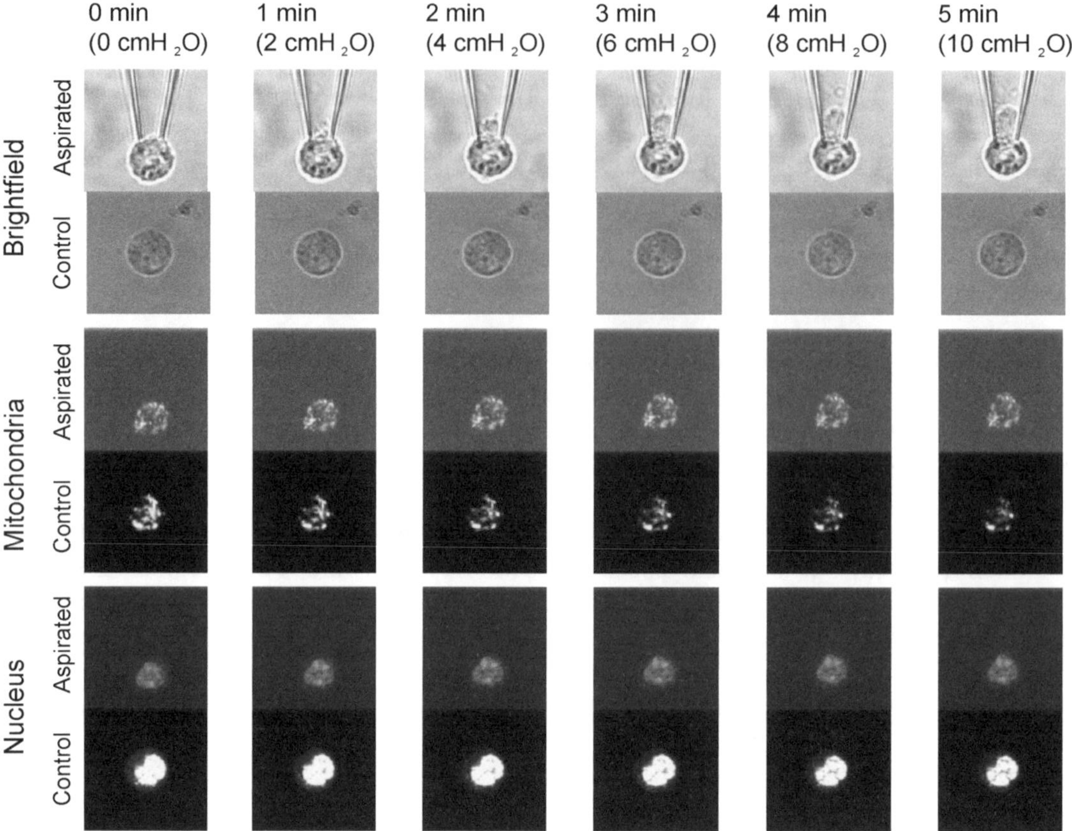



Fig. 3. Brightfield and confocal images of 2 representative cells, one subjected to incremental aspiration and the other a non-aspirated control cell visualised over the same time period. For confocal imaging, cells were incubated in JC-1 and Syto-16 to label the mitochondrial and nucleus respectively. Scale bar represents 10 $\mu$m.

Young's modulus value of approximately 1 kPa (Fig. 6). Treatment with cytochalasin D or nocodazole, induced an increase in cell aspiration length associated with a statistically significant decreases in Young's modulus (Fig. 6).

Treatment with cytochalasin D resulted in statistically significant increases in mitochondrial distortion (Fig. 7a), nucleus deformation (Fig. 8a) and nucleus displacement (Fig. 8b). By contrast, nocodazole had no such effects on the movement or distortion of these intracellular organelles.

In order to determine the influence of the actin and tubulin cytoskeleton on intracellular strain transfer it was necessary to normalise the data to account for the increased deformation of the cytoskeletal disrupted cells. For each cell, the values of percentage changes in nucleus diameter and nucleus position were divided by the aspiration length. Similarly the normalised percentage mitochondrial co-localisation was calculated as follows

$$\text{Normalised mitochondrial co-localisation} = 100 - \left( \frac{100 - \text{M}_{\text{CL}}}{\text{L}_{\text{A}}} \right).$$

Where $\text{M}_{\text{CL}}$ is the percentage mitochondrial co-localisation and $\text{L}_{\text{A}}$ is the aspiration length. In addition, only data associated with aspiration lengths within the range seen in untreated cells (0–4 $\mu$m) was used. For normalised data, it was shown that treatment with both cytochalasin D and nocodazole

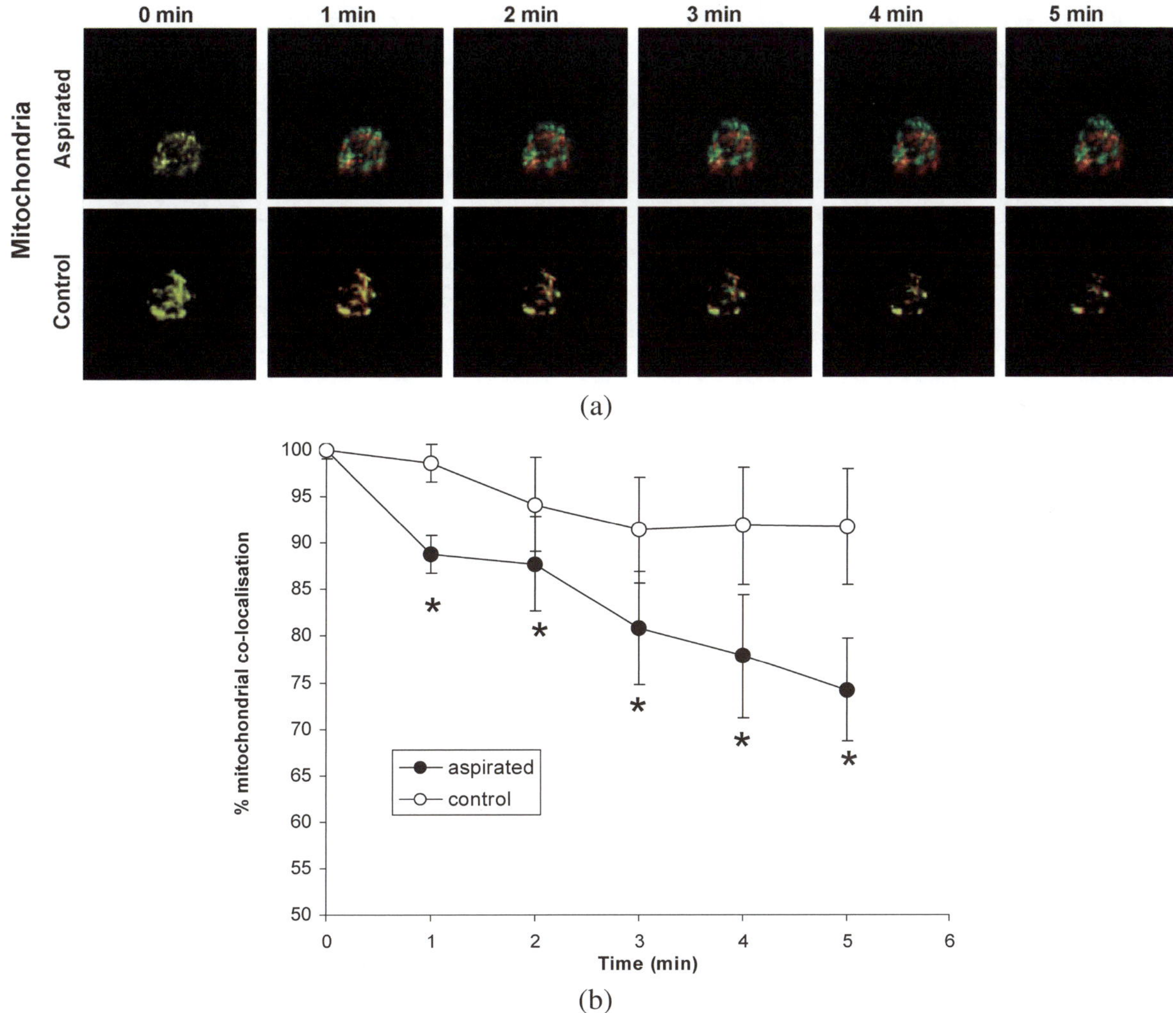

Fig. 4. Co-localisation analysis of mitochondrial distortion. Confocal co-localisation images showing 2 representative cells, one subjected to incremental aspiration and the other a non-aspirated control cell visualised over the same time period (a). The initial image at time $t = 0$ is coloured green and subsequent images are coloured red. Areas of co-localisation appear yellow. Mean percentage co-localisation for samples of aspirated and non-aspirated cells (b). Error bars indicate SEM for $n = 10$–$15$ cells. Statistically significant differences between the aspirated and time matched non-aspirated cells are indicated at $p < 0.05$ (*).

resulted in mitochondrial co-localisation values that were significantly greater than in untreated cells indicating a reduction in mitochondrial distortion (Fig. 7b). Similarly, the normalised data indicated that both cytochalasin D and nocodazole produced statistically significant reductions in nucleus deformation (Fig. 8c) and nucleus displacement (Fig. 8d).

### 3.3. Calcium signalling

Transient pipette aspiration to 5 cm of $H_2O$ induced a characteristic $Ca^{2+}$ transient in 80% of cells tested. The $Ca^{2+}$ response consisted of a rapid rise in intracellular $Ca^{2+}$ concentration followed by a slower return to baseline (Fig. 9). $Ca^{2+}$ mobilisation appeared frequently to initiate around the area of the cell closest to the mouth of the micropipette.

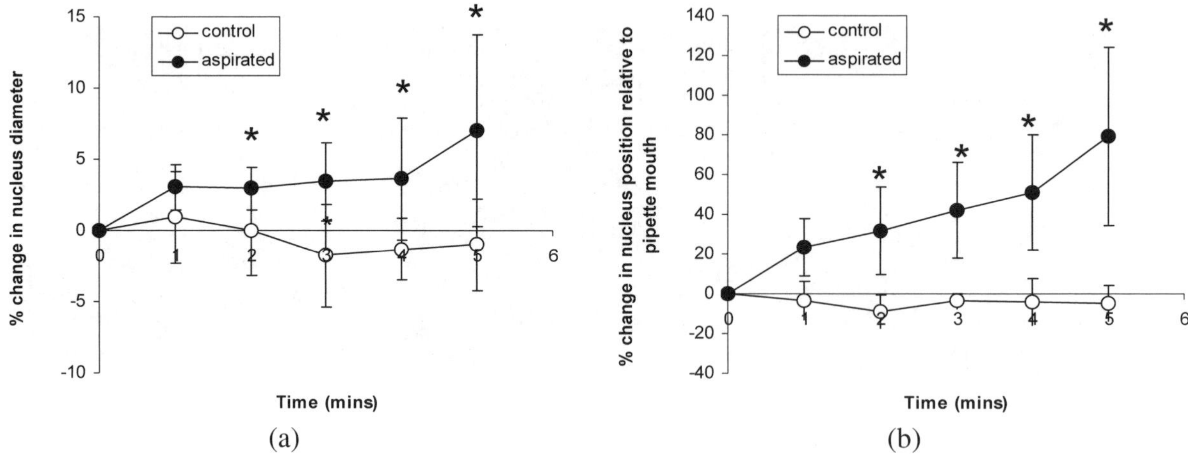

Fig. 5. Percentage change in nucleus diameter (a), and nucleus position (b), for aspirated and non-aspirated cells. A positive change in nucleus position indicates movement towards the pipette mouth. Values represent sample means for $n = 10$–15 cells. Error bars indicate SEM. Statistically significant differences between the aspirated and time matched non-aspirated cells are indicated at $p < 0.05$ ($^*$).

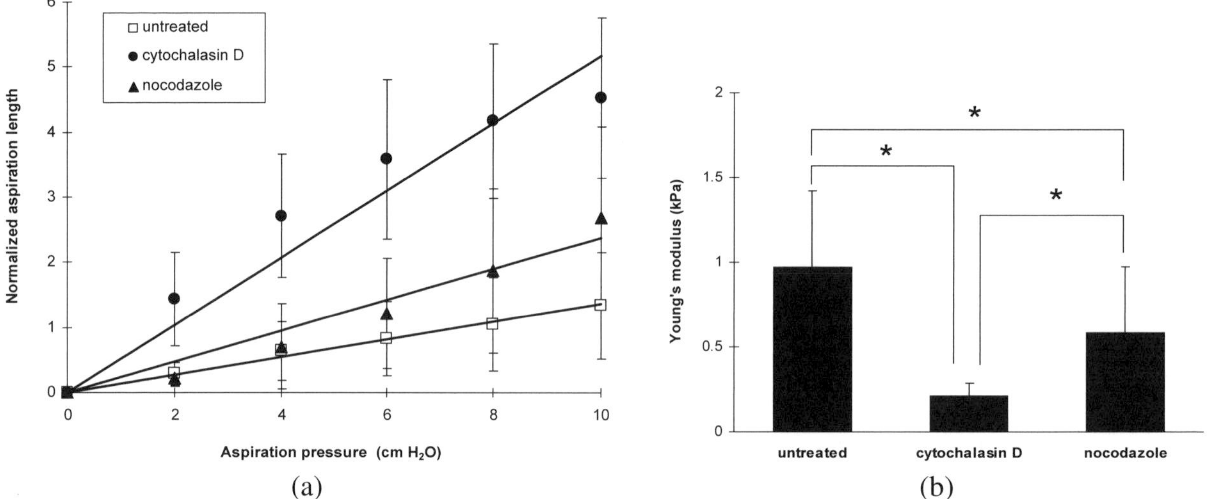

Fig. 6. Relationship between aspiration pressure and mean normalised aspiration length for untreated cells and those treated with cytochalasin D or nocodazole. Linear models have been fitted to the data. Values represent sample means with error bars showing standard deviations.

## 4. Discussion

The present study examines the influence of micropipette aspiration on intracellular mechanics and mechanotransduction events within articular chondrocytes. Pipette aspiration is a widely used experimental technique for estimating the mechanical properties, in particular, the apparent elastic Young's modulus of living cells [26]. Isolated bovine articular chondrocytes were subjected to controlled mechanical loading in the form of partial aspiration into a micropipette. This was performed on the stage of a confocal microscope system enabling simultaneous visualisation of fluorescently labelled intracellular organelles and physiological calcium imaging within living cells. Cytoskeletal disruptive agents were

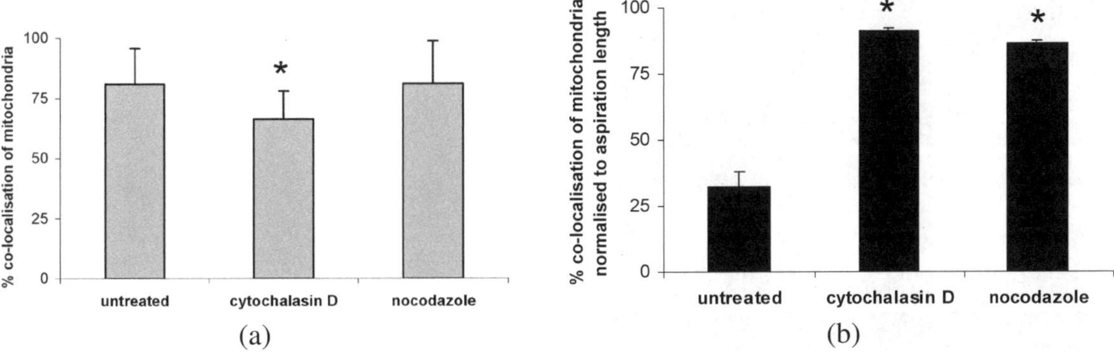

Fig. 7. Mitochondrial co-localisation for samples of cell treated with either cytochalasin D or nocodazole and subjected to pipette aspiration (a). Mitochondrial co-localisation data normalised to pipette aspiration length for aspiration lengths <4 $\mu$m (b). Values represent sample means for $n = 10$–15 cells. Error bars indicate SEM. Statistically significant differences from untreated cells are indicated at $p < 0.05$ (*).

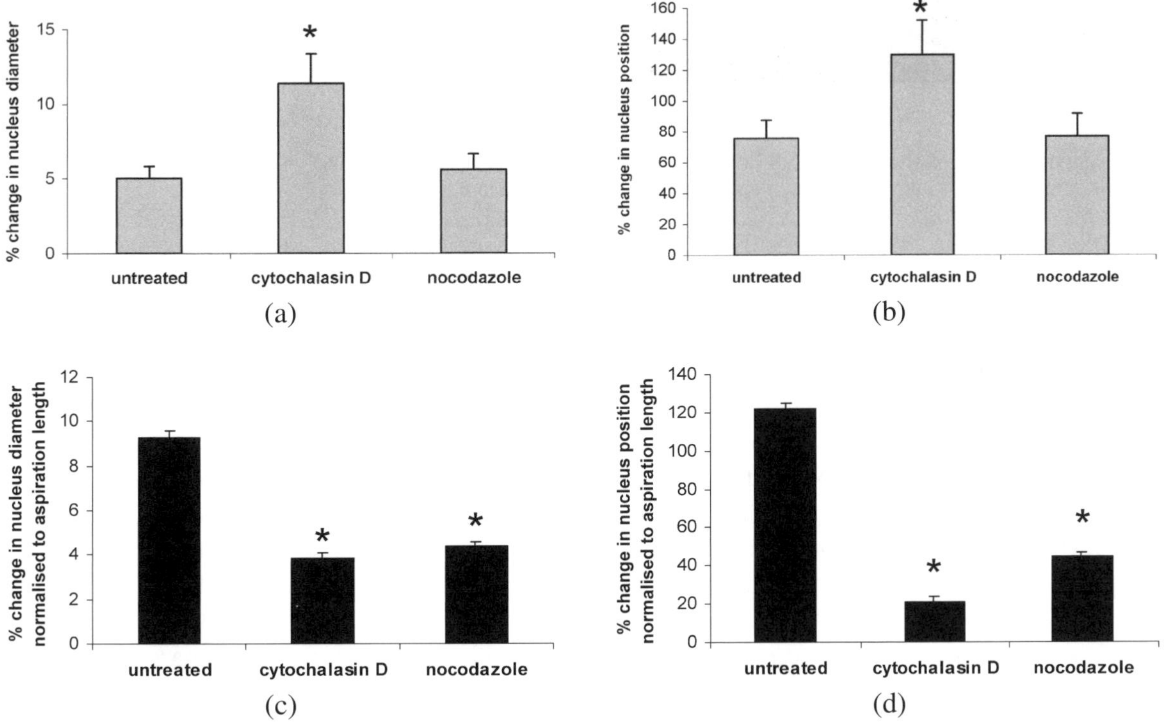

Fig. 8. Nucleus deformation (a) and displacement (c) for samples of cell treated with either cytochalasin D or nocodazole and subjected to pipette aspiration. Data for nucleus deformation and displacement has been normalised to pipette aspiration length for aspiration lengths <4 $\mu$m (b and d respectively). Values represent sample means for $n = 10$–15 cells. Error bars indicate SEM. Statistically significant differences from untreated cells are indicated at $p < 0.05$ (*).

used to examine the specific role of actin microfilaments and microtubules in intracellular mechanics and cytoplasmic strain transfer.

Measurements of cell deformation into a micropipette at aspiration pressures up to 10 cm of water were used to estimate cell Young's modulus based on the theoretical model proposed by Theret et al. [44]. The

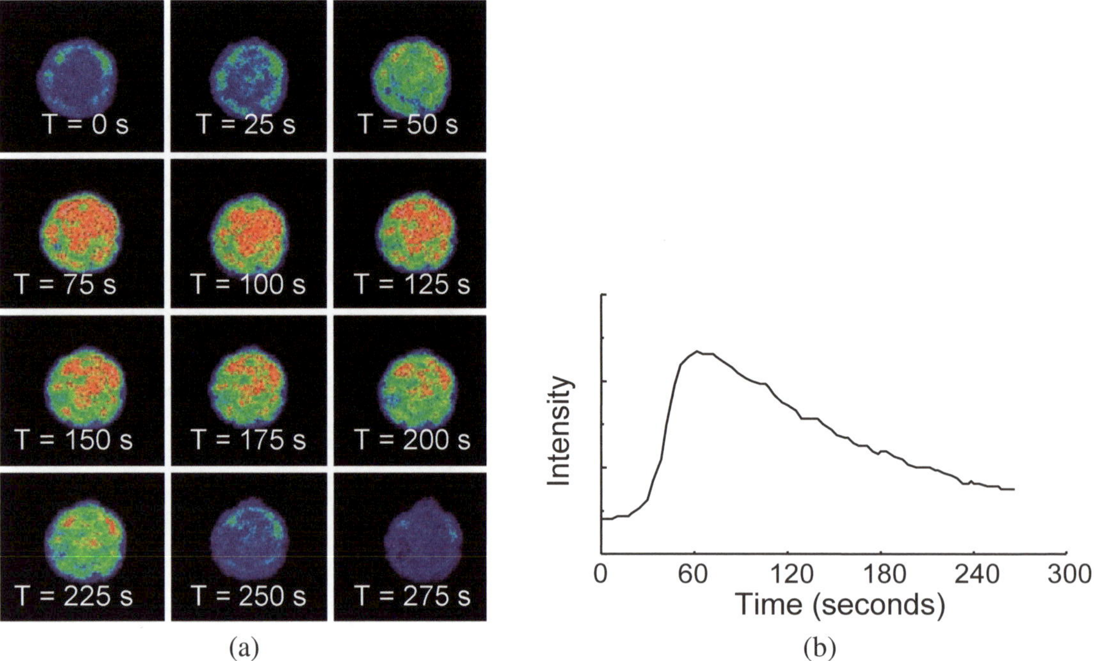



Fig. 9. Sequence of images showing Fluo-4 labelled chondrocyte following pipette aspiration (a) and the associated temporal changes in intracellular calcium indicating a characteristic calcium transient (b).

resulting mean modulus of 1 kPa is in broad agreement with that previous reported for isolated articular chondrocytes using a similar pipette aspiration technique [21,29]. However, it is noticeable less than the value of approximately 3 kPa estimated from measurements of gross cell deformation and relaxation in low modulus agarose [6] or alginate [32] gels. This difference may reflect the fact that pipette aspiration causes localised distortion of the cell cortex, which may reflect the local rigidity of this area more than the gross cell stiffness.

In the present study the mitochondria and nucleus were fluorescently labelled and visualised at increasing aspiration pressures up to 10 cm $H_2O$ over a 5 minute period. The mitochondria within articular chondrocytes formed a three-dimensional fibrous structure or reticulum similar to that observed in other cell types [12]. This mitochondrial reticulum exhibited a small degree of motility and remodelling over a 5 minute imaging period possibly associated with actin–myosin driven movement [33] as well as movement of the underlying cytoskeleton [14,25]. For this reason it was necessary to compare measurements of mitochondrial movement associated with pipette aspiration with that in non-aspirated cells visualised over the same time period. An estimate of the degree of mitochondrial movement and distortion was made from measurements of percentage co-localisation with the initial non-aspirated image at $t = 0$ minutes. It is recognised, however, that this method is not as precise or accurate as techniques, such as digital image correlation, which the authors have previously used to determine the spatial variability in mitochondrial distortion [30]. However, based on co-localisation analysis it was shown that aspiration of the cell into the micropipette resulted in substantial movement and distortion of the mitochondrial reticulum (Fig. 4a). The movement and distortion was significantly greater than that which occurred in non-aspirated cells over an identical 5 minute period (Fig. 4b). Similar mitochondrial deformation has also been reported associated with chondrocyte deformation in both cartilage explants [43] and agarose constructs [30]. Indeed mitochondrial deformation may be involved in chondrocyte mechanotransduction through the strain-medicated release of oxygen free radicals [2].

Visualisation of nucleus deformation indicated that pipette aspiration resulted in both elongation of the nucleus in the direction of aspiration and movement towards the pipette mouth (Fig. 5). Nucleus deformation has also been observed in chondrocytes compressed in cartilage explants [10,20,43] and agarose or alginate constructs [32,34] and may be involved in mechanotransduction through direct alterations in gene transcription.

Measurements of mitochondria and nucleus movement indicated that localised aspiration of the cell surface was transferred throughout the cytoplasm indicating a degree of mechanical continuity. Previous studies of cells cultured in monolayer demonstrate similar strain transfer mediated by the cytoskeleton [28]. Furthermore, the distortion of the nucleus and mitochondria reticulum suggest that these intracellular structures may be tethered in some way to the cell periphery and therefore provide a degree of mechanical resistance to localised cell deformation. These factors are largely overlooked in the models of chondrocyte mechanics, which treat the cell as either a fluid filled elastic cortical shell [18], a solid elastic or viscoelastic sphere [44], or more recently a bi- or tri-phasic poroviscoelastic sphere [5].

Within isolated non-adherent articular chondrocytes, the actin microfilaments form a dense cortical network without obvious stress fibres while the microtubules and vimentin intermediate filaments form a fibrous structure throughout the cytoplasm [27,34]. To further investigate the role of the cytoskeleton in strain transfer across the cytoplasm, cells were aspirated following treatment with cytochalasin D and nocodazole which disrupt actin microfilaments and microtubules, respectively. Disruption of these cytoskeletal networks independently resulted in increased cell aspiration length (Fig. 6a) indicating a reduction in apparent gross cell Youngs' modulus (Fig. 6b). These results are similar to those previously reported for articular chondrocytes [45]. The increased cell aspiration in cells treated with cytochalasin D was associated with increased movement and distortion of the mitochondria and nucleus (Figs 7a and 8a, b). Therefore in order to determine the influence of the cytoskeleton on strain transfer, it was necessary to normalize the data to aspiration length over the range of aspiration lengths measured in untreated cells (0–4 $\mu$m). Based on this normalised data, disruption of either the actin microfilaments or the microtubules reduced the effective strain transfer such that for both disruptive agents there was a reduction in mitochondrial (Fig. 7b) and nucleus (Fig. 8c, d) movement and distortion. Microtubules are known to co-localise with the mitochondria [3,7,30] and connect the cell cortex to the nucleus [27,34]. Thus microtubule involvement in intracellular mechanics and strain transfer appears consistent with the structural organisation of this cytoskeletal protein. However, this relationship is less clear for actin microfilaments which, in isolated chondrocytes in suspension, form a cortical mesh without cytoplasmic stress fibres [17,31,34].

The present study found that rapid transient pipette aspiration of viable chondrocytes induced a subsequent rise in intracellular $Ca^{2+}$ concentration followed by a slow return to baseline (Fig. 9). The associated transient level of cell aspiration is unlikely to exceed 0.5 $\mu$m based on the viscoelastic response of chondrocytes to a static aspiration pressure of 5 cm $H_2O$ (unpublished data). However, a study by Erickson et al. [16] reported that complete aspiration of chondrocytes, over several minutes into larger diameter micropipettes, did not induce $Ca^{2+}$ influx. This suggests that the activation of $Ca^{2+}$ signalling in response to mechanical deformation may be dependent on strain rate and/or strain magnitude. The response to pipette aspiration was abolished by treatment with either thapsigargin or EGTA (unpublished data) suggesting that $Ca^{2+}$ signalling requires the release of $Ca^{2+}$ from the intracellular stores as well as influx of extracellular $Ca^{2+}$.

Intracellular $Ca^{2+}$ is a ubiquitous second messenger widely involved in mechanotransduction in a variety of cell types [8,9]. Previous studies on isolated chondrocytes have reported $Ca^{2+}$ transients in response to a variety of mechanical stimuli including hydrostatic pressure [37], fluid flow [49], osmotic challenge [15,39,48], cell indentation [13,22] and compression in agarose [40]. However, the percentage of cells responding and the temporal characteristics of any response varies between different stimuli. For example localised deformation resulting from pipette aspiration, as in this study, or pipette indentation [13,22], produced an instantaneous single $Ca^{2+}$ transient in 80 to 100% of cells whereas gross deformation in agarose up-regulated $Ca^{2+}$ signalling in a smaller percentage of cells [40]. Thus it is likely that different signalling pathways are activated by different mechanical stimuli. It has been proposed that the rapid $Ca^{2+}$ response associated with pipette aspiration or indentation is caused by localised rupture of the cell membrane due to the very high local strains. However, this is perhaps unlikely given the characteristic transient nature of the response and the return to baseline (Fig. 9) rather than a monotonic $Ca^{2+}$ increase as expected for permeabilised cells. Thus $Ca^{2+}$ signalling may be part of a mechanotransduction pathways leading to alterations in cell function. Alternatively, $Ca^{2+}$ mobilisation may initiate alterations in actin cytoskeleton through the action of calmodulin and other calcium sensitive cytoskeletal-associated proteins [4,17]. Changes in actin polymerisation may lead to a reduction in cell modulus, as demonstrated by the response to cytochalasin D (Fig. 6). Such mechanically-induced changes in cell mechanics may be necessary to enable the extensive localized deformation of the actin-rich cell cortex which occurs during pipette aspiration.

In conclusion, the present study demonstrates that localised pipette aspiration of viable chondrocytes was associated with movement and distortion of the mitochondrial reticulum and the nucleus. This indicates, not only significant biomechanical connectivity within the cell, but also the possibility of mechanotransduction events associated with deformation of these organelles. Both actin microfilaments and microtubules played a significant role in gross cell mechanics as well as intracellular mechanics and strain transfer. In addition pipette aspiration triggered a transient rise in intracellular $Ca^{2+}$, possibly as part of a mechanotransduction pathway or as a means of regulating actin-dependent cell mechanics in response to the changing mechanical environment. Hence pipette aspiration provides a useful tool for examining cellular and intracellular mechanics and mechanotransduction at a single cell level. However, it is difficult to separate the biomechanical and physiological properties of the cell such that the cell mechanical properties may be altered by the pipette aspiration technique used to estimate them.

## Acknowledgements

Dr. Knight is funded on an EPSRC Advanced Research Fellowship. The authors wish to thank the meat inspectors at Dawn Cardington, UK, for the supply of bovine metacarpal–phalangeal joints. The authors would also like to thank Messrs. Kazushi Ito, Junichi Yamazaki, and Yousuke Ueki for their assistance in analyzing data. This work was conducted as part of an academic fellowship program for Dr Ohashi, at Queen Mary University of London, supported from the Ministry of Education, Culture, Sports, Science, and Technology, Japan.

## References

[1] A.I. Alford, C.E. Yellowley, C.R. Jacobs and H.J. Donahue, Increases in cytosolic calcium, but not fluid flow, affect aggrecan mRNA levels in articular chondrocytes, *J. Cell. Biochem.* **90** (2003), 938–944.

[2] M.H. Ali, D.P. Pearlstein, C.E. Mathieu and P.T. Schumacker, Mitochondrial requirement for endothelial responses to cyclic strain: implications for mechanotransduction, *Am. J. Physiol. Lung Cell Mol. Physiol.* **287** (2004), L486–L496.

[3] F. Appaix, A.V. Kuznetsov, Y. Usson, L. Kay, T. Andrienko, J. Olivares, T. Kaambre, P. Sikk, R. Margreiter and V. Saks, Possible role of cytoskeleton in intracellular arrangement and regulation of mitochondria, *Exp. Physiol.* **88** (2003), 175–190.

[4] A. Arai, K. Kyozuka and T. Nakazawa, Cytoplasmic $Ca^{2+}$ oscillation coordinates the formation of actin filaments in the sea urchin eggs activated with phorbol ester, *Cell Motility and the Cytoskeleton* **42** (1999), 27–35.

[5] F.P. Baaijens, W.R. Trickey, T.A. Laursen and F. Guilak, Large deformation finite element analysis of micropipette aspiration to determine the mechanical properties of the chondrocyte, *Ann. Biomed. Eng.* **33** (2005), 494–501.

[6] D.L. Bader, T. Ohashi, M.M. Knight, D.A. Lee and M. Sato, Deformation properties of articular chondrocytes: a critique of three separate techniques, *Biorheology* **39** (2002), 69–78.

[7] E.H. Ball and S.J. Singer, Mitochondria are associated with microtubules and not with intermediate filaments in cultured fibroblasts, *Proc. Natl. Acad. Sci. USA* **79** (1982), 123–126.

[8] M.J. Berridge, M.D. Bootman and P. Lipp, Calcium – a life and death signal, *Nature* **395** (1998), 645–648.

[9] M.J. Berridge, P. Lipp and M.D. Bootman, The versatility and universality of calcium signalling, *Nat. Rev. Mol. Cell Biol.* **1** (2000).

[10] M.D. Buschmann, E.B. Hunziker, Y.J. Kim and A.J. Grodzinsky, Altered aggrecan synthesis correlates with cell and nucleus structure in statically compressed cartilage, *J. Cell Sci.* **1996** (1996), 499–508.

[11] T.T. Chowdhury, D.M. Salter, D.L. Bader and D.A. Lee, Integrin-mediated mechanotransduction processes in TGFbeta-stimulated monolayer-expanded chondrocytes, *Biochem. Biophys. Res. Commun.* **318** (2004), 873–881.

[12] T.J. Collins, M.J. Berridge, P. Lipp and M.D. Bootman, Mitochondria are morphologically and functionally heterogeneous within cells, *EMBO J.* **21** (2002), 1616–1627.

[13] P. D'Andrea, A. Calabrese, I. Capozzi, M. Grandolfo, R. Tonon and F. Vittur, Intercellular $Ca^{2+}$ waves in mechanically stimulated articular chondrocytes, *Biorheology* **37** (2000), 75–83.

[14] T. Delhaas, S. Van Engeland, J. Broers, C. Bouten, N. Kuijpers, F. Ramaekers and L.H. Snoeckx, Quantification of cytoskeletal deformation in living cells based on hierarchical feature vector matching, *Am. J. Physiol. Cell Physiol.* **283** (2002), C639–C645.

[15] G.R. Erickson, L.G. Alexopoulos and F. Guilak, Hyper-osmotic stress induces volume change and calcium transients in chondrocytes by transmembrane, phospholipid, and G-protein pathways, *J. Biomech.* **34** (2001), 1527–1535.

[16] G.R. Erickson, R. Minkhorst and F. Guilak, The calcium transients elicited in chondrocytes as a result of exposure to osmotic pressure change are not due to simple membrane deformation, *Trans. Orthop. Res. Soc.* **555** (2001).

[17] G.R. Erickson, D.L. Northrup and F. Guilak, Hypo-osmotic stress induces calcium-dependent actin reorganization in articular chondrocytes, *Osteoarthritis Cartilage* **11** (2003), 187–197.

[18] E. Evans and B. Kukan, Passive material behavior of granulocytes based on large deformation and recovery after deformation tests, *Blood* **64** (1984), 1028–1035.

[19] A.J. Grodzinsky, M.E. Levenston, M. Jin and E.H. Frank, Cartilage tissue remodeling in response to mechanical forces, *Annu. Rev. Biomed. Eng.* **2** (2000), 691–713.

[20] F. Guilak, Compression-induced changes in the shape and volume of the chondrocyte nucleus, *Journal of Biomechanics* **28** (1995), 1529–1541.

[21] F. Guilak, W.R. Jones, P. Ting-Beall and G.M. Lee, The deformation behavior and mechanical properties of chondrocytes in articular cartilage, *Osteoarthritis and Cartilage* **7** (1999), 59–70.

[22] F. Guilak, R.A. Zell, G.R. Erickson, D.A. Grande, C.T. Rubin, K.J. McLeod and H.J. Donahue, Mechanically induced calcium waves in articular chondrocytes are inhibited by gadolinium and amiloride, *J. Orthop. Res.* **17** (1999), 421–429.

[23] M.A. Haider and F. Guilak, An axisymmetric boundary integral model for incompressible linear viscoelasticity: application to the micropipette aspiration contact problem, *J. Biomech. Eng.* **122** (2000).

[24] B.P. Helmke, D.B. Thakker, R.D. Goldman and P.F. Davies, Spatiotemporal analysis of flow-induced intermediate filament displacement in living endothelial cells, *Biophys. J.* **80** (2001), 184–194.

[25] C.L. Ho, J.L. Martys, A. Mikhailov, G.G. Gundersen and R.K. Liem, Novel features of intermediate filament dynamics revealed by green fluorescent protein chimeras, *J. Cell Sci.* **111** (1998), 1767–1778.

[26] R.M. Hochmuth, Micropipette aspiration of living cells, *J. Biomech.* **33** (2000), 15–22.

[27] B.D. Idowu, M.M. Knight, D.L. Bader and D.A. Lee, Confocal analysis of cytoskeletal organisation within isolated chondrocyte sub-populations cultured in agarose, *Histochem. J.* **32** (2000), 165–174.

[28] P.A. Janmey, The cytoskeleton and cell signalling: Component localization and mechanical coupling, *Physiological Reviews* **78** (1998), 763–781.

[29] W.R. Jones, H.P. Ting-Beall, G.M. Lee, S.S. Kelley, R.M. Hochmuth and F. Guilak, Alterations in the Young's modulus and volumetric properties of chondrocytes isolated from normal and osteoarthric human cartilage, *J. Biomech.* **32** (1999), 119–127.

[30] M.M. Knight, Z. Bonzom, D.A. Lee, E. Kimmel and D.L. Bader, Heterogeneity of deformation of cytoplasmic networks in compressed chondrocytes, *Trans. Orthop. Res. Soc.* (2004).

[31] M.M. Knight, B.D. Idowu, D.A. Lee and D.L. Bader, Temporal changes in cytoskeletal organisation within isolated chondrocytes quantified using a novel image analysis technique, *Med. Biol. Eng. Comput.* **39** (2001), 397–404.

[32] M.M. Knight, B.J. van de Breevaart, D.A. Lee, G.J. van Osch, H. Weinans and D.L. Bader, Cell and nucleus deformation in compressed chondrocyte-alginate constructs: temporal changes and calculation of cell modulus, *Biochim. Biophys. Acta* **1570** (2002), 1–8.

[33] M. Krendel, G. Sgourdas and E.M. Bonder, Disassembly of actin filaments leads to increased rate and frequency of mitochondrial movement along microtubules, *Cell Motil. Cytoskeleton* **40** (1998), 368–378.

[34] D.A. Lee, M.M. Knight, J.F. Bolton, B.D. Idowu, M.V. Kayser and D.L. Bader, Chondrocyte deformation within compressed agarose constructs at the cellular and sub-cellular levels, *J. Biomech.* **33** (2000), 81–95.

[35] R.L. Mauck, S.L. Seyhan, G.A. Ateshian and C.T. Hung, Influence of seeding density and dynamic deformational loading on the developing structure/function relationships of chondrocyte-seeded agarose hydrogels, *Ann. Biomed. Eng.* **30** (2002), 1046–1056.

[36] S.J. Millward-Sadler and D.M. Salter, Integrin-dependent signal cascades in chondrocyte mechanotransduction, *Ann. Biomed. Eng.* **32** (2004), 435–446.

[37] S. Mizuno, A novel method for assessing effects of hydrostatic fluid pressure on intracellular calcium: a study with bovine articular chondrocytes, *Am. J. Physiol. Cell Physiol.* **288** (2005), C329–C337.

[38] A. Mobasheri, S.D. Carter, P. Martin-Vasallo and M. Shakibaei, Integrins and stretch activated ion channels; putative components of functional cell surface mechanoreceptors in articular chondrocytes, *Cell Biol. Int.* **26** (2002), 1–18.

[39] S. Pritchard and F. Guilak, The role of F-actin in hypo-osmotically induced cell volume change and calcium signaling in anulus fibrosus cells, *Ann. Biomed. Eng.* **32** (2004), 103–111.

[40] S.R. Roberts, M.M. Knight, D.A. Lee and D.L. Bader, Mechanical compression influences intracellular $Ca^{2+}$ signaling in chondrocytes seeded in agarose constructs, *J. Appl. Physiol.* **90** (2001), 1385–1391.

[41] S. Saini and T.M. Wick, Concentric cylinder bioreactor for production of tissue engineered cartilage: effect of seeding density and hydrodynamic loading on construct development, *Biotechnol. Prog.* **19** (2003), 510–521.

[42] M. Sato, M.J. Levesque and R.M. Nerem, An application of the micropipette technique to the measurement of the mechanical properties of cultured bovine aortic endothelial cells, *J. Biomech. Eng.* **109** (1987), 27–34.

[43] J.D. Szafranski, A.J. Grodzinsky, E. Burger, V. Gaschen, H.H. Hung and E.B. Hunziker, Chondrocyte mechanotransduction: effects of compression on deformation of intracellular organelles and relevance to cellular biosynthesis, *Osteoarthritis Cartilage* **12** (2004), 937–946.

[44] D.P. Theret, M.J. Levesque, M. Sato, R.M. Nerem and L.T. Wheeler, The application of a homogeneous half-space model in the analysis of endothelial cell micropipette measurements, *J. Biomech. Eng.* **110** (1988), 190–199.

[45] W.R. Trickey, T.P. Vail and F. Guilak, The role of the cytoskeleton in the viscoelastic properties of human articular chondrocytes, *J. Orthop. Res.* **22** (2004), 131–139.

[46] J.P.G. Urban, The chondrocyte: A cell under pressure, *Br. J. Rheumatol.* **33** (1994), 901–908.

[47] G. Vunjak-Novakovic, B. Obradovic, S. Treppo, A.J. Grodzinsky, R. Langer and L.E. Freed, Bioreactor cultivation conditions modulate the composition and mechanical properties of tissue engineered cartilage, *J. Orthop. Res.* **17** (1999), 130–138.

[48] R.J. Wilkins, T.P. Fairfax, M.E. Davies, M.C. Muzyamba and J.S. Gibson, Homeostasis of intracellular $Ca^{2+}$ in equine chondrocytes: response to hypotonic shock, *Equine Vet. J.* **35** (2003), 439–443.

[49] C.E. Yellowley, C.R. Jacobs and H.J. Donahue, Mechanisms contributing to fluid-flow-induced $Ca^{2+}$ mobilization in articular chondrocytes, *J. Cell Physiol.* **180** (1999), 402–408.

Biorheology 43 (2006) 215–222
IOS Press



# The effect of hydrodynamic shear on 3D engineered chondrocyte systems subject to direct perfusion

Manuela T. Raimondi [a,*], Matteo Moretti [a], Margherita Cioffi [a], Carmen Giordano [b],
Federica Boschetti [a], Katia Laganà [a] and Riccardo Pietrabissa [a]

[a] *Laboratory of Biological Structure Mechanics, Department of Structural Engineering,
Politecnico di Milano, Milano, Italy*
[b] *Laboratory of Biocompatibility and Cell Cultures, Department of Chemistry, Materials
and Chemical Engineering "Giulio Natta", Politecnico di Milano, Milano, Italy*

**Abstract.** Bioreactors allowing direct-perfusion of culture medium through tissue-engineered constructs may overcome diffusion limitations associated with static culturing, and may provide flow-mediated mechanical stimuli. The hydrodynamic stress imposed on cells within scaffolds is directly dependent on scaffold microstructure and on bioreactor configuration. Aim of this study is to investigate optimal shear stress ranges and to quantitatively predict the levels of hydrodynamic shear imposed to cells during the experiments. Bovine articular chondrocytes were seeded on polyestherurethane foams and cultured for 2 weeks in a direct perfusion bioreactor designed to impose 4 different values of shear level at a single flow rate (0.5 ml/min). Computational fluid dynamics (CFD) simulations were carried out on reconstructions of the scaffold obtained from micro-computed tomography images. Biochemistry analyses for DNA and sGAG were performed, along with electron microscopy. The hydrodynamic shear induced on cells within constructs, as estimated by CFD simulations, ranged from 4.6 to 56 mPa. This 12-fold increase in the level of applied shear stress determined a 1.7-fold increase in the mean content in DNA and a 2.9-fold increase in the mean content in sGAG. In contrast, the mean sGAG/DNA ratio showed a tendency to decrease for increasing shear levels. Our results suggest that the optimal condition to favour sGAG synthesis in engineered constructs, at least at the beginning of culture, is direct perfusion at the lowest level of hydrodynamic shear. In conclusion, the presented results represent a first attempt to quantitatively correlate the imposed hydrodynamic shear level and the invoked biosynthetic response in 3D engineered chondrocyte systems.

Keywords: Tissue engineering, cartilage, mechanobiology, biosynthesis, computational fluid dynamics, simulation, porous biomaterials

## 1. Introduction

To date, tissue-engineered constructs are the first step towards the restoring of injured cartilage by cellular approaches. Mass transfer is a critical issue when culturing cells in porous scaffolds, to ensure that adequate nutrient and pH levels are maintained and to eliminate catabolites from the construct. *Ex vivo* tissue engineering processes involving articular cartilage are particularly dependent on mechanical stimuli for differentiation [9]. Direct perfusion bioreactors have been developed to culture cell/scaffold constructs under improved mass transfer and under hydrodynamic loading [3,5,8,10]. In such systems, the

*Address for correspondence: Manuela Teresa Raimondi, PhD, Dipartimento di Ingegneria Strutturale, Politecnico Di Milano, Piazza Leonardo da Vinci, 32, 20133 Milano, Italy. Tel.: +39 02 66214939; Fax: +39 02 66214939; E-mail: manuela.raimondi@polimi.it.

entire construct is directly perfused with culture medium, and the transport of oxygen and nutrients from the medium to cells occurs via both diffusion and convection.

Increasing the flow rate increases the shear stresses exerted on cells. This complexity presents serious hurdles in determining adequate flow rates and medium solute concentrations for maintaining cell viability and modulating matrix synthesis [2,7]. Flow determines a range of local fluid velocities in the scaffold, which themselves are impossible to measure. Most available models of flow through porous structures are homogeneous models, which implicitly average out microscopic details of the porous geometry to arrive at average permeability, velocity, shear stress, etc. These models have important limitations in situations involving spatially inhomogeneous micro architectures. Therefore, more detailed pore-scale computer simulations of fluid transport in tissue-engineering scaffolds populated with living cells have been recently proposed [11,12].

In this work, we first established a direct-perfusion bioreactor to engineer cartilaginous tissue starting from primary bovine chondrocytes seeded into a 3D polymer scaffold. The bioreactor is conceived in order to impose varying levels of hydrodynamic loading to constructs as the only culture variable. We, then, developed a 3D computational model to quantitatively predict the levels of hydrodynamic shear imposed to cells during the experiments. We used this system as a model to establish a quantitative relationship between the applied shear level and the stage of development of constructs, as defined by the sulfated glycosaminoglycan (sGAG) and DNA content, and by their ratio.

## 2. Materials and methods

### 2.1. Direct-perfusion bioreactor

In a direct perfusion system, the culture medium is conveyed through the porous structure of cellular constructs, thus exposing the cells located in the entire scaffold thickness both to convective solute transport and to a flow-induced mechanical stimulus. Our system is composed of 4 culture chambers, each carrying constructs shaped as discs, 1 mm in thickness, with a perfused section of 2, 3, 4 or 7 mm in diameter. Each chamber carries 3 constructs of identical size (Fig. 1). The bioreactor consists of two independent circuits. Each circuit (Fig. 2) consists of two chambers connected in series by gas permeable silicone tubing and of a de-bubbler placed between them, along with a medium reservoir. The priming volume in each circuit is 20 ml. The two circuits share a peristaltic pump (Watson-Marlow Bredel Inc., Wilmington, MA, USA). The pump is remotely controlled by a timer which automatically reverses the direction of flow at constant time intervals, to avoid cell migration and to guarantee a homogeneous nutrient distribution on both sides. At a constant value of flow rate imposed by the pump, the average fluid velocity induced within the constructs will be different for each of the four different diameter scaffold groups, i.e. higher in smaller constructs and lower in bigger constructs, depending on their section area. This experimental set-up allows to apply four different levels of hydrodynamic stimulus as the only culture variable between the four chambers.

### 2.2. Cell culture and assays

Articular cartilage was harvested aseptically from the metacarpophalangeal joints of 8-month old calves slaughtered at the abattoir. Chondrocyte isolation was carried out as described elsewhere [4]. The reagents used were from Sigma-Aldrich S.r.l. (Italy). The complete medium contained Dulbecco's Modified Eagle Medium (DMEM), sodium pyruvate, Hepes buffer, antibiotics (penicillin/streptomycin),

Fig. 1. Bioreactor chamber. (a) The chamber contains three scaffolds by means of a scaffold holder enclosed between shells which are sealed by simple pressure. (b) The chamber mounted inside the circuit.

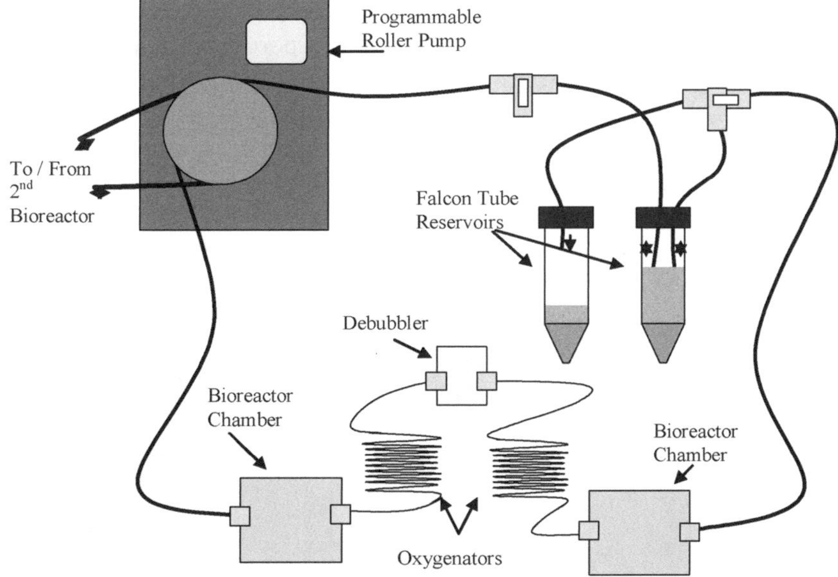



Fig. 2. Scheme of half of the hydraulic circuit. The complete bioreactor consists of two circuits identical to the one represented, mounted in parallel.

L-glutamine, L-ascorbic acid, insulin and a 10% aliquot of foetal calf serum (FCS). Viable cells were counted with trypan blue exclusion and used for scaffold seeding.

The scaffold used is DegraPol®, a biodegradable polyestherurethane foam developed at the Swiss Federal Institute of Technology (ETH) in Zurich [14]. The scaffolds employed in our study were sterile discs, 1 mm in thickness and 3, 4, 5 and 8 mm in diameter. The discs were positioned in a culture cluster and seeded with cells at a density of about 80,000 cells/mm$^3$. The discs were incubated for 24 hours at 37°C to allow for cell adhesion. The seeded constructs were, then, either placed in the bioreactor to be subjected to direct perfusion or in culture plates as static controls. The flow rate was regulated at 0.5 ml/min, the flow direction was cyclically inverted every 40 minutes. Constructs were incubated for 2 weeks, with medium completely replaced every 5 days. Three independent experiments were carried out.

At 2 weeks of culture, constructs were prepared for biochemical assays and for electron microscopy. Portions used for biochemical analysis were digested with papain and assayed for DNA content by Hoechst 33258 dye binding [13] and for sGAG content by 1.9-dimethylmethylene blue chloride [6]. Other portions of the constructs were prepared for scanning electron microscopy as previously described [12] and observed at 10 kV using a Stereoscan S260 electron microscope (Leica Cambridge Ltd., Cambridge, UK).

## 2.3. Computational model

This part of the study allowed to characterise the local fluid dynamics to which cells were exposed within the perfused constructs. We developed a simulation, at the pore scale, of culture medium flow through the scaffold used in the experiments, using computational fluid dynamics (CFD). The calculations and theory below are described in detail elsewhere [1]. Briefly, a 3D solid model of the scaffold micro-geometry was reconstructed from 250 micro-computed tomography ($\mu$CT) images (Fig. 3). This scaffold is characterised by an interconnected porosity of roughly 77% and nominal pore size of 100 $\mu$m. Cell dimension was assumed to be negligible if compared to pore size. The fluid domain was created by meshing the void volume of the 3D solid model. The commercial finite-volume code Fluent (Fluent Inc., Lebanon, NH, USA) was used to set up and solve the CFD problem. In order to quantify the fluid-dynamic environment imposed to cells in each culture chamber, simulations were run by imposing the calculated inlet velocities as boundary conditions. The inlet velocity specific to each scaffold size was calculated from the flow rate adopted in the experiments (0.5 ml/min for each bioreactor chamber

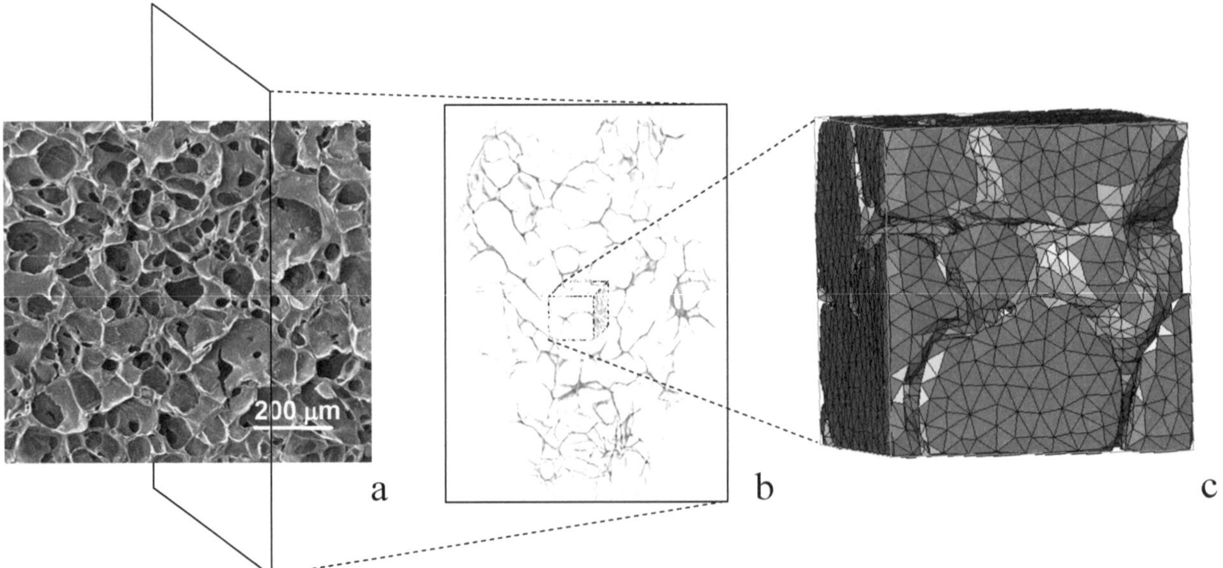


Fig. 3. Method employed to set up the CFD model. (a) SEM image of the scaffold microstructure, (b) micro-computed tomography ($\mu$CT) image acquired on the scaffold, (c) mesh of the fluid domain reconstructed for a cubic scaffold portion, 400 $\mu$m in side.

holding three scaffolds). The fluid-induced shear stresses acting at the wall of the pores were calculated and assumed as an estimate of the shear stresses acting on the membranes of the cells. The mean, mode and median shear stress values were determined.

## 3. Results

The results of the fluid dynamic simulations were analyzed at the central walls of the reconstructed fluid domain, to avoid boundary effects. The average, median and mode shear stress values calculated at the scaffold walls are shown in Table 1 for each size of perfused construct. Both the inlet fluid velocity and the median shear stresses showed a 12-fold increase between the 7 mm and the 2 mm constructs, consistently with the 12-fold reduction in perfused section area. The predicted median shear within constructs ranged from 4.6 mPa (0.046 dyn/cm$^2$) for the 7 mm constructs to 56 mPa (0.56 dyn/cm$^2$) for the 2 mm constructs.

Figure 4 shows the results of the biochemical assays at 2 weeks of culture, in terms of content in DNA and sGAG specific to construct volume, and in terms of sGAG/DNA. Values are plotted for increasing values of median shear stress and are given as the mean and standard deviation of 3 independent experiments. The mean content in DNA and in sGAG increased at increasing levels of applied shear stress. The total increase, correspondent to the 12-fold increase in the level of applied shear, was in the order of 1.7-fold and 2.9-fold for DNA and sGAG content, respectively. In contrast, the mean sGAG/DNA ratio, indicative of the biosynthesis of sGAG specific to cells, showed a tendency to decrease for increasing shear levels, although the standard deviation calculated for this parameter was quite high. The content in DNA and sGAG estimated for the non-perfused controls was comparable to that estimated for constructs subjected to a moderate regimen of hydrodynamic stimulus (i.e. a shear stress level of 25 mPa).

At the SEM, chondrocytes showed a phenotypic spherical morphology with smooth membranes in all culture conditions (Fig. 5). Cells were dispersed within large aggregates of neo-formed matrix. Surface morphology of the synthesised matrix was rougher in constructs non-perfused and in constructs cultured at lower shear level (Fig. 5a), and smoother in constructs cultured at higher shear levels (Figs 5b,c).

## 4. Discussion

The bioreactor for direct perfusion presented in this study was designed with the aim to establish a quantitative relationship between the level of hydrodynamic shear applied to 3D engineered chondrocyte systems and parameters related to the invoked biosynthetic response. Our main finding was that

Table 1

Results of the CFD simulations

| Diameter of perfused construct [mm] | Experimental flow rate [ml/min] | Construct inlet velocity [$\mu$m/sec] | Mean shear stress [mPa] | Median shear stress [mPa] | Mode shear stress [mPa] |
|---|---|---|---|---|---|
| 2 | 0.5 | 884 | 63 | 56 | 38 |
| 3 | 0.5 | 393 | 28 | 25 | 17 |
| 4 | 0.5 | 221 | 16 | 14 | 9.4 |
| 7 | 0.5 | 72 | 5.2 | 4.6 | 3.1 |

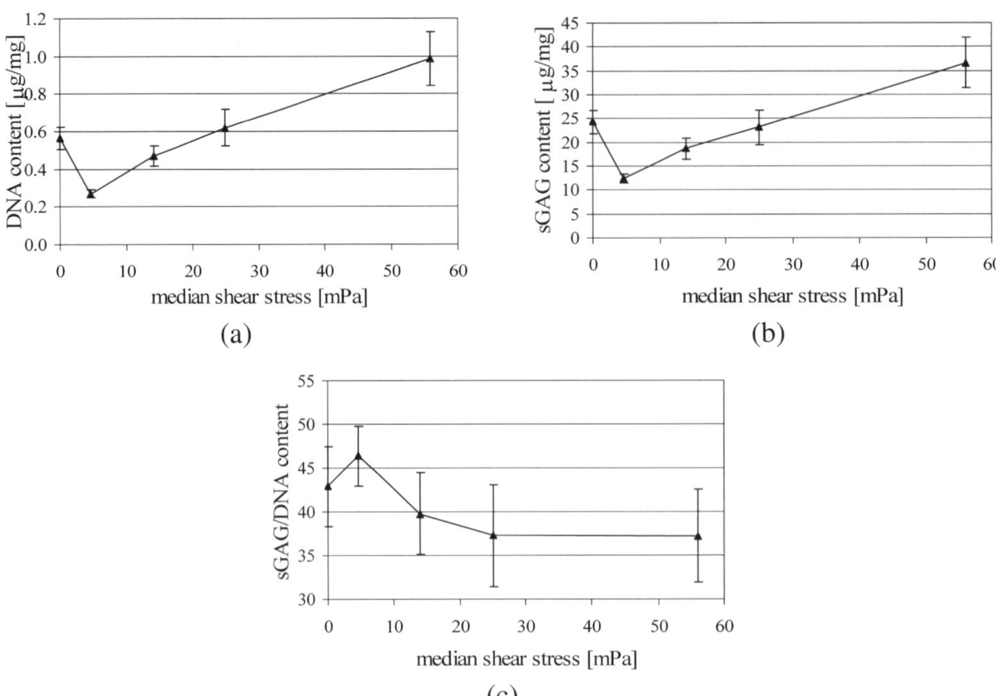

Fig. 4. Results of the biochemical assays for the chondrocyte-seeded constructs at 2 weeks of direct-perfusion under varying hydrodynamic conditions. (a) DNA content, (b) sGAG content and (c) sGAG/DNA content. Values are reported for increasing values of median shear stress (zero stress refers to non-perfused controls) and are given as mean ± standard deviation of 3 independent experiments.

at increasing levels of hydrodynamic loading, in the range 5.6 to 56 mPa, sGAG and DNA content increased after two weeks of culture, whereas sGAG specific to cells showed a tendency to decrease. Few comparisons with literature data exist. Our results are in agreement with those obtained by Saini and Wick [15], both in terms of sGAG and sGAG/cell content as a function of increasing levels of hydrodynamic loading. In the mentioned work, hydrodynamic shear was imposed at higher levels and only at the construct surface, thus only 'qualitative agreement may be expected. In most previous studies of direct-perfusion [5,8,10], constructs were cultured under a fixed fluid velocity and the level of shear induced on cells was never quantified. These studies have demonstrated an increase in cell content and matrix synthesis in perfused constructs, as compared to static controls, again in general agreement with our results. Only one study investigated the effect of direct perfusion at two levels of velocity [3]. Here, a higher fluid velocity was found to decrease cell content, as compared to a lower fluid velocity, when applied at early time points in culture. As a matter of fact, matrix content in cellular constructs is minimal at the early period of culture. It is possible that perfusion enables fluid shear to wash out cells and matrix from the scaffold, consistently with smoothening of the matrix surface observed in our study. This hypothesis could explain both the findings by Davisson [3], and our finding that content in DNA and sGAG estimated for the non-perfused controls was higher with respect to constructs subject to the lowest regimen of hydrodynamic loading (i.e. 4.6 mPa). However, cell and sGAG content in the medium were not measured so this hypothesis remains to be verified.

The bioreactor system was able to induce significantly different fluid dynamic environments within constructs cultured in the four different chambers. The magnitude of medium velocity imposed at the

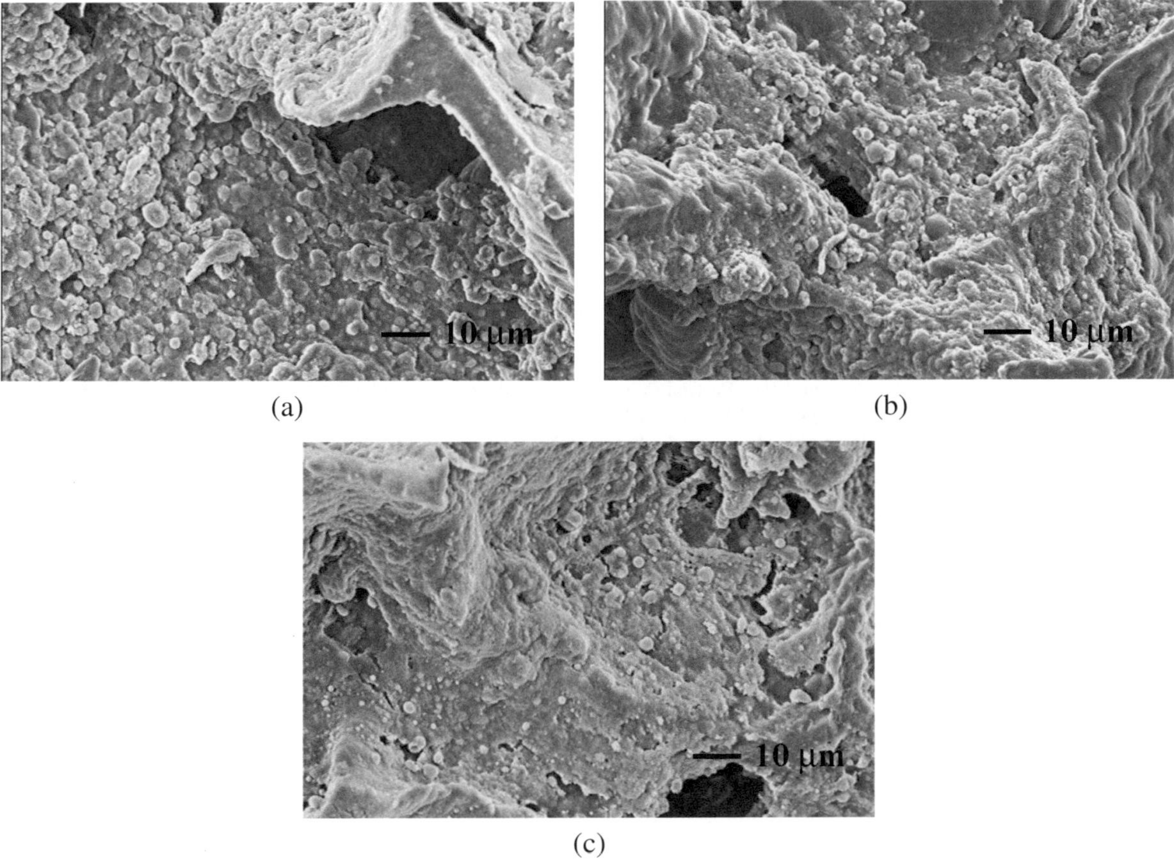

(a)          (b)

(c)

Fig. 5. SEM images of chondrocyte-seeded constructs at 2 weeks of culture under direct-perfusion. Surface morphology of the synthesised matrix varies in constructs subject to increasing levels of median hydrodynamic shear of: (a) 4.6 mPa, (b) 14 mPa and (c) 56 mPa.

construct inlet ranged from around 70 $\mu$m/sec in 7 mm constructs to around 900 $\mu$m/sec in 2 mm constructs. The level of hydrodynamic shear induced on cells within constructs, as estimated by CFD simulation, ranged from 4.6 mPa in 7 mm constructs to 56 mPa in 2 mm constructs. This level of shear is of the same order of magnitude as the level to which chondrocytes are exposed within natural articular cartilage (100 mPa for a 10 MPa joint contact pressure, as estimated by Schinagl et al. [16]).

The magnitude of shear stresses in this study was calculated from a numerical analysis of flow, which is yet to be validated. Unfortunately, local hydrodynamic shear cannot be measured within constructs. Our predicted levels of shear stress agree with other theoretical predictions [11] referred to a different 3D scaffold. Again, only a general agreement may be expected with literature data in this respect, as the local shear stresses experienced by cells can be significantly different, even for the same inlet fluid velocity, as a function of the scaffold microarchitecture.

In conclusion, the results presented here represent the first attempt to quantitatively correlate the imposed level of hydrodynamic shear and the invoked biosynthetic response in 3D engineered chondrocyte systems. Our results suggest that the optimal condition to favour sGAG synthesis in engineered constructs, at least at the beginning of culture, is direct perfusion at the lowest level of hydrodynamic shear technically achievable.

## Acknowledgements

This project is funded by Fondazione Cariplo (grant #2004.1148) and by Politecnico di Milano (grant CelTec). The DegraPol scaffolds were provided without charge by Dr. Peter Neuenschwander of ETH, Zurich.

## References

[1] M. Cioffi, F. Boschetti, M.T. Raimondi and G. Dubini, Modeling evaluation of the fluid-dynamic microenvironment in tissue-engineered constructs: A micro-CT based model, *Biotechnol. Bioeng.* **93**(3) (2006), 500–510.

[2] E.M. Darling and K.A. Athanasiou, Articular cartilage bioreactors and bioprocesses, *Tissue Eng.* **9**(1) (2003), 9–26.

[3] T. Davisson, R.L. Sah and A. Ratcliffe, Perfusion increases cell content and matrix synthesis in chondrocyte three-dimensional cultures, *Tissue Eng.* **8**(5) (2002), 807–816.

[4] O. Demarteau, D. Wendt, A. Braccini, M. Jakob, D. Schafer, M. Heberer and I. Martin, Dynamic compression of cartilage constructs engineered from expanded human articular chondrocytes, *Biochem. Biophys. Res. Commun.* **310**(2) (2003), 580–588.

[5] N.S. Dunkelman, M.P. Zimber, R.G. LeBaron, R. Pavelec, M. Kwan and A.F. Purchio, Cartilage production by rabbit articular chondrocytes on polyglycolic acid scaffolds in a closed bioreactor system, *Biotechnol. Bioeng.* **46** (1995), 299–305.

[6] R.W. Farndale, C.A. Sayers and A.J. Barrett, A direct spectrophotometric microassay for sulphated glycosaminoglycans in cartilage cultures, *Connective Tissue Research* **9** (1982), 247–248.

[7] I. Martin, D. Wendt and M. Heberer, The role of bioreactors in tissue engineering, *Trends Biotechnol.* **22**(2) (2004), 80–86.

[8] S. Mizuno, F. Allemann and J. Glowacki, Effects of medium perfusion on matrix production by bovine chondrocytes in three-dimensional collagen sponges, *J. Biomed. Mater. Res.* **56**(3) (2001), 368–375.

[9] V.C. Mow and C.C. Wang, Some bioengineering considerations for tissue engineering of articular cartilage, *Clin. Orthop.* **367** (1999), S204–S223.

[10] D. Pazzano, K.A. Mercier, J.M. Moran, S.S. Fong, D.D. DiBiasio, J.X. Rulfs, S.S. Kohles and L.J. Bonassar, Comparison of chondrogensis in static and perfused bioreactor culture, *Biotechnol. Prog.* **16**(5) (2000), 893–896.

[11] B. Porter, R. Zauel, H. Stockman, R. Guldberg and D. Fyhrie, 3-D computational modeling of media flow through scaffolds in a perfusion bioreactor, *J. Biomech.* **38**(3) (2005), 543–549.

[12] M.T. Raimondi, F. Boschetti, L. Falcone, G.B. Fiore, A. Remuzzi, M. Marazzi, E. Marinoni and R. Pietrabissa, Mechanobiology of engineered cartilage cultured under a quantified fluid dynamic environment, *Biomech. Modell. Mechanobiol.* **1** (2002), 69–82.

[13] J. Rao and W.R. Otto, Fluorimetric assay for cell growth estimation, *Anal. Biochem.* **207** (1992), 186–192.

[14] B. Saad, Y. Kuboki, M. Welti, G.K. Uhlschmid, P. Neuenschwander and U.W. Suter, DegraPol-foam: a degradable and highly porous polyesterurethane foam as a new substrate for bone formation, *Artif. Organs* **24**(12) (2000), 939–945.

[15] S. Saini and T.M. Wick, Concentric cylinder bioreactor for production of tissue engineered cartilage: effect of seeding density and hydrodynamic loading on construct development, *Biotechnol. Prog.* **19**(2) (2003), 510–521.

[16] R.M. Schinagl, M.S. Kurtis, K.D. Ellis, S. Chien and R.L. Sah, Effect of seeding duration on the strength of chondrocyte adhesion to articular cartilage, *J. Orthop. Res.* **17**(1) (1999), 121–129.

Biorheology 43 (2006) 223–233
IOS Press



# Calcium/calmodulin-dependent protein kinase II in human articular chondrocytes

A. Shimazaki [*], M.O. Wright, K. Elliot, D.M. Salter and S.J. Millward-Sadler [**],[***]
*Osteoarticular Research Group, Division of Pathology, School of Molecular and Clinical Medicine, College of Medicine and Veterinary Medicine, Edinburgh University, Medical School, Teviot Place, Edinburgh, EH8 9AG, UK*

**Abstract.** Mechanical stimuli are known to have major influences on chondrocyte function. The molecular events that regulate chondrocyte responses to mechanical stimulation have been the subject of much study. Using an *in vitro* experimental system we have identified mechanotransduction pathways that control molecular and biochemical responses of human articular chondrocytes to cyclical mechanical stimulation, and how these responses differ in cells isolated from diseased cartilage. We have previously shown that mechanical stimulation of normal articular chondrocytes leads to a cell membrane hyperpolarisation. Within 1 hour following mechanical stimulation there is an increase in aggrecan mRNA levels. These responses are mediated via $\alpha 5 \beta 1$ integrins, the neuropeptides substance P and NMDA, and the cytokine interleukin-4. In OA chondrocytes mechanical stimulation leads to cell membrane depolarisation, but no change in aggrecan mRNA at 1 hour. The depolarisation response is mediated via $\alpha 5 \beta 1$ integrins, substance P and interleukin-4, but the cells show an altered response to NMDA.

Having identified that the NMDA receptor is present in human articular cartilage and may play an important role in a chondroprotective mechanotransduction pathway, we were interested in whether other components associated with NMDA signalling may be involved in the chondrocyte mechanotransduction pathways. One such component is calcium/calmodulin-dependent protein kinase II (CaMKII).

CaMKII mediates many cellular responses to elevated $Ca^{2+}$ in a wide variety of cells and tissues. It is involved in the regulation of ion channels, cytoskeletal dynamics, gene transcription, neurotransmitter synthesis, insulin secretion, and cell division. CaMKII also shows a broad substrate specificity and is abundant in brain tissue, indicating that this kinase may play a number of roles in the functioning of the central nervous system. This kinase has been studied extensively in brain, but there is only a limited understanding of CaMKII in other tissues. CAMKII has four subunit isoforms ($\alpha, \beta, \gamma, \delta$). The $\alpha$- and $\beta$-isoforms have narrow distributions restricted mainly to neuronal tissues, but the $\gamma$- and $\delta$-isoforms are ubiquitously expressed within neuronal and non-neuronal tissues.

The aim of this study was to investigate the expression of CaMKII in normal and OA cartilage and chondrocytes, and whether this enzyme is involved in the response of chondrocytes to cyclical mechanical stimuli.

Reverse transcriptase–polymerase chain reaction (RT–PCR), using primers specific for the different CaMKII isoforms, was carried out to assess which isoforms are expressed in human articular chondrocytes. To assess whether CaMKII is expressed in human articular chondrocytes at the protein level, cultured chondrocytes were extracted and analysed by Western blotting using a pan-CaMKII antibody. Immunohistochemistry was carried out to investigate whether CaMKII is expressed by human articular chondrocytes *in vivo*. Frozen sections of normal, OA and ankle cartilage were incubated for one hour with CaMKII antibody and visualised using ABC and DAB.

To assess the role of CaMKII in the mechanotransduction responses of normal and OA chondrocytes, human normal and OA articular chondrocytes were mechanically stimulated at 0.33 Hz, or by addition of recombinant IL-4 for 20 minutes. Cell responses to these stimuli, in the absence or presence of an inhibitor of CaMKII were assessed by measuring changes in cell membrane potential or changes in relative levels of aggrecan mRNA compared with the housekeeping gene GAPDH.

---

[*]Current address: Department of Orthopaedic Surgery, Osaka City University Graduate School of Medicine, 1-4-3 Asahimachi Abeno-ku, Osaka 545-8585, Japan.

[**]Current address: Division of Laboratory and Regenerative Medicine, School of Medicine, University of Manchester, Stopford Building, Oxford Road, Manchester, M13 9PT, UK.

[***] Address for correspondence: Dr. SJ Millward-Sadler, Tel.: +44161 275 1818; Fax: +44161 275 5289; E-mail: Jane.Sadler@manchester.ac.uk.

Normal, OA, and ankle chondrocytes expressed the $\gamma$ and $\delta$ isoforms of CaMKII mRNA, but not the $\alpha$ and $\beta$ isoforms as demonstrated by RT–PCR. Western blotting showed a band at $\sim$60 kDa consistent with the expression of CaMKII. Immunohistochemistry revealed the positive staining in the middle and deep zones, but not the superficial zone, of normal, OA, and ankle cartilage.

The presence of a CaMKII inhibitor inhibits the membrane hyperpolarisation response and upregulation of aggrecan mRNA in normal chondrocytes following mechanical stimulation, but has no effect on the hyperpolarisation response to recombinant IL4. The depolarisation response of OA chondrocytes to mechanical stimulation is unaffected by the presence of the CaMKII inhibitor.

The CaMKII isoforms $\gamma$ and $\delta$ are expressed in both normal and OA chondrocytes, both in vitro and *in vivo*, but are only involved in the response of normal chondrocytes to mechanical stimulation. This response is upstream of the effect of IL4. These findings are consistent with previous findings for the NMDA receptor, and suggest that dysregulation of NMDA-CaMKII signalling may be important in onset and progression of osteoarthritis.

Keywords: Calcium/calmodulin-dependent protein kinase II, chondrocyte, cartilage, NMDA receptor, osteoarthritis, mechanotransduction

## 1. Introduction

Mechanical loading in articular joints is critically important for the maintenance of normal matrix integrity [7,10]. Animal studies have demonstrated that abnormal loading, either under-or overloading, results in cartilage loss and the development and progression of osteoarthritis [2,17].

The molecular events that regulate the chondrocytes response to mechanical stimulation have been the subject of much study. Research in our laboratory has identified an integrin-mediated mechanotransduction pathway activated in normal human articular chondrocytes following stimulation at 0.33 Hz [19,20]. This pathway involves the release of interleukin-4 from the chondrocyte that acts in an autocrine/paracrine manner, leading to cell membrane hyperpolarisation and gene regulation, including an upregulation of aggrecan [11,13]. We have shown that this pathway is altered in chondrocytes from OA cartilage, with the same stimulus leading to a cell membrane depolarisation, and loss of the anabolic gene regulation [12,14]. Recent studies have further identified the involvement of substance P and NMDA receptor signalling in this mechanotransduction pathway [14,16]. While substance P, acting via its NK1 receptor, is involved in both pathways, one of the differences identified between normal and OA responses is that the NMDA receptor signalling is altered [14,16]; despite the presence of functional NMDA receptor in both normal and OA chondrocytes [16], NMDA receptor signalling is only activated by 0.33 Hz stimulus in normal chondrocytes, and addition of exogenous NMDA/glycine is insufficient to upregulate aggrecan mRNA in normal chondrocytes (unpublished observation).

Following the observation that the NMDA receptor is present in articular cartilage, and activated when normal chondrocytes are mechanically stimulated, we were interested in identifiying whether other components of the NMDA signalling complex that have been identified in neuronal cells were present in chondrocytes and involved in the mechanotransduction pathways. One such component is calcium/calmodulin-dependent kinase II (CamKII). CamKII is one of a family of calcium/calmodulin-dependent protein kinases that play a central role in the transduction of calcium signals within the cell. CamKII is a ubiquitously expressed serine/threonine kinase that has a broad tissue distribution and has been implicated in a range of cellular functions including synaptic transmission, gene regulation and cell growth [3]. There are four isoenzymes, encoded by separate genes ($\alpha$, $\beta$, $\gamma$, and $\delta$), with alternative splicing within the variable domain of the protein generating additional diversity [1,18]. The $\alpha$- and $\beta$-isoforms are restricted to neuronal and endocrine tissues, whereas the $\gamma$- and $\delta$-isoforms have been identified within neuronal and non-neuronal tissues [6].

The aim of this study was to investigate whether CaMKII is expressed in human normal and OA articular cartilage and involved in chondrocytes response to mechanical stimulation.

## 2. Materials and methods

### 2.1. Isolation and culture of chondrocytes

Articular cartilage was obtained from human adult knee and ankle joints following arthroplasty or lower limb resection for peripheral vascular disease with patients' or relatives' consent. In all cases cartilage was assessed and graded macroscopically for the presence or absence of osteoarthritis (OA) using the Collins/McElligott system [5]. Articular cartilage was sampled from 12 males (mean age, 72 years; range 52–85 years) and 15 females (mean age, 70 years; range 50–88 years). Chondrocytes were isolated by sequential enzyme digestion and cultured as described previously [15]. Primary, nonconfluent cultures of chondrocytes were used in all experiments. Morphologically the cells from both normal and OA cartilage were typically flattened with a polygonal cell shape and did not show the fibroblastic appearance of dedifferentiated chondrocytes. In addition, reverse transcriptase–polymerase chain reaction (RT–PCR) and immunohistochemistry showed expression of cartilage-specific molecules, including type II collagen and aggrecan [19,20].

### 2.2. Immunohistochemistry

Frozen sections of normal articular cartilage (Collins grade 0) ($n = 5$) and OA cartilage (Collins grades 2, 3) ($n = 5$) of knee joint and normal articular cartilage (Collins grade 0) ($n = 5$) of ankle joint were incubated overnight at 4°C with rabbit polyclonal anti-CaMKII antibody (Santa Cruz) at a dilution of 1 : 50 followed by swine anti-rabbit biotinylated secondary antibody (Dako) at a dilution of 1 : 100 for one hour at room temperature, then visualized using 3,3′-diaminobenzidine tetrahydrochloride (DAB) (Sigma). Staining specificity was assessed by replacing the primary antibody with the IgG fraction derived from nonimmune rabbit serum.

### 2.3. RNA extraction

Total RNA was extracted from cultured chondrocytes according to the protocol of Chomczynski and Sacchi 1987 [4] using a denaturing buffer of 4 M guanidine thiocyanate, 0.75 M sodium citrate, 10% (w/v) lauroyl sarcosine and 7.2 $\mu$l/ml $\beta$-mercaptoethanol. The quantity of RNA isolated was determined by the absorbance reading at 260 nm on a spectrophotometer (GeneQuant; Pharmacia, St. Albans, UK).

### 2.4. RT–PCR and gel analysis

Template complementary DNA (cDNA) was synthesized using 0.5 $\mu$g of RNA, SuperScript II (Invitrogen), and oligo(dT)$_{12-18}$ (Amersham Biosciences) according to the manufacturer's instructions, and PCR was carried out as described previously [13]. The primers and conditions used for the PCR reactions are given in Table 1. PCR products were analyzed by electrophoresis using a 1% (w/v) agarose gel stained with ethidium bromide.

Semiquantitative PCR was performed and analysed as described previously [13] in order to examine the quantity of aggrecan expression in human normal and OA articular chondrocytes following the addition of KN-93, an inhibitor of CaMKII. Each donor was examined in duplicate, and 3 donors were used for each experiment.

Table 1
PCR primer sequences and conditions required for amplification

| Name | Sense primer | Antisense primer | Annealing temperature (°C) | dNTP concentration ($\mu$M) | Mg concentration (mM) |
|---|---|---|---|---|---|
| GAPDH | CCACCCATGGCAA ATTCCATGGGCA | TCTAGACGGCA GGTCAGGTCCACC | 60 | 100 | 2.5 |
| Aggrecan | TGAGGAGGGC TGGAACAAGTACC | GGAGGTGGTAAT TGCAGGGAACA | 60 | 100 | 1.5 |
| CamKII $\alpha$ | AATGCCAGGAG GAAACTGAAG | GCCTGGTCCT TCAATGGGGCA | 50 | 100 | 1.5 |
| CamKII $\beta$ | AGACTGTGGAG TGTCTGAAAAAGT | CGAAACCAGGGCGC AGCTCTTCACTGCAG | 45 | 100 | 1 |
| CamKII $\gamma$ | GTGGCATCC ATGATGCATCG | AAGTCCCCAT TGTTGATGGC | 52 | 100 | 2 |
| CamKII $\delta$ | ACACAAAAAT CTGTGACCCA | TGCCACAAA GAGGTGCCTCC | 52 | 100 | 1.5 |

## 2.5. Western blot analysis

Total protein extracts were prepared from cultured chondrocytes from human normal and OA articular cartilage as described previously [9]. The concentration of protein within lysates was determined using the Folin–Lowry assay method with Dynatech MR 5000 (Dynatech, Alexandria, VA). Equal quantities of chondrocyte protein in the total cell lysates were electrophoresed on 7.5% sodium dodecyl sulfate-polyacrylamide gels (SDS-PAGE) under reducing conditions. Following electrophoresis, the proteins were transferred onto polyvinylidene fluoride (PVDF) membranes (VWR). Membranes were blocked for 1 hour at room temperature with 5% nonfat dried milk in TBST (12.5 mM Tris HCl [pH 7.6], 137 mM NaCl, 0.1% Tween 20), then incubated overnight at 4°C with anti-CaMKII rabbit polyclonal antibody (Santa Cruz) at a dilution of 1 : 1000 in TBST. After washing 6 times with TBST, the blots were incubated for 1 hour at room temperature with goat anti-rabbit horseradish peroxidase (HRP)-labeled secondary antibody (Dako, Cambridge, UK) at a dilution of 1 : 2000 in TBST. Membranes were rewashed extensively, and the binding was detected using the Enhanced Chemiluminescence (ECL) Western blotting detection system (Amersham, Little Chalfont, UK) according to the manufacturer's instructions. Membranes were stripped and reprobed with mouse monoclonal anti-GAPDH antibody (Abcam) at a dilution of 1 : 5000 using 5% nonfat dried milk in TBST for 1 hour at room temperature, followed by sheep anti-mouse HRP-labeled secondary antibody at a dilution of 1 : 2000 in TBST for 1 hour at room temperature.

## 2.6. Mechanical stimulation

The technique and apparatus used for mechanical stimulation of primary human normal and OA articular chondrocytes have been previously described in detail [19,20]. The standard stimulation regimen used was a frequency of 0.33 Hz (2 seconds on, 1 second off) for 20 minutes at 37°C, at a pressure of 16 kPa above atmospheric pressure. This system produces 3700 microstrain on the base of the culture dish.

## 2.7. Experimental protocol

Human normal and OA articular chondrocytes were mechanically stimulated at 0.33 Hz with 3700 microstrain for 20 minutes at 37°C, or by addition of 10 pg/ml of recombinant IL-4 for 20 minutes. Cell responses to these stimuli, in the absence or presence of 10 $\mu$M of KN-93, an inhibitor of CaMKII, were assessed by measuring changes in cell membrane potential or changes in relative levels of aggrecan mRNA compared with the housekeeping gene GAPDH mRNA. All experiments were performed at least 3 times, using cells from different donors.

## 2.8. Electrophysiological measurements

Membrane potentials of human normal and OA articular chondrocytes were recorded using a single electrode bridge circuit and calibrator as previously described [19,20]. In each set of experiments, a control dish of chondrocytes was examined whose culture medium did not contain any agent to be tested. Resting membrane potentials of 5–10 cells were measured and, following addition of the reagent to be tested to the culture medium, membrane potentials were assessed in a further 5–10 cells to establish whether the reagent itself had an effect on the resting membrane potential of the cells. Then, following a 20-minute period of cyclical stimulation, membrane potentials were recorded in a further 5–10 cells. Reagents being studied were in contact with cells throughout the period of mechanical stimulation and afterwards when post-stimulated membrane potentials were measured. At least three experiments with cells from three separate donors were performed with each reagent.

## 2.9. Statistical analysis

The mean, standard deviation, and standard error of the mean were calculated for each experiment. For statistical comparisons, the nonparametric Mann–Whitney $U$ test was used when the $F$ ratio of the 2 variances reached significance. When the ratio did not reach significance, the Student's $t$-test was used.

# 3. Results

## 3.1. Endogenous expression of CaMKII in human normal and OA articular chondrocytes

Immunohistochemical staining of cryostat sections of snap-frozen human normal articular cartilage from knee and ankle joints and OA articular cartilage from knee joint showed extensive positivity for CaMKII in the large majority of chondrocytes in middle and deep zones of all samples (Fig. 1).

Western blot analysis of protein extracts from primary cultures of chondrocytes extracted from both normal and OA articular cartilage using specific antibody against CaMKII are shown in Fig. 2. Rat brain lysates were used as positive control for CaMKII expression. The anti-CaMKII antibody resulted in positive staining of protein band at $\sim$60 kDa in normal and OA knee cartilage and normal ankle cartilage lysates, which is consistent with the known molecular weight of CaMKII.

Isoform-specific primer pairs that anneal to sequences common to each reported isoform were used in RT–PCR analysis to determine which mRNAs are present for $\alpha$, $\beta$, $\gamma$, and $\delta$ CaMKII in human normal and OA articular chondrocytes from knee joints. As shown in Fig. 3, PCR products were obtained from the $\gamma$ and $\delta$ CaMKII primer pairs, not the $\alpha$ and $\beta$ CaMKII primer pairs. Sequencing confirmed that the PCR products obtained corresponded to $\gamma$ and $\delta$ CaMKII, indicating that mRNAs for $\gamma$ and $\delta$ CaMKII isoforms, not the $\alpha$ and $\beta$ CaMKII isoforms, are present in both human normal and OA articular chondrocytes from knee joints.

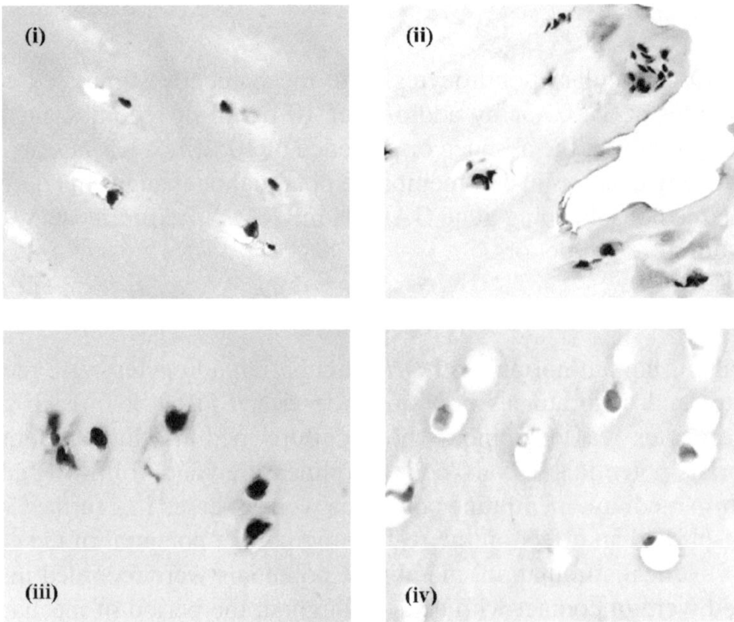

Fig. 1. Expression of CaMKII in human articular chondrocytes *in vivo*. Immunohistochemistry was performed on frozen sections of (i) normal, (ii) OA and (iii) ankle cartilage using a pan-CaMKII antibody; (iv) negative control, with IgG substituted for antibody.

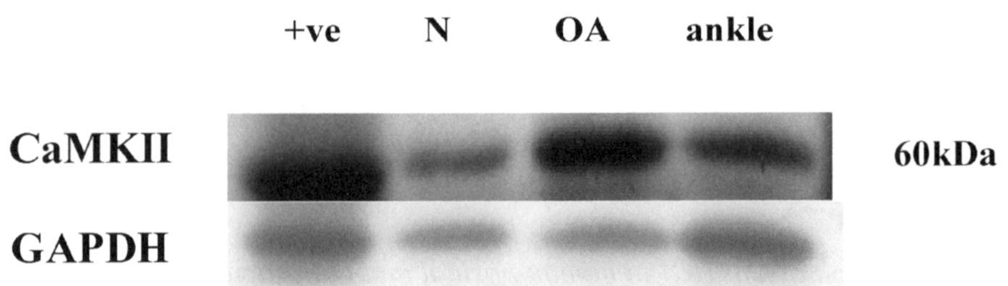

Fig. 2. Protein expression of CaMKII in cultured human articular chondrocytes. Total protein lysates extracted from (i) rat brain (positive control), (ii) cultured normal articular chondrocytes, (iii) cultured OA articular chondrocytes and, (iv) cultured ankle chondrocytes were immunoblotted with pan-CaMKII antibody.

### 3.2. *CaMKII-dependent electrophysiological response of human normal and OA articular chondrocytes to mechanical stimulation*

Having identified that CamKII is expressed in human articular cartilage, we investigated whether CamKII is involved in the mechanotransduction pathways activated following mechanical stimulation. Stimulation of human chondrocytes from normal and OA articular cartilage at a frequency of 0.33 Hz results in changes in membrane potential; chondrocytes from normal cartilage show a membrane hyperpolarisation, while cells isolated from OA cartilage depolarise in response to the same stimulus. Preincubation of chondrocytes with KN-93, an inhibitor of CaMKII, blocked the membrane hyperpolarisation of normal chondrocytes (Fig. 4a). In contrast, the presence of KN-93 had no effect on the depolarisation response of OA chondrocytes to the same stimulus (Fig. 4b).

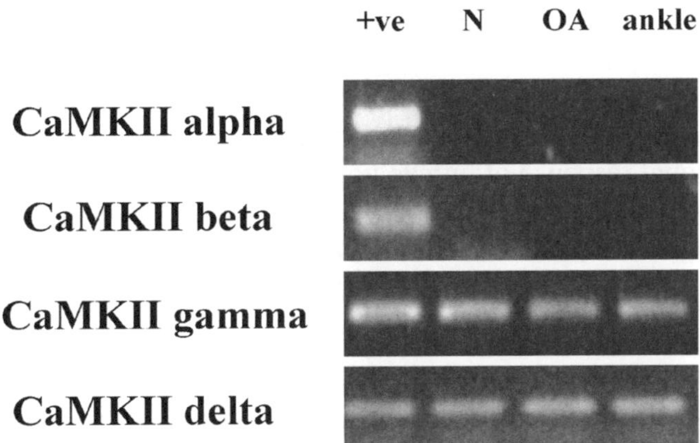



Fig. 3. Isoform expression of CaMKII in human articular chondrocytes. RT–PCR was carried out on RNA extracted from cultured human articular chondrocytes isolated from normal, OA, and ankle cartilage.

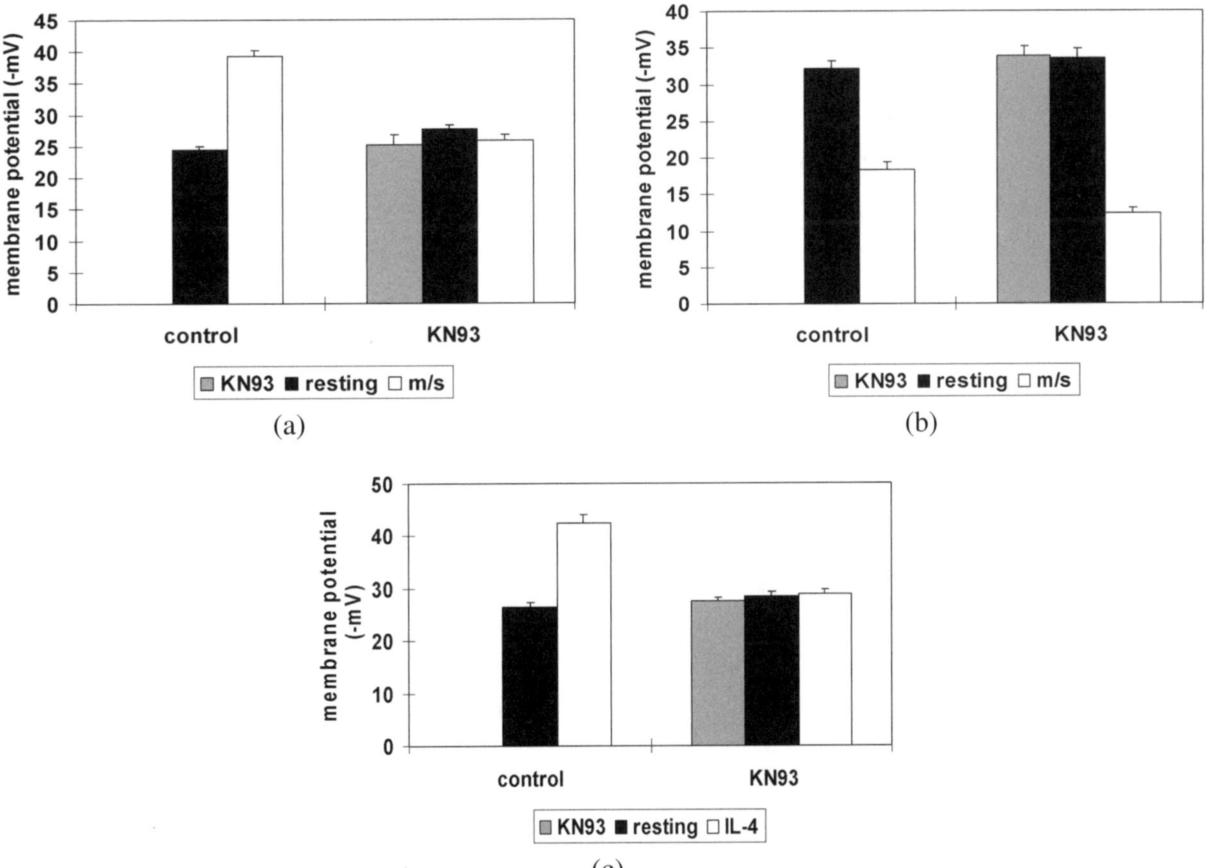

Fig. 4. Effect of a CaMKII inhibitor on the electrophysiological responses of human articular chondrocytes. (a) Normal chondrocytes mechanically stimulated in the presence or absence of KN93 (10 $\mu$M). (b) OA chondrocytes mechanically stimulated in the presence or absence of KN93 (10 $\mu$M). (c) Normal chondrocytes stimulated by the addition of recombinant human IL4 (10 pg/ml) in the presence or absence of KN93 (10 $\mu$M).

### 3.3. CaMKII–IL-4 relationship in the mechanotransduction pathway in human normal articular chondrocytes

In an attempt to identify the position of CaMKII in relation to IL-4 in the mechanotransduction pathway, the membrane potential response of human normal articular chondrocytes to IL-4 was assessed in the presence or absence of KN-93, an inhibitor of CaMKII. KN-93 inhibits the cell membrane hyperpolarisation induced by addition of IL-4 to human normal articular chondrocytes (Fig. 4c).

### 3.4. CaMKII-dependent up-regulation of aggrecan gene expression by mechanical stimulation

To determine whether CaMKII is involved in the regulation of aggrecan mRNA following mechanical stimulation, chondrocytes from human normal and OA articular cartilage were stimulated at 0.33 Hz for 20 minutes in the presence or absence of KN-93, and total RNA was extracted immediately (time point 0), 1, 3, or 6 hours following mechanical stimulation, and aggrecan mRNA levels were analysed by semiquantitative RT–PCR (Fig. 5a). Relative levels of aggrecan mRNA compared with the housekeeping

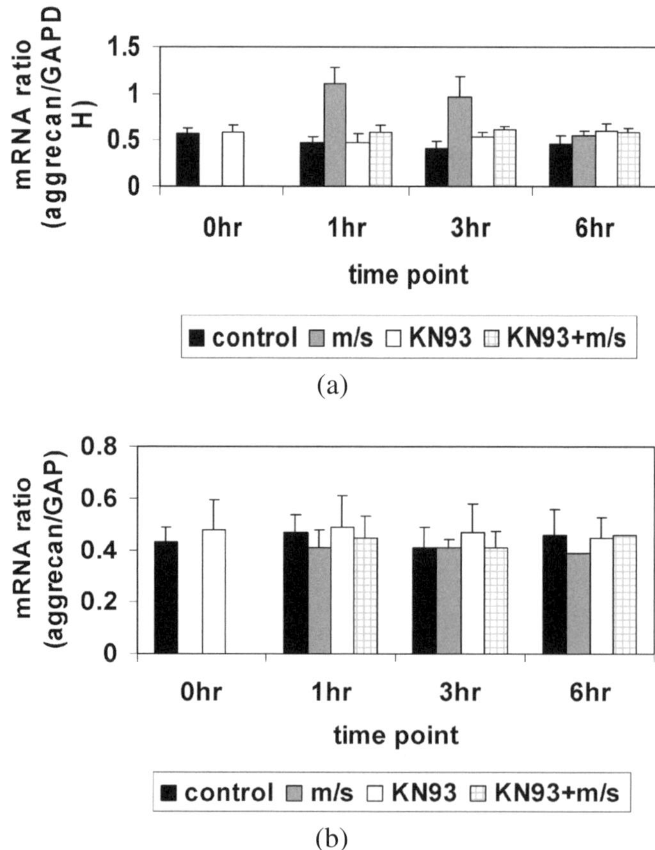


Fig. 5. Effect of a CaMKII inhibitor on aggrecan production in human articular chondrocytes, as determined by semi-quantitative RT–PCR. (a) Normal chondrocytes were mechanically stimulated in the presence or absence of KN93 (10 $\mu$M), RNA extracted at 0–6 hours post stimulation, and semi-quantitative RT–PCR carried out for aggrecan, relative to the housekeeping gene GAPDH. (b) OA chondrocytes were mechanically stimulated in the presence or absence of KN93 (10 $\mu$M), RNA extracted at 0–6 hours post stimulation, and semi-quantitative RT–PCR carried out for aggrecan, relative to the housekeeping gene GAPDH.

gene GAPDH in normal articular chondrocytes in the absence of KN-93 were significantly elevated at 1 and 3 hours post-stimulation and returned to the baseline level at 6 hours. However, in the presence of KN-93, an inhibitor of CaMKII, no changes in the aggrecan mRNA ratios were seen following mechanical stimulation. In the absence of mechanical stimulation, KN-93 had no effect on the relative levels of aggrecan mRNA. In contrast, no changes in the aggrecan mRNA ratios relative to the housekeeping gene GAPDH in human OA articular chondrocytes were seen following mechanical stimulation in the presence or absence of KN-93 (Fig. 5b).

## 4. Discussion

We have previously shown that the NMDA receptor is present in normal and OA articular cartilage [16]. In neuronal cells the cytoplasmic domains of the NMDA receptor subunits interact with the actin cytoskeleton and form large signalling complexes with a number of proteins including post-synaptic density protein-95 (PSD-95), neuronal nitric oxide and CaMKII. Activation of the NMDA receptor leads to an influx of calcium into the cell, which results in the activation of this signalling complex, including CaMKII. This study shows that CaMKII as well as the NMDA receptor is present in normal and OA knee and normal ankle articular cartilage. RT–PCR was used to investigate the isoforms expressed, and found that only the $\gamma$ and $\delta$ isoforms were expressed in the cultured chondrocytes. This is consistent with findings in other tissues, which have shown that the $\alpha$ and $\beta$ forms are restricted to neuronal cell types [1,18]. Further subtyping of splice variants was not carried out in these cells.

Previous studies have shown that the mechanotransduction pathways, including NMDA activation leading to hyperpolarisation, are the result of activation of small conductance calcium-activated potassium (SK) channels [11,16]. Studies in mouse myocytes have shown CaMKII to be involved in the regulation of SK channel activation [8]. The inhibition of the chondrocyte hyperpolarisation response in the presence of KN-93, a CaMKII inhibitor, is consistent with the involvement of CaMKII in the mechanotransduction response and regulation of SK channels in human chondrocytes. Furthermore, CaMKII is involved in the regulation of aggrecan mRNA in the normal chondrocyte, as the presence of KN-93 blocks the upregulation usually seen following mechanical stimulation.

Although both NMDA receptors and SK channels have been shown to be present on OA chondrocytes, they are not activated following the same mechanical stimulus [12,15]. The presence of KN-93 had no effect on the cell membrane response of OA chondrocytes to 0.33 Hz, indicating that CamKII does not appear to be activated in OA cells following mechanical stimulation. This suggests that the divergence of the signalling pathways in normal and OA chondrocytes to the same mechanical stimulus is upstream of NMDA receptor signalling and subsequent activation of CaMKII.

The mechanotransduction pathway in OA chondrocytes involves the release and autocrine/paracrine action of interleukin-1 beta upstream of the release and action of IL4 [11]. The fact that the presence of KN93 inhibits the hyperpolarisation response usually induced in normal chondrocytes by the addition of recombinant IL4 indicates that CaMKII acts downstream of the effects of IL4. There is substantial evidence suggesting that IL1b can modulate NMDA receptor signalling in a number of different ways, including through the activation of tyrosine kinases and phosphatases that regulate the phosphorylation status of the receptor subunits [3]. The phosphorylation status of the receptors will then affect the downstream signalling, including the activity of CamKII. This raises the possibility that the involvement of IL1b in OA signal transduction is responsible for the differences in NMDA/CamKII responses seen between normal and OA chondrocytes following mechanical stimulation.

This study indicates that CamKII is involved in the NMDA receptor signalling that is activated as part of the mechanotransduction response of normal articular chondrocytes to mechanical stimulation. However, despite being present in OA cartilage, the same signal transduction pathway is not activated in OA cells following stimulation. This suggests that dysregulation of the NMDA–CaMKII signalling may be important in the onset and progression of OA.

## Acknowledgement

This work was supported by a grant from Action Medical Research.

## References

[1] C.M. Beaman-Hall, M.J. Hozza and M.L. Vallano, Detection of mRNAs encoding distinct isoenzymes of type II calcium/calmodulin-dependent protein kinase using the polymerase chain reaction, *J. Neurochem.* **58** (1992), 1259–1267.

[2] K.D. Brandt, S.L. Myers, D. Burr and M. Albrecht, Osteoarthritic changes in canine articular cartilage, subchondral bone, and synovium fifty-four months after transection of the anterior cruciate ligament, *Arthritis Rheum.* **34** (1991), 1560–1570.

[3] A.P. Braun and H. Schulman, The multifunctional calcium/calmodulin-dependent protein kinase: from form to function, *Annu. Rev. Physiol.* **57** (1995), 417–445.

[4] P. Chomczynski and N. Sacchi, Single-step method of RNA isolation by acid guanidinium thiocyanate–phenol–chloroform extraction, *Anal. Biochem.* **162** (1987), 156–159.

[5] D.H. Collins and T.F. McElligott, Sulphate uptake by chondrocytes in relation to histological changes in osteoarthritic human articular cartilage, *Ann. Rheum. Dis.* **19** (1960), 318–330.

[6] R.A. Easom, CaM kinase II: a protein kinase with extraordinary talents germane to insulin exocytosis, *Diabetes* **48** (1999), 675–684.

[7] H.J. Helminen, A.M. Saamanen, J. Jurvelin, I. Kiviranta, J.J. Parkkinen, M.J. Lammi et al., The effect of loading on articular cartilage, *Duodecim* **108** (1992), 1097–1107.

[8] I.D. Kong, S.D. Koh, O. Bayguinov and K.M. Sanders, Small conductance $Ca^{2+}$-activated $K^+$ channels are regulated by $Ca^{2+}$-calmodulin-dependent protein kinase II in murine colonic myocytes, *J. Physiol.* **524**(2) (2000), 331–337.

[9] H.-S. Lee, S.J. Millward-Sadler, M.O. Wright, G. Nuki and D.M. Salter, Integrin and mechanosensitive ion channel-dependent tyrosine phosphorylation of focal adhesion proteins and $\beta$-catenin in human articular chondrocytes after mechanical stimulation, *J. Bone Min. Res.* **15** (2000), 1501–1509.

[10] G. Meachim, Ways of cartilage breakdown in human and experimental osteoarthrosis, in: *The Aetiopathogenesis of Osteoarthrosis*, G. Nuki, ed., Pitman Medical, London, 1980, pp. 16–28.

[11] S.J. Millward-Sadler, M.O. Wright, K. Nishida, H. Caldwell, G. Nuki and D.M. Salter, Integrin-regulated secretion of interleukin-4: a novel pathway of mechanotransduction in human articular chondrocytes, *J. Cell Biol.* **145** (1999), 183–189.

[12] S.J. Millward-Sadler, M.O. Wright, H.-S. Lee, H. Caldwell, G. Nuki and D.M. Salter, Chondrocytes derived from osteoarthritic cartilage show an altered electrophysiological response to mechanical stimulation, *OA Cart.* **8** (2000), 272–278.

[13] S.J. Millward-Sadler, M.O. Wright, L.W. Davies, G. Nuki and D.M. Salter, Cyclical mechanotransduction regulates aggrecan and matrix metalloproteinase 3 mRNA levels in normal but not osteoarthritic human articular chondrocytes, *Arthritis Rheum.* **43** (2000), 2091–2099.

[14] S.J. Millward-Sadler, A. Mackenzie, M.O. Wright, H.-S. Lee, K. Elliot, L. Gerrard, C.E. Fiskerstrand, D.M. Salter and J.P. Quinn, Tachykinin expression in cartilage and function in human articular chondrocyte mechanotransduction, *Arthritis Rheum.* **48** (2003), 146–156.

[15] D.M. Salter, J.E. Robb and M.O. Wright, Electrophysiological responses of human bone cells to mechanical stimulation: evidence for specific integrin function in mechanotransduction, *J. Bone Miner. Res.* **12** (1997), 1133–1141.

[16] D.M. Salter, M.O. Wright and S.J. Millward-Sadler, NMDA receptor expression and roles in human articular chondrocyte mechanotransduction, *Biorheology* (2004).

[17] M. Tammi, A.M. Saamanen, A. Jauhiainen, O. Malminen, I. Kiviranta and H. Helminen, Proteoglycan alterations in rabbit knee articular cartilage following physical exercise and immobilization, *Connect. Tiss. Res.* **11** (1983), 45–55.

[18] R.M. Tombes and G.W. Krystal, Identification of novel human tumour cell-specific CamK-II variants, *Biochim. Biophys. Acta* **1355** (1997), 281–292.

[19] M.O. Wright, K. Nishida, C. Bavington, J.L. Godolphin, E. Dunne, S. Walmsley et al., Hyperpolarisation of cultured human chondrocytes follows cyclical pressure-induced strain: evidence of a role for $\alpha5\beta1$ integrin as a chondrocyte mechanoreceptor, *J. Orthop. Res.* **15** (1997), 742–747.
[20] M.O. Wright, P. Jobanputra, C. Bavington, D.M Salter and G. Nuki, The effects of intermittent pressurisation on the electrophysiology of cultured human articular chondrocytes: evidence for the presence of stretch-activated membrane ion channels, *Clin. Sci.* **90** (1996), 61–71.

Biorheology 43 (2006) 235–247
IOS Press



# Poroelastic numerical modelling of natural and engineered cartilage based on *in vitro* tests

Federica Boschetti [a,*], Francesca Gervaso [a], Giancarlo Pennati [a], Giuseppe M. Peretti [b], Pasquale Vena [a] and Gabriele Dubini [a]

[a] *Laboratory of Biological Structure Mechanics, Department of Structural Engineering, Politecnico di Milano, Italy*
[b] *Orthopaedic Department, San Raffaele Hospital, Milan, Italy*

**Abstract.** The mechanisms underlying the ability of articular cartilage to withstand and distribute the loads applied across diarthrodial joints have been widely studied. Experimental tests have been done under several configurations to reveal the tissue response to mechanical stimuli, and theoretical models have been developed for the interpretation of the experimental results. The experiments demonstrated that the tissue is non-linear with strain, both in tension and in compression, non-linear with direction of stimulus, anisotropic in tension and compression, non-homogeneous with depth, resulting in depth dependent mechanical properties, and presents fluid dependent and fluid independent viscoelasticity. None of the models up to now developed is able to describe the whole set of responses of such a complex tissue. The purpose of this study was to develop a combined experimental-numerical approach for the proper description of the cartilage response under confined and unconfined compression. We defined a series of experimental tests to be performed on disks of natural and engineered cartilage and we developed a numerical model for cartilage, based on the biphasic theory, which potentially includes the tension–compression non-linearity, the strain non-linearity and the fluid independent viscoelasticity. The model successfully simulated the confined and unconfined compression experiments performed on disks of natural and engineered cartilage, and was also used to identify parameters of difficult experimental evaluation, such as the collagen stiffness and the permeability. In conclusion, the use of our model in combination with biomechanical experimental testing seems a valuable tool to analyze the mechanical properties of natural cartilage and the biofunctionality of tissue engineered cartilage.

Keywords: Articular cartilage, tissue engineered cartilage, poroelastic modeling, compression tests, permeation

## 1. Introduction

The mechanisms underlying the ability of articular cartilage to withstand and distribute the loads applied across diarthrodial joints have been widely studied. Experimental tests have been done under several configurations to reveal the tissue response to mechanical stimuli, and theoretical models have been developed for the interpretation of the experimental results. Most experiments consist of *in vitro* compression tests performed under confined or unconfined geometry, or indentation, under load or displacement control, using static or dynamic stimuli. The experiments demonstrated that the tissue is non-linear with strain, both in tension [33,39] and in compression [9], non-linear with direction of stimulus,

---

*Address for correspondence: Federica Boschetti, PhD, LaBS, Dipartimento di Ingegneria Strutturale, Politecnico di Milano, Piazza L. da Vinci 32, 20133 Milano, Italy. E-mail federica.boschetti@polimi.it.

known as tension–compression non-linearity, [11,16,35], anisotropic in tension [8,19] and compression [18], non-homogeneous with depth [17,28], resulting in depth dependent mechanical properties [10,13, 34] and presents fluid dependent and fluid independent viscoelasticity [15,20,30].

From a theoretical point of view, the homogeneous isotropic biphasic theory [30] has been the most widely used. This theory can satisfactorily explain the fluid dependent transient response of cartilage during compression. Models based on this theory have been able to adequately simulate confined compression creep and stress-relaxation experiments [1,5,30]. For such cases, the assumption of a constant permeability seems the most limiting aspect of the model [5]. The linear biphasic theory have failed to reproduce the high peak to relaxation ratio observed during unconfined compression tests [2]. The early response of cartilage during unconfined compression is thought as governed by the collagen fibrils, which are put in tension during these tests. A few models have then been developed to overcome this limitation. The linear transversely isotropic biphasic model [11] could successfully curve-fit confined and unconfined compression but could not properly predict the radial stress measured during confined compression [7]. The Conewise Linear Elasticity (CLV) model [35] could successfully curve-fit the stress-relaxation response of cartilage in confined compression, unconfined compression and shear. The model accounts for the tension–compression non-linearity of the tissue, but does not take into account the strain non-linearity and the fluid independent viscoelasticity. The fiber reinforced model [36] was successful in reproducing unconfined compression tests performed on healthy natural cartilage [26], degenerated natural cartilage [22], and tissue engineered cartilage [6]. As the CLE model and the linear transversely isotropic biphasic model, neither the fiber reinforced model takes into account the strain non-linearity and the fluid independent viscoelasticity. Furthermore, the axial-symmetric geometry allows modeling only the radial fibers while the circumferential ones are neglected. Finally, the spring elements have to be inserted into the model by the user connecting each pair of nodes, therefore the geometry discretization is the same regardless of the sample geometry.

The purpose of this study was then to develop a combined experimental-numerical approach for the proper description of the cartilage response under confined and unconfined compression. We defined a series of experimental tests to be performed on the same cartilage samples and we developed a numerical model for cartilage, based on the biphasic theory, which includes the tension–compression non-linearity, the strain non-linearity and the fluid independent viscoelasticity.

The experimental tests are based on step-wise stress relaxation confined and unconfined compression, and on permeation studies. The step-wise compression can put in evidence the strain dependence of the solid matrix properties as well as of the fluid–matrix interaction (through the strain dependent permeability). The comparison between confined and unconfined compression underline the tension–compression non-linearity. Finally the permeation studies, besides giving a direct measurement of the permeability, can elucidate about possible mechanisms of fluid independent viscoelasticity.

In this paper we present the essential formulation of the model, some results of the experiments, and a few examples of the ability of the model to simulate the stress-relaxation response of natural and engineered cartilage under unconfined compression. The fluid independent viscoelasticity effect was not considered in the present simulations. It will be the aim of future papers to show the whole set of experimental results and the capabilities of the model applied to a more wide set of experiments.

## 2. Materials and methods

### 2.1. Experimental tests

Experimental tests were performed on disks of natural and engineered cartilage.

The natural cartilage samples were harvested from a total of 6 patients (age $86 \pm 3$ years), who suffered from hip fracture and underwent hip endoprosthesis as treatment. Only those samples visually free from osteoarthritis were utilized for the study. Nineteen cylindrical plugs of bone and cartilage of approximately 10 mm in diameter were drilled from the femoral heads, then cut into three slices, of approximately 700 $\mu$m, using a custom made device. The layers obtained, though consecutive, were not representative of the real anatomical superficial, middle and deep layer. They were identified as "layer 1" ("superficial"), layer 2 ("middle"), and layer 3 ("deep"). Each slice was punched to obtain 5 or 9 mm disks, used in the tests. Cartilage disks were equilibrated in phosphate buffered saline (PBS: 0.15 mmol/l NaCl; 1.54 mmol/l $NaH_2PO_4$; 2.71 mmol/l $Na_2HPO_4$, purchased from Sigma-Aldrich Srl, Milan, Italy) containing protease inhibitors, PI (PI: 1 mmol/l phenylmethyl-sulfonyl-fluoride; 2 mmol/l EDTA; 5 mmol/l benzamidine; 10 mmol/l N-ethylmaleimide, purchased from Sigma-Aldrich Srl, Milan, Italy) and stored at $-26°C$ prior to testing. On the day of testing, cartilage slices were thawed at room temperature and equilibrated in PBS containing PI for about 30 minutes.

A first set of samples from 3 patients (Group A), 5 mm in diameter, underwent stress-relaxation confined compression and then stress-relaxation unconfined compression tests. A second set of samples from 4 patients (Group B), 9 mm in diameter, underwent step-wise stress-relaxation confined compression tests and then, after re-equilibration, permeation studies. A third set of samples from 1 patient (Group C), 9 mm in diameter, underwent permeation studies, then step-wise stress-relaxation confined tests, and finally step-wise stress-relaxation unconfined tests. For Group C, the compression tests were performed on 5 mm disks punched from the central part of the 9 mm disks after re-equilibration in PBS + PI for 24 hours.

Samples from Group A were subjected to a 10% followed by 5% strain compression, at a velocity of 1 $\mu$m/s, each strain compression being followed by stress relaxation to equilibrium. The biomechanical parameters were calculated with respect to the last 5% ramp compression. For Group B and C, the step-wise compression was defined as a series of 6 strain ramps, each 3% of the sample thickness, at a velocity of 1 $\mu$m/s, followed by stress relaxation to equilibrium. For each level of strain, the axial permeability was calculated by fitting the confined compression stress curve to the analytical solution [30].

All the compression tests were performed using an electromagnetic testing machine (Enduratec ELF3200, Enduratec-Bose, Minnetonka, MN, USA), equipped with a load cell of 22 N, under displacement control. Different chambers and indenters were used with the machine depending on the test configuration – confined or unconfined – and on the sample diameter.

The confined tests were performed by placing the specimen in a confining chamber over a stainless steel porous filter allowing free fluid flow. Compressive strain was applied to the cartilage specimen by a non-porous stainless steel indenter. The unconfined tests were performed by placing the specimen in a Plexiglas chamber and applying compressive strain by a non-porous stainless steel indenter.

The permeation studies were done using a custom made device consisting of two coaxial cylinders, a capillary flow-meter and a pressure regulator. A pressure transducer (EW-68950-02 digital pressure gauge, Cole-Parmer, Vernon Hills, IL, USA) was connected to the high pressure chamber to measure the fluid pressure during the tests. Fluid pressures ranging from 0.01 to 0.5 MPa were obtained using compressed air. The cartilage sample was placed on a polyethylene porous filter and bonded at the periphery to the surface of the inferior testing chamber with cyanoacrylate glue. The hydraulic permeability, $k$, was calculated by applying the Darcy law to the direct flow measurements. The cartilage strain level associated with each level of applied pressure was calculated for each sample using the stress–strain data derived from the confined compression stress relaxation studies performed on the same sample.

Using the equilibrium stress–strain data from confined and unconfined compression, we calculated the aggregate modulus $H_A$ and the Young modulus $E$, respectively, under the hypothesis of linear poroelastic behavior; the values of $H_A$ and $E$ were used to calculate the Poisson coefficient, as previously done by other authors [21]. The permeability, $k$, was both identified by fitting the experimental stress-relaxation curve to the analytical solution [30], and measured by the permeation studies.

The tissue engineered cartilage samples were obtained by seeding swine articular chondrocytes onto biological scaffolds and then culturing the constructs *in vitro* for up to 40 days. Swine articular chondrocytes were enzymatically isolated from pig joints and expanded in monolayer culture. When confluence was reached, cells were re-suspended and seeded onto biological collagen type I and III scaffolds *in vitro*. Samples were cultured in Ham's F-12 medium supplemented with 10% fetal calf serum, 1% glutamine, 1% antibiotic-antimycotic solution and 50 $\mu$g/ml ascorbic acid. The specimens were retrieved from culture after 18 and 39 days for analyses. Samples were biomechanically tested by unconfined compression and permeation. A step-wise compression was used, for a total of three ramps, 4% each, at a velocity of 10 $\mu$m/s. The stress relaxation data of the second ramp were used to calculate the Young modulus, $E$, and the peak to equilibrium stress ratio, $R$. The permeability, $k$, was measured using the apparatus described above and applying a hydraulic pressure difference of approximately 0.8 kPa.

The thickness of all the samples was measured from the position of the testing machine actuator, after imposing a preload of approximately 10 or 0.4 kPa for natural or tissue engineered cartilage, respectively.

Throughout the measurements, all the samples were immersed in PBS containing PI.

The statistical significance of the parameters variation among layers was assessed by $t$-test.

### 2.2. Numerical model

Cartilage was modeled as a composite material consisting of three phases, a solid matrix representing the proteoglycans, a fibril network corresponding to the collagen fibers, and a fluid. The solid phase, composed of both solid matrix and collagen fibers, was assumed to behave as a transversely isotropic and incompressible material. A continuum based formulation [12,38] was adopted.

The existence of a strain energy function that depends on the two invariants ($I_1, I_2$) of the right Cauchy–Green tensor and on a pseudo invariant ($I_4$) that accounts for a preferential direction of the fibers was postulated.

The strain energy function can be defined according to the following relationship:

$$W_0 = W_{\mathrm{vol}}(J) + W_{\mathrm{iso}}(\overline{I}_1, \overline{I}_2, \overline{I}_4), \tag{1}$$

where $J$ is the determinant of the deformation gradient $F_{ij}$, $W_{\mathrm{vol}}$ is the elastic energy stored in the continuum subject to the volumetric strain, and $W_{\mathrm{iso}}$ refers to the isochoric part of the deformation.

The term $W_{\mathrm{vol}}$ is assumed as:

$$W_{\mathrm{vol}} = \frac{1}{2}B(J-1)^2, \tag{2}$$

where $B$ is the bulk modulus of the material.

The isochoric part of the strain energy function was assumed as the sum of the strain energy stored in the matrix and the strain energy stored within the reinforcing fibers:

$$W_{\mathrm{iso}} = W_{\mathrm{iso}}^m(\overline{I}_1, \overline{I}_2) + W_{\mathrm{iso}}^f(I_4) \tag{3}$$

which means that no fiber–matrix interaction was considered.

For the matrix, the following formulation was adopted [32]:

$$W_{\text{iso}}^m = \frac{c_1}{2}\left(\overline{I}_1 - 3\right) + \frac{c_2}{2}\left(\overline{I}_2 - 3\right). \tag{4}$$

The coefficient $c_2$ was set equal to zero hence $c_1$ results to be equal to the shear modulus $G$.

The strain energy stored within the collagen fibers was split into two parts [32]: an exponential form [12], and a linear one accounting for the linear behavior of collagen for stretch greater than a given threshold, $\lambda_f^0 = \sqrt{I_4^0}$:

$$\frac{\partial W_{\text{iso}}^f}{\partial I_4} = E_f\, e^{b_2(I_4-1)^2}(I_4 - 1); \quad I_4 \leqslant I_4^0, \tag{5}$$

$$\frac{\partial W_{\text{iso}}^f}{\partial I_4} = \frac{C_f^1}{2\sqrt{I_4}} + \frac{C_f^2}{I_4}; \quad I_4 > I_4^0. \tag{6}$$

Here, $b_2$ rules the rate of collagen uncrimping and $E_f$ is a tuning factor related to the initial slope of collagen fiber stress–strain curve. In the second part, $C_f^1$ and $C_f^2$ are two constants the values of which enforce the continuity of $\partial W_{\text{iso}}^f/\partial I_4$ and $\partial^2 W_{\text{iso}}^f/\partial I_4$ at $I_4 = I_4^0$.

The fluid is incompressible and its interaction with the solid matrix was represented by the material permeability $k$ and the porosity $\phi$. The hydraulic permeability was assumed strain-dependent, according to [25]: $k = k_0 \cdot e^{-M\varepsilon}$.

Summarizing, the model is described by seven constitutive parameters: $B$, $G$, $E_f$, $b_2$, $k_0$, $M$ and $\phi$. Furthermore, for isotropic conditions the shear modulus and the bulk modulus can be related to the engineering constants $E$ and $\nu$ according to the following relationships:

$$G = \frac{E_m}{2(1 + \nu_m)}, \qquad B = \frac{E_m}{3(1 - 2\nu_m)}, \tag{7}$$

where $E_m$ and $\nu_m$ represent, respectively, the elastic modulus and the Poisson coefficient of the proteoglycan matrix. These relationships allow the calculation of the parameters $E_m$ and $\nu_m$ of the solid matrix in order to compare with the experimental values.

The described material was then implemented in a commercial Finite Element code (ABAQUS, Hibbit Karlsson and Sorenses, Inc. Pawtucket, RI, USA), via a user defined FORTRAN subroutine and a Non-Linear Composite Poro-Elastic (*NLCPE*) model was developed in order to obtain the numerical solution. The poroelastic theory in finite deformations was employed.

A sector of five degrees was considered and a uniform mesh consisting of 1600 elements (type C3D8P) was adopted.

The boundary conditions were set in order to simulate a stress-relaxation unconfined compression experiment and were as follows: the vertical displacement of the nodes at the bottom of the disk was constrained in order to model the presence of the steel plate supporting the specimen; at the nodes lying at distance $r = 2.5$ mm from the z-axis the pore pressure was set equal to zero, meaning the lateral surface was permeable to the fluid. The consolidation problem was simulated: a vertical displacement was imposed to the reference node of an analytical surface representing the compression plate. The final position of the compression plate was then kept constant for the relaxation time.

The porosity of each sample in the uncompressed state was determined by the difference between the weights of the hydrated and dried cartilage sample. The remaining six material parameters were evaluated by comparing the experimental curves to the ones simulated by the *NLCPE* model.

## 3. Results

### 3.1. Experimental tests

Results for the first set of tested cartilage samples, Group A, those subjected to confined and unconfined compression, are shown in Fig. 1 for consecutive layers of cartilage. A total of 21 cartilage disks, 7 for each layer, was tested by confined and then unconfined compression. The mean values of $H_A$ and $E$ increased with depth, $k$ decreased with depth, $\nu$ remained fairly constant. These differences were statistically significant only for $E$ and $H_A$ ($p < 0.05$) between layer 1 and layer 3.

Figures 2 and 3 show the results for the second set of tested cartilage samples, Group B, those subjected to step-wise stress relaxation tests and then permeation studies. A total of 24 cartilage disks, 8

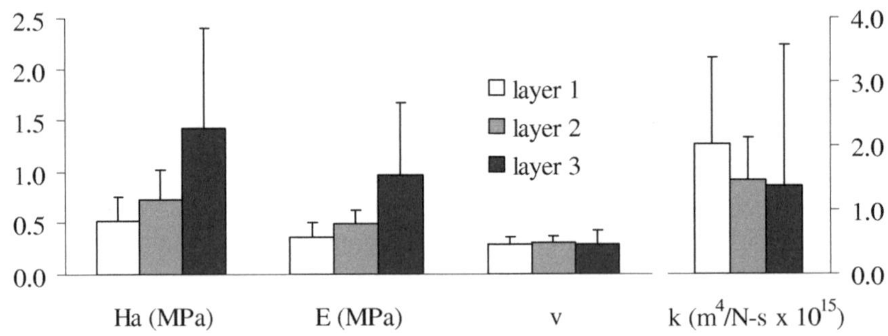


Fig. 1. Experimental results for Group A from confined ($H_A$ and $k$) and unconfined ($E$, $\nu$) compression for consecutive layers of cartilage. Mean values with standard deviations are reported for the tested disks ($n = 21$, total number of disk tested, seven for each layer). Differences were significant ($p < 0.05$) between layer 1 and layer 3 for $E$ and $H_A$. All other differences were not significant.

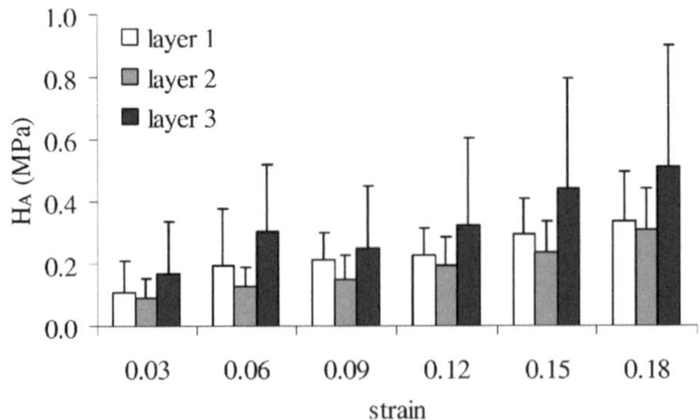


Fig. 2. Aggregate modulus ($H_A$) for Group B from confined compression for consecutive layers of cartilage and for increasing strain levels. Mean values with standard deviations are reported for the tested disks ($n = 24$, total number of disk tested, eight for each layer). All differences were non-significant.

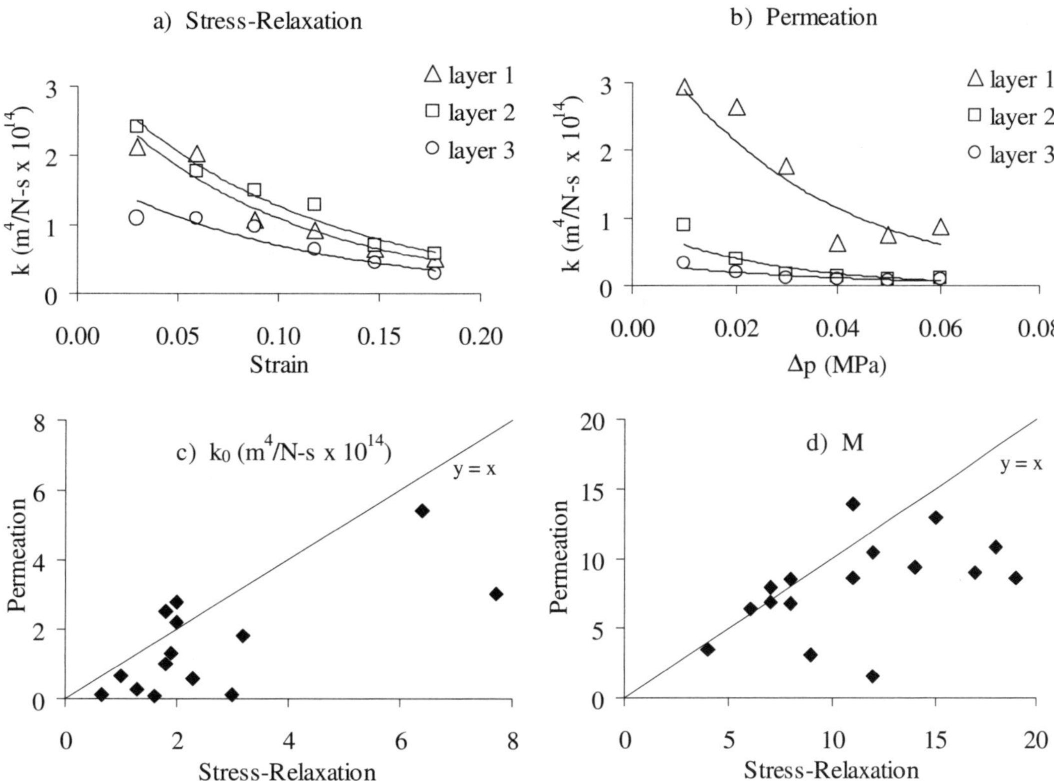



Fig. 3. Permeability results from stress relaxation confined compression and permeation studies for Group B. (a) Permeability ($k$) for consecutive cartilage layers, and for increasing levels of strain, calculated by fitting the confined compression stress relaxation curves to the analytical solution [30]. Each point is the average of the values obtained for several disks ($n = 24$, total number of disk tested, eight for each layer). (b) Permeability ($k$) for consecutive cartilage layers, and for increasing levels of applied pressure, measured from the permeation studies. Each point is the average of the values obtained for several disks ($n = 19$, total number of disks tested, eight for layer 1, 6 for layer 2, 4 for layer 3). The cartilage strain level associated with each level of applied pressure was calculated using the stress–strain data derived from the confined compression stress relaxation studies. (c), (d) Comparison between $k_0$ and $M$ derived by fitting the permeation measured permeabilities and the stress relaxation identified permeabilities to the classical equation, $k = k_0 \cdot e^{-M\varepsilon}$. Differences were non-significant for $k_0$ but significant for $M$ ($p < 0.01$).

for each layer, was tested by confined compression. Nineteen of these twenty-four disks were then subjected to the permeation study. The aggregate modulus, $H_A$, is shown in Fig. 2 for consecutive layers of cartilage and for increasing strain levels. Although the mean values show an increase with depth and strain, all differences were non-significant. The permeability values identified from the stress relaxation curves and those measured from the permeation studies are shown in Figs 3a and 3b, respectively. The values show a decrease with depth, and with strain (Fig. 3a) or applied pressure (Fig. 3b). The values of $k_0$ and $M$ obtained by curve fitting the data to the classical equation $k = k_0e^{-M\varepsilon}$ are compared in Figs 3c and 3d, and are lower for those obtained from the permeation studies than from the stress-relaxation tests. The differences were non-significant for $k_0$ and significant ($p < 0.01$) for $M$.

Results for the tissue engineered cartilage samples are shown in Fig. 4. The Young modulus $E$ decreased after 18 days ($p < 0.05$) of culture and then remained constant. The peak to equilibrium stress ratio, $R$, increased and the permeability, $k$, decreased between 0 and 18 days and between 18 and 39 days ($p < 0.05$).

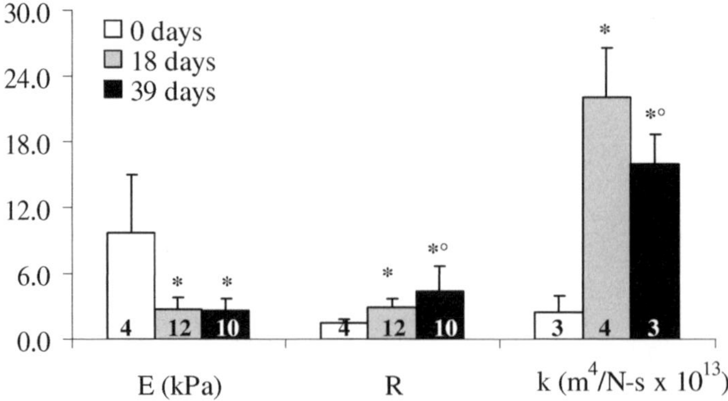


Fig. 4. Experimental results for the tissue engineered cartilage samples from unconfined compression ($E$, Young modulus, and $R$, peak to equilibrium stress ratio) and from permeation studies ($k$, permeability) for different times of culture (0 days, pure membrane, 18 and 39 days, constructs). Mean values with standard deviations are reported for the tested disks (each bar shows the number of samples tested, the total number is $n = 36$). $^*$Significant differences vs. 0 days ($p < 0.05$), $^\circ$significant differences vs. 18 days ($p < 0.05$).

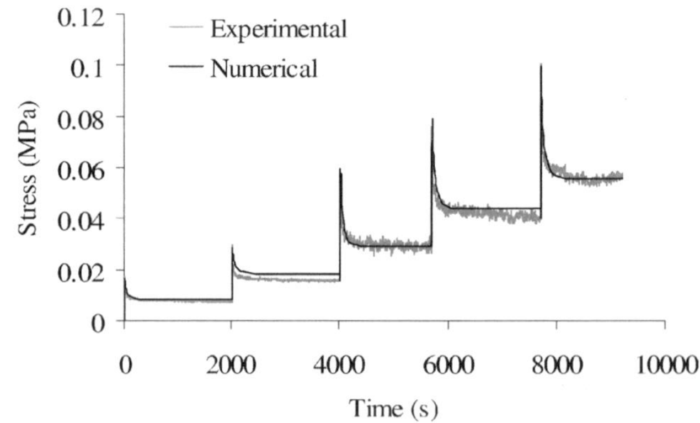


| strain | $B$ (MPa) | $G$ (MPa) | $E_f$ (MPa) | $k_r$ (m⁴/N-s) | $\phi$ (%) |
|---|---|---|---|---|---|
| 0.06 | 0.200 | 0.067 | 0.17 | 3.50E-14 | 73.5 |
| 0.09 | 0.233 | 0.078 | 0.25 | 1.99E-14 | 72.8 |
| 0.12 | 0.311 | 0.104 | 0.60 | 1.41E-14 | 72.0 |
| 0.15 | 0.333 | 0.111 | 0.70 | 1.02E-14 | 71.3 |
| 0.18 | 0.371 | 0.124 | 0.90 | 7.14E-15 | 70.5 |

Fig. 5. Comparison between the experimental step-wise unconfined compression stress relaxation curves and the simulated ones for one sample of Group C. The figure also shows the values assigned to the model parameters for each strain level in order to obtain a good curve-fit. The parameter $b_2$ was set to zero, the porosity, $\phi$, of the cartilage sample in the uncompressed state was 77%. See text for more details.

## 3.2. Numerical model

An example of the use of the *NLCPE* model is given in Fig. 5, which shows a comparison between the experimental step-wise unconfined compression stress relaxation curves and the simulated ones, for one sample of the experimental Group C, subjected to step-wise confined compression and then, after re-equilibration, to step-wise unconfined compression. Although the model allows the use of strain dependent parameters, for this example each ramp was simulated separately with constant parameters

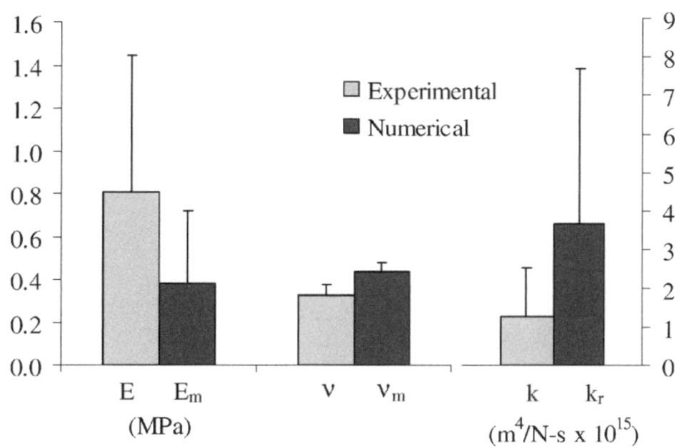


Fig. 6. Comparison between biomechanical parameters for Group A derived from the compression studies and the parameter values obtained by fitting the unconfined compression experimental curves to the numerical simulated ones ($n = 9$, total number of disks tested and simulated, three for each layer). The experimental values $E$ and $\nu$ represent the contribute of the solid phase made of collagen and proteoglycans. The numerical values $E_\mathrm{m}$ and $\nu_\mathrm{m}$ represent the contribute of the proteoglycan mesh only. The experimental values of the permeability are those derived from the confined compression tests and thus represent the axial permeability ($k$), while the numerical ones represent the radial permeability, $k_\mathrm{r}$. Differences between experimental and numerical were significant ($p < 0.05$) for all the parameters.

for each ramp. A good curve-fit could be obtained by assigning the values shown at the bottom of Fig. 5 to the model parameters. The values of $E_\mathrm{m}$ and $\nu_\mathrm{m}$ derived from the model parameters (Eqs (7)) are always lower and higher, respectively, than the experimental measured $E$ and $\nu$, since the latter represent the response of both the collagen fibers and the proteoglycans, whereas the former represent the proteoglycans contribute only. The experimental permeability is often greater than the one assigned to the model in order to obtain a good curve-fit. The experimental permeability is the one derived from the confined compression tests and is therefore related to flow in the axial direction, while the permeability used in the model for the simulation of the unconfined compression ($k_\mathrm{r}$) is related to flow in the radial direction. A comparison between model parameters and experimental values is shown in Fig. 6 for 9 samples from Group A. Differences were significant ($p < 0.05$) for all the parameters.

The *NLCPE* model was also used to simulate unconfined compression stress-relaxation curves of tissue engineered cartilage samples, shown in Fig. 7a. The biomechanical experimental values are shown in Fig. 7b. Values used in the model to obtain a good fit between experimental and numerical curves are shown in Fig. 7c for the pure membrane and for constructs retrieved from culture at 18 and 39 days ($n = 3$ for each group). The experimental and numerical Young modulus, $E$ and $E_\mathrm{m}$, significantly decreased from 0 days to 18 days and then significantly increased from 18 to 39 days ($p < 0.05$); the modulus of the collagen fibers, $E_\mathrm{f}$, was significantly higher at 0 days with respect to 39 days ($p < 0.05$); the radial permeability, $k_\mathrm{r}$, significantly decreased from 18 to 39 days ($p < 0.05$).

## 4. Discussion

The experimental tests provided values for the elastic parameters of the human femoral head. The average values were 0.6 and 0.9 MPa for $E$ and $H_\mathrm{A}$ respectively and are comparable with those measured by other authors [3,14]. The Poisson ratio, calculated indirectly from the equilibrium data of confined and unconfined compression, averaged 0.3. Values in the literature for the Poisson ratio show a great

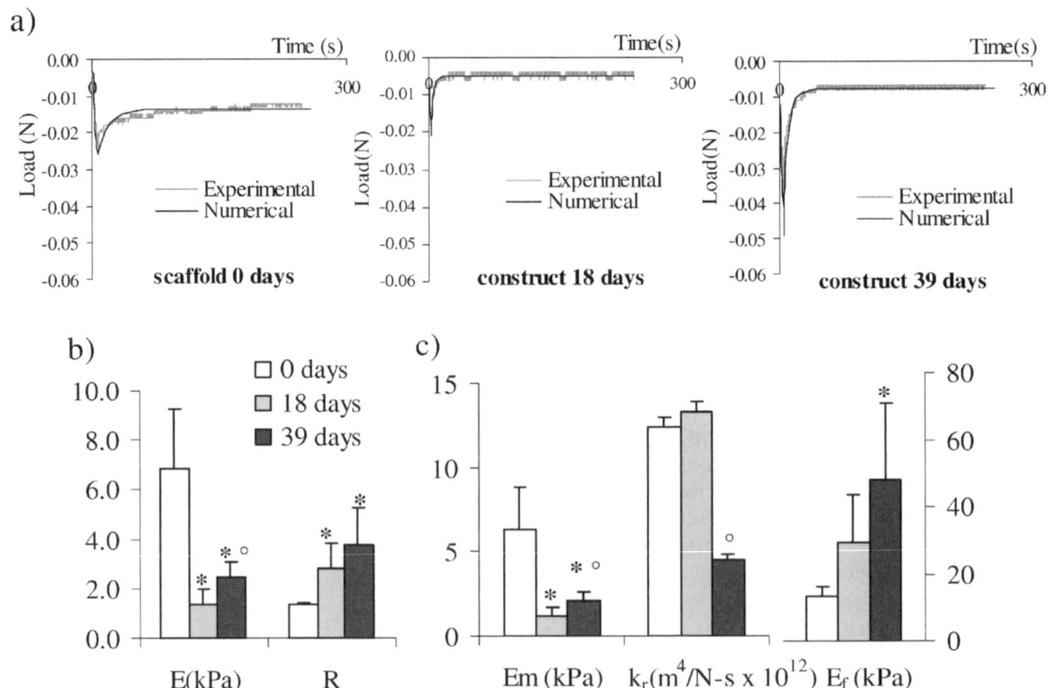

Fig. 7. (a) An example of stress relaxation curves and simulations by the *NLCPE* model for the collagen membrane (scaffold 0 days) and two constructs (at 18 and 39 days of static culture); (b) experimental values obtained from the stress-relaxation tests, $E$, Young modulus and $R$, peak to equilibrium stress ratio; (c) values used in the model to obtain a good fit between experimental and numerical curves, for the Young modulus of the matrix, $E_m$, the radial permeability, $k_r$, and the elastic modulus of the collagen fibers, $E_f$ ($n = 3$ for each group). *Significant differences vs. 0 days ($p < 0.05$), °significant differences vs. 18 days ($p < 0.05$).

variability between 0 and 0.4. Further studies, also using optical methods, are underway to determine this coefficient for the human femoral head.

The step-wise confined compression tests showed an increase of $H_A$ with strain. Although differences were not significant, the same trend has been observed by other authors [9].

The permeability values, $k$, as determined by fitting the stress relaxation confined compression data to the analytical solution, showed values comparable to those reported in the literature for the human femoral head [3], but higher than those directly measured by our permeation studies. For the latter, a comparison with the literature is not possible since the few permeation studies about human articular cartilage are not exhaustive (see for example, [27]). The values of permeability showed a decrease with increasing strain and were fitted to the classical equation $k = k_0 \cdot e^{-M\varepsilon}$. The comparison between stress relaxation and permeation suggests that both $k_0$ and $M$ are over-estimated from stress-relaxation experiments. This may be due to the well-known difficulty in assuring a perfect confinement between the confining chamber and the cartilage disk. Furthermore, the presence of a gap between the chamber and the indenter, required in order to guarantee the indenter movement, allows flow exudation and tissue extrusion around the indenter, leading to over-estimate the permeability, up to 30% [5]. On the other hand, permeation studies may underestimate the permeability, especially at low pressures when the flow is extremely low and difficult to be detected. More data are needed to understand which of the two methods is more reliable and to allow for statistical significance of the results.

With regard to tissue in-homogeneity, the parameters $H_A$, and $E$, increased while $k$ decreased with depth, in agreement with previous measurements [23,37], and reflecting the well-known variability in tissue composition and structure with depth.

The unconfined compression tests and the permeation studies performed on the tissue engineered cartilage tissues resulted in the quantification of the Young modulus and the permeability at different culture times. The Young modulus dramatically decreased and the permeability increased after 18 days of static culture with respect to the pure collagen membrane. This may indicate a faster scaffold degradation than extracellular matrix (ECM) build up by chondrocytes. However, the Young modulus was higher and the permeability lower at 39 days than at 18 days, possibly revealing a compensation between scaffold degradation and ECM production, related to glycosaminoglycan (GAG) content. The cellular ECM production with regard to collagen content may be related to the increase in peak to equilibrium stress ratio ($R$), shown by the constructs both at 18 and 39 days with respect to the pure membrane. For the natural cartilage, the equilibrium compressive stiffness is related to the GAG content [31]. The dense GAG mesh provides most of the resistance to fluid flow [29] and is therefore also linked to the permeability, $k$. The collagen network mainly determines the instantaneous compressive response of the tissue [4,31]. Hypothesizing the same behavior for tissue engineered cartilage, one can relate the Young modulus and the permeability to the GAG content, and the peak to equilibrium stress ratio ($R$) to the collagen content. Preliminary quantitative measurements of GAG content (results not shown) revealed, indeed, an increase of GAGs with culture time, which correlates with the increase of $E$ and decrease of $k$ at 39 days of culture. The increase in collagen content suggested by the increase of $R$ during culture can be correlated to the biomechanical parameter $E_f$, evaluated by fitting the experimental stress-relaxation curves to the simulations of the *NLCPE* model, which increased with culture time. Such an increase is consistent with the production of type II collagen, as confirmed by histological evaluation after immunostaining (results not shown). Quantitative evaluation of type II collagen by biochemical assays remains to be done in the future.

The *NLCPE* model could successfully curve-fit unconfined compression stress-relaxation data for natural and tissue engineered cartilage. In particular, results from this study showed the model capability for the description of the tension–compression non-linearity and for the identification of the contribute of different components of the solid phase, i.e. collagen and proteoglycans, to the equilibrium and transient tissue response to compression. The latter characteristic is particularly interesting when applied to tissue engineered cartilage, since it can help to evaluate the changing function of the constructs during culture, although this hypothesis needs to be validated by further biomechanical tests and model simulations, coupled with quantitative biochemical assays regarding collagen and GAG production. Furthermore, the model can be used in the future to elucidate about possible mechanisms of directional in-homogeneity, revealed by the difference between radial and axial permeability resulting from the fitting procedure. In particular, the radial permeability was higher than the axial one, a result in agreement with the literature [24].

The model is potentially able to describe the strain dependence of the biomechanical parameters, and the viscoelasticity of the solid matrix, by introducing the proper relations into the material properties, which can easily be modified by a user defined subroutine coupled to the ABAQUS code. To this aim, further experimental tests, based on the protocols described here, are underway and hopefully will allow to test the whole model capabilities.

In conclusion, the use of our NLCPE model in combination with biomechanical experimental testing seems a valuable tool to analyze the mechanical properties of natural cartilage and the biofunctionality of tissue engineered cartilage.

# References

[1] C.G. Armstrong and V.C. Mow, Variations in the intrinsic mechanical properties of human articular cartilage with age, degeneration, and water content, *J. Bone Joint Surg. Am.* **64** (1982), 88–94.

[2] C.G. Armstrong, W.M. Lai and V.C. Mow, An analysis of the unconfined compression of articular cartilage, *J. Biomech. Eng.* **106** (1984), 165–173.

[3] K.A. Athanasiou, A. Agarwal and F.J. Dzida, Comparative study of the intrinsic mechanical properties of the human acetabular and femoral head cartilage, *J. Orthopaedic Res.* **12** (1994), 340–349.

[4] D.L. Bader, G.E. Kempson, J. Egan, W. Gilbey and A.J. Barrett, The effects of selective matrix degradation on the short-term compressive properties of adult human articular cartilage, *Biochimica et Biophysica Acta* **1116** (1992), 147–154.

[5] F. Boschetti, G. Pennati, F. Gervaso, G.M. Peretti and G. Dubini, Biomechanical properties of human articular cartilage under compressive loads, *Biorheology* **41** (2004), 159–166.

[6] F. Boschetti, F. Gervaso, G. Pennati, C. Sosio, G.M. Peretti and G. Dubini, Use of a fiber reinforced poroelastic model to predict the mechanical properties of tissue engineered cartilage, *Trans. Orthop. Res. Soc.* **30** (2005), 1769.

[7] P.M. Bursac, T.W. Obitz, S.R. Eisenberg and D. Stamenovic, Confined and unconfined stress relaxation of cartilage: appropriateness of a transversely isotropic analysis, *J. Biomech.* **32** (1999), 1125–1130.

[8] N.O. Chahine, C.C. Wang, C.T. Hung and G.A. Ateshian, Anisotropic strain-dependent material properties of bovine articular cartilage in the transitional range from tension to compression, *J. Biomech.* **37** (2004), 1251–1261.

[9] A.C. Chen, W.C. Bae, R.M. Schinagl and R.L. Sah, Depth- and strain-dependent mechanical and electromechanical properties of full-thickness bovine articular cartilage in confined compression, *J. Biomech.* **34** (2001), 1–12.

[10] S.S. Chen, Y.H. Falcovitz, R. Schneiderman, A. Maroudas and R.L. Sah, Depth-dependent compressive properties of normal aged human femoral head articular cartilage: relationship to fixed charge density, *Osteoarthritis Cartilage* **9** (2001), 561–569.

[11] B. Cohen, W.M. Lai and V.C. Mow, A transversely isotropic biphasic model for unconfined compression of growth plate and chondroepiphysis, *J. Biomech. Eng.* **120** (1998), 491–496.

[12] R. Eberlain, G. Holzapfel and C.J. Shultze-Bauer, An anisotropic constitutive model for annulus tissue and enhanced finite element analyses of intact lumbar disc bodies, *Comput. Methods Biomech. Biomed. Eng.* **190** (2000), 4379–4403.

[13] O.K. Erne, J.B. Reid, L.W. Ehmke, M.B. Sommers, S.M. Madey and M. Bottlang, Depth-dependent strain of patellofemoral articular cartilage in unconfined compression, *J. Biomech.* **38** (2005), 667–672.

[14] T.R. Gardner, E.J. Balaguer, G.A. Athesian and V.C. Mow, Comparison of isotropic and transversely isotropic material properties from confined and unconfined compression and indentation, *ASME BED* **50** (2001).

[15] W.C. Hayes and A.J. Bodine, Flow-independent viscoelastic properties of articular cartilage matrix, *J. Biomech.* **11** (1978), 407–419.

[16] C.Y. Huang, A. Stankiewicz, G.A. Ateshian and V.C. Mow, Anisotropy, inhomogeneity, and tension–compression non-linearity of human glenohumeral cartilage in finite deformation, *J. Biomech.* **38** (2005), 799–809.

[17] E.B. Hunziker, Articular structure in humans and experimental animals, in: *Articular Cartilage and Osteoarthritis*, K.E. Kuettner, R. Schleyerbach, J.G. Peyron and V.C. Hascall, eds, Raven Press, New York, 1992, pp. 183–199.

[18] J.S. Jurvelin, M.D. Buschmann and E.B. Hunziker, Mechanical anisotropy of the human knee articular cartilage in compression, *Proc. Inst. Mech. Eng. [H]* **217** (2003), 215–219.

[19] G.E. Kempson, M.A. Freeman and S.A. Swanson, Tensile properties of articular cartilage, *Nature* **220** (1968), 1127–1128.

[20] G.E. Kempson, M.A. Freeman and S.A. Swanson, The determination of a creep modulus for articular cartilage from indentation tests of the human femoral head, *J. Biomech.* **4** (1971), 239–250.

[21] R.K. Korhonen, M.S. Laasanen, J. Toyras, J. Rieppo, J. Hirvonen, H.J. Helminen and J.S. Jurvelin, Comparison of the equilibrium response of articular cartilage in unconfined compression, confined compression and indentation, *J. Biomech.* **35** (2002), 903–909.

[22] R.K. Korhonen, M.S. Laasanen, J. Toyras, R. Lappalainen, H.J. Helminen and J.S. Jurvelin, Fibril reinforced poroelastic model predicts specifically mechanical behavior of normal, proteoglycan depleted and collagen degraded articular cartilage, *J. Biomech.* **36** (2003), 1373–1379.

[23] R. Krishnan, S. Park, M.A. Soltz, R.J. Pawluk and G.A. Ateshian, Depth-dependent tensile and compressive properties of human patellofemoral joint cartilage, *ASME BED* **50** (2001).

[24] R. Krishnan, S. Park, F. Eckstein and G.A. Ateshian, Inhomogeneous cartilage properties enhance superficial interstitial fluid support and frictional properties, but do not provide a homogeneous state of stress, *J. Biomech. Eng.* **125** (2003), 569–577.

[25] W.M. Lai and V.C. Mow, Drag-induced compression of articular cartilage during a permeation experiment, *Biorheology* **17** (1980), 111–123.

[26] L.P. Li, M.D. Buschmann and A. Shirazi-Adl, A fibril reinforced non-homogeneous poroelastic model for articular cartilage: inhomogeneous response in unconfined compression, *J. Biomech.* **33** (2000), 1533–1541.

[27] A. Maroudas and P. Bullough, Permeability of articular cartilage, *Nature* **219** (1968), 1260–1261.

[28] A. Maroudas, Physico-chemical properties of articular cartilage, in: *Adult Articular Cartilage*, M.A.R. Freeman, ed., Pitman Medical, Tunbridge Wells, England, 1979, pp. 215–290.

[29] A. Maroudas, J. Mizrahi, E. BenHaim and I. Ziv, Swelling pressure in cartilage, *Advances in Microcirculation* **13** (1987), 203–212.

[30] V.C. Mow, S.C. Kwei, W.M. Lai and C.G. Armstrong, Biphasic creep and stress relaxation of articular cartilage in compression: theory and experiments, *J. Biomech. Eng.* **102** (1980), 73–84.

[31] V.C. Mow, D.C. Fithian and M.A. Kelly, Fundamentals of articular cartilage and meniscus biomechanics, in: *Articular cartilage and knee joint function: basic science and arthroscopy*, J.W. Ewing, ed., Raven Press, New York, USA, 1990, pp. 1–18.

[32] K.M. Quapp and J.A. Weiss, Material characterization of human medial collateral ligament, *J. Biomech. Eng.* **120** (1998), 757–763.

[33] V. Roth and V.C. Mow, The intrinsic tensile behavior of the matrix of bovine articular cartilage and its variation with age, *J. Bone Joint Surg. Am.* **62** (1980), 1102–1117.

[34] R.M. Schinagl, D. Gurskis, A.C. Chen and R.L. Sah, Depth-dependent confined compression modulus of full-thickness bovine articular cartilage, *J. Orthop. Res.* **15** (1997), 499–506.

[35] M.A. Soltz and G.A. Ateshian, A conewise linear elasticity mixture model for the analysis of tension–compression nonlinearity in articular cartilage, *J. Biomech. Eng.* **122** (2000), 576–586.

[36] J. Soulhat, M.D. Buschmann and A. Shirazi-Adl, A fibril-network-reinforced biphasic model of cartilage in unconfined compression, *J. Biomech. Eng.* **121** (1999), 340–347.

[37] S. Treppo, H. Koepp, E.C. Quan, A.A. Cole, K.E. Kuettner and A.J. Grodzinsky, Comparison of biomechanical and biochemical properties of cartilege from human knee and ankle pairs, *J. Orthopaedic Res.* **18** (2000), 739–748.

[38] J.A. Weiss, B.N. Maker and S. Govindjee, Finite element implementation in incompressible, transversely isotropic hyperelasticity, *Comput. Methods Biomech. Biomed. Eng.* **135** (1996), 107–128.

[39] S.L. Woo, P. Lubock, M.A. Gomez, G.F. Jemmott, S.C. Kuei and W.H. Akeson, Large deformation non-homogeneous and directional properties of articular cartilage in uniaxial tension, *J. Biomech.* **12** (1979), 437–446.

Biorheology 43 (2006) 249–258
IOS Press



# Mechanical responses and integrin associated protein expression by human ankle chondrocytes

M. Orazizadeh [a], C. Cartlidge [b], M.O. Wright [b], S.J. Millward-Sadler [b], J. Nieman [b],
B.P. Halliday [b], H.-S. Lee [c] and D.M. Salter [b,*]
[a] *Department of Histology, School of Medicine, Ahwaz University of Medical Sciences, Ahwaz, Iran*
[b] *Division of Pathology, School of Molecular and Clinical Medicine, College of Medicine and Veterinary Medicine, Edinburgh University, UK*
[c] *Department of Pathology, Tri-Service General Hospital and National Defense Medical Center, Taipei, Taiwan*

**Abstract.** Metabolic, biochemical and biomechanical differences between ankle and knee joint cartilage and chondrocytes including resistance to the effects of catabolic cytokines and fibronectin fragments may be relevant to differences in prevalence of OA in these joints. Although there is increasing information available on how chondrocytes from knee and hip joint cartilage recognise and respond to mechanical stimuli, knowledge of mechanotransduction in ankle joint chondrocytes is limited. This study was undertaken to (i) establish whether the response of normal ankle joint derived chondrocytes to mechanical stimulation *in vitro* was similar to that of normal and osteoarthritic knee joint derived chondrocytes and (ii) to investigate whether these chondrocytes showed differences in expression of integrin associated regulatory and signalling molecules. Unlike normal knee joint chondrocytes, ankle joint chondrocytes did not show an increase in relative levels of aggrecan mRNA when mechanically stimulated. No obvious change in protein tyrosine phosphorylation was seen in ankle chondrocytes subsequent to mechanical stimulation but these cells expressed elevated levels of tyrosine phosphorylated proteins at rest when compared to normal knee joint chondrocytes. Ankle joint chondrocytes showed an increase in protein kinase B phosphorylation following 1 min 0.33 Hz stimulation which was inhibited by the presence of antibodies to $\alpha5\beta1$ integrin. Ankle joint chondrocytes appeared to show significant differences in levels of the integrin-associated proteins CD98, CD147 and galectin 3, PKC$\gamma$ and differences in responses to glutamate were seen. Chondrocytes from ankle and knee joint cartilage respond differently to 0.33 Hz mechanical stimulation. This may be related to modified integrin-dependent mechanotransduction as a result of changes in expression of integrin regulatory molecules such as CD98 or differential expression and function of downstream components of the mechanotransduction pathway such as PKC or NMDA receptors.

Keywords: Cartilage, PKB, integrin, aggrecan

## 1. Introduction

Human knee and ankle joint differ in their prevalence of osteoarthritis (OA) or chondro-degeneration *in vivo* [16]. Approximately 6% of the adult population are affected by symptomatic OA of the knee, this percentage increases to almost 10% in individuals over 65 years of age [9]. Symptomatic OA does develop in the ankle, although rarely ($<1\%$), and the prevalence does not increase with age [23]. A number

---

*Address for correspondence: D.M. Salter, Pathology, Edinburgh Royal Infirmary, Little France Crescent, Edinburgh, EH16 4AS, UK. Tel.: +44 131 242 7125; Fax: +44 131 242 7169; E-mail: donald.salter@ed.ac.uk.

of reports have shown differences in biochemical and biomechanical properties of ankle joint cartilage when compared to cartilage of knees [17]. It has been suggested that differences in susceptibility to OA might be a result of an imbalance between degenerative and repair processes, an idea that has been supported by recent work demonstrating differences in matrix turnover in early OA lesions of ankle and knee cartilage [4].

Proteoglycan (PG) synthesis [9] and PG content [32] of the ankle are higher than that of the knee. Ankle joint chondrocytes are less responsive to IL-1$\beta$ than knee joint chondrocytes [9]. With IL-1$\beta$ stimulation, there is an increase in proteolytic activity in both the knee and ankle, as evidenced by increased neopeptides for both aggrecanase and matrix metalloproteinases (MMPs), however levels of neopeptides are higher in the knee than ankle. Ankle cartilage also shows a higher rebound response to anabolic factors such as bone morphogenetic protein 7 [9]. Significant differences are also seen between on exposure of knee and ankle cartilage to fibronectin-fragment (Fn-fs), ankle joint cartilage being less responsive [15].

The extracellular matrix of ankle cartilage appears to be biomechanically different from knee joint cartilage. Higher GAG content, lower water content, higher equilibrium modulus and dynamic stiffness and lower hydraulic permeability [4] providing characteristics which will allow increased resistance to loading and mechanical damage [26,32]. Ankle joint chondrocytes are therefore likely to be in a mechanical environment which is dissimilar to that of knee joint cells.

As with responses to catabolic and anabolic cytokines and peptides, it is possible that mechanotransduction pathways and events subsequent to mechanical stimulation will differ between ankle joint and knee joint chondrocytes. Such variations could contribute significantly to differences in cartilage structure. This study was therefore undertaken to ascertain whether ankle joint chondrocytes showed differences in mechanical responses to knee joint chondrocytes in an *in vitro* model system and to investigate possible mechanisms for any differences identified.

## 2. Materials and methods

### 2.1. Source of tissue and chondrocyte culture

Human articular cartilage was obtained, with ethical approval and patients' consent, from knee joint arthroplasty specimens and amputations for peripheral vascular disease. Chondrocytes were isolated by sequential enzyme. Cells were seeded in Iscove's modified Dulbecco's medium (Gibco) supplemented with 10% fetal calf serum (Sigma), 100 IU/ml penicillin (Gibco) and 100 $\mu$g/ml streptomycin (Gibco) to a final density of $5 \times 10^5$/ml in 58 mm plastic petri dishes (Nunc). Primary, non-confluent, 1–2 week cultures of chondrocytes were used in all experiments. The day before mechanical stimulation was to be carried out culture media containing serum was replaced by serum-free media.

### 2.2. Mechanical stimulation

The technique has previously been described in detail [29,33]. Briefly, flexible 58 mm plastic culture dishes (Nunc) are placed into a sealed pressure chamber with inlet and outlet ports. The chamber is pressurised using helium gas from a cylinder, at a standard stimulation regime of 0.33 Hz (2 sec on, 1 sec off) for 20 min, 37°C, with a pressure of 16 kPa above atmospheric pressure. This system produces 3700 microstrain on the base of the culture dish. Non-stimulated controls were also placed inside the apparatus, and thus exposed to the same environmental conditions.

## 2.3. Protein extraction, western blotting and immunoprecipitation

The methods for protein extraction, immunoprecipitation and western blotting used have been described previously [18]. In brief, cells at rest or following mechanical stimulation were washed with ice-cold PBS containing 100 $\mu$M $Na_3VO_4$ (Sigma) and lysed *in situ* with ice-cold lysis buffer at 4°C for 15 min. Lysis buffer contained 1% Igepal (Sigma), 100 $\mu$M $Na_3VO_4$, and protease inhibitor cocktail tablet (Boehringer Mannheim). Supernatants were collected after centrifugation at 13,000 rpm for 15 min. Concentration of protein within lysates was determined using Folin–Lowry assay method with Dynatech MR5000 and equivalent amounts loaded and separated on a 7.5% SDS-PAGE under reducing conditions. Following electrophoresis, immunoprecipitated proteins or whole cell lysates were transferred onto polyvinylidene fluoride (PVDF) membranes (Millipore Immobilon-P, Sigma). Membranes were blocked overnight at 4°C with 2% BSA in TBST (12.5 mM Tris/HCl, pH 7.6, 137 mM NaCl, 0.1% Tween 20). After washing with TBST, blots were incubated for 1 h at room temperature with primary antibodies and then HRP labelled secondary antibodies.

## 2.4. Electrophysiological measurements

Membrane potentials of cells were recorded using a single electrode bridge circuit and calibrator [33]. Microelectrodes with tip resistances of 40 to 60 meg ohms and tip potentials of approximately 3 mV were used to impale the cells. Membrane potentials of isolated cells were measured, and results were accepted if, on cell impalement, there was a rapid change in voltage to the membrane potential level that remained constant for at least 20 s. The membrane potentials of 5–10 cells were measured prior to and 10 min following addition of 50 $\mu$M glutamate in the presence of 20 $\mu$M glycine.

## 2.5. RT-PCR and gel analysis

Total RNA was extracted using a denaturing buffer of 4 M guanidine thiocyanate, 0.75 M sodium citrate, 10% (w/v) lauroyl sarcosine and 7.2 $\mu$l/ml $\beta$-mercaptoethanol. Prior to cDNA synthesis, all RNA samples were incubated with DNAse I (Life Technologies, Paisley, UK) for 15 min in the presence of an RNAse inhibitor (Pharmacia, St. Albans, UK). Template cDNA was synthesised using 0.5 $\mu$g RNA, Superscript II and oligo dT (12–18) (Life Technologies, Paisley, UK) according to the manufacturers instructions. The primers used for the PCR reactions were (upstream/downstream):

GAPDH    5′-CCACCCATGGCAAATTCCATGGCA-3′
             5′-TCTAGACGGCAGGTCAGGTCCACC-3′;

Aggrecan    5′-TGAGGAGGGCTGGAACAAGTACC-3′
             5′-GGAGGTGGTAATTGCAGGGAACA-3′.

A typical 20 $\mu$l PCR reaction contained 20 mM ammonium sulphate, 75 mM Tris.HCl, pH 8.8, 0.01% (v/v) Tween-20, 1 $\mu$M each primer, 2 $\mu$l cDNA, 100 $\mu$M dNTPs, 0.1% (w/v) bovine serum albumin, 0.25 U Taq polymerase (Biogene). The magnesium chloride concentrations for each primer pair were: GAPDH – 2.5 mM and aggrecan – 1.25 mM and the following programme was used for aggrecan reactions: 94°C for 3 min; 24 cycles of 94°C for 1 min, 60°C for 1 min, 72°C for 1 min 30 s. PCR products were analysed by electrophoresis using a 1% (w/v) agarose gel stained with ethidium bromide, and the intensity of each band measured under UV fluorescence using EASY Image analysis software

(Scotlab, Coatbridge, UK). The ratio of intensities of the bands for the aggrecan product compared to the housekeeping gene GAPDH was calculated and compared. Each donor was tested in duplicate, and a minimum of three donors was used for each experiment.

## 2.6. Statistics

The mean, standard deviation, and standard error of the mean were calculated for each experimental point where appropriate. The student $t$-test was used to evaluate whether there was statistical significance between each of the points calculated.

## 3. Results

### 3.1. Responses of human ankle articular chondrocytes to mechanical stimulation

#### 3.1.1. Gene expression

Aggrecan is the major proteoglycan of cartilage and relative levels of aggrecan gene mRNA have been shown previously to be increased in knee joint chondrocytes following 20 min mechanical stimulation at 0.33 Hz [21]. Ankle joint derived chondrocytes were stimulated under identical conditions and changes in aggrecan gene expression assessed following a further 1, 3, 6 and 24 hours incubation. The results are shown in Fig. 1. There was a steady decrease in the relative levels of aggrecan mRNA in mechanically stimulated ankle chondrocytes from 1 to 6 hours post stimulation although this did not reach statistical significance, $p < 0.01$, until the 6 hour time point. 24 hours after 0.33 Hz stimulation aggrecan mRNA levels remained significantly decreased in comparison to control levels, $p < 0.02$.

#### 3.1.2. Activation of cell signalling pathways

*Tyrosine phosphorylation*   The results are shown in Fig. 2A. Ankle joint chondrocytes in monolayer culture showed basal levels of protein tyrosine phosphorylation. A prominent band at approximately

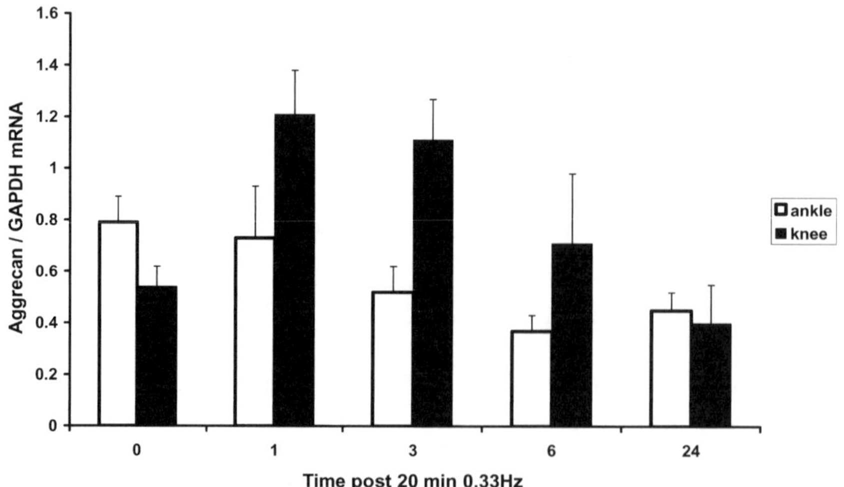


Fig. 1. Effect of 0.33 Hz mechanical stimulation on aggrecan gene expression by normal human ankle and knee joint derived articular chondrocytes. Cells were mechanically stimulated for 20 min and incubated for a further period of 1, 3, 6 or 24 hours before extraction of RNA and semiquantitative RTPCR.

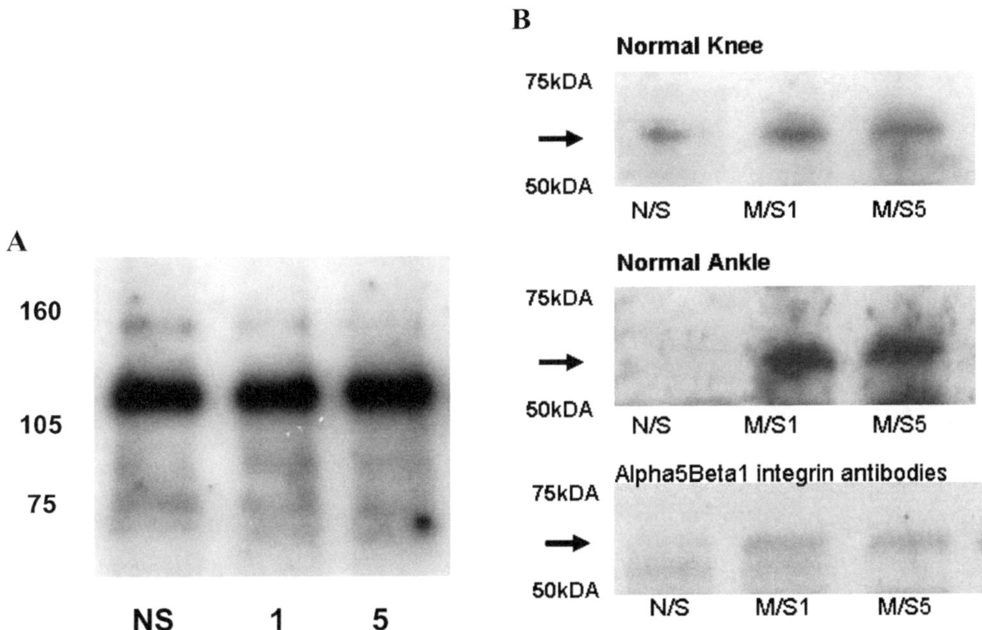

Fig. 2. Effect of 0.33 Hz mechanical stimulation on protein tyrosine phosphorylation (A) and phosphorylation of PKB (B) in human ankle and knee joint derived articular chondrocytes. Protein was extracted from unstimulated cells (NS) or following 1 (M/S1) or 5 (M/S5) min stimulation and western blotting with specific phospho-tyrosine (A) and phospho-PKB (B) antibodies. In separate experiments ankle chondrocytes were pre-incubated with anti $\alpha5\beta1$ integrin antibodies prior to mechanical stimulation and assessment of PKB phosphorylation.

125 kD was clearly seen, with slightly weaker bands being visible at around 160, 90 and 75 kD. There was no evident change in tyrosine phosphorylation following 1 and 5 minutes mechanical stimulation.

*Protein kinase B (PKB) phosphorylation*　　Ankle joint chondrocytes from normal cartilage showed very low basal levels of PKB phosphorylation, similar to that of normal knee joint chondrocytes (Fig. 2B). Following mechanical stimulation for 1 and 5 minutes ankle joint chondrocytes show an increase in PKB phosphorylation. Similar results are seen when chondrocytes from normal knee joint cartilage are stimulated. In contrast osteoarthritic knee joint chondrocytes typically showed high basal levels of phospho-PKB and following mechanical stimulation no convincing differences from unstimulated control could be seen (results not shown).

Addition of the anti-$\alpha5\beta1$ JBS5 integrin antibody for 30 minutes prior to mechanical stimulation had no effect on basal levels of PKB phosphorylation. In the presence of this antibody the increase in PKB phosphorylation following 1 and 5 minutes mechanical stimulation at 0.33 Hz stimulus was greatly reduced (Fig. 2B).

## 3.2. Expression of integrin-associated proteins and PKC isozymes

Previous studies by our group have shown that the biochemical, molecular and electrophysiological responses of normal and OA knee joint chondrocytes activated by 0.33 Hz mechanical stimulation is integrin and PKC dependent with $\alpha5\beta1$ integrin being the major mechanoreceptor. The effects of an anti-$\alpha5\beta1$ antibody on PKB phosphorylation (see above) suggests that this integrin is also involved in ankle joint chondrocyte mechanotransduction.

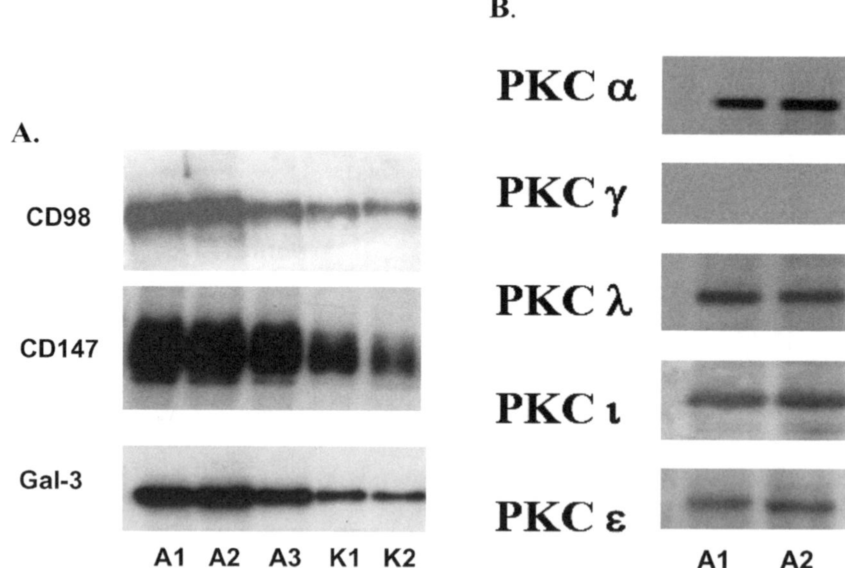

Fig. 3. Expression of CD98, CD147 and galectin 3 (A) and PKC isozymes (B) by human articular chondrocytes extracted from normal ankle (A1-3) or normal knee (Kn 1 and 2) joint articular cartilage. Extracted proteins were western blotted under reducing conditions and probed with specific antibodies against CD98, CD147, galectin 3 and PKC isozymes.

To investigate the possibility that the differential response of ankle chondrocytes is a result of differences in expression of integrin regulatory molecules, protein was extracted from unstimulated cells and expression of the integrin associated molecules CD98, galectin 3 and CD147 (Extracellular Matrix Metallo Proteinase Inducer, EMMPRIN) and PKC isozymes was assessed. The results are shown in Fig. 3. Reproducible differences in detectable levels of CD98, galectin 3 and CD147 were seen with bands of higher intensity in ankle chondrocyte extracts than the bands in lanes loaded with identical amounts of protein from normal and OA knee chondrocytes (Fig. 3A).

Chondrocytes from ankle cartilage expressed 4 of the 5 PKC isozymes studied (Fig. 3B). PKC $\alpha$, $\lambda$, $\iota$ and $\varepsilon$ were seen in protein extracts from ankle joint cartilage whilst no expression of PKC$\gamma$ was seen. Knee joint chondrocytes are known to express PKC $\alpha$, $\lambda$, $\iota$, $\varepsilon$ and $\gamma$ [19].

### 3.3. Electrophysiological response to NMDAR activation

Recently we have demonstrated that both normal and OA knee joint chondrocytes express NMDAR and that this member of the metabotropic glutamate receptors has potential roles in mechanotransduction [30]. To investigate whether ankle joint chondrocytes expressed functional NMDAR we studied the changes in membrane receptors of ankle joint chondrocytes subsequent to the addition of NMDA and compared these with the results of similar experiments carried out on knee joint chondrocytes from the same individual. The results are shown in Fig. 4. Ankle joint chondrocytes showed no significant change in membrane potential in the presence of glutamate. This is in contrast to the observations with normal and osteoarthritic knee joint derived chondrocytes which show membrane hyperpolarisation and membrane depolarisation respectively following exposure to 50 $\mu$M glutamate in the presence of 20 $\mu$M glycine.

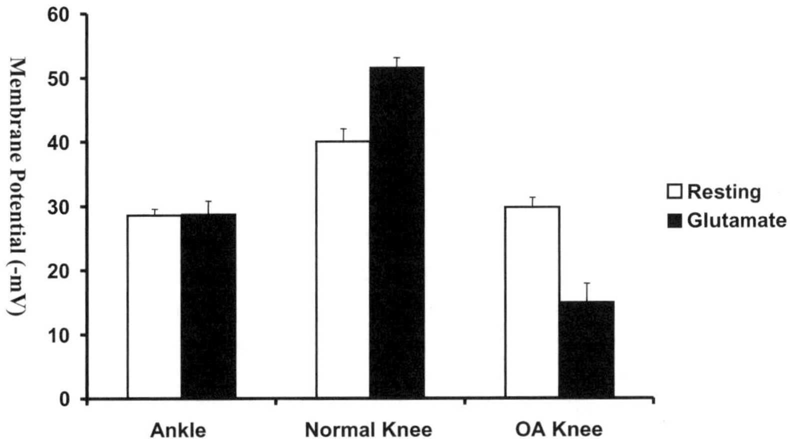


Fig. 4. Effect of 50 $\mu$M glutamate (in the presence of 20 $\mu$M glycine) on the membrane potential of ankle joint and knee joint chondrocytes derived from a single patient. Membrane potentials of cells were measured before and 5 minutes following addition of glutamate with glycine.

## 4. Discussion

Clinical observations on the incidence of osteoarthritis in the ankle joint suggest that this joint may be a privileged site when compared to knee and hip joints and it was suspected that cartilage composition and chondrocyte phenotype may be important in restricting disease in this joint. Subsequent work, primarily from Prof. Kuettner's group has shown a number of metabolic and biochemical differences that support the idea that ankle joint cartilage and resident chondrocytes significantly differ from cartilage and chondrocytes of the knee joint [17]. The results of the present study from our group would support this idea. We show that ankle joint chondrocytes, under identical experimental conditions show differences in response to mechanical and pharmacological activation when compared to knee joint chondrocytes and, additionally appear to show differences in PKC isozyme expression and levels of expression of the integrin associated proteins CD98, CD147 and galectin-3.

In the *in vitro*, model system used in the present study, 0.33 Hz mechanical stimulation of chondrocytes derived from normal knee and hip joint chondrocytes induces a chondroprotective response which is $\alpha 5\beta 1$ integrin and IL4 dependent resulting in an increase levels of aggrecan mRNA and proteoglycan synthesis. OA chondrocytes from knee joint cartilage do not show this anabolic response. When stimulated at 0.33 Hz in this model system OA chondrocytes produce IL1$\beta$ and show no changes in aggrecan gene or proteoglycan expression despite signalling through $\alpha 5\beta 1$ integrin [18,19,21,33]. In contrast, when stimulated at 0.33 Hz ankle joint chondrocytes show a decrease in relative levels of aggrecan mRNA.

Experiments have not undertaken to ascertain whether the aggrecan response in ankle chondrocytes was integrin mediated however the experiments looking at PKB phosphorylation provide evidence of integrin activation in ankle chondrocytes by mechanical stimulation. PKB is a 57 kDa serine/threonine kinase, originally identified as an inactivator of glycogen synthase (GSK3$\beta$) in response to insulin-like growth factor [5]. PKB, when activated by phosphorylation on amino acids Thr308 and Ser473 has several important effects including inhibition of apoptosis by phosphorylation and inactivation of pro-apoptotic factors Bad and caspase-9 [7]. The observations that phosphorylation of PKB in mechanically stimulated ankle chondrocytes was blocked by antibodies to $\alpha 5\beta 1$ integrin suggests that this receptor is also involved in ankle joint mechanotransduction, with at least some downstream pathways being differentially activated. Studies in other systems have demonstrated a link between the $\beta_1$ integrin signalling and PKB activation via ILK, and it is plausible that this is the method of PKB phosphorylation following mechanical stimulation [8,11]. Differences seen in protein tyrosine phosphorylation subsequent to mechanical stimulation of ankle joint chondrocytes also suggests that integrin dependent signalling is dissimilar in the cells from the different joints.

In addition to interacting with extracellular matrix (*trans*), integrins can also form associations with other receptors on the same cell (*cis*) to form multireceptor complexes. These complexes recruit signalling molecules to sites of cell–matrix adhesion, such as focal complexes or focal adhesions. So far, there is little evidence that integrins can signal through exclusive, integrin specific pathways. Instead, they appear to cooperate with other cell surface receptors to influence a variety of signalling pathways. These complexes of integrins and partner receptors can be formed either in response to or independent of integrin activation and ligation. The classes of receptors known to be associated both physically and functionally with integrins include CD47, CD98, CD147 and galectin-3.

CD47 is a unique Ig superfamily member which is known to associate with a number of different integrins and modify cell signalling through these molecules [3]. Interestingly CD47 has also been shown to be involved in bone cell responses to 0.33 Hz mechanical stimulation [29]. No clear differences in expression of this molecule and a known ligand SIRP$\alpha$ have been shown in parallel studies in our laboratory (M. Orazizadeh unpublished observations, PhD thesis, Edinburgh University). CD98, CD147 and galectin-3, however each appear to be expressed by chondrocytes extracted from normal ankle cartilage at higher levels than in chondrocytes extracted from normal and OA cartilage from knee joints. CD98 is a disulfide-linked $\sim$125 kD heterodimer composed of a glycosylated $\sim$85-kD heavy chain (designated CD98) and a nonglycosylated 40-kD light chain. CD98 is expressed at low levels on the surface of quiescent cells but expression is rapidly up-regulated after cellular activation. Functional and physical associations between CD98 and $\beta1$ integrins through which CD98 may modify integrin function have been described [22]. CD147/EMMPRIN is a highly glycosylated transmembrane glycoprotein of 50–60 kD having typical features of an integral membrane protein of the immunoglobulin superfamily. CD147 associates physically with $\beta1$ integrins in the cell membrane [2]. Galectin-3 is a non-glycosylated protein which, as a lectin, is specific for $\beta$-galactosides, has also been shown to interact with and regulate cell function through $\beta1$ integrins [24]. It is interesting to speculate that differences in expression of these molecules may be important in regulating integrin-dependent responses of chondrocytes in ankle cartilage including mechanotransduction. Alternatively integrin independent functions of these molecules may also be important in defining the distinctive characteristics of ankle cartilage. CD147 was initially identified on the surface of human cancer cells and shown to stimulate adjacent stromal cells to produce and activate several MMPs, including MMP-1, MMP-2, MMP-3, membrane type 1MMP (MT1-MMP), and MT-2-MMP [13,31]. Galelctin-3 has been described in osteoarthritic articular cartilage [12] where it may function in chondrocyte–matrix interactions and through matrix metalloproteinase-9 [25] regulate tissue remodelling. Nevertheless additional work is clearly necessary to establish whether direct or indirect (via integrins) effects of these molecules on chondrocyte function may be important in the recognised unique properties of ankle articular cartilage.

The significance of the differences in expression of PKC $\gamma$ expression is not clear. PKC signalling has diverse roles in regulation of chondrocyte function and differentiation. Less is known at present about roles of specific isozymes. PKC $\gamma$ is a member of the classical PKC subfamily which is activated by calcium and diacylglycerol in the presence of phosphatidylserine. Unlike the other classical PKC isozymes, PKC $\alpha$ and $\beta$, PKC $\gamma$ shows restricted distribution in certain tissues such as the central neural system [28].

Addition of NMDA/glutamate to normal human articular chondrocytes has previously been shown to induce membrane hyperpolarisation and blockade of NMDAR prevents the hyperpolarisation response of normal human articular chondrocytes to 0.33 Hz mechanical stimulation [30]. In the present study we show that, unlike knee joint chondrocytes ankle joint chondrocytes do not show a membrane potential response when exposed to glutamate. The reasons for these differences remain to be elucidated. It may be that ankle joint chondrocytes do not express appropriate glutamate/NMDA receptors. Alternatively ankle joint chondrocytes may express NMDA receptors but signalling through this receptor family differs in the different cell populations. Interestingly PKC $\gamma$ is known to be associated with NMDA signalling in neuronal cells [20]. In view of the observations that NMDA receptors appear to be involved in the mechanical response of normal knee joint chondrocytes [30] additional work requires to be undertaken to ascertain potential roles for this receptor in ankle joint responses to mechanical stimulation.

The current work adds to the growing body of information in the literature indicating that ankle joint cartilage and resident chondrocytes are significantly different from their counterparts in knee cartilage. We have identified evidence that ankle chondrocytes respond to mechanical stimulation but stimuli which induce an anabolic response in knee joint chondrocytes appear to have different effects on ankle chondrocytes. The reasons for this difference is unclear but probably relate to the different and biochemical and biomechanical properties of ankle joint cartilage which may lead to resident cells perceiving different mechanical forces from those seen in knee joint cartilage.

## Acknowledgements

This work was supported by grants from Action Medical Research and Arthritis Research Campaign, Ministry of Health and Medical Education, Iran (M. Orazizadeh) and National Defence Medical Center and Tri-Service General Hospital, Taiwan (H.-S. Lee, NSC 91-2314-B-016-034 and TGSH-C91-58).

## References

[1] M. Aurich, A.R. Poole, A. Reiner, C. Mollenhauer, A. Margulis, K.E. Kuettner et al., Matrix homeostasis in aging normal human ankle cartilage, *Arthritis Rheum.* **46** (2002), 2903–2910.

[2] F. Berditchevski, S. Chang, J. Bodorova and M.E. Hemler, Generation of monoclonal antibodies to integrin-associated proteins. Evidence that alpha3beta1 complexes with EMMPRIN/basigin/OX47/M6, *J. Biol. Chem.* **272** (1997), 29174–29180.

[3] E.J. Brown and W.A. Frazier, Integrin-associated protein (CD47) and its ligands, *Trends Cell Biol.* **11** (2001), 130–135.

[4] A.A. Cole, A. Margulis and K.E. Kuettner, Distinguishing ankle and knee articular cartilage, *Foot Ankle Clin.* **8** (2003), 305–316.

[5] D.A.E. Cross, D.R. Alessi, P. Cohen, M. Andjelkovic and B.A. Hemmings, The poto-oncogene PKB/Akt mediates the inhibiton of glycogen synthase kinase-3 by insulin in the skeletal muscle cell line L6, *Nature* **378** (1995), 785–789.

[6] Y. Dang, A.A. Cole and G.A. Homandberg, Comparison of the catabolic effects of fibronectin fragments in human knee and ankle cartilages, *Osteoarthritis Cartilage* **11** (2003), 538–547.

[7] S.R. Datta, A. Brunet and M.E. Greenberg, Cellular survival: a play in three Akts, *Genes Dev.* **13** (1999), 2903–2927.

[8] S.R. Delcommene, C. Tan, V. Gray, L. Rue, J. Woodgett and S. Dedhar, Phosphoinositide-3-OH kinase-dependent regulation of glycogen synthase kinase 3 and protein kinase B/Akt by the integrin-linked kinase, *Proc. Natl. Acad. Sci. USA* **95** (1998), 11211–11216.

[9] W. Eger, B.L. Schumacher, J. Mollenhauer, K.E. Kuettner and A.A. Cole, Human knee and ankle cartilage explants: catabolic differences, *J. Orthop. Res.* **20** (2002), 526–534.

[10] D.T. Felson, A. Naimark, J. Anderson, L. Kazis, W. Castelli and R.F. Meenan, The prevalence of knee osteoarthritis in the elderly: The Framingham Osteoarthritis Study, *Arthritis Rheum.* **30** (1987), 914–918.

[11] D. Gary, O. Milhavet, S. Camandola and M.P. Mattson, Essential role for integrin linked kinase in Akt-medicated integrin survival signalling in hippocampal neurons, *J. Neurochem.* **84** (2003), 878–890.

[12] M. Guévremont, J. Martel-Pelletier, C. Boileau, F.-T. Liu, M. Richard, J.-C. Fernandes, J.-P. Pelletier and P. Reboul, Galectin-3 surface expression on human adult chondrocytes: a potential substrate for collagenase-3, *Ann. Rheum. Dis.* **63** (2004), 636–643.

[13] H. Guo, S. Zucker, M.K. Gordon, B.P. Toole and C. Biswas, Stimulation of matrix metalloproteinase production by recombinant extracellular matrix metalloproteinase inducer from transfected Chinese hamster ovary cells, *J. Biol. Chem.* **272** (1997), 24–27.

[14] Huch, K.E. Kuettner and P. Dieppe, Osteoarthritis in ankle and knee joints [review], *Semin. Arthritis Rheum.* **26** (1997), 667–674.

[15] Y. Kang, H. Koepp, A.A. Cole, K.E. Kuettner and G.A. Homandberg, Cultured human ankle and knee cartilage differ in susceptibility to damage mediated by fibronectin fragments, *J. Orthop. Res.* **16** (1998), 551–556.

[16] H. Koepp, W. Eger, C. Muehleman, A. Valdellon, J.A. Buckwalter, K.E. Kuettner and A.A. Cole, Prevalence of articular cartilage degeneration in the ankle and knee joints of human organ donors, *J. Orthop. Sci.* **4** (1999), 407–412.

[17] K.E. Kuettner and A.A. Cole, Cartilage degeneration in different human joints, *Osteoarthritis Cartilage* **13** (2005), 93–103.

[18] H.-S. Lee, S.J. Millward-Sadler, M.O. Wright, G. Nuki and D.M. Salter, Integrin and mechanosensitive ion channel-dependent tyrosine phosphorylation of focal adhesion proteins and beta-catenin in human articular chondrocytes after mechanical stimulation, *J. Bone Miner. Res.* **15** (2000), 1501–1509.

[19] H.-S. Lee, S.J. Millward-Sadler, M.O. Wright, G. Nuki, R. Al-Jamal and D.M. Salter, Activation of Integrin-RACK1/PKCα signalling in human articular chondrocyte mechanotransduction, *Osteoarthritis Cartilage* **10** (2002), 890–897.

[20] W.J. Martin, A.B. Malmberg and A.I. Basbaum, PKCgamma contributes to a subset of the NMDA-dependent spinal circuits that underlie injury-induced persistent pain, *J. Neurosci.* **21** (2001), 5321–5327.

[21] S.J. Millward-Sadler, M.O. Wright, L.W. Davies, G. Nuki and D.M. Salter, Mechanotransduction via integrins and interleukin-4 results in altered aggrecan and matrix metalloproteinase 3 gene expression in normal but not osteoarthritic human articular chondrocytes, *Arthritis Rheum.* **43** (2000), 2091–2099.

[22] Y.J. Miyamoto, J.S. Mitchell and B.W. McIntyre, Physical association and functional interaction between beta1 integrin and CD98 on human T lymphocytes, *Mol. Immunol.* **39** (2003), 739–751.

[23] C. Muehleman, D. Bareither, K. Huch, A.A. Cole and K.E. Kuettner, Prevalence of degenerative morphological changes in the joints of the lower extremity, *Osteoarthritis Cartilage* **5** (1997), 23–37.

[24] J. Ochieng, M.L. Leite-Browning and P. Warfield, Regulation of cellular adhesion to extracellular matrix proteins by galectin-3, *Biochem. Biophys. Res. Commun.* **246** (1998), 788–791.

[25] N. Ortega, D.J. Behonick, C. Colnot, D.N. Cooper and Z. Werb, Galectin-3 is a downstream regulator of matrix metalloproteinase-9 function during endochondral bone formation, *Mol. Biol. Cell.* (2005) [Epub ahead of print].

[26] P. Patwari, M.N. Cook, M.A. DiMicco, S.M. Blake, I.E. James, S. Kumar et al., Proteoglycan degradation after injurious compression of bovine and human articular cartilage *in vitro*: interaction with exogenous cytokines, *Arthritis Rheum.* **48** (2003), 1292–1301.

[27] A.R. Poole, M. Ionescu, A. Swan and P.A. Dieppe, Changes in cartilage metabolism in arthritis are reflected by altered serum and synovial fluid levels of the cartilage proteoglycan aggrecan: implications for pathogenesis, *J. Clin. Invest.* **94** (1994), 25–33.

[28] N. Saito and Y. Shirai, Protein kinase C gamma (PKC gamma): function of neuron specific isotype, *J. Biochem. (Tokyo)* **132** (2002), 683–687.

[29] D.M. Salter, J.E. Robb and M.O. Wright, Electrophysiological responses of human bone cells to mechanical stimulation: Evidence for specific integrin function in mechanotransduction, *J. Bone Miner. Res.* **12** (1997), 1133–1141.

[30] D.M. Salter, M.O. Wright and S.J. Millward-Sadler, NMDA receptor expression and roles in human articular chondrocyte mechanotransduction, *Biorheology* **41** (2004), 273–281.

[31] T. Sameshima, K. Nabeshima, B.P. Toole, K. Yokogami, Y. Okada, T. Goya et al., Glioma cell extracellular matrix metalloproteinase inducer (EMMPRIN) (CD147) stimulates production of membrane-type matrix metalloproteinases and activated gelatinase A in co-cultures with brain-derived fibroblasts, *Cancer Lett.* **157** (2000), 177–184.

[32] S. Treppo, H. Koepp, E.C. Quan, A.A. Cole, K.E. Kuettner and A.J. Grodzinsky, Comparison of biomechanical and biochemical properties of cartilage from human knee and ankle pairs, *J. Orthop. Res.* **18** (2000), 739–748.

[33] M.O. Wright, K. Nishida, C. Bavington, J.L. Godolphin, E. Dunne, S. Walmsley, P. Jobanputra, G. Nuki and D.M. Salter, Hyperpolarisation of cultured human chondrocytes following cyclical pressure-induced strain: Evidence of a role for α5β1 integrin as a chondrocyte mechanoreceptor, *J. Orthop. Res.* **15** (1997), 742–747.

Biorheology 43 (2006) 259–269
IOS Press



# Chondrocyte gene expression under applied surface motion

Sibylle Grad [a,*], Cynthia R. Lee [a], Markus A. Wimmer [b] and Mauro Alini [a]

[a] *Biomaterials and Tissue Engineering Program, AO Research Institute, Davos, Switzerland*
[b] *Department of Orthopedics, Rush University Medical Center, Chicago, IL, USA*

**Abstract.** A cartilage bioreactor has been designed that is intended to approximate the kinematics of natural joints and allows for functional cartilage tissue engineering studies. In particular, interface motion can be generated by oscillation of a ball over the surface of a construct. The present study investigated the specific effect of applied articular motion on the gene expression of chondrocytes cultured in 3D scaffolds, with a particular emphasis on different superficial zone protein (SZP)/lubricin transcripts. Cylindrical porous polyurethane scaffolds were seeded with bovine articular chondrocytes and subjected to dynamic compression, with or without articulation against a ceramic hip ball. Articular motion markedly up-regulated the mRNA expression of the four previously described and two newly identified SZP/lubricin isoforms and of cartilage oligomeric matrix protein (COMP), and, to a lesser extent, aggrecan, type II collagen and TIMPs, while axial compression alone had no effect on the chondrocytes' gene expression levels. These results demonstrate the beneficial effect of articular motion not only for stimulation of important lubricating molecules, but also for the preservation of the chondrocytic phenotype.

Keywords: Cartilage tissue engineering, surface motion, gene expression, superficial zone protein (SZP)

## 1. Introduction

With increasing awareness that mechanical stimuli play a fundamental role in the natural development and maintenance of healthy articular cartilage, the field of functional cartilage tissue engineering has grown substantially in recent years. In an attempt to generate functional, biomechanically competent tissue, controlled physical forces have been applied to chondrocytic cells in various cell culture systems. Classically, mechanical stimuli have included tissue level dynamic compression and the associated phenomena of fluid-flow induced shear, tissue shear, and hydrostatic pressure. In particular, numerous studies have applied unconfined dynamic compression protocols to tissue engineering systems using hydrogels or macroporous scaffolds to stimulate cell differentiation, proliferation, biosynthetic activity, and the development of a functional extracellular matrix [2,4,19,23,24,33]. Spinner flasks, rotating wall and perfusion culture systems have been used to investigate the effects of fluid flow-induced convective transport and shear on tissue engineered constructs [28,36]. Direct tissue shear strain has also been applied to cell-scaffold constructs to improve the development of cartilaginous constructs [37]. In addition, the benefits of cyclic hydrostatic pressure on the accumulation of extracellular matrix components have been reported in different systems [3,27]. However, the loading experienced by natural joints is more complicated than simple compression, shear or pressurization. Thus, articular cartilage *in vivo* is exposed to a complex interaction of physical forces. To address this combined interplay of

---

*Address for correspondence: Sibylle Grad, PhD, AO Research Institute, Clavadelerstrasse, 7270 Davos Platz, Switzerland. Tel.: +41 81 414 23 90; Fax: +41 81 414 22 88; E-mail: sibylle.grad@aofoundation.org.

joint-level compression, shear, and articular motion, we have developed a cartilage bioreactor that allows for simultaneous compression, shear, and articular fluid transport of developing constructs [12,38]. This concept of construct loading is motivated by an attempt to reproduce the movements and contact pressures that are relevant during normal joint loading *in vivo*. One characteristic feature of our bioreactor is that articular surface motion can be simulated by oscillation of a ceramic hip ball over the surface of a tissue-engineered construct. This type of motion should approximate joint movement during human locomotion and is believed to enhance the transport of fluid, helping to build up fluid flow at the ball–specimen interface. Most notably, we have shown that applied surface motion, generated by intermittent articulation of an alumina hip ball over the surface of chondrocyte-seeded porous polyurethane scaffolds, resulted in an increase in both mRNA expression and release of superficial zone protein (SZP) [12]. SZP is a product of the proteoglycan 4 (PRG4) gene that is specifically expressed by superficial zone chondrocytes at the cartilage surface [31,32]. PRG4 also encodes the related gene products lubricin [16,17], megakaryocyte-stimulating factor (MSF) [26] and hemangiopoietin (HAPO) [25]. Lubricin is a major component of synovial fluid that participates in the boundary lubrication of synovial joints [16]. Previous reports have shown that four SZP/lubricin mRNA transcripts were expressed by both human synovial fibroblasts and articular chondrocytes through alternatively splicing the exons 2, 4, and/or 5 of the 12 exons out of the precursor gene [18]. The specific roles of the different isoforms are still not known. To address the question if applied surface motion preferably up-regulates the expression of distinct isoforms, the mRNA expression levels of different SZP/lubricin splice variants of stimulated compared to free swelling chondrocytes-seeded polyurethane constructs were analyzed. In addition, the effect of applied dynamic compressive loading with and without articular surface motion on the gene expression of cartilage matrix proteins, matrix degrading enzymes and their inhibitors were determined to evaluate the potential of our bioreactor concept for promoting the development of functional constructs.

## 2. Materials and methods

### 2.1. Polyurethane scaffolds

Cylindrical (8 mm × 4 mm) porous polyurethane scaffolds were prepared as described elsewhere [10]. The scaffolds with interconnected pores had an average pore size of 200–400 $\mu$m. The polymers used for scaffold preparation were synthesized using hexamethylene diisocyanate, poly($\varepsilon$-caprolactone) diol with a molecular weight of 530 daltons and isosorbide diol (1,4:3,6-dianhydro-D-sorbitol) as chain extender [11]. The scaffolds were attached to the specimen holders [38] using bone cement (Norian SRS, Norian Corp., Cupertino, CA). Then they were sterilized in a cold-cycle (37°C) ethylene oxide process and subsequently evacuated at 45°C and 150 mbar for 3–4 days.

### 2.2. Chondrocyte isolation, seeding and culture conditions

Chondrocytes were isolated from full thickness metacarpal joint cartilage of young (3–4 months old) calves using sequential pronase and collagenase digestion [13]. Isolated chondrocytes ($10 \times 10^6$ cells/scaffold) were suspended in fibrinogen solution and then combined with thrombin solution immediately prior to seeding them into the polyurethane scaffold [22]. The fibrin components were provided by Baxter Biosurgery (Vienna, Austria). The final concentrations of the fibrin gel were 17 mg/ml fibrinogen, 0.5 U/ml thrombin, and 665 KIE/ml aprotinin [22]. Constructs were incubated for 45 min (37°C, 5% $CO_2$, 95% humidity) to permit fibrin gelation before adding growth medium (DMEM supplemented

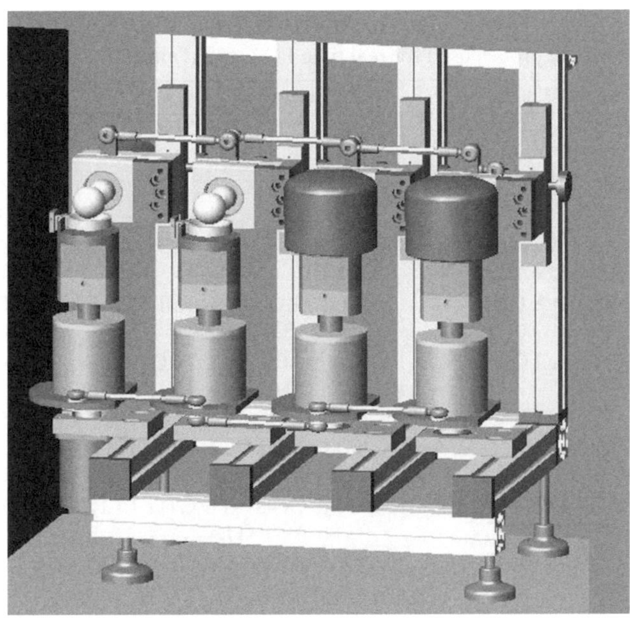

Fig. 1. Four-station cartilage bioreactor that allows for application of joint specific biomechanical stimuli.

with antibiotics, 10% FCS, 50 $\mu$g/ml ascorbic acid, 40 $\mu$g/ml L-proline, non-essential amino acids, and 500 kIU/ml aprotinin). Since cells are particularly susceptible to oxidative damage directly after enzymatic treatment for cell isolation [35], the addition of ascorbic acid was delayed until two days post-seeding. After 5 days in culture, cell-scaffold constructs were exposed to mechanical loading as described below.

### 2.3. Mechanical loading

Mechanical conditioning of cell-scaffold constructs was performed using our bioreactor system, which was installed into a $CO_2$ incubator held at 37°C, 5% $CO_2$, and 85% humidity [38] (Fig. 1). Briefly, a commercially available ceramic hip ball 32 mm in diameter was pressed onto the cell-seeded scaffolds. Interface motion was generated by oscillation of the ball over the construct surface. Compressive strains were applied simultaneously along the cylindrical axis of the construct.

One hour of mechanical loading was performed twice a day over 5 consecutive days. Dynamic compression was performed at a frequency of 0.1 Hz, with 10% sinusoidal strain, superimposed on a 10% static offset strain, resulting in actual strain amplitude of 10–20%. Simultaneously, the ball oscillated at 0.1 Hz with amplitude of $\pm30°$. Between the loading periods, the constructs were kept free swelling, without any contact to the ball. Unloaded constructs served as controls. Results of six independent experiments run in duplicates are presented.

### 2.4. Gene expression analysis

After 10 one-hour loading cycles, constructs were pulverized under liquid nitrogen, and total RNA was extracted with TRI Reagent (Molecular Research Center, Cincinnati, OH) according to the manufacturer's specifications, using the modified precipitation method with a high salt precipitation solution

Table 1

Oligonucleotide primers and probes used for real time PCR. (a)–(f) in SZP isoforms correspond to Fig. 2

| Gene | Primer fw ($5'$–$3'$) | Primer rev ($5'$–$3'$) | Probe ($5'$FAM/$3'$TAMRA) |
|---|---|---|---|
| Aggrecan | CCA ACG AAA CCT ATG ACG TGT ACT | GCA CTC GTT GGC TGC CTC | ATG TTG CAT AGA AGA CCT CGC CCT CCA T |
| Procollagen 1A2 | TGC AGT AAC TTC GTG CCT AGC A | CGC GTG GTC CTC TAT CTC CA | CAT GCC AAT CCT TAC AAG AGG CAA CTG C |
| Procollagen 2A1 | AAG AAA CAC ATC TGG TTT GGA GAA A | TGG GAG CCA GGT TGT CAT C | CAA CGG TGG CTT CCA CTT CAG CTA TGG |
| COMP | CCA GAA GAA CGA CGA CCA GAA | TCT GAT CTG AGT TGG GCA CCT T | ACG GCG ACC GGA TCC GCA A |
| MMP-3 | GGC TGC AAG GGA CAA GGA A | CAA ACT GTT TCG TAT CCT TTG CAA | CAC CAT GGA GCT TGT TCA GCA ATA TCT AGA AAA C |
| MMP-13 | CCA TCT ACA CCT ACA CTG GCA AAA G | GTC TGG CGT TTT GGG ATG TT | TCT CTC TAT GGT CCA GGA GAT GAA GAC CCC |
| SZP (all isoforms) | GAG CAG ACC TGA ATC CGT GTA TT | GGT GGG TTC CTG TTT GTA AGT GTA | CTG AAC GCT GCC ACC TCT CTT GAA A |
| TIMP1 | GGA CCG CAG AAG TCA ATG AAA | GGG TGT AGA TGA ACC GGA TGT C | TAC CAG CGT TAT GAG ATC AAG ATG ACT AAG ATG |
| TIMP3 | CCT TTG GCA CGA TGG TCT ACA | TTA AGG CCA CAG AGA CTT TCA GAA G | AAG CAG ATG AAG ATG TAC CGA GGA TTC A |
| SZP V0 | TGCACTGTGGAGCTTTCC TGTA (**a**) | GAATGTTCTTCTGTTATTT CCTCTGATTC (**c**) | TCTTGCAATCTGAGTCGC AGTCACACTCC (**f**) |
| SZP V1 | | GTTATCTTTTACTTCTGTTA TTTCCTCTGATTC (**d**) | |
| SZP V2 | | TGTTCTTCTTGTTATCTTTT ACTGCTTCAC (**e**) | |
| SZP V3 | GTTTCATCTCAAGAGCTTT CCTGTA (**b**) | | |
| SZP V0a | | GAATGTTCTTCTGTTATTT CCTCTGATTC (**c**) | |
| SZP V1a | | GTTATCTTTTACTTCTGTTA TTTCCTCTGATTC (**d**) | |

(Molecular Research Center). Reverse transcription was performed with TaqMan reverse transcription reagents (Applied Biosystems, Foster City, CA) using random hexamers and 1 $\mu$g of total RNA sample.

PCR was performed on a 7500 Real Time PCR System (Applied Biosystems, Foster City, CA). Oligonucleotide primers and TaqMan probes (Table 1; all from Microsynth, Balgach, CH) were designed with Primer Express Oligo Design software, versions 1.5/2.0 (Applied Biosystems). Figure 2 shows a schematic overview of the four phenotypic isoforms of SZP/lubricin expression identified in human synovial fibroblasts and articular chondrocytes [18], the two additional variants analyzed, and the location of the primers and probe specified in Table 1. Probes were labelled with the reporter dye molecule FAM (6-carboxyfluorescein) at the $5'$ end and with the quencher dye TAMRA (6-carboxy-N, N, N$'$, N$'$-tetramethylrhodamine) at the $3'$ end. The nucleotide sequences were taken from the GenBank database. Regions where bovine sequences were not available were substituted by the human sequence. To exclude amplification of genomic DNA, the probe or one of the primers were selected to overlap an exon-exon junction. Primers and probe for amplification of 18S ribosomal RNA for use as endogenous

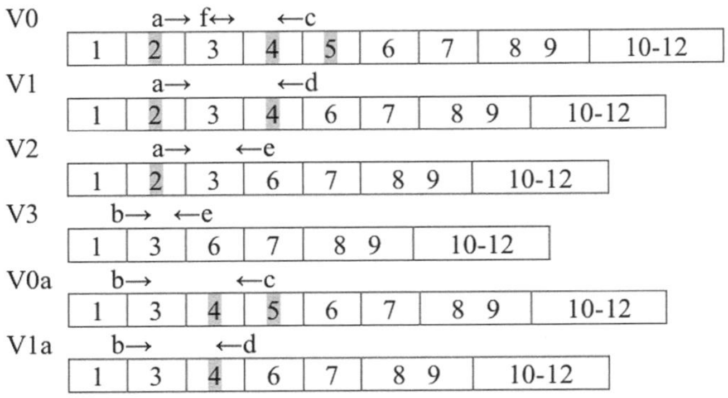

V0    a→ f↔    ←c
| 1 | 2 | 3 | 4 | 5 | 6 | 7 | 8 | 9 | 10-12 |

V1    a→    ←d
| 1 | 2 | 3 | 4 | 6 | 7 | 8 | 9 | 10-12 |

V2    a→  ←e
| 1 | 2 | 3 | 6 | 7 | 8 | 9 | 10-12 |

V3    b→ ←e
| 1 | 3 | 6 | 7 | 8 | 9 | 10-12 |

V0a   b→    ←c
| 1 | 3 | 4 | 5 | 6 | 7 | 8 | 9 | 10-12 |

V1a   b→  ←d
| 1 | 3 | 4 | 6 | 7 | 8 | 9 | 10-12 |

Fig. 2. Scheme of the four SZP/lubricin isoforms described in [18] and the two additional variants analyzed, and location of the primers/probe specified in Table 1.

control were from Applied Biosystems. The PCR was carried out under standard thermal conditions with TaqMan Universal PCR master mix (Applied Biosystems), 900 nM primers (forward and reverse), and 250 nM TaqMan probe. Relative quantification of target mRNA was performed according to the comparative $C_T$ method with 18S ribosomal RNA as endogenous control [1].

## 2.5. ELISA for MSF/SZP/lubricin in conditioned media

Conditioned culture media of the 5 days of mechanical loading were pooled for each construct and were analyzed by ELISA using a method adapted from Klein et al. [20]. Medium was diluted 1 : 2 with PBS containing 0.2 M guanidine-HCl, pH 7.4, and was allowed to bind to wells of 96-well ELISA plates overnight at 4°C. Recombinant human MSF (kindly provided by C.R. Flannery, Wyeth Research, Cambridge, MA) was used to prepare a standard curve. Plates were washed three times with PBS containing 0.05% Tween 20 between all steps. Non-specific binding was blocked with PBS containing 1% bovine serum albumin (BSA) for 1 hour. Plates were then incubated for 1 hour with a 1 : 1000 dilution of rabbit anti-human MSF antibody 06A10 (kindly provided by C.R. Flannery, Wyeth, MA) in PBS containing 1% BSA and 0.05% Tween 20, followed by an additional 1 hour incubation with a 1 : 5000 dilution of a horseradish peroxidase-linked donkey anti-rabbit IgG (Amersham Biosciences, Buckinghamshire, UK). TMB substrate (Sigma-Aldrich, St. Louis, MO) was applied for 30 minutes, and the reaction was stopped by the addition of 1 M sulfuric acid. Optical density was measured at 450 nm and used to calculate sample MSF/SZP/lubricin concentrations, expressed as MSF equivalent, using the standard curve. The DNA content of each construct was determined using Hoechst 33258 dye assay [21] after proteinase K digestion, and MSF/SZP/lubricin immunoreactivity was normalized to the DNA amount of the samples.

## 2.6. Statistical analysis

The mRNA expression levels of mechanically stimulated constructs were normalized to the expression levels of unloaded controls, and normalized values were analyzed by one-way ANOVA and Fisher PLSD post-hoc testing (STATVIEW statistical software, SAS Institute, Inc.) to compare loading groups (significance at $p < 0.05$).

## 3. Results

As expected, the mRNA expression of SZP/lubricin (not distinguishing isoforms) was increased (4.1 ± 0.6 times) in cell-scaffold constructs exposed to surface motion, superimposed onto cyclic compression, compared to free swelling controls [12]. The mRNA expression of COMP was also enhanced (6.6 ± 1.0 times) by applied surface motion, which is consistent with previous observations [38]. The gene expression levels of aggrecan, procollagen type II and of TIMP-1 and TIMP-3 were also up-regulated in constructs subjected to compression and surface motion; however, these genes showed a mean increase of between 1.6 and 2.2 times compared to unloaded control samples, which, although statistically significant, may not be of biological relevance. Cyclic axial compression alone had a slight effect on the mRNA expression of TIMP-1, but did not affect the expression levels of any of the other genes analyzed (Fig. 3).

Separate gene expression analysis of the four known SZP/lubricin isoforms and of the two additionally analyzed variants showed that all transcripts were enhanced in constructs exposed to surface motion superimposed onto cyclic compression. The variants V0, V1, V3, and the new forms V0a and V1a were up-regulated to a similar extent, whereas the up-regulation of variant V2 was less, but still significant (Fig. 4).

The MSF/SZP/lubricin immunoreactivity, expressed as MSF equivalent and normalized to the DNA content of the samples, was not altered in conditioned culture media of constructs exposed to dynamic compression compared to media of unloaded control samples. However, significantly elevated MSF/SZP/lubricin immunoreactivity was observed in media of constructs exposed to dynamic compression and superimposed surface motion (Fig. 5).

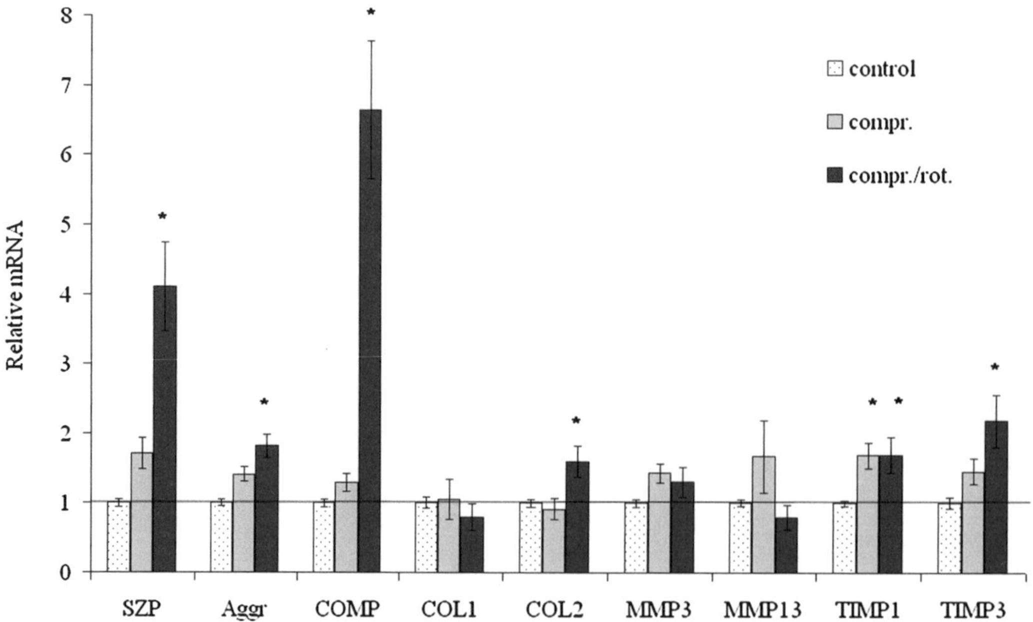


Fig. 3. Relative mRNA expression of chondrocyte-polyurethane constructs exposed to dynamic compression (compr.) or dynamic compression and superimposed surface motion (compr./rot.), normalized to the mRNA levels of unloaded constructs (control). Mean ± SEM, $n = 12$; $^*p < 0.05$ vs. unloaded control.

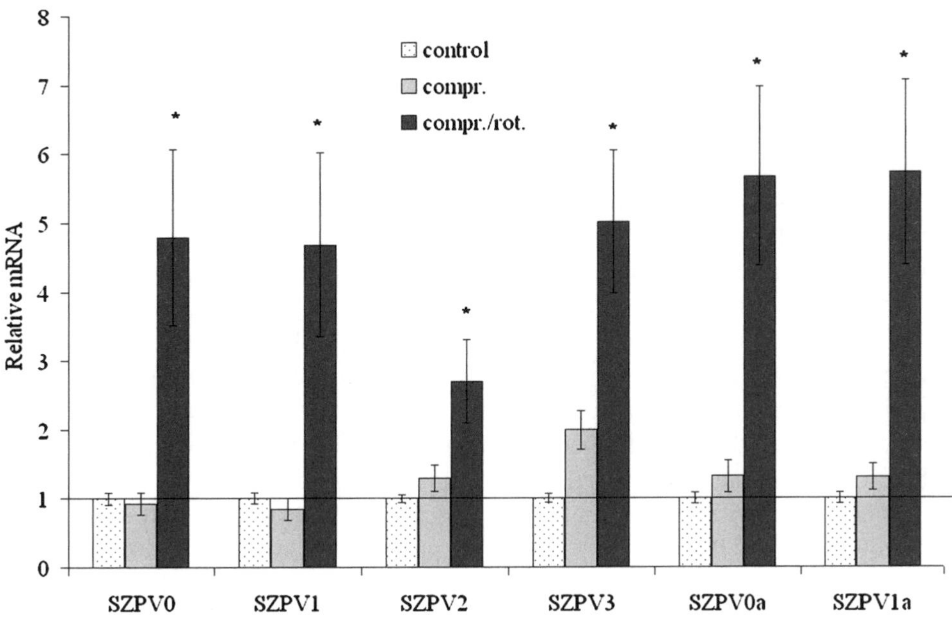

Fig. 4. Relative mRNA expression of different SZP/lubricin isoforms in chondrocyte–polyurethane constructs exposed to dynamic compression (compr.) or dynamic compression and superimposed surface motion (compr./rot.), normalized to the mRNA levels of unloaded constructs (control). Mean ± SEM, $n = 12$; $^*p < 0.05$ vs. unloaded control.

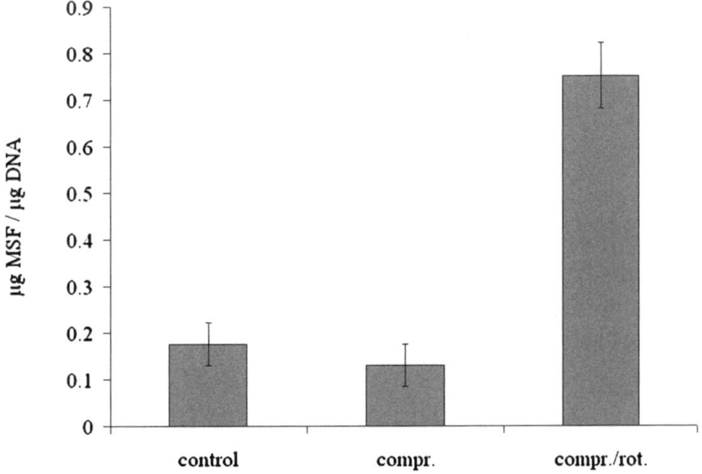

Fig. 5. MSF/SZP/lubricin immunoreactivity, expressed as MSF equivalent and normalized to the DNA content of the samples, in conditioned culture media of unloaded chondrocyte–polyurethane constructs (control), constructs exposed to dynamic compression (compr.), and constructs exposed to dynamic compression and superimposed surface motion (compr./rot.). Mean ± SEM, $n = 12$; $^*p < 0.0001$ vs. unloaded control.

## 4. Discussion

The present study investigated the effects of applied articular motion and dynamic compression on the gene expression of primary articular chondrocytes cultured in biodegradable porous polymer scaffolds. As observed in a previous study, applied surface motion significantly up-regulated SZP/lubricin expression [12]. The slightly lower response observed in this study may be explained by the implementation of a fibrin gel, which embedded the cells and therefore might have changed the responsiveness of the cells at the surface of the constructs. The lower seeding density compared to our previous study might also contribute to the decreased responsiveness. Furthermore, measurements of mRNA levels only represent a current state of the cells, and the previous study determined the mRNA levels after 3 days of loading. The results of this study demonstrate that after 5 days of mechanical loading, the SZP/lubricin expression was still clearly enhanced by surface motion stimulation. Importantly, analysis of MSF/SZP/lubricin immunoreactivity of conditioned media also confirmed that the respective protein was synthesized at enhanced levels, and increased amounts were released by constructs exposed to surface motion.

Six different transcripts, that can be generated by alternative splicing of exons 2, 4, and/or 5 of the PRG4 gene were identified in primary bovine articular chondrocytes and were all found to be up-regulated by applied surface motion. The roles of the different isoforms at the articular cartilage-synovial interface still need to be elucidated. It has been reported [6,30] that PRG4, which is known to contribute to cartilage lubrication, also can reduce cartilage–cartilage integration and thus inhibit integrative carti-lage repair. In this context it will be of interest to investigate if all the identified transcripts are translated into functional proteins and if the different functional properties (desirable and undesirable effects) of this glycoprotein can be attributed to specific isoforms or to the expression of distinct exons of the PRG4 gene.

Relative surface motion also clearly promoted the gene expression level of the cartilage oligomeric matrix protein (COMP). COMP is one of the major noncollagenous proteins in cartilage and plays a role in the organization and formation of the extracellular network [14,29,34]. Importantly, COMP is known as a sensitive marker of chondrocyte differentiation [41]. Unlike SZP/lubricin expression, the incorporation of a fibrin gel and the longer loading period resulted in a more pronounced increase in COMP mRNA levels compared to a previous study [38]. This may indicate that repetitive loading has a cumulative effect (5 days of loading in the current study versus 3 days in the previous study) on COMP mRNA levels and/or that fibrin enhances the effect by facilitating the transduction of the mechanical signal to the cells distributed in the construct. Increases in COMP and SZP/lubricin mRNA levels were also observed in chondrocyte-seeded alginate constructs exposed to cyclic tension for a period of 3 hours per day over 3 days [40]. This suggests that similar mechanotransduction pathways may be activated by cyclic tensile strain and the motion trajectories applied by articulation of the ceramic ball. COMP mRNA expression has also been reported to be enhanced by dynamic compressive loading of both cartilage explants and chondrocyte-alginate constructs [7]. In that study, the presence of a well-developed pericellular matrix was found to be a precondition for COMP up-regulation, which might be one reason for the non-responsiveness of COMP mRNA levels to dynamic compression in our system.

The slight, but significant increases in aggrecan and type II procollagen gene expression indicate that the applied articular motion may contribute to the prevention of the dedifferentiation of the cells with longer time in culture. Although the implementation of the fibrin gel improved the maintenance of the chondrocytic phenotype compared to cell–polyurethane constructs cultured without addition of fibrin, certain signs of dedifferentiation were noticeable also in fibrin-containing constructs after a 4-week culture period [22]. Thus, surface motion may promote the preservation of the articular phenotype not only with respect to SZP/lubricin and COMP expression, but also to the expression of other cartilage specific matrix molecules.

Cyclic axial compression without surface motion did not affect the gene expression of the cartilage matrix molecules and enzymes analyzed. This observation is consistent with previous studies reporting no change in gene expression of matrix molecules with the same loading regimes that are known to stimulate protein biosynthesis. In both bovine chondrocytes cultured in collagen gels [15] and human chondrocytes cultured in PEGT/PBT foams [5], no changes in aggrecan, collagen I or collagen II mRNA levels could be observed by dynamic compression. Similarly, investigating the effect of mechanical compression on chondrocytes-seeded polyurethane scaffolds cultured with or without the addition of fibrin gel, we found little effects on the gene expression levels of the constructs [22].

The obvious non-responsiveness of the chondrocyte gene expression to the dynamic compressive load applied in this study may be related to the fact that the loading conditions are not appropriately adapted to the polymer scaffold properties. Finite element analysis has shown that, due to the high permeability of the porous scaffold, no meaningful pressure gradients are generated in an unconfined setup [39]. On-going work applying the FEA model to predict pore pressures and the spatial distribution of the pressure gradients in the polyurethane scaffold will be used to optimize the boundary conditions and mechanical loading parameters [8,9].

The present study describes the gene expression response of chondrocytes-scaffold constructs to mechanical loading over a relatively short period of 5 days. Longer-term experiments under optimized conditions will be required to evaluate the effect of mechanical conditioning on the biochemical composition and biomechanical properties of the tissue-engineered constructs.

## Acknowledgements

We would like to thank Dr. Carl Flannery, Wyeth Research, Cambridge, MA, for providing anti-MSF antibody and rhMSF standard, Dr. Andreas Goessl, Baxter Bioscience, Vienna, for providing the fibrinogen and thrombin solutions, and Robert Peter for technical assistance with cell culture and biochemical analysis. This work was supported by the Swiss National Science Foundation grant #3200B0-104083.

## References

[1] *ABI PRISM 7700 Sequence Detection System User Bulletin (2)*, Perkin Elmer Applied Biosystems, 1997, pp. 11–15.

[2] M.D. Buschmann, Y.A. Gluzband, A.J. Grodzinsky and E.B. Hunziker, Mechanical compression modulates matrix biosynthesis in chondrocyte/agarose culture, *J. Cell Sci.* **108**(Pt 4) (1995), 1497–1508.

[3] S.E. Carver and C.A. Heath, Increasing extracellular matrix production in regenerating cartilage with intermittent physiological pressure, *Biotechnol. Bioeng.* **62** (1999), 166–174.

[4] T. Davisson, S. Kunig, A. Chen, R. Sah and A. Ratcliffe, Static and dynamic compression modulate matrix metabolism in tissue engineered cartilage, *J. Orthop. Res.* **20** (2002), 842–848.

[5] O. Demarteau, D. Wendt, A. Braccini, M. Jakob, D. Schafer, M. Heberer and I. Martin, Dynamic compression of cartilage constructs engineered from expanded human articular chondrocytes, *Biochem. Biophys. Res. Commun.* **310** (2003), 580–588.

[6] C. Englert, K.B. McGowan, T.J. Klein, A. Giurea, B.L. Schumacher and R.L. Sah, Inhibition of integrative cartilage repair by proteoglycan 4 in synovial fluid, *Arthritis Rheum.* **52** (2005), 1091–1099.

[7] P. Giannoni, M. Siegrist, E.B. Hunziker and M. Wong, The mechanosensitivity of cartilage oligomeric matrix protein (COMP), *Biorheology* **40** (2003), 101–109.

[8] U.J. Goerke, H. Guenther and M.A. Wimmer, Multiscale FE-modeling of native and engineered articular cartilage tissue, in: *European congress on computational methods in applied sciences and engineering. ECCOMAS*, P. Neittaanmäki, T. Rossi, K. Majava and O. Pironneau, eds, Jyväskylä, 2004, pp. 1–20.

[9] U.J. Goerke, M.A. Wimmer, S. Grad, C. Lee, E. Schneider, M. Alini and H. Guenther, Experimental and Numerical Simulation of Biological Activities in Engineered Tissue Constructs for Articular Cartilage Repair, *European Society of Biomechanics Congress*, 'S-Hertogenbosch NL, 2004, p. 68.

[10] K. Gorna and S. Gogolewski, Novel biodegradable polyurethanes for mediccal applications, in: *Synthetic Bioresorbable Polymers for Implants*, C.M. Agrawal, J.E. Parr and S.T. Lin, eds, ASTM, West Conshohocken, PA, 2000, pp. 39–57.

[11] K. Gorna and S. Gogolewski, Biodegradable polyurethanes for implants. II. In vitro degradation and calcification of materials from poly(epsilon-caprolactone)-poly(ethylene oxide) diols and various chain extenders, *J. Biomed. Mater. Res.* **60** (2002), 592–606.

[12] S. Grad, C.R. Lee, K. Gorna, S. Gogolewski, M.A. Wimmer and M. Alini, Surface motion upregulates superficial zone protein and hyaluronan production in chondrocyte-seeded three-dimensional scaffolds, *Tissue Eng.* **11** (2005), 249–256.

[13] S. Grad, L. Zhou, S. Gogolewski and M. Alini, Chondrocytes seeded onto poly (L/DL-lactide) 80%/20% porous scaffolds: a biochemical evaluation, *J. Biomed. Mater. Res. A* **66** (2003), 571–579.

[14] E. Hedbom, P. Antonsson, A. Hjerpe, D. Aeschlimann, M. Paulsson, E. Rosa-Pimentel, Y. Sommarin, M. Wendel, A. Oldberg and D. Heinegard, Cartilage matrix proteins. An acidic oligomeric protein (COMP) detected only in cartilage, *J. Biol. Chem.* **267** (1992), 6132–6136.

[15] C.J. Hunter, S.M. Imler, P. Malaviya, R.M. Nerem and M.E. Levenston, Mechanical compression alters gene expression and extracellular matrix synthesis by chondrocytes cultured in collagen I gels, *Biomaterials* **23** (2002), 1249–1259.

[16] G.D. Jay, Characterization of a bovine synovial fluid lubricating factor. I. Chemical, surface activity and lubricating properties, *Connect. Tissue Res.* **28** (1992), 71–88.

[17] G.D. Jay, D.E. Britt and C.J. Cha, Lubricin is a product of megakaryocyte stimulating factor gene expression by human synovial fibroblasts, *J. Rheumatol.* **27** (2000), 594–600.

[18] G.D. Jay, U. Tantravahi, D.E. Britt, H.J. Barrach and C.J. Cha, Homology of lubricin and superficial zone protein (SZP): products of megakaryocyte stimulating factor (MSF) gene expression by human synovial fibroblasts and articular chondrocytes localized to chromosome 1q25, *J. Orthop. Res.* **19** (2001), 677–687.

[19] J.D. Kisiday, M. Jin, M.A. DiMicco, B. Kurz and A.J. Grodzinsky, Effects of dynamic compressive loading on chondrocyte biosynthesis in self-assembling peptide scaffolds, *J. Biomech.* **37** (2004), 595–604.

[20] T.J. Klein, B.L. Schumacher, T.A. Schmidt, K.W. Li, M.S. Voegtline, K. Masuda, E.J. Thonar and R.L. Sah, Tissue engineering of stratified articular cartilage from chondrocyte subpopulations, *Osteoarthritis Cartilage* **11** (2003), 595–602.

[21] C. Labarca and K. Paigen, A simple, rapid, and sensitive DNA assay procedure, *Anal. Biochem.* **102** (1980), 344–352.

[22] C.R. Lee, S. Grad, K. Gorna, S. Gogolewski, A. Goessl and M. Alini, Fibrin-polyurethane composites for articular cartilage tissue engineering: A preliminary analysis, *Tissue Eng.* **11** (2005), 1562–1573.

[23] C.R. Lee, A.J. Grodzinsky and M. Spector, Biosynthetic response of passaged chondrocytes in a type II collagen scaffold to mechanical compression, *J. Biomed. Mater. Res. A* **64** (2003), 560–569.

[24] D.A. Lee, T. Noguchi, S.P. Frean, P. Lees and D.L. Bader, The influence of mechanical loading on isolated chondrocytes seeded in agarose constructs, *Biorheology* **37** (2000), 149–161.

[25] Y.J. Liu, S.H. Lu, B. Xu, R.C. Yang, Q. Ren, B. Liu, B. Li, M. Lu, F.Y. Yan, Z.B. Han and Z.C. Han, Hemangiopoietin, a novel human growth factor for the primitive cells of both hematopoietic and endothelial cell lineages, *Blood* **103** (2004), 4449–4456.

[26] D.M. Merberg, L.J. Fitz and P. Temple, A comparison of vitronectin and megakaryocyte stimulating facotor, in: *Biology of Vitronectins and Their Receptors*, K. Preissner, S. Rosenblatt, C. Kost, J. Wegerhoff and D. Mosher, eds, Elsevier Science Publishers BV, Amsterdam, 1993, pp. 45–54.

[27] S. Mizuno, T. Tateishi, T. Ushida and J. Glowacki, Hydrostatic fluid pressure enhances matrix synthesis and accumulation by bovine chondrocytes in three-dimensional culture, *J. Cell Physiol.* **193** (2002), 319–327.

[28] M.T. Raimondi, F. Boschetti, L. Falcone, G.B. Fiore, A. Remuzzi, E. Marinoni, M. Marazzi and R. Pietrabissa, Mechanobiology of engineered cartilage cultured under a quantified fluid-dynamic environment, *Biomech. Model. Mechanobiol.* **1** (2002), 69–82.

[29] K. Rosenberg, H. Olsson, M. Morgelin and D. Heinegard, Cartilage oligomeric matrix protein shows high affinity zinc-dependent interaction with triple helical collagen, *J. Biol. Chem.* **273** (1998), 20397–20403.

[30] D.B. Schaefer, D. Wendt, M. Moretti, M. Jakob, G.D. Jay, M. Heberer and I. Martin, Lubricin reduces cartilage–cartilage integration, *Biorheology* **41** (2004), 503–508.

[31] B.L. Schumacher, J.A. Block, T.M. Schmid, M.B. Aydelotte and K.E. Kuettner, A novel proteoglycan synthesized and secreted by chondrocytes of the superficial zone of articular cartilage, *Arch. Biochem. Biophys.* **311** (1994), 144–152.

[32] B.L. Schumacher, C.E. Hughes, K.E. Kuettner, B. Caterson and M.B. Aydelotte, Immunodetection and partial cDNA sequence of the proteoglycan, superficial zone protein, synthesized by cells lining synovial joints, *J. Orthop. Res.* **17** (1999), 110–120.

[33] J.C. Shelton, D.L. Bader and D.A. Lee, Mechanical conditioning influences the metabolic response of cell-seeded constructs, *Cells Tissues Organs* **175** (2003), 140–150.

[34] J. Thur, K. Rosenberg, D.P. Nitsche, T. Pihlajamaa, L. Ala-Kokko, D. Heinegard, M. Paulsson and P. Maurer, Mutations in cartilage oligomeric matrix protein causing pseudoachondroplasia and multiple epiphyseal dysplasia affect binding of calcium and collagen I, II, and IX, *J. Biol. Chem.* **276** (2001), 6083–6092.

[35] T. Tschan, I. Hoerler, Y. Houze, K.H. Winterhalter, C. Richter and P. Bruckner, Resting chondrocytes in culture survive without growth factors, but are sensitive to toxic oxygen metabolites, *J. Cell Biol.* **111** (1990), 257–260.

[36] G. Vunjak-Novakovic, I. Martin, B. Obradovic, S. Treppo, A.J. Grodzinsky, R. Langer and L.E. Freed, Bioreactor cultivation conditions modulate the composition and mechanical properties of tissue-engineered cartilage, *J. Orthop. Res.* **17** (1999), 130–138.

[37] S.D. Waldman, C.G. Spiteri, M.D. Grynpas, R.M. Pilliar and R.A. Kandel, Long-term intermittent shear deformation improves the quality of cartilaginous tissue formed in vitro, *J. Orthop. Res.* **21** (2003), 590–596.

[38] M.A. Wimmer, S. Grad, T. Kaup, M. Hanni, E. Schneider, S. Gogolewski and M. Alini, Tribology approach to the engineering and study of articular cartilage, *Tissue Eng.* **10** (2004), 1436–1445.

[39] M.A. Wimmer, H. Guenther, S. Grad, C. Lee, M. Haenni, S. Gogolewski and M. Alini, Numerical simulation as a tool to evaluate loading regimes for tissue engineering, in: *Transactions of the 49th Annual Meeting of the Orthopaedic Research Society*, New Orleans, 2003, 0992.

[40] M. Wong, M. Siegrist and K. Goodwin, Cyclic tensile strain and cyclic hydrostatic pressure differentially regulate expression of hypertrophic markers in primary chondrocytes, *Bone* **33** (2003), 685–693.

[41] F. Zaucke, R. Dinser, P. Maurer and M. Paulsson, Cartilage oligomeric matrix protein (COMP) and collagen IX are sensitive markers for the differentiation state of articular primary chondrocytes, *Biochem. J.* **358** (2001), 17–24.

Biorheology 43 (2006) 271–282
IOS Press



# Ultrasound stimulates proteoglycan synthesis in bovine primary chondrocytes

Milla Kopakkala-Tani [a,1], Jarkko J. Leskinen [b,1], Hannu M. Karjalainen [a], Tero Karjalainen [b], Kullervo Hynynen [b], Juha Töyräs [c], Jukka S. Jurvelin [b,d] and Mikko J. Lammi [a,*]

[a] *Department of Anatomy, Institute of Biomedicine, University of Kuopio, Kuopio, Finland*
[b] *Department of Physics, University of Kuopio, Kuopio, Finland*
[c] *Department of Clinical Neurophysiology, Kuopio University Hospital and University of Kuopio, Kuopio, Finland*
[d] *Department of Clinical Physiology and Nuclear Medicine, Kuopio University Hospital and University of Kuopio, Kuopio, Finland*

**Abstract.** Mechanical forces can stimulate the production of extracellular matrix molecules. We tested the efficacy of ultrasound to increase proteoglycan synthesis in bovine primary chondrocytes. The ultrasound-induced temperature rise was measured and its contribution to the synthesis was investigated using bare heat stimulus. Chondrocytes from five cellular isolations were exposed in triplicate to ultrasound (1 MHz, duty cycle 20%, pulse repetition frequency 1 kHz) at average intensity of 580 mW/cm$^2$ for 10 minutes daily for 1–5 days. Temperature evolution was recorded during the sonication and corresponding temperature history was created using a controllable water bath. This exposure profile was used in 10-minute-long heat treatments of chondrocytes. Heat shock protein 70 (Hsp70) levels after one-time treatment to ultrasound and heat was analyzed by Western blotting, and proteoglycan synthesis was evaluated by $^{35}$S-sulfate incorporation. Ultrasound treatment did not induce Hsp70, while heat treatment caused a slight heat stress response. Proteoglycan synthesis was increased approximately 2-fold after 3–4 daily ultrasound stimulations, and remained at that level until day 5 in responsive cell isolates. However, chondrocytes from one donor cell isolation out of five remained non-responsive. Heat treatment alone did not increase proteoglycan synthesis. In conclusion, our study confirms that pulsed ultrasound stimulation can induce proteoglycan synthesis in chondrocytes.

Keywords: Ultrasound treatment, heat, proteoglycans, chondrocyte

## 1. Introduction

Osteoarthritis and bone fractures are common health problems. New techniques are needed for better treatment of patients suffering from these conditions. An idea of using ultrasound to stimulate joint metabolism was tested already more than twenty years ago [35]. Bone has a much better capacity for repair than cartilage, and more effort has been focused on techniques that utilize ultrasound stimulation for bone repair. Low intensity pulsed ultrasound has enhanced the healing of fracture callus in animal models [1,30,36,37] and in clinical studies [9,17,20]. Ultrasound stimulation during endochondral ossification has been shown to affect cell differentiation by increasing the percentage of calcified cartilage, but not cell proliferation in fetal mouse long bones [16]. In diabetic BB Wistar rats, low-intensity pulsed ultrasound increased the fracture callus strength without any effects in cellular proliferation [8].

---

[1] Equal contribution to the work.

[*] Address for correspondence: Mikko Lammi, Department of Anatomy, Institute of Biomedicine, University of Kuopio, PO Box 1627, 70211 Kuopio, Finland. Fax: +358 17 163032; E-mail: mikko.lammi@uku.fi.

Effects of ultrasound treatment have been investigated also in cell cultures. It increased insulin-like growth factor in bone marrow-derived stromal cells and osteoblasts [24,25], and prostaglandin $E_2$ production was elevated via the induction of cyclooxygenase-2 mRNA in osteoblasts [15]. The growth of osteoblasts occurred, at least partly, due to increase in the synthesis and secretion of prostaglandin $E_2$ [22]. The optimum results were achieved with the intensity of 600 mW/cm$^2$ (spatial-average temporal-peak, duty cycle 20%) at 1 MHz frequency [22].

Experiments with an *in vitro* endochondral ossification model of fetal mouse long bones showed that low intensity ultrasound (30 mW/cm$^2$, 1.5 MHz) significantly increased the length of the calcified diaphysis within few days of application of ultrasound for 20 min/day, but did not influence the total length of the bone rudiments [26]. Also, significant responses in meniscus tears [23] and congenital pseudoarthrosis of tibia [27] treated with low intensity ultrasound stimulation have been reported, too.

Surgical or other clinical operations provide a costly way to repair cartilage lesions, and new strategies for prevention and treatment of these diseases are of great interest. Low intensity ultrasound treatment increased proteoglycan expression in cartilaginous fracture callus during bone fracture healing [37], and proteoglycan synthesis in rat chondrocytes *in vitro* [28,29]. The signalling mechanisms and the optimal sonication parameters are still only partly known. However, these results indicate that ultrasound may be a practical method to increase the synthesis of cartilage matrix also in various scaffolds used for tissue engineering.

The purpose of this study was to test the capacity of a custom-made ultrasound device to stimulate proteoglycan production in bovine primary chondrocyte cultures, and to explore whether the ultrasound-induced temperature increase could explain the changes in proteoglycan production.

## 2. Materials and methods

### 2.1. Cell cultures

Primary bovine chondrocytes were isolated from five one- to two-year-old animals. The cells from different donors were separately used for the experiments. The articular cartilage from patellar surface of the femur was cut into small pieces, washed twice with phosphate-buffered saline (PBS), and digested with hyaluronidase (0.5 mg/ml, Sigma, St. Louis, MO, USA) supplemented with gentamycin (0.1 mg/ml, PAA, Linz, Austria), fungizone (2.5 $\mu$g/ml, PAA), and ascorbic acid (0.5 $\mu$g/ml, Sigma) in high glucose Dulbecco's modified Eagle's medium (DMEM, Gibco, Paisley, UK) at 37°C for 30 min with continuous shaking. The medium was then changed to collagenase (3 mg/ml, Sigma) digestion solution, supplemented with 10% fetal calf serum (FCS, PAA), DNase (0.2 mg/ml, Sigma), 2 mM glutamine (PAA), 2.5 $\mu$g/ml of fungizone, 50 units/ml of penicillin (PAA), 50 units/ml of streptomycin (PAA), and 0.5 $\mu$g/ml of ascorbic acid (Sigma) in high glucose DMEM, and kept at 37°C in periodical magnetic stirring overnight. Next morning, the cells were divided into six-well culture plates (1 million cells/well) and incubated in culture medium at 37°C under a gas mixture of 95% air and 5% of carbon dioxide. The chondrocytes were cultured in high glucose DMEM including 10% FCS, penicillin, streptomycin, and glutamine. The cells were used for ultrasound exposure experiments nine days after plating.

### 2.2. Ultrasound device and calibration measurements

Ultrasound exposures were applied using a home-made ultrasound device (Fig. 1A). The device consisted of six piezoceramic discs (PZT26, Ferroperm Piezoceramics A/S, Denmark) with the diameter

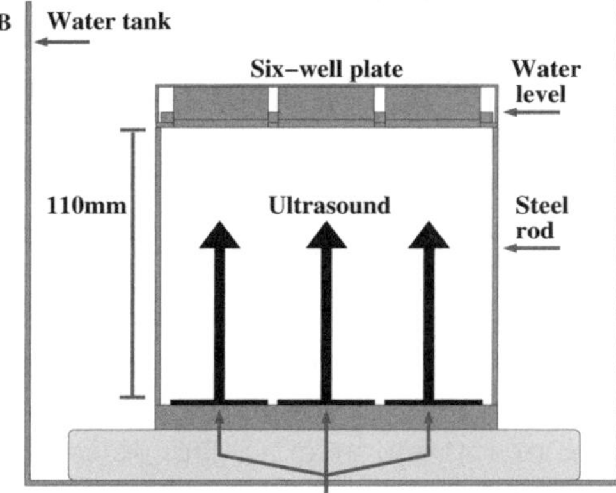


Fig. 1. Instrument used for ultrasound stimulations (A). The six-well plate sonication device with six transducer elements. (B) The cells are inside the plate and are located 110 mm above the transducers' surfaces.

of 38 mm and resonance frequency of 1 MHz. The air-backed transducers were covered using a thin layer of protecting rubber (thickness 0.1 mm). Discs were mounted in a form of six-well plate, in a way that another six-well plate with cells could be positioned parallel to the transducers at the distance of 110 mm from the transducers' surfaces using supporting steel rods (Fig. 1B). An instrument with ultrasound transducers placed only 3 mm away from the bottom of six-well culture plate was previously used to investigate the possibility to stimulate proteoglycan synthesis in cultured chondrocytes [29]. For these conditions it was estimated that the temperature rise in the culture wells would not exceed 1°C. Our first experimental data obtained with a set-up where the transducers are placed right under the plate showed a substantial temperature rise. Therefore, we changed the set-up and placed the transducers further away from the culture plate with an idea that a larger water mass between the culture plate and ultrasound transducers would result in a more efficient balancing of the temperature change. Three of the elements were active during the sonications to give the ultrasound treatment for the cells inside the wells, while the remaining three wells acted as controls. The device was immersed in a water container filled with distilled and degassed water. The water level of the container was adjusted so that the cell plate was maximally covered with water without the risk of contamination. The plate was protected with the lid supplied by the manufacturer. The water-filled container with the ultrasound apparatus was placed into the incubator kept at constant 37°C temperature and 100% relative humidity.

The ultrasound device was calibrated and characterized using radiation force balance and hydrophone measurements [31]. All the measurements were performed at room temperature in distilled and degassed water, except the plate attenuation measurement which was measured at 37°C in distilled and degassed water. During ultrasound exposures of cell cultures, the ultrasound waves travel through 110 mm of water path, through the bottom of the culture plate well, and after traveling through the cellular layer and cell culture medium, the waves reflect back from the culture medium–air interface. Thus, the cells are exposed to standing waves. The phase of the standing waves formed due to ultrasound reflections depends on the cell culture volume (height of the cell culture medium) and can thus vary from experiment to experiment. Weaker but more stable reflections are also formed between the transducer and bottom of the plastic well. When these reflected acoustic waves strike the transducer's face they interfere with the new outgoing waves, but also have an impact on the transducer's electric properties and thus on its performance [2]. Free-field calibrations do not include these factors and do not present the actual acoustic exposures of the cells. However, they will allow the experiments to be repeated in a similar set up.

The acoustic power values that were measured in free-field circumstances using radiation force balance were corrected with the attenuation of the plastic plate. The six-well plates (Nalge Nunc Int., NY, USA) with 1.2 mm thick polystyrene bottom induced a 5% insertion loss to a free field at 1 MHz frequency. This was measured in water bath at 37°C using Panametrics V303 probe (center frequency 1 MHz), Panametrics 500PR pulser-receiver (Panametrics-NDT, MA, USA), and 1.0 mm polyvinylidene fluoride (PVDF) needle hydrophone (Precision Acoustics Ltd., UK).

The hydrophone measurements were performed through the six-well plate immersed in the tank. The relative intensity distributions were measured with 0.5 mm PVDF needle hydrophone (Precision Acoustics Ltd., UK) at the distance of 110 mm in an acrylic, water-filled measurement tank lined with rubber absorbers to minimize reflections. In radiation force measurements, intensity field measurements, and exposures, the waveforms were generated with Agilent 33120A function generators (Agilent Technologies Inc., CO, USA) and amplified with ENI 240L 50dB RF amplifiers (Electronic Navigation Industries, NY, USA). The ultrasound exposures were administered with the frequency of 1 MHz, duty cycle 20%, and pulse repetition frequency of 1 kHz (Fig. 2).

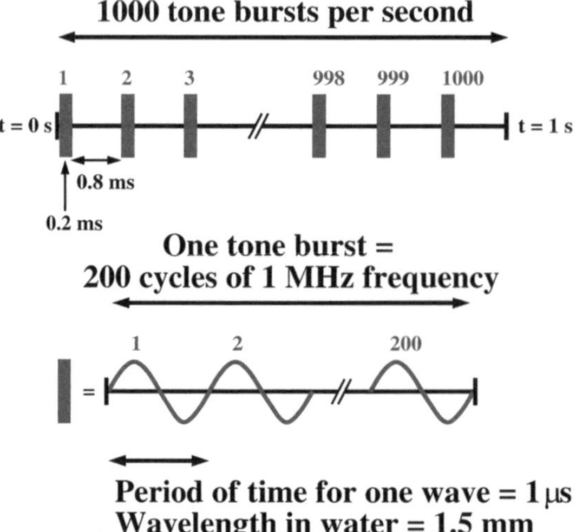


Fig. 2. Temporal parameters and their relations in the ultrasound exposure used in this study. The pulse repetition frequency controls when the stimulus on cells is turned on and off (upper image) while the effects during the on-phase are specified by the operating frequency (lower image). Total treatment time is 600 seconds.

## 2.3. Temperature measurements and exposures

The ultrasound-induced temperature rise was recorded using a copper–constantan thermocouple (T-type, single wire diameter 0.2 mm) connected to a Keithley 2000 multimeter (Keithley Instruments Inc., OH, USA). The thermocouple was placed parallel to the ultrasound field and the insulation on the tip of the probe was removed to decrease the absorption of the ultrasound waves and overestimation of temperature [11,12,18]. Since the sound absorbing plastic is probably the hottest place in the well, the bare tip of the temperature probe was put in contact with the bottom of the plastic at the center of the well. The wells were filled with 2.5 ml of the cell culture medium. The six-well plates were covered with the lid during the measurements, and all the measurements ($n = 3$) were performed in the same incubation conditions present during the actual ultrasound exposures of chondrocytes.

To observe the possible temperature peaks within the culture well, temperatures were recorded also at room temperature at nine different spots inside the culture well (single measurement). These measurement points were equally spaced to cover the whole bowl area (well diameter 35 mm). The well plates were covered with the lids as in actual exposures since it was observed that the temperature rise is much smaller without the lid. The lid prevents the convection through air and, thereby, increases the temperature inside the wells.

On the basis of thermocouple measurements described above, the temperature profile was mimicked using water bath exposures. The purpose of these experiments was to investigate whether the temperature rise observed during ultrasound treatments could alone explain the effects on the cellular metabolism. In these measurements, six-well plates, filled with 2.5 ml of the cell culture medium, were immersed in the thermostat-adjustable water bath (Thermomix®BU, B. Braun AG, Germany) similarly as in the ultrasound exposures. An identical T-type thermocouple probe was positioned at the center of the culture well touching the plastic bottom of the culture plate. The warm water heats the whole culture plate area equally because the water is circulated in the bath. The temperature of the bath was adjusted to get similar time-temperature profile as with the ultrasound exposures. The duration of ultrasound and heat exposures was ten minutes.

## 2.4. Exposure of primary chondrocytes to ultrasound and heat

Bovine primary chondrocytes were cultured as monolayer cultures in 6-well plates. Three wells of each 6-well plate were subjected to ultrasound or heat once a day for up to five days. For ultrasound treatments, three wells that were not subjected to ultrasound exposure served as controls, while for heat exposure experiments untreated 6-well plates were used as controls. The daily exposure time was ten minutes. The attenuation corrected temporal average intensity averaged over the well area was 580 mW/cm$^2$. The six-well plate with the cover was immersed into the water container inside the incubator. The container was filled with distilled and degassed water similar to calibration measurements, and the water was allowed to balance to the incubator's temperature. Possible air bubbles at the water–plate interface were removed. Small weight was placed at the top of the culture plate lid to eliminate the buoyancy. After the ultrasound exposure, the plates were taken out of the container and left inside the incubator. During the ultrasound exposures, the control cells were on the same plate as the treated cells, and they were also immersed in a water bath inside the incubator, but were not exposed to ultrasound.

The heat-treated cells had a similar temporal protocol as the ultrasound-treated cells. Covered culture plates were taken out of the incubator and immersed in warm water using a special holder. Buoyancy was again removed with the weight. After ten-minute-period in the water bath, the plate was placed inside the incubator. Control culture plates were taken out of the incubator to room temperature for few seconds, and put then back into the incubator to experience the same potential effects caused by the movement of the cells, or the small temperature drop which both occur in heat-treatments before the plates are immersed in the water bath.

## 2.5. $^{35}$Sulfate incorporation assay

After the last treatment (at maximum for five days), labeling in medium containing 10 $\mu$Ci/ml of $^{35}$S-sulfate was performed for 24 h. Incorporated and free sulfate were separated with PD-10 desalting columns and incorporation rates were determined after liquid scintillation analysis [19].

## 2.6. Western blotting

After one-time ultrasound or heat treatment, the six-well plates were kept inside the incubator for 3, 6 and 24 hours. The cells were detached from the culture plate and centrifuged for 4 min at 10,000 $\times$ g, and the cell pellets were stored at $-70°$C until used for further analysis. The samples were then resuspended in buffer containing 1 $\times$ PBS, 1% Nonidet P-40, 0.5% sodium deoxycholate, 0.1% sodium dodecyl sulfate (SDS), 3 $\mu$g/ml of phenyl methyl sulfonyl fluoride, and 1 mM sodium orthovanadate. The protein concentrations of the samples were measured with Bradford's assay [3] before electrophoresis.

Cellular samples, containing 20 $\mu$g of protein, were mixed with the electrophoresis sample buffer, boiled for 5 minutes at 100°C, and the proteins were separated in 10% SDS-polyacrylamide gel. The proteins were then transferred onto nitrocellulose membrane (Protran, Schleicher&Schuell, Dassel, Germany). Ponceau S solution (Sigma-Aldrich, St. Louis, MO) was used to check for the presence of equal amounts of transferred proteins on the membrane. The blotted membranes were placed in blocking buffer (Tris-buffered saline, pH 8.0, supplemented with 0.3% Tween 20 and 5% nonfat dry milk) for 1 h followed by 1 h incubation with heat shock protein 70 (Hsp70) primary antibody (StressGen Biotechnologies, Victoria, BC, Canada) diluted 1 : 10,000 in Tris-buffered saline (pH 8.0) containing 0.3% Tween 20 and 1% nonfat dry milk and overnight incubation with $\beta$-tubulin primary antibody (Boehringer Mannheim Biochemica, Mannheim, Germany) diluted 1 : 1000 in Tris-buffered saline (pH 8.0) containing 0.1% Tween 20 and 1% nonfat dry milk. Secondary antibody (goat anti-mouse IgG horseradish peroxidase, Santa Cruz Biotechnology, San Francisco, CA, USA) was diluted 1 : 1000 in Tris-buffered saline (pH 8.0) containing 0.1% Tween 20 and 1% nonfat dry milk, and membranes were incubated for 1 h. The membranes were developed with the Supersignal West Pico Chemiluminescent Substrate detection kit (Pierce, Rockford, IL, USA).

## 3. Results

The ultrasound intensity distribution showed a clear peak at the center of the field when measured through the bottom of the plate (Fig. 3). However, no similar temperature peaks were observed between nine different spots in the culture well (Fig. 4). Temperature measurements showed relatively constant heat rise on the whole well area. The longer distance did not prevent the gradual heating of the culture plate, since the average temperature rise in the plastic bottom was found to be 6.9°C at the end of ten-minute-exposure (Fig. 5).

The rise in temperature during ultrasound exposure may affect the general metabolism of the cell cultures. To estimate the effect of plain heat we experimentally determined the protocol for heating

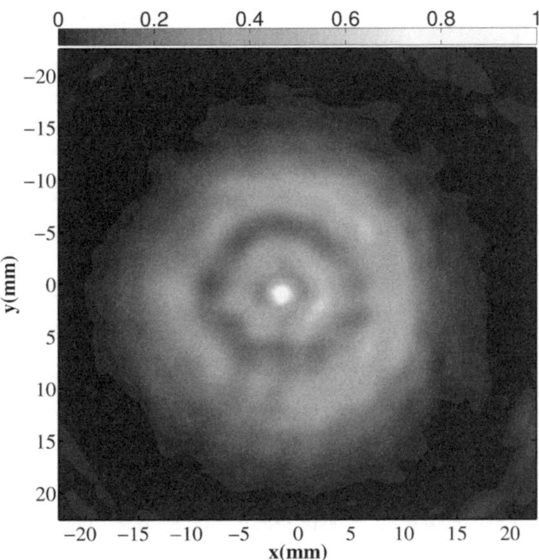

Fig. 3. Normalized intensity distribution through the culture plate for the single element on the radial plane at the distance of 110 mm of the transducer's surface.

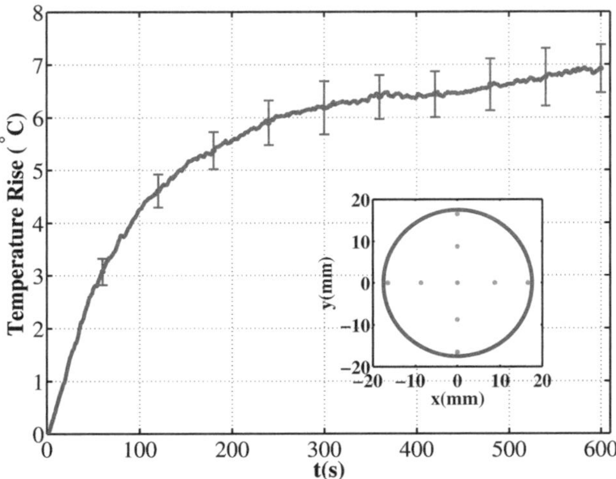

Fig. 4. Measured average ultrasound-induced temperature rise inside the well. Error bars (60-s intervals) show the standard deviation between the different measurement points. The small figure presents the locations of the nine measurements points inside the well used in this measurement.

of the culture plate that would lead to a similar temperature rise as observed during the ultrasound treatment. The chosen protocol for heat treatment closely followed the temperature change observed during ultrasound exposure (Fig. 5).

Stressful temperature rise in cells leads easily to induction of Hsp70. Although the temperature rise caused by ultrasound was gradual in our present study, and did not last for very long time, the recorded maximal temperature was relatively high at the end of the ultrasound exposure. Therefore, we determined the level of Hsp70 in ultrasound-treated chondrocyte cultures to estimate the severity of the heat stress for the cells. Western blot analysis of cellular proteins isolated after one-time sonication showed

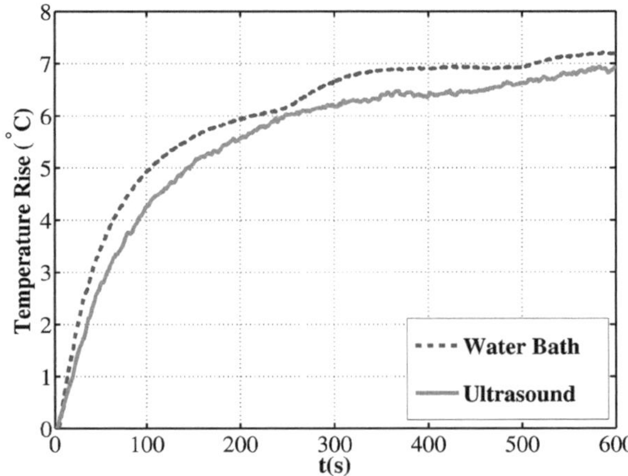

Fig. 5. Measured average temperature rise. Recordings (average of three measurements) from the bottom of the culture well during ultrasound (–) and heat exposure (- -). In both measurements, the culture plates were covered and the bare thermocouple touched the plastic bottom of the well.

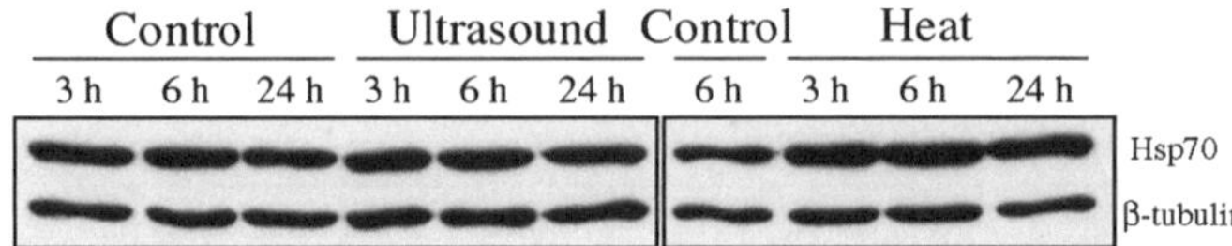

Fig. 6. Accumulation of Hsp70 after ultrasound treatment. Western blot analysis of cellular proteins isolated after one-time exposure showed increase in Hsp70 after heat treatment, but not after ten-minute-sonication. $\beta$-tubulin was used as a loading control. Numerical markings of the lanes represent the duration of follow-up after treatments.

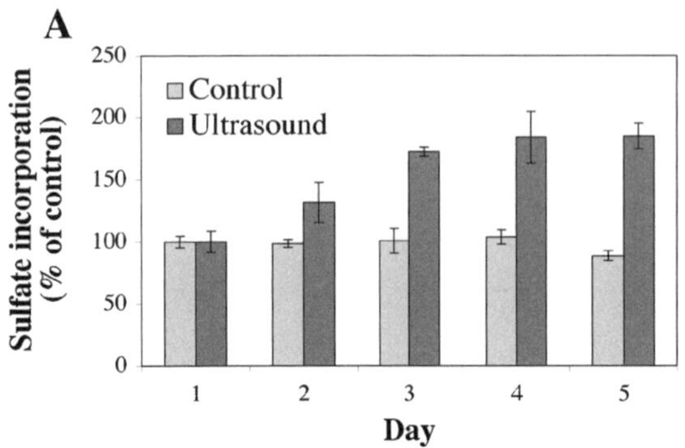

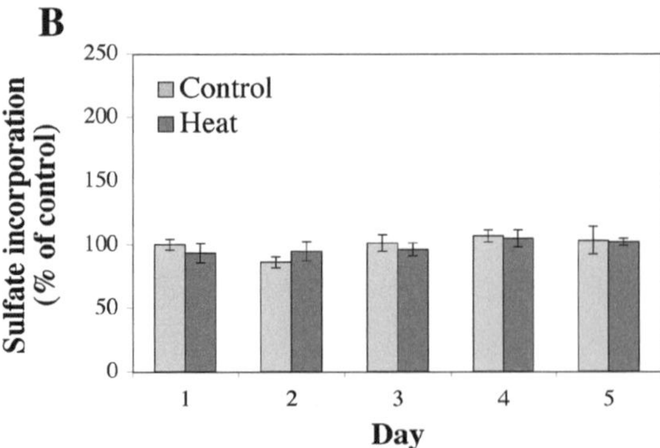

Fig. 7. [35]S-sulfate incorporation rates after ultrasound and heat treatments. (A) Incorporations rate was increased from day 2 by daily ten-minute-sonication. (B) The heat treatment alone did not increase the incorporation rate. The figure shows representative results of triplicate samples for chondrocyte cultures responsive to ultrasound (four isolations out of five) and effect of heat treatment on the cells from the same cellular isolation. The values show mean $\pm$ SD.

no induction of Hsp70 within 24 hours after the ultrasound treatment (Fig. 6). However, a slight accumulation of Hsp70 was seen for the cellular proteins isolated after the heat treatment. Obviously, the heat treatment transmitted the heat more effectively on the cell cultures, perhaps because the whole culture plate was subjected to the heat instead of only well areas affected by the ultrasound treatment. Nevertheless, the chondrocytes appear to tolerate the ultrasound-induced heat production rather well.

Daily ultrasound treatment for ten minutes increased proteoglycan synthesis in bovine primary chondrocytes starting generally from the second treatment and reaching approximately two-fold induction after three-four days (Fig. 7A). The experiment was repeated five times in triplicate samples. The heat treatment alone did not increase the incorporation rates (Fig. 7B). Proteoglycan synthesis in chondrocytes from one donor remained at control level after ultrasound treatments.

## 4. Discussion

Our present study with bovine primary chondrocytes subjected to ultrasound-treatment implies that ultrasound may be used to increase extracellular matrix production in cultured chondrocytes. Under the conditions that we used, a considerable heat production was observed. Yet, no induction of Hsp70 was obvious after ultrasound treatment, while temperature stimulation mimicking the temperature change caused by ultrasound slightly increased Hsp70. Previously, we have observed that primary chondrocytes can resist a stress response induced by high hydrostatic pressure in conditions which in most other cell types so far investigated lead to the accumulation of Hsp70 [6,7,13,14]. This may explain the lack of Hsp70 induction by the ultrasound treatment. Since the heat treatment alone did not increase proteoglycan synthesis in chondrocytes, the temperature rise as such cannot explain the stimulatory effect of ultrasound treatment. For functional tissue engineering purposes this technique may combine the benefits of extra heat on general metabolism and directional force created by ultrasound to produce implants with high matrix content and correct cartilage structure.

There are several studies concerning the combined effect of heat and ultrasound [21,32–34]. These studies revealed the non-thermal nature of ultrasound exposure by increasing the cell death when heat is combined with ultrasound. Remarkably, similar results have been obtained by exposing cells to both shear stresses and heat [4]. This suggests that heat and mechanical factors, such as shear stress due to fluid transport (streaming), can co-operate in the cellular responses. Our results indicate that in order to increase the proteoglycan synthesis with ultrasound treatment some non-thermal effect of ultrasound is required.

Recent studies have indicated that the level of the heat exposure can be very crucial for the proteoglycan synthesis. One-time heat treatment at 48°C for ten minutes caused chondrocyte apoptosis and suppression of proteoglycan synthesis in rat articular cartilage [38]. Temperature in this case was approximately 4°C higher than the maximum in our present study. Interestingly, immersion of HCS-2/8 chondrosarcoma cells at 41°C for fifteen or thirty minutes was found to have positive effect on both the viability and proteoglycan synthesis rate [10]. However, the viability and the metabolism were decreased when the cells were exposed to 43°C or higher for thirty minutes [10]. In the present study, the cells experienced temperatures of 43°C or higher for only five minutes.

The measured intensity distribution through the well was found to be non-uniform. However, temperature values inside the well were relatively constant and no clear peaks were found. It is likely that the multiple reflections inside the well smooth the ultrasound exposure and also the temperature rise.

The ultrasound-induced temperature rise is relatively fast inside the well, and similar temperature profile or history is difficult to create and control, for example, using water bath. In this study, the water bath temperature profile was higher in all time points, but the difference was always under 1°C. In the bath exposures the culture medium was not mixed, so that the whole well volume may warm up slower than in ultrasound exposures which involve mixing through the acoustic streaming. On the other hand, cells are attached to the well bottom that heats up most rapidly in the both exposure methods. The difference in Hsp70 response between heat and ultrasound treatments suggests that the temperature recordings during ultrasound treatment may be overestimated.

The major difference between these two exposure methods is the lack of fluid streaming or some other mechanical stimulus at the heat exposures. On the other hand, ultrasound exposure can involve many different non-thermal effects [5]. Ultrasound violently shakes and mixes the culture medium, and the attached cells experience this motion of fluid or streaming. Stable cavitation is another possible mechanism affecting the cell responses, especially at *in vitro systems*. This mechanism involves a response of different sizes of gas bubbles present in the culture medium to ultrasound waves. In addition, this bubble response is a source for acoustic microstreaming.

Parvizi et al. speculated that the possible key factor in ultrasound exposures could be the pulse repetition frequency [28]. They argued that the wavelength of the acoustic wave (mm) is much longer than the physical dimensions of the cell ($\mu$m) even at megahertz range sound waves and the cells are exposed to relatively static pressure. Thereby, no gradients are formed inside the cells. The pulse repetition frequency, on the other hand, controls the periods when the pressure stimulus is turned on and off, and it can create vibrations to the cells. In addition, the operating frequency plays a vital role with the pulse duration in the occurrence of thermal and non-thermal effects. Higher frequency means higher absorption and higher temperature rise. Lower frequency is more likely to cause cavitation than the higher one. Pulse duration can have an influence on the occurrence of inertial cavitation, but also as a means to decrease temperature rise while maintaining the same duty cycle.

The design of our ultrasound device and the equipment previously used to investigate ultrasound effects on rat chondrocytes [29] has certain limitations. The major disadvantage of the system is the formation of standing waves inside the wells between the bottom of the well and the medium–air interface. This means that the stimulating acoustic field is very sensitive to the height of the culture medium. Also, the rippling fluid and the reflections of sound waves make the system unstable. Therefore, we are currently building a modified multi-element system for *in vitro* studies. In this modified set-up there is no liquid-air interface and, thus, no standing waves are formed. This system allows more accurate control of the sonication parameters which is important for optimization of the method for tissue engineering and clinical purposes.

## Acknowledgements

The authors acknowledge Mrs Elma Sorsa for the valuable help with the analyses. This study was supported by the grants from Academy of Finland (#54054, #206113), and Heikki and Hilma Honkanen Foundation.

## References

[1] Y. Azuma, M. Ito, Y. Harada, H. Takagi, T. Ohta and S. Jingushi, Low-intensity pulsed ultrasound accelerates rat femoral fracture healing by acting on the various cellular reactions in the fracture callus, *J. Bone Miner. Res.* **16** (2001), 671–680.

[2] K. Beissner, The influence of membrane reflections on ultrasonic power measurements, *Acustica* **50** (1982), 194–200.

[3] M.M. Bradford, A rapid and sensitive method for the quantitation of microgram quantities of protein utilizing the principle of protein-dye binding, *Anal. Biochem.* **72** (1976), 248–254.

[4] F. Dunn, Cellular inactivation by heat and shear, *Radiat. Environ. Biophys.* **24** (1985), 131–139.

[5] M. Dyson, Non-thermal cellular effects of ultrasound, *Br. J. Cancer Suppl.* **45** (1982), 165–171.

[6] M.A. Elo, R.K. Sironen, K. Kaarniranta, S. Auriola, H.J. Helminen and M.J. Lammi, Differential regulation of stress proteins by high hydrostatic pressure, heat shock, and unbalanced calcium homeostasis in chondrocytic cells, *J. Cell. Biochem.* **79** (2000), 610–619.

[7] M.A. Elo, R.K. Sironen, H.M. Karjalainen, K. Kaarniranta, H.J. Helminen and M.J. Lammi, Specific induction of heat shock protein 90beta by high hydrostatic pressure, *Biorheology* **40** (2003), 141–146.

[8] G.P. Gebauer, S.S. Lin, H.A. Beam, P. Vieira and J.R. Parsons, Low-intensity pulsed ultrasound increases the fracture callus strength in diabetic BB Wistar rats but does not affect cellular proliferation, *J. Orthop. Res.* **20** (2002), 587–592.

[9] J.D. Heckman, J.P. Ryaby, J. McCabe, J.J. Frey and R.F. Kilcoyne, Acceleration of tibial fracture-healing by non-invasive, low-intensity pulsed ultrasound, *J. Bone Joint Surg. Am.* **76** (1994), 26–34.

[10] T. Hojo, M. Fujioka, G. Otsuka, S. Inoue, U. Kim and T. Kubo, Effect of heat stimulation on viability and proteoglycan metabolism of cultured chondrocytes: preliminary report, *J. Orthop. Sci.* **8** (2003), 396–399.

[11] K. Hynynen, C.J. Martin, D.J. Watmough and J.R. Mallard, Errors in temperature measurements by thermocouple probes during ultrasound induced hyperthermia, *Br. J. Radiol.* **56** (1983), 969–970.

[12] K. Hynynen and D.K. Edwards, Temperature measurements during ultrasound hyperthermia, *Med. Phys.* **16** (1989), 618–626.

[13] K. Kaarniranta, M. Elo, R. Sironen, M.J. Lammi, M.B. Goldring, J.E. Eriksson, L. Sistonen and H.J. Helminen, Hsp70 accumulation in chondrocytic cells exposed to high continuous hydrostatic pressure coincides with mRNA stabilization rather than transcriptional activation, *Proc. Natl. Acad. Sci. USA* **95** (1998), 2319–2324.

[14] K. Kaarniranta, C.I. Holmberg, M.J. Lammi, J.E. Eriksson, L. Sistonen and H.J. Helminen, Primary chondrocytes resist hydrostatic pressure-induced stress while primary synovial cells and fibroblasts show modified Hsp70 response, *Osteoarthritis Cartilage* **9** (2001), 7–13.

[15] T. Kokubu, N. Matsui, H. Fujioka, M. Tsunoda and K. Mizuno, Low intensity pulsed ultrasound exposure increases prostaglandin E2 production via the induction of cyclooxygenase-2 mRNA in mouse osteoblasts, *Biochem. Biophys. Res. Commun.* **256** (1999), 284–287.

[16] C.M. Korstjens, P.A. Nolte, E.H. Burger, G.H. Albers, C.M. Semeins, I.H. Aartman, S.W. Goei and J. Klein-Nulend, Stimulation of bone cell differentiation by low-intensity ultrasound – a histomorphometric in vitro study, *J. Orthop. Res.* **22** (2004), 495–500.

[17] T.K. Kristiansen, J.P. Ryaby, J. McCabe, J.J. Frey and L.R. Roe, Accelerated healing of distal radial fractures with the use of specific, low-intensity ultrasound. A multicenter, prospective, randomized, double-blind, placebo-controlled study, *J. Bone Joint Surg. Am.* **79** (1997), 961–973.

[18] P.K. Kuhn and D.A. Christensen, Influence of temperature probe sheathing materials during ultrasonic heating, *IEEE Trans. Biomed. Eng.* **33** (1986), 536–538.

[19] M.J. Lammi, T.P. Häkkinen, J.J. Parkkinen, M.M. Hyttinen, M. Jortikka, H.J. Helminen and M.I. Tammi, Adaptation of canine femoral head articular cartilage to long distance running exercise in young beagles, *Ann. Rheum. Dis.* **52** (1993), 369–377.

[20] K.S. Leung, W.S. Lee, H.F. Tsui, P.P. Liu and W.H. Cheung, Complex tibial fracture outcomes following treatment with low-intensity pulsed ultrasound, *Ultrasound Med. Biol.* **30** (2004), 389–395.

[21] G.C. Li, G.M. Hahn and L.J. Tolmach, Cellular inactivation by ultrasound, *Nature* **267** (1977), 163–165.

[22] J.G. Li, W.H. Chang, J.C. Lin and J.S. Sun, Optimum intensities of ultrasound for PGE(2) secretion and growth of osteoblasts, *Ultrasound Med. Biol.* **28** (2002), 683–690.

[23] J.A. Muche, Efficacy of therapeutic ultrasound treatment of a meniscus tear in a severely disabled patient: a case report, *Arch. Phys. Med. Rehabil.* **84** (2003), 1558–1559.

[24] K. Naruse, Y. Mikuni-Takagaki, Y. Azuma, M. Ito, T. Oota, K. Kameyama and M. Itoman, Anabolic response of mouse bone-marrow-derived stromal cell clone ST2 cells to low-intensity pulsed ultrasound, *Biochem. Biophys. Res. Commun.* **268** (2000), 216–220.

[25] K. Naruse, A. Miyauchi, M. Itoman and Y. Mikuni-Takagaki, Distinct anabolic response of osteoblast to low-intensity pulsed ultrasound, *J. Bone Miner. Res.* **18** (2003), 360–369.

[26] P.A. Nolte, J. Klein-Nulend, G.H. Albers, R.K. Marti, C.M. Semeins, S.W. Goei and E.H. Burger, Low-intensity ultrasound stimulates endochondral ossification *in vitro*, *J. Orthop. Res.* **19** (2001), 301–307.

[27] K. Okada, N. Miyakoshi, S. Takahashi, S. Ishigaki, J. Nishida and E. Itoi, Congenital pseudoarthrosis of the tibia treated with low-intensity pulsed ultrasound stimulation (LIPUS), *Ultrasound Med. Biol.* **29** (2003), 1061–1064.

[28] J. Parvizi, V. Parpura, J.F. Greenleaf and M.E. Bolander, Calcium signaling is required for ultrasound-stimulated aggrecan synthesis by rat chondrocytes, *J. Orthop. Res.* **20** (2002), 51–57.

[29] J. Parvizi, C.C. Wu, D.G. Lewallen, J.F. Greenleaf and M.E. Bolander, Low-intensity ultrasound stimulates proteoglycan synthesis in rat chondrocytes by increasing aggrecan gene expression, *J. Orthop. Res.* **17** (1999), 488–494.

[30] A.A. Pilla, M.A. Mont, P.R. Nasser, S.A. Khan, M. Figueiredo, J.J. Kaufman and R.S. Siffert, Non-invasive low-intensity pulsed ultrasound accelerates bone healing in the rabbit, *J. Orthop. Trauma* **4** (1990), 246–253.

[31] H.F. Stewart, Ultrasonic measurement techniques and equipment output levels, in: *Essentials of Medical Ultrasound. A Practical Introduction to the Principles, Techniques and Biomedical Applications*, M.H. Repacholi and D.A. Benwell, eds, Humana Press, Clifton, NJ, 1982, pp. 77–116.

[32] G. Ter Haar, I.J. Stratford and C.R. Hill, Ultrasonic irradiation of mammalian cells in vitro at hyperthermic temperatures, *Br. J. Radiol.* **53** (1980), 784–789.

[33] G.R. Ter Haar and I.J. Stratford, Evidence for a non-thermal effect of ultrasound, *Br. J. Cancer Suppl.* **45** (1982), 172–175.

[34] G.T. Ter Haar, J. Walling, P. Loverock and S. Townsend, The effect of combined heat and ultrasound on multicellular tumour spheroids, *Int. J. Radiat. Biol. Relat. Stud. Phys. Chem. Med.* **53** (1988), 813–827.

[35] H. Vanharanta, I. Eronen and T. Videman, Effect of ultrasound on glycosaminoglycan metabolism in the rabbit knee, *Am. J. Phys. Med.* **61** (1982), 221–228.

[36] S.J. Wang, D.G. Lewallen, M.E. Bolander, E.Y. Chao, D.M. Ilstrup and J.F. Greenleaf, Low intensity ultrasound treatment increases strength in a rat femoral fracture model, *J. Orthop. Res.* **12** (1994), 40–47.

[37] K.H. Yang, J. Parvizi, S.J. Wang, D.G. Lewallen, R.R. Kinnick, J.F. Greenleaf and M.E. Bolander, Exposure to low-intensity ultrasound increases aggrecan gene expression in a rat femur fracture model, *J. Orthop. Res.* **14** (1996), 802–809.

[38] J. Ye, H. Haro, M. Takahashi, H. Kuroda and K. Shinomiya, Induction of apoptosis of articular chondrocytes and suppression of articular cartilage proteoglycan synthesis by heat shock, *J. Orthop. Sci.* **8** (2003), 387–395.

Biorheology 43 (2006) 283–291
IOS Press



# Diurnal fluid expression and activity of intervertebral disc cells

S. Sivan [a], C. Neidlinger-Wilke [b], K. Würtz [b], A. Maroudas [a] and J.P.G. Urban [c,*]

[a] *Department of Biomedical Engineering, Israel Technion, Haifa, Israel*
[b] *Institute of Orthopaedic Research and Biomechanics, University of Ulm, Germany*
[c] *Physiology Laboratory, Parks Rd, Oxford, OX1 3PT, UK*

**Abstract.** The intervertebral discs are large cartilaginous structures situated between the vertebral bodies, occupying around one third of the length of the spinal column. They act as the joints of the spine and carry mechanical load arising from body weight and muscle activity. Loads change with every alteration of posture and activity and the discs thus undergo a diurnal loading pattern with high loads on the discs during the day's activity and low loads on it at night during rest. As the disc is an osmotic system, around 25% of the disc's fluid is expressed and re-imbibed during each diurnal cycle with consequent changes in the osmotic environment of the disc cells. Here, present information on the effect of osmotic changes in disc cell metabolism is reviewed; results indicate that prevailing osmolarity is a powerful regulator of disc cell activity.

## 1. Introduction

The intervertebral discs are large cartilaginous structures between the vertebral bodies; they occupy around one third of the length of the spinal column. Their predominant role is mechanical. Like articular cartilage, they support compressive loads arising from body weight and muscle tension. They also anchor one vertebral body to the next. Most importantly they are the joints of the spine, enabling it to bend, flex and twist [52]. The ability of the discs to perform its biomechanical role depends on the organisation and composition of the complex extracellular matrix which make up this tissue and thus ultimately on a sparse population of cells which are responsible for synthesising and maintaining the matrix. These cells are always under loads which vary with every change of posture and activity; as the disc is an osmotic system, changes in load affect the disc's fluid content significantly. Here we will review how matrix production by these cells is affected by disc fluid content under varying applied loads.

## 2. Structure and composition of the disc

### 2.1. Disc morphology

Morphologically the disc consists of two distinct regions, the outer tough annulus fibrosus which encloses the softer more flexible nucleus pulposus (Fig. 1). These lamellae of the annulus consist of concentric sheets of dense collagen bundles running obliquely from one vertebral body to the next with the

---

*Address for correspondence: Jill P.G. Urban, Physiology Laboratory, Oxford University, Parks Rd, Oxford, OX1 3PT, UK. Tel.: +44 1865 272509; Fax: +44 1865 272469; E-mail: jpgu@physiol.ox.ac.uk.

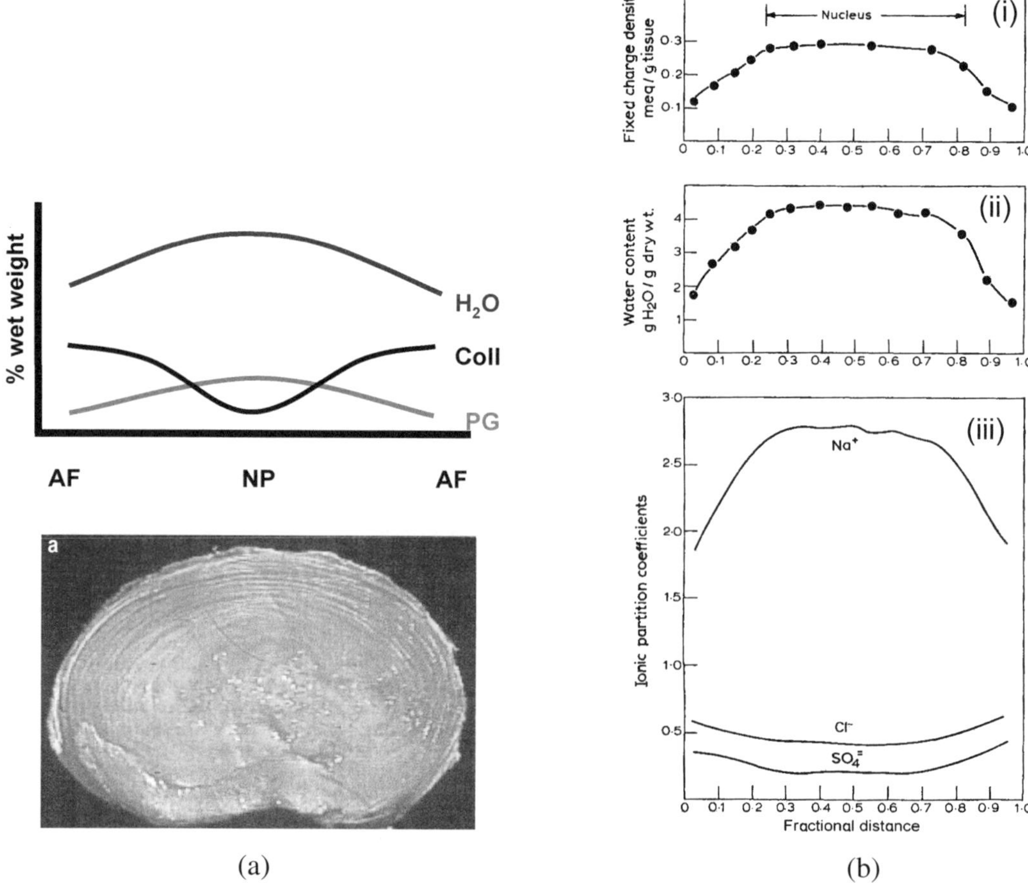


Fig. 1. Profiles of composition across the disc from annulus to annulus. (a) A schematic view of the concentrations of the main constituents of an adult lumbar intervertebral disc in the tissue. (b) Profiles of (i) FCD, (ii) water, (iii) content and ion concentrations adapted from [50].

angle relative to the vertebral bodies alternating from one lamella to the next [47]. The collagen bundles of adjacent lamellae appear only loosely interconnected which is thought to allow relative movement during flexion, torsion and extension. An organised elastic fibre network, concentrated in the interlamellar space is thought to also play a mechanical role in restoring annulus organisation after deformation [54]. By contrast, the mature nucleus has only a loose collagen network, is highly hydrated and, when non-degenerate, behaves hydrostatically under load [34].

## 2.2. Disc cells

The disc is distinguished by a very low cell density 1–5 million cells/ml tissue; each cell thus has to maintain 200 times its own volume of matrix. It contains at least three distinct cell populations, which differ in morphology and in amount and type of matrix they produce [21] and which reflect their distinct embryological origins [51]. The annulus arises from the mesenchyme. The cells of the outer annulus are long and fibroblast-like with long cellular processes; three different cellular morphologies have been described [6] but whether the cellular activity varies with morphology is unknown at present. The cells of the inner annulus are more chondrocyte-like being round and surrounded by a capsule. The nucleus arises from the notochord. In animal species where notochordal cells persist (such as rodents, cats, non-chondrodystrophoid dogs [8,22]) the nucleus is translucent, very highly hydrated and has a very low collagen content. In other animals species (in humans between 4–10 yrs, in chondrodystrophoid dogs at c. 1 yr and in ruminants before birth) the notochordal cells disappear. The nucleus becomes populated by cells of uncertain origin [16,19]; mature nucleus cells resemble articular chondrocytes as they are round and surrounded by a capsule but unlike chondrocytes they have long cellular processes extending into the matrix for at least 3–4 cell diameters. With loss of notochordal cells the nucleus loses its translucent appearance, becomes firmer and more collagenous.

## 2.3. Disc composition

As in all load-bearing cartilages, the three major constituents of the disc are water, collagen and aggrecan, the large aggregating proteoglycan with its high concentration of sulphated glycosaminoglycans. The proportion and organisation of these components varies considerably with position across the disc. In general the proportion of aggrecan and water are highest in the nucleus and lowest in the outer annulus and endplate, while collagen shows the reverse profile.

Collagen is the major structural macromolecule of the disc with the major fibrils of the annulus and nucleus formed from the fibrillar collagens, collagen I and collagen II. Collagen I is the major collagen in the outer annulus but its proportion decreases radially towards the nucleus as the proportion of collagen II increases [14]. Type II collagen is the major fibrillar collagen in the disc nucleus which normally contains no type I collagen [14]. Other collagen types, viz. collagens V, VI, IX, XI, XII and XIV are all reported to be present in the matrix [13]; some appear localized to the pericellular space while others appear predominantly within the matrix with their distribution varying with disc region [43]. The role of these collagens, present in minor amounts, is unclear, but it appears important in some respects as type IX collagen polymorphisms are strongly associated with disc degeneration in a Finnish population [1].

The large aggregating proteoglycan, aggrecan is the other major macromolecule of the disc [25]. It is similar to aggrecan of articular cartilage, consisting of a protein core to which up to 100 highly sulphated GAG chains, principally chondroitin and keratan sulphate are covalently attached [46], and like cartilage aggrecan is able to form aggregates with hyaluronan when newly synthesized. Soon after synthesis, disc aggrecan is attacked by proteases, loses its ability to form aggregates and breaks down into smaller fragments [3]. Nevertheless, on account of the disc's size these fragments remain in the tissue and are still functional [48].

Apart from these major species, there are many other macromolecules present in the disc in minor amounts. Proteoglycans such as versican and the small leucine-rich proteoglycans such as decorin, biglycan and fibromodulin have been found in the disc [12,26,36] and are present mostly in the annulus. The disc also contains other proteins and glycoproteins such as fibronectin and tenascin [17,38] and a well organised elastic fibre network [54]. The role of these minor components in maintaining matrix structure and function is in many cases unclear, but appears important; polymorphisms of the minor protein CILP are, for example, associated with disc degeneration [44].

As well as these macromolecular components, the disc also contains several different classes of proteases and their inhibitors [35,42] active in normal turnover as is shown by the fragmentation of aggrecan soon after synthesis [33]; the level of active proteases present in the disc increases with age and pathology as does the concentration of degraded macromolecules [2,11].

## 3.  Disc water content and load

### 3.1.  Swelling pressure

Aggrecan, through the negatively charged groups on its constituent glycosaminoglycan (GAG) chains, contributes a high fixed negative charge to the disc matrix [31]. In order to maintain charge equilibria, ions distribute themselves between plasma and the aggrecan-rich disc in a manner consistent with Gibbs–Donnan equilibria [30]. The concentration of cations in the disc is thus high relative to that in the plasma, and varies directly with GAG concentration i.e. with FCD (Fig. 1b); the concentration of anions however, is lower than in the plasma as these are excluded by the negatively charged matrix. As FCD varies directly with concentration of the constituent GAGs [50], concentrations of ions such as sodium and potassium follow those of GAG, with cation concentrations highest in the nucleus where they may be around 3–4 times as high as in the plasma [28,50] and lowest in the outer annulus, whereas anion concentrations show a reverse profile (Fig. 1b).

This charge distribution is a major contributor to the osmotic pressure of aggrecan. The osmotic pressure of aggrecan solutions increases steeply with increase in FCD; measured values are, however, lower than those predicted from an ideal Donnan distribution. In the tissue, aggrecan is also osmotically active and exerts a similar concentration-dependent osmotic pressure. However as aggrecan is excluded from within the collagen fibril [27] and is thus only present in the extrafibrillar space, the osmotically active FCD is higher than that shown in Fig. 1b. The excluded volume fraction thus varies with collagen concentration. In the disc nucleus, the intrafibrillar fluid content has been estimated to be around 1.3 g water/g dry collagen [49] and is around 1.0 g/g dry collagen in the annulus [45]. However as collagen fibril water content varies with extrafibrillar osmotic pressure [27], excluded volume fraction also depends on FCD. In the collagen-rich annulus fibrosus, which has a relatively low fluid content, up to 30% of the total fluid may be intrafibrillar and unavailable to GAGs whose swelling pressure is thus correspondingly greater than calculated on a total volume basis.

### 3.2.  Diurnal loading pattern on the disc and disc hydration

Disc osmotic pressure has an important mechanical function as it maintains disc turgor under load. Since the osmotic pressure aggrecan in the disc is considerably greater than the osmotic pressure of the plasma, the disc tends to imbibe water. This imbibition is resisted by external loads. *In vivo*, these arise from body weight and muscle and ligament tensions [49] and vary with posture and activity as shown in Fig. 2; it is evident that the disc undergoes a diurnal loading pattern with loads lowest when lying down during rest at night and increasing 5–10 fold during the day's activities. As the disc is an osmotic system, fluid content varies with load (Fig. 3a); fluid is expressed when the applied pressure exceeds that of the existing aggrecan osmotic pressure and collagen tension while fluid is imbibed if the applied pressure falls. Around 20–25% of the disc's fluid is expressed during the high loads experienced during the day's activities; the fluid is re-imbibed at night under low loads during rest [5,37]. As a result of such changes in fluid content, concentrations of GAG and other macromolecules also undergo a diurnal cycle; the consequent changes in FCD and hence in ion concentrations affect extracellular osmolality, which in a normal human adult, cycles between c. 400 when the water content is highest and c. 550 mOsm after maximum daily fluid expression.

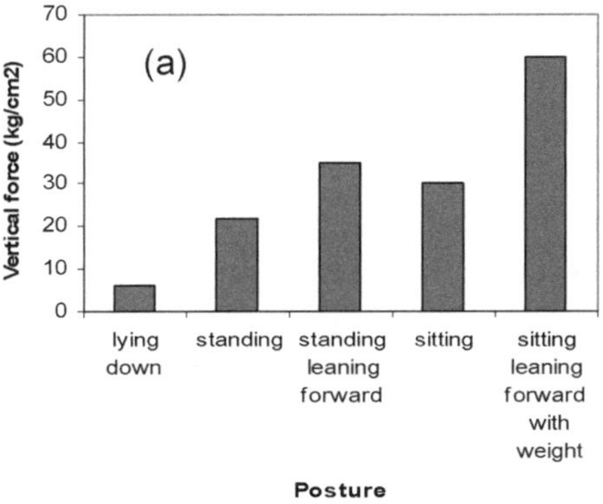
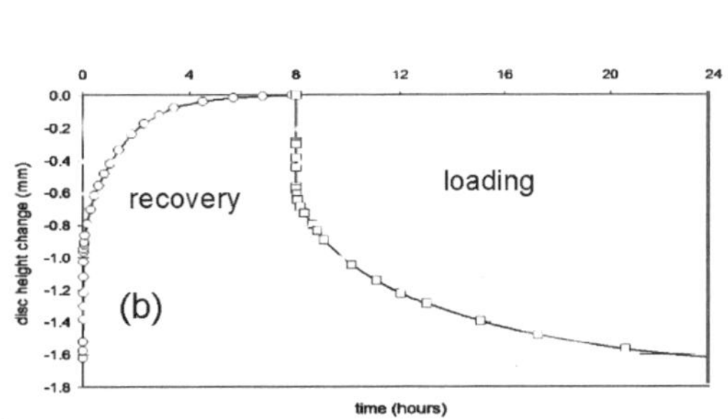



Fig. 2. Effect of posture on disc loading and consequent changes in disc height. (a) Pressures in the L3–L4 disc measured *in vivo* under different postures adapted from [37]. (b) Change in height in response to diurnal loading pattern (adapted from [15]).

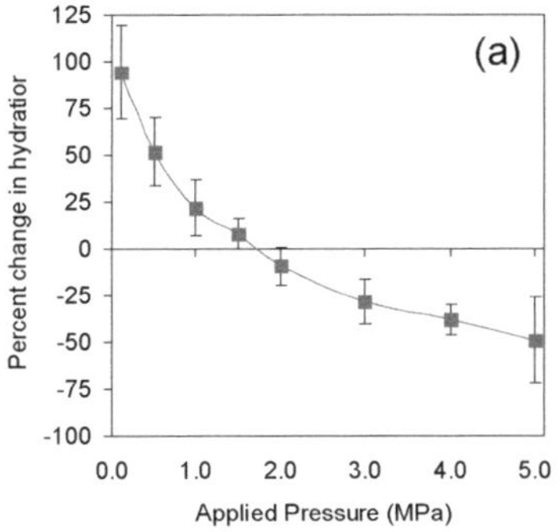
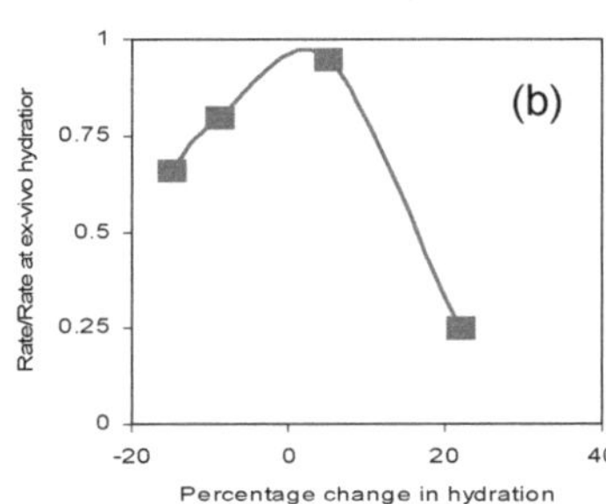
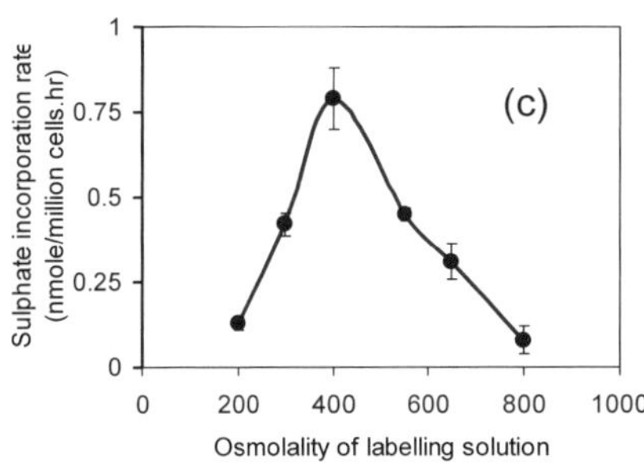



Fig. 3. Effect of changes in water content and osmolarity on synthesis of sulphated GAGs. (a) Changes of hydration of bovine disc nucleus with load (adapted from [39]). (b) Changes in sulphate incorporation in response to load and hence to hydration (adapted from [38]). (c) Effect of change in osmolarity on synthesis of sulphated GAGs by isolated nucleus cells. Cells were isolated from the tissue by enzyme digestion, encapsulated in alginate beads, incubated in DMEM with 6% serum for 48 hours before the osmolarity of the medium was altered by NaCl addition. Two hours after osmotic shock $^{35}$S-sulphate was added as a tracer for measuring GAG synthesis and incorporation measured over the following 4 hours.

## 4. Effect of load-induced changes in fluid content on disc cell activity

The fluid expression which is associated with increase in loading, concentrates GAGs and hence cations in the extracellular matrix and raises its osmolarity (i.e. the concentration of osmotically active solutes). These changes have powerful effects on disc cells. Firstly, the load-induced increase in extracellular osmolarity leads to cell shrinkage; as the plasma membrane of mammalian cells is permeable to water but not solutes, the disc cells decrease in volume to maintain osmotic equilibrium with the matrix. As a consequence of change in volume, intracellular ions and other solutes increase in concentration, affecting gene expression [7]. Change in volume can also activate transporters and channels able to move ions across the plasma membrane allowing many cell types to 'volume regulate'; the little information available on disc cells suggests that mature disc cells do not have this capacity [24].

### 4.1. Matrix synthesis and gene expression

The effect of hydration on disc cells appears to be mediated to a large extent by the resulting changes to extracellular osmolarity and hence to cell volume; the consequent changes in intracellular ion concentrations affect activity of enzymes involved in protein synthesis and can also trigger gene expression [7].

A number of *in vitro* studies have shown that fluid expression affects disc cell metabolism significantly. Tests on disc explants demonstrated that sulphate incorporation rate, a marker of the rate of GAG synthesis, was maximal if disc hydration was maintained at *ex vivo* levels [4,25]. Similar results were seen in whole bovine discs loaded in a perfusion system and then exposed to a range of different loads [39]. Synthesis rates appeared maximal at loads which maintained disc hydration, and fell in a dose-dependent manner if the disc hydration decreased or increased (Fig. 3). Similar responses of isolated mature nucleus and annulus cells to changes in osmolarity have been seen after short term and long-term exposure to solutions over a range of different osmolarities (Fig. 3). The extent of change in synthesis rate is marked, with cells synthesizing 40–100 percent more sulphated GAG at osmolarities seen *in vivo* (400–500 mOsm) than in standard tissue culture medium [24].

The effect of change in osmolarity on gene expression has been examined in notochordal and inner annulus (transition zone) cells. Notochordal cells appear rather insensitive to osmotic change, but several genes were upregulated by exposure to hypo-osmolarity in transition zone cells, notably collagen II and biglycan; biglycan was also upregulated in hyperosmotic medium [9].

### 4.2. Response to mechanical signals

There have been a number of studies on the responses of disc explants [18,29] cells to a variety of mechanical signals such as hydrostatic pressure [23], stretch [32,40,41] and compression [10]. Results show that both adult nucleus (but not notochordal nucleus) and annulus cells are very sensitive to mechanical stimuli and that *in vitro* at least, the level and type of mechanical signal influences both matrix synthesis and the level of protease expression and activity. However, virtually all measurements on isolated cells have been made in standard medium whose osmolarity is around 300 mOsmol (depending on medium formulation), well below the osmolarity seen by disc cells *in vivo*. Recent work however, shows that response of disc cells to mechanical signals is very sensitive to the extracellular osmotic environment; gene expression to the same signal can be inhibited rather than stimulated if the osmotic environment is altered (Fig. 4) [53].

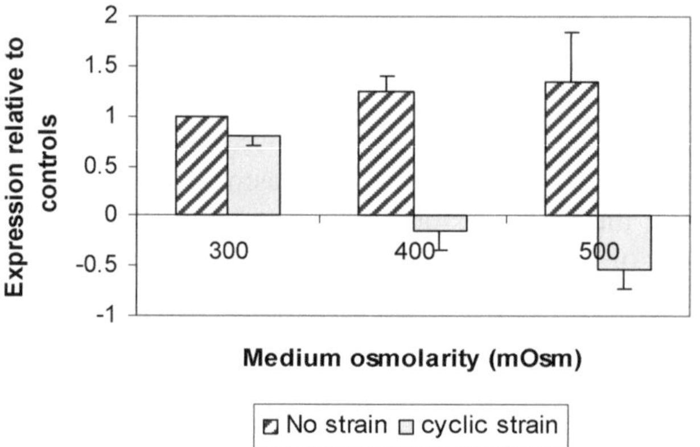


Fig. 4. Effect of osmolarity on response of outer annulus cells to cyclic strain (adapted from [53]). The effect of osmolarity on collagen-2 gene expression with and without application of cyclic strain (4%, 1 Hz for 30 minutes, results obtained after 24 hours).

## 5. Conclusions

It is apparent that the cells of the disc, *in vitro* at least, appear very sensitive to their prevailing osmotic environment which has a powerful effect on matrix synthesis and hence ultimately on the structure and composition of the disc. It is thus surprising that so little information is available on how the diurnal changes in osmolarity affect disc cells and the maintenance and turnover of the complex extracellular environment they produce. Virtually all results available come from data which have not simulated the diurnal extracellular osmotic but have provided a constant osmotic environment after an initial 'osmotic shock'. Since cartilaginous cells adapt to the prevailing osmolarity over a period of hours [20], the response of disc cells *in vivo* may be rather different to that measured to date and is worth further investigation.

## Acknowledgements

This study was funded by the EU (Eurodisc QLK6-CT-2002-02582): PGU is an Arthritis Research Campaign Senior fellow.

## References

[1] L. Ala-Kokko, Genetic risk factors for lumbar disc disease, *Ann. Med.* **34** (2002), 42–47.
[2] J. Antoniou, T. Steffen, F. Nelson, N. Winterbottom, A.P. Hollander, R.A. Poole, M. Aebi and M. Alini, The human lumbar intervertebral disc: evidence for changes in the biosynthesis and denaturation of the extracellular matrix with growth, maturation, ageing, and degeneration, *J. Clin. Invest.* **98** (1996), 996–1003.
[3] M.T. Bayliss, B. Johnstone and J.P. O'Brien, 1988 Volvo award in basic science. Proteoglycan synthesis in the human intervertebral disc. Variation with age, region and pathology, *Spine* **13** (1988), 972–981.
[4] M.T. Bayliss, J.P.G. Urban, B. Johnstone and S. Holm, In vitro method for measuring synthesis rates in the intervertebral disc, *J. Orthop. Res.* **4** (1986), 10–17.
[5] N. Boos, A. Wallin, T. Gbedegbegnon, M. Aebi and C. Boesch, Quantitative MR imaging of lumbar intervertebral disks and vertebral bodies: influence of diurnal water content variations, *Radiology* **188** (1993), 351–354.
[6] S.B. Bruehlmann, J.B. Rattner, J.R. Matyas and N.A. Duncan, Regional variations in the cellular matrix of the annulus fibrosus of the intervertebral disc, *J. Anat.* **201** (2002), 159–171.
[7] M.B. Burg, E.D. Kwon and D. Kultz, Osmotic regulation of gene expression, *FASEB J.* **10** (1996), 1598–1606.
[8] W.F. Butler, Comparative anatomy and development of the mammalian disc, in: *The Biology of the Intervertebral Disc*, P. Ghost, ed., CRC Press, Boca Raton, 1989, pp. 84–108.
[9] J. Chen, A.E. Baer, P.Y. Paik, W. Yan and L.A. Setton, Matrix protein gene expression in intervertebral disc cells subjected to altered osmolarity, *Biochem. Biophys. Res. Commun.* **293** (2002), 932–938.
[10] J. Chen, W. Yan and L.A. Setton, Static compression induces zonal-specific changes in gene expression for extracellular matrix and cytoskeletal proteins in intervertebral disc cells in vitro, *Matrix Biol.* **22** (2004), 573–583.
[11] J.K. Crean, S. Roberts, D.C. Jaffray, S.M. Eisenstein and V.C. Duance, Matrix metalloproteinases in the human intervertebral disc: role in disc degeneration and scoliosis, *Spine* **22** (1997), 2877–2884.
[12] G. Cs-Szabo, D. Ragasa-San Juan, V. Turumella, K. Masuda, E.J. Thonar and H.S. An, Changes in mRNA and protein levels of proteoglycans of the anulus fibrosus and nucleus pulposus during intervertebral disc degeneration, *Spine* **27** (2002), 2212–2219.
[13] D.R. Eyre, Y. Matsui and J.J. Wu, Collagen polymorphisms of the intervertebral disc, *Biochem. Soc. Trans.* **30** (2001), 844–848.
[14] D.R. Eyre and H. Muir, Quantitative analysis of types I and II collagens in the human intervertebral disc at various ages, *Biochimica et Biophysica Acta* **492** (1977), 29–42.
[15] S.J. Ferguson, K. Ito and L.P. Nolte, Fluid flow and convective transport of solutes within the intervertebral disc, *J. Biomech.* **37** (2004), 213–221.
[16] D. Gottschalk, M. Fehn, S. Patt, W. Saeger, T. Kirchner and T. Aigner, Matrix gene expression analysis and cellular phenotyping in chordoma reveals focal differentiation pattern of neoplastic cells mimicking nucleus pulposus development, *Am. J. Pathol.* **158** (2001), 1571–1578.

[17] H.E. Gruber, J.A. Ingram and E.N. Hanley, Jr., Tenascin in the human intervertebral disc: alterations with aging and disc degeneration, *Biotech. Histochem.* **77** (2002), 37–41.

[18] T. Handa, H. Ishihara, H. Ohshima, R. Osada, H. Tsuji and K. Obata, Effects of hydrostatic pressure on matrix synthesis and matrix metalloproteinase production in the human lumbar intervertebral disc, *Spine* **22** (1997), 1085–1091.

[19] J.M. Heaton and D.R. Turner, Reflections on notochordal differentiation arising from a study of chordomas, *Histopathology* **9** (1985), 543–550.

[20] B. Hopewell and J.P. Urban, Adaptation of articular chondrocytes to changes in osmolality, *Biorheology* **40** (2003), 73–77.

[21] H.A. Horner, S. Roberts, R.C. Bielby, J. Menage, H. Evans and J.P. Urban, Cells from different regions of the intervertebral disc: effect of culture system on matrix expression and cell phenotype, *Spine* **27** (2002), 1018–1028.

[22] C.J. Hunter, J.R. Matyas and N.A. Duncan, Cytomorphology of notochordal and chondrocytic cells from the nucleus pulposus: a species comparison, *J. Anat.* **205** (2004), 357–362.

[23] W.C. Hutton, W.A. Elmer, S.D. Boden, S. Hyon, Y. Toribatake, K. Tomita and G.A. Hair, The effect of hydrostatic pressure on intervertebral disc metabolism, *Spine* **24** (1999), 1507–1515.

[24] H. Ishihara, K. Warensjo, S. Roberts and J.P. Urban, Proteoglycan synthesis in the intervertebral disk nucleus: the role of extracellular osmolality, *Am. J. Physiol.* **272** (1997), C1499–C1506.

[25] B. Johnstone and M.T. Bayliss, The large proteoglycans of the human intervertebral disc. Changes in their biosynthesis and structure with age, topography, and pathology, *Spine* **20** (1995), 674–684.

[26] B. Johnstone, M. Markopolous, P. Neame and B. Caterson, Identification and characterization of glycanated and non-glycanated forms of biglycan and decorin in the human intervertebral disc, *Biochem. J.* **292** (1993), 661–666.

[27] E.P. Katz, E.J. Wachtel and A. Maroudas, Extrafibrillar proteoglycans osmotically regulate the molecular packing of collagen in cartilage, *Biochim. Biophys. Acta* **882** (1986), 136–139.

[28] J. Kraemer, D. Kolditz and R. Gowin, Water and electrolyte content of human intervertebral discs under variable load, *Spine* **10** (1985), 69–71.

[29] G.Z. Liu, H. Ishihara, R. Osada, T. Kimura and H. Tsuji, Nitric oxide mediates the change of proteoglycan synthesis in the human lumbar intervertebral disc in response to hydrostatic pressure, *Spine* **26** (2001), 134–141.

[30] A. Maroudas, Distribution and diffusion of solutes in articular cartilage, *Biophys. J.* **10** (1970), 365–379.

[31] A. Maroudas, H. Muir and J. Wingham, The correlation of fixed negative charge with glycosaminoglycan content of human articular cartilage, *Biochim. Biophys. Acta* **177** (1969), 492–500.

[32] T. Matsumoto, M. Kawakami, K. Kuribayashi, T. Takenaka and T. Tamaki, Cyclic mechanical stretch stress increases the growth rate and collagen synthesis of nucleus pulposus cells in vitro, *Spine* **24** (1999), 315–319.

[33] C.A. McDevitt, Proteoglycans of the intervertebral disc, in: *The Biology of the Intervertebral Disc*, P. Ghost, ed., CRC Press, Inc., Boca Raton, FL, 1988, pp. 151–170.

[34] D.S. McNally, Biomechanics of the intervertebral disc–disc pressure measurements and significance, in: *Lunbar Spine Disorders*, R.M. Aspden and R.W. Parter, eds, World Scientific Publishing Co., Singapore, 1955, pp. 42–50.

[35] J. Melrose, P. Ghosh and T.K. Taylor, Neutral proteinases of the human intervertebral disc, *Biochim. Biophys. Acta* **923** (1987), 483–495.

[36] J. Melrose, P. Ghosh and T.K. Taylor, A comparative analysis of the differential spatial and temporal distributions of the large (aggrecan, versican) and small (decorin, biglycan, fibromodulin) proteoglycans of the intervertebral disc, *J. Anat.* **198** (2001), 3–15.

[37] A. Nachemson and J.M. Morris, In vivo measurements of interdiscal pressure, *J. Bone Jt. Surg.* **46A** (1964), 1077–1092.

[38] T.R. Oegema, Jr., S.L. Johnson, D.J. Aguiar and J.W. Ogilvie, Fibronectin and its fragments increase with degeneration in the human intervertebral disc, *Spine* **25** (2000), 2742–2747.

[39] H. Ohshima, J.P.G. Urban and D.H. Bergel, The effect of static load on matrix synthesis rates in the intervertebral disc measured in vitro by a new perfusion technique, *J. Orthop. Res.* **13** (1995), 22–29.

[40] F. Rannou, S. Poiraudeau, V. Foltz, M. Boiteux, M. Corvol and M. Revel, Monolayer anulus fibrosus cell cultures in a mechanically active environment: local culture condition adaptations and cell phenotype study, *J. Lab. Clin. Med.* **136** (2000), 412–421.

[41] F. Rannou, P. Richette, M. Benallaoua, M. Francois, V. Genries, C. Korwin-Zmijowska, M. Revel, M. Corvol and S. Poiraudeau, Cyclic tensile stretch modulates proteoglycan production by intervertebral disc annulus fibrosus cells through production of nitric oxide, *J. Cell Biochem.* **90** (2003), 148–157.

[42] S. Roberts, B. Caterson, J. Menage, E.H. Evans, D.C. Jaffray and S.M. Eisenstein, Matrix metalloproteinases and aggrecanase: their role in disorders of the human intervertebral disc, *Spine* **25** (2000), 3005–3013.

[43] S. Roberts, J. Menage, V. Duance, S. Wotton and S. Ayad, 1991 Volvo award in basic sciences: Collagen types around the cells of the intervertebral disc and cartilage end plate: An immunolocalization study, *Spine* **16** (1991), 1030–1038.

[44] S. Seki, Y. Kawaguchi, K. Chiba, Y. Mikami, H. Kizawa, T. Oya, F. Mio, M. Mori, Y. Miyamoto, I. Masuda, T. Tsunoda, M. Kamata, T. Kubo, Y. Toyama, T. Kimura, Y. Nakamura and S. Ikegawa, A functional SNP in CILP, encoding cartilage intermediate layer protein, is associated with susceptibility to lumbar disc disease, *Nat. Genet.* **37** (2005), 607–612.

[45] S. Sivan, E. Wachtel, E. Merkher and A. Maroudas, Correlation of swelling pressure and intra-fibrillar water in young and aged annuli of the intervertebral disc, *Journal of Orthopaedic Research*, in press (2005).

[46] R.L. Stevens, R.J. Ewins, P.A. Revell and H. Muir, Proteoglycans of the intervertebral disc. Homology of structure with laryngeal proteoglycans, *Biochem. J.* **179** (1979), 561–572.

[47] T. Takeda, Three-dimensional observations of collagen framework of human lumbar discs, *J. Japan Orthop. Assoc.* **49** (1975), 45–57.

[48] J.P. Urban, A. Maroudas, M.T. Bayliss and J. Dillon, Swelling pressures of proteoglycans at the concentrations found in cartilaginous tissues, *Biorheology* **16** (1979), 447–464.

[49] J.P. Urban and J.F. McMullin, Swelling pressure of the inervertebral disc: influence of proteoglycan and collagen contents, *Biorheology* **22** (1985), 145–157.

[50] J.P.G. Urban and A. Maroudas, Measurement of fixed charge density and partition coefficients in the intervertebral disc, *Biochim. Biophys. Acta* **586** (1979), 166–178.

[51] R. Walmsley, The development and growth of the intervertebral disc, *Edinburgh Medical Journal* **60** (1953), 341–364.

[52] A.A. White and M.M. Panjabi, Clinical Biomechanics of the Spine, 1978.

[53] K. Würtz, J.P.G. Urban and C. Neidlinger-Wilke, Effects of osmolarity on responses of disc cells to hydrostatic pressure, in: *ISSLS Proceedings*, New York Meeting, 2005.

[54] J. Yu, C. Peter, S. Roberts and J.P. Urban, Elastic fibre organization in the intervertebral discs of the bovine tail, *J. Anat.* **201** (2002), 465–475.

Biorheology 43 (2006) 293–302
IOS Press



# A mandibular propulsive appliance modulates collagen-binding integrins distribution in the young rat condylar cartilage

Mara Rúbia Marques, Denise Hajjar, Virgínia Oliveira Crema, Edna Teruko Kimura and
Marinilce Fagundes Santos *
*Cell and Developmental Biology Department, Biomedical Sciences Institute, University of São Paulo,
São Paulo, Brazil*

**Abstract.** We have previously shown that a mandibular propulsive appliance (MPA) stimulates cell proliferation and the synthesis of growth factors in the rat condylar cartilage. The aim of this study was to evaluate the effects of a MPA in the distribution of the integrin subunits $\alpha 1$ and $\alpha 2$ in this cartilage. Twenty eight days-old male Wistar rats were divided into treated (T) and age-matched control groups (C). Treated rats wore the appliance during 3, 5, 7, 9, 11, 15, 20, 30 and 35 days. The condyles were fixed, decalcified and paraffin-embedded. The distribution of $\alpha 1$ and $\alpha 2$ was studied by immunohistochemistry. Alpha1 distribution was uniform along the cartilage, increasing in 48 days-old rats (C20). Treated animals anticipated this increase to the age of 36 days (T9). The number of $\alpha 2$-positive cells was increased in C9 in the anterior condylar region, in C9 and C20 in the middle region and showed no differences in the posterior region. The MPA apparently abolished all variations, leading to a single increase at T30 in all regions. These results suggest that integrins containing the $\alpha 1$ and $\alpha 2$ subunits are modulated by forces promoted by the MPA, participating of the biological response to this therapy.

Keywords: Functional orthopedics, condyle, alpha1, alpha2, integrin

## 1. Introduction

The modulation of mandibular growth by the application of mechanical forces has been studied to improve the correction of discrepancies between the maxilla and the mandible, as well as dental malocclusions [19]. It is believed that the use of functional orthopedic appliances by young individuals generate tension in orofacial muscles, which are transmitted to the mandibular condyle, an important growth center of the mandible, generating adaptation. Clinical results obtained with this therapy are well characterized, although the molecular mechanisms underlying this response still remain unclear.

The mandibular condyle is part of the temporomandibular joint, a bilateral synovial joint between the mandible and the temporal bones. The young condylar cartilage is a hyaline cartilage covered by a fibrous layer. Proliferative undifferentiated cells, located immediately below this layer, differentiate into chondroblasts, which produce extracellular matrix (ECM), mostly collagen type II and proteoglycans.

---

*Address for correspondence: M.F. Santos, Cell and Developmental Biology Department, Av. Prof. Lineu Prestes 1524 sala 434, São Paulo, SP, Brazil, CEP 05508-900.

Later on, these cells become entrapped within this matrix, becoming chondrocytes. Hypertrophy of chondrocytes is followed by endochondral ossification. The proliferative compartment in the condylar cartilage is formed by undifferentiated cells and chondroblasts [19,24].

It is possible to modulate growth of the condylar cartilage by the application of forces [6]. Small load alterations such as those promoted by food hardness, for example, affect growth [1]. In cultured rat condylar cartilage hydrostatic compressive forces stimulated glycosaminoglycan (GAG) and DNA synthesis in a magnitude-dependent manner [22], and intermittent compressive loading increased collagen I and fibronectin synthesis [17].

We have previously observed that an intermittent indirect force generated by a mandibular propulsive orthopedic appliance was able to increase the expression of Insulin-like Growth Factors I and II (IGF-I and IGF-II), and cell proliferation in the young rat condylar cartilage [7]. Charlier et al. (1969) also observed an increase in the size of the undifferentiated layer after using this kind of appliance [2]. In our previous study, if this appliance was used for a short period of time (up to 9 days), the effect on mandibular repositioning was reversible. After 11 days of use, the mandibular repositioning was permanent, even after the appliance's withdrawn (unpublished observation).

The transmission of forces applied to the condylar chondrocytes is probably mediated by integrins, heterodimeric transmembrane receptors for ECM proteins. Integrins contain two non-covalently associated subunits, $\alpha$ and $\beta$, which bind ECM proteins through their extracellular domains [5,9]. Through their cytoplasmic domains integrins usually interact with the cytoskeleton. Although integrins have no intrinsic enzymatic activity, they trigger signaling pathways upon binding to ECM, which might regulate several chondrocyte functions such as differentiation, matrix remodeling, gene expression, responses to mechanical stimulation and cell survival [13]. In normal human articular chondrocytes, integrin activation, consequent to mechanical stimulation *in vitro*, results in tyrosine phosphorylation of regulatory proteins and subsequent secretion of autocrine and paracrine factors [15].

The expression of integrins in human femoral condyles has been shown to be modulated by health condition and developmental stage [8,15,18,20]. During human growth, $\alpha 1$ and $\alpha 2$ integrin subunits were observed in the proliferative and hypertrophic layers [8], determined mostly by the ECM composition [11]. Condrossarcoma cells, as well as fetal chondrocytes, showed high levels of the collagen-binding $\alpha 2 \beta 1$ integrin [12].

We believe that an improved comprehension of how integrins mediate chondrocyte responses to mechanical stimulation, and how cross talk between integrins, ECM, and autocrine/paracrine signaling molecules (such as IGF-I and IGF-II) regulate mechanotransduction are important for further understanding how functional orthopedic appliances promote cartilage remodeling.

## 2. Materials and methods

### 2.1. Animals and treatment

All experiments were conducted in accordance with the NIH guidelines, and the protocols were approved by the Biomedical Sciences Institute/USP Ethical Committee for Animal Research. Seventy-two male Wistar rats were used, beginning at 28 days of age. They were divided into treated (T) and control age-matched groups (C). Treated animals wore the appliance for 3, 5, 7, 9, 11, 15, 20, 30 and 35 days, from 8:00 a.m. to 6:00 p.m. The appliance was changed when necessary, according to the animal's development. Food and water were available at night, *ad libitum*. At the time of sacrifice, animals were anaesthetized with chloral hydrate (30 mg/100 g of body weight) and perfused with saline and with fixative solution (4% formaldehyde, Sigma, St. Louis, MO, USA). The mandibular condyles were dissected out, post-fixed in 4% formaldehyde at 4°C during 24 h, decalcified with 10% EDTA for 20 days and embedded in Paraplast®. Sagittal 5 $\mu$m-thick sections were placed on Poly-L-Lysine-coated slides (Sigma, St. Louis, MO, USA).

## 2.2. Orthopedic appliance

The appliance was designed as described by Hajjar et al. [7]. Briefly, it consisted of an inclined copper plane that raised the anterior displacement of the mandible every time the animals attempted to close their mouths at the normal position. The appliance was banded 90° at the anterior portion and a rubber tube was glued close, to fit in the upper incisor of the animal. All edges were rounded to avoid mucosal irritation. A soft leather collar was used to prevent the removal of the appliance; control animals wore the collar alone.

## 2.3. Immunohistochemistry

Integrin subunits were investigated by immunohistochemistry, using the avidin–biotin peroxidase technique. Rabbit polyclonal anti-$\alpha$1 (Chemicon, AB1934, Temecula, CA, USA) and anti-$\alpha$2 (Chemicon, AB1936, Temecula, CA, USA) were used at a 1 : 100 dilution in PBS, overnight at room temperature (RT). After washing in PBS, sections were incubated with the biotin-conjugated goat anti-rabbit secondary antibody (Jackson Immunoresearch Laboratories, West Grove, PA, USA) at a 1 : 250 dilution during 2 hours at RT. The sections were then revealed with 3,3′diamino-benzidine (DAB) in the presence of 0.02% $H_2O_2$ for 30 min, mounted with Permount® (Fisher Chemicals) and examined on a Leitz microscope (Aristoplan). As a negative control the omission of the primary antibody was used, and no labeling was observed under these conditions. Pictures were taken using a CCD camera and the Scion Image software (NIH, Bethesda, USA).

## 2.4. Counting of labeled cells and statistical analysis

The quantification of labeled cells was performed by the evaluation of 900 cells in 3 randomly chosen microscopic fields at the anterior, central and posterior regions of the condylar cartilage per section. The number of labeled cells was expressed as a percentage of the total number of cells observed. Data were submitted to analysis of variance (ANOVA) followed by the Tukey's post-test for multiple comparisons. Main comparisons were made within the control group (between different ages) and between the treated groups and their age-matched controls. A $P$ value of less than 0.05 was considered to be significant. Results were expressed as mean $\pm$ standard deviation (SD).

# 3. Results

## 3.1. The propulsor appliance modulates $\alpha$1 integrin subunit distribution

During normal development, $\alpha$1 distribution was almost restricted to the hypertrophic layer in C3 (Fig. 1). From C5 on, $\alpha$1 staining was distributed throughout the cartilage, including the proliferative

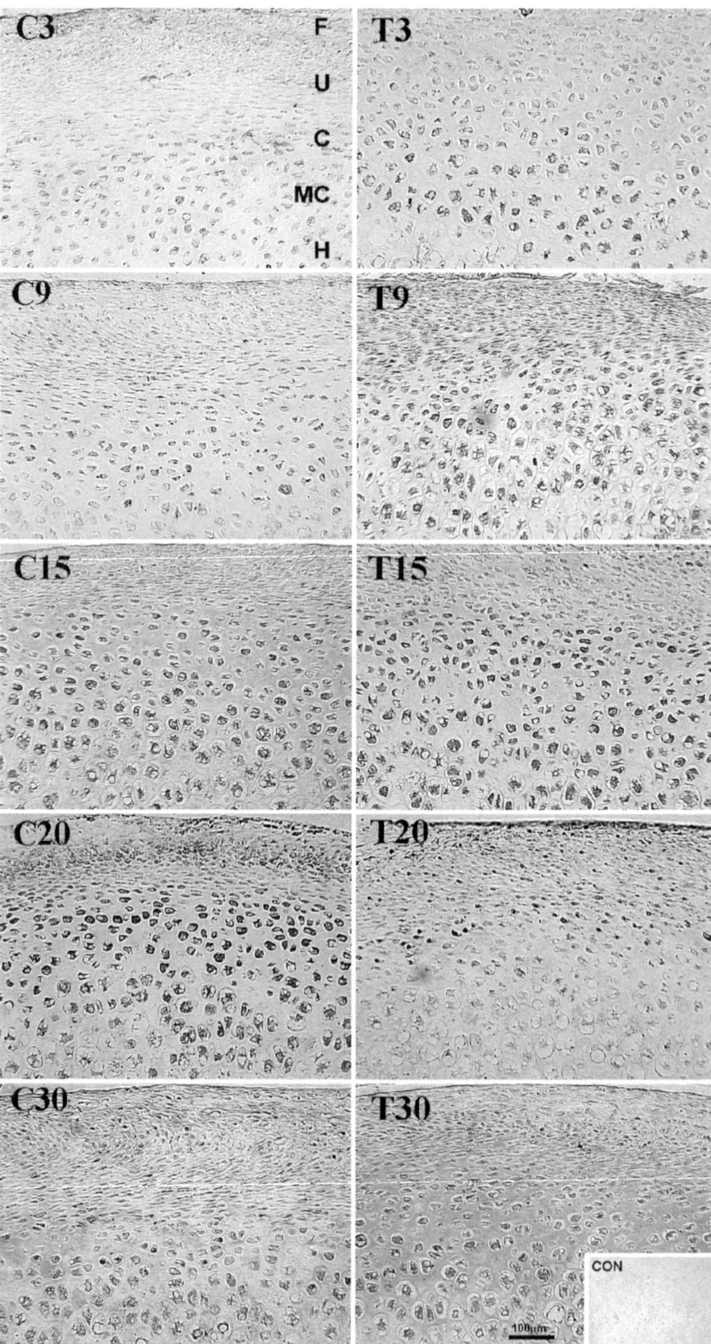


Fig. 1. Distribution of the $\alpha$1 integrin subunit in the rat condylar cartilage during normal growth (C3–C30) and under therapy with a mandibular propulsive appliance (T3–T30). The $\alpha$1 subunit was studied by immunohistochemistry using the avidin–biotin-peroxidase technique in sagittal sections of the cartilage. T3, T9, T15, T20 and T30 correspond to young animals that wore the appliance for 3, 9, 15, 20 and 30 days, respectively. Each group was sacrificed with their age-matched control groups. In C3, all layers are shown: fibrous (F), undifferentiated (U), chondroblasts (C), mature chondrocytes (MC) and Hypertrophic (H). In detail, the negative control consisting of the omission of the primary antibody (CON). Magnification bar = 100 $\mu$m.

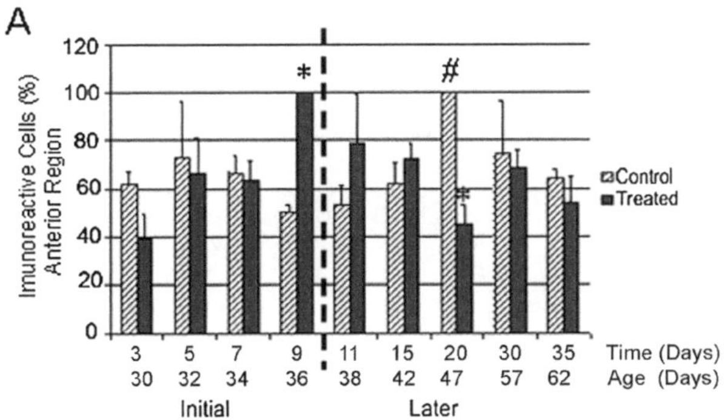
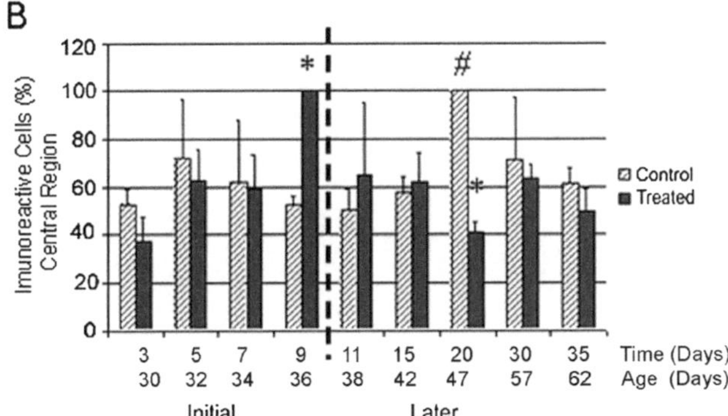
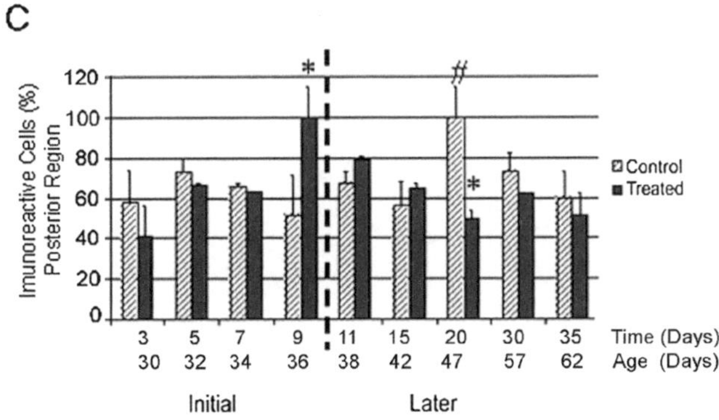

Fig. 2. Quantification of $\alpha$1-labeled cells in different regions of the rat condylar cartilage: (A) anterior region; (B) central region; and (C) posterior region. The dotted line separates the early and late phases of the therapy. Data are presented as percentage of total cells counted, mean $\pm$ SD. $^{\#}P < 0.001$ when compared to the previous group; $^{*}P < 0.001$ when compared to the age-matched control group, according to ANOVA and Tukey's post-test.

layer. Labeling intensity, however, was variable, mainly in the proliferative compartment and the chondrocyte layer (Fig. 1).

The number of $\alpha$1-positive cells was uniform among the anterior, central and posterior regions of the cartilage, varying from 50–70% (Fig. 2). In the control group, there was an increase in $\alpha$1-positively labeled cells in the C20 group (100% labeled cells compared to 58.52% in C15). Labeling intensity was increased as well. Both the number of labeled cells and labeling intensity decreased afterwards, returning to initial levels (Figs 1 and 2).

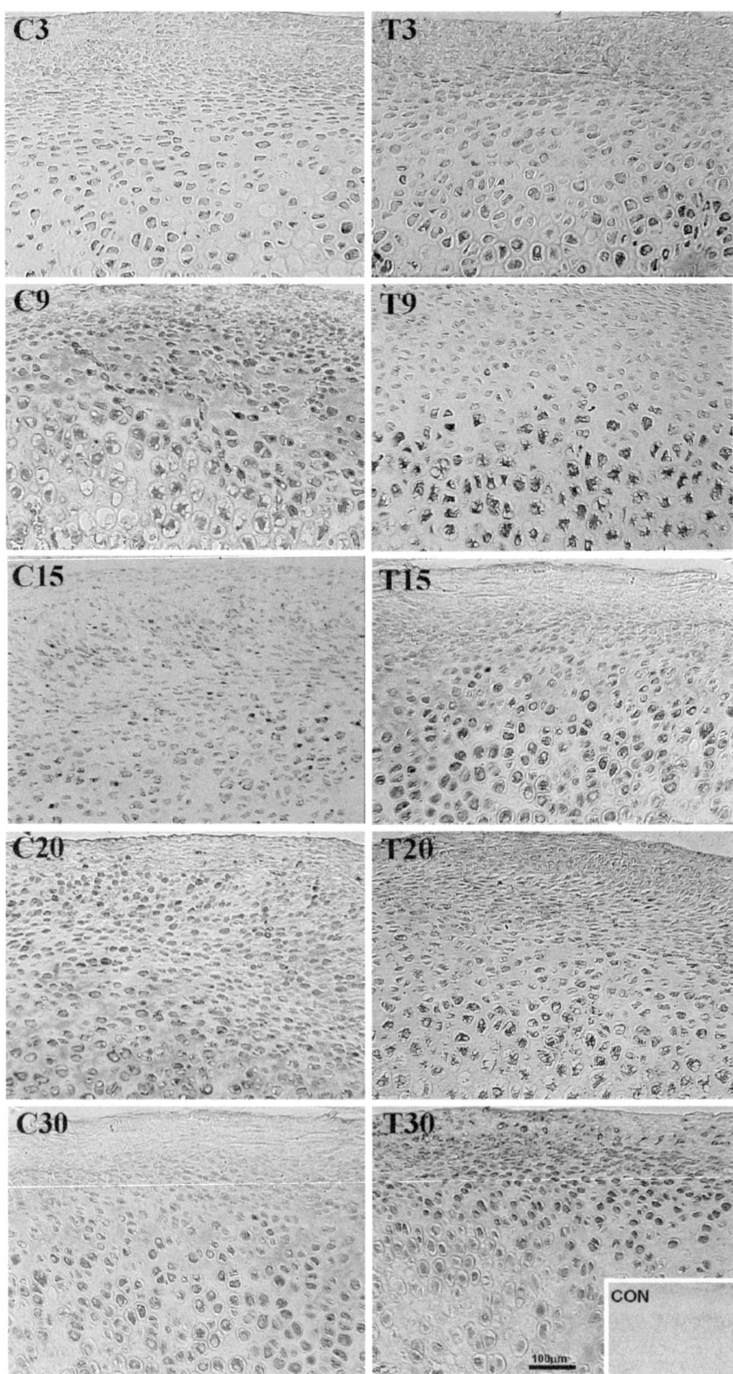


Fig. 3. Distribution of the $\alpha 2$ integrin subunit in the rat condylar cartilage during normal growth (C3–C30) and under therapy with a mandibular propulsive appliance (T3–T30). The $\alpha 2$ subunit was studied by immunohistochemistry using the avidin–biotin-peroxidase technique in sagittal sections of the cartilage. T3, T9, T15, T20 and T30 correspond to young animals that wore the appliance for 3, 9, 15, 20 and 30 days, respectively. Each group was sacrificed with their age-matched control groups. In detail, the negative control consisting of the omission of the primary antibody (CON). Magnification bar = 100 $\mu$m.

The appliance promoted a slight increase in labeling intensity at T3, mostly in the hypertrophic layer (Fig. 1). At T9, however, both the number of positive cells and the intensity of the staining were enhanced, comparing to C9 (100% labeled cells vs. 51.33% in C9) (Figs 1 and 2). From T9 on, the labeling was similar to that observed in the control group, except for T20, which did not show the increase in $\alpha1$ observed in C20.

## 3.2. *The $\alpha2$ integrin subunit distribution varies according to the cartilage region and is modulated by the propulsor appliance*

At C3, $\alpha2$-positive cells were observed mostly in the hypertrophic layer (Fig. 3). At C9 there was a marked increase in the number of positive cells in the anterior and middle regions of the cartilage, but

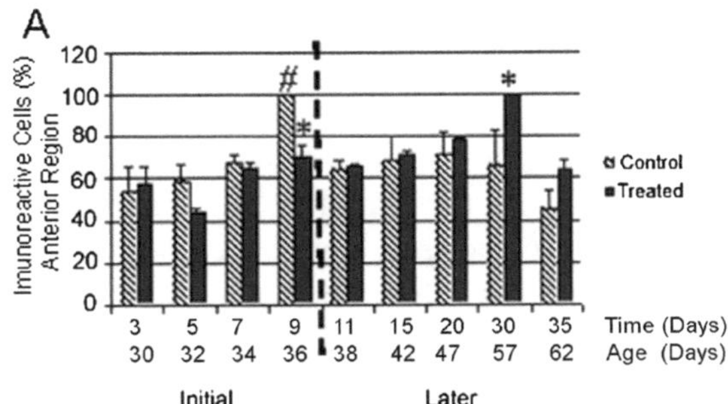

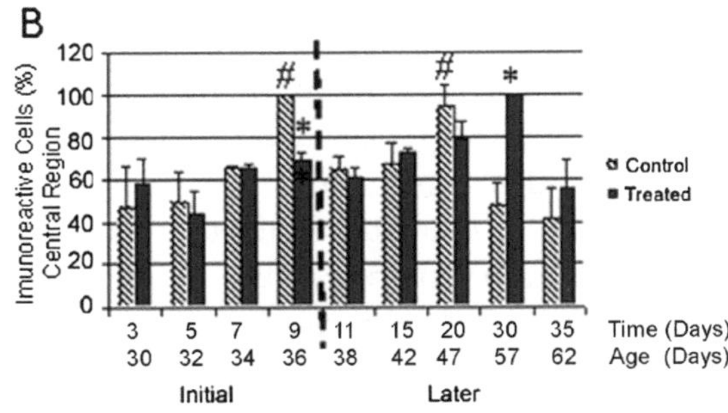

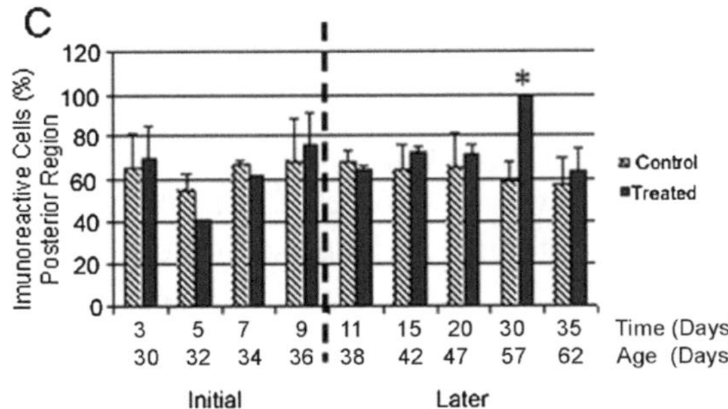

Fig. 4. Quantification of $\alpha2$-labeled cells in different regions of the rat condylar cartilage: (A) anterior region; (B) central region; and (C) posterior region. The dotted line separates the early and late phases of the therapy. Data are presented as percentage of total cells counted, mean $\pm$ SD. [#]$P < 0.001$ when compared to the previous group; [*]$P < 0.001$ when compared to the age-matched control group, according to ANOVA and Tukey's post-test.

not in the posterior region (100% labeled cells vs. 67% in C7, Fig. 4). We observed an enhanced labeling intensity as well, and the presence of $\alpha2$ in all cartilage layers (Fig. 3). This increase in the number of labeled cells was observed at C20 again, but only in the central cartilage region (77% of labeled cells). From C20 on, the number of marked cells gradually returned to the initial levels, being mostly restricted to the hypertrophic layer (Figs 3 and 4). No variations were observed for $\alpha2$ labeling in the posterior region (Fig. 4).

As observed for $\alpha1$, the staining in T3 was slightly more intense than that observed in the control group (Fig. 3). Curiously, the appliance abolished variations described above in the number of labeled cells, leading to a single increase in number and staining intensity at T30, in the anterior, central and posterior regions (Figs 3 and 4). This difference, observed mainly in the proliferative compartment, was not observed in T35.

## 4. Discussion

The modulation of the mandibular condylar cartilage growth and adaptation is considered an effective way to correct growth discrepancies between the mandible and maxilla. In ortodonthics, the application of mechanical forces through the use of orthopedic appliances promotes very good clinical results. In this study, we showed that a mandibular propulsor appliance changed the distribution of the integrin subunits $\alpha1$ and $\alpha2$, probably corresponding to the collagen-binding integrins $\alpha1\beta1$ and $\alpha2\beta1$, in the rat condylar cartilage.

For both $\alpha1$ and $\alpha2$ there was variation in expression and distribution during growth. In control animals, the $\alpha1$ subunit raised uniformly in all cartilage regions in C20, when the rats were 48 days-old, around puberty. The $\alpha2$ subunit, on the other hand, was uniformly expressed along the condylar cartilage, but showed some differential distribution with growth. Specifically, there was an increase in C9, but only in the anterior and posterior regions of the cartilage, and another peak in C20, similar to $\alpha1$, but observed only in the central region. An interesting idea is that while $\alpha1$ is modulated by hormones and growth factors present in the cartilage during growth, $\alpha2$ could be related to force, because its expression was higher close to the muscle insertion. However, maybe contrary to this hypothesis, the appliance, which we consider as an intermittent indirect force applied to the cartilage, anticipated the $\alpha1$ peak (previously in C20) to 9 days of treatment (T9). On the other hand, all $\alpha2$ variations were abolished by the appliance's use, and a peak of expression appeared at T30, only in the treated group. In general, expression peaks mean a broader distribution, with all layers marked, including the undifferentiated proliferative layer.

It is tempting to think that $\alpha2$ was being replaced by $\alpha1$ in T9, and that there was a late effect of this therapy after 30 days of treatment, leading to an increase in $\alpha2$. After 9 days of treatment, it is known that there is a marked increase in cell proliferation, as observed by PCNA (Proliferating Cell Nuclear Antigen) labeling [7]. However, we believe that in order to consolidate this effect on growth, the process of endochondral ossification must be continued, and the use of the appliance ensures that those new cells will differentiate and synthesize ECM proteins, such as collagen II and proteoglycans. If the appliance is removed after 9 days, the mandible does not return to its original position.

Several studies showed that the condylar cartilage responds to mechanical stimuli by altering cell proliferation and ECM proteins expression. For example, intermittent compressive forces increased sulfated glycosaminoglycans and collagen synthesis, while continuous compressive forces reduced ECM synthesis [3,4]. Another study showed that compressive forces applied uniformly *in vitro* to the whole condyle using a pneumatic loading system increased collagen I and fibronectin [17]. One has to consider, however, the force direction in these experiments, since it appears to be determinant for the effects on cartilage. Opposite forces such as those exerted by a protrusive appliance compress the anterior portion of the condyle against the temporal articular eminence, decreasing cartilage thickness and aggrecan immunostaining in rats [23]. However, in the same anterior region there are considerable stretching forces compared to the rest of the condyle, due to the muscle intermittent activity.

Apparently, ECM composition defines the expression of different integrins in chondrocytes. For example, *in vitro*, $\alpha 1$ and $\alpha 2$ expression were upregulated by collagens I and II, respectively, in porcine knee articular cartilage [11]. It is possible that, in our study, the increase in $\alpha 2$ after 30 days of treatment was related to a higher amount of collagen type II produced by chondrocytes, after cell proliferation and differentiation. In monkeys and humans, the distribution and expression of collagens I and II vary with age [10,16], suggesting that the normal variation of $\alpha 1$ and $\alpha 2$ in rats could also be associated with ECM composition variations.

Corroborating this hypothesis, stretching strain *in vitro* upregulated cartilaginous collagen II and aggrecan expression, as well as $\alpha 2$ and $\alpha 5$ (fibronectin-binding) integrin subunits [14]. Similar observation was made for the $\beta 1$ integrin subunit, *in vivo* [21].

We have shown previously that this mandibular propulsive appliance stimulates the synthesis of IGF-I and IGF-II in the condylar cartilage, closely related to an increase in PCNA labeling [7]. After 9 days of treatment, for example, the expression of IGF-I and IGF-II was increased in the cartilage, which is coincident with the increased expression of $\alpha 1$ observed in the present study. Even in control animals, the expression of IGFs was slightly higher during puberty [7]. It is possible that IGFs affected $\alpha 1$ expression in the condylar cartilage in our study, and this hypothesis deserves further investigation.

Alternatively, $\alpha 1$-containing integrins, functioning as mechanoreceptors, might activate signal transduction pathways that regulate gene expression, leading to an increase in IGFs production. In human articular chondrocytes, for example, the fibronectin receptor $\alpha 5 \beta 1$ functions as a mechanoreceptor. After mechanical stimulation, there was activation of a signal cascade involving stretch-activated ion channels, the actin cytoskeleton and the tyrosine phosphorylation of several components of focal adhesions. Subsequently, there was secretion of interleukin-4, which acted in an autocrine manner to regulate several cell functions and gene expression [20].

In summary, we have shown that integrin subunits $\alpha 1$ and $\alpha 2$ are expressed in the rat condylar cartilage during normal growth, with varying distribution according to the animal's age. A mandibular propulsive appliance, probably acting as an intermittent force, altered their expression and distribution, suggesting that this modulation may be part of the appliance's effect on a molecular level. The understanding of how forces exerted by orthopedic appliances modulate chondrocyte function is very important to improve our knowledge about this widely employed therapy.

## Acknowledgements

The authors are grateful to Emília Ribeiro for the excellent technical support. This study was supported by the São Paulo State Research Foundation (FAPESP), grants # 97/09507-6 and 01/09047-2. Mara R. Marques is the recipient of a scholarship from FAPESP (02/11920-9).

## References

[1] M. Bouvier and M.L. Zimny, Effects of loads on surface morphology of the condylar cartilage of the mandible in rats, *Acta Anat.* **129** (1997), 293–300.

[2] J.P. Charlier, A. Petrovic and J. Herrmann-Stutzmann, Effects of mandibular hyperpropulsion on the prechondroblastic zone of young rat condyle, *Am. J. Orthod.* **55** (1969), 71–74.

[3] J.C. Copray and W. Jansen, Cyclic nucleotides and growth regulation of the mandibular condylar cartilage of the rat in vitro, *Archs. Oral Biol.* **30** (1985), 299–304.

[4] J.C. Copray, H.W. Jansen and H.S. Duterloo, An in-vitro system for studying the effect of variable compressive forces on the mandibular condylar cartilage of the rat, *Arch. Oral Biol.* **30** (1985), 305–311.

[5] E.H. Danen, Integrins: regulators of tissue function and cancer progression, *Curr. Pharm. Des.* **11** (2005), 881–891.

[6] A.J. Grodzinsky, M.E. Leventston, J. Moonsoo and E.H. Frank, Cartilage tissue remodeling in response to mechanical forces, *Annu. Rev. Biomed. Eng.* **02** (2000), 691–713.

[7] D. Hajjar, M.F. Santos and E.T. Kimura, Propulsive appliance stimulates the synthesis of insulin-like growth factors I and II in the mandibular condylar cartilage of young rats, *Arch. Oral Biol.* **48** (2003), 635–642.

[8] G. Hausler, M. Helmreich, S. Marlovits and M. Egerbacher, Integrins and extracellular matrix proteins in the human childhood and adolescent growth plate, *Calcif. Tissue Int.* **71** (2002), 212–218.

[9] J. Iqbal and M. Zaidi, Molecular regulation of mechanotransduction, *Biochem. Biophys. Res. Commun.* **328** (2005), 751–755, Review.

[10] H. Ishibashi, Y. Takenoshita, K. Ishibashi and M. Oka, Expression of extracellular matrix in human mandibular condyle, *Oral Surg. Oral Med. Oral Pathol. Oral Radiol. Endod.* **81** (1996), 402–414.

[11] S.J. Kim, E.J. Kim, Y.H. Kim, S.B. Hahn and J.W. Lee, The modulation of integrin expression by the extracellular matrix in articular chondrocytes, *Yonsei Med. J.* **44** (2003), 493–501.

[12] W. Knudson and R.F. Loeser, CD44 and integrin matrix receptors participate in cartilage homeostasis, *Cell Mol. Life Sci.* **59** (2002), 36–44.

[13] R.F. Loeser, Chondrocyte integrin expression and function, *Biorheology* **37** (2002), 109–116.

[14] K. Lahiji, A. Polotsky, D.S. Hungerford and C.G. Frondoza, Cyclic strain stimulates proliferative capacity, alpha2 and alpha5 integrin, gene marker expression by human articular chondrocytes propagated on flexible silicone membranes, *In Vitro Cell Dev. Biol. Anim.* **40** (2004), 138–142.

[15] S.J. Millward-Sadler and D.M. Salter, Integrin-dependent signal cascades in chondrocytes mechanotransduction, *Ann. Biom. Eng.* **32** (2004), 435–446.

[16] I. Mizoguchi, I. Takahashi, Y. Sasano, M. Kagayama, Y. Kuboki and H. Mitani, Localization of types I, II and X collagen and osteocalcin in intramembranous, endochondral and chondroid bone of rats, *Anat. Embryol.* **195** (1997), 127–135.

[17] H. Nakai, A. Niimi and M. Ueda, The influence of compressive loading on growth of cartilage of the mandibular condyle in vitro, *Arch. Oral Biol.* **43** (1998), 505–515.

[18] K. Ostergaard, D.M. Salter, J. Petersen, K. Bendtzen, J. Hvolris and C.B. Andersen, Expression of alpha and beta subunits of the integrin superfamily in articular cartilage from macroscopically normal and osteoarthritic human femoral heads, *Ann. Rheum. Dis.* **57** (1998), 303–308.

[19] A.G. Petrovic, J. Stutzmann and C. Oudet, Control processes in the postnatal growth of the condylar cartilage of the mandible, in: *Determinants of Mandibular form and Growth*, J.A. McNamara, Jr., ed., Center for Human Growth and Development, Michigan, 1975, pp. 101–153.

[20] D.M. Salter, S.J. Millward-Sadler, G. Nuki and M.O. Wright, Integrin–interleukin-4 mechanotransduction pathways in human chondrocytes, *Clin. Orthop. Relat. Res.* **391**(Suppl.) (2001), 49–60, Review.

[21] I. Takahashi, K. Onodera, Y. Sasano, I. Mizoguchi, J.W. Bae, H. Mitani, M. Kagayama and H. Mitani, Effect of stretching on gene expression of beta1 integrin and focal adhesion kinase and on chondrogenesis through cell–extracellular matrix interactions, *Eur. J. Cell Biol.* **82** (2003), 182–192.

[22] T. Takano-Yamamoto, S. Soma, K. Nakagawa, Y. Kobayashi, M. Kawakami and M. Sakuda, Comparison of the effects of hydrostatic compressive force on glycosaminoglycan synthesis and proliferation in rabbit chondrocytes from mandibular condylar cartilage, nasal septum, and spheno-occipital synchondrosis *in vitro*, *Am. J. Orthod. Dentofacial Orthop.* **99** (1991), 448–455.

[23] M. Teramoto, S. Kaneko, S. Shibata, M. Yanagishita and K. Soma, Effect of compressive forces on extracellular matrix in rat mandibular condylar cartilage, *J. Bone Miner. Metab.* **21** (2003), 276–286.

[24] V. Visnapuu, T. Peltomaki, K. Isotupa, T. Kantomaa and H. Helenius, Distribution and characterization of proliferative cells in the rat mandibular condyle during growth, *Eur. J. Orthod.* **22** (2000), 631–638.

Biorheology 43 (2006) 303–310
IOS Press



# Modulation of proteoglycan production by cyclic tensile stretch in intervertebral disc cells through a post-translational mechanism

Mourad Benallaoua [a,b], Pascal Richette [a], Mathias François [a], Lydia Tsagris [a], Michel Revel [b], Maité Corvol [a], Serge Poiraudeau [a,b], Jean-François Savouret [a] and François Rannou [a,b,*]

[a] *Institut National de la Santé Et de la Recherche Médicale (INSERM) UMR-S-747, Université Paris 5, UFR Biomédicale, 45 Rue des Saints-Pères, 75006 Paris, France*
[b] *Service de rééducation et de réadaptation de l'appareil locomoteur et des pathologies du rachis, Réseau Fédératif de Recherche sur le Handicap INSERM, Hôpital Cochin, Assistance Publique-Hôpitaux de Paris, Université Paris 5, 27 rue du faubourg Saint-Jacques, 75679 Paris, France*

**Abstract.** Proteoglycan production is one of the major extracellular matrix components implicated in the dynamic process of intervertebral disc degeneration. Mechanical stress is an important modulator of the degeneration, but the underlying molecular mechanism at the proteoglycan level remains unclear. The aim of this work was to study the regulation of proteoglycan production by cyclic tensile stretch applied to intervertebral disc annulus fibrosus cells. Matrix metalloproteinases do not seem to be implicated in the regulation of proteoglycan production. By contrast, nitrite oxide production is induced by cyclic tensile stretch, in a time, intensity, and frequency dependant manner. Using a non-specific nitric oxide synthases inhibitor [$N^G$-methyl-L-arginine (L-NMA)], we suppress totally the inhibition of proteoglycan production induced by cyclic tensile stretch suggesting the implication of nitric oxide synthases in the observed phenomenon. Introducing the transcriptional inhibitor 5,6-dichloro-1-$\beta$-D-ribofuranosylbenzimidazole or a more specific inhibitor of nitric oxide synthases II [N-iminoethyl-L-lysine (L-NIL)] did not affect the decreased proteoglycan production, which suggests a post-translational regulation. In contrast, N-omega nitro-L-arginine (L-NNA) a more specific inhibitor of NOS I and III abrogated the cyclic tensile stretch-dependant inhibition of proteoglycan production. These results suggest that cyclic tensile stretch regulates proteoglycan production through a post-translational mechanism involving nitrite oxide. This result could be of interest in the development of local therapeutic strategies aimed at controlling intervertebral disc degeneration.

Keywords: Intervertebral disc, mechanical stretch, annulus fibrosus, proteoglycan, chondrocyte, nitrite oxide

## 1. Introduction

In developed countries, low back pain is a major health problem and accounts for high medical expenses, absenteeism from work, disability and handicap [28]. Recent studies suggest that the main anatomical structure implicated in low back pain is the intervertebral disc (IVD) and the pathogenic process is its degeneration that predisposes disc herniation [7,9,18,27].

---

*Address for correspondence: Francois Rannou, MD, PhD, Service de rééducation et de réadaptation de l'appareil locomoteur et des pathologies du rachis, Hôpital Cochin, Assistance Publique-Hôpitaux de Paris, Université Paris 5, 27 rue du faubourg Saint Jacques, 75679 Paris, France. Tel.: +33 1 58 41 25 35; Fax: +33 1 58 41 25 45; E-mail: francois.rannou@cch.aphp.fr.

*In vivo*, the homeostasis of the extracellular-matrix (ECM) in the IVD is regulated by mechanical forces [12,17], but the molecular mechanisms are not yet known. Some evidence indicates that intervertebral disc degeneration begins with a progressive decrease in proteoglycan content leading to dehydration of the disc and modification of the load-bearing ability of this tissue [2,4,20]. Evidence exists that intervertebral disc cells are implicated in proteoglycan degradation and synthesis [21,22,25]. The major proteoglycan of the disc is aggrecan, which forms high molecular weight complexes with hyaluronan to produce a large swelling pressure that is important for the intervertebral disc to respond to mechanical stimulation [13]. Aggrecan is thought to play a role in maintening the collagen network and in collagen fibrillogenesis [26,29]. Although the turnover of collagen within the disc is estimated to be very slow ($>$100 years), that of aggrecan is more rapid, with a half-life of 8–300 days in rabbits [19]. Because of this relatively rapid turnover, the decreased production of aggrecan could significantly induce intervertebral disc degeneration. Finally, it has been shown than an autosomal recessive mutation in the aggrecan gene leads to intervertebral disc degeneration in mice [29]. Taken together, these results point out the main role of proteoglycans in intervertebral disc degeneration.

In a previous study, we have shown that cyclic tensile stretch (CTS) is involved in the post-translational regulation of proteoglycan production by IVD annulus fibrosus (AF) cells [25]. The aim of this work was to characterize the post-translational regulation of proteoglycan production by cyclic tensile stretch applied to AF cells.

## 2. Materials and methods

### 2.1. Cell culture and cyclic tensile stretch experiments

Rabbit IVD AF cell culture was performed as previously described [21,23–25]. Confluent AF cells cultured in 6-well plates with flexible bottoms were subjected (stretched cells) or not (static control) to CTS. The transcriptional inhibitor 5,6-dichloro-1-$\beta$-D-ribofuranosylbenzimidazole (10 $\mu$g/ml) was added or not 1 h before stretching. In some experiments, a non-specific inhibitor of nitric oxide synthases (NOSs), $N^G$-methyl-L-arginine (L-NMA, 1 mM), or a specific inhibitor of inducible NOS (iNOS), N-iminoethyl-L-lysine (L-NIL, 10 $\mu$M), or a relative specific inhibitor of constitutive NOS (cNOS), N-omega nitro-L-arginine (L-NNA, 5 $\mu$M), was added during stretching. CTS was applied by use of a Flexercell$^{TM}$ stress unit [3]. The cells were maintained in a Ham F12 medium containing 10% fetal bovine serum (FBS), 100 IU/ml penicillin, and 100 $\mu$g/ml streptomycin. The experimental protocol delivered 5% stretch at a frequency from 0.05 to 1 Hz, for 8 to 24 h.

### 2.2. Radiolabeling studies

The incorporation of $^{35}SO_4$ into proteoglycans by stretched and static control AF cells was assessed in the culture medium as previously described [5,25]. After CTS application, stretched and static control AF cells were incubated in serum and sulfate-free DMEM plus 1.5 $\mu$Ci/ml $Na_2[^{35}SO_4]$ for an additional 20 h. Proteoglycans were extracted from the culture media with use of 3 M guanidinium chloride in 0.05 M Tris-HCl, pH 7.4, in the presence of protease inhibitors. Aliquots of the guanidinium extract were then spotted on a set of Whatman 3MM paper and precipitated with 1% cetylpyridinium chloride (CPC) in 0.3 M NaCl. Scintillation liquid was added to each strip, and were counted with use of a Packard Tricarb $\beta$-spectrometer. Each measurement was made in triplicate. For each experiment, results were calculated as mean total medium dpm per well of three similarly treated wells.

## 2.3. Nitrite determination

The nitrite levels of the culture medium were determined through indirect measurement of nitrite oxide production by the method of Griess, the results are expressed in $\mu$M.

## 2.4. Zymography

Zymographic analysis of gelatinase activities was performed as previously described by using cell culture medium (10 $\mu$l) [8]. Briefly, the cell culture medium, mixed with 4× sample buffer was subjected to electrophoresis on a 0.1% SDS-polyacrylamide gel containing 1 mg/ml gelatin. After electrophoresis, the gel was washing and then incubated at 37°C for 24 h in incubation buffer. Gels were stained with 0.5% Coomassie Blue, 10% acetic acid, 30% methanol buffer and destained with sequential discoloration solutions until the digested bands were clearly separated.

## 2.5. Statistical analysis

Analysis of variance was performed with variable duration. When an $F$ value was found to be significant, the analysis of variance was followed by multiple comparisons with use of the Tukey test. $P$ values above 0.05 were not considered statistically significant.

## 3. Results

### 3.1. Cyclic tensile stretch increases nitrite production and not matrix metalloproteinases activities

Nitrite oxide (NO) and matrix metalloproteinases (MMPs) have been shown to be involved in the post-translational regulation of proteoglycan production. Thus, we hypothesize than NO and MMPs could be involved in the regulation by CTS of proteoglycan production. The basal production of nitrites observed in the culture medium of static AF cells was at the limit of detection and evaluated at a mean of $3.90 \pm 0.35$ $\mu$M after 8 h (Fig. 1A). Nitrite concentration remained stable for 8 to 24 h in static controls. CTS at 5% elongation and 1 Hz had a significant effect on nitrite concentration, a 3.8-fold increase was observed after 8 h at 5% CTS ($F = 29.22$, $p < 0.001$) (Fig. 1A). There was no difference in effect between 8 and 24 h of 5% CTS. When the frequency decreased to 0.1 or 0.05 Hz, we observed a concomitant decrease of the nitrite production (Fig. 1A). Gelatinolytic activities at an apparent molecular mass of 55 kDa, 72 kDa, and 92 kDa was not significantly increased by 5% CTS, 1 Hz, and 24 h (Fig. 1B). These molecular mass presumably correspond to MMP-1, -3, MMP-2, and MMP-9 respectively. Treatment with IL-1$\beta$ (1 ng/ml) was used as a positive control. These results suggest than NOS activation could be more implicated with the 5% CTS-dependant post-translational regulation of proteoglycan production. than MMPs activation.

### 3.2. The non-specific inhibition of nitrite oxide production suppresses the cyclic tensile stretch-dependant inhibition of proteoglycans secreted by annulus fibrosus cells

To study the implication of the CTS-induced NO production on the CTS-induced decreased production of $^{35}$S-labeled proteoglycans, AF cells were incubated with a non-specific inhibitor of nitric oxide synthases (1 mM L-NMA) during stretching. $^{35}$SO$_4$ incorporation into secreted neosynthesized proteoglycans was measured after 24 h of 5% CTS, and 1 Hz. CTS induce a 3.6-fold increase in nitrite

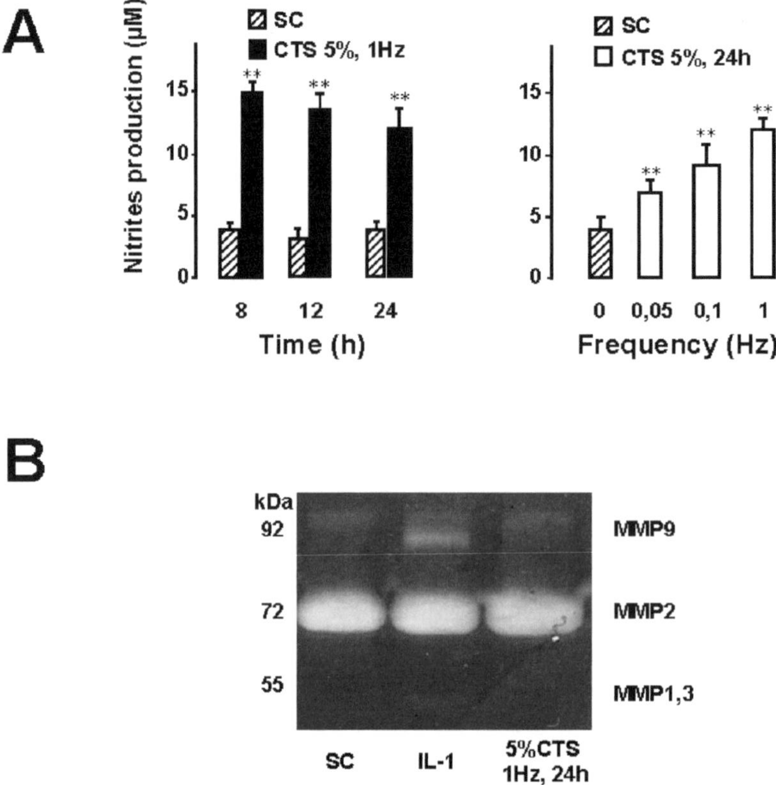



Fig. 1. Effect of CTS on NO production and MMPs activities. (A) Accumulation of NO in the culture supernatant was assessed by Griess reaction in various conditions: static control AF cells (SC), AF cells subjected to 5% CTS, 1 Hz (CTS 5%, 1 Hz) from 8 to 24 h, or AF cells subjected to 5% CTS, 24 h (CTS 5%, 24 h) at a frequency from 0.05 to 1 Hz. Results are presented as mean $\pm$ standard deviation of three independent experiments. $^{**}p < 0.01$ relative to SC. (B) Culture medium was collected and subjected to gelatin zymography as described in materials and methods section. AF cells were subjected to various conditions: SC, IL-1$\beta$, or CTS 5%, 1 Hz, 24 h (CTS 5%, 1 Hz, 24 h). This panel is representative of one of three separate experiments.

production. This effect was totally abolished in presence of L-NMA (Fig. 2A). Concomitantly, CTS-induced decreased secretion of $^{35}$S-labeled proteoglycans was totally inhibited after 24 h in the presence of L-NMA (Fig. 2B). These results suggest that the CTS-induced inhibition of proteoglycan production requires NO production mainly through NOS activation for long periods of stretch (24 h).

### 3.3. Constitutive form of nitric oxide synthases are more implicated in the nitric oxide production induced by cyclic tensile stretch than inducible form

The 5% CTS-induced nitrite production totally abolished by the addition of L-NMA (Fig. 2A) was partially abolished by the addition of L-NNA (Fig. 3A). By contrast, the introduction of L-NIL did not abolish the 5% CTS-induced nitrite production after 24 h. AF cells pretreatment by the transcriptional inhibitor 5,6-dichloro-1-$\beta$-D-ribofuranosylbenzimidazole (DRB) did not affect the nitrite production induced by CTS (Fig. 3A). Lastly, the proteoglycan production was recovered in the presence of L-NNA and not L-NIL (Fig. 3B). These results suggest that the CTS-induced inhibition of proteoglycan production requires NO production mainly through the activation of cNOS.

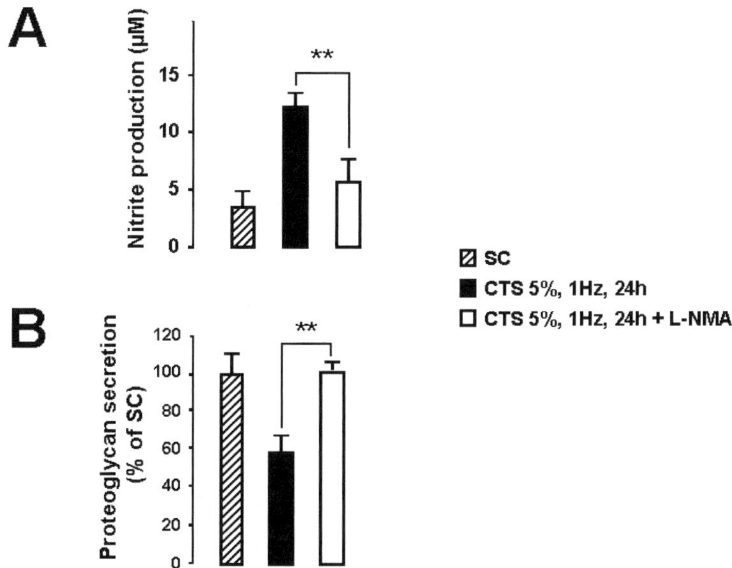



Fig. 2. Effect of a non-specific NOS inhibitor on NO production and proteoglycans secreted by AF cells subjected to 5% CTS. (A) Accumulation of NO in the culture supernatant was assessed by Griess reaction in various conditions: static control AF cells (SC), AF cells subjected to 5% CTS, 1 Hz, 24 h (CTS 5%, 1 Hz, 24 h), and AF cells subjected to 5% CTS, 1 Hz, 24 h in the presence of a non-specific inhibitor of NOS at 1 mM (CTS 5%, 1 Hz, 24 h + L-NMA). (B) After stretching or static period, AF cells were labeled with $^{35}SO_4$ for 20 h. Proteoglycans were extracted with 1% cetylpyridinium chloride in 0.3 M NaCl in the presence of protease inhibitors. Results are presented as mean ± standard deviation of three independent experiments.

## 4. Discussion

Our results show that NO is involved in the regulation of proteoglycan production by CTS and that this phenomenon depends on the amount, duration, and frequency of stretch. Constitutive NOS activation seems to be more implicated with the 5% CTS-dependant post-translational regulation of proteoglycan production, than MMPs or iNOS activation.

According to Kang et al., degenerated human discs produce spontaneously more NO than do normal discs [14]. Liu et al. showed that hydrostatic pressure-induced NO production decreases proteoglycan synthesis in degenerated human lumbar discs [16]. These results point out a close connection between mechanical stress, proteoglycans, and NO in the degeneration process of the discs. Our results suggest than IVD AF cells can produce NO in an autocrine–paracrine manner leading to the modulation of proteoglycan production.

The mechanism implicated in the mechanical modulation of NO production remains unclear. CTS applied to articular chondrocytes and fibrochondrocytes is accompanied by a decrease of NO production, which depends on the downregulation of the iNOS gene [1,10]. This phenomenon was observed only with the presence of IL-1$\beta$ and for a higher intensity of stimulation (20% and 6%, respectively) than that used in the present study. By contrast, in our work, cNOS activity seems more involved that iNOS. This observation is in accordance with previous results showing than compressive stress applied to chondrocytes might induce a modulation of NO production through modulation of cNOS activity without transcriptional regulation of NOS genes [15]. Taken together, these data emphasize the importance of the cell type and the inflammatory state of the mechanical-stimulated cells.

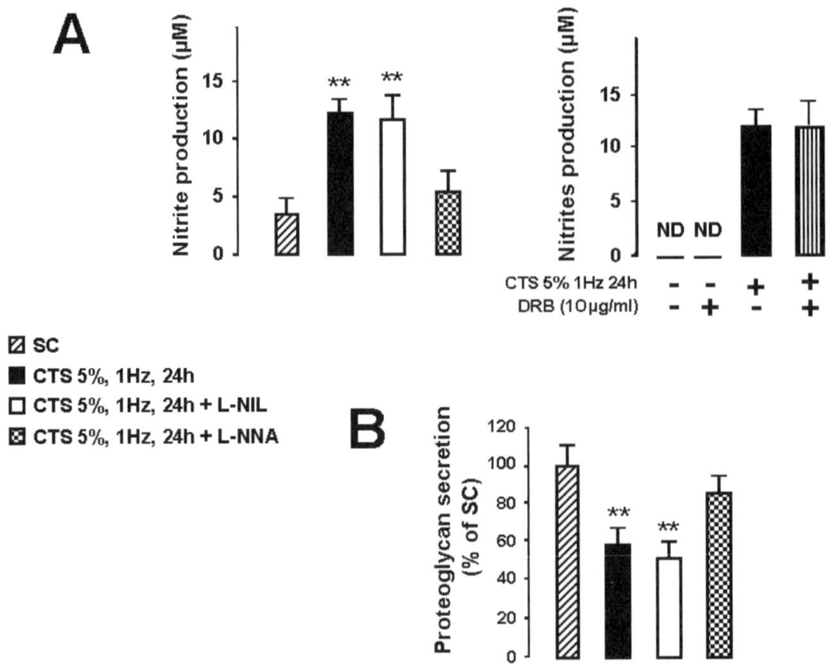


Fig. 3. Effect of a specific inhibition of constitutive NOS on NO production and proteoglycans secreted by AF cells subjected to 5% CTS. (A) Accumulation of NO in the culture supernatant was assessed by Griess reaction in various conditions: static control AF cells (SC), AF cells subjected to 5% CTS, 1 Hz, 24 h (CTS 5%, 1 Hz, 24 h), AF cells subjected to 5% CTS, 1 Hz, 24 h in the presence of a specific inhibitor of inducible NOS at 10 $\mu$M (CTS 5%, 1 Hz, 24 h + L-NIL), and AF cells subjected to 5% CTS, 1 Hz, 24 h in the presence of a specific inhibitor of constitutive NOS at 5 $\mu$M (CTS 5%, 1 Hz, 24 h + L-NNA). In some experiments AF cells were pretreated with the transcriptional inhibitor 5,6-dichloro-1-$\beta$-D-ribofuranosylbenzimidazole at 10 $\mu$g/ml (DRB). (B) After stretching or static period, AF cells were labeled with $^{35}$SO$_4$ for 20 h. Proteoglycans were extracted with 1% cetylpyridinium chloride in 0.3 M NaCl in the presence of protease inhibitors. Results are presented as mean $\pm$ standard deviation of three independent experiments. $^{**}p < 0.01$ relative to SC.

Our results suggest that IVD cells can sense and convert applied loads to biological signals that result in tissue responses. The mechanosensors are not yet identified but could be the stretch-activated Ca$^{2+}$-channels. Elfervig et al. has shown the involvement of stretch-activated Ca$^{2+}$-channels in the signaling pathway of mechanical stress in IVD cells [6]. With our results, we could hypothesize that in AF cells, mechanical stress via stretch-activated Ca$^{2+}$-channels modulates a Ca$^{2+}$-dependant enzyme like cNOS but not iNOS, which is known to be Ca$^{2+}$-independant.

How NO modulates proteoglycan production by AF cells subjected to mechanical stress requires further investigation. One hypothesis could be the induction of IVD cells apoptosis by CTS. This hypothesis seems not relevant because we have recently shown that IVD cell apoptosis was induced by CTS only for higher stretch intensity (15%) and not at 5% [23]. An other hypothesis could be than NO directly inhibits glycosaminoglycan sulfatation, the last step of aggrecan intracellular synthesis, as suggested by Hickery and Bayliss [11]. Lastly, NO could act more like a second messenger in the mechanotransduction process, as is seen in other cell types.

In conclusion, CTS can participate in the regulation of the intervertebral disc matrix by decreasing proteoglycan production through NO regulation. This regulation seems to be related to cNOS activation. This result could be of interest in the development of local therapeutic strategies aimed at controlling intervertebral disc degeneration.

## Acknowledgements

Grant sponsor: Fondation de l'Avenir, Société Française de Rhumatologie and Bonus Qualité Recherche (Université Paris 5).

## References

[1] S. Agarwal, P. Long, R. Gassner, N.P. Piesco and M.J. Buckley, Cyclic tensile strain suppresses catabolic effects of interleukin-1beta in fibrochondrocytes from the temporomandibular joint, *Arthritis Rheum.* **44** (2001), 608–617.

[2] J. Antoniou, T. Steffen, F. Nelson, N. Winterbottom, A.P. Hollander, R.A. Poole, M. Aebi and M. Alini, The human lumbar intervertebral disc, *J. Clin. Invest.* **98** (1996), 996–1003.

[3] A.J. Banes, J. Gilbert, D. Taylor and O. Monbureau, A new vacuum-operated stress providing instrument that applies static or variable duration cyclic tension or compression to cells in vitro, *J. Cell. Sci.* **75** (1985), 35–42.

[4] J.A. Buckwalter, Aging and degeneration of the human intervertebral disc, *Spine* **20** (1995), 1307–1314.

[5] M.T. Corvol, M.F. Dumontier, R. Rappaport, H. Guyda and B.I. Posner, The effect of a slightly acidic somatomedin peptide (ILAs) on the sulphation of proteoglycans from articular and growth plate chondrocytes in culture, *Acta Endocrinol.* **89** (1978), 263–275.

[6] M.K. Elfervig, J.T. Minchew, E. Francke, M. Tsuzaki and A.J. Banes, IL-1$\beta$ sensitizes intervertebral disc annulus cells to fluid-induced shear stress, *J. Cell. Biochem.* **82** (2001), 290–298.

[7] J. Fairbank, Clinical importance of the intervertebral disc, or back pain for biochemists, *Biochem. Soc. Trans.* **30** (2002), 829–831.

[8] M. Francois, P. Richette, L. Tsagris, M. Raymondjean, M.C. Fulchignoni-Lataud, C. Forest, J.F. Savouret and M.T. Corvol, Peroxisome proliferator-activated receptor-gamma down-regulates chondrocyte matrix metalloproteinase-1 via a novel composite element, *J. Biol. Chem.* **279** (2004), 28411–28418.

[9] A.J. Freemont, A.J. Watkins, A. Watkins, C. Le Maitre, M. Jeziorska and J.A. Hoyland, Current understanding of cellular and molecular events in intervertebral disc degeneration: implications for therapy, *J. Pathol.* **196** (2002), 374–379.

[10] R. Gassner, M.J. Buckley, H. Georgescu, R. Studer, M. Stefanovic-Racic, N.P. Piesco, C.H. Evans and S. Agarwal, Cyclic tensile stress exerts antiinflammatory actions on chondrocytes by inhibiting inducible nitric oxide synthase, *J. Immunol.* **163** (1999), 2187–2192.

[11] M.S. Hickery and M.T. Bayliss, Interleukin-1 induced nitric oxide inhibits sulphation of glycosaminoglycan chains in human articular chondrocytes, *Biochim. Biophys. Acta* **1425** (1998), 282–290.

[12] W.C. Hutton, Y. Toribatake, W.A. Elmer, T.M. Ganey, K. Tomita and T.E. Whitesides, The effect of compressive force applied to the intervertebral disc in vivo, *Spine* **23** (1998), 2524–2537.

[13] B. Johnstone and M. Bayliss, The large proteoglycans of the intervertebral disc. Changes in their biosynthesis and structure with age, topography and pathology, *Spine* **20** (1995), 674–684.

[14] J.D. Kang, H.I. Goergescu, L. McIntyre-Larkin, M. Stefanovic-Racic, W. Donaldson and C.H. Evans, Herniated lumbar intervertebral discs spontaneously produce matrix metalloproteinases, nitric oxide, interleukin-6, and prostaglandin $E_2$, *Spine* **21** (1996), 271–277.

[15] D.A. Lee, T. Noguchi, S.P. Frean, P. Lees and D.L. Bader, The influence of mechanical loading on isolated chondrocytes seeded in agarose constructs, *Biorheology* **37** (2000), 149–161.

[16] G.Z. Liu, H. Ishihara, R. Osada, T. Kimura and H. Tsuji, Nitric Oxide mediates the change of proteoglycan synthesis in the human lumbar intervertebral disc in response to hydrostatic pressure, *Spine* **26** (2001), 134–141.

[17] J.C. Lotz, O.K. Colliou, J.R. Chin, N.A. Duncan and E. Liebenberg, Compression-induced degeneration of the intervertebral disc: an in vivo mouse model and finite-element study, *Spine* **23** (1998), 2493–2506.

[18] K. Luoma, H. Riihimaki, R. Luukkonen, R. Raininko, E. Viikari-Juntura and A. Lamminen, Low back pain in relation to lumbar disc degeneration, *Spine* **25** (2000), 487–492.

[19] H.J. Mankin and L. Lippiello, The turnover of adult rabbit articular cartilage, *J. Bone Joint Surg. Am.* **51** (1969), 1591–1600.

[20] A.G. Nerlich, E.D. Schleicher and N. Boos, Immunohistologic markers for age-related changes of human intervertebral discs, *Spine* **22** (1997), 2781–2795.

[21] S. Poiraudeau, I. Monteiro, P. Anract, O. Blanchard, M. Revel and M.T. Corvol, Phenotypic characteristics of rabbit intervertebral disc cells: comparison with cartilage cells from the same animals, *Spine* **24** (1999), 837–844.

[22] F. Rannou, M.T. Corvol, C. Hudry, P. Anract, O. Blanchard, L. Tsagris, M. Revel and S. Poiraudeau, Sensitivity of anulus fibrosus cells to interleukin 1$\beta$: comparison with articular chondrocytes, *Spine* **25** (2000), 17–23.

[23] F. Rannou, T.S. Lee, R.H. Zhou, J. Chin, J.C. Lotz, M.A. Mayoux-Benhamou, J.P. Barbet, A. Chevrot and J.Y.J. Shyy, Intervertebral disc degeneration: the role of the mitochondrial pathway in annulus fibrosus cell apoptosis induced by overload, *Am. J. Pathol.* **164** (2004), 915–924.

[24] F. Rannou, S. Poiraudeau, V. Foltz, M. Boiteux, M. Corvol and M. Revel, Monolayer anulus fibrosus cell cultures in a mechanically active environment: local culture condition adaptations and cell phenotype study, *J. Lab. Clin. Med.* **136** (2000), 412–421.

[25] F. Rannou, P. Richette, M. Benallaoua, M. Francois, V. Genries, C. Korwin-Zimjowska, M. Revel, M. Corvol and S. Poiraudeau, Cyclic tensile stretch modulates proteoglycan production by intervertebral disc annulus fibrosus cells through production of nitrite oxide, *J. Cell. Biochem.* **90** (2003), 148–157.

[26] M.B. Schmidt, V.C. Mow, L.E. Chun and D.R. Eyre, Effects of proteoglycan extraction on the tensile behavior of articular cartilage, *J. Orthop. Res.* **8** (1990), 353–363.

[27] J.P. Urban and S. Robert, Degeneration of the intervertebral disc, *Arthritis Res.* **5** (2003), 120–130.

[28] M.W. Van Tulder, B.W. Koes and L.M. Bouter, A cost-of-illness study of back pain in the Netherlands, *Pain* **62** (1995), 233–240.

[29] H. Watanabe, K. Nakata, K. Kimata, I. Nakanishi and Y. Yamada, Dwarfism and age-associated spinal degeneration of heterozygote cmd mice defective in aggrecan, *Proc. Natl. Acad. Sci. USA* **94** (1997), 6943–6947.

Biorheology 43 (2006) 311–321
IOS Press



# Mandibular repositioning modulates IGFBP-3, -4, -5 and -6 expression in the mandibular condylar cartilage of young rats

Denise Hajjar, Marinilce F. Santos and Edna Teruko Kimura [*]
*Department of Cell and Developmental Biology, Institute of Biomedical Sciences, University of São Paulo, Av. Prof. Lineu Prestes 1524, São Paulo 05508-900, SP, Brazil*

**Abstract.** Functional orthopedic appliances correct dental malocclusion partially by exerting indirect mechanical stimulus on the condylar cartilage, modulating growth and the adaptation of orofacial structures. However, the exact nature of the biological responses to this therapy is not well understood. Insulin-like growth factors I and II (IGF-I and II) are important local factors during growth and differentiation in the condylar cartilage [D. Hajjar, M.F. Santos and E.T. Kimura, Propulsive appliance stimulates the synthesis of insulin-like growth factors I and II in the mandibular condylar cartilage of young rats, *Arch. Oral Biol.* **48** (2003), 635–642]. The bioefficacy of IGFs at the cellular level is modulated by IGF binding proteins (IGFBP). The aim of this study was to verify the mRNA and protein expression of IGFBP-3, IGFBP-4, IGFBP-5 and IGFBP-6 in the condylar cartilage of young male Wistar rats that used a mandibular propulsive appliance for 3, 9, 15, 20, 30 or 35 days. For this purpose, sagittal sections of decalcified and paraffin-embedded condyles were submitted to immunohistochemistry and the condylar cartilage to RT–PCR. The control group showed a gradual increase in the protein expression of all IGFBPs, except IGFBP-4. Following use of the appliance, IGFBP-3 and IGFBP-6 expression decreased in the early stage of the treatment. At 20 days of treatment there was a decline in the IGFs and IGFBP-3, IGFBP-4 and IGFBP-5 expression and at 30 days there was a peak in the IGFs and all IGFBPs expression except for IGFBP-3 where the peak was observed in the control animals. The expression patterns of all IGFBPs in the condylar cartilage were similar. The modulation of IGFBP-3, -4, -5 and -6 expression in the condylar cartilage in response to the propulsive appliance suggests that those peptides are involved in the mandibular adaptation during this therapy.

Keywords: Insulin-like growth factor binding protein, mechanical stress, mandibular condyle, cartilage, functional orthopedics

## 1. Introduction

Among all therapeutic methods used to correct dental malocclusions, the functional orthopedic appliance is a clinical procedure used mainly to change the mandibular posture, reducing anterior–posterior bone discrepancies between the maxillaries. Orthopedic appliances promote mandibular repositioning in relation to the maxilla, generating mechanical forces through muscular tension in the mandibular condylar cartilage, an important growth site at the temporomandibular joint (TMJ) [21,37]. Although the clinical result obtained with the use of these appliances is undeniable, the molecular mechanisms through which they act remain to be fully characterized [28,37].

For that purpose, we have developed an experimental animal model using a mandibular appliance, based on the one developed by Petrovic et al. [37]. The device was able to induce an advancement of

*Address for correspondence: Edna Teruko Kimura, Cell and Developmental Biology Department, Av. Prof. Lineu Prestes, 1524, São Paulo, SP, Brazil, CEP 05508-900. Tel.: +55 11 3091 7304; Fax: +55 11 3091 7402; E-mail: etkimura@usp.br.

the animal's mandible, mimicking the effect produced by the functional orthopedic appliances in the patients' mouth. This model provides an opportunity to study the physical stimuli and the resulting cellular responses at the molecular, cellular and tissue levels on the condylar cartilage.

During the past decade, increasing attention has been focused on the ability of cells and tissues to respond to mechanical forces and other physical stimuli in their environment. Similar to bone and skeletal muscle, the condylar cartilage responds and adapts to changes in loading state via mechanisms that appear to be intrinsic to the cartilage [37]. It is now well accepted that mechanical stimuli in the microenvironment of the chondrocytes can significantly affect the proliferation and surrounding matrix biology.

Mandibular condylar cartilage reportedly exhibits a different pattern of growth and differentiation from the growth plate and the articular cartilage of the long bones [28,37]. There is increasing evidence that chondrocyte proliferation is modulated by mechanical stress [13,44]. During joint movement, articular cartilage is exposed to a range of mechanical forces that are involved in the regulation of cartilage metabolism and integrity [16].

Recent data suggest that there are multiple regulatory pathways by which chondrocytes sense and respond to mechanical stimuli, including upstream signaling pathways and mechanisms that may lead to direct changes at the level of transcription, translation and post-translation modifications and cell-mediated extracellular assembly and degradation of matrix. However, the cellular mechanisms that govern mandibular cartilage response to mechanical stimuli are not yet fully clarified.

One of the mechanisms modulating the condylar cartilage adaptation involves the autocrine or paracrine production of the growth factors. We have focused on understanding the role of insulin like growth factors I and II (IGF-I and IGF-II) and we have observed its increased expression *in vivo* in rat condylar cartilage, submitted to mechanical stimulus by mandibular advancement, promoted with a functional orthopedic appliance [17]. Also, Itoh et al. [18] showed that the local administration of IGF-I to the rat mandibular articular cavity, stimulated the condylar cartilage to continue growing, even after normal growth was complete.

The IGF-I and IGF-II are the major growth promoting factors present in several tissues. However, the bioefficacy of IGFs at the cellular level is modulated by IGF binding proteins (IGFBP) [10]. Six distinct, yet structurally homologous, IGFBPs have been characterized and designated IGFBP-1 through IGFBP-6 [36]. The initial interpretation of the function of such binding proteins was that they would prolong the half-life of the IGFs in the circulation and inhibit their metabolic effects by preventing them from binding to receptors. These highly conserved proteins can also modulate IGF bioactivity positively or negatively and also exert IGF-independent effects [10,39].

Despite their structural similarity, each IGFBP has unique characteristics and function being expressed by specific gene locus and distributed in different chromosomes. The bioavailability and function of IGFs are therefore thought to be modulated by binding proteins and, in particular, by IGFBP-3. It can be a potent inhibitor of cell growth by virtue of its ability to bind IGF with high affinity, thereby preventing its interaction with membrane receptors [6]. Furthermore, IGFBP-3 can potentialize the IGF action in a number of cell systems [3]. Moreover, there is evidence that IGFBP-3 and IGFBP-5 may exert an IGF-independent action by binding itself to the cell surface [9]. IGFBP-5 is considered to be rather a stimulatory IGFBP that appears to counteract the inhibitory action of IGFBP-4 in systems such as bone.

IGFBP-6 has markedly higher affinity for IGF-II than for IGF-I, whereas the other IGFBPs bind the two IGFs with relatively similar affinities [3]. It has not been shown to potentialize IGF actions and its expression is associated with inhibition of growth of tumor cells *in vitro* and *in vivo* [2].

Although the IGF-IGFBP action is better understood in other tissues, we do not know its role in the developing condylar cartilage and much less is known about IGFBPs and their properties under mechanical stress.

## 2. Material and methods

### 2.1. Animals and experimental protocol

Ninety-six (96) 21-day-old male Wistar rats were divided into six groups ($n = 6$). In each group, 3 rats were fitted with a mandibular propulsor appliance for periods denominated early stage (ES): days 3 (T3), 9 (T9) and 15 (T15); and later stage (LS): days 20 (T20), 30 (T30) and 35 (T35), together with a soft leather collar to prevent removal of the appliance. The remaining 3 rats in each group used the collar alone to serve as controls (C), and were sacrificed together with the rats in the respective experimental groups. At the time of sacrifice, the animals were anesthetized and perfused with 0.9% saline and 4% formaldehyde solution. After surgical removal, the mandibular condyles were post-fixed in 4% formaldehyde solution at 4°C for 24 h, followed by decalcification with 10% EDTA for 60 days and embedded in Paraplast®. Sagittal sections, 5 $\mu$m thick, were placed on 2% 3-amino-propyltriethoxy-silano-coated slides (A3648 Sigma Chem. Co., St Louis, USA).

### 2.2. Orthopedic appliance

The appliance design was based on the work by Petrovic et al. [37] as described elsewhere [16]. Briefly, on the basis of functional orthopedics, exploiting the anterior displacement of the mandible, each time the animals attempted to close their mouths. The propulsor was a $4 \times 9$ mm rectangle cut from a 55.88 mm thick copper plate banded in an angle of 90 degree, 2 mm from the anterior end. All the edges were rounded to avoid irritation and a 3 mm diameter rubber tube was glued close to the appliance vertical portion. The rubber tube was fitted into the superior incisors of the animal. The animals used the appliance daily for 10 h (8:00 AM to 6:00 PM). Food and water were made available at night *ad libitum*.

### 2.3. Condylar cartilage morphology and growth

The slides were stained with hematoxylin and eosin for routine histology to study cartilage morphology by analyzing the cellular layers (Fig. 1). The anterior and posterior segments of the cartilage were also identified to facilitate further analysis of the IHQ expression studies.

The tissue growth, spontaneous or induced by the mechanical stress promoted by orthopedic appliances, was evaluated by proliferating cell nuclear antigen (PCNA) immunohistochemistry in the condylar cartilage [11].

### 2.4. Immunohistochemistry: IGFBP-3, IGFBP-4, IGFBP-5, IGFBP-6 and PCNA

Forty-eight (48) animals were used for immunohistochemistry method. Goat anti-rat IGFBP-3, IGFBP-4, IGFBP-5 and IGFBP-6 polyclonal primary antibodies (sc 6004, sc 6005, sc 6006 and sc 6008, Santa Cruz Biotech., Sta. Cruz, USA) and were used at a dilution of 1 : 100. Mouse monoclonal antibody PC-10 against PCNA (M0879 Dakopatts, Copenhagen, Denmark) was used at a concentration

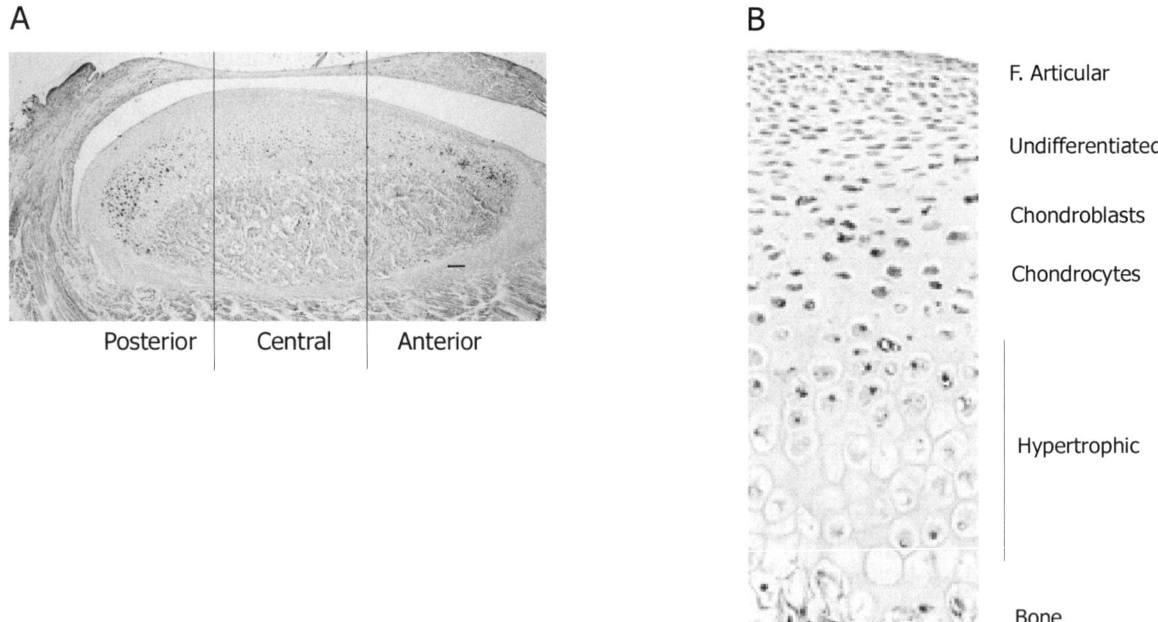


Fig. 1. Mandibular condylar regions and cell layers. Sagittal histological section of the rat mandibular condyle showing (A) the posterior, central and anterior regions ($10\times$) and (B) the cartilage cell layers ($40\times$). Immunohistochemistry for IGFBP-4 at the posterior region of the condylar cartilage staining at the undifferentiated, chondroblast and chondrocyte cell layers (proliferative zone) and some cells from the hyperthropic cell layer.

of 1 : 250. The immunohistochemistry reaction for IGFBP-3, IGFBP-4, IGFBP-5 and IGFBP-6 were detected by extrAvidin® conjugated with peroxidase (Sigma Chem. Co., St Louis, USA) and with alkaline phosphatase for PCNA. The peroxidase activity was visualized using diaminobenzidine (DAB) while alkaline phosphatase activity was detected using 5-bromo-4-chloro-3-indolyl phosphate/nitroblue tetrazolium (BCIP/NBT) (Sigma Chem. Co., St Louis, USA). As a negative control, the replacement of the primary antibody with non-immune serum was used. The quantification was performed at the condylar cartilage, where each segment was analyzed under a light microscope and the percentage of positive cells was observed.

## 2.5. RNA extraction and semi-quantitative RT–PCR

Another 48 rats were used in the same experimental treatment as described above for RT–PCR method. Mandibular condylar cartilages were isolated and immersed into liquid nitrogen. The oligonucleotides and all reagents used in RT–PCR were purchased from Life Technologies (Long Island, NY). The RNA was extracted using TRIZOL® reagent as recommended by manufacturer (Life Technologies, Long Island, NY). The first strand of complementary DNA (cDNA) was generated from 3 $\mu$g of condylar cartilage total RNA, using 200 ng random hexamer primer, 10 U RNase inhibitor, 1 mM dNTP MIX and 400 U M-MLV reverse transcriptase, in a total volume of 20 $\mu$l. The reaction was incubated at $21°C$ for 10 minutes, at $42°C$ for 30 minutes and at $99°C$ for 10 minutes in a Cyclogene® thermal cycler (Thecne, Cambridge, UK). IGFBPs' cDNA was amplified in a 50 $\mu$l volume using 2.5 U Taq DNA polymerase, 1.5 mM MgCl$_2$, 0.2 mM dNTP mix and 30 pmol of specific primers designed based on the BLASTN 2.2.1 program (Table 1). Rat RPL 19 (ribosomal protein L19) gene was amplified as mRNA integrity

Table 1

| Gene | Accession No. | Primer (5′–3′) | Product size (bp) |
|---|---|---|---|
| IGFBP-3 | M33300 | (f) GGT CCC TCG CGC AGA GAA A | 210 |
| | | (r) ACG TCG TCT TTC CCC TTG G | |
| IGFBP-4 | CK364123 | (f) GAG CCG TAC CCA CGA AGA C | 310 |
| [38] | X81582 | (r) GAC TCA GGC CAA GAC TCC AT | |
| | AY686592 | | |
| IGFBP-5 | M62781 | (f) GGG GTT TGC CTC AAC GAA | 309 |
| | | (r) CTG GAG GGA AGC TTC CAT | |
| IGFBP-6 | NM 013104 | (f) CGA GAG AAC GAA GAG ACA CCT | 410 |
| | | (r) CTG GCC ATC TGG AGA CAC T | |
| RPL | JO2650 | (f) AGT ATG CTT AGG CTA CAG AA | 500 |
| | | (r) TTC CTT GGT CTT AGA CCT GC | |

Specific forward (f) and reverse (r) primers designed for IGFBP-3, IGFBP-4 [38], IGFBP-5 and IGFBP-6.

control, using 20 pmol of each primer. Each cycle of PCR amplification involved denaturation for 35 seconds at 94°C, annealing for 1 minute at a specific temperature for each primer, and extension for 1 minute at 72°C, and was performed in a GeneAmp® PCR System 9700 thermal cycler (PE Applied Biosystems, Foster City, CA). To ensure the exponential phase of amplification, the number of PCR cycles was determined and optimized. After every 30 cycles, the PCR reaction was repeated at least twice. Aliquots of the amplified PCR products were fractionated on 1.2 % agarose gel containing ethidium bromide and the image of PCR product was captured by Typhoon 8600® (Molecular Dynamics, Sunnyvale, CA).

## 3. Results

### 3.1. IGFBPs protein expression in the condylar cartilage

Immunohistochemistry showed IGFBP-3, -4, -5 and -6 positivity in the cells' cytoplasm of the condylar cartilage. The positivity varied according to the condylar cartilage region and cell layers. In general in the animals of the control group and mainly among those treated with the propulsor appliance, the IGFBPs expression was more intense at the anterior and posterior regions of the condylar cartilage, close to the lateral pterygoid muscle and the bilaminar zone respectively (Fig. 1A). The animals treated with the appliance showed an even larger number of positive cells in those regions. The more superficial layer of the cartilage showed some positivity for IGFBP-4 and none for the other IGFBPs. The immunohistochemistry stain was not homogeneous in all the cartilage cell layers. The IGFBPs were predominantly present at the chondroblast and chondrocyte cell layers, with a lower number of positive cells at the undifferentiated and hypertrophic cell layers (Fig. 1B). The observation of the cell proliferation with PCNA immunoreaction showed positivity in the undifferentiated, chondroblast and chondrocyte cell layers which correspond to the proliferative zone of the cartilage.

### 3.2. IGFBPs expression during the animals' development

The IGFBPs expression showed an increase and progressive rise on the number of positive cells until 35 days of age. After that the IGFBPs expression, especially IGFBP-4, showed a tendency to stabilize

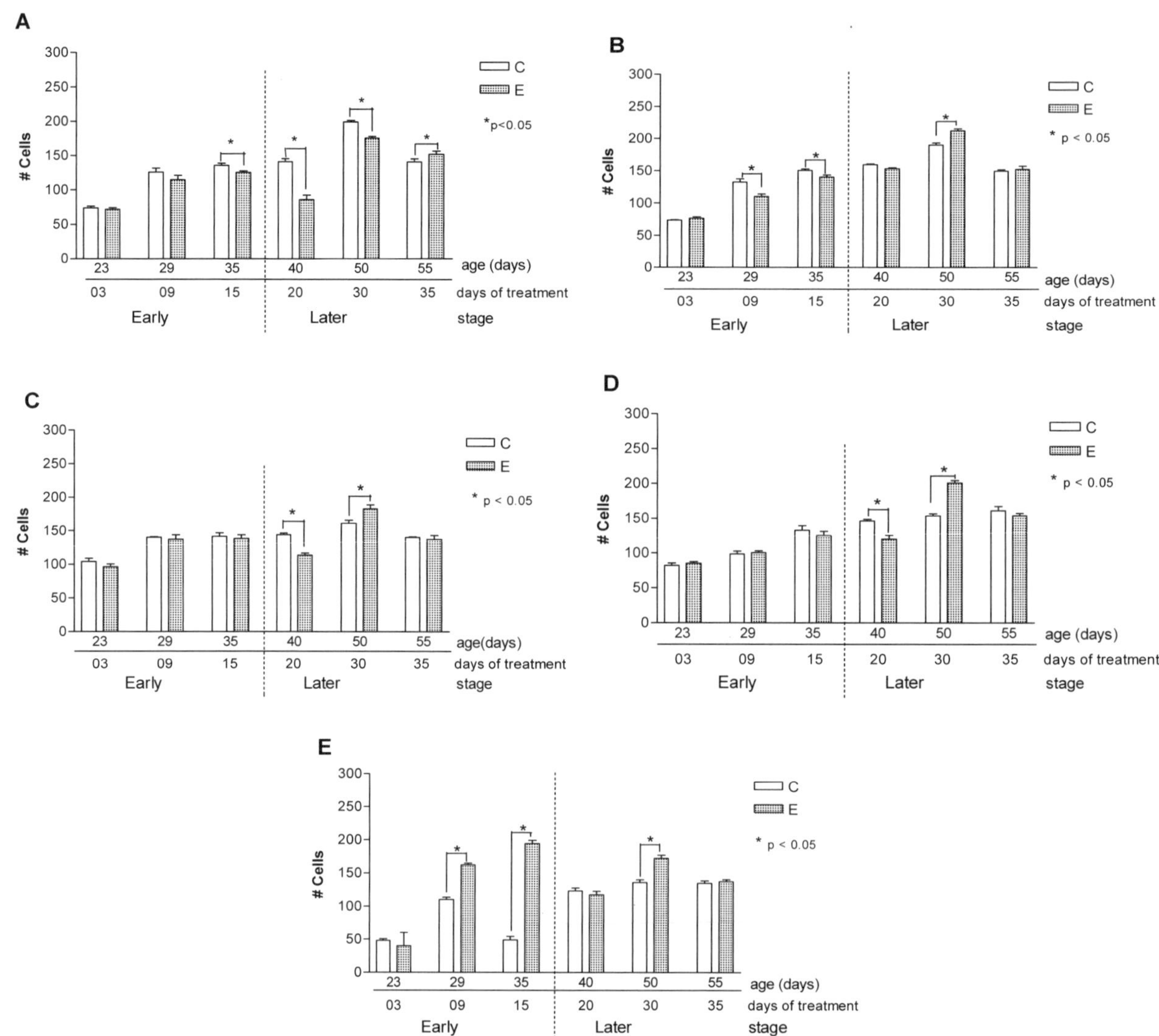

Fig. 2. Quantification of cells labeled by immunohistochemistry for IGFBP-3, IGFBP-6, IGFBP-4, IGFBP-5 and PCNA. Number of immunohistochemistry positively-labeled cells for IGFBP-3 (A), IGFBP-6 (B), IGFBP-4 (C), IGFBP-5 (D) and PCNA (D) in the condylar cartilage at the control and treated animal groups. Each value represents the mean $\pm$ SD for four replicates ($n = 3$). Variance analysis followed by Student Newman–Keuls test indicates differences between treated and control values as follows: $^*p < 0.05$. The early and later stages of the experiment are denoted by doted lines.

until the end of the experiment (Fig. 2C). Only the expression of IGFBP-3 and IGFBP-6 showed a slight peak at 50 days of age (Fig. 2A,B). The cell proliferation marker, PCNA, was present during these developmental periods, though expressed less at 23 and 25 days of age (Fig. 2E).

## 3.3. IGFBPs expression with the mandibular propulsor appliance

During the early phase of the treatment, the IGFBPs showed variation in the number of stained cells, with less expression among the animals that used the appliance when compared with age-matched controls. However it followed the same pattern of expression of the control groups with an increase from 23

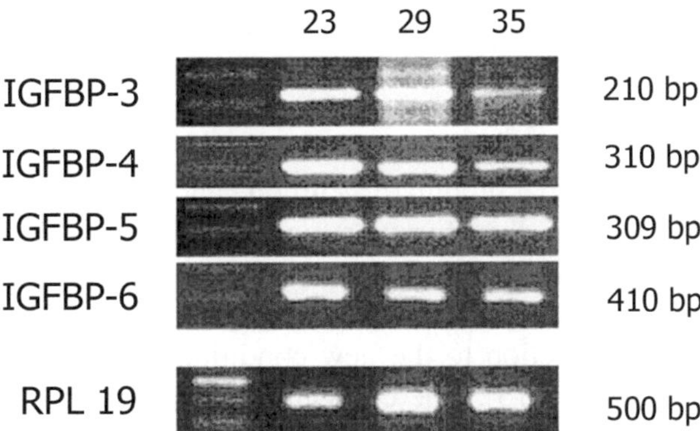



Fig. 3. IGFBPs mRNA expression in the condylar cartilage by RT–PCR. Representative photo from gel shows IGFBP-3 (210 bp), IGFBP-4 (310 bp), IGFBP-5 (309 bp), IGFBP-6 (410 bp) and RPL 19 (500 bp) RT–PCR amplification products, from control young rats of 23, 29 and 35 days of age.

to 25 days of age. The expression of IGFBP-3 and IGFBP-6 after 9 and 15 days of treatment (T9 and T15) were significantly lower than the respective controls (Fig. 2A,B).

At the beginning of the later stage, it was interesting to observe a decrease in the number of positive cells for IGFBP-3, -4 and -5 in the animals treated with the propulsor appliance when compared with the respective controls. All IGFBPs showed a peak at 30 days of appliance use although the IGFBP-3 expression was still lower than its respective control (Fig. 2A–D). IGFBP-4 and IGFBP-5 showed a similar pattern of expression and seems to be influenced by the appliance on these later stages (Fig. 2C,D). The PCNA expression was almost the same as the controls, differing only by a rising number of positive cells after 30 days of treatment (T30) ($p < 0.05$).

### 3.4. Genic expression of IGFBP-3, IGFBP-4, IGFBP-5 and IGFBP-6

The condylar cartilage expresses IGFBP-3, IGFBP-4, IGFBP-5 and IGFBP-6 mRNA indicating its local synthesis by the chondrocyte cells (Fig. 3).

## 4. Discussion

Modification of the mandibular posture is one of the basic principals of functional orthopedics to correct dental malocclusions. In this paper we have demonstrated that there is a modulation of the physiological levels of IGFBP-3, -4, -5 and -6 in rat condylar cartilage during growth. Furthermore use of a functional orthopedic appliance changed the IGFBPs expression in the condylar tissue, as a consequence of the mechanical forces generated indirectly on this cartilage.

Previous studies have demonstrated that the forward displacement of the mandible in rats promotes an increase in the proliferative cells and increment of the cartilage width while the posterior displacement promotes the opposite effect [37]. In our study, the propulsor appliance used in this experimental model presented two distinct phases during treatment: an early stage where clinically there was a change at the dental occlusion and biologically a higher expression of IGF-I and IGF-II [14]; followed by a stable phase, but indispensable for the consolidation of the clinical results previously obtained, where we observed a higher variation on the IGFBPs expression. These results suggest that the early stage is related with cell proliferation and the later stage with the tissue maturation and differentiation that is necessary for the consolidation of the remodeling process of the condylar cartilage.

The modification in the mandibular posture of orthodontic patients in the examining room as well as in the animals comprising this experimental model occurs due to changes in the neuromuscular tonus, which is considered one of the principal modulators of bone and temporomandibular joint growth [40]. In the initial process the alteration in the muscular function, generated by the change in mandibular posture, creates an acute response to the stimulus in the condylar cartilage which generates a rapid growth response through the increased expression of IGF-I and IGF-II [17]. After a certain time there is an adaptation of the muscular function to the new condition, when we observed a decrease in the IGFs and modulation of the IGFBPs [15]. The clinical results obtained in this study corroborate that observed in functional orthopedic human treatment, where after the active treatment a contain period was necessary to consolidate the results obtained. The contention period, stipulated in clinical practice to avoid malocclusion relapse is dictated by the time necessary to perform the initial alteration of the occlusion.

We failed to alter the occlusion of young adult rats that used the appliance for a prolonged period of 30 days (data not shown), suggesting that the efficiency of the clinical treatment depends not only on the tissue responsiveness but also on factors that are present in the tissue during the various growth phases.

Data obtained from different methods and experiments *in vivo* and *in vitro* indicate other functions for the IGFBPs besides being mere binding proteins. In the cartilage tissue the IGFBPs expression has been studied predominately on the epiphyseal cartilage of the long bones. The presence of IGFBP-3, -4, and -5 was observed in different types of animals in the proliferative zone and hypertrophic chondrocytes, demonstrating that IGF-I is able to stimulate theses IGFBPs expression [4,7,34,35,41]. In the condylar cartilage, only some cells in the hypertrophic zone are positive for IGFBP-3, -4, -5 and -6, especially in the marginal zones of the cartilage, suggesting a different chondrocyte activity in different regions. The same was observed in the cartilage of the long bones although their growth occurs in longitudinal direction [20,29].

On the epiphyseal cartilage of different animal species the elevation in IGF-I stimulates an increase in IGFBPs expression [4,7,34,35,41]. Interestingly there were almost no variations in the IGFs expression during the animal development. The condylar cartilage does not appear to suffer the influence of endogenous sexual hormones to stimulate the cartilage IGFs expression, at least during the early stage. In the condylar cartilage of young rats the paracrine production of IGF-I, is independent from GH and responsible for the newborn rats condylar development [24,25]. However the same authors have also demonstrated that testosterone raises the genic and protein expression of IGF-I and IGF-IR in an organ culture system of young rats. On the other hand, we observed a gradual increase in IGFBP-3 and IGFBP-6 expression among the control animals, suggesting that maybe these IGFBPs are sensitive to the elevation in sexual hormones and GH as well as the IGFs in the later stage of our experiment.

It is interesting to observe the general pattern of expression of the IGFs and IGFBPs in the condylar cartilage of the animals that used the propulsor. In the early stage there was a high elevation in the IGFs expression [17], not followed by the IGFBPs, which had the tendency to a lower expression. On the other hand at the later stage, a period that coincides with the animals' puberty, we observed a less important increase in the IGFs, followed by a larger increase the IGFBPs. The second intensification in the expression of growth factors at 30 days-treatment occurred 15 days after the first one and may indicate a consolidation of the remodeling process.

The mechanical force applied on the cartilage generates a particular situation regarding IGF/IGFBP, where on the physiological and pathological state (arthritis rheumatoid and osteoarthritis for example) the increase in IGF-I is followed by an increase in the expression of IGFBP-3 [8,9,12,26,32]. An opposite pattern is observed at the condylar cartilage regarding the IGF-I/IGFBP-3 expression. Low expression of IGFBP-3, the most important BP, when a high expression of IGF-I was observed, may suggest that the mechanical forces are able to induce more release of IGF-I in the tissue, promoting greater cell proliferation, as shown in our previous study [17]. Other studies have shown a mitogenic and differentiation effect with an increase in the extracellular matrix of cultivated cells of condylar, articular and epiphyseal cartilage [23,33,42,43]. Continuous treatment with IGF-I sustains the cartilage cells phenotype *in vitro*, while its removal rapidly leads the cells to a hypertrophic stage [5]. There is evidence that the chondrocyte evolution to the hyperthropic stage is followed by a decrease in the IGF-I and IGF-II mRNA as has been demonstrated in bovine epiphyseal cartilage [34]. IGF independent effects of the IGFBPs have been reported. IGFBP-3 may exert selective effects on cell growth and death through a direct connection with the cell surface [9,22].

IGFBP-6 differs from the other IGFBPs by presenting a higher affinity for IGF-II than IGF-I [3, 32]. In our previous study we showed a high expression of IGF-II in the condylar cartilage [17]. The lower expression of this IGFBP at the early stage of the experiment suggests that there may be more IGF-II available in cartilage submitted to mechanical stress and that this growth factor may also play an important role on these type of cartilage growth. On the other hand, at the later stage IGFBP-6 and IGF-II exhibit the same expression.

IGFBP-4 is considered a negative regulator of the local action of IGF by sequestering this growth factor, thereby inhibiting its binding to the receptor [27,36]. If we consider that IGFBP-4 expression in this cartilage has the same effect observed in bone tissue, we can affirm that the 30-day treatment group is the reflex not only of elevated IGF-I expression [15] and its higher viability by the lower expression of IGFBP-3, but also of its capture by IGFBP-4 which at this time presents its highest expression observed during the whole period of treatment. IGFBP-3 and IGFBP-4 may play an important role in the modulation of accessibility of the IGFs to their receptors with antagonist effects, and that this function may be influenced by the mechanical stress in the cartilage.

IGFBP-5 is the most abundant protein stored in bone tissue, presenting high affinity binding with hydroxyapatite and is therefore capable of maintaining IGF-I within the bone tissue, where it stimulates the proliferation of osteoblasts *in vitro* [31]. However opposite effects have been described in relation to IGFBP-5 [30]. IGFBP-5 has a role on long-term muscular adaptation to changes in load [1] and its presence in the skeletal muscle of mice was three times lower in relation to the controls in the presence of load. Whereas absence of load was followed by a higher expression of this factor. Recent *in vitro* studies on the epiphyseal cartilage have suggested that IGFBP-5 is able to stimulate chondrocyte growth only in the presence of IGF-I [19]. IGFBP-5 binds to IGF-I with less affinity than IGFBP-3; however the higher expression of this factor after 30 days of treatment could also potentialize IGF-I action and simultaneously oppose the inhibitory effect of IGFBP-4, as described in bone tissue and leaving more free IGF-I, probably released from the lower expression of IGFBP-3, to bind to its receptor.

In conclusion, we can emphasize that IGFBPs showed a varying expression according to the animal's age, thereby influencing the IGF-I and IGF-II viability in the system. The IGFBPs expression in different chondrocyte layers indicates a role for the IGFBPs in chondrocyte proliferation as well as in cell maturation. This experimental protocol with application of an appliance in rats demonstrates the same clinical efficacy as is observed in human treatment. The variation of the IGF-IGFBP system protein expression during growth and under the orthopedic appliance treatment suggests an important role in the condylar cartilage remodeling.

# References

[1] B. Awede, J. Thissen, P. Gailly and J. Lebacq, Regulation of IGF-I, IGFBP-4 and IGFBP-5 gene expression by loading in mouse skeletal muscle, *FEBS Lett.* **19** (1999), 263–267.

[2] L.A. Bach, The insulin-like growth factor system: basic and clinical aspects, *Aust. N. Z. J. Med.* **29** (1999), 355–361.

[3] R.C. Baxter, Insulin-like growth factor (IGF)-binding proteins: interactions with IGFs and intrinsic bioactivities, *Am. J. Physiol. Endocrinol. Metab.* **278** (2000), E967–976.

[4] B. Bhaumick, Insulin-like growth factor (IGF) binding proteins and insulin-like growth factor secretion by cultured chondrocyte cells: identification, characterization and ontogeny during cell differentiation, *Regul. Pept.* **20** (1993), 113–122.

[5] K. Bohme, M. Conscience-Egli, T. Tschan, K.H. Winterhalter and P. Bruckner, Induction of proliferation or hypertrophy of chondrocytes in serum-free culture: the role of insulin-like growth factor-I, insulin, or thyroxine, *J. Cell Biol.* **116** (1992), 1035–1042.

[6] C.A. Conover and D.D. De Leon, Acid-activated insulin-like growth factor-binding protein-3 proteolysis in normal and transformed cells, Role of cathepsin D, *J. Biol. Chem.* **269** (1994), 7076–7080.

[7] P. De Los Rios and D.J. Hill, Cellular localization and expression of insulin-like growth factors (IGFs) and IGF binding proteins within the epiphyseal growth plate of the ovine fetus: possible functional implications, *Can. J. Physiol. Pharmacol.* **77** (1999), 235–249.

[8] P. De Los Rios and D.J. Hill, Expression and release of insulin-like growth factor binding proteins in isolated epiphyseal growth plate chondrocytes from the ovine fetus, *J. Cell Physiol.* **183** (2000), 172–181.

[9] T. Eviatar, H. Kauffman and A. Maroudas, Synthesis of insulin-like growth factor binding protein 3 in vitro in human articular cartilage cultures, *Arthritis Rheum.* **48** (2003), 410–417.

[10] S.M. Firth and R.C. Baxter, Cellular actions of the insulin-like growth factor binding proteins, *Endocr. Rev.* **23** (2002), 824–854.

[11] P. Galand and C. Degraef, Cyclin/PCNA immunostaining as an alternative to H$^3$ pulse labeling for marking S phase cells in paraffin sections from animal and human tissues, *Cell Tissue Kinet.* **22** (1989), 383–392.

[12] A.M. Garcia, N. Szasz, S.B. Trippel, T.I. Morales, A.J. Grodzinsky and E.H. Frank, Transport and binding of insulin-like growth factor I through articular cartilage, *Arch. Biochem. Biophys.* **415** (2003), 69–79.

[13] A.J. Grodzinsky, M.E. Levenston, M. Jin and E.H. Frank, Cartilage tissue remodeling in response to mechanical forces, *Annu. Rev. Biomed. Eng.* **2** (2000), 691–713.

[14] D. Hajjar, Efeito da utilização do aparelho propulsor da mandíbula: expressão de IGF-I e IGF-II na cartilagem do côndilo mandibular de ratos, Tese de Mestrado, 2001.

[15] D. Hajjar, J.C. Ricarte-Filho, S.E. Matsuo, M.F. Santos and E.T. Kimura, Mudança de postura da mandíbula altera expressão de IGFBP-3, 4, 5 e 6 na cartilagem condylar de ratos, *Brazilian Oral Research* **18**(Suppl.) (2004), 203.

[16] D. Hajjar, M.F. Santos and E.T. Kimura, Mandibular propulsive appliance for rats. An experimental model for functional orthopedics, *J. Dent. Res.* **79** (2000), 1117.

[17] D. Hajjar, M.F. Santos and E.T. Kimura, Propulsive appliance stimulates the synthesis of insulin-like growth factors I and II in the mandibular condylar cartilage of young rats, *Arch. Oral Biol.* **48** (2003), 635–642.

[18] K. Itoh, S. Suzuki and T. Kuroda, Effects of local administration of insulin-like growth factor-I on mandibular condylar growth in rats, *J. Med. Dent. Sci.* **50** (2003), 79–85.

[19] D. Kiepe, D.L. Andress, S. Mohan, L. Standker, T. Ulinski, R. Himmele, O. Mehls and B. Tonshoff, Intact IGF-binding protein-4 and -5 and their respective fragments isolated from chronic renal failure serum differentially modulate IGF-I actions in cultured growth plate chondrocytes, *J. Am. Soc. Nephrol.* **12** (2001), 2400–2410.

[20] D.A. Lazowski, L.J. Fraher, A. Hodsman, B. Steer, D. Modrowski and V.K. Han, Regional variation of IGF-I gene expression in mature rat bone and cartilage, *Bone* **15** (1994), 563–576.

[21] E. Livne, C. Oliver and M. Silbermann, Further characterization of the chondroprogenitor zone in mandibular condyles of suckling mice. An ultrastructural and cytochemical study, *Acta Anat.* **129** (1987), 231–237.

[22] L. Longobardi, M. Torello, C. Buckway, L. O'Rear, W.A. Horton, V. Hwa, C.T. Roberts, Jr., F. Chiarelli, R.G. Rosenfeld and A. Spagnoli, A novel insulin-like growth factor (IGF)-independent role for IGF binding protein-3 in mesenchymal chondroprogenitor cell apoptosis, *Endocrinology* **144** (2003), 1695–1702.

[23] G. Maor, Z. Hochberg and M. Silbermann, IGF-I accelerates proliferation and differentiation of cartilage progenitor cells in cultures of neonatal mandibular condyles, *Acta Endocrinol. (Copenh.)* **128** (1993), 56–64.

[24] G. Maor, Z. Laron, R. Eshet and M. Silbermann, The early postnatal development of the murine mandibular condyle is regulated by endogenous IGF-I, *J. Endocrinol.* **137** (1993), 21–26.

[25] G. Maor, Y. Segev and M. Phillip, Testosterone stimulates IGF-I and IGF-I receptor gene expression in the mandibular condyle-a model of endochondral ossification, *Endocrinology* **140** (1999), 1901–1910.

[26] T. Matsumoto, S.E. Gargosky, K. Iwasaki and R.G. Rosenfeld, Identification and characterization of insulin-like growth factors (IGFs), IGF-binding proteins (IGFBPs), and IGFBP proteases in human synovial fluid, *J. Clin. Endocrinol. Metab.* **81** (1996), 150–155.

[27] S. Mazerbourg, I. Callebaut, J. Zapf, S. Mohan, M. Overgaard and P. Monget, Up date on IGFBP-4: regulation of IGFBP-4 levels and functions, in vitro and in vivo, *Growth Horm. IGF Res.* **14** (2004), 71–84.

[28] J.A. McNamara, R.J. Hinton and D.L. Hoffman, Histologic analysis of temporo-mandibular joint adaptation to protrusive function in young adult rhesus monkeys, *Am. J. Orthod.* **82** (1982), 288–298.

[29] C. Miralles-Flores and E. Delgado-Baeza, Histomorphometric differences between the lateral region and central region of the growth plate in fifteen-day-old rats, *Acta Anat. (Basel)* **139** (1990), 209–213.

[30] N. Miyakoshi, C. Richman, Y. Kasukawa, T.A. Linkhart, D.J. Baylink and S. Mohan, Evidence that IGF-binding protein-5 functions as a growth factor, *J. Clin. Invest.* **107** (2001), 73–81

[31] S. Mohan, Y. Nakao, Y. Honda, E. Landale, U. Leser, C. Dony, K. Lang and D.J. Baylink, Studies on the mechanisms by which insulin-like growth factor (IGF) binding protein-4 (IGFBP-4) and IGFBP-5 modulate IGF actions in bone cells, *J. Biol. Chem.* **1** (1995), 20424–20431.

[32] T.I. Morales, The insulin-like growth factor binding proteins in uncultured human cartilage: increases in insulin-like growth factor binding protein 3 during osteoarthritis, *Arthritis Rheum.* **46** (2002), 2358–2367.

[33] R.J. O'keefe, I.D. Crabb, J.E. Puzas and R.N. Rosier, Effects of transforming growth factor-beta 1 and fibroblast growth factor on DNA synthesis in growth plate chondrocytes are enhanced by insulin-like growth factor-I, *J. Orthop. Res.* **12** (1994), 299–310.

[34] R.C. Olney and E.B. Mougey, Expression of the components of the insulin-like growth factor axis across the growth-plate, *Mol. Cell Endocrinol.* **25** (1999), 63–71.

[35] R.C. Olney, R.L. Smith, Y. Kee and D.M. Wilson, Production and hormonal regulation of insulin-like growth factor binding proteins in bovine chondrocytes, *Endocrinology* **133** (1993), 563–570.

[36] C.O. Ortiz, B.K. Chen, L.K. Bale, M.T. Overgaard, C. Oxvig and C.A. Conover, Transforming growth factor-beta regulation of the insulin-like growth factor binding protein-4 protease system in cultured human osteoblasts, *J. Bone Miner. Res.* **18** (2003), 1066–1072.

[37] A.G. Petrovic, J. Stutzmannand, C. Oudet, Control processes in the postnatal growth of the condylar cartilage of the mandible, in: *Determinants of Mandibular Form and Growt.*, J.A. McNamara, Jr., eds, Ann Arbor, MI, 1975, pp. 101–153.

[38] J.C.M. Ricarte-Filho, D. Hajjar and E.T. Kimura, Rattus norvegicus insulin-like growth factor binding 4 (Igfbp4) mRNA, complete cds. LOCUS AY686592. Submitted (15-Jul-2004) *Cellular Biology and Developmental*, Institute of Biomedical Sciences, University of Sao Paulo, Professor Lineu Prestes, 1524, Sao Paulo, SP 05508-900, Brazil.

[39] M.R. Schneider, E. Wolf, A. Hoeflich and H. Lahm, IGF-binding protein-5: flexible player in the IGF system and effector on its own, *J. Endocrinol.* **172** (2002), 423–440.

[40] W.A. Simões, Ortopedia funcional dos maxilares-vista através da reabilitação neuro-oclusal. 1 Ediçăo, São Paulo, 1985, pp. 69–85.

[41] D. Sunic, D.A. Belford, J.D. Mcneil and O.W. Wiebkin, Insulin-like growth factor binding proteins (IGF-BPs) in bovine articular and ovine growth-plate chondrocyte cultures: their regulation by IGFs and modulation of proteoglycan synthesis, *Biochim. Biophys. Acta* **17** (1995), 43–48.

[42] S.B. Trippel, M.T. CorvoL, M.F. Dumontier, R. Rappaport, H.H. Hung and H.J. Mankin, Effect of somatomedin-C/insulin-like growth factor I and growth hormone on cultured growth plate and articular chondrocytes, *Pediatr. Res.* **25** (1989), 76–82.

[43] U. Vetter, J. Zapf, W. Heit, G. Helbing, E. Heinze, E.R. Froesch and W.M. Teller, Human fetal and adult chondrocytes. Effect of insulinlike growth factors I and II, insulin, and growth hormone on clonal growth, *J. Clin. Invest.* **77** (1986), 1903–1908.

[44] X. Wang and J.J. Mao, Accelerated chondrogenesis of the rabbit cranial base growth plate by oscillatory mechanical stimuli, *J. Bone Miner. Res.* **17** (2002), 1843–1850.

Biorheology 43 (2006) 323–335
IOS Press



# Physical signals and solute transport in human intervertebral disc during compressive stress relaxation: 3D finite element analysis

Hai Yao [a] and Wei Yong Gu [b,*]

[a] *Department of Bioengineering, Clemson University, Clemson, SC, USA*
[b] *Tissue Biomechanics Laboratory, Department of Biomedical Engineering, University of Miami, Coral Gables, FL, USA*

**Abstract.** A 3D finite element model for charged hydrated soft tissues containing charged/uncharged solutes was developed based on the multi-phasic mechano-electrochemical mixture theory (Lai et al., *J. Biomech. Eng.* **113** (1991), 245–258; Gu et al., *J. Biomech. Eng.* **120** (1998), 169–180). This model was applied to analyze the mechanical, chemical and electrical signals within the human intervertebral disc during an unconfined compressive stress relaxation test. The effects of tissue composition [e.g., water content and fixed charge density (FCD)] on the physical signals and the transport rate of fluid, ions and nutrients were investigated. The numerical simulation showed that, during disc compression, the fluid pressurization was more pronounced at the center (nucleus) region of the disc while the effective (von Mises) stress was higher at the outer (annulus) region. Parametric analyses revealed that the decrease in initial tissue water content (0.7–0.8) increased the peak stress and relaxation time due to the reduction of permeability, causing greater fluid pressurization effect. The electrical signals within the disc were more sensitive to FCD than tissue porosity, and mechanical loading affected the large solute (e.g., growth factor) transport significantly, but not for small solute (e.g., glucose). Moreover, this study confirmed that the interstitial fluid pressurization plays an important role in the load support mechanism of IVD by sharing more than 40% of the total load during disc compression. This study is important for understanding disc biomechanics, disc nutrition and disc mechanobiology.

Keywords: Intervertebral disc, triphasic theory, finite element method, solute transport, soft tissue mechanics

## 1. Introduction

The intervertebral disc (IVD) is the largest, avascular cartilaginous structure in human body that contributes to flexibility and load support in the spine. Knowledge of mechanical, chemical and electrical signals within the tissue is important for understanding mechanobiology of IVD [40]. Numerous works have been done to study the biomechanical behavior of spinal motion segments as well as isolated IVD samples [1,14,15,17,19,21–23,28–32,35,36,38]. Most of those studies were based on the model in which the nucleus pulposus (NP) was assumed to be an incompressible fluid and the annulus fibrosus (AF) was a linear or nonlinear, elastic or viscoelastic, single phase or composite material. Simon et al. and Laible et al. used the poroelastic (biphasic) theory for their finite element analysis (FEA) of a spinal motion segment, which included fluid transport and swelling effects [19,30,31]. A two-dimensional poroelastic

---

*Address for correspondence: Weiyong Gu, Ph.D., Department of Biomedical Engineering, College of Engineering, University of Miami, P.O. Box 248294, Coral Gables, FL 33124-0621, USA. Tel.: +1 305 284 5434; Fax: +1 305 284 4720; E-mail: wgu@miami.edu.

axisymmetric finite element model (FEM) was also presented by Whyne et al. (2001) [41]. The influence of load-induced fluid flow on solute transport within the IVD was investigated by Ferguson et al. (2004) using poroelastic FEM [4]. However, none of above models considered the charged nature of the IVD and the mechano-electrochemical coupling effects on the solute transport, and thus lacking the power to provide the comprehensive picture of the mechano-electrochemical environment within the disc under mechanical loading condition.

More recently, mechano-electrochemical theories [12,18] have been used to study the biomechanical behavior of disc under 1D [5] and 2D assumptions [13]. In order to quantitatively determine physical signals within the disc, however, it is necessary to develop a 3D triphasic finite element model of the entire IVD.

Thus, the objectives of this study were (1) to develop a 3D finite element model for charged hydrated soft tissues containing charged/uncharged solutes based on the mechano-electrochemical triphasic theory [18] and (2) to apply this model to analyze the mechanical, chemical and electrical signals within the human intervertebral disc under mechanical loading. The effects of tissue composition [e.g., water content and fixed charge density (FCD)] and transport properties (e.g., hydraulic permeability and solute diffusivity) on the physical signals and the transport rate of fluid, ions and nutrients were investigated. The role of the fluid pressurization in the load support mechanism was also clarified. This study is important for understanding disc biomechanics, disc nutrition and disc mechanobiology.

## 2. Methods

Responses of physical signals and solute transport in the human lumbar disc (Fig. 1a) to unconfined compression (stress-relaxation test, Fig. 1) were analyzed in this study. The size and geometry of a representative disc was shown in Fig. 1a [13]. The thickness of the disc was $h = 10$ mm. The disc

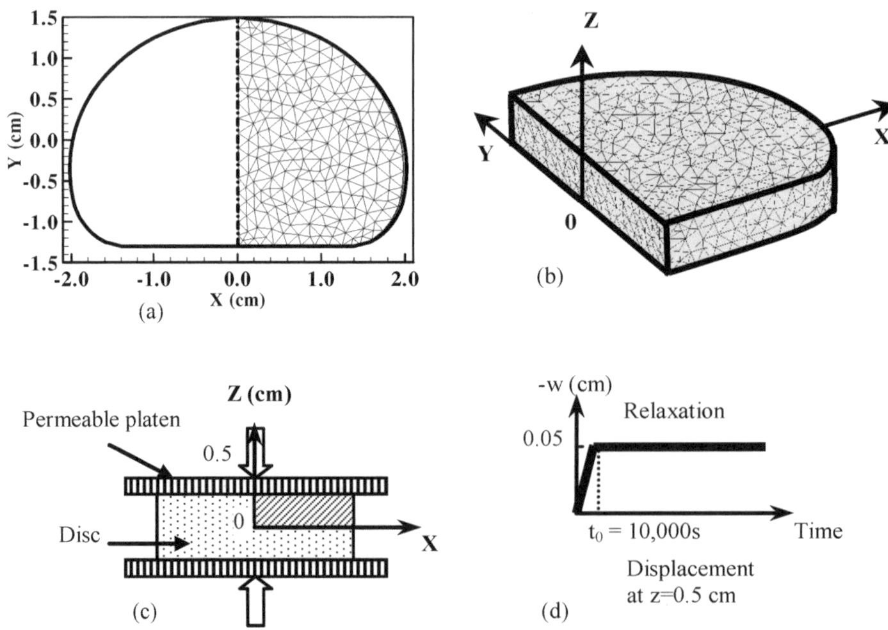


Fig. 1. (a) Disc geometry, (b) mesh, (c) test configuration, and (d) testing protocol.

sample was initially equilibrated with a bathing solution of 0.15 M NaCl. An uncharged solute was introduced into the bathing solution at $t = 0$, and the disc was subjected to a ramp compression (10% strain in 10,000 s) via two perfectly adhesive porous platens (Fig. 1d). In order to investigate the effect of mechanical loading on the transport of uncharged solute, a diffusion process of uncharged solute into the disc without ramp compression was also simulated. In this study, the human disc was modeled as an isotropic homogeneous mixture consisting of an intrinsically incompressible elastic solid (with fixed charge), water, ion ($Na^+$ and $Cl^-$) and uncharged solute (glucose or uncharged growth factor) phases.

## 2.1. Governing equations

An equivalent formulation of the triphasic theory was used for the finite element implementation [37]. In this formulation, the solid displacement $\mathbf{u}$ and the modified (electro)chemical potentials $\varepsilon^\alpha$ ($\alpha = w, +, -, o$) of water, cation, anion, and uncharged solute were chosen as the primary degrees of freedom. Thus, this formulation has the advantage that the jump condition between elements and across the interface boundaries are satisfied automatically [37].

The balance of linear momentum for the mixture and the conservation of mass for each of phases or species led to the following governing equations [7,18,37]:

$$\nabla \cdot \boldsymbol{\sigma} = 0, \tag{1}$$

$$\nabla \cdot (\mathbf{v}^s + \mathbf{J}^w) = 0, \tag{2}$$

$$\partial(\phi^w c^+)/\partial t + \nabla \cdot (\mathbf{J}^+ + \phi^w c^+ \mathbf{v}^s) = 0, \tag{3}$$

$$\partial(\phi^w c^-)/\partial t + \nabla \cdot (\mathbf{J}^- + \phi^w c^- \mathbf{v}^s) = 0, \tag{4}$$

$$\partial(\phi^w c^o)/\partial t + \nabla \cdot (\mathbf{J}^o + \phi^w c^o \mathbf{v}^s) = 0, \tag{5}$$

where $\phi^w$ is tissue porosity defined as a ratio of water volume to total tissue volume. The total stress tensor of the mixture $\boldsymbol{\sigma}$, the velocity of the solid phase $\mathbf{v}^s$, the fluxes relative to the solid phase $\mathbf{J}^\alpha$ ($\alpha = w, +, -, o$) and concentrations (per unit water volume) $c^\alpha$ ($\alpha = +, -, o$) were related to solid displacement $\mathbf{u}$, modified (electro)chemical potentials of water $\varepsilon^w$, cation $\varepsilon^+$, anion $\varepsilon^-$, and uncharged solute $\varepsilon^o$ [37]:

$$\mathbf{v}^s = \frac{\partial \mathbf{u}}{\partial t}, \tag{6}$$

$$\boldsymbol{\sigma} = -\lfloor RT\varepsilon^w + RT\phi(c^+ + c^- + c^o) - p_o \rfloor + (\lambda + B_w)\nabla \cdot \mathbf{u} + \mu[\nabla\mathbf{u} + (\nabla\mathbf{u})^T], \tag{7}$$

$$\mathbf{J}^w = -RTk\left(\nabla\varepsilon^w + \frac{c^+}{\varepsilon^+}H^+\nabla\varepsilon^+ + \frac{c^-}{\varepsilon^-}H^-\nabla\varepsilon^- + \frac{c^o}{\varepsilon^o}H^o\nabla\varepsilon^o\right), \tag{8}$$

$$\mathbf{J}^+ = H^+c^+\mathbf{J}^w - \frac{\phi^w c^+ D^+}{\varepsilon^+}\nabla\varepsilon^+, \tag{9}$$

$$\mathbf{J}^- = H^-c^-\mathbf{J}^w - \frac{\phi^w c^- D^-}{\varepsilon^-}\nabla\varepsilon^-, \tag{10}$$

$$\mathbf{J}^o = H^o c^o \mathbf{J}^w - \frac{\phi^w c^o D^o}{\varepsilon^o}\nabla\varepsilon^o, \tag{11}$$

where $R$ is the universal gas constant; $T$ is the absolute temperature; $\phi$ is the osmotic coefficient; $p_o$ is the pre-stress [see Eq. (12) below]; $B_w$ is the interphase coupling coefficient; $\lambda$ and $\mu$ are Lame coefficients of solid matrix; $k$ is the intrinsic (i.e., closed-circuit) permeability; $D^+$, $D^-$ and $D^o$ are the intra-tissue diffusivities of cation, anion and uncharged solute, respectively. The parameters $H^\alpha$ ($\alpha = +,-,o$) in Eqs (9)–(11) are the convective coefficients since the convective velocity in a mixture is in general not equal to the water velocity. The value of the convective coefficients represent the hindrance factor of the solute for convection that incorporates hydrodynamic and steric interactions between solute and the matrix [2]. They are functions of drag coefficients between a solute and other phases. Note that the mechanical interactions among solutes were assumed to be negligible.

In arriving at Eqs (6) and (7), infinitesimal deformation assumption was made [18,37]. The value of pre-stress ($p_o$) in Eq. (7) depends on the choice of the reference configuration for strain. In this study, the free-swelling state of tissue equilibrated with the bathing (NaCl) solution of concentration $c^*$ (no other uncharged solute) was chosen as the reference configuration for strain. Thus, the pre-stress ($p_o$) was given as

$$p_o = RT\left[\phi(c_o^+ + c_o^-) - 2\phi^*c^*\right] - B_we_o, \tag{12}$$

where $c_o^+$ and $c_o^-$ are the ion concentrations within the tissue at equilibrium (i.e., the initial ion concentrations), and $e_o$ is the dilatation of tissue at equilibrium (relative to the hypertonic state). If one chooses the tissue configuration at the hypertonic state (load-free) as the reference configuration, then $p_o = 0$ and $e_o = 0$. In the following calculations, all strains were relative to the free-swelling state of tissue equilibrated with 0.15 M NaCl bathing solution [8].

In Eqs (7)–(10), the modified (electro)chemical potentials ($\varepsilon^w$, $\varepsilon^+$, $\varepsilon^-$, $\varepsilon^o$) were related to fluid pressure ($p$), electrical potential ($\psi$), and solute concentrations ($c^+$, $c^-$, $c^o$) by [37]:

$$\varepsilon^w = \frac{p}{RT} - \phi(c^+ + c^- + c^o) + \frac{B_w}{RT}e, \tag{13}$$

$$\varepsilon^+ = \gamma_+c^+ \exp\left(\frac{F_c\psi}{RT}\right), \tag{14}$$

$$\varepsilon^- = \gamma_-c^- \exp\left(-\frac{F_c\psi}{RT}\right), \tag{15}$$

$$\varepsilon^o = \gamma_oc^o, \tag{16}$$

where $e$ is the dilatation; $F_c$ is the Faraday constant; $\gamma_+, \gamma_-$ and $\gamma_o$ are the activity coefficients of cation, anion and uncharged solute, respectively. At equilibrium, the activity coefficient of the uncharged solute within the tissue ($\gamma_o$) was related to the partition coefficient ($\Phi$) (i.e., the ratio of solute concentration in tissue to solute concentration in bathing solute at equilibrium) and the activity coefficient of the uncharged solute in bathing solution ($\gamma_o^*$) by

$$\gamma_o = \gamma_o^*/\Phi. \tag{17}$$

The ion concentrations were related to the value of negatively fixed charged density ($c^F$) through the electroneutrality condition in this model [18]:

$$c^+ = c^- + c^F, \tag{18}$$

where $c^F$ is related to the tissue porosity, the reference FCD ($c_0^F$) and reference porosity ($\phi_0^w$) by

$$c^F = \frac{c_0^F(1 - \phi^w)\phi_0^w}{(1 - \phi_0^w)\phi^w}.$$ (19)

In this study, the following constitutive relationship was used for the intrinsic permeability:

$$k = a\left(\frac{\phi^w}{\phi^s}\right)^n,$$ (20)

where $a$ and $n$ are parameters of which the values depend on the structure and composition of the porous media [10]. The dependence of ion diffusivities on the tissue porosity was estimated using the Mackie and Meares model [25]:

$$\frac{D^\alpha}{D_0^\alpha} = \frac{(\phi^w)^2}{(2 - \phi^w)^2}, \quad \text{for } \alpha = +, -,$$ (21)

where $D_0^\alpha$ is the ion diffusivity in aqueous solution. However, in this study, the diffusivity of the neutral solute was assumed to be constant within the tissue.

Since tissue porosity ($\phi^w$) was related to tissue dilatation and the porosity ($\phi_o^w$) at the reference configuration (i.e., $e = 0$) [18]

$$\phi^w = \frac{\phi_o^w + e}{1 + e},$$ (22)

the intrinsic permeability, ion diffusivity, and FCD were all strain-dependent.

For this 3D problem of interest, only the upper quadrant of the sample was modeled due to the symmetry with respect to plane $X = 0$ and plane $Z = 0$.

*Initial condition:*

$$t = 0: \quad \mathbf{u} = 0, \quad \varepsilon^w = \varepsilon^{w*}, \quad \varepsilon^+ = \varepsilon^{+*}, \quad \varepsilon^- = \varepsilon^{-*}, \quad \varepsilon^o = 0,$$ (23)

where superscript * stands for the quantities in the bathing solution.

*Boundary condition:*

$$\text{Plane } X = 0: \quad u = \sigma_{xy} = \sigma_{xz} = 0, \quad J_x^w = J_x^\alpha = J_x^o = 0;$$ (24a)

$$\text{Plane } Z = 0: \quad w = \sigma_{zx} = \sigma_{zy} = 0, \quad J_z^w = J_z^\alpha = J_z^o = 0;$$ (24b)

$$\text{Plane } Z = 0.5: \quad w = w(t), \quad u = v = 0, \quad \varepsilon^w = \varepsilon^{w*}, \quad \varepsilon^\alpha = \varepsilon^{\alpha*}, \quad \varepsilon^o = \varepsilon^{o*}H(t);$$ (24c)

$$\text{Lateral surface:} \quad \boldsymbol{\sigma} \cdot \vec{n} = 0, \quad \varepsilon^w = \varepsilon^{w*}, \quad \varepsilon^+ = \varepsilon^{+*}, \quad \varepsilon^- = \varepsilon^{-*}, \quad \varepsilon^o = \varepsilon^{o*}H(t).$$ (24d)

Here $u$ and $w$ are the components of the solid displacement $\mathbf{u}$ in $X$ and $Z$ directions, respectively. $H(t)$ is the Heaviside step function, and $w(t)$ is the pre-described displacement history (Fig. 1d). In the case of a pure diffusion process, $w(t)$ was set to zero.

## 2.2. Numerical implementation

The numerical technique for solving this problem was similar to the procedure provided by Sun et al. [37]. The modified Newton–Raphson iterative procedure was employed to handle the nonlinear terms in the above set of equations. The resulting first-order ordinary differential equations with respect to time were solved using the implicit Euler backward scheme. The primary degrees of freedom, i.e., the solid displacement $\mathbf{u}$ and the modified (electro)chemical potentials $\varepsilon^\alpha$ ($\alpha = w, +, -, o$) of water, cation, anion and uncharged solute, were independently interpolated. Since the highest order of derivatives was the same for each degree of freedom, the same order shape functions were used. In this study, the weak formulation of these 3D initial- and boundary-value problems was solved using FEMLAB software (FEMLAB3.0, COSMOL Inc., Burlington, MA).

The upper quadrant of the disc was modeled with a mesh of 3975 second-order, tetrahedral Lagrange elements (Fig. 1b). The maximum time-step of 100 s was used during the ramp phase, and variable maximum time-steps from 5 s to 1000 s were used during the relaxation phase. The convergence of the numerical model was examined by refining the mesh and tightening the tolerance. The accuracy of the numerical method was checked with the results of a 2D case published in the literature [42].

In this study, the parameters used were $c^* = 0.15$ M NaCl, $\lambda = 0.2$ MPa, $\mu = 0.1$ MPa, $T = 298$ K, $D_o^+ = 1.28 \times 10^{-9}$ m$^2$/s, $D_o^- = 1.77 \times 10^{-9}$ m$^2$/s, $\gamma_+ = \gamma_- = 1$, $\phi = 1$, and $B_w = 0$ [26]. The values of the diffusivity and partition coefficient for a neutral solute were: $D^o = 2.15 \times 10^{-10}$ m$^2$/s and $\Phi = 0.85$ for glucose, and $D^o = 2.23 \times 10^{-11}$ m$^2$/s and $\Phi = 0.1$ for large solute. The mean values of $a = 0.00044$ nm$^2$ and $n = 7.193$ [9] for porcine AF were used in the constitutive equation of the permeability [i.e., Eq. (20)].

The parameters $H^\alpha$ ($\alpha = +, -, o$) in Eqs (8)–(11) have not been determined for IVD tissues. In this study, the values of these parameters were assumed to be unity.

## 3. Results

All mechanical, chemical and electrical signals within the disc during stress–relaxation test were calculated. The following are the descriptions of the major results.

### 3.1. Profiles of physical signals

The distributions of FCD (Fig. 2), fluid pressure (Fig. 3), and electrical potential (Fig. 4), and effect (von Mises) stress (Fig. 5) during stress relaxation test were non homogeneous. While the fluid pressure was higher at the NP region, the effective (von Mises) stress was found to be higher at the outer region, indicating the annulus may serve chiefly to constrain the nucleus and share the major part of the load (Fig. 5).

### 3.2. Effects of porosity and FCD on signals, load support and fluid transport pathway

The increase in initial tissue water content (0.7–0.8) slightly increased the peak electrical potential, but not the equilibrium potential, at the center point of the disc (Fig. 6a), while the increase in initial FCD (0.05–0.2 mEq/ml) significantly decreased both potentials (Fig. 6b). It is evident that the electrical signal is more sensitive to FCD than tissue porosity.

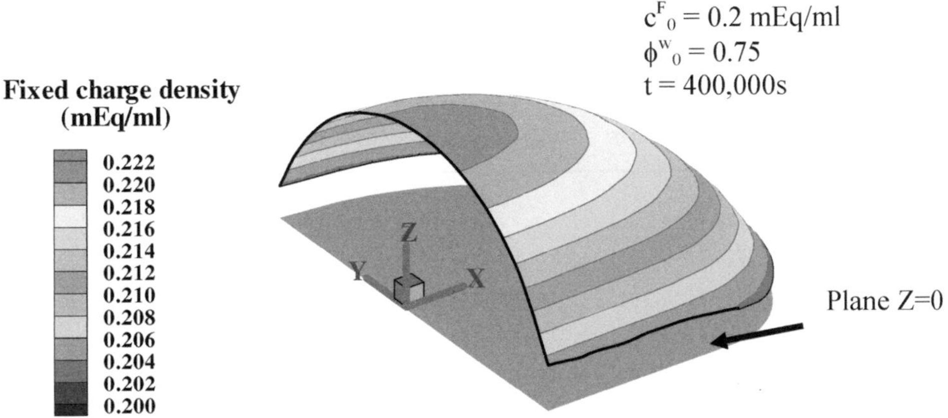



Fig. 2. Distribution of the fixed charge density on plane $Z = 0$ at $t = 400,000$ s (i.e., at equilibrium). Before the disc was subjected to a ramp compression, the initial water content was $\phi^w_0 = 0.75$ and the initial fixed charge density was $c^F_0 = 0.2$ mEq/ml.

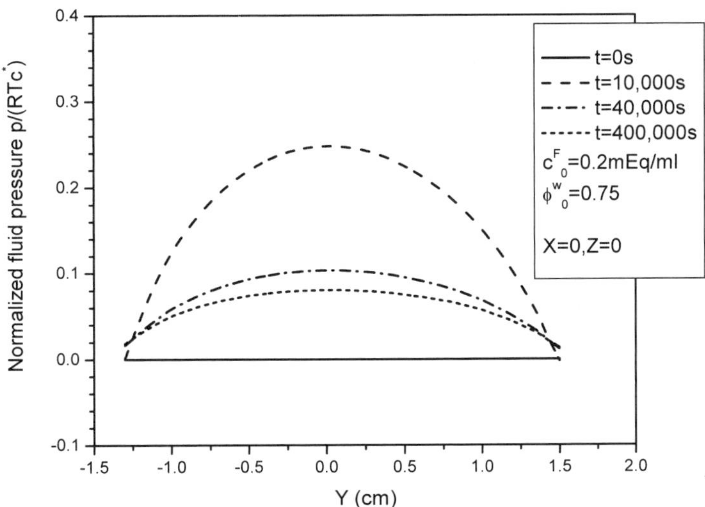



Fig. 3. Distributions of the fluid pressure in the axial direction during stress relaxation test. The pressure was normalized by $RTc^*(= 0.372$ MPa) at room temperature ($T = 298$ K).

A decrease in initial tissue water content increased the peak stress and relaxation time at the center point of the disc (Fig. 7a). This is due to the decrease in permeability as water content decreases, causing greater fluid pressurization effect (Fig. 8a). The equilibrium stress increased with the increase of initial FCD (Fig. 7b) due to the increase in osmotic pressure effect (Fig. 8b). It is apparent that the fluid plays an important role in the load support mechanism since the fluid supported more than 40% of the total load during the ramp compression (Fig. 9).

### 3.3. Effect of mechanical loading on nutrient transport

The compression-induced fluid flow retarded the large solute (e.g., growth factor) transport from the outer annulus boundary to the center of the disc (Fig. 10a) due to the opposite direction of the fluid flow

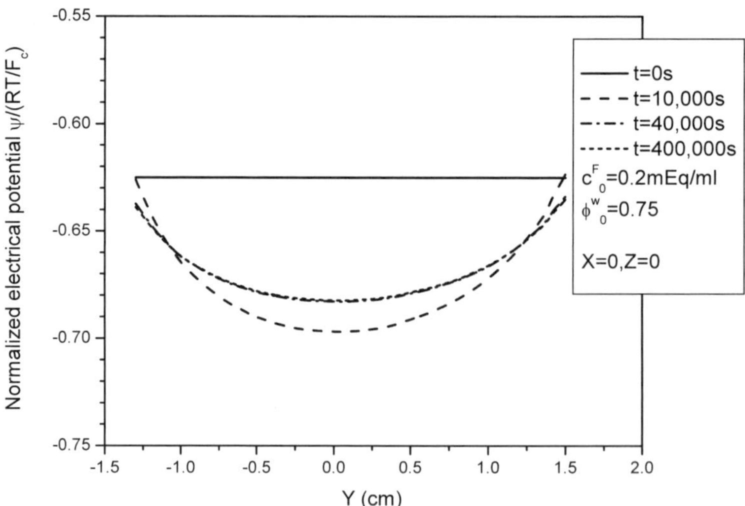

Fig. 4. Distributions of the electrical potential in the axial direction during the stress relaxation test. The electrical potential was normalized by $RT/F_c$ ($= 25.7$ mV).

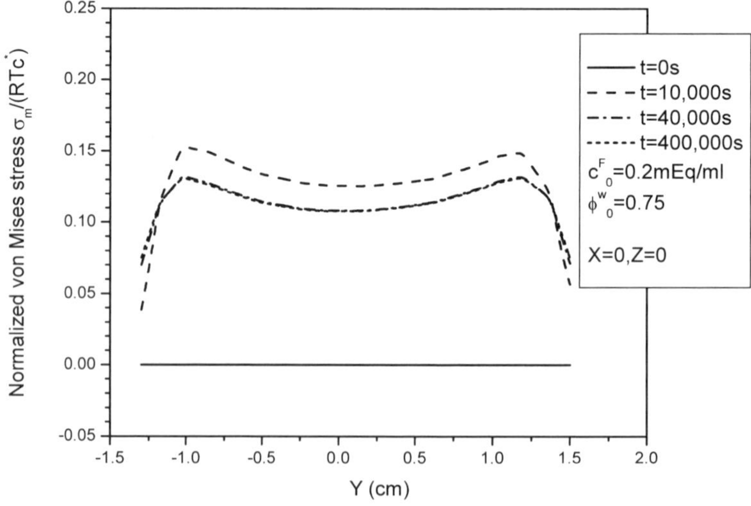

Fig. 5. Distributions of the von Mises effective stress in the radial direction during the stress relaxation test.

to the solute flux, while the fluid flow with the same direction of the solute flux enhanced the large solute transport from the endplate to the center of the disc (Fig. 10b). For small solute transport (e.g., glucose), the mechanical loading effect was not significant (not show).

## 4. Discussion

The most significant biochemical change seen in a degenerated disc is the loss of the proteoglycan, resulting in the decrease in FCD [24]. Our analysis shows that the electrical signal is very sensitive to FCD. Thus, the measurement of this electrical signal might be used as an indicator for disc degeneration or for tissue growth in tissue engineering applications.

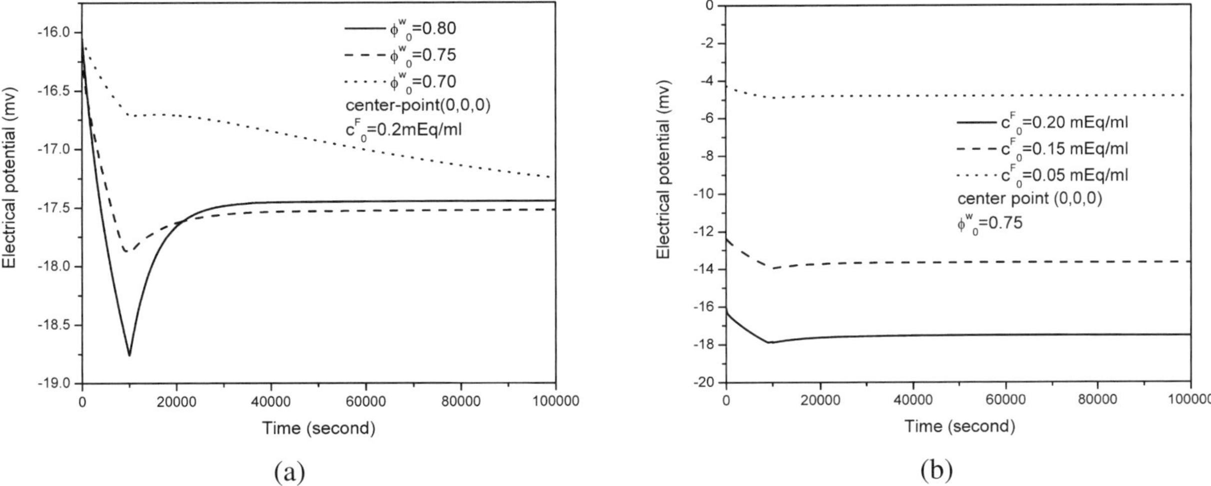

Fig. 6. Effects of (a) initial water content and (b) initial fixed charge density on the electrical potential at the center of the disc (relative to the potential in the bathing solution).

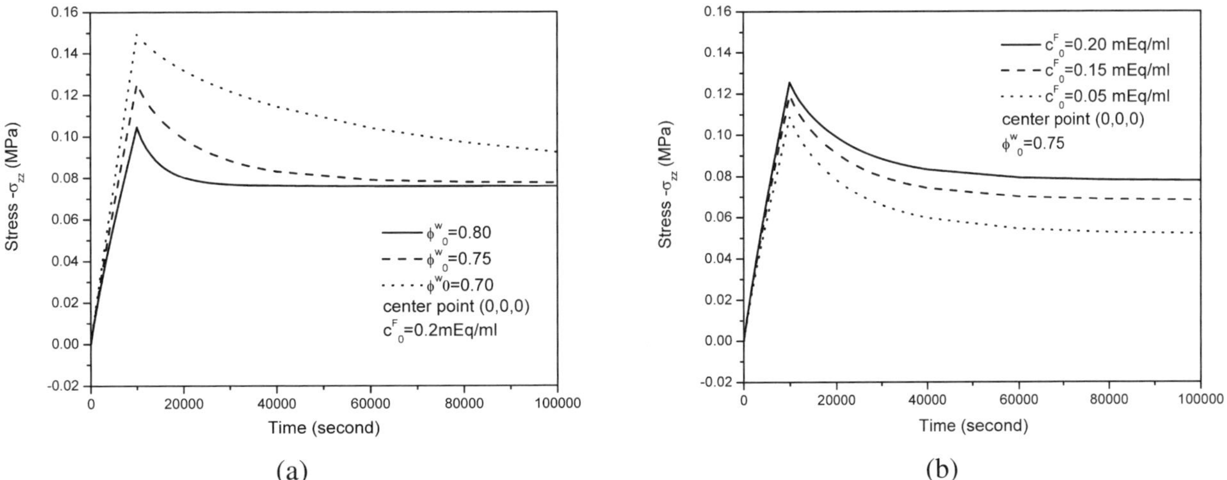

Fig. 7. Effects of (a) initial water content and (b) initial fixed charge density on the normal stress component $(-\sigma_{zz})$ at the center of the disc.

Interstitial fluid pressurization has long been hypothesized to play a fundamental role in the load-support mechanism of IVD. The results of this study showed that the fluid pressure within the disc, produced by the compression, could exceed 90% of the total stress, and the interstitial fluid shared more than 40% of the total load during the ramp compression. These results are consistent with the experimental findings obtained for articular cartilage by Soltz et al. [33,34].

The effect of mechanical loading on the solute transport was investigated in this study, which provided new insights into the mechanism of solute transport within the IVD. The results of this study show that the mechanical loading has little effect on small solute transport (e.g., glucose), but significantly affects large solute transport (e.g., growth factor). These results confirm that diffusion is the major mechanism for transport of small solutes in IVD, while convection plays a significant role in large solute transport.

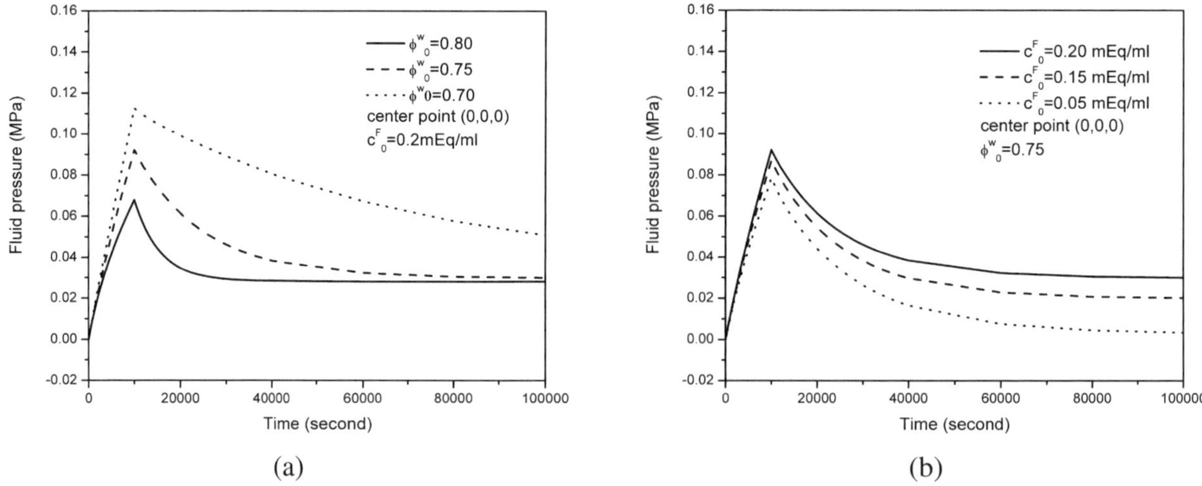

Fig. 8. Effects of (a) initial water content and (b) initial fixed charge density on the fluid pressure at the center of the disc.

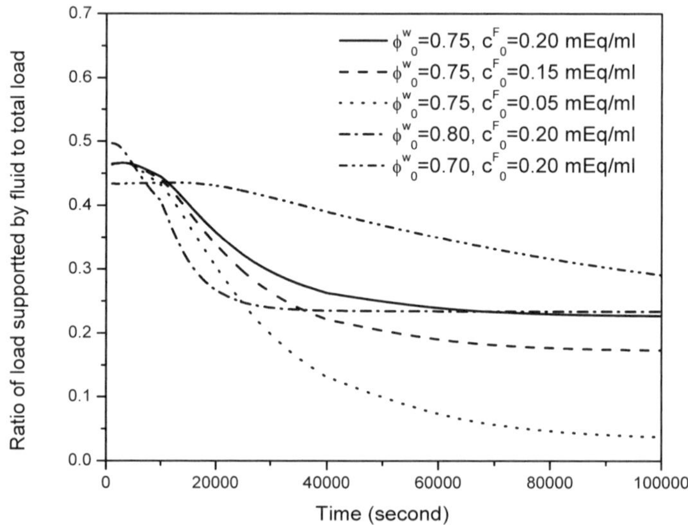

Fig. 9. Effects of the initial water content and the initial fixed charge density on the ratio of the load supported by the fluid (i.e., $\iint_{Z=0} p\phi^w \, d\Omega$) to that by the tissue (i.e., $\iint_{Z=0} -\sigma_{zz} \, d\Omega$).

This conclusion is consistent with the early findings from an *in vivo* animal model [39] and an *in vitro* experiment [16].

The mechanical interaction between the solute and the matrix may restrict solute transport in cartilaginous tissue for both diffusion and convection [6]. For diffusion, the intra-tissue diffusivities $D^\alpha$ in Eqs (9)–(11) were always smaller than their corresponding values in free solution $D_0^\alpha$, and its value is compression-dependent [20,27]. Recently, a new model for strain-dependent diffusivity of solutes in porous fibrous tissues was proposed [11]. However, we did not use this model in the present study simply because the strain is relatively small in our analysis. For convection, the restriction effect is considered by the convective coefficient $H^\alpha$ in Eqs (8)–(11). The values of $H^\alpha$ should be in the range of $(D^\alpha/D_0^\alpha) \leqslant H^\alpha \leqslant 1$ (a separate paper). Note that $H^\alpha$ might be significantly less than 1 for large

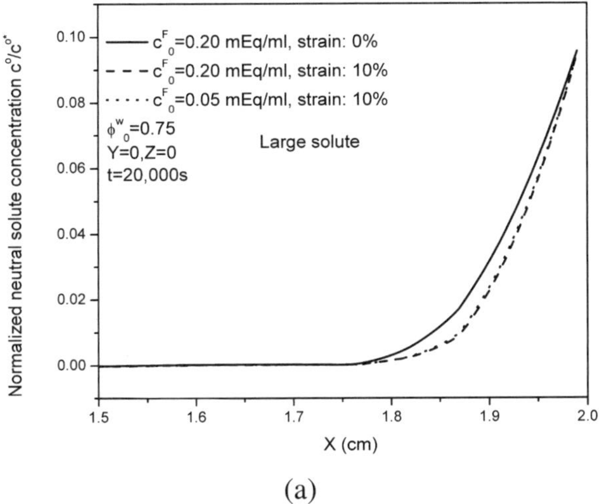


(a)

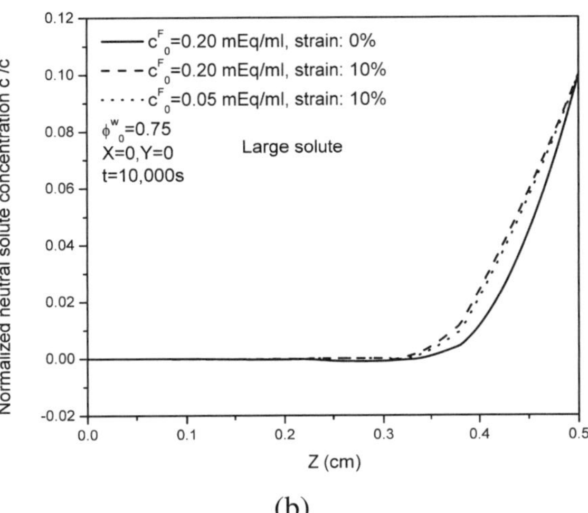


(b)

Fig. 10. Effect of the mechanical loading on the transport of large solute into the disc (a) in the radial direction and (b) in the axial direction. The concentration of large solute in the bathing solution was $c^{o*} = 4 \times 10^{-5}$ mol/m$^3$.

solutes, while its value is close to unity for small solutes. In this study, the values of $H^\alpha = 1$ were used for both small and large solutes, since there is no data available for IVD tissue. Consequently, the convective flux of large solute might be overestimated in our results. This is one of the limitations of this study.

The other limitations of this study include that the IVD was assumed to be isotropic, homogeneous mixture without chemical reaction. Theses assumptions do not reflect the real material behavior of AF and ignores the variation in material properties from region to region [3]. In the future, this finite element model needs to be improved to address these issues.

In summary, this is the first report of 3D finite element analysis of mechanical, chemical and electrical signals within the human disc under stress relaxation test. The effects of tissue porosity and FCD on physical signals and fluid transport were investigated parametrically. The transport of uncharged solute into IVD was also elucidated. The results of this study provide additional insights into load-supporting mechanisms and physical signals within the human disc under mechanical loading.

## Acknowledgements

This study was supported by a grant from the NIH (AR46860) and by General Research Support Award from the University of Miami.

## References


[1] T. Belytschko, R.F. Kulak, A.B. Schultz and J.O. Galante, Finite element stress analysis of an intervertebral disc, *J. Biomech.* **7** (1974), 277–285.
[2] W.M. Deen, Hindered transport of large molecules in liquid-filled pores, *AIChE Journal* **33** (1987), 1409–1425.
[3] D.M. Elliott and L.A. Setton, Anisotropic and inhomogeneous tensile behavior of the human anulus fibrosus: experimental measurement and material model predictions, *J. Biomech. Eng.* **123** (2001), 256–263.
[4] S.J. Ferguson, K. Ito and L.P. Nolte, Fluid flow and convective transport of solutes within the intervertebral disc, *J. Biomech.* **37** (2004), 213–221.

[5] A.J.H. Frijns, J.M. Huyghe and J.D. Janssen, A validation of the quadriphasic mixture theory for intervertebral disc tissue, *International Journal of Engineering Science* **35** (1997), 1419–1429.

[6] A.M. Garcia, E.H. Frank, P.E. Grimshaw and A.J. Grodzinsky, Contributions of fluid convection and electrical migration to transport in cartilage: relevance to loading, *Arch. Biochem. Biophys.* **333** (1996), 317–325.

[7] W.Y. Gu, W.M. Lai and V.C. Mow, A mixture theory for charged-hydrated soft tissues containing multi-electrolytes: passive transport and swelling behaviors, *Journal of Biomechanical Engineering* **120** (1998), 169–180.

[8] W.Y. Gu, W.M. Lai and V.C. Mow, Transport of fluid and ions through a porous-permeable charged-hydrated tissue, and streaming potential data on normal bovine articular cartilage, *J. Biomech.* **26** (1993), 709–723.

[9] W.Y. Gu and H. Yao, Effects of hydration and fixed charge density on fluid transport in charged hydrated soft tissue, *Annals of Biomedical Engineering* **31** (2003), 1162–1170.

[10] W.Y. Gu, H. Yao, C.-Y. Huang and H.S. Cheung, New insight into deformation-dependent hydraulic permeability of gels and cartilage, and dynamic behavior of agarose gels in confined compression, *J. Biomech.* **36** (2003), 593–598.

[11] W.Y. Gu, H. Yao, A.L. Vega and D. Flagler, Diffusivity of ions in agarose gels and intervertebral disc: Effect of porosity, *Annals of Biomedical Engineering* **32** (2004), 1710–1717.

[12] J.M. Huyghe and J.D. Janssen, Quadriphasic mechanics of swelling incompressible porous media, *Int. J. Engng. Sci.* **35** (1997), 793–802.

[13] J.C. Iatridis, J.P. Laible and M.H. Krag, Influence of fixed charge density magnitude and distribution on the intervertebral disc: applications of a poroelastic and chemical electric (PEACE) model, *J. Biomech. Eng.* **125** (2003), 12–24.

[14] M. Kanayama et al., A cineradiographic study on the lumbar disc deformation during flexion and extension of the trunk, *Clinical Biomech.* **10** (1995), 193–199.

[15] M. Kasra, A. Shirazi-Adl and G. Drouin, Dynamics of human lumbar intervertebral joints. Experimental and finite-element investigations, *Spine* **17** (1992), 93–102.

[16] M.M. Katz, A.R. Hargens and S.R. Garfin, Intervertebral disc nutrition. Diffusion versus convection, *Clin. Orthop.* (1986), 243–245.

[17] R.F. Kulak, T.B. Belyschko and A.B. Schultz, Nonlinear behavior of the human intervertebral disc under axial load, *J. Biomech.* **9** (1976), 377–386.

[18] W.M. Lai, J.S. Hou and V.C. Mow, A triphasic theory for the swelling and deformation behaviors of articular cartilage, *J. Biomech. Eng.* **113** (1991), 245–258.

[19] J.P. Laible, D.S. Pflaster, M.H. Krag, B.R. Simon and L.D. Haugh, A poroelastic-swelling finite element model with application to the intervertebral disc, *Spine* **18** (1993), 659–670.

[20] H.A. Leddy and F. Guilak, Site-specific molecular diffusion in articular cartilage measured using fluorescence recovery after photobleaching, *Annals of Biomedical Engineering* **31** (2003), 753–760.

[21] H.S. Lin, Y.K. Liu, G. Ray and P. Nikiavesh, Systems identification for material properties of the intervertebral joint, *J. Biomech.* **11** (1978), 1–14.

[22] Y.M. Lu, W.C. Hutton and V.M. Gharpuray, Do bending, twisting, and diurnal fluid changes in the disc affect the propensity to prolapse? A viscoelastic finite element model, *Spine* **21** (1996), 2570–2579.

[23] Y.M. Lu, W.C. Hutton and V.M. Gharpuray, The effect of fluid loss on the viscoelastic behavior of the lumbar intervertebral disc in compression, *J. Biomech. Eng.* **120** (1998), 48–54.

[24] G. Lyons, S.M. Eisenstein and M.B. Sweet, Biochemical changes in intervertebral disc degeneration, *Biochim. Biophys. Acta* **673** (1981), 443–453.

[25] J.S. Mackie and P. Meares, The diffusion of electrolytes in a cation-exchange resin. I. Theoretical, *Proc. Roy. Soc. London A* **232** (1955), 498–509.

[26] V.C. Mow, D.N. Sun, X.E. Guo, M. Likhitpanichkul and W.M. Lai, Fixed negative charges modulate mechanical behavior and electrical signals in articular cartilage under unconfined compression – a triphasic paradigm, in: *Porous Media: Theory, Experiments and Numerical Application*, W. Ehlers and J. Bluhm, eds, Springer, Berlin, 2002, pp. 227–247.

[27] T.M. Quinn, V. Morel and J.J. Meister, Static compression of articular cartilage can reduce solute diffusivity and partitioning: implications for the chondrocyte biological response, *J. Biomech.* **34** (2001), 1463–1469.

[28] S.A. Shirazi-Adl, A.M. Ahmed and S.C. Shrivastava, A finite element study of a lumbar motion segment subjected to pure sagittal plane moments, *J. Biomech.* **19** (1986), 331–350.

[29] S.A. Shirazi-Adl, S.C. Shrivastava and A.M. Ahmed, Stress analysis of the lumbar disc-body unit in compression: A three dimensional nonlinear finite element study, *Spine* **9** (1984), 120–134.

[30] B.R. Simon, J.P. Liable, D. Pflaster, Y. Yuan and M.H. Krag, A poroelastic finite element formulation including transport and swelling in soft tissue structures, *J. Biomech. Eng.* **118** (1996), 1–9.

[31] B.R. Simon, J.S.S. Wu, M.W. Carlton, J.H. Evans and L.E. Kazarian, Structural models for human spinal motion segments based on a poroelastic view of the intervertebral disk, *J. Biomech. Engng.* **107** (1985), 327–335.

[32] H. Snijders, M.R. Drost, P. Willems, J.M. Huyghe, P. van Dijk and J.D. Janssen, Triphasic material parameters of canine anulus fibrosus, in: *Computer Methods in Biomechanics and Biomedical Engineering*, J. Middleton, G.N. Pande and K.R. Williams, eds, Books and Journals International, Ltd., Swansea, UK, 1992, pp. 220–229.

[33] M.A. Soltz and G.A. Ateshian, Experimental verification and theoretical prediction of cartilage interstitial fluid pressurization at an impermeable contact interface in confined compression, *J. Biomech.* **31** (1998), 927–934.

[34] M.A. Soltz and G.A. Ateshian, Interstitial fluid pressurization during confined compression cyclical loading of articular cartilage, *Ann. Biomed. Eng.* **28** (2000), 150–159.

[35] R.L. Spilker, D.M. Daugirda and A.B. Schultz, Mechanical response of a simple finite element model of the intervertebral disc under complex loading, *J. Biomech.* **17** (1984), 103–112.

[36] R.L. Spilker, Mechanical behavior of a simple model of an intervertebral disk under compressive loading, *J. Biomech.* **13** (1980), 895–901.

[37] D.N. Sun, W.Y. Gu, X.E. Guo, W.M. Lai and V.C. Mow, A mixed finite element formulation of triphasic mechano-electrochemical theory for charged, hydrated biological soft tissues, *International Journal for Numerical Methods in Engineering* **45** (1999), 1375–1402.

[38] K. Ueno and Y.K. Liu, A three-dimensional nonlinear finite element model of lumbar intervertebral joint in torsion, *J. Biomech. Engng.* **109** (1987), 1987.

[39] J.P. Urban, S. Holm and A. Maroudas, Diffusion of small solutes into the intervertebral disc: as in vivo study, *Biorheology* **15** (1978), 203–221.

[40] J.P. Urban, The role of the physicochemical environment in determining idsc cell behaviour, *Biochem. Soc. Trans.* **30** (2002), 858–864.

[41] C.M. Whyne, S.S. Hu and J.C. Lotz, Parametric finite element analysis of vertebral bodies affected by tumors, *J. Biomech.* **34** (2001), 1317–1324.

[42] H. Yao and W.Y. Gu, Physical signals and solute transport in cartilage under dynamic unconfined compression: finite element analysis, *Annals of Biomed. Engng.* **32** (2004), 380–390.

Biorheology 43 (2006) 337–345
IOS Press



# Contraction of collagen gels seeded with tendon cells

M.-Y. Chen, L. Jeng, Y.-L. Sun, C.-F. Zhao, M.E. Zobitz, S.L. Moran, P.C. Amadio and
K.-N. An[*]
*Biomechanics Laboratory, Division of Orthopedic Research, Mayo Clinic College of Medicine,
200 First Street S.W., Rochester, MN, USA*

**Abstract.** Knowledge of the adaptation of the soft tissue to mechanical factors and biomolecules would be essential to better understand the mechanism of tendon injury and to improve the outcome of tendon repair. The responses to these factors could be different for the distinct types of cells in the tendon: cells from the tendon sheath, fibroblasts from the epitenon surface, or fibroblasts from the internal endotenon. In this study, we examined the mechanical and histological characteristics of the rate of contraction of the collagen gel seeded with epitenon and endotenon fibroblasts. The rate of contraction and the mechanical property of the contracted construct depend on the gel concentration and also the treatment of TGF-$\beta_1$.

Keywords: Gel concentration, TGF-$\beta$, fibroblast, tendon, cell culture, biomechanics

## 1. Introduction

Fibroblasts have been grown and observed in hydrated collagen lattices by Ehrmann and Gey [11] and Elsdale and Bard [12]. However, it wasn't until 1979 that Bell et al. first reported the lattice contraction and formation of tissue-like structure as a consequence of the seeded-cell contraction [4]. Varying the protein content and cell concentration could regulate the rate and extent of lattice contraction. Subsequently, the model of cell seeded collage gels has been studied extensively to better understand the mechanism of contraction. The effects of mechanical stimulation and growth factors on the contraction and the mechanical properties of the contracted construct were also examined.

Tendons are fibrous connective tissues that join muscle to bone. They are subjected to tension when transmitting forces from muscles to bones during contraction and to shear stress as a result of gliding over other tissues. Tendons are composed of three distinct types of cells: cells from the tendon sheath, fibroblast from the epitenon surface, and fibroblasts from the internal endotenon. Each cell type has unique properties and behavior [3,6,17]. After tendon wounding, fibrosis and adhesions may develop, which prevents the tissue from healing to its original quality [17]. Fibroblasts are heavily involved in the wound healing process. They proliferate, reorganize structural proteins, such as collagen, synthesize and degrade the extracellular matrix, and are important in wound contraction [19]. The gel contraction process is similar to the wound contraction that occurs during tendon wound healing *in vivo*, and the

---

[*]Address for correspondence: Kai-Nan An, PhD, Mayo Clinic College of Medicine, Biomechanics Laboratory, Division of Orthopedic Research, 200 First Street SW, Rochester, MN, USA. Tel.: +1 507 538 1717; Fax: +1 507 284 5392; E-mail: an@mayo.edu.

use of a collagen gel provides a convenient method for quantifying the contractile ability of the cells *in vitro* [4,9,16,20].

Injured and surgically repaired tendons heal with the formation of scar tissue. Scar tissue represents one of the most unpredictable factors contributing to postoperative morbidity. The main cell involved in scar formation is the fibroblast. Khan et al. [17] examined the relative activity of fibroblasts from the fibro-osseous sheath and the endotenon with respect to proliferation and the ability to contract a collagen lattice. The fibroblasts derived from the fibro-osseous sheath were more active in both these respects. In addition, the amount of matrix metalloproteinase activity was found to be greater for the fibro-osseous sheath fibroblasts, implying a greater capacity to degrade and disorganize connective tissue and thus migrate.

In order to improve tendon healing, it is necessary to study the healing process and wound contraction at the cellular level. Our ultimate goal is to further understand the specific similarity and difference in behavior of these tendon fibroblasts. Knowledge of the biochemical and biomechanical properties of tendons and understanding of the wound healing process are important in order to improve the tendon healing process and to develop engineered tendon graft to replace damaged tissue in the body. In this study, we seeded epitenon and endotenon fibroblasts in collagen gels and allowed the gels to contract around an inner ring. The rate of gel contraction and the material properties of the fibroblast gel contracted constructs were evaluated.

## 2. Materials and methods

### 2.1. Harvesting and culturing of tendon cells

The flexor digitorum profundus tendon of a dog was dissected and cells were harvested as previously described [2]. The tendon was tied into a loop and suspended in a sterile solution of 0.25% trypsin in 20 mM HEPES buffer solution for 20 minutes at $37°C$ in a 5% $CO_2$ humidified incubator to harvest the epitenon cells. The cut ends of the tendon were kept above the trypsin in order to avoid collecting endotenon cells with the epitenon cells. The supernatant was collected and centrifuged at $1132 \times g$ for 5 minutes to remove the trypsin solution, and the remaining epitenon cells were plated in 100 mm culture dishes. The entire tendon was then placed in a sterile solution of 0.5% collagenase in 20 mM HEPES buffer solution (Invitrogen Corporation, Grand Island, NY, USA) for 20 minutes at room temperature to harvest the endotenon cells. The supernatant was collected and centrifuged at $1132 \times g$ for 5 minutes to remove the collagenase solution, and the remaining endotenon cells were plated in 100 mm culture dishes.

The cells were cultured in growth medium consisting of minimum essential medium, 10% (v/v) fetal bovine serum (FBS), penicillin (100 units/ml)/streptomycin (100 $\mu$g/ml), and amphotericin B (2.5 $\mu$g/ml). The cells were incubated at $37°C$ in a 5% $CO_2$ humidified incubator, and media was changed every other day. At confluency, the cells were passaged by washing with $1\times$ sterile phosphate-buffered saline (PBS) and incubating in trypsin for 5 minutes. Cells were used between passages 2 and 4.

### 2.2. Collagen gel ring formation

Vitrogen bovine dermal collagen (Cohesion Technologies, Palo Alto, CA, USA) were prepared following the company's instructions. Briefly, 10 ml of sterile, chilled Vitrogen collagen were mixed with 3 ml of sterile $5\times$ MEM and 1.05 ml of sterile 0.1 67 M NaOH and 0.95 ml distilled $H_2O$ to adjust pH to $7.4 \pm 0.2$, to make 15 ml temporary collagen/MEM solution on ice, and the resulting collagen concentration was 2.0 mg/ml.Then, the collagen/MEM solution was diluted in 2.0 mg/ml with $1 \times$ MEM to make final collagen concentrations of 0.5, 1.0, 1.5, and 2.0 mg/ml. The solution in various collagen concentrations were stored at 4–6°C for no longer than 1 hour until use.

To prepare the wells, the bottoms of sterile small cloning rings, outer diameter of 8 mm and height of 8 mm, were coated with sterile vacuum grease. Each of the rings was affixed to the center of a well in a six-well culture plate.

Confluent plates of endotenon cells were washed with sterile PBS and trypsinized. The cells were counted with a hemacytometer and centrifuged to remove the media and leave behind a cell pellet with a known number of cells. The cells were seeded in the Vitrogen collagen at an initial density of $1 \times 10^6$ cells/ml. Then, 2 ml aliquots of the cell-seeded collagen solution were added to each well of the six-well plates in the area enclosed by the well and the inner cloning ring. The collagen rings were incubated at 37°C in a 5% $CO_2$ humidified incubator for 2 hours to allow for gelation. Additional collagen rings were also seeded using epitenon cells.

To prepare the supplemented media, recombinant human TGF-$\beta_1$ (R&D Systems, Minneapolis, MN, USA) was divided into concentrations of 2.0, 5.0, and 10.0 ng/ml in media under sterile conditions. Following gelation, 5 ml of the supplemented media was added to each of the collagen gels. A negative control of 5 ml of unsupplemented media was also tested. The gels were incubated at 37°C in a 5% $CO_2$ humidified incubator for 14 days, and media was changed every other day. The fibroblasts were regularly examined using a light microscope to ensure cell viability.

### 2.3. Quantification of gel contraction

The progress of the gel contraction was monitored by measuring the surface area of the gels at days 0, 2, 4, 6, 8, 10, 12, and 14 after treatment. Fibroblast-seeded gels were placed on a transparent plastic table with a ruler and photographed from below using a digital camera. The digital images were analyzed and the surface areas of the gels were calculated using Scion Image Beta 4.02 (Scion Corporation, Frederick, MD, USA). The surface areas of the gels were expressed as the averages of six replicated experiments ($n = 6$) plus or minus the standard deviations. Statistical significance was determined using a t-test with a significance level of $p < 0.005$.

### 2.4. Mechanical testing

At the end of the 14-day contraction period, the collagen rings underwent mechanical testing. Stepwise stretch-relaxation tests were carried out on each of the rings using a custom-built micro-tester system to determine the material properties of the rings. The collagen rings were removed from the wells and carefully set up on two 0.6 mm glass posts. The ramp test was carried out at a velocity of 0.05 mm/s until the ring was stretched to a length of 11.4 mm, a length at which the ring fit snugly around the hooks. The force output by the computer at the end of the ramp test was recorded as the preload force. The stepwise stretch-relaxation test to failure was then carried out with the preload force at a velocity of 5 mm/s with step sizes of 0.5 mm and a holding time of 30 seconds for each step.

## 3. Results

The endotenon cells were observed to contract collagen rings over a 14-day period, with the rate of contraction depending on the gel concentration (Fig. 1). At low gel concentrations, the contraction was

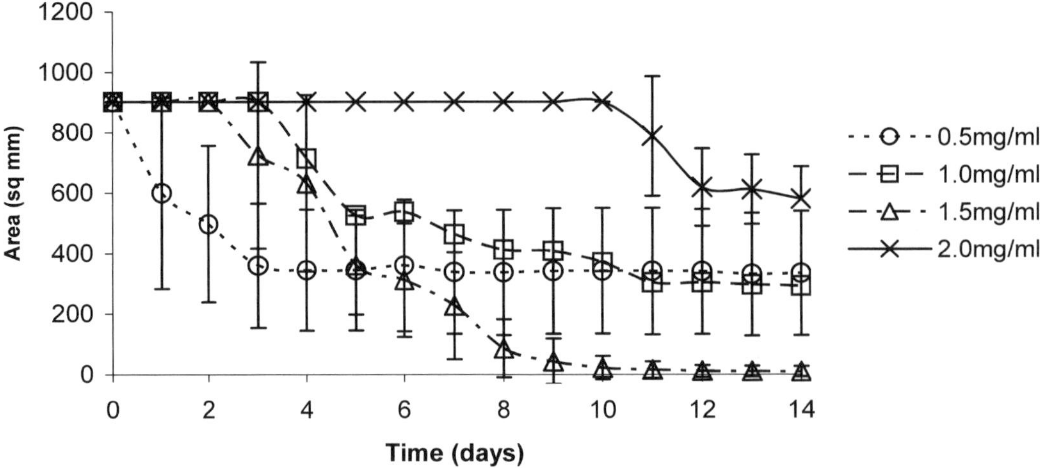



Fig. 1. The rate of contraction (area changes) as function of the gel concentration.

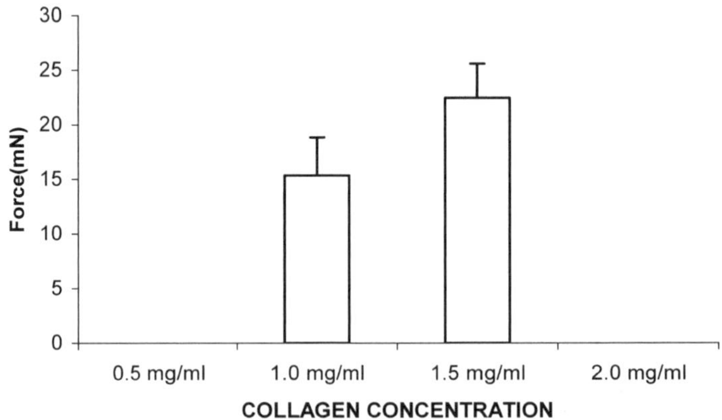



Fig. 2. Failure load of the construct with various gel concentrations.

either not completed (0.5 mg/ml) or was variable (1.0 mg/ml) within the period of observation. On the other hand, at a high gel concentration (2.0 mg/ml), the gel contraction was more consistent but at a slower rate. In this study a gel concentration of 1.5 mg/ml seemed to be optimal for gel contraction. At 10 days all the gels completely contracted to the inner ring.

For the gel constructs that completely contracted, mechanical testing was performed. The failure load was higher for a gel concentration of 1.5 mg/ml compared to 1.0 mg/ml (Fig. 2). However, the stiffness for both concentrations was similar (Fig. 3). Figure 4 shows the extent of contraction of gels treated with TGF-$\beta_1$ and untreated control gels. The surface area of the gels treated with TGF-$\beta_1$ was significantly smaller than the untreated gels on days 6, 8, 12, and 14 ($p < 0.005$ using a $t$-test), indicating that TGF-$\beta_1$ increased the rate of gel contraction in comparison to the untreated controls. However, there was not a dose-dependent effect of TGF-$\beta_1$ on the rate of endotenon-seeded collagen gel contraction. The failure loads (Fig. 5) and the stiffness (Fig. 6) of the constructs treated with TGF-$\beta_1$ are higher than the control.

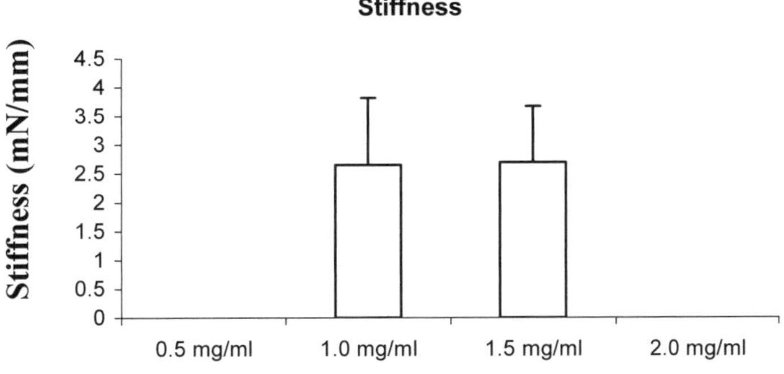


Fig. 3. The stiffness of the construct as function of the gel concentration.

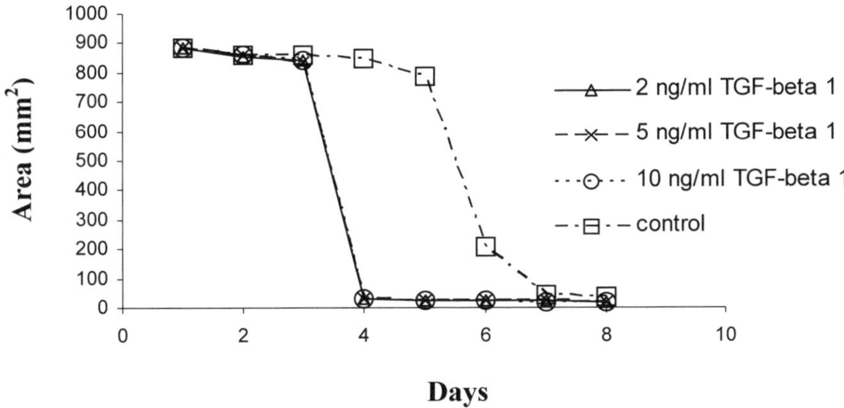


Fig. 4. The rate of contraction (area changes) by TGF-$\beta_1$ concentration with each point representing the average of six replicated trials.

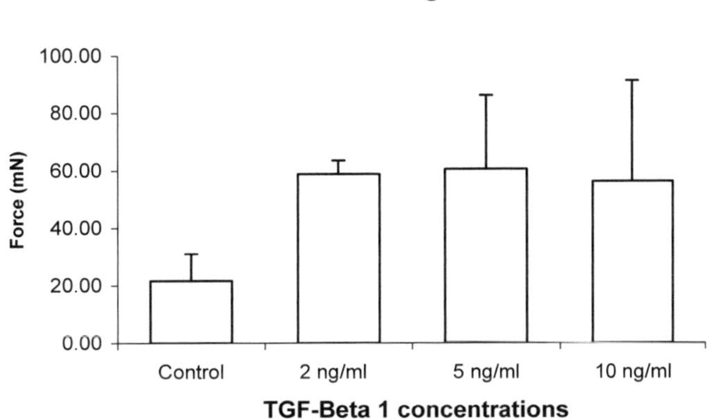


Fig. 5. Failure strength of the constructs with contracted gel at different concentration of TGF-$\beta_1$.

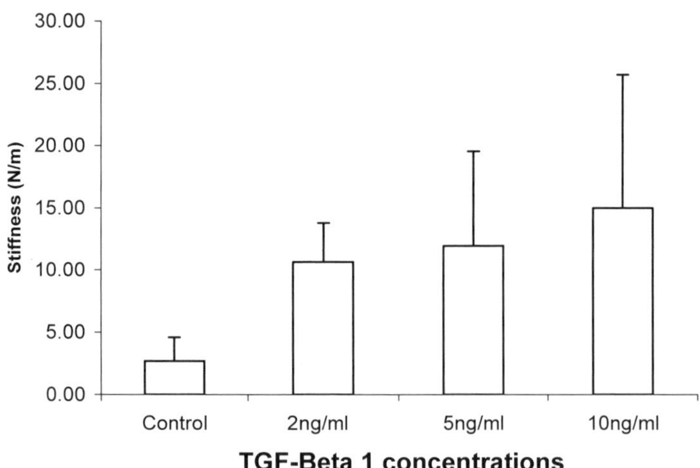


Fig. 6. Stiffness of the constructs with contracted gel at different concentration of TGF-$\beta_1$.

## 4. Discussion

Previous studies have reported a phenomenon of contraction of fibroblast-seeded collagen gels *in vitro* [4,8–10,13,17]. Fibroblasts are capable of synthesizing and degrading collagen and when they are cultivated in collagen gels, they cause the gel to reorganize [4,9,13,17]. Specifically, it is believed that the cells attach to the collagen fibers, elongate, and then contract the gels [9,13].

Two possible mechanisms had been considered in association with the tissue contraction [15]. One was related to the myofibroblast that caused contraction by making contact through gap junctions and then the aggregate could function like a muscle. The other mechanism was the consequence of migratory cells moving through a scaffold or on top of the substrate which applied traction through adhesion between cells and the extracellular matrix.

The gel contraction would involve the degradation of the matrix as well as collagen synthesis. Fibroblast migration is an integral component of the processes resulting in the formation of restrictive adhesions in the injured tendon. Prerequisites for cell migration are an intact cytoskeleton and the ability to biochemically degrading the extra-cellular matrix. Ragoowansi et al. [21] examined the relative characteristics of fibroblasts from the fibro-osseus sheath and the endotenon with respect to morphology, cytoskeletal structure and ability to produce matrix metalloproteinases (MMPs) 2 and 9. They found that fibro-osseus sheath cells were larger in size and demonstrated greater amounts of intra-cellular alpha-smooth muscle actin (alpha-SMA) and intra-membranous vinculin. Filamentous actin (F-actin) fibres in fibro-osseus sheath cells were more densely packed and concentrated, resulting in stress fibres. The fibro-osseus sheath cells also produce greater amounts of MMP-2 and MMP-9.

Collagen synthesis was detected as early as the second day in an *in vitro* fibrin gel culture model of human dermal fibroblasts [22]. Type I collagen was the major collagen synthesized by fibroblasts with small amounts of type V and type III collagen. Although collagen was biochemically detected early, the organized collagen fibrils were only in the later stage of cultures in transmission electron micrographs. Measurement of the rate of collagen synthesis within collagen-based systems is problematic. Methods which assess total collagen are not appropriate due to the pre-existing collagen. Lee et al. [18] (1998) developed a novel method using sirius red F3B precipitation for the assessment of collagen synthesis by cells in collagen-based materials. Using this method, collagen synthesis by dermal fibroblasts was measured for up to 6 days. It was found that collagen synthesis started at day one and then were down-regulated with time.

The rate of contraction is influenced by such factors as cell density, protein content, and concentration of matrix metalloproteinase, growth factors, and other biomolecules [4,8,10]. Bell et al. [4] first reported that fibroblast incorporated in a collagen lattice induced a progressive contraction. The rate of contraction of collagen lattices can be regulated by the number of cells embedded in the gel and inversely related to the protein concentration.

Montesano and Orci [20] first noted that TGF-$\beta$ dramatically increased the rate of contraction of the collagen gels by the fibroblasts. They speculated the mechanism by which TGF-$\beta$ stimulated fibroblasts to contact collagen gel was either increased force generated by the fibroblast contractile machinery, or improved transmission of the force to the matrix. In addition to $\alpha$-smooth muscle actin expression, Vaughan et al. [23] demonstrated that TGF-$\beta_1$ enhanced the formation of the structural elements important in myofibroblast contractile force generation and transmission, including stress fibers, vinculin-containing fibronexus adhesion complexes, and fibronectin fibrils.

Brown et al. [7] measured the effects of TGF-$\beta_1$ and TGF-$\beta_3$ on contractile forces generated by human dermal fibroblasts using culture force monitor instrument. Maximal stimulation was between 7.5 and 15 ng/ml of TGF-$\beta_1$, and higher doses were inhibitory. They suggested that in cytomechanical responses the initial generation of force is due to fibroblast traction on the matrix attachment and cell motility, while later stages are dominated by active contraction of cells on the collagen structures with the appearance of stress fibers. TGF-$\beta$ can stimulate force output from the cytoskeletal motor of actinomyosin elements as well as enhance the expression of appropriate integrins which allows greater transmission of force to the matrix. It was therefore concluded that TGF-$\beta_1$ and TGF-$\beta_3$ behave primarily as mechanoregulatory growth factors and that stimulation of integrin expression may be a consequence of the altered cell stress.

The induction of the $\alpha$-smooth muscle actin ($\alpha$-SMA) by the transforming growth factor-$\beta_1$ (TGF-$\beta_1$), and the effect of this process by the compliance of collagen substrates were studied [1]. Three different types of collagen gels with wide variations of mechanical compliance were used for the cell culture. These included that cells plated on thin films of collagen-coated plastic representing minimal compliance and maximal intracellular tension; cells in anchored collagen gels representing moderate compliance and tension; and cells in floating collagen gels representing high compliance and low tension. It was found that TGF-$\beta_1$-induced increases of $\alpha$-SMA contents are dependent on the resistance of the substrate to deformation. Furthermore, as the intracellular tension on collagen substrates was reduced by the blocking antibodies to the $\alpha_2$ and $\beta_1$ integrin subunits, TGF-$\beta_1$ failed to increase $\alpha$-SMA protein content in all three types of collagen matrices.

Grinnel and Ho [14] further noted that the TGF-$\beta$ stimulates fibroblast contraction of collagen matrix by a different mechanism depending on the mechanical loading condition of the cells in the matrix. Under mechanically unloaded condition, the TGF-$\beta$ stimulated the contraction directly as an agonist and indirectly by differentiating the cell to myofibroblast phenotype. On the other hand, under mechanical loaded condition, fibroblasts preactivated to become myofibroblasts to transfer tension to the matrices.

The effects of external uniaxial cyclic strain on oriented fibroblast-seeded collagen gels were performed [5]. In that study, the constructs were preloaded to either 2 or 10 mN and then subsequently subjected to a further 10% cyclic strain at 1 Hz. They found that cell proliferation was enhanced; collagen synthesis was enhanced by cyclic strain within constructs preloaded at 2 mN only. The constructs that were preloaded at 10 mN and subjected to cyclic strain expressed enhanced levels of staining for latent MMP-1, latent MMP-9 and active MMP-2. Interestingly, the structural stiffness of constructs preloaded at 2 mN and subjected to cyclic strain was enhanced compared with control specimens, reflecting the increase in collagen synthesis. By contrast, the initial failure loads for cyclically strained constructs preloaded at 10 mN were reduced, potentially because of enhanced catabolic activity. There seems to be synergy of mechanical loading and the growth factor stimulation.

The product of the gel contraction is a tissue-like structure that can undergo further testing, such as analysis of mechanical strength and cellular metabolic processes [4,8,9]. Previous studies have also tested the mechanical strength of fibroblast-seeded collagen constructs. Recent evidence found that the tissue-like products have a lower stiffness than that of native tissue [8]. Cacou et al. [8] tested human neonatal dermal fibroblasts in the collagen ring system and observed dynamic structural stiffnesses between 0.2 and 22 mN/mm and average failure loads between 70 and 90 mN. Chapuis et al. [9] tested human fibroblasts in the collagen ring system and observed stiffness moduli of 0.06 to 1 MPa and failure loads between 150 and 450 mN.

It is known that the ability of the tendon to heal after injury is limited, and the quality of the healing tendon remains inferior to that of native, healthy tendon [16]. Fibroblasts play a key role in the wound healing process. These cells have been shown to contract collagen gel lattices *in vitro*, a phenomenon similar to wound contraction *in vivo*, but the mechanisms behind the fibroblast contraction of the collagen gels need to be further understood. The fibroblast-seeded collagen ring can be used as an effective model to determine the effects of various growth factors on wound contraction *in vitro* and to examine the material properties of the resultant contracted structures. Knowledge of the biochemical and biomechanical properties of tendons and understanding of the wound healing process are important in order to improve the tendon healing process and to develop engineered tendon grafts to replace damaged tissue in the body.

# References

[1] P.D. Arora, N. Narani and C.A.G. McCulloch, The compliance of collagen gels regulates transforming growth factor-$\beta$ induction of $\alpha$-smooth muscle actin in fibroblasts, *Am. J. Path.* **154** (1999), 871–882.

[2] A.J. Banes, K. Donlon, G.W. Link, Y. Gillespie, A.G. Bevin, H.D. Peterson, D. Bynum, S. Watts and L. Dahners, Cell populations of tendon: A simplified method for isolation of synovial cells and internal fibroblasts: Confirmation of origin and biologic properties, *J. Orthop. Res.* **6**(1) (1988), 83–94.

[3] A.J. Banes, G. Horesovsky, C. Larson, M. Tsuzaki, S. Judex, J. Archambault, R. Zernicke, W. Herzog, S. Kelley and L. Miller, Mechanical load stimulates expression of novel genes *in vivo* and *in vitro* in avian flexor tendon cells, *Osteoarthritis Cartilage* **7**(1) (1999), 141–153.

[4] E. Bell, B. Ivarsson and C. Merrill, Production of a tissue-like structure by contraction of collagen lattices by human fibroblasts of different proliferative potential *in vitro*, *Proc. Natl. Acad. Sci. USA* **76**(3) (1979), 1274–1278.

[5] C.C. Berry, J.C. Shelton, D.L. Bader and D.A. Lee, Influence of external uniaxial cyclic strain on oriented fibroblast-seeded collagen gels, *Tissue Eng.* **9**(4) (2003), 613–624.

[6] B.E. Brigman, P. Hu, H. Yin, M. Tsuzaki, W.T. Lawrence and A.J. Banes, Fibronectin in the tendon-synovial complex: Quantitation *in vivo* and *in vitro* by ELISA and relative mRNA levels by polymerase chain reaction and northern blot, *J. Orthop. Res.* **12**(2) (1994), 253–261.

[7] R.A. Brown, K.K. Sethi, I. Gwanmesia, D. Raemdonck, M. Eastwood and V. Mudera, Enhanced fibroblast contraction of 3D collagen lattices and integrin expression by TGF-$\beta_1$ and -$\beta_3$: mechanoregulatory growth factors?, *Exp. Cell Res.* **274**(2) (2002), 310–322.

[8] C. Cacou, D. Palmer, D.A. Lee, K.L. Bader and J.C. Shelton, A system for monitoring the response of uniaxial strain on cell seeded collagen gels, *Med. Eng. Phys.* **22**(5) (2000), 327–333.

[9] J.F. Chapuis and P. Agache, A new technique to study the mechanical properties of collagen lattices, *J. Biomech.* **25**(1) (1992), 115–120.

[10] M. Eastwood, R. Porter, U. Khan, G. McGrouther and R. Brown, Quantitative analysis of collagen gel contractile forces generated by dermal fibroblasts and the relationship to cell morphology, *J. Cell. Physiol.* **166**(1) (1996), 33–42.

[11] R.L. Ehrmann and G.O. Gey, The use of cell colonies on glass for evaluating nutrition and growth in roller-tube cultures, *J. Natl. Cancer Inst.* **13**(5) (1953), 1099–1121.

[12] T. Elsdale and J. Bard, Collagen substrata for studies on cell behavior, *J. Cell Biol.* **54**(3) (1972), 626–637.

[13] Z. Feng, T. Matsumoto and T. Nakamura, Measurements of the mechanical properties of contracted collagen gels populated with rat fibroblasts or cardiomyocytes, *J. Artificial Organs* **6**(3) (2003), 192–196.

[14] F. Grinnell and C.H. Ho, Transforming growth factor beta stimulates fibroblast–collagen matrix contraction by different mechanisms in mechanically loaded and unloaded matrices, *Exp. Cell Res.* **273**(2) (2002), 248–255.

[15] I.M. Khouw, P.B. van Wachem, J.A. Plantinga, Z. Vujaskovic, J.J. Wissink, LF. De Leiu and M.J. van Luyn, TGF-$\beta$ and bFGF affect the differentiation of proliferating porcine fibroblasts into myofibroblasts in vitro, *Biomat.* **20** (1999), 1815–1822.

[16] U. Khan, N.L. Occleston, P.T. Khaw and D.A. McGrouther, Single exposures to 5-fluorouracil: A possible mode of targeted therapy to reduce contractile scarring in the injured tendon, *Plas. Reconstr. Surg.* **99**(2) (1997), 465–471.

[17] U. Khan, N.L. Occleston, P.T. Khaw and D.A. McGrouther, Differences in proliferative rate and collagen lattice contraction between endotenon and synovial fibroblasts, *J. Hand Surg.* **23**(2) (1998), 266–273.

[18] D.A. Lee, E.E. Assoku and V. Doyle, A specific quantitative assay for collagen synthesis by cells seeded in collagen-based biomaterials using sirius red F3B precipitation, *J. Mat. Sci.* **9** (1998), 47–51.

[19] T. Molloy, Y. Wang and G.A.C. Murrell, The roles of growth factors in tendon and ligament healing, *Sports Med.* **33**(5) (2003), 381–394.

[20] R. Montesano and L. Orci, Transforming growth factor $\beta$ stimulates collagen–matrix contraction by fibroblasts: Implications for wound healing, *Proc. Natl. Acad. Sci. USA* **85** (1988), 4894–4897.

[21] R. Ragoowansi, U. Khan, R.A. Brown and D.A. McGrouther, Differences in morphology, cytoskeletal architecture and protease production between zone II tendon and synovial fibroblasts in vitro, *J. Hand Surg. (Br.)* **28**(5) (2003), 465–470.

[22] T.L. Tuan, A. Song, S. Chang, S. Younai and M.E. Nimni, In vitro fibroplasia: matrix contraction, cell growth, and collagen production of fibroblasts cultured in fibrin gels, *Exp. Cell Res.* **223**(1) (1996), 127–134.

[23] M.B. Vaughan, E.W. Howard and J.J. Tomasek, Transforming growth factor-beta1 promotes the morphological and functional differentiation of the myofibroblast, *Exp. Cell Res.* **257**(1) (2000), 180–189.

Biorheology 43 (2006) 347–354
IOS Press



# Beneficial actions of hyaluronan (HA) on arthritic joints: Effects of molecular weight of HA on elasticity of cartilage matrix

Yukio Kato [a,*], Shigeo Nakamura [b] and Masahiro Nishimura [c]

[a] *Department of Dental and Medical Biochemistry, Graduate School of Biomedical Sciences, Hiroshima University, Hiroshima 734-8553, Japan*
[b] *Department of Periodontal Medicine, Graduate School of Biomedical Sciences, Hiroshima University, Hiroshima 734-8553, Japan*
[c] *Department of Prosthetic Dentistry, Graduate School of Biomedical Sciences, Hiroshima University, Hiroshima 734-8553, Japan*

**Abstract.** Hyaluronan (HA) has viscoelastic, anti-inflammatory and protective actions in joint tissues, and is being widely used for treatment of OA and RA patients. However, the mechanisms underlying the pharmacological action of HA on OA and RA have not been fully understood. In this article, we review the molecular weight-dependent, anti-inflammatory actions of HA preparations – produced in Japan – in joint tissues, and show that the molecular weight of HA, but not its concentration, is crucial for maintenance of cartilage elasticity.

Keywords: Hyaluronan, aggrecan, cartilage, OA, RA, elasticity

## 1. Introduction

Hyaluronan/hyaluronic acid (HA) – the longest glycosaminoglycan – is an unbranched glycosamino-glycan composed of repeating disaccharide units of [D-glucuronic acid-N-acetyl-D-glucosamine]. It is widely distributed from some microorganisms to all animals, being conserved throughout the evolution, perhaps because HA is required for the protection of cells/tissues. Aggregates of HA and proteogly-cans or HA alone form viscoelastic matrices. In addition, HA has anti-inflammatory actions in various tissues, and these actions are partly mediated by activation of HA receptor CD44 and Rhamm. In this article, we review studies on the benefits of high molecular weight HA preparations, produced in Japan for arthritic joints, and our recent studies on the effect of the molecular weight of HA on elasticity of cartilage matrix.

---

*Address for correspondence: Yukio Kato, Department of Dental and Medical Biochemistry, Graduate School of Biomedical Sciences, Hiroshima University, Hiroshima 734-8553, Japan. Tel.: +81 82 257 5628; Fax: +81 82 257 5629; E-mail: ykato@hiroshima-u.ac.jp.

## 2. Beneficial actions of high molecular weight HA on arthritic joints

### 2.1. Prehistory of high molecular weight HA preparations

In Japan, progress in biotechnology has led to a method for producing natural-type relatively-high-molecular-weight HA preparations, using microorganism fermentation technology, to allow clinical applications of the preparations. The molecular weight of the HA preparations, measured using Mark–Houwink parameters (intrinsic viscosity) or MALLS (multi-angle laser light scattering, weight average), was $>1.9 \times 10^6$ Da or $>3.3 \times 10^6$ Da[#], respectively. The former method has been used widely to determine the molecular weight of HA, but the latter[#] is more accurate.

In early 1990, basic pharmacology studies were conducted on an HA preparation (Suvenyl, Chugai Pharmaceutical/Denki Kagaku Kogyo, Tokyo, Japan) with a molecular weight of $1.9 \times 10^6$ Da / $3.3 \times 10^6$ Da[#], purified from the *Streptococcus equi* culture supernatant. The studies demonstrated that HA preparations with a higher molecular weight were more effective in suppressing pain (including arthralgia), inflammation and cartilage degeneration. This led to high hopes for the potential of HA in the treatment of rheumatoid arthritis (RA), along with osteoarthritis (OA) of the knee and periarthritis scapulohumeralis (PS).

In 1991, clinical trials with Suvenyl were initiated in patients with OA of the knee and PS [13,14], and Suvenyl proved to be effective for the treatment of RA as well [12]. In 2000, Suvenyl was first approved for the treatment of RA in Japan, in addition to OA of the knee and PS.

Table 1 summarizes the history of studies on the beneficial actions of HA on arthritis – including the prehistory of Suvenyl. Commercially available HA preparations are listed in Table 2.

### 2.2. Viscoelastic properties and joint-lubricating effect of HA

HA appears to exert its beneficial effects on knee pain in OA and RA by acting as a lubricant and shock absorber for joints, thus protecting the superficial layer of cartilage, based on HA's viscoelastic properties. Oka et al. reported that HA of $1.9 \times 10^6$ Da maintains a liquid film longer than does HA of $0.8 \times 10^6$ Da [7], which indicates that relatively heavier HA provides better conditions for liquid film formation. The researchers also reported that, in OA and RA synovial fluids, the reduction in the coefficient of friction by $1.9 \times 10^6$ Da HA is greater than that by $0.8 \times 10^6$ Da HA [8]. The heavy HA is therefore superior to the relatively-low-molecular-weight HA in terms of both fluid and boundary lubrications. HA binds to chondroitin sulfate glycosaminoglycans (GAG) via sugar-sugar interactions and

Table 1

History of the HA research and the development of Suvenyl

| | |
|---|---|
| 1934: | Karl Meyer discovered HA from bovine vitreous body. |
| 1960: | HA with a molecular weight of 150–200 kDa containing about 10% protein was first clinically used in Austria. |
| 1967: | E.A. Balazs reported that molecular weight of HA in human joint fluids exceed 2000 kDa. |
| 1970: | N.W. Rydell reported the efficacy of HA on the clinical symptoms of arthritis in track horses. |
| 1978: | J.G. Peyron first reported the efficacy of HA on human osteoarthritis (OA). |
| 1987: | HAs with a molecular weight of approximately 800 kDa were marketed in Italy and Japan. |
| 1991: | Clinical trials of Suvenyl were initiated in patients with OA, periarthritis scapulohumeralis (PS) and rheumatoid arthritis (RA) in Japan. |
| 1998: | Cross-linked HA named Synvisc were marketed in U.S. |
| 2000: | Suvenyl was first approved for knee RA in addition to knee OA and PS in Japan. |

Table 2
ACR Subcommittee on Osteoarthritis Guideline 2000

| Commercial name | Molecular weight (kDa) | Indication |
| --- | --- | --- |
| Artz (1987, Japan) | 600–1200 | OA, PS |
| Hyalgan (1998, USA) | 500–730 | OA |
| Synvisc (1998, USA) | Cross-linked | OA |
| Suvenyl (2000, Japan) | 1900–2500 (intrinsic viscosity[*]) | OA, PS, RA |
| | 2700–3700 (weight average[**]) | |

[*]MW determined using Mark–Houwink parameter (intrinsic viscosity).
[**]MW determined using MALLS (multi-angle laser light scattering) (weight average).
Ref.: T. Yomota, *Bull. Natl. Inst. Health Sci.* **121** (2003), 30–33.

aggrecan monomers via sugar-protein interactions [6]; the HA–GAG or HA–proteoglycan interactions may be involved in boundary lubrication [6]. The effects of the molecular weight of HA on cartilage elasticity are described in Section 3 (see below).

## 2.3. Analgesic and anti-inflammatory effects of HA

Fujii et al. injected HA (Suvenyl, $1.9 \times 10^6$ Da) into the arthritic knees of monkeys as an animal model of RA to examine the effectiveness of Suvenyl on knee pain: The injection reduced cartilage degeneration and inflammatory proliferation of synovial membranes, maintained cartilage matrices, and blocked the expression of inflammatory cytokines and matrix metalloproteinase-3 in cartilages as an immunohistochemical feature, when the treated knees were compared with non-treated knees.

Tamoto et al. studied the inhibitory effect of 0.3, 0.8 and $1.9 \times 10^6$ Da HA preparations on prostaglandin $E_2$ production by RA synovial cells [10]: HA of $1.9 \times 10^6$ Da had the most potent inhibitory effect on the production in the presence or absence of interleukin-1. The researchers also investigated the mechanism of the inhibitory effect, and HA of $1.9 \times 10^6$ Da again had the most potent inhibitory effect on the expression of cyclooxygenase-2 induced by RA synovial cells in the presence of interleukin-1. Since cyclooxygenase-2 expression is blocked by p38 MAP kinase inhibitors, HA of $1.9 \times 10^6$ Da appears to exert its anti-inflammatory effects by (a) binding firmly with HA receptors on synovial cells, (b) inhibiting activation of p38 MAP kinase, and (c) blocking transcription of the cycloxygenase-2 gene.

The most prominent clinical effect of HA injected into OA joints appears to be local pain relief. Oka et al. injected either 0.8 or $1.9 \times 10^6$ Da HA into the joint cavity in a beagle model of urate crystal-induced arthritis. The analgesic effect of HA became stronger in a molecular weight- and concentration-dependent manner.

## 2.4. Effects of HA on cartilages

Kikuchi et al. investigated the *in vivo* effect of HA on cartilage degeneration in a partial menisectomy model of OA in rabbit knee [3]. Intra-articular injection of HA or PBS was initiated immediately after surgery and repeated twice weekly. The injection of HA suppressed fibrosis and fibrillation of the cartilage matrix surface, which is often observed in early arthritis, in both the femurs and tibias. Protection against cartilage degeneration was again more remarkable with $1.9 \times 10^6$ Da HA than with $0.8 \times 10^6$ Da HA or PBS.

Table 3

The results of phase III studies (RA, OA, PS) overall improvement

|           | Drug    | Markedly improved | Moderately improved | Mildly improved | Unchanged / aggravated | Total | Comparison between groups[*] |
|-----------|---------|-------------------|---------------------|-----------------|------------------------|-------|------------------------------|
| RA[1)]    | Suvenyl | 13 (19.1)         | 31 (64.7)           | 13              | 11                     | 68    | $z = 6.780$                  |
|           | Placebo | 0                 | 4 (5.7)             | 29              | 37                     | 70    | $p = 0.0001$                 |
| OA[2)]    | Suvenyl | 20 (21.1)         | 49 (72.6)           | 18              | 8                      | 95    | $z = 2.521$                  |
|           | Artz    | 8 (9.2)           | 43 (58.6)           | 24              | 12                     | 87    | $p = 0.0117$                 |
| PS[3)]    | Suvenyl | 10 (10.1)         | 55 (65.7)           | 26              | 8                      | 99    | $z = 2.047$                  |
|           | Artz    | 8 (8.1)           | 42 (50.5)           | 36              | 13                     | 99    | $p = 0.041$                  |

( ): cumulative %, [*]: Wilcoxon's rank sum test. MW (Suvenyl) = 2700–3700 kDa (weight average); MW (Artz) = 600–1200 kDa; RA: rheumatoid arthritis; OA: osteoarthritis; PS: periarthritis scapulohumeralis.
[1)] S. Tanaka et al., *Clin. Rheumatol.* **5** (1994).
[2)] M. Yamamoto et al., *Jpn. Pharmacol. Ther.* **22** (1994).
[3)] R. Yamamoto et al., *Jpn. Pharmacol. Ther.* **22** (1994).

Tanaka et al. demonstrated that $1.9 \times 10^6$ Da HA inhibited the synthesis of MMP-1 and RANTES by chondrocytes, which were stimulated by IL-1beta, through CD44, in a molecular weight-dependent manner [11]. MMP-1 plays a key role in cartilage destruction in OA, and RANTES is a C-C chemokine family member that has been shown to increase MMP synthesis and reduce proteoglycan production.

These results suggest that heavy HA suppresses the activity of cartilage destruction factors more markedly and is more effective for the inhibition of cartilage destruction in OA and RA.

## 2.5. Effects of HA on subchondral bones

Pelletier et al. reported that HA inhibited, in a molecular weight-dependent manner, production of IL-6, prostaglandin $E_2$, and urokinase-type plasminogen activator by osteoblasts in subchondral bones in OA [4]. Abnormal metabolism by osteoblasts in subchondral bones appears to be related closely to spur formation and osteosclerotic changes; heavy HA probably delays the progress of OA.

## 2.6. Clinical effects of HA

In a phase III double-blind comparative study of HA in RA patients, 2.5 ml of 1% Suvenyl was injected into arthritic joints once weekly for a total of five consecutive injections, based on the results of a phase II dose-finding study [12]. Comparison of 1% Suvenyl with 0.01% Suvenyl revealed that the improvement rate for 1% Suvenyl was higher (65%) than for 0.01% Suvenyl (Table 3). In addition, Suvenyl showed greater efficacy on OA and periarthritis scapulohumeralis (PS) than did Artz (smaller HA, MW $0.8 \times 10^6$ Da). Few adverse reactions were reported throughout the study. These are the first results to demonstrate the clinical efficacy of HA with a molecular weight of $1.9 \times 10^6$ Da in the treatment of knee pain in RA as well as OA and PS.

## 2.7. Conclusion and future of HA

HA of $1.9 \times 10^6$ Da is the first HA product (Suvenyl) that has demonstrated efficacy in RA. Markedly increased synovial fluid in RA has been declining drastically since HA preparations of $1.9 \times 10^6$ Da became commercially available, so there is no longer any doubt about the inhibitory effect of HA on local inflammation of joints.

Thus, HA preparations will be able to reduce or eliminate the necessity of corticosteroid therapy, which can cause adverse reactions associated with frequent or long-term use of these steroids.

HA preparation of $1.9 \times 10^6$ Da has been used as a conservative treatment in some advanced OA and RA patients who are candidates for arthroplasty – joint replacement – but who wish to avoid this surgery. HA preparation is an option for the treatment of advanced OA or RA by delaying arthroplasty as long as possible for OA or RA patients who cannot undergo revision surgery because of aging. In addition, HA preparation may be effective as an adjuvant for physiotherapy to prevent joint deformity and contracture.

Although the use of $1.9 \times 10^6$ Da HA in RA is at present limited to knee pain, Sohen et al. have demonstrated the efficacy of HA in the treatment of pain in shoulders, elbows, finger joints, and ankles, as well as knee [9]. This finding may lead to expanded use of HA for the treatment of various joints in RA.

## 3. Effects of the molecular weight of HA on elasticity of cartilage matrix

Cartilage elasticity often decreases with age and during inflammation, and this decrease in elasticity accelerates cartilage degeneration. While many matrix components are involved in cartilage elasticity, the elasticity of HA and aggrecan complexes is basic to understanding the biomechanics and pathology of cartilage.

The HA molecular weight in articular cartilage markedly decreases with age from $2 \times 10^6$ to $3 \times 10^5$ Da between the ages of 2.5 and 86 years, although the HA content increases over the same age range [1,2]; HA size may also decrease during OA and RA. On the other hand, aggrecan content decreases via activation of MMPs and ADAMTSs (aggrecanases) during inflammation. These changes in HA size, HA concentration and aggrecan concentration may reduce cartilage elasticity.

To address this issue, we examined the effects of HA size, HA concentration and aggrecan concentration on elasticity of HA–aggrecan solutions under dynamic conditions, using a controlled-stress rheometer. Since normal young cartilage contains ~0.5 mg/ml HA of ~$3 \times 10^6$ Da and ~50 mg/ml aggrecan, solutions containing $3.3 \times 10^6$ Da[#] HA (Suvenyl) at 0.001–5 mg/ml and/or aggrecan at 10–50 mg/ml were used. At the same time, we determined the elasticity and viscosity of HA–aggrecan solutions as a function of frequency using a cone-on-plate rheometer. The frequencies of 0.5 and 2.5 Hz correspond to those of joint movement during walking and running.

At a constant level of $3.3 \times 10^6$ Da[#] HA (0.5 mg/ml), the elasticity of the aggrecan–HA complex was (a) very low at an aggrecan concentration of 10 mg/ml, (b) become detectable at 20–30 mg/ml, and (c) further increased until 50 mg/ml at 0.1, 1 and 10 Hz (Fig. 1). The aggrecan concentrations had far less effect on viscosity of the solution [5], which suggests that only a 50% decrease in aggrecan content – compared with that in normal cartilage – can almost abolish the elasticity of cartilage.

At 20–30 mg/ml, aggrecan and HA (0.5 mg/ml) ratio was $1:40 - 1:60$, which indicates that $>10\%$ saturation of HA with aggrecan monomers can initiate elastic network formation, since HA was saturated with aggrecan at $1:500$, as described below.

Figure 2 shows the effects of increasing concentrations of heavy HA ($3.3 \times 10^6$ Da[#]) on the elasticity of HA–aggrecan solutions. At a constant level of aggrecan monomers (50 mg/ml), the elasticity of HA–aggrecan solution sharply increased HA concentration-dependently at 0.001 to 0.1 mg/ml, but did not further increase at higher HA concentrations (0.1–1 mg/ml): The HA–aggrecan ratio was optimal at $1:500$ at all Hz. Since even aged cartilage contains HA at $>0.5$ mg/ml, HA concentration may not be as important in determining cartilage elasticity as aggrecan concentration is.

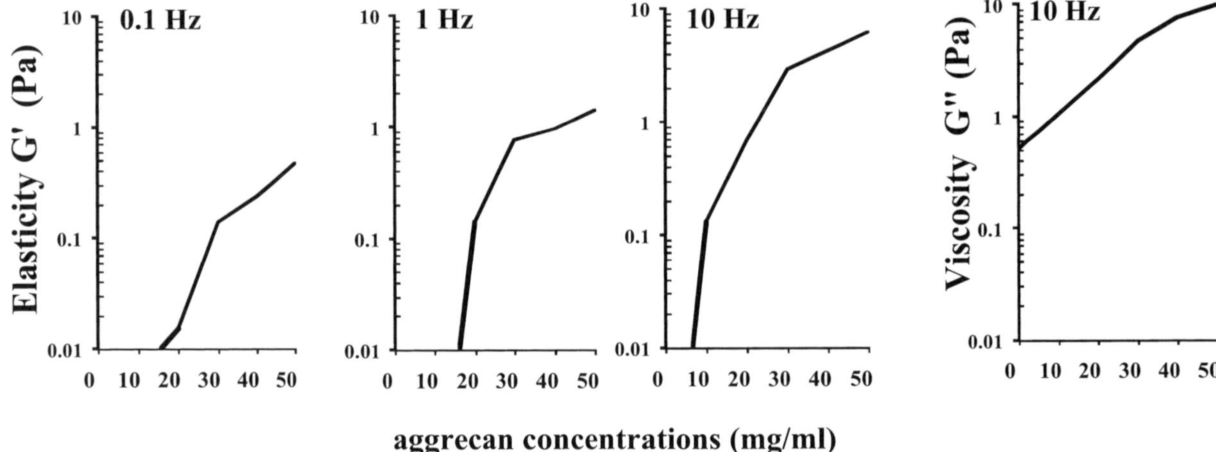


Fig. 1. Elasticity and viscosity of aggrecan–HA (0.5 mg/ml) complex at various aggrecan concentrations. Elasticity and viscosity of aggrecan–HA solutions containing a constant level of HA (0.5 mg/ml) and various concentrations of aggrecan were determined at 0.1, 1 and 10 Hz, using a cone-on-plate rheometer.

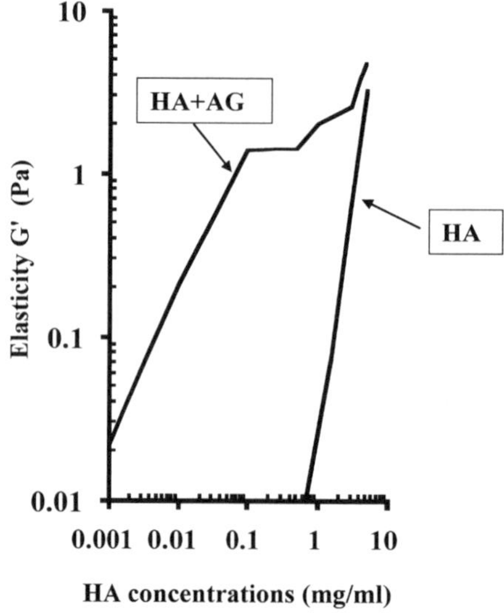


Fig. 2. Elasticity of aggrecan (50 mg/ml)–HA (3300 kDa) complex at various HA concentrations. Elasticity of the solutions was determined at 1 Hz.

The HA chain saturated with aggrecan monomers becomes a platform for close aggrecan–aggrecan interactions, forming the elastic structure. Since HA size in articular cartilage markedly decreases with age, we examined the effect of HA size on the elasticity of HA–aggrecan solutions. Aggrecan at 50 mg/ml alone had very low elasticity at any Hz, but heavy HA ($3.3 \times 10^6$ Da[#]) (0.5 mg/ml)-aggrecan solution showed greater elasticity than smaller HA ($1.0 \times 10^6$ Da[#])–aggrecan solution at 0.1 and 1 Hz (Fig. 3). On the other hand, HA size had little effect on viscosity at all Hz [5]. Large aggregates formed with high molecular weight HA force close aggrecan–aggrecan interactions even with slow motion, but such interactions are fewer with small aggregates. Since the HA size in articular cartilage markedly decreases with age, the failure of small aggregates to form an elastic network must therefore contribute to age-

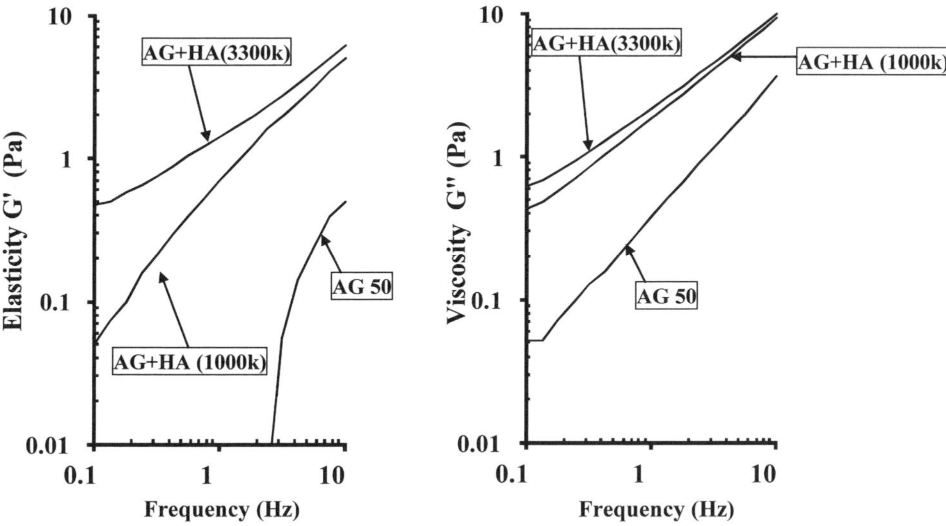


Fig. 3. Elasticity and viscosity of aggrecan (50 mg/ml)–HA (0.5 mg/ml) complex at various HA sizes. Elasticity and viscosity of the solutions were determined as a function of frequency.

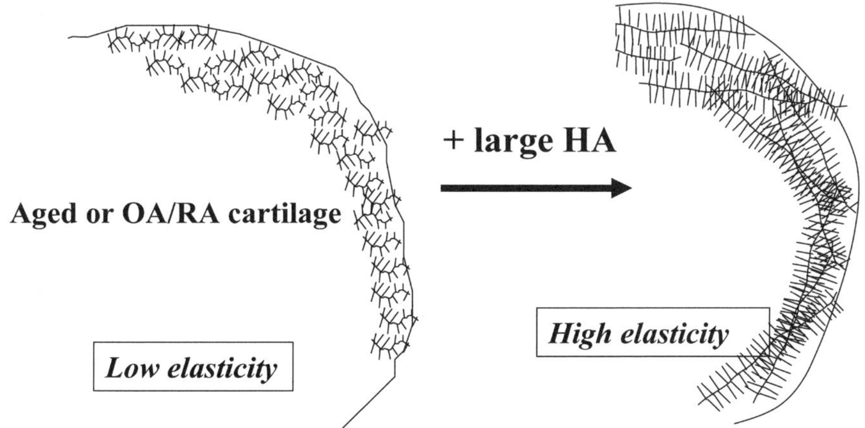


Fig. 4. Intra-articular injections of large HA may recover the elasticity of cartilage matrix.

related weakness of the cartilage. Intra-articular injection of heavy HA may help to recover the elasticity of cartilage matrix (Fig. 4), since HA can penetrate into the surface portion of arthritic cartilage.

## References


[1] M.W. Holmes, M.T. Bayliss and H. Muir, Hyaluronic acid in human articular cartilage. Age-related changes in content and size, *Biochem. J.* **250** (1988), 435–441.
[2] H. Kamada, K. Masuda, A.L. D'Souza, M.E. Lenz, D. Pietryla, L. Otten and E.J.-M.A. Thonar, Age-related differences in the accumulation and size of hyaluronan in alginate culture, *Arch. Biochem. Biophys.* **408** (2002), 192–199.
[3] T. Kikuchi, H. Yamada and K. Fujikawa, Effects of high molecular weight hyaluronan on the distribtion and movement of proteoglycan around chondrocytes cultured in alginate beads, *Osteoarthritis Cartilage* **4** (1996), 99–110.
[4] D. Lajeunesse, A. Delalandre, J. Martel-Pelletier and J.P. Pelletier, *Bone* **33** (2003), 703–710.

[5]  M. Nishimura, M. Kawata, W. Yan, A. Okamoto, H. Nishimura, Y. Ozaki, T. Hamada and Y. Kato, Quantitative analysis of the effects of hyaluronan and aggrecan concentration and hyaluronan size on the elasticity of hyaluronan-aggrecan solutions, *Biorheology* **41** (2004), 629–639.

[6]  M. Nishimura, W. Yan, Y. Mukudai, S. Nakamura, K. Nakamasu, M. Kawata, T. Kawamoto, M. Noshiro, T. Hamada and Y. Kato, Toe of chondroitin sulfate–hyaluronan interactions in the viscoelastic properties of extracellular matrices and fluids, *Biochim. Biophys. Acta* **1380** (1998), 1–9.

[7]  M. Oka, T. Nakamura and T. Kitsuji, Effect of high molecular weight hyaluronic acid on joint lubrication (in Japanese), *Nihon Riumach, Kansetu Geka, Gakkai-Zassi* **12** (1993), 259–266.

[8]  M. Oka, T. Nakamura, Y. Matsusue, M. Akagi and M. Horiguchi, Effects of high-molecular weight hyaluronate on joint disorders (in Japanese), *Seikei Saigai Geka* **40** (1997), 77–84.

[9]  S. Sohen, K. Kikuchi, T. Nonaka and C. Hamanishi, Clinical and experimental evaluation of hyaluronic acid for rheumatoid arthritis (in Japanese), *Clin. Rheumatol.* **15** (2003), 70–75.

[10]  K. Tamoto, H. Nochi and Y. Tokumitsu, High molecular weight hyaluronic acids inhibit interleukin-1-induced prostaglandin E2 generation and prostaglandin E2-elicited cyclic AMP accumulation in human rheumatoid arthritic synovial cells, *Japanese J. Rheum.* **5** (1994), 227–236.

[11]  M. Tanaka, K. Fujii, M. Tsuji and T. Sawai, Autoimmune reaction to type II collagen and cartilage degeneration in MRL/Mp-lpr/lpr mouse, *Rheumatol. Int.* **24** (2004), 84–92.

[12]  S. Tanaka, M. Yamamoto, Y. Komatsubara, S. Sugawara, S. Soen, T. Matsubara, S. Kashiwazaki and M. Nakashima, Clinical study of high molecular hyaluronic acid (NRD101) on rheumatoid arthritis. A multi-center, joint phase III comparative study (in Japanese), *Clin. Rheumatol.* **5** (1994), 304–332.

[13]  M. Yamamoto, S. Sugawara, Y. Tsukamoto, M. Motegi, H. Iwata, J. Ryu, K. Takagishi and M. Nakashima, Clinical evaluation of high molecular weight sodium hyaluronate (NRD101) on osteoarthritis of the knee (in Japanese), *Jpn. Pharmacol. Ther.* **22** (1994), 4059–4087.

[14]  R. Yamamoto, S. Tabata, M. Mikasa, K. Takagishi and M. Nakashima, Phase III comparative clinical study of high molecular weight sodium hyaluronate (NRD101) with ARTZ® on periarthritis scapulohumeralis (in Japanese), *Jpn. Pharmacol. Ther.* **22** (1994), 4029–4057.

Biorheology 43 (2006) 355–370
IOS Press



# Mechanical regulation of secondary chondrogenesis

Charles W. Archer [a], Paul Buxton [b], Brian K. Hall [c] and Philippa Francis-West [b]

[a] *Connective Tissue Biology Laboratories, School of Biosciences, Cardiff University, Museum Avenue, Cardiff, Wales, UK*
[b] *Department Craniofacial Development, Guy's, King's and St. Thomas' School of Dentistry, Guy's Hospital, London Bridge, London, SE1 9RT, UK*
[c] *Department of Biology, Dalhousie University, Halifax, Nova Scotia, Canada, NJ B3 4J1*

**Abstract.** The development of the skull is characterised by its dependence upon epigenetic influences. One of the most important of these is secondary chondrogenesis, which occurs following ossification within certain membrane bone periostea, as a result of biomechanical articulation. We have studied the genesis, character and function of the secondary chondrocytes of the quadratojugal of the chick between embryonic days 11 and 14. Analysis of gene expression revealed that secondary chondrocytes formed coincident with *Sox9* upregulation from a precursor population expressing *Cbfa1/Runx2*: a reversal of the normal sequence. Such secondary chondrocytes rapidly acquired a phenotype that is a compound of prehypertrophic and hypertrophic chondrocytes, exited from the cell cycle and upregulated *Ihh*. Pulse and pulse/chase experiments with BrdU confirmed the germinal region as the highly proliferative source of the secondary chondrocytes, which formed by division of chondrocyte-committed precursors. By blocking Hh signalling in explant cultures we show that the enhanced proliferation of the germinal region surrounding the secondary chondrocytes derives from this *Ihh* source. Additionally, *in vitro* studies on membrane bone periosteal cells (nongerminal region) demonstrated that these cells can also respond to Ihh, and do so both by enhanced proliferation and precocious osteogenesis. Despite the pro-osteogenic effects of Ihh on periosteal cell differentiation, mechanical articulation of the quadratojugal/quadrate joint in explant culture revealed a negative role for articulation in the regulation of *osteocalcin* by germinal region descendants. Thus, the mechanical stimulus that is the spur to secondary chondrocyte formation appears able to override the osteogenic influence of Ihh on the periosteum, but does not interfere with the cell cycle-promoting component of Hh signalling.

Keywords: *Cbfa-1*, *Ihh*, *Sox9*, hypertrophy, chondrocyte, membrane bone, proliferation, chick

## 1. Introduction

During development of the vertebrate skeleton, cartilaginous structures can be classified as either primary as in the cartilage of the developing limbs whereby the cartilage acts as a morphogenetic template for the future bones and secondary cartilage where bone acts as a template for cartilage. Examples of this type of cartilage are those found in the temporo-mandibular joint and the clavicular joints. In the joints that comprise primary cartilage, joint initiation is independent of movement but subsequent development and maintenance requires movement. In contrast, during secondary joint formation, both the initiation of joint formation and development both require movement [4,12,13]. The bases of such development lie in the periosteum that surrounds the original bony elements in the location of where the joint will form, responds to mechanical cues generated across the joint by the surrounding musculature and that the cells differentiating from the progenitor cells within the periosteum form chondrocytes rather than osteoblasts.

Using the chick embryo as a model, secondary cartilage develops within the quadratojugal joint comprising two membrane bones [33]. Because this joint is very small, it can be organ cultured and mechanical articulation of the joint leads to chondrogenic differentiation [17].

While it is known that all connective tissue cells can respond to mechanical cues and are essential for the functioning of many tissues including cartilage, the requirement for particular cues in determining cell fate selection is less well appreciated. Pivotal in this lineage selection process, is the expression of two transcription factors, *Runx2* (*CBFA1*) and *Sox9*. *Runx2* is central to the differentiation of bone and appears to be regulated by the canonical Wnt pathway via $\beta$-catenin [8,22,27] but is also involved in chondrocyte hypertrophy [26]. Indeed, inhibition of Wnt signalling can lead to cells that would normally differentiate into osteoblasts becoming chondrocytes. Sox9, however, is required for chondrocyte differentiation [3,31]. Wnt signalling in chondrogenic precursors often leads to down-regulation of the phenotype [20]. It is not known how mechanical cues interact and regulate these pathways or why only certain precursor cells can respond. It may even be the case that other pathways are involved in this particular precursor population. Here, we investigate the factors that contribute to this lineage selection and those that regulate the proliferation of the progenitor cells within the surrounding periosteum.

## 2. Materials and methods

### 2.1. In situ hybridisation and probe details

Radioactive *in situ* hybridisation was performed as described [15]. In brief, embryos were fixed in 4% paraformaldehyde and processed through to wax, sectioned at 8 mm and hybridised with [35S]-labelled riboprobes at 55°C. Chick probes used were as follows: *Cbfa1/Runx2–Xho*I (T3; gift from Dr J. Helms), *Sox9* [24], *Col2* [7], *Col10* [28], *Ihh* [40], *Ptc2* [35], *Bmp7* [23], *Frzb* [29], and *Osteocalcin* and *Osteopontin* PCR fragments (gift from Dr V. Church). Sections were viewed on a Zeiss Axioskop and photographed using a Sony DSC75 digital camera. Montages were arranged using Adobe Photoshop. In some cases bright field images were overlayed with inverted dark field images.

### 2.2. Manipulation of the quadratojugal/quadrate joint

Fertilised wild-type chicken eggs (Ross White) were obtained from Henry Stewart and Co., Lincolnshire, UK. Eggs were incubated at $\sim$38°C for the stated number of days. Embryos at stage e14 were used for the *ex vivo* experiments, as the joint was less liable to dislocation when manipulated. The quadratojugal/quadrate (QJ/Q) was dissected from either side of the head and stripped of most attendant connective tissue, except that holding the two elements together. The Q and QJ were trimmed and then placed on a filter (Millipore; 0.2 $\mu$M) upon a metal grid at the air/medium interface. Explants were cultured in DMEM (Gibco) plus 15% chick serum (Sigma), at 37°C/5% $CO_2$. Articulation regimes involved operating the joint through its normal movement (10 times on the hour) in order to mimic the sporadic movements *in ovo*, for the specified time [17].

### 2.3. Treatment with blocking antibody

Anti-hedgehog antibody, 5E1 [11], was obtained from the DSHB, University of Iowa, IA. Partially purified antibody was obtained and used at a dilution of 1 : 10.

## 2.4. Cell proliferation studies

Explants were labelled in $10 \times$ BrdU diluted in culture medium (as above), according to the manufacturer's protocol (Roche), immediately following dissection (for 0 hour timepoint) or at the specified time post-dissection for 90 minutes. Explants were then fixed in 4% paraformaldehyde, processed through to wax and sectioned. BrdU incorporation was revealed using a biotin-conjugated anti-BrdU antibody (MD-5215; Caltag, USA). ABC kit from Vectastain (Vector Labs, Peterborough, UK) was then used, followed by DAB (Vector Labs) for the colour reaction. Sections were counterstained with Toluidine Blue (0.02%) for cell counting, or Alcian Blue (prior to immunochemistry) and Chlorantine Fast Red (post-immunochemistry) to assess differentiation. For pulse-chase experiments the QJ/Q joint was rinsed in medium, placed on a fresh filter in fresh medium and re-incubated. Parathyroid hormone (PTH) was used at a concentration of 10 nM.

# 3. Results

## 3.1. Genetic basis of secondary chondrocyte initiation

Previous studies in the chick by Murray, and by Fang and Hall have demonstrated that secondary chondrocytes arise in the quadratojugal (QJ) between embryonic days (e) 10 and 11 [13,33]. In order to characterize the formation of secondary chondrocytes in this location, we studied initially the expression of *Cbfa1* and *Sox9*. Using radioactive *in situ* hybridisation of adjacent sections at e10 (Fig. 1A), we found *Cbfa1* (Fig. 1B) and *Sox9* (Fig. 1C) to be expressed in exclusive tissues: *Cbfa1* was expressed in membrane bone periostea and osteoblasts, and *Sox9* was expressed in the primary cartilaginous elements (e.g. the quadrate; Q). *Cbfa1* was expressed throughout the periosteum (Fig. 1E), but at the tip of the QJ at e10, where the secondary chondrocytes will arise, there was neither Alcian Blue staining (Fig. 1D), nor expression of *Sox9* (Fig. 1C), *type II collagen* (*Col2*; Fig. 1G) or *type X collagen* (*Col10*; data not shown). Analysis of *Gdf5* expression, a potent pro-chondrogenic growth factor of the TGF$\beta$ superfamily [16], revealed it to be expressed only in the perichondrium of the Q (Fig. 1F). Alcian Blue staining can first be detected within the boundary of the periosteum of the membrane bone at e11, on the inner face of the QJ-hook (Fig. 1H), the earliest histological evidence of secondary chondrocytes (SCs). Observed simultaneously with this change, (though necessarily slightly in advance of it; data not shown) *Sox9* transcripts could be detected within the same region of the QJ (Fig. 1J). The *Sox9* expression domain was confined to the inner region of the QJ tip (the secondary chondrocytes) and did not extend to the adjacent periosteum, which, from hereon, we shall call the germinal region [18]. *Cbfa1* continued to be expressed unaltered in the periosteum and germinal region, and was also expressed in the differentiating chondrocytes (Fig. 1I). Detection of the same genes at e13 (Fig. 1K,L,M) show clearly that *Cbfa1* was expressed in secondary chondrocytes (Fig. 1L), the periosteum and the germinal region. *Sox9* was also expressed throughout the SCs at this time point, but was excluded from the germinal region (Fig. 1M).

We further characterised the secondary chondrocytes by examining the expression of *Col2*, an early chondrocyte matrix constituent; *Ihh*, expressed by pre-hypertrophic chondrocytes [40]; *Col10*, a marker for terminally differentiated chondrocytes [7]; and *Frzb*, expressed by condensing mesenchyme and early chondrocytes [29]. *Col2* was expressed in a domain coincident with that of Alcian Blue staining (Fig. 2A,B); in this example, both in anterior and posterior SC domains of the tip of the QJ, but not in the germinal region. *Col10* expression on an adjacent section (Fig. 2C) revealed that although the domain of *Col10* was contained within that of *Col2*, it had been rapidly upregulated by the secondary chondrocytes.

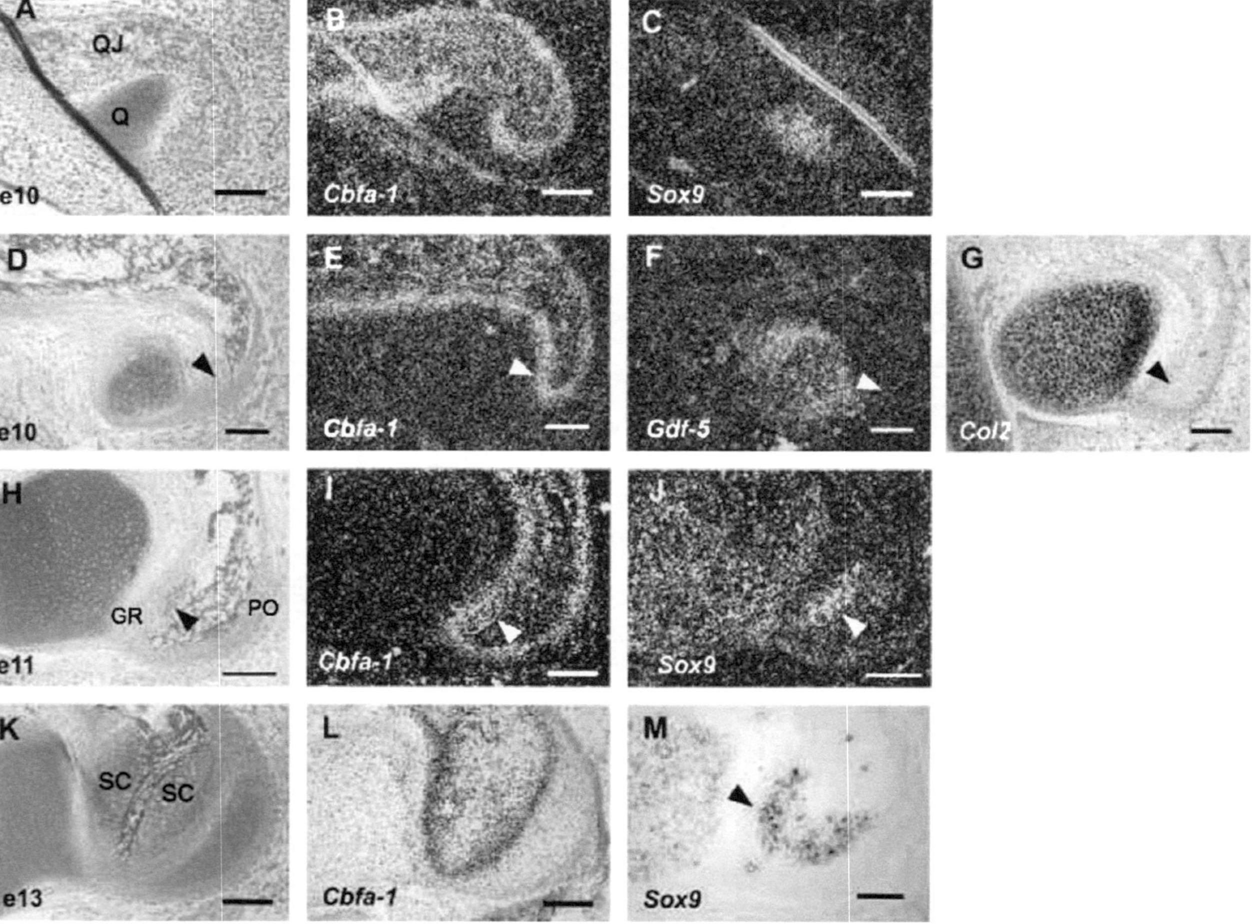

Fig. 1. Initiation of secondary chondrocyte formation. (A–M) quadratojugal/quadrate joints at stage: e10 (A–G), e11 (H–J) and e13 (K–M). (A,D,H,K) Alcian Blue/Chlorantine Fast Red stained controls. (B) *Cbfa1* expression is found only in the membranous periosteum of the quadratojugal (QJ). (C) *Sox9* expression is only found in the cartilaginous quadrate (Q). (D–G) Comparison of *Cbfa1*, *Gdf5* and*Col2* expression. Arrowheads (D–J) indicate future sites of secondary chondrocytes (SCs). (E) *Cbfa1* expression is maintained in the periosteum prior to SC formation. *Gdf5* (F) is expressed around the quadrate and *Col2* (G) is expressed within the quadrate, but neither is expressed within the quadratojugal. (H–J) *Cbfa1* and *Sox9* expression on adjacent sections after formation of the first SCs, adjacent to the germinal region (GR). (I) *Cbfa1* is expressed by the periosteum, and by the nascent SCs (arrowed and circled). (J) *Sox9* is expressed only by the SCs of the QJ (arrowhead). (K–M) *Cbfa1* and *Sox9* expression at e13. (L) *Cbfa1* continues to be expressed in the periosteum and is also expressed by SCs (asterisk). (M) *Sox9* is expressed only by SCs (arrowed), not in the periosteum. GR, germinal region; PO, periosteum; Q, quadrate; QJ, quadratojugal; SC, secondary chondrocytes. Scale bars: 100 $\mu$m.

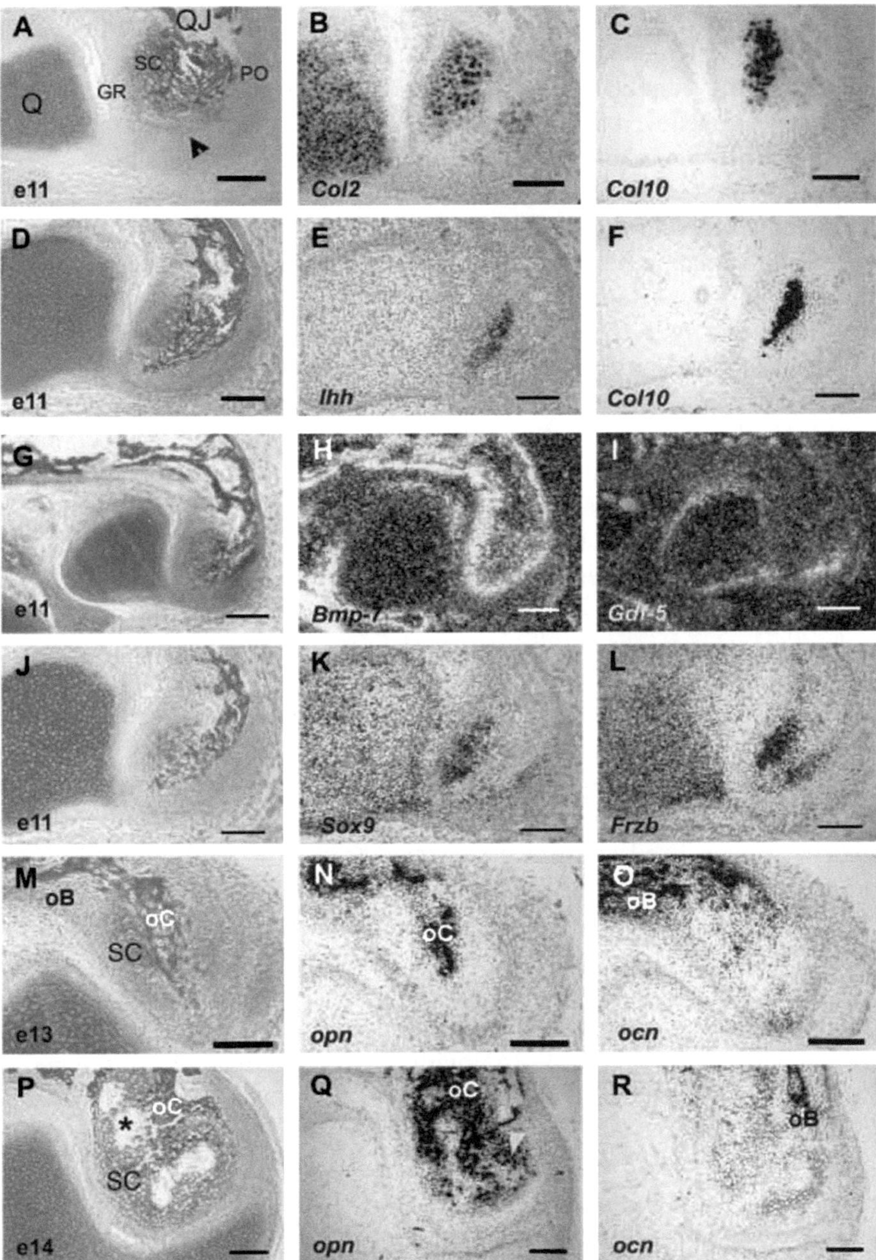

Fig. 2. Differentiation of secondary chondrocytes. (A–L) e11, (M–O) e13, (P–R) e14. (A,D,G,J,M,P) Alcian Blue/Chlorantine Fast Red stained controls. (A) SCs forming on the anterior and posterior (arrowhead) aspect of the QJ. (B,C) Adjacent sections to A, showing that *Col2* (B) is upregulated as SCs form, and that *Col10* (C) upregulation is delayed. (E,F) Adjacent sections to D, showing *Ihh* expression (E) in the same domain as *Col10* (F). (H,I) Adjacent sections to G. (H) *Bmp7* expression is observed throughout the periosteum, and (I) *Gdf5* expression is restricted to the perichondrium of the Q; neither reveal a change in the germinal region. (K,L) Adjacent sections to J, which show that *Sox9* (K) expression is confined to the SCs of the QJ, and that *Frzb* (L) expression is upregulated by SCs. (N,O) Adjacent sections to M, showing *opn* expression (N) in osteocytes; *ocn* expression (O) in osteoblasts adjacent to the PO precedes *opn* expression. (P,Q,R) At e14, chondroclast activity is evident (asterisk in P), and *opn* expression (Q) is additionally upregulated by late SCs (yellow arrowhead); *ocn* expression (R) remains specific for osteoblasts. oC, osteocytes; oB, osteoblasts. Scale bars: 100 $\mu$m.

This apparent rapid hypertrophy was borne out by the co-expression of *Ihh* (Fig. 2E) and *Col10* (Fig. 2F), and was consistent with the enlarged appearance of the secondary chondrocytes. Additionally, a negative regulator of chondrocytes differentiation, *Pthrp*, was undetectable at e11 and at e14 (data not shown). As secondary chondrocytes are believed to arise from the germinal region, we examined the expression of known perichondrial/periosteal markers to see whether changes in the expression of these genes presaged or reflected the formation of SCs. We found that *Bmp7* was expressed in the periosteum of the membrane bones (Fig. 2H), including around the SCs. *Gdf5* expression was not detected in the germinal region (Fig. 2I), even after SC formation, nor was it expressed in the remaining periosteum, but it could be detected in the quadrate perichondrium. We also looked at further markers of chondrocytes, for example, the pre-chondrogenic marker *Frzb* (Fig. 2L), which was expressed coincident with *Sox9* upregulation (Fig. 2K) and in advance of Alcian Blue staining. For markers of definitive osteoblasts and late stage hypertrophic chondrocytes we also compared the expression of *osteocalcin* (*ocn*) and *osteopontin* (*opn*) in the quadratojugal. Neither gene was expressed in SCs up to e13 (Fig. 2M,N,O; data not shown). In the membrane bone, *opn* was expressed by the osteocytes of the trabecular bone (Fig. 2N), whereas *ocn* was expressed earlier, in the osteoblasts adjacent to the periosteum (Fig. 2O). *Opn* was upregulated in SCs at e14 (Fig. 2P,Q), coinciding with the erosion of secondary cartilage by chondroclasts. By contrast, *ocn* expression remained osteoblast specific (Fig. 2R).

## 3.2. Secondary chondrocytes rapidly exit the cell cycle

Exit from the cell cycle is a key event in the maturation of chondrocytes, occurring at the transition to prehypertrophic chondrocyte. To determine how, or whether, this transition occurred in secondary chondrocytes, we explanted the quadratojugal/quadrate joints (QJ/Q), from e14 embryos, onto filters and labelled them with bromo-2′-deoxyuridine (BrdU) for 90 minutes whilst culturing at the air/medium interface. Explants were then fixed and processed for detection of BrdU. Labelling revealed striking differences in proliferation between the SCs, the germinal region and the periosteum – very few SCs were found to be in S-phase. Comparison of the percentage of BrdU labelled cells in the three regions confirmed the clear variation in proliferation rates (Fig. 3): periosteum (5.2%; 114/2175), germinal region (15.9%; 563/3531) and secondary chondrocytes (0.51%; 10/1978) (nine sections from three QJs were assessed for each result). Histologically, these results represented a marked cut-off at the boundary between the SCs and germinal region (Fig. 4A), with only occasional cells at the periphery of the SC region being labelled (Fig. 4C; $n = 6$, 36 sections).

## 3.3. Pulse and chase confirms origin of secondary chondrocytes

Given the distinction between proliferation and non-proliferation, and the parallel contrast between undifferentiated germinal region and differentiated secondary chondrocytes (as assayed by genetic markers), we realised that it would be possible to use BrdU incorporation to label the germinal cells and then follow their fate. The immediate labelling by BrdU of S-phase cells would therefore assess commitment to the chondrocytic pathway under the *in vivo* conditions of articulation. Subsequently, however, in the absence of mechanical stimulation *ex vivo*, the germinal region should start to give rise to osteoblasts, not chondrocytes [18] (see also Fig. 7). All the following experiments were carried out in the absence of mechanical stimulation, unless otherwise stated. In order to follow the fate of proliferating cells, paired QJ/Q joints were incubated in BrdU, as described above, and one was immediately fixed (time 0). The contralateral side was rinsed of BrdU, moved to a new well with a fresh filter/medium

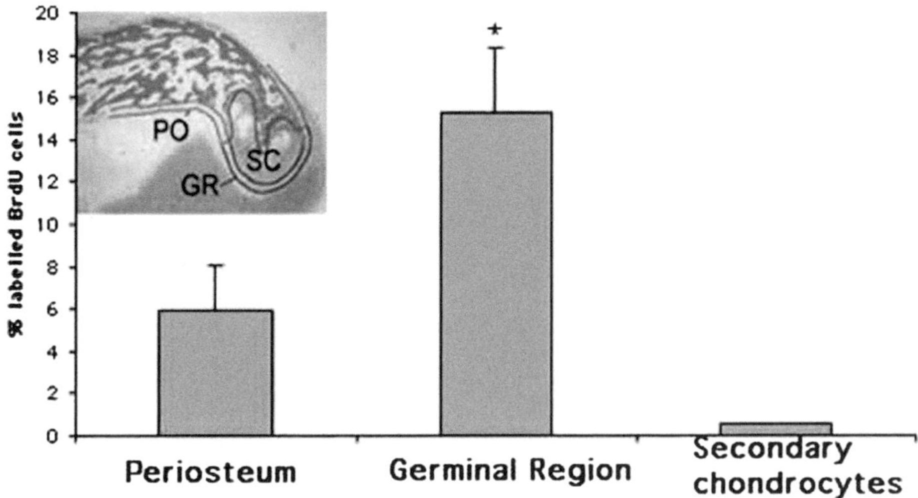


Fig. 3. Proliferation in the QJ at e14, as shown by the percentage of S-phase cells after BrdU labelling for 90 minutes at 0 hours. Figure (inset) indicates regions of the QJ counted for each value. Values shown are mean ± s.d. from three QJs, a total of 6 sections. Asterisk indicates a significant difference between the periosteum and the germinal region ($P < 0.001$, Student's $t$-test).

and re-incubated for 24 hours (the chase) prior to fixation. By following the pulse/chase protocol, we found that pairs of BrdU-labelled nuclei could be detected within the chondrocyte zone (Fig. 4B,D). The daughter cells shared a lacuna within the hypertrophic-like domain of the secondary chondrocytes, where at time 0 BrdU data showed proliferation to be very low or undetectable. To confirm that self-renewing pre-chondrocytes were not present, we incubated the explanted QJ/Q joint for 16 hours prior to labelling with BrdU, and followed this by a 24-hour chase in the absence of mechanical stimulation. If there was an amplification zone of prechondrocytes these would be labelled at the second division and labelled chondrocytes would be observed. We found that pulsing at 16 hours, followed by immediate fixation, labelled the germinal region (Fig. 4E), but the chase of the contralateral side failed to label chondrocytes (Fig. 4F; labelling osteoblasts instead), even though conditions were clearly permissive for chondrocyte proliferation as judged by that detected in the quadrate.

To explore the role of cell division in differentiation, we assessed the expression of *Col2* and *Col10* in explanted QJ/Q joints labelled with BrdU, on adjacent sections. As at e11 and e13 (see Fig. 2B,C; data not shown), *Col2* expression (Fig. 4H) slightly preceded that of *Col10* (Fig. 4H), although BrdU labelling within the *Col2*-expressing area was rare (Fig. 4G). When we compared this to the situation after a 24 hour chase (Fig. 4I, BrdU labelling at 0 hours), we found that the *Col2* and *Col10* expression domains at the germinal region/SC boundary were now coincident (Fig. 4J). Also, cell doublets could now be observed within the *Col10* expression domain (Fig. 4I). This observation was consistent with a cell committing to the chondrocyte lineage, dividing once, and then up-regulating *Col10* ($n = 3$). The rapid hypertrophy indicated by this result, led us to assess whether this was an intrinsic property of SCs, or was related to the absence of *Pthrp* expression. We attempted to delay hypertrophy by adding PTH peptide to cultured QJ/Q explants. We found no significant difference between PTH-treated and untreated explants when we compared *Col2* with *Col10* expression on adjacent sections after 24 hours culture ($n = 4$; data not shown).

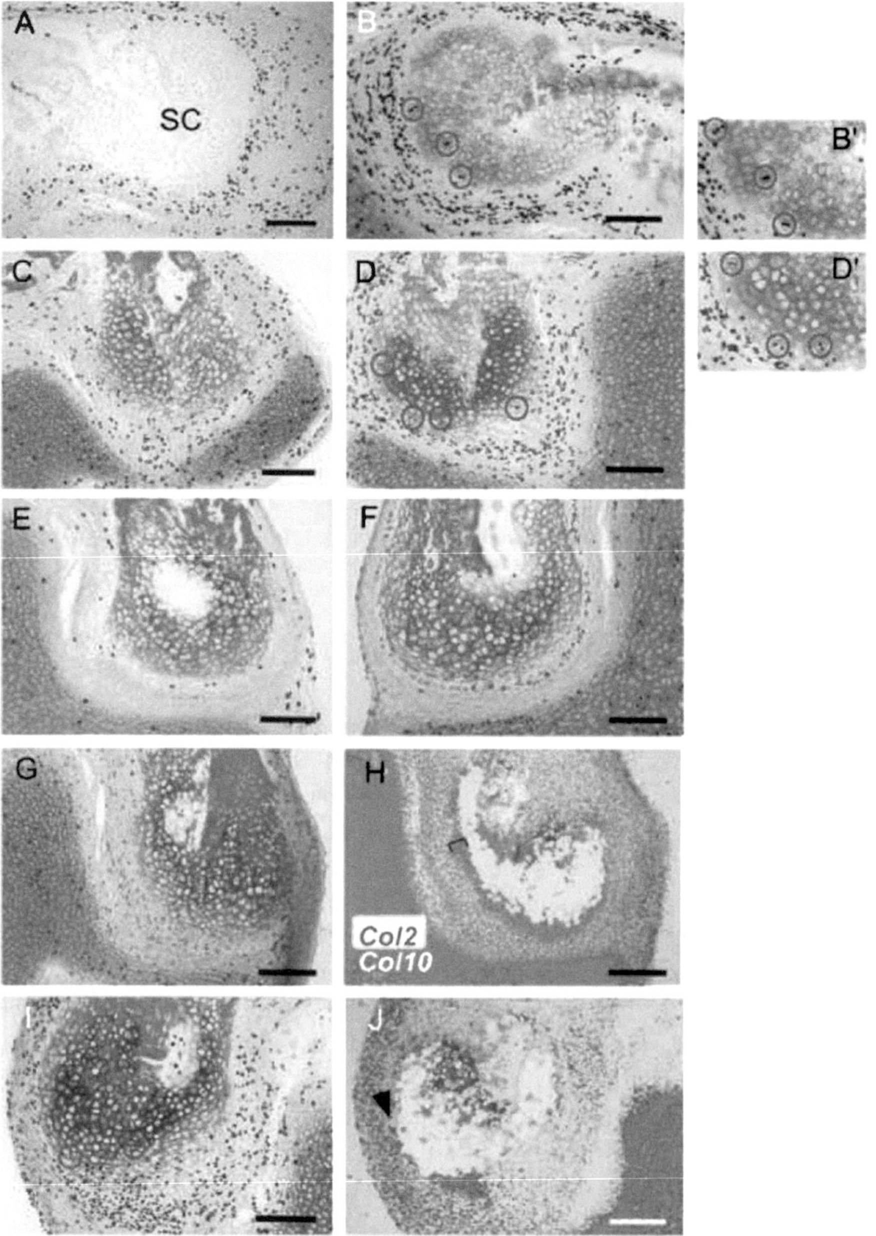

Fig. 4. BrdU labelling and the origin of secondary chondrocytes at e14. Counterstains: Toluidine Blue (A); (B–G,I) Alcian Blue/Chlorantine Fast Red. BrdU labelled cells are black. (A) Pulse and immediate fix shows cycling cells in the germinal region, and a relative absence of BrdU label within the SCs. (B) Contralateral side to A, with a 24-hour chase following pulse. Doublets found within the SC domain are highlighted with circles. (B′) Higher magnification of image in B. (C) BrdU labelled cells at the boundary between the germinal region and the SCs. (D) Contralateral side to C shows labelled doublets after a 24-hour chase (circled). (D′) Higher magnification of image in D. (E) BrdU labelling at 16 hours after explant and immediate fixation labels cells in the germinal region. (F) Contralateral side to E after a 24-hour chase shows no labelled chondrocytes in the SC domain. (G) BrdU label at 0 hour. (H) Adjacent section to G, which shows *Col10* expression superimposed on, and contained within, *Col2* expression. Bracket indicates delay in *Col10* upregulation. (I) 24-hour chase on contralateral side to G,H results in labelled doublets. (J) Adjacent section to I, showing that the *Col10* domain has expanded to coincide with *Col2* domain, black arrowhead. Scale bars: 100 μm.

### 3.4. Ihh signalling during secondary chondrogenesis

The proliferation rate in the germinal region was almost three times that of the periosteum in other regions, indicating that the germinal region was subject to influences unique to this location. In order to assess whether hedgehog (Hh) signalling from SCs was responsible, we analysed the expression of *Ptc2* in relation to *Ihh*. As an inducible transcriptional target, *Ptc2* is a sensitive marker for receipt of Hh signal [20,35]. At e11, although *Ihh* was readily detectable (Fig. 2E, Fig. 5B), *Ptc2* expression was at the limit of detection (Fig. 5C). However, at e12 and e14 *Ptc2* expression was evident within the germinal layer (Fig. 5F,I), adjacent to the *Ihh* domain in the SCs (Fig. 5E,H). *Ptc2* was not expressed in the periosteum surrounding the membrane bone, and we were unable to detect *Ptc1* expression at either e11 or e14 (data not shown).

We addressed the functional significance of the *Ihh* domain by using the Hh blocking antibody 5E1 to ablate Ihh function in our explant system [11,43]. We then assessed proliferation after 16 hours of incubation by BrdU labelling as before. We found that proliferation in the contralateral joint in 5E1-treated

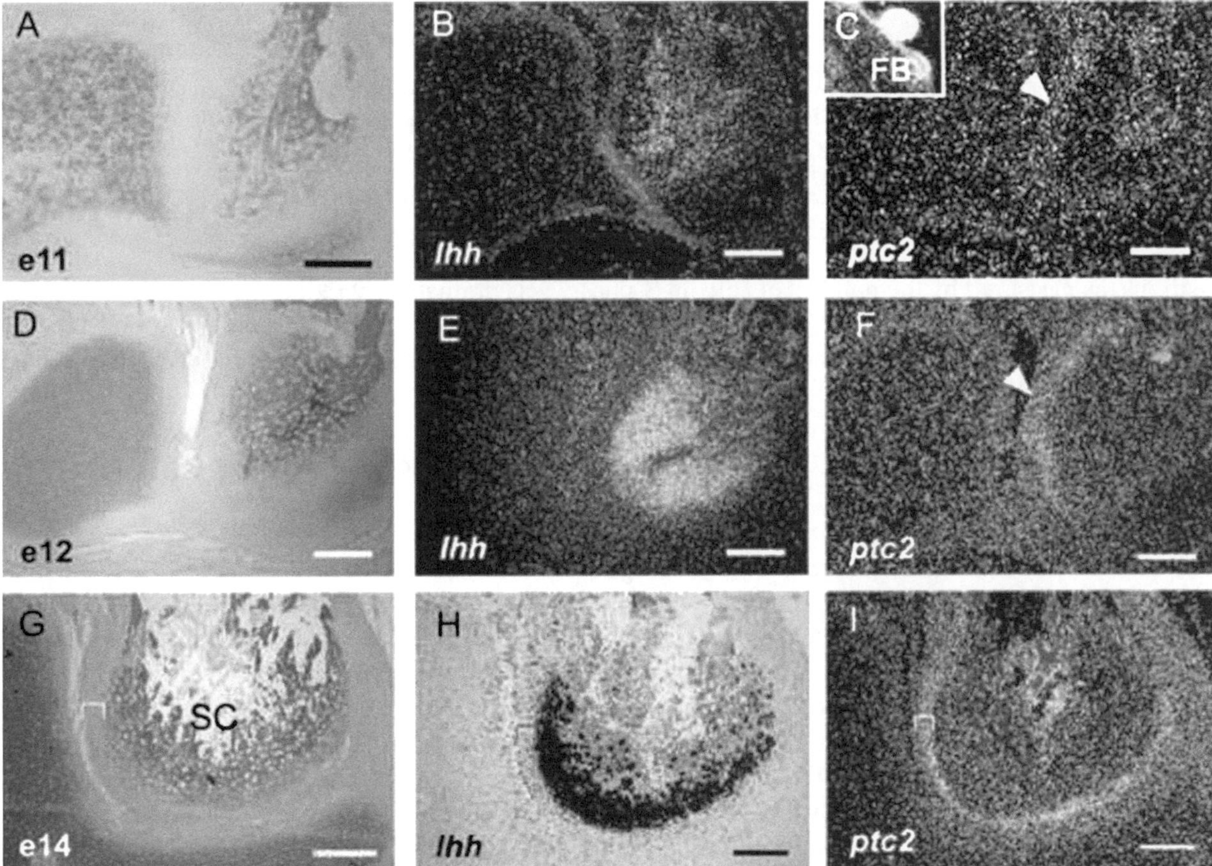


Fig. 5. Ihh signalling from secondary chondrocytes. (A,D,G,) Alcian Blue/Chlorantine Fast Red stained controls. (A–C) At e11, *Ihh* is expressed by SCs (B), but, as shown on an adjacent section (C), *ptc2* expression is at the limit of detection in the germinal region (white arrowhead); however, it is expressed in the feather buds (FB; inset in C). (D–F) At e12, *Ihh* is expressed by SCs (E), and *ptc2* (F) is expressed in the germinal region (arrowhead). (G–I) At e14, the relationship between *Ihh* (H) and *ptc2* (I) expression is maintained, brackets indicate extent of the germinal region. Scale bars: 100 $\mu$m.

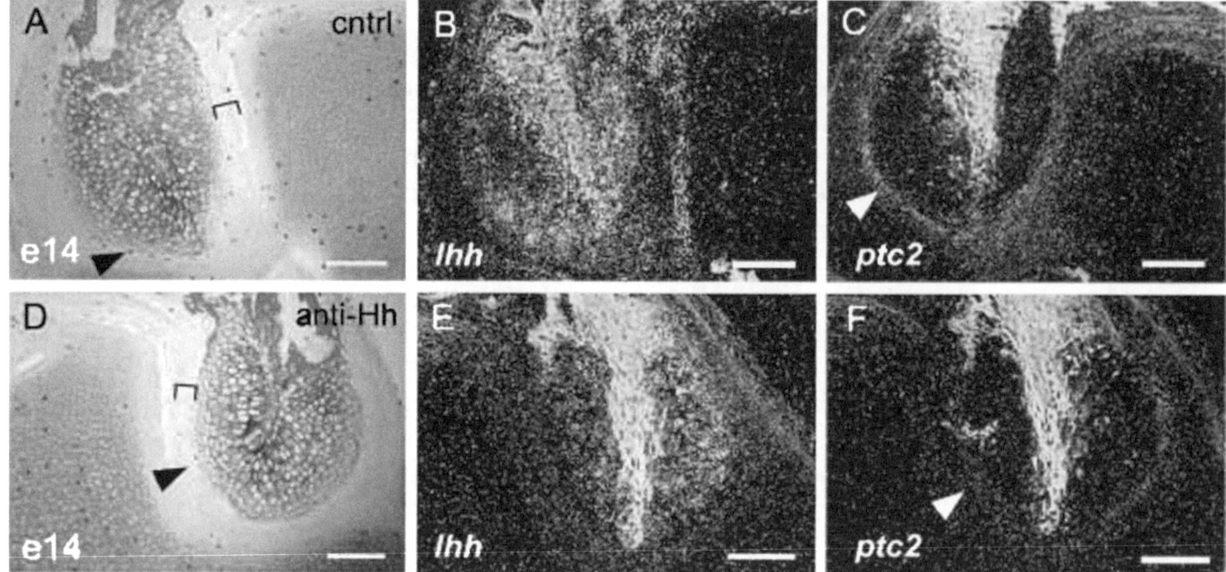

Fig. 6. Inhibition of Ihh signalling reduces proliferation in the germinal region at e14. (A) control Q/QJ cultures cultured for 16 hours and labelled with BrdU. Bracket demarcates the germinal region, black arrowhead indicates proliferating cells. (B) *Ihh* expression in an adjacent section. (C) *ptc2* is expressed in the germinal region (white arrowhead). (D–F) Hh blocking-antibody Treated contralateral side. (D) The reduction in number of BrdU labelled cells is indicated by the black arrowhead. (E) An adjacent section showing unaffected *Ihh* expression. (F) An adjacent section showing downregulation of *ptc2* expression in the germinal region (arrowhead). Scale bars: 100 $\mu$m.

sections, as measured by the number of cells in S-phase, was dramatically reduced when compared with non-Hh antibody treated, or no antibody treated controls (Fig. 6A,D). $T$-test analyses on scores for four pairs of QJ/Qs revealed significant differences in each case ($P < 0.05$): on average the total number of BrdU-labelled cells in the germinal region of 5E1-treated samples was reduced to a third of that in the controls. To ensure the effect was specific, we analysed adjacent sections for *Ihh* and *Ptc2* expression: *Ihh* was unaffected (Fig. 6B,E), whereas *Ptc2* was downregulated in the treated side relative to the control (Fig. 6C,F).

### 3.5. Mechanical regulation of secondary chondrogenesis

In order to explore the role of mechanical cues in the genesis of SCs, we used the transcriptional regulation of *osteocalcin* as an assay; this would also provide evidence on whether the *Cbfa1*-expressing nature of the germinal cells was truly indicative of pre-osteoblastic commitment. We utilised an hourly articulation regime, based on the methods of Hall [18], in which the joint is manually flexed whilst in explant culture to recapitulate the effects of jaw opening and closing as occurs *in ovo*. We coupled analysis of *ocn* expression with BrdU labelling in order to determine potential effects on the cell cycle simultaneously. Pulsing at 0 hours, followed by hourly articulation for 9 hours and analysis at 24 hours, revealed labelling of chondrocyte pairs (Fig. 7A) and an absence of *ocn* expression (Fig. 7B). The control joint possessed similar labelling of chondrocyte pairs (Fig. 7C) after 24 hours, but in contrast to the articulated joint, *ocn* expression was readily detected at the periphery of the SC domain (Fig. 7D; 0/4 in articulated joints; 3/4 in non-articulated joints). Thus, in the absence of mechanical stress, the cells of the germinal region rapidly give rise to definitive osteoblasts. Having established that mechanical stress

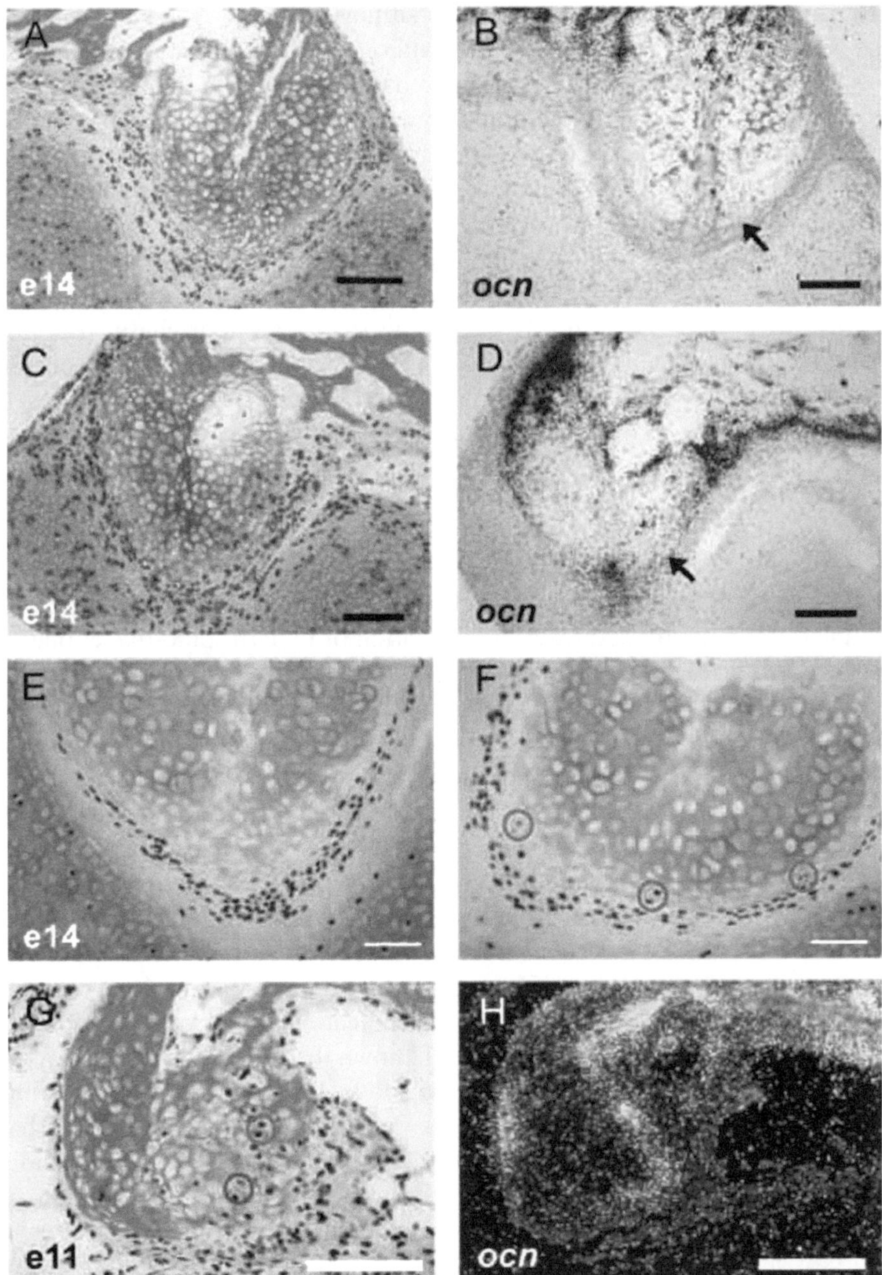

Fig. 7. Mechanical stimulation in secondary chondrogenesis. (A–F) e14, (G–H) e11. (A) BrdU labelled at 0 hours, with a 24-hour chase with mechanical stimulation; labelled doublets are observed. (B) *osteocalcin (ocn)* expression in an adjacent section: *ocn* is not seen around the SCs (black arrow). (C) BrdU labelled at 0 hours, with a 24-hour chase without mechanical stimulation; again labelled doublets are found. (D) Upregulation of *ocn* expression around the SCs on an adjacent section (black arrow). (E) BrdU labelled at 16 hours followed by 24-hour chase without mechanical stimulation; labelled cells are found in the germinal region. (F) As E, but with mechanical stimulation; BrdU labelled cells are found in the chondrocyte zone (circled). (G) e11 BrdU pulse at 0 hours followed by a chase for 48 hours without mechanical stimulation, both chondrocytes (blue circle) and osteoblasts (red circle) are labelled. (H) Adjacent section to G, showing *ocn* expression in BrdU-labelled cells. Scale bars: E, 100 $\mu$m; F, 50 $\mu$m.

resulting from the articulation of the QJ/Q joint was sufficient to repress osteoblast differentiation we wanted to assess whether it was also promoting chondrogenic differentiation. Using the same regime but with a delayed labelling strategy incorporating a chase (i.e. pulse at 24 hours, chase for 24 hours), we could follow the fate of cycling cells. By doing so we found that, as before, in the absence of articulation chondrocytes were not labelled (Fig. 7E), but after the articulation regime labelled chondrocytes could be detected (Fig. 7F; $n = 3$). This indicated that germinal region cells had been committed to the chondrocyte pathway. The rapid shift in cell-type differentiation, dependent upon the presence or absence of mechanical stimulation, indicated that the germinal region was extremely plastic, with the fate decision being a late event in germinal region expansion. In order to determine whether this was the case, we labelled e11 explanted QJ/Qs at 0 hours followed by a 48 hour chase ($n = 3$). The resultant BrdU-labelling of both chondrocytes and osteoblasts (Fig. 7G) indicated that precursor cells capable of giving rise to osteoblasts, as well as cells committed to giving rise to chondrocytes, had both been proliferating under the influence of mechanical stimulation (Fig. 7H).

## 4. Discussion

Our study of the events underlying secondary chondrogenesis has revealed the details and potential function of a unique differentiation pathway: the formation of hypertrophic-like chondrocytes from osteoblast precursors. In essence, this transition reflects the transcriptional upregulation of the *Sox9* gene within a precursor population that already expresses *Cbfa1*. Subsequently, a raft of genetic markers, normally associated with the hypertrophic chondrocyte phenotype become rapidly expressed within these cells; *Ihh, Col10, Wnt11* (data not shown) and, later, *osteopontin*. While there is a temporal sequence of gene expression which reflects differentiative status (i.e. *Col2* expression precedes *Col110* and *osteopontin* expression, the differentiative process is very much more rapid than primary chondrocytes which may be related to the lack of *PTHrp* expression in secondary chondrocytes which is known to be a check point for progression to hypertrophy. Thus the hypertrophic character that Murray observed [33] and used to distinguish secondary chondrocytes from their primary counterparts is largely true in common with mammalian secondary chondrocytes [37]. In itself, the evidence that *Sox9* expression is upregulated is unsurprising given its pivotal role in chondrocyte specification and gene regulation [5,6] what is unusual is the succession of gene upregulation. It is well known that immature chondrocytes of the growth plate gradually acquire the range of markers outlined above, but start from a population of precursors expressing *Sox9*. Misexpression of *Cbfa1* in immature, *Sox9*-expressing chondrocytes has shown that it is sufficient to promote hypertrophy and exit from the cell cycle [10,39]. What our study of SCs shows is that the reverse is also true: mechanical evocation of *Sox9 in ovo*, in a population expressing *Cbfa1*, leads to the same array of gene targets becoming upregulated.

This reversibility of succession is evidence for a simple mechanism underlying hypertrophy and the control of chondrocyte maturation. It is also noteworthy that although the endpoint for pre-osteoblasts and chondrocytes can be the same, the genes expressed by (pre-) hypertrophic chondrocytes are not merely the sum of Sox9 targets [6] plus Cbfa1 targets [9]. We characterise the germinal cells as pre-osteoblastic because they express *Cbfa1*, and, if not diverted into chondrogenesis, they will upregulate *osteocalcin* and produce matrix that binds Chlorantine Fast Red when they differentiated i.e. a bone-like matrix. Furthermore, prior to the evocation of secondary chondrocytes, the periosteum at the tip of the QJ gives rise to osteoblasts and no changes in the expression of peripheral markers occurred coincident with the differentiation of secondary chondrocytes. Nevertheless, we have yet to demonstrate conclusively that the secondary chondrocytes do derive from *Cbfa1* expressing cells at the single cell level.

### 4.1. Exit from the cell cycle follows commitment to the chondrogenic lineage

The transition from perichondrium to periosteum during endochondral ossification follows exit from the cell cycle and pre-hypertrophy in the underlying chondrocytes. This process of chondrocyte maturation is meticulously regulated and deficiency of any of a number of genes that affect proliferation or cell cycle control causes its perturbation, for example, *Pthrp* [25], *Pthrp-r* [30], *Ihh* (St. Jacques et al., 1999), *p57* [44] and *Cyclin D1* [2]. When considering secondary chondrogenesis, the periosteum is the source of the chondrocytes as well as the responding tissue. However, the chondrocytes formed rapidly exit the cell cycle and undergo hypertrophy. Thus, commitment to the chondrocyte lineage appears not to be accompanied by the formation of a self-renewing proliferative sub-population that is evident in the epiphyses of the developing long bone elements or the progenitor layer seen at the top of growth plates and is independent of movement. This conclusion is also supported by *in ovo* experiments that show following paralysis of the embryo, proliferating secondary chondrocyte-committed cells are not found [19]. This mechanism achieves rapid hypertrophy through continued mechanical articulation of the joint to maintain commitment to the chondrocyte lineage [17–19,34]. Our finding that *Col10* expression expanded to coincide with the *Col2* domain after 24 hours in culture indicated that commitment, differentiation and exit from the cell cycle are intimately correlated in SCs. Thus the initial spur to chondrogenesis of compression and upregulation of *Sox9* [38], appears to be followed by a single round of division and hypertrophy.

### 4.2. Ihh drives proliferation of the germinal region

The mouse chondrodysplasias mentioned above result in either delayed or accelerated hypertrophy. These changes then feed into ossification via *Ihh* signalling, resulting in reduced or increased bone formation, respectively. As in endochondral ossification, accelerated exit from the cell cycle in SCs is accompanied by *Ihh* upregulation, which then signals to the germinal region but not to the chondrocytes. As a result, cycling cells in the germinal region express *Ptc2*, but, as they become chondrocytes, the cells exit the cell cycle, downregulate *Ptc2* and upregulate *Ihh*. Hence, the cells progressed from Hh-signal receiving to Hh-signal producing at the germinal region/SC boundary, superficially resembling an extreme form of the growth plate of the *Pthrp-r* knockout mouse [29].

The threefold enhanced proliferation evident in the germinal region around the SCs was quenched by blocking Ihh function with the 5E1 antibody [11,43]. This finding was consistent with the demonstrated role of Ihh signalling to immature chondrocytes to drive proliferation [32], but was transposed to the germinal region. Given the pivotal roles of Ihh in chondrogenesis and osteogenesis of the long bones, it might also be expected to affect differentiation of the germinal cells. However, we did not find precocious or delayed expression of the osteogenic marker *osteocalcin* in 5E1 treated samples (data not shown). It was also apparent from the *osteocalcin* expression data at 24 and 48 hours that Ihh alone, in the absence of mechanical stimulation, was not sufficient to cause chondrogenic differentiation in germinal region descendants. This is consistent with the fact that *Ihh* is expressed only after SCs have formed, excluding it from any role in the initiation of SCs themselves. Thus the principal, and perhaps sole, function of Ihh signalling in secondary chondrogenesis is to drive proliferation of the germinal region and thereby increase indirectly recruitment to the chondrogenic fate. In turn, exit from the cell cycle appears to be a prerequisite for the synthesis of Ihh.

### 4.3. Mechano-transduction regulates gene expression and cell fate in the germinal region

One of the most striking features of the SCs described is that their formation is dependent upon articulation of the QJ/Q joint. *In ovo* this is achieved by muscular contraction that opens and closes the beak, which can be mimicked in explant by the manual operation of the joint ([18] and this paper). Whereas evocation of chondrogenesis has certainly been described, we found that the discreet nature of the germinal region as proliferative centre allowed us to explore the short-term response to mechanical articulation. The ability of mechanical articulation to prevent the default osteogenic pathway, as determined by *ocn* upregulation, indicates a direct role for mechanical stimulation in the regulation of gene expression. *Ocn* is upregulated by osteoblasts as they exit the cell cycle [1] and in the absence of mechanical stimulation the germinal cells follow this route. Thus, the cells of the germinal region that give rise to secondary chondrocytes would normally exit the cell cycle at this juncture, but the acquisition of chondrogenic fate diverts them and appears to allow them at least one further division. The fact that no self-renewing chondrocyte-committed precursors appear to be established, as evidenced by gene expression and BrdU labelling (see also [20]), accentuates this as a defining feature of secondary chondrogenesis in the chick. The chick quadratojugal system appears a useful system for the further analysis of the regulation of Sox9 and associated signalling pathways by mechanical cues.

## 5. Conclusion

The unusual differentiation undergone by SCs, and their transient nature, alludes to a role for SCs that is not structural, but is rather as a growth and signalling centre; our study demonstrates that a molecular basis for this function is Ihh production. The clear inference is that the pre-hypertrophic chondrocyte is of such utility as a source of extracellular signals for bone morphogenesis that, even when no cartilage scaffold exists, pre-hypertrophic chondrocytes are evoked. The finding that Ihh stimulates proliferation of the germinal region highlights the reciprocity between epigenetic and genetic pathways that characterize the development of the cranial tissues, and that underpin their evolution. Such a scenario resonates with other phenomena, such as fracture repair, where a short-lived, *Ihh*-synthesising soft callus forms from the periosteum under the influence of movement [14,41]. Conversely, the common characteristics that we have found between endochondral and secondary chondrogenesis indicate that the subordinate role of bone to cartilage [and other epigenetic influences [21] is a skeleton-wide phenomenon. Moreover, although requiring confirmation in mammals, our findings provide additional mechanistic evidence in support of Scott's view of cartilage as a pacemaker of cranial bone growth [36].

## Acknowledgements

We would like to thank Stalin Kariyawasam for help with histology, and all those who provided reagents for this study: C. Devlin, B. Houston, A. Kwan, C. Healy, J. Helms, A. Sinclair, C. Tabin and A. Vortkamp. This work was funded by the Arthritis Research Campaign.

## References

[1] J.E. Aubin and F. Liu, The osteoblast lineage, in: *Principles of Bone Biology*, Vol. 1, J.P. Bilezikian, L.G. Raisz and G.A. Rodan, eds, Academic Press, San Diego, 1996, pp. 51–67.

[2] F. Beier, Z. Ali, D. Mok, A.C. Taylor, T. Leask, C. Albanese, R.G. Pestell and P. LuValle, TGFbeta and PTHrP control chondrocyte proliferation by activating cyclin D1 expression, *Mol. Biol. Cell* **12** (2001), 3852–3863.

[3] D.M. Bell, K.K. Leung, S.C. Wheatley, L.J. Ng, S. Zhou, K.W. Ling, M.H. Sham, P. Koopman, P.P. Tam and K.S. Cheah, SOX9 directly regulates the type-II collagen gene, *Nat. Genet.* **16** (1997), 174–178.

[4] W.A. Beresford, *Chondroid Bone, Secondary Cartilage and Metaplasia*, Urban and Schwarzenberg, Baltimore, 1981.

[5] W. Bi, J.M. Deng, Z. Zhang, R.R. Behringer and B. de Crombrugghe, Sox9 is required for cartilage formation, *Nat. Genet.* **22** (1999), 85–89.

[6] B. de Crombrugghe, V. Lefebvre, R.R. Behringer, W. Bi, S. Murakami and W. Huang, Transcriptional mechanisms of chondrocyte differentiation, *Matrix Biol.* **19** (2000), 389–394.

[7] C.J. Devlin, P.M. Brickell, E.R. Taylor, A. Hornbruch, R.K. Craig and L. Wolpert, *In situ* hybridization reveals differential spatial distribution of mRNAs for type I and type II collagen in the chick limb bud, *Development* **103** (1988), 111–118.

[8] P. Ducy, R. Zhang, V. Geoffroy, A.L. Ridall and G. Karsenty, Osf2/Cbfa1: a transcriptional activator of osteoblast differentiation, *Cell* **89** (1997), 747–754.

[9] P. Ducy, Cbfa1: a molecular switch in osteoblast biology, *Dev. Dyn.* **219** (2000), 461–471.

[10] H. Enomoto, M. Enomoto-Iwamoto, M. Iwamoto, S. Nomura, M. Himeno, Y. Kitamura, T. Kishimoto and T. Komori, Cbfa1 is a positive regulatory factor in chondrocyte maturation, *J. Biol. Chem.* **275** (2000), 8695–8702.

[11] J. Ericson, S. Morton, A. Kawakami, H. Roelink and T.M. Jessell, Two critical periods of Sonic Hedgehog signaling required for the specification of motor neuron identity, *Cell* **87** (1996), 661–673.

[12] J. Fang and B.K. Hall, In vitro differentiation potential of the periosteal cells from a membrane bone, the quadratojugal of the embryonic chick. *Dev. Biol.* **180** (1996), 701–712.

[13] J. Fang and B.K. Hall, Chondrogenic cell differentiation from membrane bone periostea, *Anat. Embryol. (Berl.)* **196** (1997), 349–362.

[14] C. Ferguson, E. Alpern, T. Miclau and J.A. Helms, Does adult fracture repair recapitulate embryonic skeletal formation?, *Mech. Dev.* **87** (1999), 57–66.

[15] P.H. Francis, M.K. Richardson, P.M. Brickell and C. Tickle, Bone morphogenetic proteins and a signalling pathway that controls patterning in the developing chick limb, *Development* **120** (1994), 209–218.

[16] P.H. Francis-West, A. Abdelfattah, P. Chen, C. Allen, J. Parish, R. Ladher, S. Allen, S. MacPherson, F.P. Luyten and C.W. Archer, Mechanisms of GDF-5 action during skeletal development, *Development* **126** (1999), 1305–1315.

[17] B.K. Hall, The formation of adventitious cartilage by membrane bones under the influence of mechanical stimulation applied *in vitro*, *Life Sciences* **6** (1967), 663–667.

[18] B.K. Hall, *In vitro* studies on the mechanical evocation of adventitious cartilage in the chick, *J. Exp. Zool.* **168** (1968), 283–306.

[19] B.K. Hall, Immobilization and cartilage transformation into bone in the embryonic chick, *Anat. Rec.* **173** (1972), 391–403.

[20] B.K. Hall, Selective proliferation and accumulation of chondroprogenitor cells as the mode of action of biomechanical factors during secondary chondrogenesis, *Teratology* **20** (1979), 81–92.

[21] C. Hartmann and C.J. Tabin, Wnt-14 plays a pivotal role in inducing synovial joint formation in the developing appendicular skeleton, *Cell* **104** (2001), 341–351.

[22] S.W. Herring, Epigenetic and functional influences on skull growth, in: *The Skull*, Vol. 1, B.K. Hall and J. Hanken, eds, University of Chicago Press, Chicago and London, 1993, pp. 153–206.

[23] T.P. Hill, D. Spater, M.M. Taketo, W. Birchmeier and C. Hartmann, Canonical Wnt/$\beta$-catenin signalling prevents osteoblasts from differentiating into chondrocytes, *Developmental Cell* **8** (2005), 727–738.

[24] B. Houston, B.H. Thorp and D.W. Burt, Molecular cloning and expression of bone morphogenetic protein-7 in the chick epiphyseal growth plate, *J. Mol. Endocrinol.* **13** (1994), 289–301.

[25] A.C. Karaplis, A. Luz, J. Glowacki, R.T. Bronson, V.L. Tybulewicz, H.M. Kronenberg and R.C. Mulligan, Lethal skeletal dysplasia from targeted disruption of the parathyroid hormone-related peptide gene, *Genes Dev.* **8** (1994), 277–289.

[26] J. Kent, S.C. Wheatley, J.E. Andrews, A.H. Sinclair and P. Koopman, A male-specific role for SOX9 in vertebrate sex determination, *Development* **122** (1996), 2813–2822.

[27] I.S. Kim, F. Otto, B. Zabel and S. Mundlos, Regulation of chondrocyte differentiation by Cbfa1, *Mech. Dev.* **80** (1999), 159–170.

[28] T. Komori, H. Yagi, S. Nomura, A. Yamaguchi, K. Sasaki, K. Deguchi, Y. Shimizu, R.T. Bronson, Y.H. Gao, M. Inada, et al., Targeted disruption of Cbfa1 results in a complete lack of bone formation owing to maturational arrest of osteoblasts, *Cell* **89** (1997), 755–764.

[29] A.P. Kwan, I.R. Dickson, A.J. Freemont and M.E. Grant, Comparative studies of type X collagen expression in normal and rachitic chicken epiphyseal cartilage, *J. Cell Biol.* **109** (1989), 1849–1856.

[30] R.K. Ladher, V.L. Church, S. Allen, L. Robson, A. Abdelfattah, N.A. Brown, G. Hattersley, V. Rosen, F.P. Luyten, L. Dale, et al., Cloning and expression of the Wnt antagonists Sfrp-2 and Frzb during chick development, *Dev. Biol.* **218** (2000), 183–198.

[31] B. Lanske, A.C. Karaplis, K. Lee, A. Luz, A. Vortkamp, A. Pirro, M. Karperien, L.H. Defize, C. Ho, R.C. Mulligan et al., PTH/PTHrP receptor in early development and Indian hedgehog-regulated bone growth, *Science* **273** (1996), 663–666.

[32] V. Lefebvre, W. Huang, V.R. Harley, P.N. Goodfellow and B. de Crombrugghe, SOX9 is a potent activator of the chondrocytespecific enhancer of the pro alpha1(II) collagen gene, *Mol. Cell. Biol.* **17** (1997), 2336–2346.

[33] F. Long, X.M. Zhang, S. Karp, Y. Yang and A.P. McMahon, Genetic manipulation of hedgehog signaling in the endochondral skeleton reveals a direct role in the regulation of chondrocyte proliferation, *Development* **128** (2001), 5099–5108.

[34] P.D.F. Murray, Adventitious (secondary) cartilage in the chick embryo, and the development of certain bones and articulations in the chick skull, *Aust. J. Zool.* **11** (1963), 368–430.

[35] P.D.F. Murray and M. Smiles, Factors in the evocation of adventitious (secondary) cartilage in the chick embryo, *Aust. J. Zool.* **13** (1965), 351–381.

[36] R.V. Pearse 2nd, K.J. Vogan and C.J. Tabin, Ptc1 and Ptc2 transcripts provide distinct readouts of Hedgehog signaling activity during chick embryogenesis, *Dev. Biol.* **239**, (2001), 15–29.

[37] J.H. Scott, The growth of the human face, *Proc. Royal Soc. Med.* **47** (1954), 91–100.

[38] S. Shibata, K. Fukada, S. Suzuki and Y. Yamashita, Immunohistochemistry of collagen types II and X, and enzymehistochemistry of alkaline phosphatase in the developing condylar cartilage of the fetal mouse mandible, *J. Anat.* **191** (1997), 561–570.

[39] I. Takahashi, G.H. Nuckolls, K. Takahashi, O. Tanaka, I. Semba, R. Dashner, L. Shum, L. and H.C. Slavkin, Compressive force promotes sox9, type II collagen and aggrecan and inhibits IL-1beta expression resulting in chondrogenesis in mouse embryonic limb bud mesenchymal cells, *J. Cell Sci.* **111** (1998), 2067–2076.

[40] S. Takeda, J.P. Bonnamy, M.J. Owen, P. Ducy and G. Karsenty, Continuous expression of Cbfa1 in nonhypertrophic chondrocytes uncovers is ability to induce hypertrophic chondrocyte differentiation and partially rescues Cbfa1-deficient mice, *Genes Dev.* **15** (2001), 467–481.

[41] A. Vortkamp, K. Lee, B. Lanske, G.V. Segre, H.M. Kronenberg and C.J. Tabin, Regulation of rate of cartilage differentiation by Indian hedgehog and PTH-related protein, *Science* **273** (1996), 613–622.

[42] A. Vortkamp, S. Pathi, G.M. Peretti, E.M. Caruso, D.J. Zaleske and C.J. Tabin, Recapitulation of signals regulating embryonic bone formation during postnatal growth and in fracture repair, *Mech. Dev.* **71** (1998), 65–76.

[43] Q. Wu, Y. Zhang and Q. Chen, Indian hedgehog is an essential component of mechanotransduction complex to stimulate chondrocyte proliferation, *J. Biol. Chem.* **276** (2001), 35290–35296.

[44] P. Zhang, N.J. Liegeois, C. Wong, M. Finegold, H. Hou, J.C. Thompson, A. Silverman, J.W. Harper, R.A. DePinho and S.J. Elledge, Altered cell differentiation and proliferation in mice lacking p57KIP2 indicates a role in Beckwith-Wiedemann syndrome, *Nature* **387** (1997), 151–158.

Biorheology 43 (2006) 371–375
IOS Press



# Influence of mechanical stress on cell viability

C. Huselstein [a,1,*], N. de Isla [a,1], M.N. Kolopp-Sarda [b], H. Kerdjoudj [a], S. Muller [a] and J.F. Stoltz [a]

[a] *Mécanique et Ingénierie Cellulaire et Tissulaire, LEMTA UMR INPL CNRS 7563 et IFR 111 Bioingénierie, Faculté de Médecine, 54505 Vandoeuvre-lès-Nancy Cedex, France*
[b] *Immunologie, Faculté de Médecine et CHU - EA3443, 54505 Vandoeuvre-lès-Nancy, France*

**Abstract.** The cartilage is a hydrated connective tissue in joints that withstands and distributes mechanical forces. The chondrocytes utilize mechanical signals to regulate their metabolic activity through complex biological and biophysical interactions with the extracellular matrix (ECM). The aim of this work was to study the influence of mechanical stress on cells behavior cultured in 3D biosystems (alginate and alginate supplemented with hyaluronate). After mechanical stimulation, cell viability and cell death process were the main studied parameters. Our results indicated that viability and cell cycle progression were inhibited under mechanical stimulation, as far as the extracellular matrix was not yet synthesized. In contrast, on day 21, the mechanical stimulation had positive effect on these parameters.

Keywords: Chondrocyte, alginate hydrogel, cell viability, mechanical stimulation, extracellular matrix

## 1. Introduction

Articular cartilage restoration is problematic due to its poor capacity for repair once damaged or diseased. Tissue engineering of articular cartilage is a promising alternative for cartilage repair. Cell, biomaterial scaffolds, biochemical and physical regulatory signals can be utilized to engineer cartilage *in vitro* and *in vivo*. Mechanical loading applied on 3D biomaterials may play an important role to manufacture constructs that mimic native extracellular matrix. Although the effects of loading on extracellular matrix synthesis in cartilage tissue have been investigated a lot, there has been a lack of systematic effort to find the optimal ratio between loading and rest for the most efficient matrix synthesis.

Numerous *in vitro* studies have been performed using chondrocytes embedded in biomaterial to study how cells respond to mechanical stimulations [1,5]. These studies have shown that mechanical stimulation can affect the biosynthesis, turnover and structure of the macromolecules produced by the chondrocytes. Few investigators, however, have specifically focused on cell cycle and cell death. The focus of this study was to investigate the effect of mechanical stimulation not only on viability but also on cell death process by flow cytometry analysis.

---

[1] Both authors contributed equally to this work.

[*] Address for correspondence: Laboratoire de Mécanique et Ingénierie Cellulaire et Tissulaire, LEMTA UMR CNRS 7563, Faculté de Médecine, B.P. 184, 54505 Vandoeuvre-lès-Nancy Cedex, France. E-mail: Celine.Huselstein@medecine.uhp-nancy.fr.

## 2. Material and method

### 2.1. Isolation of chondrocytes and cell culture

Rat articular chondrocytes were seeded onto 3D biosystems. Chondrocytes were isolated enzymatically from femoral heads of male Wistar rats (Charles River, France).

### 2.2. Cell seeding into three-dimensional constructs

Sodium alginate (medium viscosity) from *Macrocystis pyrifera* (Sigma Aldrich) and HA from bacterial origin (Acros Organics, Belgium) were used. An alginate (Alg) solution (16 g/l) or a mixture containing Alg (16 g/l) and HA (4 g/l) were prepared by dissolution of autoclaved polymers in sterile 0.9% NaCl. The cell pellet was suspended at $3 \times 10^6$ cells/ml in polymer solutions. Then, the mixtures were dripped in 100 mM $CaCl_2$, using the method described by Guo et al. [3]. An average number of $10^5$ cells/bead was obtained and placed in complete medium. Cultures in beads were maintained in a humidified atmosphere of 5% $CO_2$ at 37°C for 10 and 24 days.

### 2.3. Mechanical testing

An agitator 10°/tridimensional Polymax 1040 (Fisher Bioblock Scientific, France) was used to stimulate the cells embedded in 3D constructs. Beads were deposed in tubes (Costar, France, 10 beads/tube) containing complete medium supplemented with HEPES 10 mM. Then, these tubes were all aligned parallel with the agitation plate and rotated at 30 cycles per minute. Mechanical stimulation was applied 48 h on neosynthetized matrixes of 7 and 21 days.

### 2.4. Flow cytometry analysis

Apoptosis and necrosis were measured by using Alexa Fluor 488-conjugated annexin V (1 : 10 final dilution) and propidium iodide (10 $\mu$g/ml) labeling (Molecular Probes), respectively. Cell cycle analysis was performed with DNA-Prep Stain kit (Beckman Coulter). For all the analysis, at least 5000 events were analyzed with EPICS XL flow cytometer (Beckman Coulter) and the SYSTEM II® software (Beckman Coulter). For the detection of cell cycle distribution, data were obtained with MULTI CYCLE AV® software (Phoenix Flow Systems).

### 2.5. Statistical analysis

All data are expressed as the mean ± SD. One-way analysis of variance followed by Student's comparison test were used. Significant differences were taken into account only when a probability of less than or equal to 5% was obtained.

## 3. Results

Propidium iodure and Alexa Fluor 488-conjugated annexin V used for study of cell death progression suggest that, on day 7, mechanical stimulation induced a significant decrease of cell viability and a significant increase of cell death (Fig. 1). On day 21, no variation of cell viability or cell death was observed

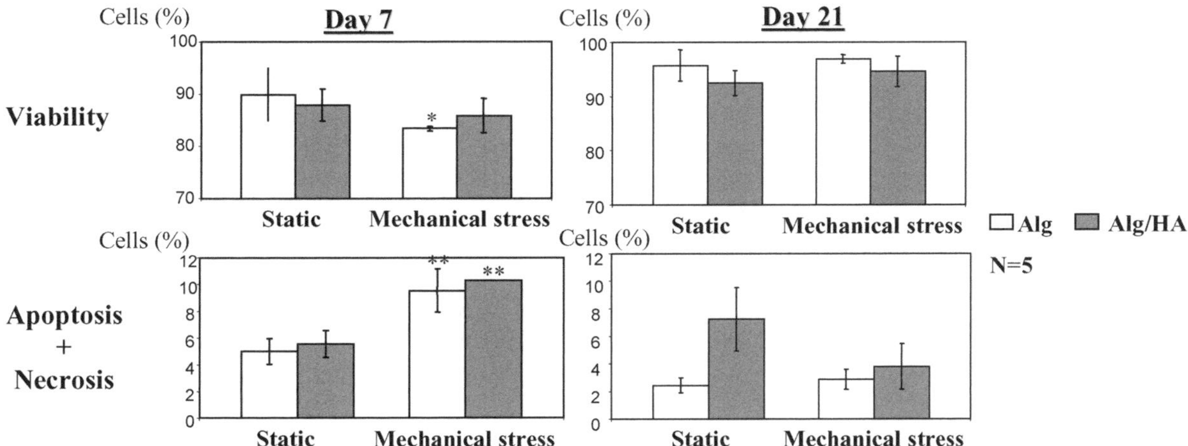

Fig. 1. Influence of mechanical stress on viability, apoptosis and necrosis. Chondrocytes were analyzed after culture of 7 and 21 days into Alg and Alg/HA biosystems and under static conditions or mechanical stress. $^{**}p < 0.025$, $^{***}p < 0.01$ (mechanical stress versus static conditions).

under mechanical stimulation as compared with static conditions. If we compared results obtained on day 21 with those obtained on day 7, we observed that the proportion of viable cells was significantly increased under mechanical stress.

Figure 2 shows that mechanical stress induced a decrease of quiescence state (G0/G1 phase) and an increase of cell proliferation (S and G2/M phases) only on biosystems with neosynthetised matrix (D21). On day 7, significant variations were observed when HA was present on biosystems (increase of G0/G1 phase). In contrast, on day 21, the presence of HA induced a significant increase of DNA synthesis.

## 4. Discussion

Cell growth, morphology and differentiation are known to be influenced by cellular environment (extracellular matrix, soluble factors, mechanical stress . . .). In this study, we investigated the influence of mechanical stimulation on chondrocyte viability and cell cycle progression. Chondrocytes were embedded onto alginate and alginate supplemented with sodium hyaluronate. Cell viability was analyzed after 7 and 21 days of culture.

Our results indicated that viability was inhibited under mechanical stimulation, as far as the extracellular matrix was not yet synthesized. In contrast, on day 21, the mechanical stimulation had positive effect on this parameter. Apoptosis is a physiological form of programmed cell death that plays an important role in many biological processes of the musculoskeletal system. It is reported that chondrocytes undergo apoptosis in the growth plate during endochondral bone formation [6]. In addition, there is an increase in apoptotic chondrocytes in degenerative cartilage disease accompanying aging that could be related to a decline in the number of chondrocytes, resulting in an inadequate cartilage formation with respect to turnover [4,6]. In our study, it seems that the quantity of neosynthetized matrix have an important effect on cell death under mechanical stress. It could be possible that, on day 7, cells are more exposed to intensity of mechanical signal due to the absence of neosynthetized matrix. Apoptosis and then necrosis of cells embedded into scaffolds may lead to a no functional biomaterial. In contrast, on day 21, the presence of a neosynthetized matrix may modulate the intensity of mechanical signal on the cells by modulating the shock wave.

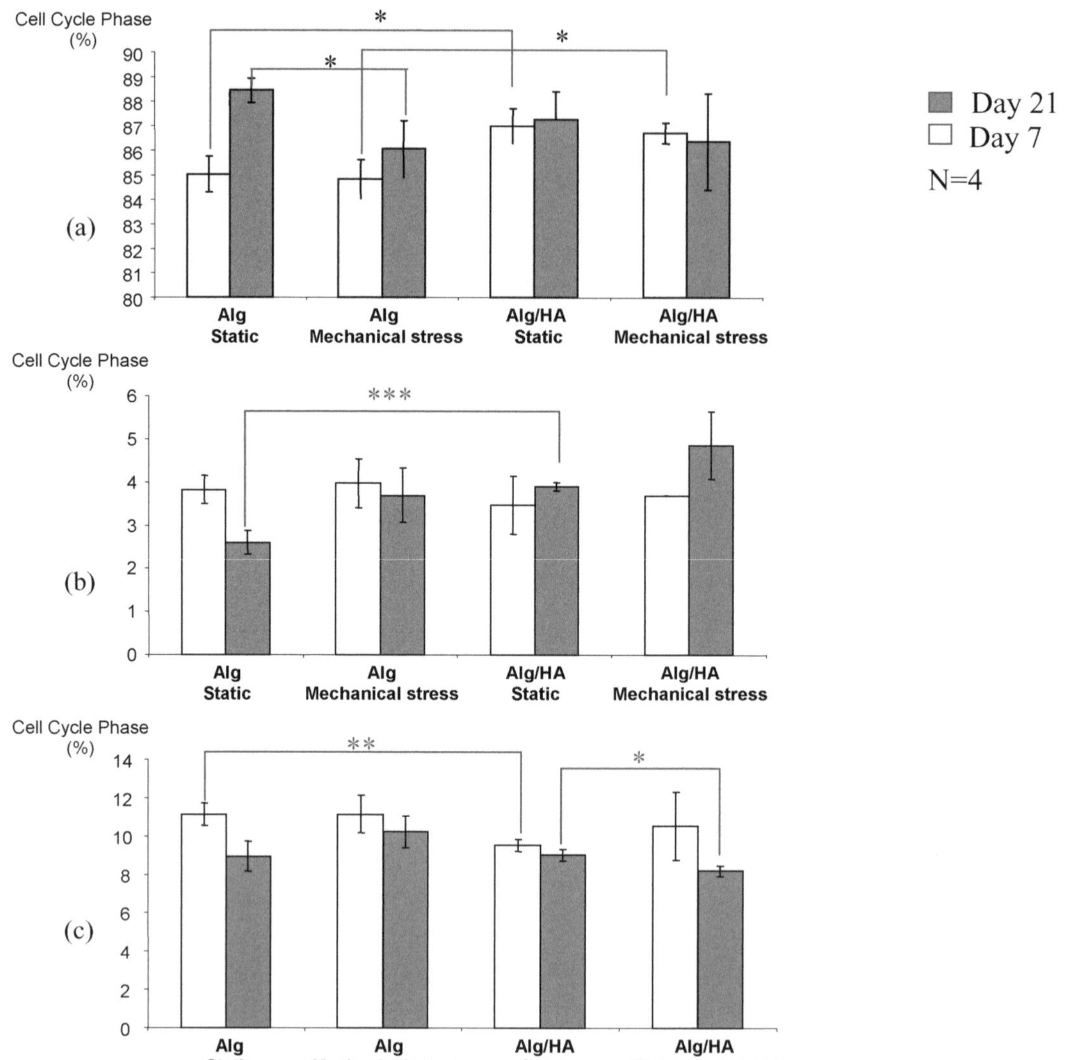



Fig. 2. Influence of mechanical stress on cell cycle progression: (a) G0/G1 phase, (b) S phase, (c) G2/M phase. Chondrocytes were analyzed after culture of 7 and 21 days into Alg and Alg/HA biosystems. $^*p < 0.05$; $^{**}p < 0.025$; $^{***}p < 0.01$.

This study demonstrated, also, that mechanical stimulation applied on 21 days neosynthesized matrixes induced DNA synthesis. In previous study, we have shown that chondrocyte phenotype was maintained in all the biomaterials and under mechanical stress [2]. The data from our studies suggest that it is very important to apply a stress like the knocking process to obtain an *in vitro* neosynthesized matrix comparable to native cartilage. However, mechanical stimulation does not be applied on biomaterials at the beginning of the culture.

## References

[1] T. Chowdhury, D.L. Bader, J.C. Shelton and D.A. Lee, Temporal regulation of chondrocyte metabolism in agarose constructs subjected to dynamic compression, *Arch. Biochem. Biophys.* **417** (2003), 105–111.

[2] C. Gigant-Huselstein, P. Hubert, D. Dumas, E. Dellacherie, P. Netter, E. Payan and J.F. Stoltz, Expression of adhesion molecules and collagen on rat chondrocyte seeded into alginate and hyaluronate based 3D biosystems. Influence of mechanical stresses, *Biorheology* **41** (2004), 423–431.

[3] J. Guo, G.W. Jourdian and D.K. McCallum, Culture and growth characteristics of chondrocytes encapsulated in alginate beads, *Connect. Tissue Res.* **19** (1989), 277–297.

[4] W.E. Horton, Jr., R. Yagi, D. Laverty and S. Weiner, Overview of studies comparing human normal cartilage with minimal and advanced osteoarthritic cartilage, *Clin. Exp. Rheumatol.* **23** (2005), 103–112.

[5] C.J. Hunter, J.K. Mouw and M.E. Levenston, Dynamic compression of chondrocyte-seeded fibrin gels: effects on matrix accumulation and mechanical stiffness, *Osteoarthritis Cartilage* **12** (2004), 117–130.

[6] M.W. Orth, The regulation of growth plate cartilage turnover, *J. Anim. Sci.* **77**(Suppl. 2) (1999), 183–189.

# Part II: Subchondral bone

Biorheology 43 (2006) 379–388
IOS Press



# Quantification of subchondral bone changes in a murine osteoarthritis model using micro-CT

S.M. Botter [a,b], G.J.V.M. van Osch [a,c], J.H. Waarsing [a], J.S. Day [a], J.A.N. Verhaar [a], H.A.P. Pols [b], J.P.T.M. van Leeuwen [b] and H. Weinans [a,*]

[a] *Department of Orthopaedics, Erasmus MC, University Medical Center, Rotterdam, The Netherlands*
[b] *Department of Internal Medicine, Erasmus MC, University Medical Center, Rotterdam, The Netherlands*
[c] *Department of Otorhinolaryngology, Erasmus MC, University Medical Center, Rotterdam, The Netherlands*

**Abstract.** In the past few years there has been a considerable interest in the role of bone in osteoarthritis. Despite the increasing evidence of the involvement of bone in osteoarthritis, it remains very difficult to attribute the cause or effect of changes in subchondral bone to the process of osteoarthritis. Although osteoarthritis in mice provides a useful model to study changes in the subchondral bone, detailed quantification of these changes is lacking. Therefore, the goal of this study was to quantify subchondral bone changes in a murine osteoarthritis model by use of micro-computed tomography (micro-CT). We induced osteoarthritis-like characteristics in the knee joints of mice using collagenase injections, and after four weeks we calculated various 3D morphometric parameters in the epiphysis of the proximal tibia. The collagenase injections caused cartilage damage, visible in histological sections, particularly on the medial tibial plateau. Micro-CT analysis revealed that the thickness of the subchondral bone plate was decreased both at the lateral and the medial side. The trabecular compartment demonstrated a small but significant reduction in bone volume fraction compared to the contralateral control joints. Trabeculae in the collagenase-injected joints were thinner but their shape remained rod-like. Furthermore, the connectivity between trabeculae was reduced and the trabecular spacing was increased. In conclusion, four weeks after induction of osteoarthritis in the murine knee subtle but significant changes in subchondral bone architecture could be detected and quantified in 3D with micro-CT analysis.

## 1. Introduction

In osteoarthritis cartilage becomes progressively damaged and many of the other joint tissues, such as synovium, ligaments and bone, are altered as well. The bone changes consist of the formation of osteophytes and sclerosis of the subchondral bone, which are both considered radiological hallmarks of osteoarthritis. In the past few years there has been a considerable interest in the role of the subchondral bone in the disease process of osteoarthritis. Despite the rising evidence of the involvement of bone in osteoarthritis (reviewed in [9,12,22,23,25], examples of experimental data in [4,8,29]), it remains very difficult to attribute the cause or effect of subchondral bone changes to the process of osteoarthritis.

---

*Address for correspondence: H. Weinans, Erasmus MC, University Medical Center, Erasmus Orthopaedic Research Lab, EE1614, P.O. Box 1738, 3000 DR Rotterdam, The Netherlands. Tel.: +31 10 40 87367/87384; Fax: +31 10 40 89415; E-mail: h.weinans@erasmusmc.nl.

The subchondral bone presumably has an important role in evenly distributing the forces resulting from joint loading, thereby protecting the cartilage from high peak stresses and damage. A change in subchondral bone architecture could in this view lead to altered loading patterns, which may play a role in osteoarthritis initiation and progression. Among the first to recognize the role of bone in the disease process were Radin and co-workers [34]: They stated that initiation of cartilage lesions probably required local stiffening of the subchondral bone, creating transverse stresses at the base of the articular cartilage, which would eventually lead to formation of cartilage lesions. These lesions would then progress as a result of stiffer subchondral bone [35]. Although this concept of osteoarthritis progression was never confirmed by experiments [11], this hypothesis led to an increasing interest in the role of subchondral bone changes in osteoarthritis [5,10,11,16,21,24]. In the laboratory, animal models are useful to further deepen our insights in the etiology of the disease. Various rodent models exist that simulate some of the osteoarthritic changes, either developing spontaneously [1,28] or post-traumatically [18,36,37]. In one of these models osteoarthritis-like characteristics can be induced by an intra-articular injection of collagenase in the knee joints of mice, creating instability by weakening the joint ligaments [37,38]. This instability eventually leads to cartilage damage and bone changes and is likely to reflect a post-traumatic osteoarthritic situation [39].

Micro-computed tomography (micro-CT) is currently applied to study bone changes in various animal models of osteoarthritis, such as dogs [14], cats [7] and guinea pigs [15]. Furthermore, it is known that micro-CT can be of use for disease monitoring in mice [42] but until now detailed quantification of the changes occuring in the subchondral bone of mice remains obscure. Therefore, the goal of this study was to quantify subchondral bone changes in a murine osteoarthritis model. To achieve this, we induced osteoarthritis-like changes in the knee joints of mice using collagenase injections and after four weeks we evaluated whether, and if so which, changes in the subchondral bone architecture could be detected. By calculating various 3D morphometric parameters we quantified changes of the subchondral bone plate and the subchondral trabeculae in the epiphysis of the proximal tibia.


## 2. Materials and methods

Following the approval of the local animal ethics committee, 16-week old male C3H/HeJ mice ($n = 13$) were used. The animals were housed in Individually Ventilated Cages (IVC's) with three brother littermates per cage and were fed *ad libitum*.

Animals were anesthetized with a 5% isoflurane/$N_2O/O_2$ mixture, and a small incision was made in the skin on the frontal side of the knee joint. Six $\mu$l containing 10 U of highly purified bacterial type VII collagenase (Sigma, St Louis, MO) was injected into the right knee intra-articular space whereas the left knee was injected with 6 $\mu$l saline [37]. As an analgesic, the animals also received a subcutaneous injection of buprenorphine (Temgesic, 0.01 mg/kg body weight). After four weeks the mice were euthanized and their knee joints were excised.

After fixation in 4% formalin for two days, the knee joints were scanned using a micro-computed tomography (micro-CT) scanner (Skyscan 1072, Skyscan, Aartselaar, Belgium) with a voxelsize of 8 $\mu$m. In order to distinguish bone tissue from non-bone tissue, the reconstructed greyscale images were segmented, using an automated tresholding algorithm [40]. Using 3D data analysis software (CTAnalyzer, Skyscan) the epiphysis of the tibia was selected as region of interest (ROI) for further analysis. Since we were only interested in bone changes occurring underneath the articular cartilage layer, care was taken not to include any outgrowing osteophytes. Furthermore, to verify whether changes in bone only occurred in the epiphysis or also at remote sites of the joint, a sample of the distal metaphyseal cortex was analyzed as well. The epiphysis was further divided in a cortical part (i.e. the subchondral bone plate) and a trabecular part, which were analyzed separately for differences in bone structure using the freely available software package 3D-Calculator (*www.eur.nl/fgg/orthopaedics/Downloads.html*). The following 3D morphometric parameters were calculated to describe the bone structure of the trabecular compartment: bone volume fraction, which describes the ratio of bone volume over tissue volume (BV/TV), structure model index (SMI), describing whether a bone structure is rod-like or plate-like [20], connectivity density (CD), which calculates the number of trabecular connections in a given volume [32], and bone thickness [19], which could be applied for trabeculae (trabecular thickness, Tb.Th.), the cortex of the distal metaphysis, and the medial and lateral subchondral bone plate (both expressed as Ct.Th.). Furthermore, the bone thickness algorithm could also be used to calculate the size of the marrow cavities, reflecting trabecular spacing (Tb.Sp.). The above-described procedure for the analysis of the epiphysis is depicted schematically in Fig. 1. The differences found between collagenase-injected joints and controls were expressed as percental changes from controls and Student's paired $t$-test was applied for statistical analysis, where $p < 0.05$ was considered significantly different.

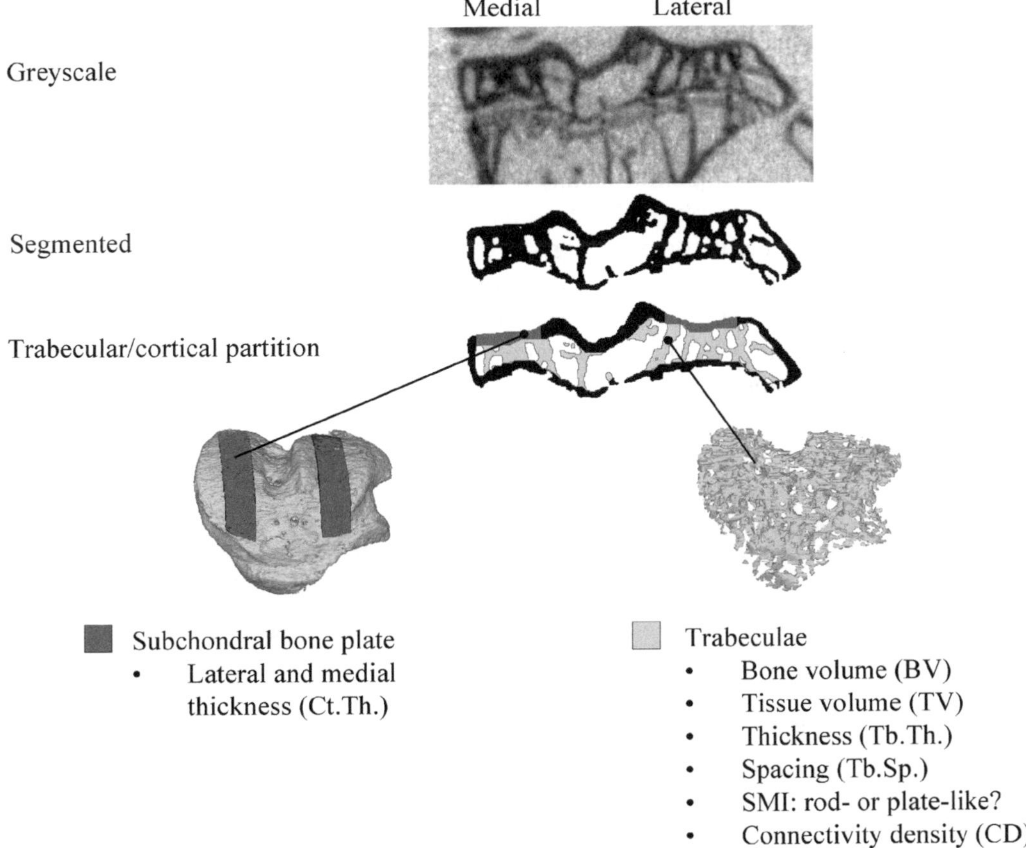


Fig. 1. Stepwise display of the methods used to analyse subchondral bone architecture with micro-CT. The micro-CT scans of the knee joints were reconstructed into greyscale images and segmented, i.e. binarized, to discriminate bone from non-bone. Following segmentation, the epiphysis of the proximal tibia was selected and the trabecular and cortical compartment were isolated. Each compartment was analyzed separately for the 3D morphometric parameters shown in the lower part of the scheme. BV = bone volume, TV = tissue volume, SMI = structure model index, CD = connectivity density, Tb.Th. = trabecular thickness, Tb.Sp. = trabecular spacing, Ct.Th. = cortical thickness.

Following the scanning procedure, knee joints were embedded in methyl methacrylate (MMA) and sectioned in coronal orientation to verify the presence of osteoarthritis characteristics such as cartilage damage and osteophytes. Per knee joint 8–14 undecalcified sections of 6 $\mu$m thickness were obtained, with 100 $\mu$m interspacing. Goldner staining was performed and all sections were scored for cartilage damage on the medial and lateral side of the tibia. A semi-quantitative scoring system was applied as follows: (0) no damage, (1) mild disruption of the cartilage surface, (2) moderate disruption of the cartilage surface accompanied by fissures, and (3) exposure of the calcified zone or subchondral bone. The average cartilage damage score was calculated, ranging from 0 (no damage) to 3 (maximal damage in medial or lateral side of the joint, in every section). In this case, the non-parametric Mann–Whitney test was applied for statistical analysis, where $p < 0.05$ was considered significantly different.

## 3. Results

The animals resumed normal activity within a few hours after surgery and limping of the collagenase-treated limb was observed only in the first few days after the operation. There was no significant decline in body weight during the experiment.

After four weeks typical osteoarthritis characteristics were found, such as osteophytes and cartilage damage (Fig. 2). The cartilage damage score in the collagenase-injected knees was 1.2 at the medial side vs. 0.2 at the lateral side, whereas in the contralateral control joints virtually no damage was observed (Fig. 3). The collagenase-injection changed the bone architecture of the epiphysis. The bone volume fraction (BV/TV) of the epiphyseal trabecular bone was consistently lower in all osteoarthritic knees compared to the contralateral controls (Fig. 4). On average a decrease of 11% was found, mainly as a result of increased TV (i.e. trabecular bone volume + marrow cavity volume). Although the general shape of the trabeculae remained rod-like as indicated by the unchanged SMI, a small decrease in trabecular thickness (Tb.Th.) was observed, accompanied by a corresponding increase in trabecular spacing (Tb.Sp.). Furthermore, the number of trabecular connections (CD) was found to be lower

Saline (left knee)        Collagenase (right knee)

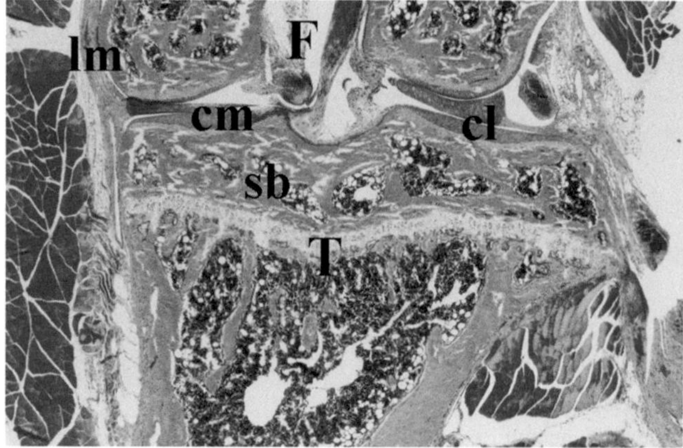

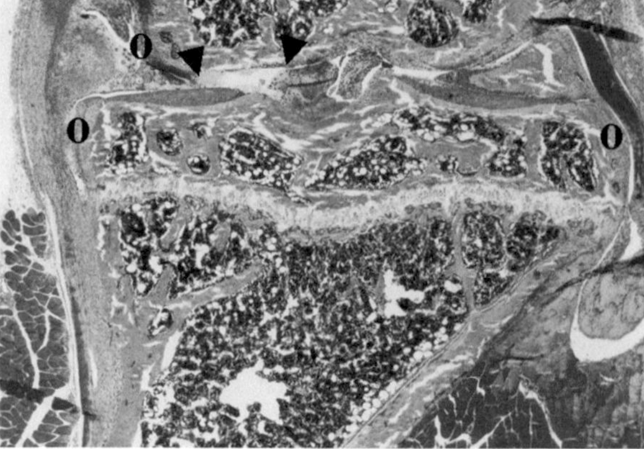


Fig. 2. Histology of a saline-injected and collagenase-injected murine knee joint. Knee joints were fixed in formalin, embedded in MMA, sectioned and Goldner staining was performed. Both cartilage damage (on medial femur between arrowheads) and osteophytes (o) were observed. The cartilage damage in the tibia was scored and is presented in Fig. 3. Original magnification ×40. F = femur, T = tibia, cm = cartilage medial, cl = cartilage lateral, lm = ligament, sb = subchondral bone in the epiphysis.

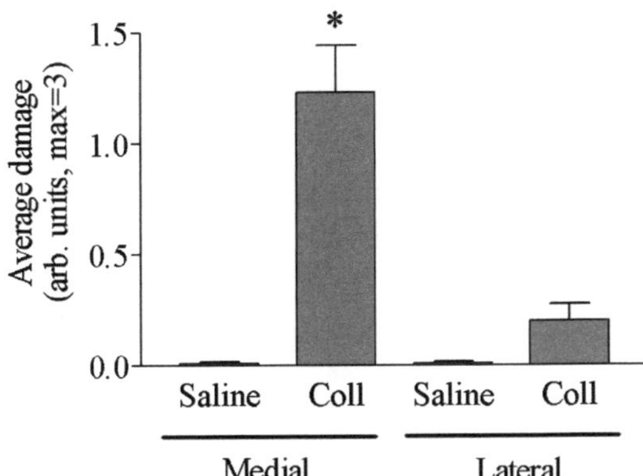

Fig. 3. Cartilage damage in the tibia at the lateral and medial side. For each collagenase or saline-injected contralateral joint (both $n = 13$), microscopic scoring of cartilage damage in the tibia was performed, using a semi-quantitative scale. This resulted in an average lateral or medial damage value with a maximum value of 3 (i.e. maximum damage observed in every analyzed section, on both the medial and lateral side). Values are mean $\pm$ SEM, $^*p < 0.05$. Coll = collagenase.

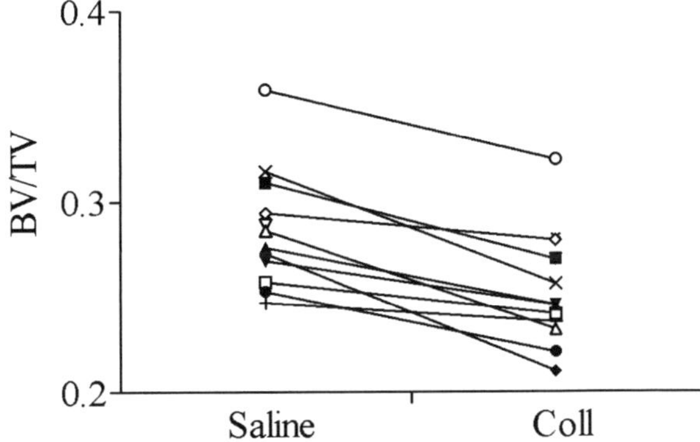

Fig. 4. Consistently lower bone volume fraction in the epiphysis. The crude BV/TV value of each individual knee joint is shown, and a connection is made between the joints of one animal to indicate the direction of change. In all cases, the value was lower in the collagenase-injected knee joint (average decrease: 11%, see Fig. 5).

in collagenase-injected knee joints than in controls (Fig. 5). These bone changes were apparently restricted to the epiphyseal area close to the joint, since the cortical thicknesses of the distal metaphysis remained unchanged (mean $\pm$ standard deviation saline-injected: $272.7 \pm 13.6$ $\mu$m, collagenase-injected: $268.8 \pm 13.6$ $\mu$m, $p = 0.312$).

The cortex of the epiphysis also changed in osteoarthritic joints: The subchondral bone plate was thinner both at the lateral and the medial side (Fig. 6a). This finding, derived from micro-CT analysis, was qualitatively confirmed by histology (Fig. 6b).

## 4. Discussion

This study showed that micro-CT can be used to detect subtle changes in subchondral bone in a murine model at a relatively early stage of osteoarthritis. Since the most prominent bone changes were to be expected in close vicinity of the diseased joint, we concentrated on the epiphysis of murine tibiae, and analyzed both the subchondral bone plate as well as the subchondral trabecular compartment.

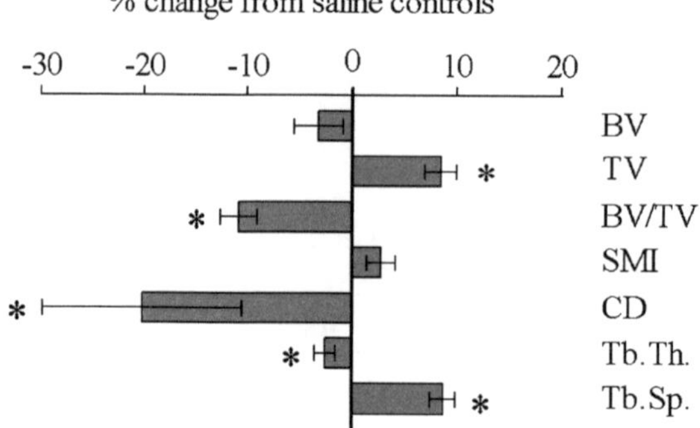


Fig. 5. Changes in 3D bone morphometric parameters. The isolated trabeculae (see Fig. 1) were analyzed separately for the indicated morphometric parameters. Of each parameter, the percental change from the saline-injected control knee joints was calculated. Values are mean ± SEM, $^{*}p < 0.05$. BV = bone volume, TV = tissue volume, SMI = structure model index, CD = connectivity density, Tb.Th. = trabecular thickness, Tb.Sp. = trabecular spacing, Ct.Th. = cortical thickness.

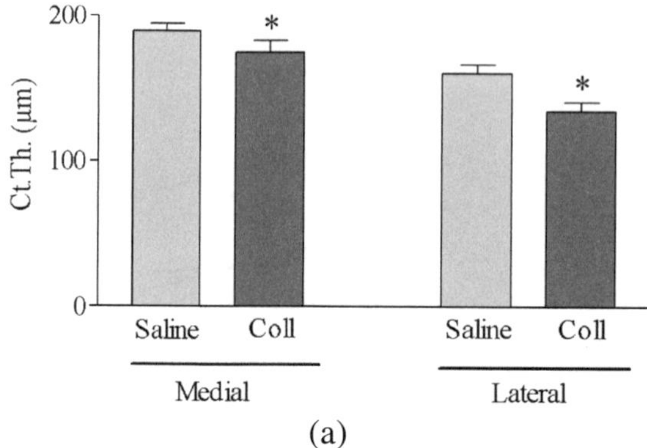


Fig. 6. Decreased thickness of the subchondral bone plate. (a) The thickness of the subchondral plate at the medial and lateral side was calculated from the segmented images, values are mean ± SEM, $^{*}p < 0.05$; (b) segmented image from a saline and collagenase-injected joint, with the subchondral bone plate shown in grey. Note the decrease in thickness on both the medial and lateral side. Below each segmented image are the corresponding histological sections, with the thickness of the subchondral bone plate indicated between dotted lines. Original magnification ×200. ucc = uncalcified cartilage, cc = calcified cartilage, sbp = subchondral bone plate. (*Figure continued on next page.*)

Opposite to what has been reported in studies using clinical samples with late-stage osteoarthritis [4,9,17], our results show that the subchondral bone plate became thinner. As a result, the volume of the trabecular compartment became slightly larger, leading to a decrease in bone volume fraction (BV/TV). The trabecular architecture also changed: We measured a reduction in trabecular thickness contributing to the observed increase in trabecular spacing. Furthermore, connections between trabeculae were lost as indicated by the decrease in connectivity density.

The fact that instead of sclerosis, subchondral bone loss occurred might reflect that in our model the osteoarthritic process is still in a relatively early stage, and that sclerosis might occur afterwards. Other animal models confirm this hypothesis: In 1993, Dedrick et al. reported on subchondral bone changes in a canine model in which the anterior cruciate ligament was transected (ACLT). At three time points (3, 18 and 54 months) after ACLT, they found loss of subchondral trabecular bone [14]. More recently, the same model was used by Boyd et al., who also found loss of trabecular bone 12 weeks post-ACLT [6]. The latter group also applied the ACLT model in cats and reported both a decrease in trabecular bone

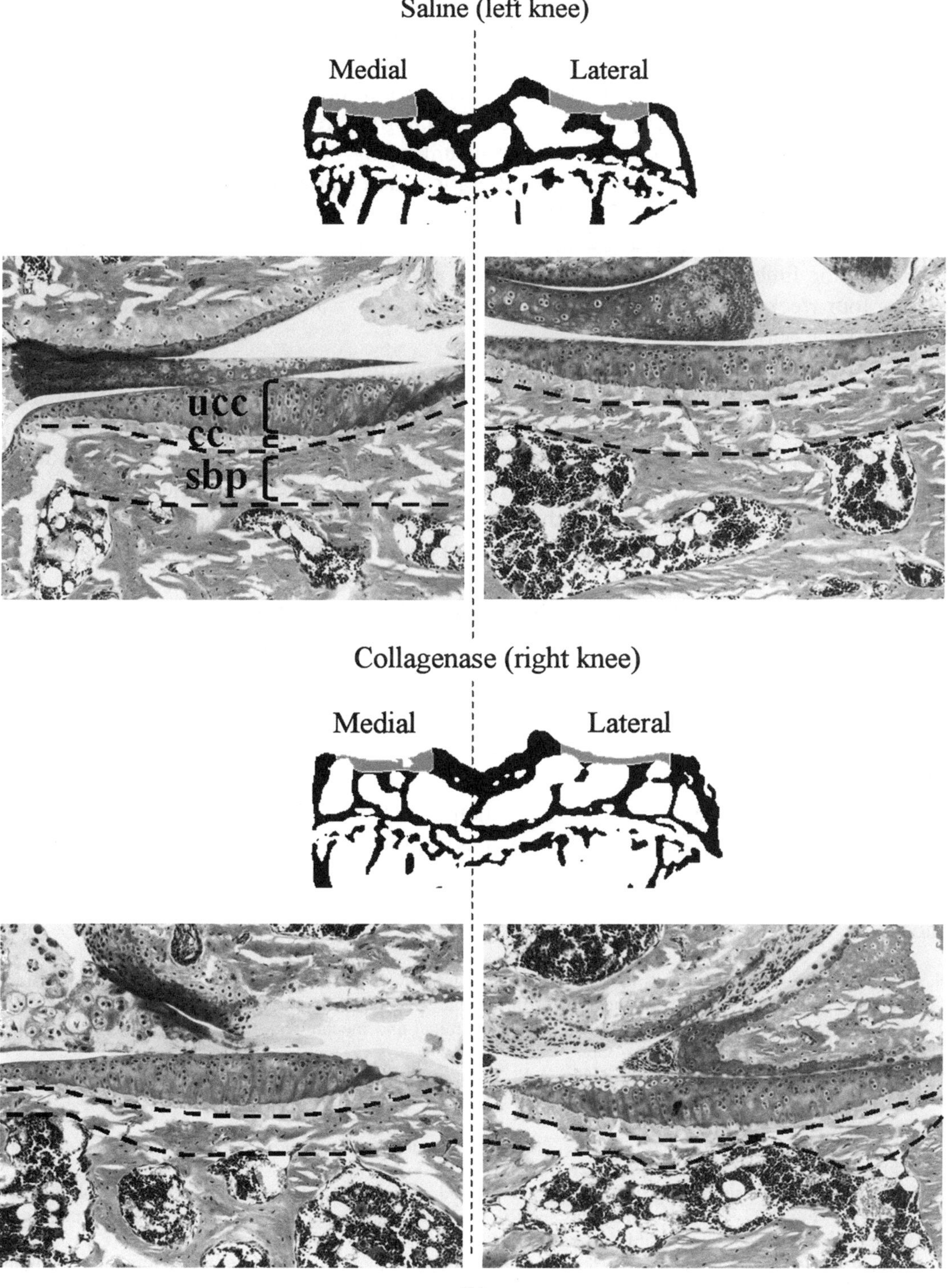



(b)

Fig. 6. Continue.

as thinning of the subchondral bone plate [7]. In the long-term though a discrepancy existed between cats [7], where the subchondral bone plate remained thin, and dogs, where the bone plate became almost twice as thick [14]. Recovery of bone loss was also observed in a rabbit [3] and rat [18] ACLT model and in a guinea pig meniscectomy model [33]. So, recovery of bone loss, as well as high turnover and anabolic effects on osteoblasts [2,26,27,29] seem to be a normal trend in a later phase of osteoarthritis.

Whether the observed bone loss might be due to unloading of the treated (collagenase-injected or ACLT) limb remains subject of discussion. Gait alterations and unloading of the osteoarthritic limb have been reported in dogs [31] and rats [13]. Although we have not quantified the degree of unloading, we did observe some limping in the first few days after the injection procedure but after this initial period normal usage of the limb was observed. Indeed, the cortical thickness of the distal metaphysis was not changed after four weeks, indicating that the observed bone loss was localized in the subchondral region of the diseased knee only.

The current study indicates that in this murine osteoarthritis model subtle subchondral bone changes occur at a relative early stage in the disease process. When these bone changes are compared to the observed cartilage damage, another interesting finding was that the decrease in subchondral bone plate thickness at the medial and lateral sides was comparable despite the fact that the lateral side had only little cartilage damage. This finding is suggestive for the fact that the changes found in the subchondral bone run parallel to or maybe even precede the cartilage degeneration. However, to confirm this hypothesis, a longitudinal study is necessary. Thus, our future plans are aimed at analysing the process in time using *in vivo* micro-CT scanning, which has already been proven useful in both a rat osteoarthritis [30] and osteoporosis model [41]. By scanning the joints of the same animal at successive time points in the disease process, we hope to further elucidate the relation between bone and cartilage changes in osteoarthritis.


## Acknowledgements

The authors would like to thank Nicole Kops for help with histology, Jan Floris de Graaf for computing a large part of the morphometric data and the Erasmus Animal Laboratory for taking care of the animals.


## References

[1] J.M. Anderson-Mackenzie, H.L. Quasnichka, R.L. Starr, E.J. Lewis, M.E. Billingham and A.J. Bailey, Fundamental subchondral bone changes in spontaneous knee osteoarthritis, *Int. J. Biochem. Cell Biol.* **37** (2005), 224–236.

[2] A.J. Bailey, J.P. Mansell, T.J. Sims and X. Banse, Biochemical and mechanical properties of subchondral bone in osteoarthritis, *Biorheology* **41** (2004), 349–358.

[3] D.L. Batiste, A. Kirkley, S. Laverty, L.M. Thain, A.R. Spouge and D.W. Holdsworth, Ex vivo characterization of articular cartilage and bone lesions in a rabbit ACL transection model of osteoarthritis using MRI and micro-CT, *Osteoarthritis Cartilage* **12** (2004), 986–996.

[4] D. Bobinac, J. Spanjol, S. Zoricic and I. Maric, Changes in articular cartilage and subchondral bone histomorphometry in osteoarthritic knee joints in humans, *Bone* **32** (2003), 284–290.

[5] C.S. Bonnet and D.A. Walsh, Osteoarthritis, angiogenesis and inflammation, *Rheumatology (Oxford)* **44** (2005), 7–16.

[6] S.K. Boyd, J.R. Matyas, G.R. Wohl, A. Kantzas and R.F. Zernicke, Early regional adaptation of periarticular bone mineral density after anterior cruciate ligament injury, *J. Appl. Physiol.* **89** (2000), 2359–2364.

[7] S.K. Boyd, R. Muller, T. Leonard and W. Herzog, Long-term periarticular bone adaptation in a feline knee injury model for post-traumatic experimental osteoarthritis, *Osteoarthritis Cartilage* **13** (2005), 235–242.

[8] O. Bruyere, C. Dardenne, E. Lejeune, B. Zegels, A. Pahaut, F. Richy, L. Seidel, O. Ethgen, Y. Henrotin and J.Y. Reginster, Subchondral tibial bone mineral density predicts future joint space narrowing at the medial femoro-tibial compartment in patients with knee osteoarthritis, *Bone* **32** (2003), 541–545.

[9] D.B. Burr, Anatomy and physiology of the mineralized tissues: role in the pathogenesis of osteoarthrosis, *Osteoarthritis Cartilage* **12** (Suppl. A) (2004), S20–S30.

[10] D.B. Burr and E.L. Radin, Microfractures and microcracks in subchondral bone: are they relevant to osteoarthrosis?, *Rheum. Dis. Clin. North Am.* **29** (2003), 675–685.

[11] D.B. Burr and M.B. Schaffler, The involvement of subchondral mineralized tissues in osteoarthrosis: quantitative microscopic evidence, *Microsc. Res. Tech.* **37** (1997), 343–357.

[12] D.B. Burr, The importance of subchondral bone in osteoarthrosis, *Curr. Opin. Rheumatol.* **10** (1998), 256–262.

[13] K.A. Clarke, S.A. Heitmeyer, A.G. Smith and Y.O. Taiwo, Gait analysis in a rat model of osteoarthrosis, *Physiol. Behav.* **62** (1997), 951–954.

[14] D.K. Dedrick, S.A. Goldstein, K.D. Brandt, B.L. O'Connor, R.W. Goulet and M. Albrecht, A longitudinal study of subchondral plate and trabecular bone in cruciate-deficient dogs with osteoarthritis followed up for 54 months, *Arthritis Rheum.* **36** (1993), 1460–1467.

[15] D.K. Dedrick, R. Goulet, L. Huston, S.A. Goldstein and G.G. Bole, Early bone changes in experimental osteoarthritis using microscopic computed tomography, *J. Rheumatol. Suppl.* **27** (1991), 44–45.

[16] M. Ding, A. Odgaard and I. Hvid, Changes in the three-dimensional microstructure of human tibial cancellous bone in early osteoarthritis, *J. Bone Joint Surg. Br.* **85** (2003), 906–912.

[17] M.D. Grynpas, B. Alpert, I. Katz, I. Lieberman and K.P.H. Pritzker, Subchondral bone in osteoarthritis, *Calcif. Tissue Int.* **49** (1991), 20–26.

[18] T. Hayami, M. Pickarski, G.A. Wesolowski, J. McLane, A. Bone, J. Destefano, G.A. Rodan and L.T. Duong, The role of subchondral bone remodeling in osteoarthritis: reduction of cartilage degeneration and prevention of osteophyte formation by alendronate in the rat anterior cruciate ligament transection model, *Arthritis Rheum.* **50** (2004), 1193–1206.

[19] T. Hildebrand and P. Ruegsegger, A new method for the model-independent assessment of thickness in three-dimensional images, *J. Microscopy* **185** (1997), 67–75.

[20] T. Hildebrand and P. Ruegsegger, Quantification of Bone Microarchitecture with the Structure Model Index, *Comput. Methods Biomech. Biomed. Engin.* **1** (1997), 15–23.

[21] H. Imhof, M. Breitenseher, F. Kainberger, T. Rand and S. Trattnig, Importance of subchondral bone to articular cartilage in health and disease, *Top. Magn. Reson. Imaging* **10** (1999), 180–192.

[22] H. Imhof, I. Sulzbacher, S. Grampp, C. Czerny, S. Youssefzadeh and F. Kainberger, Subchondral bone and cartilage disease: a rediscovered functional unit, *Invest. Radiol.* **35** (2000), 581–588.

[23] C.E. Kawcak, C.W. McIlwraith, R.W. Norrdin, R.D. Park and S.P. James, The role of subchondral bone in joint disease: a review, *Equine Vet. J.* **33** (2001), 120–126.

[24] D. Lajeunesse and P. Reboul, Subchondral bone in osteoarthritis: a biologic link with articular cartilage leading to abnormal remodeling, *Curr. Opin. Rheumatol.* **15** (2003), 628–633.

[25] D. Lajeunesse, The role of bone in the treatment of osteoarthritis, *Osteoarthritis Cartilage.* **12** (Suppl. A) (2004), S34–S38.

[26] P. Lavigne, M. Benderdour, D. Lajeunesse, P. Reboul, Q. Shi, J.P. Pelletier, J. Martel-Pelletier and J.C. Fernandes, Subchondral and trabecular bone metabolism regulation in canine experimental knee osteoarthritis, *Osteoarthritis Cartilage* **13** (2005), 310–317.

[27] J.P. Mansell and A.J. Bailey, Abnormal cancellous bone collagen metabolism in osteoarthritis, *J. Clin. Invest.* **101** (1998), 1596–1603.

[28] R.M. Mason, M.G. Chambers, J. Flannelly, J.D. Gaffen, J. Dudhia and M.T. Bayliss, The STR/ort mouse and its use as a model of osteoarthritis, *Osteoarthritis Cartilage* **9** (2001), 85–91.

[29] H. Matsui, M. Shimizu and H. Tsuji, Cartilage and subchondral bone interaction in osteoarthrosis of human knee joint: a histological and histomorphometric study, *Microsc. Res. Tech.* **37** (1997), 333–342.

[30] B.J. Morenko, S.E. Bove, L. Chen, R.E. Guzman, P. Juneau, T.M. Bocan, G.K. Peter, R. Arora and K.S. Kilgore, In vivo micro computed tomography of subchondral bone in the rat after intra-articular administration of monosodium iodoacetate, *Contemp. Top. Lab. Anim. Sci.* **43** (2004), 39–43.

[31] B.L. O'Connor, D.M. Visco, D.A. Heck, S.L. Myers and K.D. Brandt, Gait alterations in dogs after transection of the anterior cruciate ligament, *Arthritis Rheum.* **32** (1989), 1142–1147.

[32] A. Odgaard and H.J. Gundersen, Quantification of connectivity in cancellous bone, with special emphasis on 3-D reconstructions, *Bone* **14** (1993), 173–182.

[33] P.C. Pastoureau, A.C. Chomel and J. Bonnet, Evidence of early subchondral bone changes in the meniscectomized guinea pig. A densitometric study using dual-energy X-ray absorptiometry subregional analysis, *Osteoarthritis Cartilage* **7** (1999), 466–473.

[34] E.L. Radin, I.L. Paul and M.J. Tolkoff, Subchondral bone changes in patients with early degenerative joint disease, *Arthritis Rheum.* **13** (1970), 400–405.

[35] E.L. Radin and R.M. Rose, Role of subchondral bone in the initiation and progression of cartilage damage, *Clin. Orthop.* (1986), 34–40.

[36] H.M. van Beuningen, H.L. Glansbeek, P.M. Van der Kraan and W.B. Van den Berg, Osteoarthritis-like changes in the

murine knee joint resulting from intra-articular transforming growth factor-beta injections, *Osteoarthritis Cartilage* **8** (2000), 25–33.

[37] P.M. Van der Kraan, E.L. Vitters, H.M. Van Beuningen, L.B.A. Van de Putte and W.B. Van den Berg, Degenerative knee-joint lesions in mice after a single intraarticular collagenase injection. A new model of osteoarthritis, *J. Exp. Pathol. (Oxford)* **71** (1990), 19–31.

[38] G.J. van Osch, L. Blankevoort, P.M. Van der Kraan, B. Janssen, E. Hekman, R. Huiskes and W.B. Van den Berg, Laxity characteristics of normal and pathological murine knee joints in vitro, *J. Orthop. Res.* **13** (1995), 783–791.

[39] G.J. van Osch, P.M. Van der Kraan, E.L. Vitters, L. Blankevoort and W.B. Van den Berg, Induction of osteoarthritis by intra-articular injection of collagenase in mice. Strain and sex related differences, *Osteoarthritis Cartilage* **1** (1993), 171–177.

[40] J. Waarsing, J. Day and H. Weinans, An improved segmentation method for in-vivo micro-CT imaging, *J. Bone Miner. Res.* (2004).

[41] J.H. Waarsing, J.S. Day, J.C. van der Linden, A.G. Ederveen, C. Spanjers, N. De Clerck, A. Sasov, J.A. Verhaar and H. Weinans, Detecting and tracking local changes in the tibiae of individual rats: a novel method to analyse longitudinal in vivo micro-CT data, *Bone* **34** (2004), 163–169.

[42] L. Wachsmuth and K. Engelke, High-resolution imaging of osteoarthritis using microcomputed tomography, *Methods Mol. Med.* **101** (2004), 231–248.

Biorheology 43 (2006) 389–397
IOS Press



# Subchondral bone and ligament changes precede cartilage degradation in guinea pig osteoarthritis

Helen L. Quasnichka, Janet M. Anderson-MacKenzie and Allen J. Bailey *
*Collagen Research Group, University of Bristol, Bristol, UK*

**Abstract.** There is increasing recognition that osteoarthritis (OA) is a complex disease involving the whole synovial joint, rather than the articular cartilage alone, however its aetiology and pathogenesis is not understood. Our initial studies revealed elevated turnover of bone and ligament collagen in human and mouse OA, respectively. To investigate the relative appearance of pathology in cartilage, bone and ligament, we studied the progression of spontaneous OA in the Dunkin–Hartley (DH) guinea pig knee, and compared with age-matched control Bristol Strain 2 (BS2) knees. The classical radiographic OA score of the DH knees compared to BS2 knees was 2-fold higher at 24 weeks of age. The patella perimeter and subchondral bone density was significantly greater in the DHs at 24 and 36 weeks compared to BS2. The femoral intercondylar notch width was found to be significantly lower in the DHs at 24 and 36 weeks, compared to BS2, indicating bone remodelling at the cruciate ligament (CL) insertion site. We found significantly greater laxity of the DH anterior CL at 12, 16 and 20 weeks compared to BS2. This elevated laxity was associated with increased remodelling of the CLs, based on markers of collagen turnover, and occurred prior to bone and cartilage pathology. We propose that the laxity of the CL leads to remodelling of the subchondral bone, and intercondylar notch, due to a change in load through the joint. Remodelling of the CLs and bone occurs prior to and concomitant with histopathological changes in the articular cartilage respectively, demonstrating the fundamental role of the ligament and subchondral bone in the aetiology of knee OA.

Keywords: Osteoarthritis, subchondral bone, ligaments, collagen, metabolism, biomechanics, notch width

## 1. Introduction

Osteoarthritis is a common and cruelly debilitating disease of the joints. It is characterised by artic-ular cartilage destruction, local systemic increases in bone mineral density, bony outgrowths and cysts, thickening of bone tissue and narrowing of the joint space. However, based on its overall destruction, observed clinically in late-stage disease, OA is generally considered as a disorder of articular cartilage. Consequently, biochemical research has mainly focused on the mechanisms of cartilage destruction. Bio-chemical differences can undoubtedly be detected between fibrillating osteoarthritic cartilage and nor-mal cartilage; including increases in low sulphate proteoglycan (PG), COMP, matrix metalloproteinases (MMPs), aggrecanase and fibronectin [23]. These differences become more notable in late-stage disease when fibrillation is severe. Furthermore, collagen synthesis in osteoarthritic cartilage is only elevated in late-stage disease [1].

Alternatively some workers have proposed that changes in bone biochemistry and density are at the heart of OA aetiology. For example, Radin and Rose [29] proposed that bone changes in early-stage OA

*Address for correspondence: Professor Allen J. Bailey, Collagen Research Group, University of Bristol, Bristol BS40 5DU, UK. E-mail: A.J.Bailey@bristol.ac.uk.

produced a hardening (sclerosis) of the bone surface that physically challenges the cartilage on repetitive impact, initiating cartilage destruction. More recent studies have revealed that the OA-related bone changes are more complex. Dieppe et al. [12] measured bone metabolism in knee joints by scintigraphy of technetium-labelled bisphosphonate, which is incorporated into the mineral at sites of bone turnover, and found that levels correlated with the severity of the OA. We also reported an elevated turnover of subchondral bone collagen in the femoral heads of late-stage OA patients compared to controls [24]. Additionally, we found that this increase in collagen synthesis was associated with increased collagen content of the osteoarthritic femoral heads and we proposed that this indicated a fibrotic state in the bone. The elevated collagen production in osteoarthritic bone was associated with, and perhaps driven by, elevated levels of TGF-$\beta$.

More recently we reported that a proportion of osteoblasts, in osteoarthritic femoral heads, have undergone a phenotypic alteration and synthesise type I homotrimer collagen, in addition to the normal heterotrimer bone collagen [6]. The homotrimer fibre has been reported to possess a reduced mechanical strength and mineralisation in bone [25]. This altered collagen composition may account for the decrease in the mechanical strength of osteoarthritic bone as demonstrated by Li and Aspen [22]. Cultured osteoblasts from osteoarthritic bone have also been shown to secrete an enzyme capable of degrading articular cartilage [37]. Additionally, explants and primary cultured osteoblasts from osteoarthritic bone have been shown to secrete elevated levels of urokinase plasminogen activator, a serine protease capable of degrading joint ECM and cartilage [15,20]. Taken together, these observations indicate that there are fundamental biochemical alterations in osteoarthritic bone that have the potential to drive cartilage degeneration either biochemically or mechanically, or both. This clearly poses a problem for cartilage tissue engineers. Any new replacement cartilage will be placed on highly metabolic, mechanically weaker bone, and perhaps more importantly bone osteoblasts that are capable of degrading cartilage.

For many years joint instability has been known to play a critical role in the initiation of OA. Deterioration or rupture of the primary stabilisers of the knee, the cruciate ligaments, is associated with the early onset of secondary OA in humans [9,17,19,26,31]. In addition, knee laxity and a complete deterioration of the ACL have been reported in early-stage [35] and late-stage [36] idiopathic OA respectively. The integrity of the knee ligaments may therefore be a determining factor in the aetiopathogenesis and progression of primary OA.

Clearly the whole joint is affected, not just the articular cartilage alone. Hence the question arises as to whether OA is initiated in the cartilage, bone and ligaments simultaneously, or whether the initiating event occurs in a single one of these tissues. Identification of when and where the earliest osteoarthritic changes occur could lead to a redirection of the current therapies. However, the detection of pathology in these tissues at the point of initiation cannot be undertaken in humans. A suitable animal model of spontaneously occurring OA must therefore be used.

Several animal models have been employed to answer this question, most of which have obvious disadvantages [8]. The Dunkin–Hartley (DH) guinea pig, however, develops knee OA spontaneously and this model has been extensively described [7,18]. As with humans, OA in the guinea pig knee manifests primarily in the medial tibio-femoral compartment. Additionally, guinea pig OA shares similar cartilage histopathology and biochemistry with humans, including chondrocyte clustering and death, PG loss and surface fibrillation [7]. A criticism of the guinea pig model of OA in the past has been the absence of a non-OA developing strain. However, our studies of the Bristol strain 2 (BS2) guinea pig revealed that it was not susceptible to OA and that it would be a suitable control strain. This enabled us to differentiate between age-related and disease-related changes in joint tissues. We therefore used the DH and BS2 guinea pigs to determine the relative appearance of bone and ligament alterations in the knee and compared with the appearance of gross cartilage pathology.

## 1.1. Subchondral bone changes

We employed traditional radiological and histological methods to assess knee OA, and compared with changes in the bone morphology, in the guinea pig model of naturally occurring OA [5]. The DH animals (DHs) were found to have a significantly higher radiographic OA score than age matched BS2 animals (BS2s) at 24 and 36 weeks of age. The OA histological score, based on Toluidene staining, was higher in the DHs, compared to the BS2s, at 24 weeks (Fig. 1a). Similarly, we found that DHs possessed a significantly increased patella perimeter, compared to the BS2s, at 24 and 36 weeks; further demonstrating the development of OA in the DHs by 24 weeks, but not in the BS2s.

The thickness of the DH tibial plateau subchondral bone region was greater than that of the BS2s at 20 weeks. Additionally, the bone to cartilage ratio and subchondral bone area was greater in the DHs, compared to the BS2s, at 20 and 24 weeks, and 24 and 36 weeks respectively. Subchondral bone density, determined by microfocal X-radiography, was elevated in the DHs, compared to the BS2s, at 24 and 36 weeks (Fig. 1b). The shift to increased subchondral bone density in the DHs, at 24 weeks, was greater in the medial side of the joint than the lateral side. Changes in bone density, both immediately subchondral and throughout the epiphyseal region, increased with age and with OA progression in the DHs. The increase in bone density in the DHs therefore occurred concomitantly with the first appearance of changes in the cartilage pathology at 24 weeks. However, the earliest histopathological signs of OA detected in the DHs, such as increased acellularity and decreased PG staining, are generally agreed to be reversible. Irreversible fibrillation was generally only seen in the older DHs at 40 weeks. In contrast, the BS2s failed to show any histological or radiographic signs of pathology, at least up to 52 weeks of age. Importantly, bone thickness was greater in the DHs at 20 weeks, which is prior to the first detectable histological changes in the cartilage. This finding indicates that biochemical changes, which presumably drive the later morphological changes in the bone, may precede histological changes in the cartilage. Indeed, preliminary work has shown that MMP-2 and collagenase levels are elevated in the subchondral bone of the DHs, compared to BS2s, at 16 and 20 weeks [3], suggesting elevated bone turnover in the OA prone strain, which occurs prior to histological changes in the cartilage.

A previous study also reported bone changes associated with OA in the guinea pig [16], but the study was confined to two time points (2 and 12 months) and therefore bone pathology could not be related to an aetiological time frame. Carlson et al. [10] have previously reported progressive subchondral bone thickening in cynomolgus macaque OA. Importantly, the onset of articular cartilage fibrillation occurred following site specific thickening of the subchondral bone in this model. Our study has shown, by sequentially comparing an OA prone guinea pig with age matched control animals, that bone metabolism is occurring at the very earliest stage of disease progression. Our findings further support the theory that bone changes are a critical inductive component of spontaneous knee OA.

## 1.2. Cruciate ligament changes

We hypothesised that OA-related bone fibrosis might be initiated by a change in tension on the bone due to altered cruciate ligament function, since the cruciate ligaments are the primary stabilisers of the knee. Additionally, tension exerted by tendons and ligaments on bone is known to affect bone remodelling [20]. Although the precise mechanism is not yet clear, the osteocytes and oteoblasts are probably

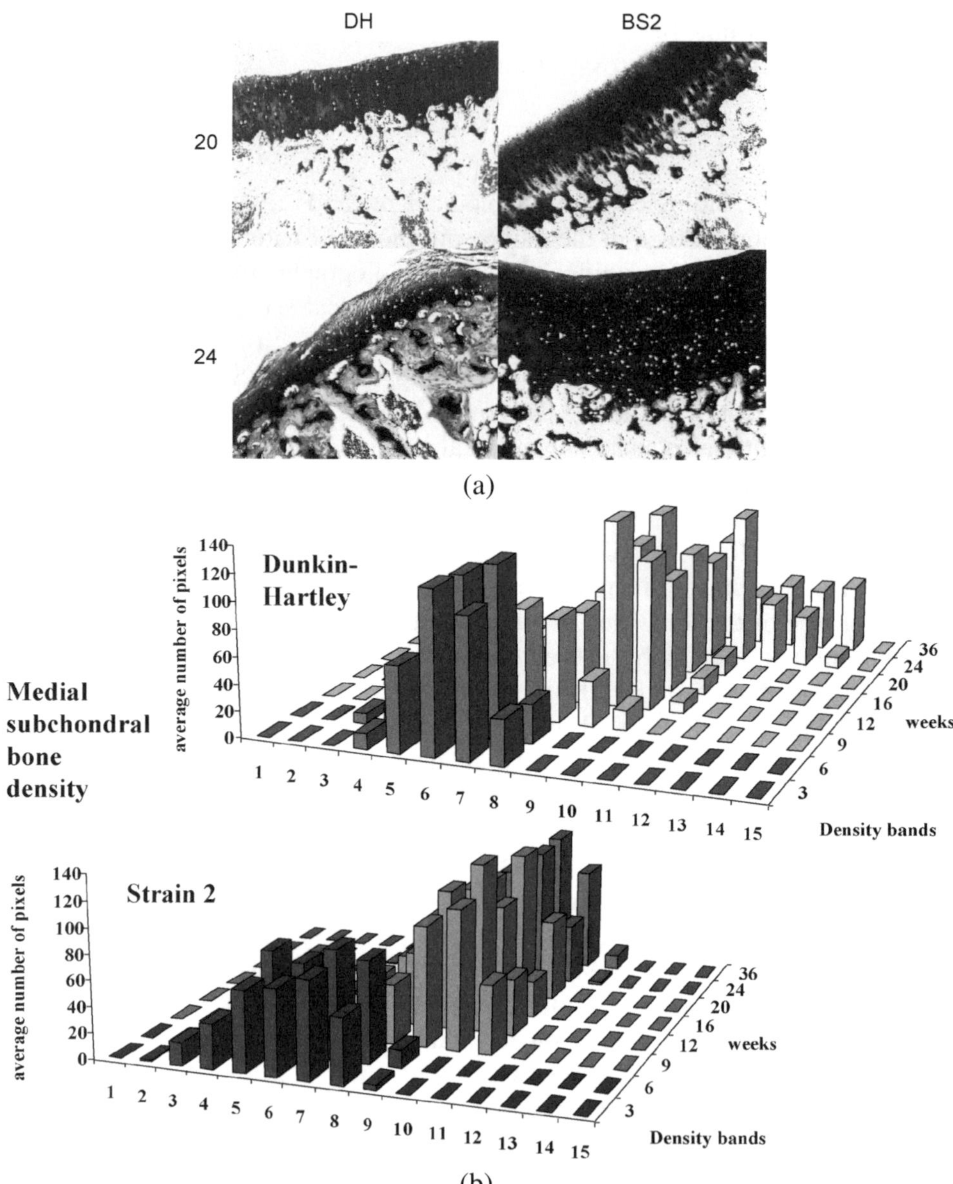



Fig. 1. (a) Typical cartilage histology from the medial tibial plateau (Toluidine Blue staining) at 24 weeks for the BS2 and DH strains, showing increased subchondral bone thickness and early signs of disruption of the DH cartilage pathology associated with OA, fibrillation, chondrocyte clustering and acellularity. These signs are not present in the BS2 strain. (b) Bone density band analysis of the immediate subchondral bone area, demonstrating an increase in bone density at the end of the growth period (12 weeks) in both DH and BS2 strains, and a significant second further increase in the DH strain from 20 weeks, but not in the case of the BS2 strain.

the mechano-responsive cells, and a change in tension may result in the release of signalling molecules and growth factors in the bone [13].

Many studies have shown a correlation between the extent of experimental or accidental cruciate ligament damage and the severity of osteoarthritic characteristics, such as cartilage destruction and os-

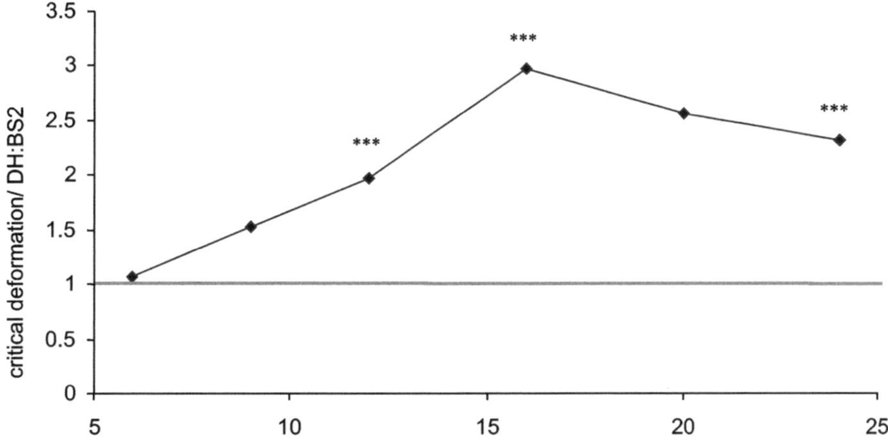

Fig. 2. The ratio of DH critical deformation to BS2 critical deformation showing significant elevated critical deformation in the DH at 12,16 and 24 weeks, compared to the BS2s ***$p < 0.001$ (2-way ANOVA).

teophyte formation [26,27,34]. However, the potential role of clinically undamaged cruciate ligaments in spontaneous OA had not been previously considered.

Our preliminary studies on the cruciate ligaments in the STR/ORT mouse model of spontaneous OA revealed an up-regulation of collagen metabolism, indicated by high MMP-2 levels, and a decrease in mechanical strength, compared to the CBA control strain [2]. Differences in the ligaments between the two strains occurred prior to radiological signs of narrowing manifested in the STR/ORT mice. Unfortunately the mouse ligaments were too small for extensive biochemical and mechanical analyses to establish the mechanism of cruciate ligament alteration. We therefore analysed the mechanics and biochemistry of the cruciate ligaments in the guinea pig model of spontaneous OA, using the same cohort of animals as the bone study.

We observed an increase in ultimate load with age, as anticipated from earlier work [11], but no difference between the DH and BS2 strains. However, examination of the toe region of the load–deformation curve, within which the ligament is easily extendible and the load physiologically relevant, revealed that the size of the toe region (defined as critical deformation) was greater in the DH ligaments [28], compared to BS2s (Fig. 2). The elevated critical deformation, which is a function of ligament laxity, in the DHs was independent of ligament size. This suggests that an alteration in the ligament matrix, of which 90% is fibrillar collagen, is driving the elevated laxity in the DH ligaments rather than an alteration in ligament morphology. Elevated turnover of collagenous tissue is known to result in poor collagen fibre alignment and increased collagen crimp periodicity and this has been associated with increased laxity [11]. In order to determine the metabolic turnover of the guinea pig cruciate ligaments we measured markers of collagen synthesis and degradation.

Collagen synthesis was found to be higher in the DH strain by quantifying the levels of procollagen C-terminal propeptide (PICP) in ligament extract [30], and by measuring levels of the immature crosslink hydroxylysino-keto-norleucine. Similarly increased collagen degradation in the DH ligament compared to the BS2 was demonstrated from the levels of MMP-2 and the tissue inhibitor of MMP-2, TIMP-2, by zymography and reverse zymography respectively [28].

In conclusion, markers of both collagen synthesis and degradation are higher in the OA–prone DH ligaments (Fig. 3a and b), which indicates an increased level rate of tissue turnover compared to the control BS2 ligaments. Elevated tissue turnover in the DHs probably accounts for the increased laxity

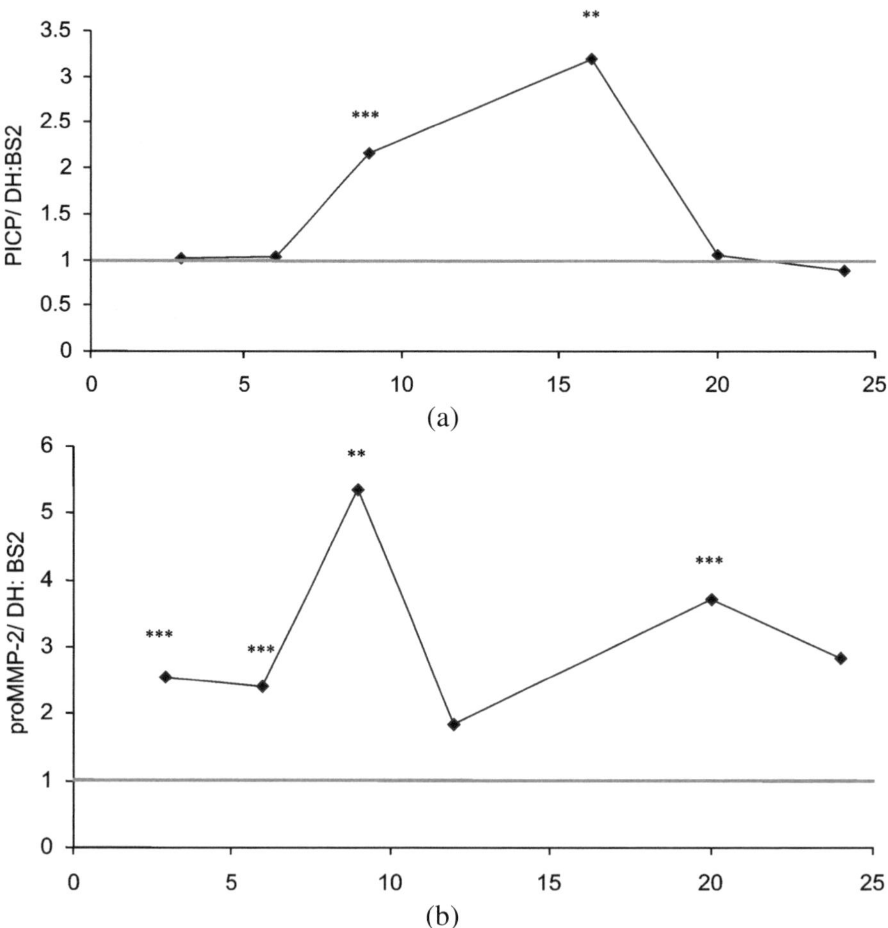

Fig. 3. Turnover of collagen in guinea pig cruciate ligaments. (a) Collagen synthesis in guinea pig cruciate ligaments. The ratio of PICP for the DH to PICP for the BS2s. PICP is significantly higher in DH ligaments at 9 and 16 weeks, compared to BS2. $^{**}p < 0.01$, $^{***}p < 0.001$ (2-way ANOVA). (b) Collagen degradation in guinea pig ligaments. The ratio of proMMP-2 for DH to proMMP-2 for BS2. ProMMP-2 is significantly higher in DH ligaments at 3, 6, 9 and 20 weeks compared to BS2. $^{**}p < 0.01$, $^{***}p < 0.001$ (2-way ANOVA).

in the DH ligaments. Importantly, these metabolic and mechanical differences between the two strains are observed well prior to the onset of histological OA in the DH cartilage. The cruciate ligaments are therefore likely to have an important, if not exclusive, role in the initiation and development of spontaneous knee OA.

### 1.3. Notch width changes

Several workers have reported an association between the integrity of the cruciate ligament and the width of the femoral intercondylar notch [14]. Others have reported a correlation between intercondylar notch width and the severity of OA [33,35]. Wada [35] proposed that new bone formation in this area was due to an attempt to increase knee stability by increasing contact between the joint surfaces, which could be a compensatory event for defective cruciate ligaments. Conversely, intercondylar notch narrowing is known to impinge the ACL insertion site and damage the ACL [32]. However it is unknown whether

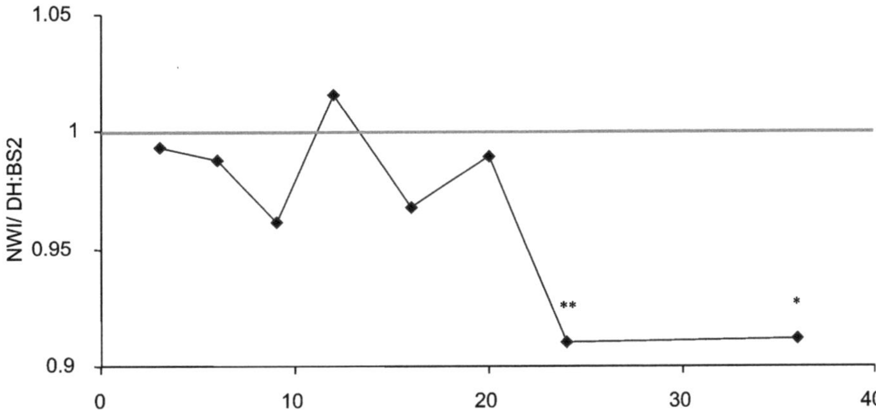

Fig. 4. Intercondylar notch width index (NWI) in guinea pig femurs. The ratio of NWI for DH to NWI for BS2. NWI is significantly lower in DH animals at 24 and 36 weeks, compared to the BS2s, indicating elevated remodelling in the ACL insertion site. $^*p < 0.05$, $^{**}p < 0.01$ (2-way ANOVA).

OA associated notch width narrowing is a cause or consequence of cruciate ligament dysfunction. We therefore determined the femoral intercondylar notch width index (NWI), the width at two thirds of the notch height relative to the condylar width of the femur, in order to determine the order of events in the guinea pig model of spontaneous OA.

NWI decreased with age for both the DHs and BS2s (Fig. 4). However the NWI of the DH femurs was significantly lower than that of the BS2 femurs at 24 and 36 weeks of age. The difference in NWI between the two strains occurred 12 weeks after a difference in ligament laxity is observed [28]. This finding clearly demonstrates that intercondylar notch narrowing is a consequence, rather than cause, of ligament dysfuction. Thus we have been able to show a time course of events involving OA development, cruciate ligament function, and notch narrowing in a way that has not been possible in humans.

## 2. Concluding remarks

We have shown that the BS2 strain of guinea pig is a viable control for the DH model of spontaneous OA, which has meant that for the first time the initiating events of spontaneous OA could be investigated. We have shown that the OA prone DH has increased bone volume and density concomitant with the first signs of disease in the cartilage, and that cruciate ligament laxity is elevated prior to disease onset. The DH strain displays a higher rate of tissue turnover in its cruciate ligaments resulting in elevated ligament laxity and joint instability. We therefore suggest that increased subchondral bone metabolism and inter-condylar notch osteophyte formation are both initiated by early ligament laxity in spontaneous OA. In addition, since elevated cruciate ligament turnover and laxity is apparent well prior to histopathological changes in the cartilage we suggest that ligament dysfuction is an initiating factor in spontaneous OA. We therefore conclude that both cruciate ligament and bone alterations are fundamental factors in the aetiology and progression of knee OA.

## Acknowledgment

This work was funded by the Medical Research Council and Roche Discovery, Welwyn, UK.

# References


[1] T. Aigner, K. Gluckert and K. von der Mark, Activation of fibrillar collagen synthesis and phenotypic modulation of chondrocytes in early human osteoarthritic cartilage lesions, *Osteoarthritis and Cartilage* **5** (1997), 183–189.

[2] J.M. Anderson-MacKenzie, M.E. Billingham and A.J. Bailey, Collagen remodelling in the anterior cruciate ligament associated with developing spontaneous murine osteoarthritis, *Biochem. Biophys. Res. Commun.* **258** (1999), 763–767.

[3] J.M. Anderson-MacKenzie, H.L. Quasnichka, R.L. Starr, E.J. Lewis, M.E.J. Billingham and A.J. Bailey, Fundamental bone remodelling in guinea pig osteoarthritis development, *Osteoarthritis and Cartilage* **10** (2002), S54.

[4] J.M. Anderson-MacKenzie, H.L. Quasnichka, R.L. Starr, E.J. Lewis, M.E.J. Billingham and A.J. Bailey, Femoral intercondylar notch shape in guinea pig osteoarthritis development, *Osteoarthritis and Cartilage* **10** (2002), S55.

[5] J.M. Anderson-MacKenzie, H. Quasnicha, R.L. Starr, E.J. Lewis, M.E.J. Billingham and A.J. Bailey, Fundamental subchondral bone changes in spontaneous guinea pig knee osteoarthritis, *Int. J. Biochem. Cell. Biol.* **37** (2005), 224–236.

[6] A.J. Bailey, T.J. Sims and L. Knott, Altered phenotypic expression of osteoblast collagen in osteoarthritic bone: production of type I homotrimer, *Int. J. Biochem. Cell Biol.* **34** (2002), 176–182.

[7] A.M. Bendele and J.F. Hulman, Effects of body weight restriction on the development and progress of spontaneous osteoarthritis in guinea pigs, *Arth. & Rheum.* **31** (1991), 561–565.

[8] A. Bendele, J. McComb, T. Gould, T. McAbee, G. Senello and E. Chlipala, Animal models of arthritis: relevance to human disease, *Toxicol. Pathol.* **27** (1999), 134–142.

[9] J.C. Buckland-Wright, J.A. Lynch and B. Dave, Early radiographic features in patients with anterior cruciate ligament rupture, *Ann. Rheum. Dis.* **59** (2000), 641–646.

[10] C.S. Carlson, R.F. Loesser, C.B. Purser, J.F. Gardin and C.P. Jerome, Osteoarthritis in cynomolgus macaques III: effects of age, gender and subchondral bone thickness on the severity of the disease, *Arth. & Rheum.* **11** (1996), 1209–1217.

[11] J. Diamant, A. Keller, E. Baer, M. Litt and R.G. Arridge, Collagen: ultrastructure and its relation to mechanical properties as a function of ageing, *Proc. Roy. Soc. Lond. B Biol. Sci.* **180** (1972), 293–315.

[12] P. Dieppe, J. Cushnaghan, P. Young and J. Kirwan, Prediction of the progression of joint space narrowing in osteoarthritis of the knee by bone scintigraphy, *Ann. Rheum. Dis.* **52** (1993), 557–563.

[13] P.J. Ehrlich and L.E. Lanyon, Mechanical strain and bone cell function: A review, *Osteoporosis. Int.* **13** (2002), 688–700.

[14] P. Hernigou and J.M. Garabedian, Intercondylar notch width and the risk for anterior cruciate ligament rupture in the osteoarthritic knee: evaluation by plain radiography and CT scan, *Knee* **9** (2000), 313–316.

[15] G. Hilal, J.F. Martell-Pelletier, P. Pelletiere, P. Ranger and D. Lajeunesse, Osteoblast-like cells from human subchondral bone demonstrate an altered phenotype in vitro: possible role of subchondral bone sclerosis, *Arth. & Rheum.* **41** (1998), 981–999.

[16] J.L. Huebner, M.A. Haines, B. Beekman, J.M. TeKoppel amd V.B. Kraus, A comparative analysis of bone and cartilage metabolism in two strains of guinea pig with varying degrees of naturally occurring osteoarthritis, *Osteoarthritis and Cartilage* **10** (2000), 758–767.

[17] K. Jacobsen, Osteoarthrosis following insufficiency of the cruciate ligaments in man. A clinical study, *Acta Orthop. Scand.* **48** (1977), 520–526.

[18] P.A. Jimenez, S.S. Glasson, O.V. Trubetsko and H.B. Haines, Spontaneous osteoarthritis in Dunkin–Hartley guinea pigs: histologic, radiologic and biochemical changes, *Lab. Animal. Sci.* **47** (1997), 598–601.

[19] P. Kannus and M. Jarvinen, Post-traumatic anterior cruciate ligament insufficiency as a cause of osteoarthtitis in a knee joint, *Clin. Rheumatol.* **8** (1989), 251–260.

[20] D. Lajeunesse, Altered subchondral osteoblast cellular metabolism in osteoarthritis: cytokines, eicosanoids and growth factors, *J. Musculoskeletal Neuronal Interact.* **2** (2002), 504–506.

[21] L.E. Lanyon, Using functional loading to influence bone mass and architecture: objectives, mechanisms and relationship with estrogen of the mechanical adaptive process in bone, *Bone* **18** (1996), 375–435.

[22] B. Li and R.M. Aspden, Composition and mechanical properties of cancellous bone from the femoral head of patients with osteporosis and osteoarthritis, *J. Bone Mineral. Res.* **12** (1997), 641–651.

[23] P. Lorenzo, M.T. Bayliss and D. Heinegard, Altered patterns and synthesis of extracellular matrix macromolecules in early osteoarthritis, *Matrix Biol.* **23** (2004), 381–391.

[24] J.P. Mansell and A.J. Bailey, Abnormal cancellous bone collagen metabolism in osteoarthritis, *J. Clin. Invest.* **101** (1998), 1596–1603.

[25] K. Misof, W.J. Landis, P. Klaushofer and P. Fratzl, Collagen from osteogenesis imperfecta mouse model (oim) shows reduced resistance against tensile stress, *J. Clin. Invest.* **100** (1997), 40–45.

[26] P. Neyret, S.T. Donell, D. DeJour and H. DeJour, Partial meniscectomy and anterior cruciate ligament rupture in soccer players. A study with a minimum 20 year follow-up, *Am. J. Sports Med.* **21** (1993), 455–460.

[27] M.J. Pond and G. Nuki, Experimentally-induced osteoarthritis in the dog, *Ann. Rheum. Dis.* **32** (1973), 387–388.

[28] H.L. Quasnichka, J.M. Anderson-MacKenzie, J.F. Tarlton, T.J. Sims, M.E. Billingham and A.J. Bailey, Cruciate ligament laxity and femoral intercondyl notch narrowing in early stage knee osteoarthritis, *Arthritis & Rheumatism* **52** (2005), 3100–3109.

[29] E. Radin and R.M. Rose, Role of subchondral bone in the initiation and progression of cartilage damage, *Clin. Orthopaed.* **213** (1986), 34–40.

[30] J. Risteli, S. Niemi, S. Kauppila, J. Melkko and L. Risteli, Collagen propeptides as indicators of collagen assembly, *Acta Orthop. Scand. Suppl.* **266** (1995), 183–188.

[31] H. Rous, T. Adalberth, L. Dahlberg and L.S. Lohmander, Osteoarthritis of the knee after injury to the anterior cruciate ligament or meniscus: the influence of time and age, *Osteoarthritis and Cartilage* **3** (1995), 261–267.

[32] K.D. Shelbourne, T.J. Davis and T.E. Klootwyk, The relationship between intercondylar notch width of the femur and the incidence of anterior cruciate ligament tears. A prospective study, *Am. J. Sports Med.* **26** (1998), 402–408.

[33] L. Shepstone, J. Rogers, J.R. Kirwan and B.W. Silverman, Shape of the condylar notch of the human femur: a comparison of osteoarthritic and non-osteoarthritic bones from a skeletal sample, *Ann. Rheum. Dis.* **60** (2001), 968–973.

[34] G.J. Van Osch, van der Kraan, L. Blankevoort, R. Huiskes and W.B. van den Berg, Relation of ligament damage with site specific cartilage loss and osteophyte formation in collagenase induced osteoarthritis in mice, *J. Rheumatol.* **23** (1996), 1227–1232.

[35] M. Wada, S. Imura, H. Baba and S. Shimeda, Knee laxity in patients with osteoarthritis and rheumatoid arthritis, *Br. J. Rheumatol.* **35** (1996), 560–563.

[36] M. Wada, H. Tatsuo, H. Baba, K. Asmoto and Y. Nojyo, Femoral intercondyl notch measurement in osteoarthritic knees, *Rheumatology (Oxford)* **38** (1999), 554–558.

[37] C.I. Westacott, G.R. Webb, M.S. Warnock, J.V. Sims and C.J. Elson, Alteration in cartilage metabolism by cells from osteoarthritic bone, *Arth. and Rheum.* **40** (1997), 1282–1291.

# Part III: Lectures of the Negma Lerards-prize

Biorheology 43 (2006) 401–402
IOS Press



# Introduction to "Negma Lerads Prize 2005" on mechanobiology of cartilage and chondrocyte

It is for me a great pleasure to present the winners of the second Negma-Lerads prize. As stated by Martine Burger, the winners were selected by an international panel from Europe and USA which evaluated the scientific merit, the scope and objectives of the research, the methods used and the potential medical interest. Among the proposals received four were selected, and it was impossible to yank these after receiving all notations.

My presentation of the winners will begin with the lady and afterwards the three men in alphabetic order.

The first winner is a young lady Astrid Watrin. After being a postdoctoral research fellow in 2001–2002 in the team of my friend P. Netter, Astrid is now a CNRS researcher. Her main research area concerns new imaging techniques for detecting microscopic cartilage damages, with special attention to tissue characterization. Since 2001 she has published 12 papers in international journals. Congratulations Astrid.

The second winner and the first man in alphabetic order is David Alan Lee. Professor Lee is well known in the field of tissue engineering. After two post docs in the Institute of Rheumatology of Hammersmith and in London, he is since 2004 Professor of Cell and Tissue Engineering at Queen Mary University of London. The research activities of David Lee concern mechanotransduction pathways and tissue engineering. During the last 7 years David Lee was 11 times invited to present a lecture in international conferences, and during the same period he published 37 papers in international journals and 7 book chapters. Congratulations David.

The third winner is Professor Marc Levenston. Marc defended his PhD thesis in 1995 in Stanford University and was a postdoctoral fellow at MIT from 1995 to 1998. Afterwards, he moved to Georgia Institute of Technology in Atlanta, where he is now Associate Professor. His research interests lie in the area of orthopaedic soft tissue biomechanics and tissue engineering, including mechanical regulation of cartilage metabolism and computational modelling of soft tissue mechanics. In the last 10 years Marc Levenston published 25 papers in international journals. Congratulations Marc.

The fourth winner is Doctor Detlef Schumann. Dr. Schumann presently works in the Department of Experimental Trauma Surgery in Regensburg (Germany). Like Astrid, Detlef is a young researcher. His PhD thesis, defended last year, was on cartilage tissue engineering, stem cells, mechanobiology and angiogenesis. In the last 4 years Detlef Schumann published 16 papers in various international journals. Congratulations Detlef.

I am glad to congratulate the four winners and I ask David Lee, Mark Levenston and Detlef Schumann each to present in 4 or 5 minutes their research (more information will be given in their talks during the symposium), and afterwards Astrid Watrin to give a 15 minute lecture on her research.

Before concluding I want to thank the members of the board for their important work and evaluations. I hope that we will receive more proposals for the third prize in two years.

Thank you.

J.F. Stoltz

Biorheology 43 (2006) 403–412
IOS Press



# Increased apoptosis in rat osteoarthritic cartilage corresponds to degenerative chondral lesions and concomitant expression of caspase-3

Astrid Watrin-Pinzano [a], Stéphanie Etienne [a], Laurent Grossin [a], Nadège Gaborit [a], Christel Cournil-Henrionnet [a], Didier Mainard [a,b], Patrick Netter [a], Pierre Gillet [a,*,1] and Laurent Galois [a,b,1]

[a] *Unité Mixte de Recherche 7561, CNRS - Université Nancy 1, Faculté de Médecine, Avenue de la Forêt de Haye, BP184, F-54505 Vandoeuvre lès Nancy, France*
[b] *Service de Chirurgie Orthopédique et Traumatologique, CHU de Nancy, Hôpital Central, CO34, Avenue de Lattre de Tassigny, 54035 Nancy Cedex, France*

Keywords: Cartilage, osteoarthritis, apoptosis, caspase, experimental model

## 1. Introduction

Articular cartilage is a specialized and uniquely structured tissue that forms the smooth, gliding surface of the diarthrodial joints. It consists of an extracellular matrix (ECM) that is synthesized by the sparsely distributed resident cells, the chondrocytes. Osteoarthritis (OA) is a typical slow degenerative disease of the joints characterized by fibrillation and erosion of hyaline cartilage, subchondral bone sclerosis and osteophyte formation at the joint margins, leading to a gradual loss of articular cartilage [20]. These progressive changes result from poorly understood events occurring over a long period that leads to both cartilage ECM degradation and inhibition of ECM components synthesis. Additionally, cartilage hypo-cellularity also contributes to the development of clinical or experimental OA, due to chondrocyte death by either apoptosis or necrosis. In fact, the progression of OA can be judged by the viability of chondrocytes and their ability to resist to apoptosis. Nevertheless, it is not clear if apoptosis is caused by senescence or is the cause of senescence [22].

Apoptosis and necrosis are the two forms of chondrocyte death. Apoptosis is the process in which each cell systematically inactivates its function and disassembles its structure. Apoptosis can occur to a

---

*Address for correspondence: Pierre Gillet, Unité Mixte de Recherche 7561, CNRS - Université Nancy 1, Faculté de Médecine, Avenue de la Forêt de Haye, BP184, F-54505 Vandoeuvre lès Nancy, France. Tel.: +33 383 683 950; Fax: +33 383 683 959; E-mail: Pierre.Gillet@medecine.uhp-nancy.fr.
[1] Equal seniorship.

single cell, even though the cells surrounding it are normal. Recently, the occurrence of apoptotic events was confirmed in the articular chondrocytes of OA patients. In parallel, numerous *in vivo* experimental studies have confirmed that apoptotic events in chondrocytes are linked with the severity of experimental OA [12]. In fact OA cartilage is heavily involved with apoptosis, resulting in the decimation of chondrocytes [5].

Understanding the interactions that promote chondrocyte apoptosis (recently called chondroptosis [19]) and cartilage hypocellularity is essential for developing appropriate targeted therapies for inhibition of chondrocyte apoptosis and subsequent treatment of OA. With this idea in mind, experimental OA models are crucial to establish the linkage between chondrocyte hypocellularity and apoptosis [25], and also to assess that controlled pharmacological inhibition of key steps in cell death process could be targeted for novel therapeutic strategies for the treatment of OA.

Recent investigations [23] have questioned the true extent of apoptosis in human OA cartilage and the specificity of the widely used TUNEL method versus expression of a key enzyme, caspase 3, that mediate the final stage of cell death by apoptosis [10]. The aims of this study were thus to determine the extent of chondrocyte apoptosis by using concomitantly TUNEL method and caspase 3 immunostaining, and to investigate the relationship between the process and the severity of chondral lesions in articular cartilage obtained at key points from rat knees with biochemical (mono-iodo acetate, MIA model [8]) and surgical (anterior cruciate ligament transection, ACLT model [4]) induced-OA and from nonarthritic controls.

## 2. Material and methods

### 2.1. Animals

Male Wistar rats (200 g; 8 weeks old) were obtained from Charles River Laboratories (St Aubin les Elbeuf, France). The maintenance and care of the experimental rats were in accordance with the guidelines of the NIH for Animal Welfare Act. Rats were kept in individual plastic cages in a 12 : 12 light–dark cycle (light-on period, 6:00 AM–6:00 PM) in a controlled temperature chamber on sawdust bedding. They were fed a standard diet and had access to tap water *ad libitum*. Body weight was recorded at regular intervals.

### 2.2. Surgically-induced rat OA

On day 0, rats underwent an ACLT under anesthesia (*i.p.* injection of a mixture of acepromazine 1.25 mg/kg + ketamine 38 mg/kg). According to Williams [26], a para-patellar skin incision was performed on the medial side of the right knee joint, and thereafter on the medial side of the patellar tendon. Patella was then dislocated laterally to provide access to the joint space and ACL was transected in the flexed knee. A positive anterior drawer test validated complete transection of the ligament. The joint was then irrigated with sterile saline to avoid ancillary inflammatory process, and a purpose-made suture was processed. A naive group (sham group) undergoing arthrotomy without ACLT was included as an internal control. Rats were killed on Day 28 [4].

### 2.3. Biochemically-induced OA

On Day 0, one dose of MIA (0.03 mg, Merck-Clevenot, Nogent-sur-Marne, France), dissolved in 50 $\mu$l of sterile physiologic saline, was injected through the infra-patellar ligament [11] in both knees.

Controls animals received an intra-articular injection of 50 $\mu$l of saline (0.9% NaCl). The animals were sacrificed at D10 and D15 [7].

## 2.4. Histological assessment

Animals were killed at various key points by cervical dislocation under anesthesia. Whole knee joints were dissected, fixed in 4% paraformaldehyde (pH 7.4), decalcified with *"Rapid Decalcifiant Osseux"* (RDO, Apex, Canada), dehydrated through a descending series of ethanol with the use of an automated tissue processing apparatus. After embedding in paraffin, serial sections with a thickness of 5 $\mu$m were prepared for histological examination and immunohistochemistry. The sections were stained with hematoxylin–eosin to observe cellularity, and toluidine blue to assess proteoglycan content.

The severity of OA lesions was graded on a scale adapted from Mankin's score [4] by two independent observers (LG and SE). This score ranged from 0 to 15 according to structure, cellularity, toluidine blue staining, thickness of hypertrophic chondrocyte layer, bone remodeling and osteolysis. Structure was graded from 0 to 5 (0 = Normal, 1 = Pannus and surface irregularities, 2 = Clefts to transitional zone, 3 = Clefts to radial zone, 4 = Clefts to calcified zone, 5 = Complete disorganization). Cellularity was graded from 0 to 3 (0 = Normal, 1 = Diffuse hypercellularity, 2 = Cloning, 3 = Hypocellularity). Toluidine blue staining was graded from 0 to 3 (0 = Normal, 1 = Slight reduction, 2 = Moderate reduction, 3 = Severe reduction). Thickness of hypertrophic chondrocyte layer graded from 0 to 2 (0 = Normal, 1 = Moderate decrease, 2 = Total decrease). Bone remodelling and bone osteolysis were graded from 0 to 1, respectively, with 0 = No and 1 = Yes.

## 2.5. TUNEL method

TUNEL is a specific immunohistochemical technique that enables sensitive and specific staining of the high concentrations of DNA 3-OH ends that are localized in apoptotic bodies. Apoptotic nuclei were stained as described in the protocol provided by the manufacturer by using the *"In situ* cell death detection, AP" kit (Boehringer Mannheim). To perform the TUNEL assay, slides were first deparaffinized, sections circled with PAP pen. Each section was incubated with ABC chondroitinase (0.25 U/ml) for 45 min at 37°C, then permeabilized respectively with Triton X-100 (0.3% PBS, 30 min) and Proteinase K (20 $\mu$g/ml, 30 min, 37°C), with PBS wash between the two steps. The slides were then flooded with TdT-enzyme and digoxigenin-dUTP reaction buffer (TUNEL) reagent for 60 min in a humidity chamber at 37°C, washed with PBS, and incubated for 30 minutes with Converter-AP solution. After 3 washes in PBS, substrate solution (NBT/BCIP/1 mM levamisole) was added to each slide and incubated at room temperature (monitoring under microscope). The slides were washed 3 times with PBS and mounted for optical observations. The stained mounted cells were examined at 10×, and 40× magnification (Olympus BH-2 microscope). Cell death was quantified by counting positively stained cells (versus total present cells) in 3–5 separate fields of view per section, and noting the percentage of apoptotic cells based on morphological appearance.

## 2.6. Detection of activated caspase 3

Immunohistochemistry was performed on the serial paraffin sections as previously described. Tissue sections were deparaffinized and rehydrated. The sections were pre-treated with chondroitinase ABC (0.25 U/ml in PBS, pH 8.0; Sigma, St. Louis, MO, USA) for 90 min at 37°C. Permeability was enhanced by using 0.3% Triton X-100 in PBS (Sigma, St. Louis, MO, USA) for 30 min. Endogenous peroxidase activity was blocked by incubating sections with freshly prepared 3% hydrogen peroxide (Sigma, St. Louis, MO, USA) for 30 min. Non-specific staining was blocked by incubation of the sections with blocking serum supplied by Novostain super ABC kit (Novocastra, Newcastle, UK) for 60 min. Sections were incubated overnight at 4°C with primary antibodies in a humidified chamber. A biotin-labelled goat anti-rabbit IgG (Novostain superABC kit) was used as a secondary antibody for 45 min. A biotin–avidin detection system (Novostain superABC kit) was used according to the manufacturer's recommendations. The peroxidase was detected using liquid diaminobenzidine substrate kit (Novocastra, Newcastle, UK). After counterstaining with methyl green, slides were dehydrated and mounted with Eukitt (Labonord, France).

The primary antibodies used in this study were: rabbit polyclonal active caspase 3 antibody (R & D Systems, Abington, UK) diluted at a ratio of $1 \times 300$. The presence of antigen was estimated by determining the number of specific chondrocytes staining positive in the superficial zone (superficial and upper intermediate cartilage layers) and in deep zone (lower intermediate and deep layers). Each zone was divided into 4 different sections. The total number of chondrocytes and the number of chondrocytes staining positive for the specific antigen were determined at $40\times$ magnification for the superficial and the deep zones, respectively.

### 2.7. Statistical analysis

Data are expressed as mean $\pm$ standard error of the mean (SEM). One-way ANOVA followed by a Student's $t$ test was used to determine the statistical significance of the differences between OA groups and control rat. $P$ values lesser than 0.05 were considered significant.

## 3. Results

### 3.1. ACLT model

Articular cartilage from the sham-operated knee joints was histologically normal. On D28, in ACLT rats, proteoglycan depletion and hypocellularity were present in all compartments. Structural alteration of cartilage consisted of clefts to radial zone, leading in some cases to bone exposure. A dramatic decrease in cellularity was observed in the whole cartilage (Fig. 1). Degenerative lesions were homogeneous in all compartments (plateaus and condyles). Mirror-image lesions were observed specially in medial compartment. Subchondral remodeling was pronounced at this stage, leading to a Mankin's score of $8.75 \pm 0.89$ vs. $1.2 \pm 0.56$ in sham operated rats.

### 3.2. MIA model

As depicted in Fig. 2, on D10 and D15 HES staining revealed a strong remodeling of subchondral bone, disappearance of hypertrophic chondrocytes layer and a decrease in cartilage cellularity. The toluidine blue staining revealed a decrease in proteoglycan content in animals exposed to MIA. Mankin's score was quite similar on D10 and D15 ($8.17 \pm 0.54$ and $8.00 \pm 1.30$, NS), while no chondral lesion was observed in saline control matched rats ($0.17 \pm 0.16$).

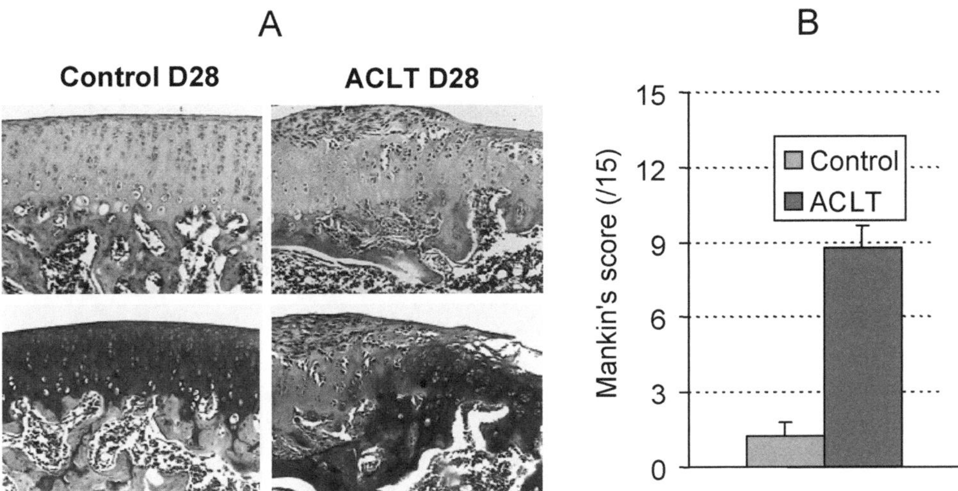



Fig. 1. A. Histological lesions (medial tibial plateaus) in ACLT rats (D28). Proteoglycan content was evaluated by toluidine blue staining (bottom images), cellularity, and surface integrity by Hematoxylin–Eosin HES staining (upper images). Loss of proteoglycans, reduced cellularity, and subchondral bone modifications are present. Clefts were pronounced at this stage. Magnification ×10. B. Mankin's score in control and ACLT groups ($n = 5$/batch, mean ± s.e.m.). Each slide was subjected to a double-blind evaluation by trained investigators and a maximal variation of only 5% was obtained.

## 3.3. TUNEL

Apoptosis (according to the TUNEL reaction) was more frequent in rat OA cartilage than in nonarthritic control cartilage in both models: 2–3-fold in ACLT-induced OA (Fig. 3) and 5–6-fold in MIA-induced OA cartilage (Fig. 4) versus age- and model-matched controls. One-way ANOVA showed that the differences between OA and control groups were highly significant ($p < 0.05$). The basal level of apoptosis in controls was inherent to the maturation process and physiological enchondral ossification in the deep layer in these young animals.

## 3.4. Caspase 3

Concomitantly, the cells expressing caspase 3 increased 3 fold in ACLT and 7–8-fold in MIA models, respectively. There was a highly significant positive correlation between TUNEL-positive cells and expression of caspase-3 in both models (Figs 5 and 6). Spearman's correlation coefficient for the correlation between the percentages of TUNEL positive and caspase 3 positive cells were excellent. Besides, TUNEL appears less specific and that apoptosis seems overestimated by TUNEL assay compared with caspase 3 detection.

## 4. Discussion

The data presented here confirm that apoptosis (TUNEL positive chondrocytes) in OA cartilage was present in ACLT and MIA models, 2 to 6 fold higher than that in nonarthritic controls, depending on the model and the severity of the degenerative lesions. Expression of caspase 3 was also higher in rat OA cartilage and correlated with apoptosis as determined by the TUNEL method. These results suggest,

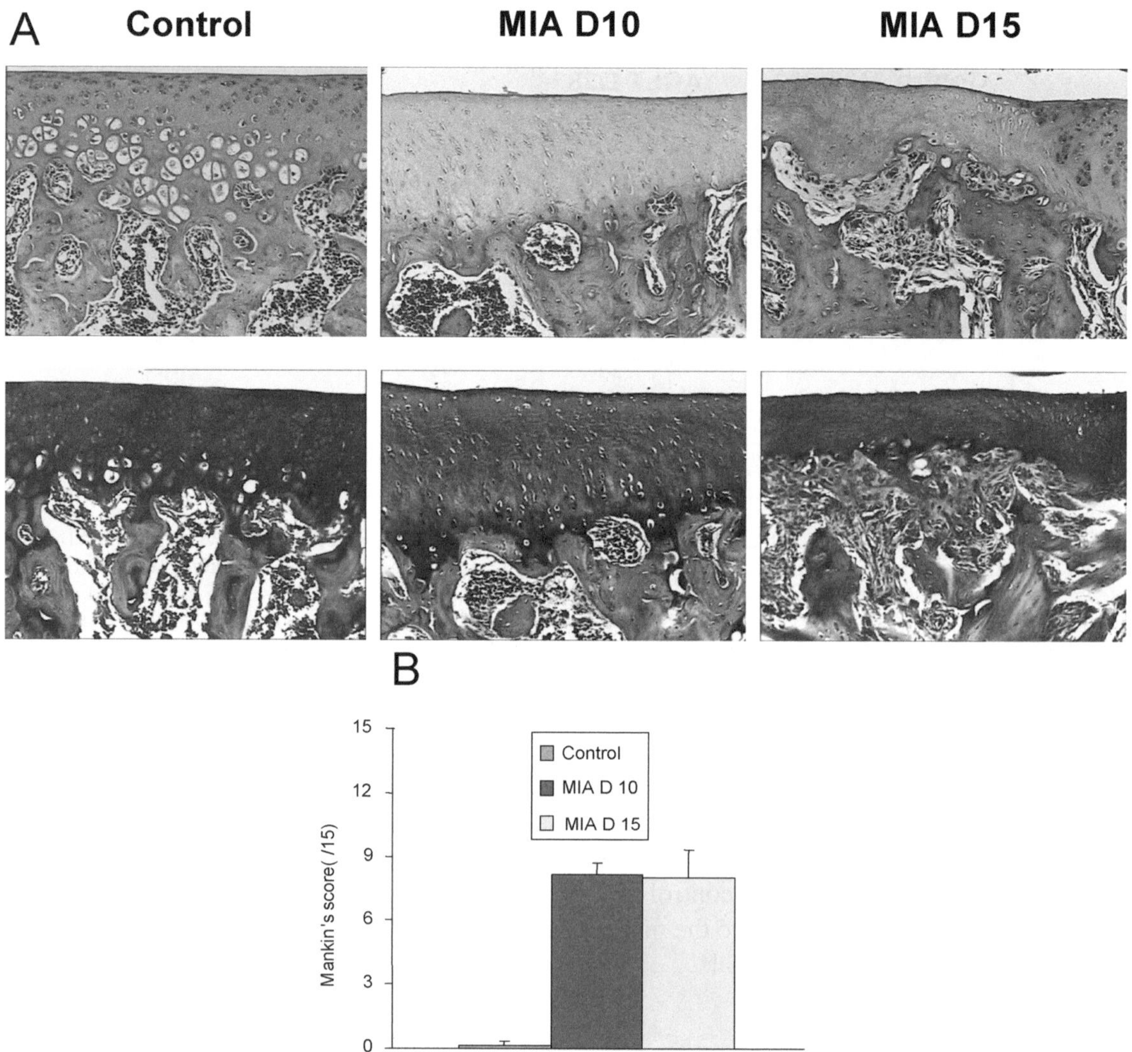



Fig. 2. A. Histological lesions (medial tibial plateaus) in MIA rats (D10 and D15): subchondral bone remodeling, hypocellularity and clusters of chondrocytes, and decrease in proteoglycans integrity are present. Magnification ×10. B. Mankin's score in control and MIA groups ($n = 5$/batch, mean ± s.e.m.). Each slide was subjected to a double-blind evaluation by trained investigators and a maximal variation of only 5% was obtained.

as in humans [23], that a significant proportion of chondrocytes die by apoptosis in experimental OA cartilage, leading to a reduced density of living cells.

Experimental OA can be induced in various ways [11]: abnormal biomechanical forces resulting from joint destabilization, displaced loading or structural alterations resulting either from a degradation of the extracellular matrix by physical, enzymatic means (papain, collagenase) or from a disturbance of chondrocyte metabolism (e.g. iodoacetate, vitamin A).

Anterior cruciate ligament transection (ACLT) model has widely been studied in various animal species (rabbits [9], dogs [17] and more recently in the rat [3,24]), thus providing new insights into pathogenic mechanisms of apoptosis in experimental OA. Biomechanical calculations suggest that damage of the surface zone leads to increased loading of the cartilage matrix and higher stress on the under-

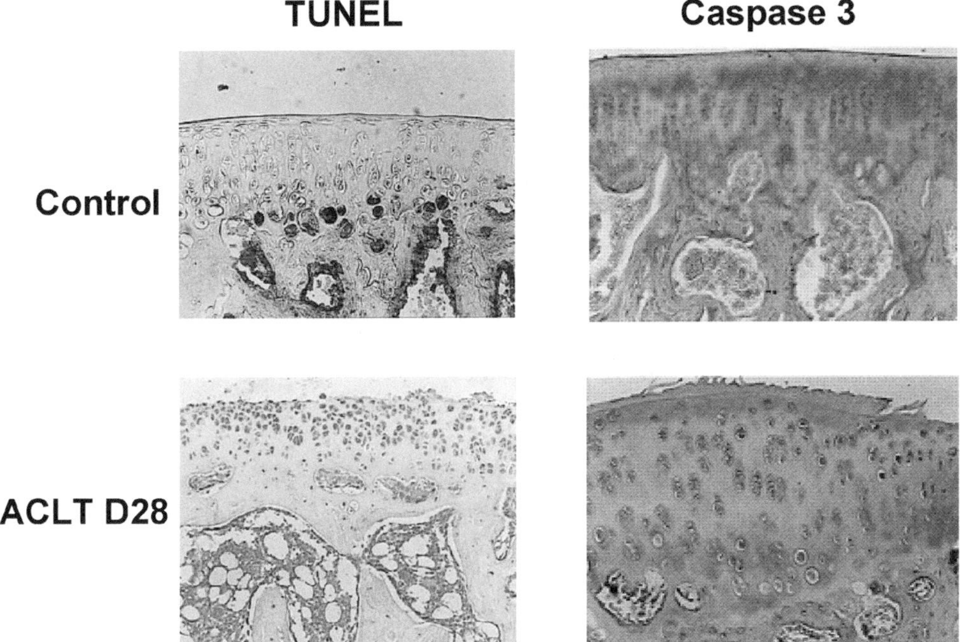


Fig. 3. Apoptotic chondrocytes (TUNEL) in the ACLT model versus age-matched controls and corresponding immunostaining for activated caspase 3. (Magnification ×40.)

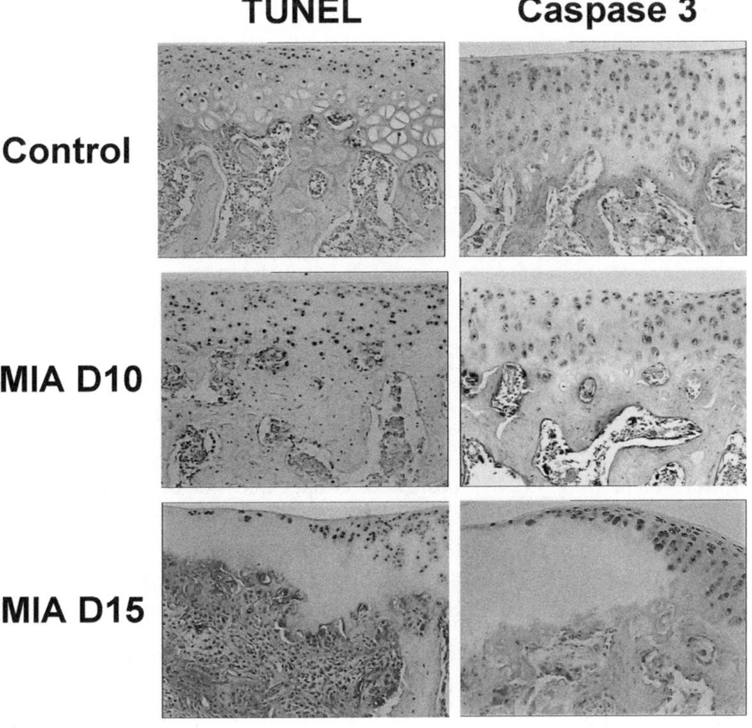


Fig. 4. Apoptotic chondrocytes (TUNEL) in the MIA model versus control rats and corresponding immunostaining for activated caspase 3. (Magnification ×40.)

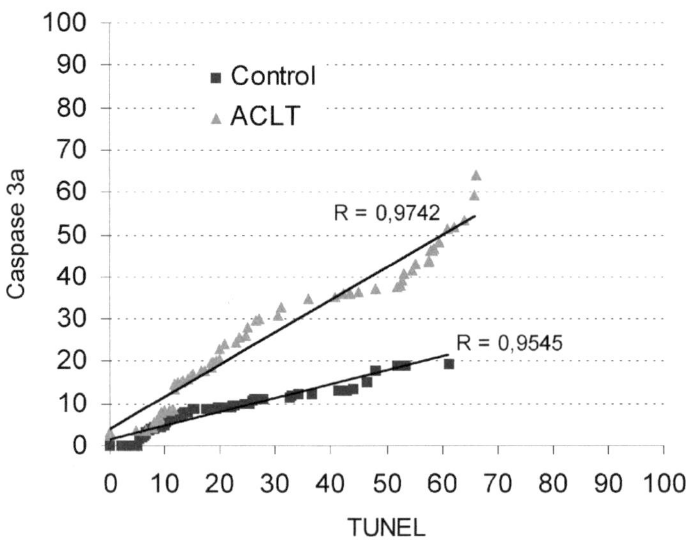

Fig. 5. Good correlation between TUNEL-positive cells and expression of caspase-3 in the ACLT model.

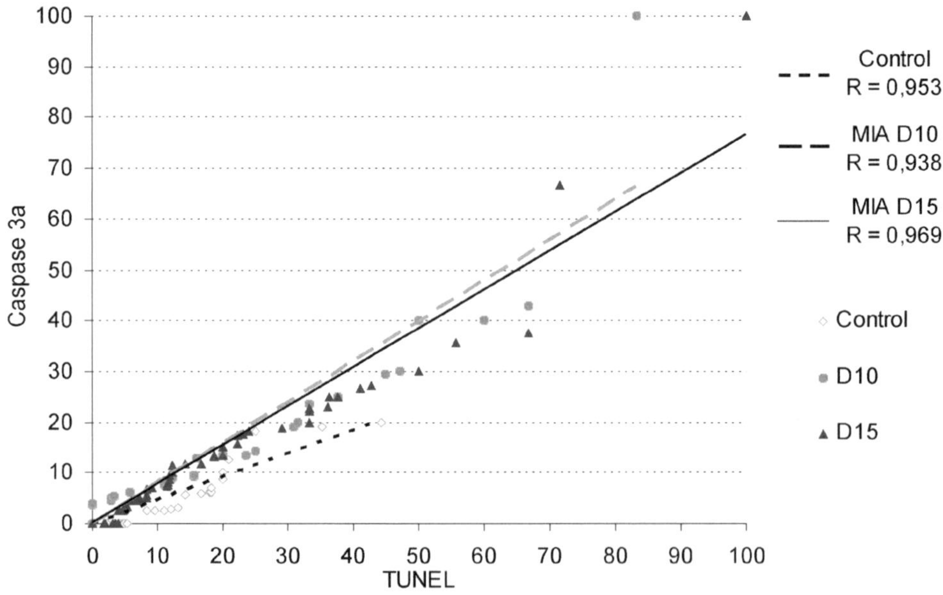

Fig. 6. Good correlation between TUNEL-positive cells and expression of caspase-3 in the MIA model.

lying cartilage, leading to a sequence of events in which the degeneration of the superficial zone develops into fibrillations of the cartilage and eventually results in erosions and ulcerations. In rat ACLT, there is an increase in the number of cells exhibiting signs of degeneration or even death [24], often related to chondrocyte apoptosis, as previously shown in the dog [17] and in the rabbit [9]. The present data confirm that caspase-dependent apoptosis of chondrocytes also occurs in ACLT-induced OA in the rat.

MIA-induced OA in the rat easily and quickly reproduces dose, time, and site-dependant OA-like lesions, and, in certain circumstances, functional impairment, similar to that observed in human disease [8]. These parameters, as well as proteoglycan metabolism and cytokine expression in weight-bearing areas [2], could serve as indicators for studying chondroprotective drugs, or for evaluating the ability of imaging techniques to detect and evaluate chondral lesions. Pro-apoptotic properties of MIA, an ATP depleting agent inhibiting glycolysis, are well established *in vitro* on various cell-lines, including chondrocytes [7] as an archetypal anaerobic cell-line being especially sensitive to MIA-induced mitochondrial dysfunction. These data confirm that caspase-dependant apoptosis also occurs in MIA-induced OA in the rat and the prevalence of caspase 3 positive cells is significantly correlated with the severity of OA.

Histologic studies of rat knees showed in both ACLT and MIA models structural changes including the characteristics of OA. Most OA chondrocytes displayed a positive TUNEL reaction whereas only a few cells in control cartilage did. To corroborate the implication of caspase 3 in the pathophysiology of OA we used 2 *in vivo* OA models, i.e., a surgical and a biochemical, to avoid any bias inherent to the induction process, as previously published by Lee et al. [14] who used the same approach for studying the expression of TRAIL in experimentally induced OA with confirmed apoptosis.

Caspases are a family of proteases that have been demonstrated to play a prominent role in determining DNA damages and ancillary caspase-dependent cell-death [13]. The initial death signals can activate the apoptotic pathway which involves a cascade of highly regulated hierarchical molecular events [5,21]. These events are mediated by a proteolytic cascade in which upstream activator caspases initiate and amplify the maturation of effector caspases that, in turn, cleave a discrete subset of cellular polypeptides to manifest the apoptotic phenotype [12]. Caspase 3 in its active form is one of the key mediators of apoptosis in its execution. It is activated by initiator caspases such as caspase 8 and 9, and cleaves vital intra-cellular proteins. As performed in clinics [15], we routinely use active caspase 3 immunostaining in evaluating experimental OA, because TUNEL is less specific [6,16,21] and apoptosis is overestimated by TUNEL assay compared with caspase 3 detection.

## 5. Conclusion

Our results suggest that positivity for caspase 3 in experimental rat OA closely correlates with OA grade and percentage of apoptosis as depicted by TUNEL. Caspase 3 expression may herald imminent apoptosis better than TUNEL assay [1,15,18], and thus, may act as a surrogate specific marker for early chondrocyte apoptosis (or chondroptosis). Finally, the key role that caspase 3 plays in executing apoptosis make it a potential target for apoptosis modulation [7,12,13].

## Acknowledgments

Authors thank Venkatesan Narayanan for his expert advices and Michel Thiery for his good care to animals. This study was supported by grants from Région Lorraine, GIP Fonds de recherches HMR AVENTIS (FR99RHU037) and CPRC CHU Nancy.

## References

[1] T. Aigner, Apoptosis, necrosis, or whatever: how to find out what really happens?, *J. Pathol.* **198** (2002), 1–4.
[2] H. Dumond, N. Presle, P. Pottie, S. Pacquelet, B. Terlain, P. Netter, A. Gepstein, E. Livne and J.Y. Jouzeau, Site specific changes in gene expression and cartilage metabolism during early experimental osteoarthritis, *Osteoarthritis Cartilage* **12** (2004), 284–295.

[3] L. Galois, S. Etienne, L. Grossin, C. Cournil, A. Pinzano, P. Netter, D. Mainard and P. Gillet, Moderate-impact exercise is associated with decreased severity of experimental osteoarthritis in rats, *Rheumatology (Oxford)* **42** (2003), 692–693; author reply 693–694.

[4] L. Galois, S. Etienne, L. Grossin, A. Watrin-Pinzano, C. Cournil-Henrionnet, D. Loeuille, P. Netter, D. Mainard and P. Gillet, Dose-response relationship for exercise on severity of experimental osteoarthritis in rats: a pilot study, *Osteoarthritis Cartilage* **12** (2004), 779–786.

[5] R. Goggs, S.D. Carter, G. Schulze-Tanzil, M. Shakibaei and A. Mobasheri, Apoptosis and the loss of chondrocyte survival signals contribute to articular cartilage degradation in osteoarthritis, *Vet. J.* **166** (2003), 140–158.

[6] S.P. Grogan, B. Aklin, M. Frenz, T. Brunner, T. Schaffner and P. Mainil-Varlet, In vitro model for the study of necrosis and apoptosis in native cartilage, *J. Pathol.* **198** (2002), 5–13.

[7] L. Grossin, S. Etienne, N. Gaborit, A. Pinzano, C. Cournil-Henrionnet, C. Gerard, E. Payan, P. Netter, B. Terlain and P. Gillet, Induction of heat shock protein 70 (Hsp70) by proteasome inhibitor MG 132 protects articular chondrocytes from cellular death in vitro and in vivo, *Biorheology* **41** (2004), 521–534.

[8] C. Guingamp, P. Gegout-Pottie, L. Philippe, B. Terlain, P. Netter and P. Gillet, Mono-iodoacetate-induced experimental osteoarthritis: a dose-response study of loss of mobility, morphology, and biochemistry, *Arthritis Rheum.* **40** (1997), 1670–1679.

[9] S. Hashimoto, K. Takahashi, D. Amiel, R.D. Coutts and M. Lotz, Chondrocyte apoptosis and nitric oxide production during experimentally induced osteoarthritis, *Arthritis Rheum.* **41** (1998), 1266–1274.

[10] R.U. Janicke, M.L. Sprengart, M.R. Wati and A.G. Porter, Caspase-3 is required for DNA fragmentation and morphological changes associated with apoptosis, *J. Biol. Chem.* **273** (1998), 9357–9360.

[11] J.Y. Jouzeau, P. Gillet and P. Netter, Interest of animal models in the preclinical screening of anti-osteoarthritic drugs, *Joint Bone Spine* **67** (2000), 565–569.

[12] K. Kuhn, D.D. D'Lima, S. Hashimoto and M. Lotz, Cell death in cartilage, *Osteoarthritis Cartilage* **12** (2004), 1–16.

[13] I.N. Lavrik, A. Golks and P.H. Krammer, Caspases: pharmacological manipulation of cell death, *J. Clin. Invest.* **115** (2005), 2665–2672.

[14] S.W. Lee, H.J. Lee, W.T. Chung, S.M. Choi, S.H. Rhyu, D.K. Kim, K.T. Kim, J.Y. Kim, J.M. Kim and Y.H. Yoo, TRAIL induces apoptosis of chondrocytes and influences the pathogenesis of experimentally induced rat osteoarthritis, *Arthritis Rheum.* **50** (2004), 534–542.

[15] M. Matsuo, K. Nishida, A. Yoshida, T. Murakami and H. Inoue, Expression of caspase-3 and -9 relevant to cartilage destruction and chondrocyte apoptosis in human osteoarthritic cartilage, *Acta Med. Okayama* **55** (2001), 333–340.

[16] D. Mistry, Y. Oue, M.G. Chambers, M.V. Kayser and R.M. Mason, Chondrocyte death during murine osteoarthritis, *Osteoarthritis Cartilage* **12** (2004), 131–141.

[17] J.P. Pelletier, D.V. Jovanovic, V. Lascau-Coman, J.C. Fernandes, P.T. Manning, J.R. Connor, M.G. Currie and J. Martel-Pelletier, Selective inhibition of inducible nitric oxide synthase reduces progression of experimental osteoarthritis in vivo: possible link with the reduction in chondrocyte apoptosis and caspase 3 level, *Arthritis Rheum.* **43** (2000), 1290–1299.

[18] A.R. Resendes, N. Majo, J. Segales, J. Espadamala, E. Mateu, F. Chianini, M. Nofrarias and M. Domingo, Apoptosis in normal lymphoid organs from healthy normal, conventional pigs at different ages detected by TUNEL and cleaved caspase-3 immunohistochemistry in paraffin-embedded tissues, *Vet. Immunol. Immunopathol.* **99** (2004), 203–213.

[19] H.I. Roach, T. Aigner and J.B. Kouri, Chondroptosis: a variant of apoptotic cell death in chondrocytes?, *Apoptosis* **9** (2004), 265–277.

[20] P. Sarzi-Puttini, M.A. Cimmino, R. Scarpa, R. Caporali, F. Parazzini, A. Zaninelli, F. Atzeni and B. Canesi, Osteoarthritis: an overview of the disease and its treatment strategies, *Semin. Arthritis Rheum.* **35** (2005), 1–10.

[21] D.R. Schultz and W.J. Harrington, Jr., Apoptosis: programmed cell death at a molecular level, *Semin. Arthritis Rheum.* **32** (2003), 345–369.

[22] D.J. Schurman and R.L. Smith, Osteoarthritis: current treatment and future prospects for surgical, medical, and biologic intervention, *Clin. Orthop. Relat. Res.* (2004), S183–189.

[23] M. Sharif, A. Whitehouse, P. Sharman, M. Perry and M. Adams, Increased apoptosis in human osteoarthritic cartilage corresponds to reduced cell density and expression of caspase-3, *Arthritis Rheum.* **50** (2004), 507–515.

[24] R. Stoop, P. Buma, P.M. van der Kraan, A.P. Hollander, R.C. Billinghurst, T.H. Meijers, A.R. Poole and W.B. van den Berg, Type II collagen degradation in articular cartilage fibrillation after anterior cruciate ligament transection in rats, *Osteoarthritis Cartilage* **9** (2001), 308–315.

[25] R. Todd Allen, C.M. Robertson, F.L. Harwood, T. Sasho, S.K. Williams, A.C. Pomerleau and D. Amiel, Characterization of mature vs aged rabbit articular cartilage: analysis of cell density, apoptosis-related gene expression and mechanisms controlling chondrocyte apoptosis, *Osteoarthritis Cartilage* **12** (2004), 917–923.

[26] J.M. Williams, D.L. Felten, R.G. Peterson and B.L. O'Connor, Effects of surgically induced instability on rat knee articular cartilage, *J. Anat.* **134** (1982), 103–109.

Biorheology 43 (2006) 413–429
IOS Press



# Dynamic compression counteracts IL-1$\beta$ induced iNOS and COX-2 activity by human chondrocytes cultured in agarose constructs

Tina T. Chowdhury *, Dan L. Bader and David A. Lee
*Medical Engineering Division and IRC in Biomedical Materials, Department of Engineering, Queen Mary, University of London, Mile End Road, London, E1 4NS, UK*

**Abstract.** ˙NO and PGE$_2$ are inflammatory mediators derived from the inducible iNOS and COX enzymes and are potentially important pharmacological targets in OA. Both mechanical loading and IL-1$\beta$ will influence the release of ˙NO and PGE$_2$. Accordingly, the current study examines the effect of dynamic compression on ˙NO and PGE$_2$ release by human chondrocytes cultured in agarose constructs in the presence and absence of selective iNOS and COX-2 inhibitors. The current data demonstrate that IL-1$\beta$ induced nitrite and PGE$_2$ release and inhibited [$^3$H]-thymidine and $^{35}$SO$_4$ incorporation. Inhibitor experiments indicate that 1400W and NS-398 either partially reversed or abolished IL-1$\beta$ induced nitrite and PGE$_2$ release. IL-1$\beta$ induced inhibition of cell proliferation and proteoglycan synthesis was partially reversed with 1400W but was not influenced by NS-398. For the dynamic loading experiments, 1400W and NS-398 either reduced or abolished the compression-induced inhibition of ˙NO and PGE$_2$ release in the presence of IL-1$\beta$. The IL-1$\beta$ induced inhibition of cell proliferation was not influenced by 1400W or NS-398 whereas strain-induced stimulation of proteoglycan synthesis in the presence of IL-1$\beta$ was enhanced by 1400W. The data obtained using human chondrocytes demonstrate that IL-1$\beta$ induced ˙NO and PGE$_2$ release via an iNOS-driven-COX-2 inter-dependent pathway. This response could be reversed by dynamic compression. These data indicate interactions exist between the NOS and COX pathways, a finding which will provide new insights in the development of pharmacological or biophysical treatments for cartilage disorders such as OA.

Keywords: Nitric oxide, PGE$_2$, mechanotransduction, osteoarthritis

## 1. Introduction

The importance of nitric oxide (˙NO) as a key mediator in the pathophysiology of osteoarthritis (OA) has been well documented [53,54]. The underlying mechanisms involve the induction of ˙NO and prostaglandin E$_2$ (PGE$_2$) by pro-inflammatory cytokines which activate a complex cascade of catabolic events that lead to the degradation of the extracellular matrix constituents. Mechanical signals will also play an integral role in maintaining cartilage homeostasis [23,48,49]. However, the activation or inhibition of the multiple pathways mediated by mechanical and biochemical stimuli, are unclear. Thus, targeting the intracellular signalling pathways triggered in mechanotransduction is an attractive concept for the treatment of OA.

Within chondrocytes, exposure to interleukin-1$\beta$ (IL-1$\beta$) will induce the release of ˙NO and PGE$_2$ via activation of inducible nitric oxide synthase (iNOS) and cyclo-oxygenase (COX-2) enzymes [5,37,46,50,

---

*Address for correspondence: Dr. Tina T. Chowdhury, Department of Engineering, Queen Mary, University of London, Mile End Road, London, E1 4NS, UK. Tel.: +44 207 882 5368; Fax: +44 208 983 3052; E-mail: T.T.Chowdhury@qmul.ac.uk.

58]. Matrix metalloproteinase (MMP) activation is downstream of the inducible signalling pathways and therefore, inhibition of the NOS or COX enzymes may provide a means to prevent the catabolic events observed in OA. For example, Celecoxib, which is a selective inhibitor of COX-2, has been shown to prevent IL-1β induced cartilage damage in normal and OA affected explants [34,35]. However, clinical studies have also shown unwanted side-effects, suggesting that the decreased levels of prostaglandins may influence the regulatory signalling mechanisms associated with cartilage matrix turnover [2,33,40, 47,57].

A potential strategy which is worthy of clinical study is the beneficial effects of physical therapy. It has been demonstrated that the application of physiological loads maintains the balance between anabolic and catabolic activities of chondrocytes *in vivo* [7,8,23,44,48]. *In vitro* studies have also shown anti-inflammatory effects of mechanical strain through inhibition of ·NO and $PGE_2$ release. Mechanical strain of physiological magnitude counteracts IL-1β induced release of ·NO and $PGE_2$ in isolated human chondrocytes and by bovine cells cultured in agarose gel [9,10,20,31,59]. It is unclear, however, whether dynamic compression acts specifically to inhibit iNOS and COX-2 activity. Accordingly, the current study aims to determine the effect of dynamic compression on ·NO and $PGE_2$ release by human chondrocytes cultured in agarose constructs, in the presence and absence of selective iNOS and COX-2 inhibitors.

## 2. Materials and methods

### 2.1. Preparation of chondrocyte/agarose constructs

Normal human cartilage was isolated from the medial femoral condyles (biopsies weighing 200–300 mg) of 11 patients (age range, 18–50 years) undergoing autologous chondrocyte implantation. Cells were isolated within two to five hours and expanded in monolayer culture for up to five passages by Verigen Transplantation Services, Leverkeusen, Germany, using proprietary methods. Following expansion, the cell number and viability of a proportion of the cells was assessed at the supply laboratory using the trypan blue exclusion test [19]. The cells were subsequently couriered by overnight delivery in 20 ml of proprietary transport medium (Verigen Transplantation Services, Leverkeusen, Germany). Upon arrival at the test laboratory, chondrocytes were washed twice in 20 ml of a defined culture medium, as previously described [11,22]. Briefly, the culture medium comprised Dulbecco's Modified Eagles Medium (DMEM), 0.1 $\mu$M dexamethasone, 0.17 mM ascorbate, 1 mM sodium pyruvate, 0.35 mM proline, 5 $\mu$g.ml$^{-1}$ penicillin, 5 $\mu$g.ml$^{-1}$ streptomycin, 2 $\mu$M L-glutamine, ITS and 1.25 mg.ml$^{-1}$ bovine serum albumin (all from Cambrex Bioscience, Wokingham, UK). Cell yield and viability was calculated using a haemocytometer and the trypan blue exclusion assay, respectively [19]. Chondrocytes were resuspended in defined medium, to give a final cell concentration of $8 \times 10^6$ cells.ml$^{-1}$. The chondrocyte suspension was added to an equal volume of molten 6% (w/v) agarose type VII (Sigma Chemical Co., Poole, UK) in Earle's Balanced Salt Solutions (EBSS, Sigma Chemical Co., Poole, UK) to yield a final cell concentration of $4 \times 10^6$ cells.ml$^{-1}$ in 3% (w/v) agarose. Using a Pasteur pipette, the chondrocyte/agarose suspension was transferred into a sterile stainless steel mould, containing holes measuring 5 mm in diameter and 5 mm in height. The chondrocyte/agarose suspension was allowed to gel at 4°C for 20 minutes to yield cylindrical constructs, which were equilibrated by culture at 37°C in 5% $CO_2$ for 24 hours in defined medium containing 10 ng.ml$^{-1}$ TGFβ$_3$ (Cambrex Bioscience, Wokingham, UK).

## 2.2. Induction and inhibition of ·NO and PGE$_2$ release in free-swelling constructs

Constructs were cultured under free-swelling conditions, for a further 48 hours in defined medium containing 10 ng.ml$^{-1}$ TGFβ$_3$ supplemented with 0, 0.1, 1, 10 and 100 ng.ml$^{-1}$ human recombinant IL-1β (PeproTech EC Ltd, London, UK). All constructs were additionally incubated in the presence of 1 μCi.ml$^{-1}$ [$^3$H]-thymidine and 10 μCi.ml$^{-1}$ $^{35}$SO$_4$ (both Amersham Biosciences Ltd, Bucks, UK), for the assessment of chondrocyte proliferation and proteoglycan synthesis, respectively. This medium is designated hereafter as radiolabelled medium. In further experiments, constructs were equilibrated in culture for 24 hours as described above and incubated for a further 48 hours in radiolabelled medium containing 10 ng.ml$^{-1}$ IL-1β + 0.1, 1 or 2 mM N-[3-(aminomethyl) benzyl]acetamidine·2HCl (1400W, iNOS inhibitor) or 0.1, 1, 10 or 100 μM N-(2-cyclohexyloxy-4-nitrophenyl)-methanesulfonamide (NS-398, COX-2 inhibitor) (both from Merck Biosciences, Nottingham, UK).

## 2.3. Application of mechanical compression

A fully characterised cell-straining system (Zwick Testing Machines Ltd, Leominster, UK) was used to apply dynamic compression to chondrocyte/agarose constructs, as detailed previously [27–29]. Constructs were equilibrated in culture for 24 hours, transferred into a 24-well culture plate (Costar, High Wycombe, UK) and mounted within the cell-straining apparatus. One millilitre of radiolabelled medium containing 0 or 10 ng.ml$^{-1}$ IL-1β, either in the presence or absence of 2 mM 1400W or 100 μM NS-398 was introduced into each well. Strained constructs were subjected to a compressive strain amplitude of 15% at 1 Hz for 48 hours. Unstrained control constructs, subjected to a tare strain of approximately 0.8% and identical boundary conditions to the strained constructs, were maintained within the cell-straining apparatus [27–29]. Both groups of constructs were incubated for 48 hour at 37°C/5% CO$_2$.

## 2.4. Biochemical analysis

At the end of the culture period, all constructs and medium were frozen at −20°C for subsequent biochemical analysis, as previously detailed by the authors [9,10,27–29]. To review briefly, constructs were digested overnight at 37°C with 10 U.ml$^{-1}$ agarase and for 1 hr at 60°C with 2.8 U.ml$^{-1}$ papain (both Sigma Chemical Co., Poole, UK). Absolute concentrations of nitrite (μM), a stable end-product of ·NO metabolism, were measured in the media using a spectrophotometric method based on the Griess assay [9,10,28]. Absorbance was measured at 550 nm and nitrite (μM) was determined by comparison with standard solutions of sodium nitrite. PGE$_2$ release was measured in the culture supernatant, using a high sensitivity enzyme immunoassay kit (Amersham Biosciences Ltd, Bucks, UK) [9,10]. The absorbance was measured at 450 nm using PGE$_2$ diluted in assay buffer as a stock standard over the concentration range 20–640 pg.ml$^{-1}$. [$^3$H]-thymidine incorporation was measured in the agarase/papain digests by 10% Tricholoroacetic acid precipitation onto filters using the Millipore Multiscreen system (Millipore, Watford, UK) [27]. Incorporation of $^{35}$SO$_4$ was determined in both medium and agarase/papain digests, using the Alcian blue precipitation method [27,36]. Total DNA, determined using the Hoescht 33258 method, was used as a baseline for [$^3$H]-thymidine and $^{35}$SO$_4$ incorporation [45].

## 2.5. Statistical analysis

Unless indicated, all data represent the mean and SEM values of between 6 to 16 replicates, from three separate experiments. Two-way ANOVA with *post hoc* Bonferroni-corrected *t*-tests were used to examine inter and intra-group differences for absolute data. Unpaired Student's $t$-tests were used to examine normalised data. In all cases, a level of 5% was considered statistically significant ($^*p < 0.05$).

## 3. Results

### 3.1. Effect of IL-1β on the metabolism of chondrocytes cultured in agarose constructs

Figure 1 illustrates nitrite and $PGE_2$ release, [$^3H$]-thymidine and $^{35}SO_4$ incorporation by constructs cultured under free-swelling conditions in medium supplemented with IL-1β at concentrations up to 100 ng.ml$^{-1}$. In the absence of the cytokine, nitrite levels measured in the media were approximately 15.5 $\mu$M (Fig. 1A). The presence of the cytokine induced a dose-dependent increase in the release

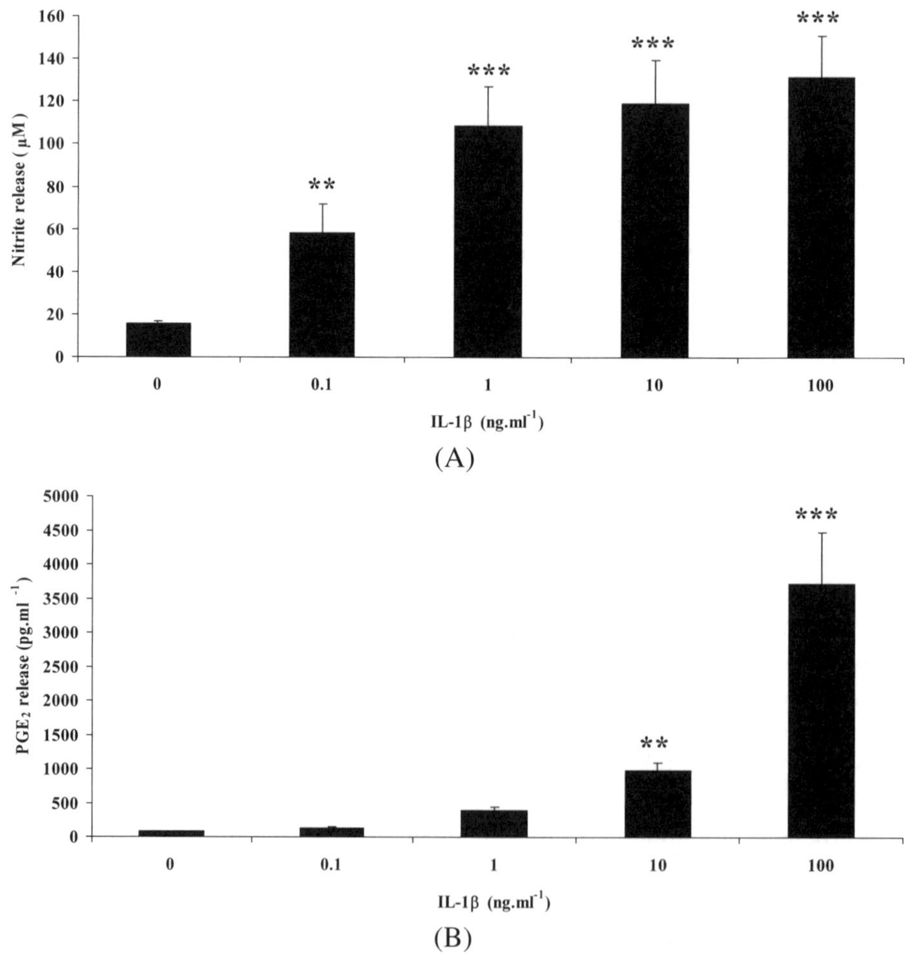

Fig. 1. Nitrite (A) and $PGE_2$ (B) release, [$^3H$]-thymidine (C) and $^{35}SO_4$ incorporation for constructs cultured for 48 hours in medium containing 0–100 ng.ml$^{-1}$ IL-1β. Bars represent the mean and SEM of 6 replicates. Two way ANOVA with post hoc Bonferroni-corrected $t$-tests was used to examine all data, where values are as follows: $^*p < 0.05$, $^{**}p < 0.01$, $^{***}p < 0.001$. Comparisons indicate significant differences between unsupplemented constructs and constructs treated with different concentrations of IL-1β. All other comparisons were not significant (not indicated).

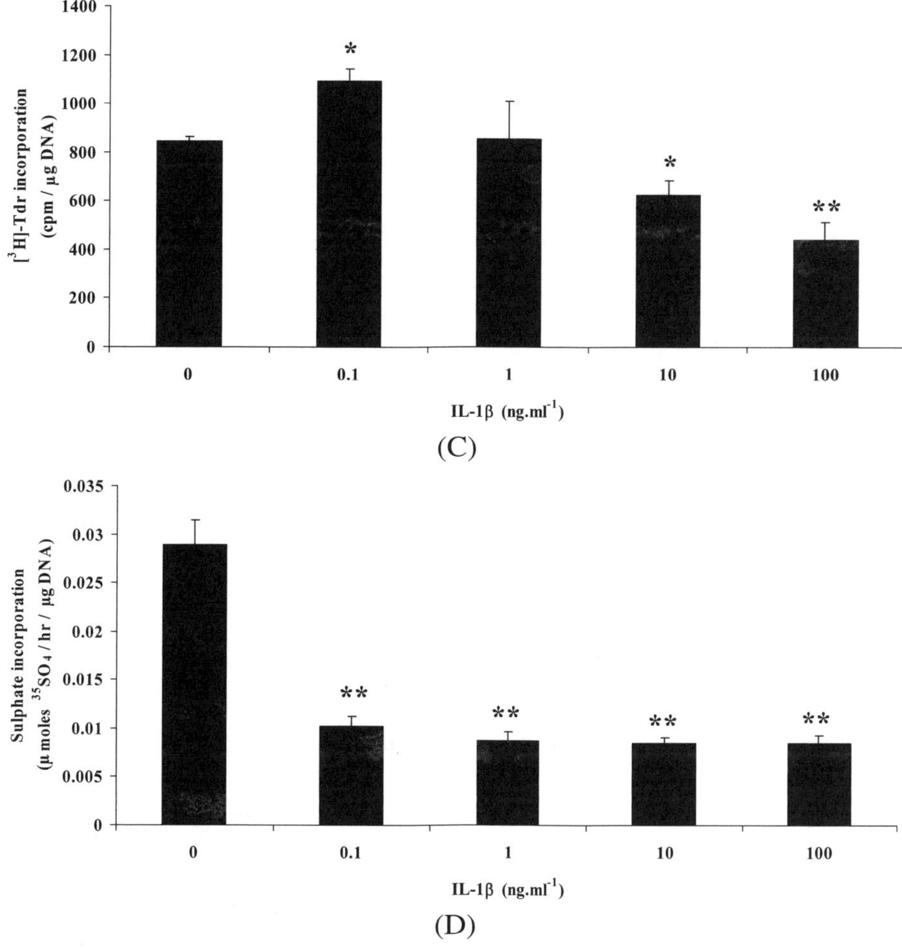

Fig. 1. (Continued).

of nitrite to a maximal level of approximately 132 $\mu$M ($0.01 < p < 0.001$). IL-1$\beta$ also induced a dose-dependent release of PGE$_2$, with maximal values of approximately 3714 pg.ml$^{-1}$ for constructs stimulated with 100 ng.ml$^{-1}$ IL-1$\beta$ ($p < 0.001$, Fig. 1B). There was a marginal upregulation of [$^3$H]-thymidine incorporation in the presence of 0.1 ng.ml$^{-1}$ IL-1$\beta$ ($p < 0.05$, Fig. 1C). However, [$^3$H]-thymidine incorporation levels were significantly inhibited with IL-1$\beta$ at concentrations of 10 ng.ml$^{-1}$ ($p < 0.05$) and 100 ng.ml$^{-1}$ ($p < 0.01$). Furthermore, IL-1$\beta$ inhibited $^{35}$SO$_4$ incorporation at all concentrations tested (all $p < 0.01$, Fig. 1D).

Figure 2 indicates the effects of the iNOS inhibitor, 1400W (0.1 to 2 mM) on nitrite and PGE$_2$ release, [$^3$H]-thymidine and $^{35}$SO$_4$ incorporation by unsupplemented constructs or constructs cultured in the presence of 10 ng.ml$^{-1}$ IL-1$\beta$. As illustrated in Fig. 2A, the cytokine-induced nitrite release was abolished by 1400W at all concentrations tested (all $p < 0.001$). IL-1$\beta$ induced PGE$_2$ release could also be partially inhibited with 0.1 to 2 mM 1400W ($0.01 < p < 0.001$, Fig. 2B). IL-1$\beta$ induced inhibition of [$^3$H]-thymidine incorporation was partially reversed with 0.1 to 2 mM 1400W (all $p < 0.05$, Fig. 2C). Similarly, 1400W (0.1 to 2 mM) partially reversed IL-1$\beta$ induced inhibition of $^{35}$SO$_4$ incorporation ($0.05 < p < 0.01$; Fig. 2D).

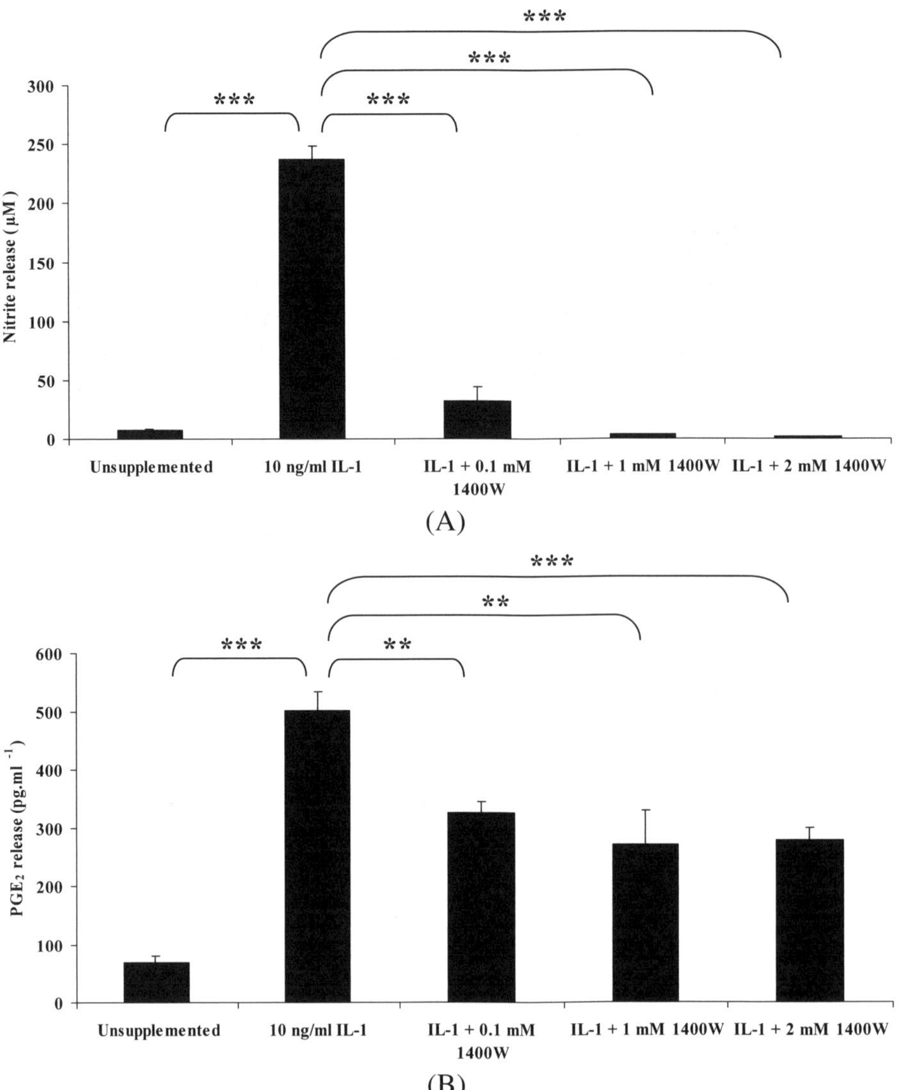

Fig. 2. Nitrite (A) and PGE$_2$ (B) release, [$^3$H]-thymidine (C) and $^{35}$SO$_4$ incorporation for constructs cultured for 48 hours in medium with 0 or 10 ng.ml$^{-1}$ IL-1β supplemented with 1 mM L-NIO or 0.1, 1 or 2 mM 1400W. Bars represent the mean and SEM of 6 to 10 replicates from two separate experiments. Two way ANOVA with post hoc Bonferroni-corrected $t$-tests was used to examine all data where values are as follows: $^*p < 0.05$, $^{**}p < 0.01$, $^{***}p < 0.001$. All other comparisons were not significant (not indicated).

Figure 3 indicates the effects of the COX-2 inhibitor, NS-398, over a concentration from 0.01 to 100 $\mu$M on constructs cultured in the presence of 10 ng.ml$^{-1}$ IL-1β. As presented in Fig. 3A, IL-1β induced nitrite release was not significantly affected by the presence of NS-398 over concentrations ranging from 0.1 to 10 $\mu$M. However, 100 $\mu$M NS-398 partially inhibited the IL-1β induced nitrite release ($p < 0.01$). By contrast, the IL-1β induced PGE$_2$ release was abolished by the presence of NS-398 at all concentrations tested (all $p < 0.01$, Fig. 3B). In addition, the IL-1β induced inhibition of [$^3$H]-thymidine and $^{35}$SO$_4$ incorporation was not significantly influenced by the presence of NS-398 (Figs 3C and 3D, respectively).

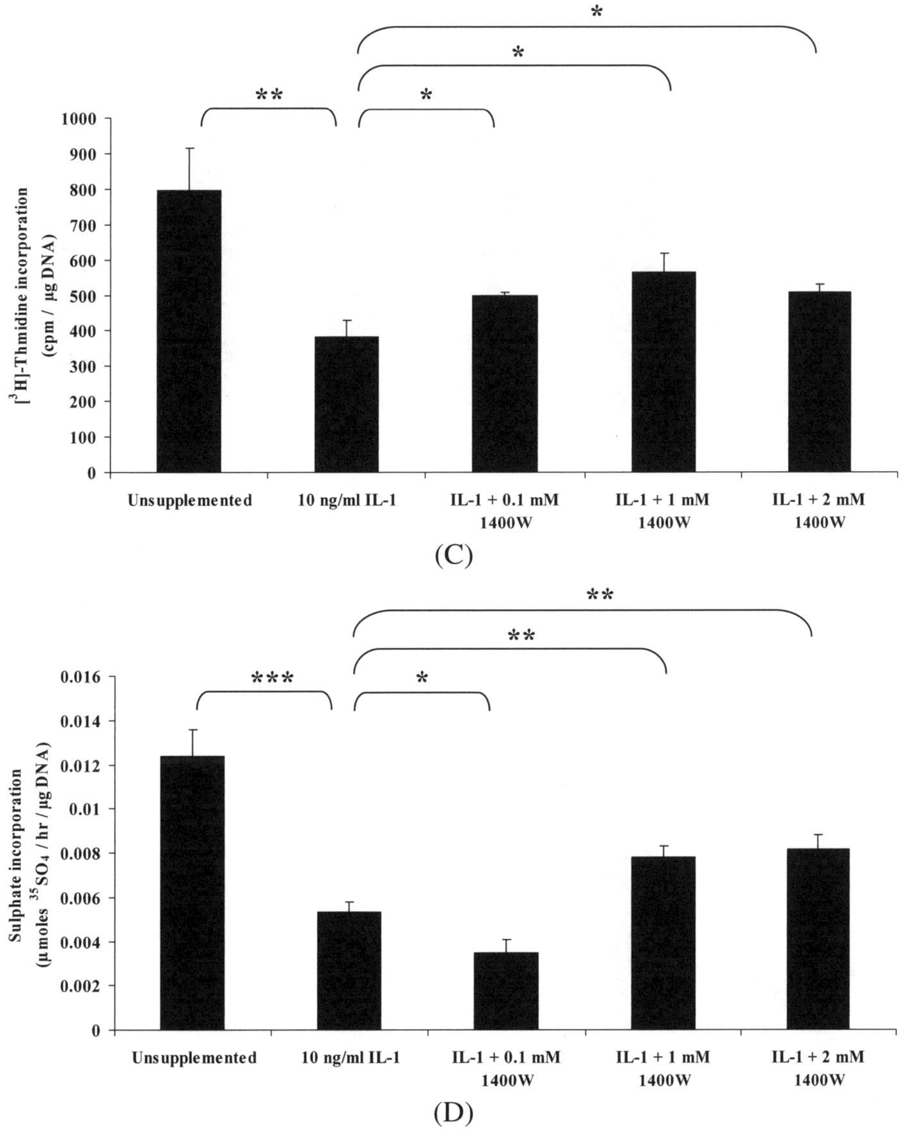

Fig. 2. (Continued).

## 3.2. Effect of iNOS and COX-2 inhibitors in mechanically stimulated constructs cultured in the presence and absence of IL-1β

Absolute and normalised values for nitrite and $PGE_2$ release, $[^3H]$-thymidine and $^{35}SO_4$ incorporation by cells cultured within unstrained constructs and constructs subjected to dynamic compression, in the presence and absence of 10 ng.ml$^{-1}$ IL-1β and/or 2 mM 1400W or 100 μM NS-398, are presented in Figs 4–7. In unstrained constructs, nitrite release increased from approximately 12 μM to 140 μM in the presence of 10 ng.ml$^{-1}$ IL-1β when compared to unsupplemented constructs ($p < 0.001$, Fig. 4A). The application of dynamic compression significantly reduced levels of nitrite release by 17% and 50%, in the absence and presence of the cytokine, respectively (both $p < 0.01$; Fig. 4A). 1400W reduced nitrite

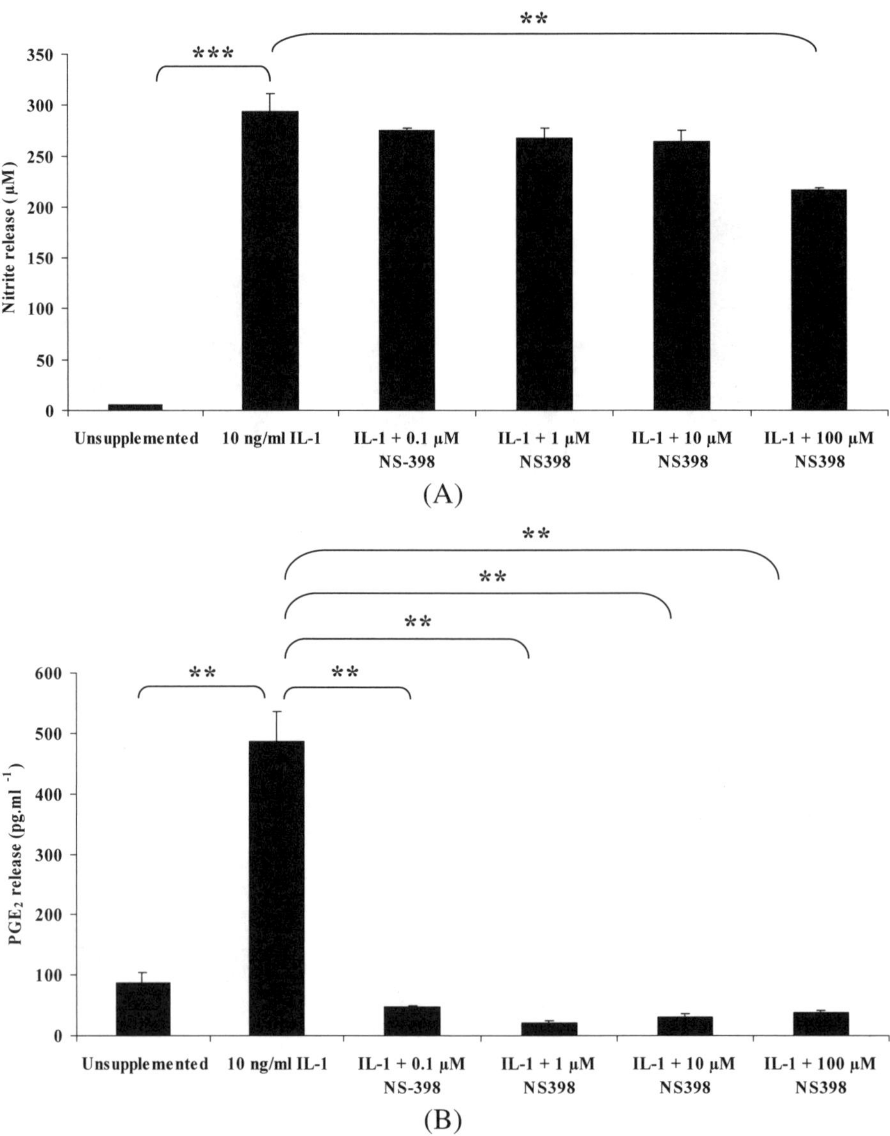


Fig. 3. Nitrite (A) and PGE$_2$ (B) release, [$^3$H]-thymidine (C) and $^{35}$SO$_4$ incorporation for constructs cultured for 48 hours in medium with 0 or 10 ng.ml$^{-1}$ IL-1β supplemented with 0.1, 1, 10 or 100 μM NS-398. Bars represent the mean and SEM of 6 replicates. Two way ANOVA with post hoc Bonferroni-corrected $t$-tests was used to examine all data where values are as follows: $^*p < 0.05$, $^{**}p < 0.01$, $^{***}p < 0.001$. All other comparisons were not significant (not indicated).

levels in unstrained and strained constructs cultured in the presence of IL-1β, such that the response to dynamic strain was abolished in the presence of the inhibitor (Fig. 4B). In addition, NS-398 partially reversed strain-induced inhibition of nitrite release in the presence of IL-1β ($p < 0.001$, Fig. 4B).

Dynamic compression did not significantly influence PGE$_2$ levels in unstrained and strained constructs cultured in unsupplemented conditions (Fig. 5A). In unstrained constructs, the presence of IL-1β enhanced PGE$_2$ release from approximately 44 pg.ml$^{-1}$ to 1202 pg.ml$^{-1}$ ($p < 0.001$; Fig. 5A). This effect was reduced by dynamic compression by approximately 46% in IL-1β stimulated constructs ($p < 0.01$; Fig. 5A). The presence of 1400W or NS-398 significantly reduced PGE$_2$ release in unstrained constructs

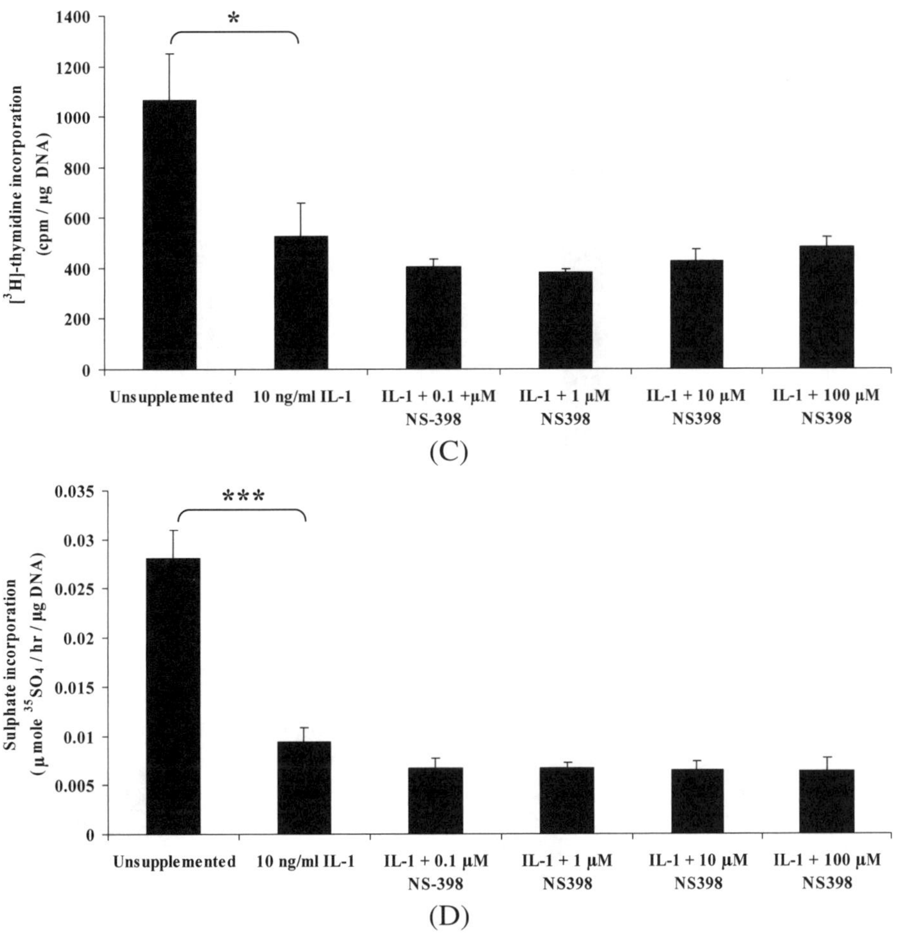



Fig. 3. (Continued).

(all $p < 0.001$; Fig. 5A). However, strain-induced inhibition of $PGE_2$ release was not significantly affected by 1400W but could be abolished with NS-398 (Fig. 5B).

In unstrained constructs, [³H]-thymidine incorporation was inhibited by the presence of IL-1β, as illustrated in Fig. 6A ($p < 0.001$). The application of dynamic compression enhanced [³H]-thymidine incorporation in the presence and absence of the cytokine ($p < 0.05$ and $p < 0.001$, respectively; Fig. 6A). The presence of 1400W or NS-398 did not significantly influence [³H]-thymidine incorporation levels in unstrained constructs cultured with IL-1β (Fig. 6A). Furthermore, the presence of 1400W or NS-398 did not significantly affect the magnitude of strain-induced stimulation of [³H]-thymidine incorporation in the presence IL-1β (Fig. 6B).

In the absence of IL-1β, dynamic compression induced $^{35}SO_4$ incorporation by approximately 88% ($p < 0.001$; Fig. 7A). The presence of IL-1β inhibited $^{35}SO_4$ incorporation levels in unstrained constructs ($p < 0.01$; Fig. 7A) and this effect was not significantly influenced by the presence of 1400W or NS-398. The magnitude of strain-induced stimulation of $^{35}SO_4$ incorporation in IL-1β stimulated constructs was not significantly influenced by NS-398 but was enhanced by 1400W when compared to unsupplemented constructs (Fig. 7B).

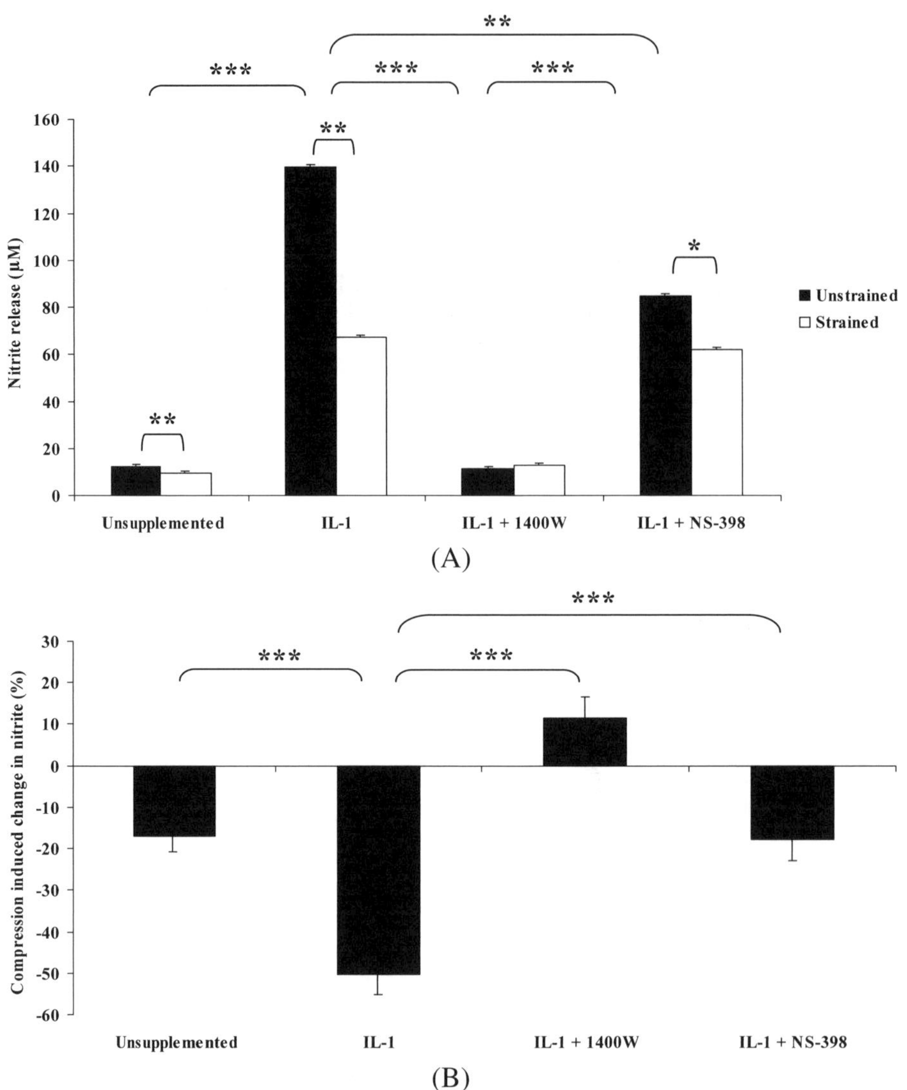


Fig. 4. Absolute (A) and normalised (B) values for nitrite release by constructs subjected to 15% dynamic compression (1 Hz) in medium supplemented with 0 or 10 ng.ml$^{-1}$ of IL-1β, either in the presence and absence of 2 mM 1400W or 100 $\mu$M NS-398. Bars represent the mean and SEM of between 6 and 16 replicates from three separate experiments. Two-way ANOVA with post hoc Bonferroni-corrected $t$-tests was used to examine absolute data. Unpaired Student's $t$-test was used to examine normalised data, where values are as follows: $^{*}p < 0.05$, $^{**}p < 0.01$, $^{***}p < 0.001$. All other comparisons were not significant (not indicated).

## 4. Discussion

IL-1 is a pro-inflammatory cytokine that contributes to the pathophysiology of OA in a number of ways. For example, IL-1 will trigger the release of ·NO and PGE$_2$ and decrease the synthesis of the extracellular matrix macromolecules by inhibiting proteoglycan synthesis and cell proliferation [5,24, 26,55,56]. Abnormally high levels of ·NO, in turn, activate MMPs and trigger chondrocyte apoptosis, resulting in the breakdown of the cartilage matrix [4,42]. These effects however, will vary across species and the experimental model systems used to examine the intracellular signaling pathways involved.

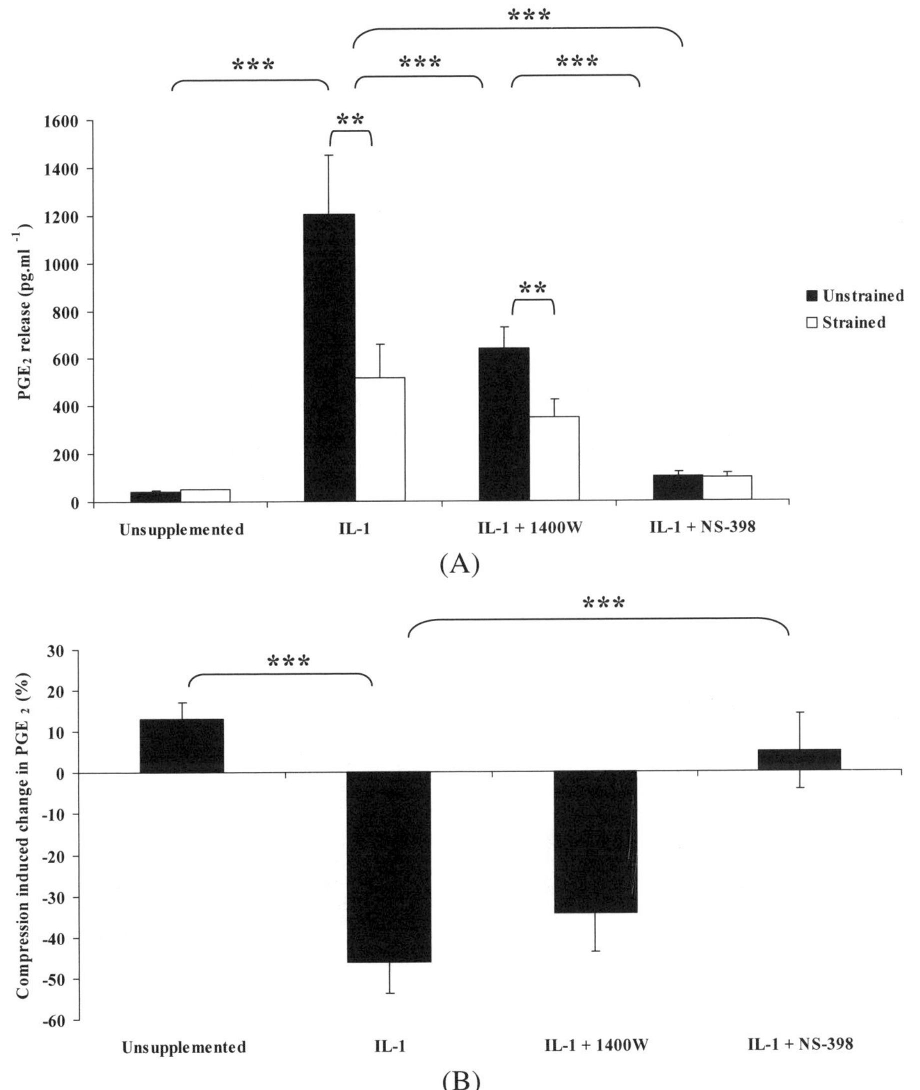

Fig. 5. Absolute (A) and normalised (B) values for $PGE_2$ release by constructs subjected to 15% dynamic compression (1 Hz) in medium supplemented with 0 or 10 ng.ml$^{-1}$ of IL-1β, either in the presence and absence of 2 mM 1400W or 100 μM NS-398. Bars represent the mean and SEM of between 6 and 16 replicates from three separate experiments. Two-way ANOVA with post hoc Bonferroni-corrected $t$-tests was used to examine absolute data. Unpaired Student's $t$-test was used to examine normalised data, where values are as follows: $^*p < 0.05$, $^{**}p < 0.01$, $^{***}p < 0.001$. All other comparisons were not significant (not indicated).

Experimental studies by the authors have shown that dynamic compression counteracts IL-1β induced ·NO and $PGE_2$ release by bovine chondrocytes seeded in agarose constructs. However, it is unclear whether dynamic compression acts to abrogate the IL-1β mediated pathways via an iNOS and/or COX-2 dependent pathway and whether these effects are replicated using human chondrocytes.

Our findings demonstrate that the presence of IL-1β induced high levels of ·NO and $PGE_2$ release by human chondrocytes, via an iNOS driven COX-2 inter-dependent pathway. Inhibition of iNOS down-regulated both ·NO and $PGE_2$ release and inhibition of COX-2 abolished $PGE_2$ release. Interestingly, the

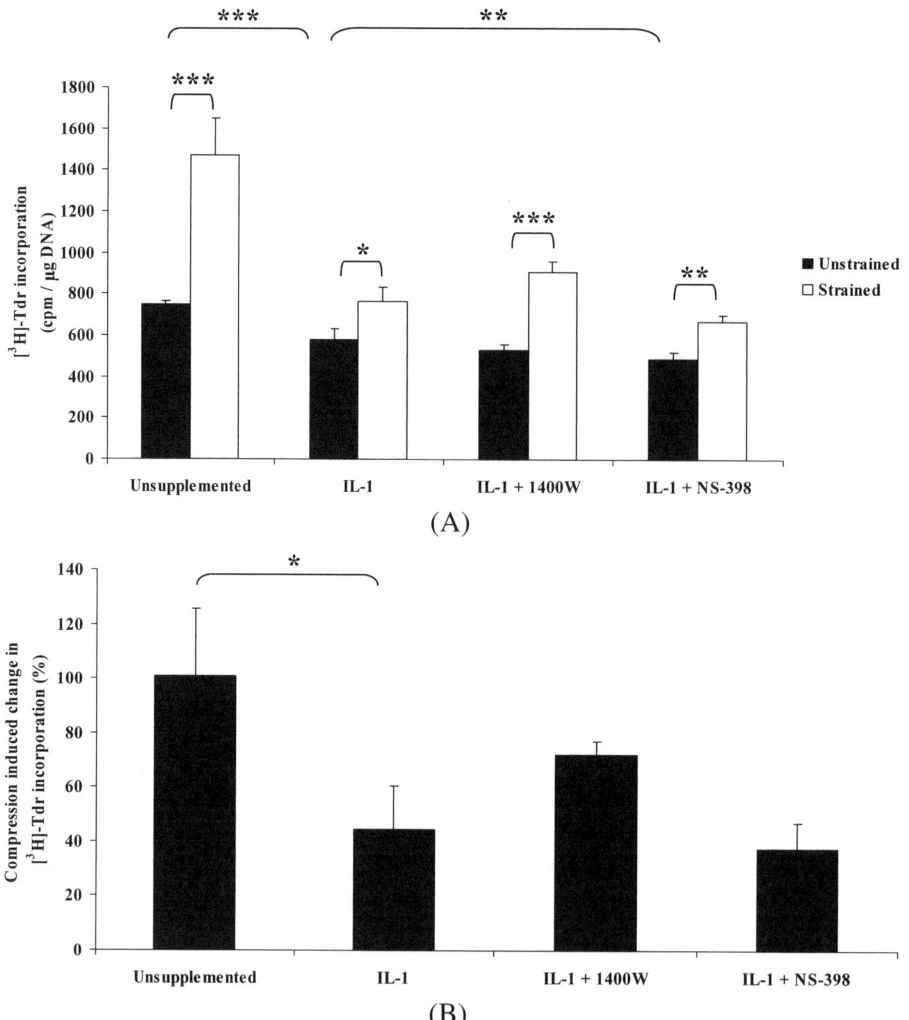

Fig. 6. Absolute (A) and normalised (B) values for $[^3H]$-thymidine incorporation by constructs subjected to 15% dynamic compression (1 Hz) in medium supplemented with 0 or 10 ng.ml$^{-1}$ of IL-1β, either in the presence and absence of 2 mM 1400W or 100 μM NS-398. Bars represent the mean and SEM of between 6 and 16 replicates from three separate experiments. Two-way ANOVA with post hoc Bonferroni-corrected $t$-tests was used to examine absolute data. Unpaired Student's $t$-test was used to examine normalised data, where values are as follows: $^*p < 0.05$, $^{**}p < 0.01$, $^{***}p < 0.001$. All other comparisons were not significant (not indicated).

COX-2 inhibitor marginally inhibited ·NO only when a high concentration was used. These data indicate a negative feedback role of COX-2 on ·NO release suggesting that PGE$_2$ could have a beneficial effect on cartilage homeostasis. Indeed, previous studies have shown that high levels of PGE$_2$, generated in response to IL-1, could downregulate ·NO release whereas low PGE$_2$ levels had the opposite effect [5, 21,38]. The biphasic response of PGE$_2$ on ·NO release may be due to the anti-inflammatory properties of the different eiconosoid end products [12]. Each product may act via a different receptor and signalling pathway and therefore have divergent effects on chondrocyte metabolism. Moreover, NOS inhibitors have been shown to augment PGE$_2$ production in osteoarthritic cartilage explants [3]. Furthermore, PGE$_2$ has been shown to stimulate cell proliferation and aggrecan synthesis in rat chondrocytes

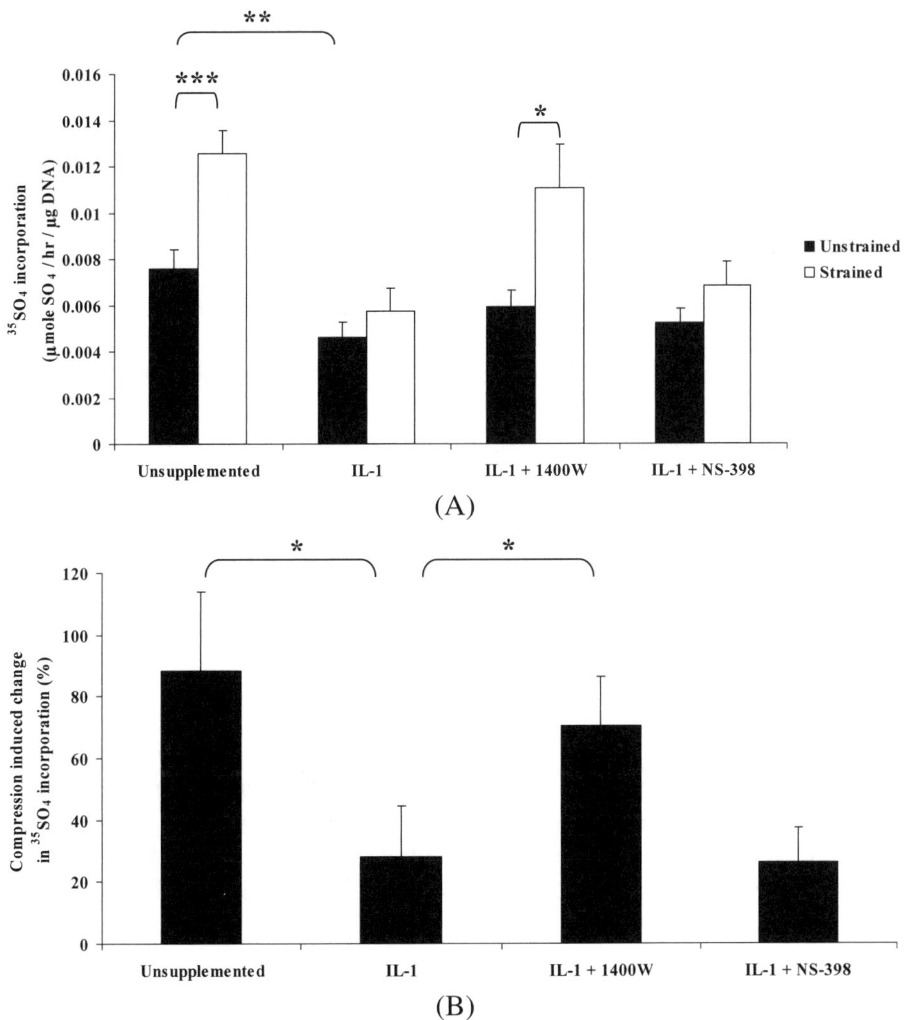

Fig. 7. Absolute (A) and normalised (B) values for $^{35}SO_4$ incorporation by constructs subjected to 15% dynamic compression (1 Hz) in medium supplemented with 0 or 10 ng.ml$^{-1}$ of IL-1β, either in the presence and absence of 2 mM 1400W or 100 μM NS-398. Bars represent the mean and SEM of between 6 and 16 replicates from three separate experiments. Two-way ANOVA with post hoc Bonferroni-corrected $t$-tests was used to examine absolute data. Unpaired Student's $t$-test was used to examine normalised data, where values are as follows: $^*p < 0.05$, $^{**}p < 0.01$, $^{***}p < 0.001$. All other comparisons were not significant (not indicated).

and to enhance proteoglycan and collagen synthesis in human cartilage through the induction of IGF [14, 32]. These data indicate significant interaction between the NOS and COX pathways and suggest a high degree of complexity of the intracellular pathways involved.

The current data shows clear inhibition of cell proliferation and proteoglycan synthesis by IL-1β. This response could be partially reversed by the iNOS inhibitor but was not influenced by the COX-2 inhibitor. These findings, however, contradicts previous studies which demonstrated a beneficial effect of PGE$_2$ on proteoglycan turnover in osteoarthritic cartilage [15,16,35]. These differences in the effects of COX-2 inhibition on chondrocyte activity may be due to the species and model system used during investigation. For instance, *in vitro* studies have shown variable effects of IL-1 on proteoglycan synthesis in lapine and bovine chondrocytes cultured in monolayer, explants or in agarose gel [9,10,51,52]. The present study demonstrates a highly significant inhibition of proteoglycan synthesis by IL-1β in the presence of the iNOS inhibitor, suggesting that factors downstream to ·NO such as oxygen free radical species including peroxynitrite or superoxide, could be responsible for the inhibition [25,43].

The current study indicates that dynamic mechanical compression can counteract IL-1β induced pro-inflammatory effects by inhibiting ·NO and $PGE_2$ release via an iNOS and COX-2-dependent pathways. Previous *in vitro* studies have shown that cyclic tensile strain of low magnitude acts as a potent antagonist of IL-1β actions and inhibits the expression of NOS and COX enzymes and the activity of MMP 1, 3, 9 and 13. The downregulation of these genes further inhibits ·NO and $PGE_2$ release as well as matrix degradation [1,20,59]. The current study involving compressive strain reveals similar antagonistic effects to those reported for tensile strain, suggesting that mechanical signals have an important anti-inflammatory role. The data also suggest that significant interactions exist between pathways induced by compressive strain and IL-1β involved in the regulation of iNOS and COX-2. Moreover, compressive strain partially reversed IL-1β induced inhibition of both cell proliferation and proteoglycan synthesis, although these effects were COX-2 independent for both parameters and iNOS independent for cell proliferation. The data are similar to studies utilizing low magnitude tensile strain which also reported that strain reversed IL-1β induced inhibition of proteoglycan synthesis [59]. Interestingly, studies by the same group suggest that high magnitude cyclic tensile strain is pro-inflammatory and acts to induce ·NO and $PGE_2$ release and inhibit proteoglycan in the absence of IL-1β [13]. Pro-inflammatory effects of mechanical loading have also been reported for explants or isolated cells subjected to static compression or high impact loads in the absence of the cytokine [6,30,41]. These data indicate the importance of physiological loads in the maintenance of cartilage homeostasis, which may be lost during application of static compression or non-physiological loading. In a further *in vitro* study utilizing cartilage explants, the application of intermittent compression in the absence of IL-1β, increased the release of ·NO and $PGE_2$ via an iNOS and COX-2 dependent pathway [17,18]. These data contradicts the current and previous studies utilizing chondrocytes in agarose gel [9,10,28,39], demonstrating the importance of the model system and loading regime when examining intracellular signalling pathways associated with mechanical loading and IL-1β.

There is now significant evidence that mechanical load will mediate the response of the chondrocytes to pro-inflammatory cytokines via interacting pathways. The current study demonstrates that both ·NO and $PGE_2$ release are influenced by mechanical loading and IL-1β and that the cytokine induced effects could be abrogated by dynamic loading via an iNOS and COX-2 inter-dependent pathway. Further studies will identify the mechanisms associated with IL-1β and mechanical load and will contribute to the design and treatment of OA therapy by targeting specific molecules known to be involved in the inflammatory process.

## Acknowledgements

This study was partly supported by a Framework *V* European grant, "Imbiotor", project number GRD1-2000-25394 and a project grant from The Wellcome Trust, project number 073972.

## References

[1] S. Agarwal, P. Long, R. Gassner, N.P. Piesco and M.J. Buckley, Cyclic tensile strain suppresses catabolic effects of interleukin-1beta in fibrochondrocytes from the temporomandibular joint, *Arthritis. Rheum.* **44**(3) (2001), 608–617.

[2] J.S. Alpert, The Vioxx debacle, *Am. J. Med.* **118**(3) (2005), 203–204.

[3] A.R. Amin, M. Attur, R.N. Patel, G.D. Thakker, P.J. Marshall, J. Rediske, S.A. Stuchin, I.R. Patel and S.B. Abramson, Superinduction of cyclooxygenase-2 activity in human osteoarthritis-affected cartilage. Influence of nitric oxide, *J. Clin. Invest.* **99**(6) (1997), 1231–1237.

[4] F.J. Blanco and M. Lotz, IL-1-induced nitric oxide inhibits chondrocyte proliferation via PGE$_2$, *Exp. Cell Res.* **218**(1) (1995), 319–325.

[5] F.J. Blanco, R.L. Ochs, H. Schwarz and M. Lotz, Chondrocyte apoptosis induced by nitric oxide, *Am. J. Pathol.* **146**(1) (1995), 75–85.

[6] J.A. Buckwalter and N.E. Lane, Athletics and osteoarthritis, *Am. J. Sports. Med.* **25**(6) (1997), 873–881.

[7] M.D. Buschmann, Y.A. Gluzband, A.J. Grodzinsky, J.H. Kimura and E.B. Hunziker, Chondrocytes in agarose culture synthesise a mechanically functional extracellular matrix, *J. Orthop. Res.* **10** (1992), 745–758.

[8] M.D. Buschmann, Y.A. Gluzband, A.J. Grodzinsky and E.B. Hunziker, Mechanical compression modulates matrix biosynthesis in chondrocyte/agarose culture, *J. Cell Sci.* **108** (1995), 1497–1508.

[9] T.T. Chowdhury, D.L. Bader and D.A. Lee, Dynamic compression inhibits the synthesis of nitric oxide and PGE$_2$ by IL-1β stimulated chondrocytes cultured in agarose constructs, *Biochem. Biophys. Res. Commun.* **285**(5) (2001), 1168–1174.

[10] T.T. Chowdhury, D.L. Bader and D.A. Lee, Dynamic compression counteracts IL-1 beta-induced release of nitric oxide and PGE2 by superficial zone chondrocytes cultured in agarose constructs, *Osteoarthritis Cartilage* **11** (2003), 688–696.

[11] T.T. Chowdhury, D.M. Salter, D.L. Bader and D.A. Lee, Integrin-mediated mechanotransduction processes in TGFβ stimulated monolayer-expanded chondrocytes, *Biochem. Biophys. Res. Commun.* **318**(4) (2004), 873–881.

[12] C.A. Clark, E.M. Schwarz, X. Zhang, N.M. Ziran, H. Drissi, R.J. O'Keefe and M.J. Zuscik, Differential regulation of EP receptor isoforms during chondrogenesis and chondrocyte maturation, *Biochem. Biophys. Res. Commun.* **328**(3) (2005), 764–776.

[13] J. Deschner, C.R. Hofman, N.P. Piesco and S. Agarwal, Signal transduction by mechanical strain in chondrocytes, *Curr. Opin. Clin. Nutr. Metab. Care* **6** (2003), 289–293.

[14] J.A. Di Battisata, S. Dore, J. Martel-Pelletier and J.P. Pelletier, Prostaglandin E2 stimulates incorporation of proline into collagenase digestible proteins in human articular chondrocytes: identification of an effector autocrine loop involving insulin-like growth factor I, *Mol. Cell. Endocrinol.* **123**(1) (1996), 27–35.

[15] J.T. Dingle, Cartilage maintenance in osteoarthritis: interaction of cytokines, NSAID and prostaglandins in articular cartilage damage and repair, *J. Rhematol. Suppl.* **28** (1991), 30–37.

[16] H. El Hajjaji, A. Marcelis, J.P. Devogelaer and D.H. Manicourt, Celecoxib has a positive effect on the overall metabolism of hyaluronan and proteoglycans in human osteoarthritic cartilage, *J. Rheumatol.* **30**(11) (2003), 2444–2451.

[17] B. Fermor, J.B. Weinberg, D.S. Pisetsky, M.A. Misukonis, A.J. Banes and F. Guilak, The effects of static and intermittent compression on nitric oxide production in articular cartilage explants, *J. Orthop. Res.* **19**(4) (2001), 729–737.

[18] B. Fermor, J.B. Weinberg, D.S. Pisetsky, M.A. Misukonis, C. Fink and F. Guilak, Induction of cyclooxygenase-2 by mechanical stress through a nitric oxide-regulated pathway, *Osteoarthritis Cartilage* **10**(10) (2002), 792–798.

[19] R. Fong and F. Kissmeyer-Nielsen, A combined dye exclusion (trypan blue) and fluorochromatic technique for the microdroplet lymphocytotoxicity test, *Tissue Antigens* **2**(1) (1972), 257–263.

[20] R. Gassner, M.J. Buckely, H. Georgescu, R. Studer, M. Stefanovic-Racic, N.P. Piesco, C.H. Evans and S. Agarwal, Cyclic tensile stress exerts antiinflammatory actions on chondrocytes by inhibiting inducible nitric oxide synthase, *J. Immunol.* **163**(4) (1999), 2187–2192.

[21] Y. Geng, F.J. Blanco, M. Cornelisson and M. Lotz, Regulation of cyclooxygenase-2 expression in normal human articular chondrocytes, *J. Immunol.* **155**(2) (1995), 796–801.

[22] A.J. Goldberg, D.A. Lee, D.L. Bader and G. Bentley, Autologous chondrocyte implantation. Culture in a TGF-beta-containing medium enhances the re-expression of a chondrocytic phenotype in passaged human chondrocytes in pellet culture, *J. Bone. Joint. Surg. Br.* **87**(1) (2005), 128–134.

[23] F. Guilak, A. Ratcliffe, N. Lane, M.P. Rosenwasser and V.C. Mow, Mechanical and biochemical changes in the superficial zone of articular cartilage in canine experimental osteoarthritis, *J. Orthop. Res.* **12**(4) (1997), 474–484.

[24] H.J. Hauselmann, L. Oppliger, B.A. Michel, M. Stefanovic-Racic and C.H. Evans, Nitric oxide and proteoglycan biosynthesis by human articular chondrocytes in alginate culture, *FEBS. Lett.* **352**(3) (1994), 361–364.

[25] Y.E. Henrotin, P. Bruckner and J.P. Pujol, The role of reactive oxygen species in homeostasis and degradation of cartilage, *Osteoarthritis Cartilage* **11**(10) (2003), 747–755.

[26] T.A. Jarvinen, T. Moilanen, T.L. Jarvinen and E. Moilanen, Endogenous nitric oxide and prostaglandin E2 do not regulate the synthesis of each other in interleukin-1 beta-stimulated rat articular cartilage, *Inflammation* **20**(6) (1996), 683–692.

[27] D.A. Lee and D.L. Bader, Compressive strains at physiological frequencies influence the metabolism of chondrocytes seeded in agarose, *J. Orthop. Res.* **15**(2) (1997), 181–188.

[28] D.A. Lee, S.P. Frean, P. Lees and D.L. Bader, Dynamic mechanical compression influences nitric oxide production by articular chondrocytes seeded in agarose, *Biochem. Biophys. Res. Commun.* **251**(2) (1998), 580–585.

[29] D.A. Lee and M.M. Knight, Mechanical loading of chondrocytes embedded in 3D constructs: in vitro methods for assessment of morphological and metabolic response to compressive strain, *Methods. Mol. Med.* **100** (2004), 307–324.

[30] M.G. Lequesne, N. Dang and N.E. Lane, Sport practice and osteoarthritis of the limbs, *Ostoarthritis Cartilage* **5**(2) (1997), 75–86.

[31] P. Long, R. Gassner and S. Agarwal, Tumor necrosis factor alpha-dependent proinflammatory gene induction is inhibited by cyclic tensile strain in articular chondrocytes in vitro, *Arthritis Rheum.* **44**(10) (2001), 2311–2319.

[32] G.N. Lowe, Y.H. Fu, S. McDougall, R. Polendo, A. Williams, P.D. Benya and T.J. Hahn, Effects of prostaglandins on deoxyribonucleic acid and aggrecan synthesis in the RCJ 3.1C5.18 chondrocyte cell line: role of second messengers, *Endrocrinology* **137**(6) (1996), 2208–2216.

[33] C.J. Malemud, N. Islam and T.M. Haqqi, Pathophysiological mechanisms in osteoarthritis lead to novel therapeutic strategies, *Cells Tissues Organs* **174**(1–2) (2003), 34–48.

[34] S.C. Mastbergen, F.P. Lafeber and J.W. Bijlsma, Selective COX-2 inhibition prevents proinflammatory cytokine-induced cartilage damage, *Rheumatology* **41**(7) (2002), 801–808.

[35] S.C. Mastbergen, J.W. Bijlsma and F.P. Lafeber, Selective COX-2 inhibition is favorable to human early and late-stage osteoarthritic cartilage: a human in vitro study, *Osteoarthritis Cartilage* **13**(6) (2005), 519–526.

[36] K. Masuda, H. Shirota and E.J. Thonar, Quantification of $^{35}$S-labeled proteoglycans complexed to alcian blue by rapid filtration in multiwell plates, *Anal. Biochem.* **317**(2) (1994), 167–175.

[37] J.A. Mengshol, M.P. Vincenti, A. Barchowsky and C.E. Brinckerhoff, Interleukin-1 induction of collagenase 3 (matrix metalloproteinase 13) gene expression in chondrocytes requires p38, c-Jun N-terminal kinase, and nuclear factor kappaB: differential regulation of collagenase 1 and collagenase 3, *Arthritis Rheum.* **43**(4) (2000), 801–811.

[38] S. Milano, F. Arcoleo, M. Dieli, R. D'Agostino, P. D'Agostino, G. De Nucci and E. Cillari, Prostaglandin E2 regulates inducible nitric oxide synthase in the murine macrophage cell line J774, *Prostaglandins* **49**(2) (1995), 105–115.

[39] K. Mio, S. Saito, T. Tomatsu and Y. Toyama, Intermittent compressive strain may reduce aggrecnase expression in cartilage, *Clin. Orthop. Relat. Res.* **433** (2005), 225–232.

[40] S. Mohammed and D.W. Croom, Gastropathy due to celecoxib, a cyclooxygenase-2 inhibitor, *N. Engl. J. Med.* **340**(25) (1999), 2005–2006.

[41] M. Murata, L.J. Bonassar, M. Wright, H.J. Mankin and C.A. Towle, A role for the interleukin-1 receptor in the pathway linking static mechanical compression to decreased proteoglycan synthesis in surface articular cartilage, *Arch. Biochem. Biophys.* **413**(2) (2003), 229–235.

[42] G.A.C. Murrell, D. Jang and R.J. Williams, Nitric oxide activates metalloprotease enzymes in articular cartilage, *Biochem. Biophys. Res. Commun.* **206**(1) (1995), 15–21.

[43] M. Oh, K. Fukuda, S. Asada, Y. Yasuda and S. Tanaka, Concurrent generation of nitric oxide and superoxide inhibits proteoglycan synthesis in bovine articular chondrocytes: involvement of peroxynitrite, *J. Rheumatol.* **25**(11) (1998), 2169–2174.

[44] M.J. Palmoski and K.D. Brandt, Effects of static and cyclic compressive loading on articular cartilage plugs in vitro, *Arthritis Rheum.* **27**(6) (1984), 675–681.

[45] J. Rao and W.R. Otto, Fluorimetric DNA assay for cell growth estimation, *Analytical Biochem.* **207** (1992), 186–192.

[46] P. Reboul, J.P. Pelletier, G. Tardiff, J.M. Cloutier and J. Martel-Pelletier, The new collagenase, collagenase-3, is expressed and synthesized by human chondrocytes but not by synoviocytes. A role in osteoarthritis, *J. Clin. Invest.* **97**(9) (1996), 2011–2019.

[47] J.M. Rusche, Tolerability of rofecoxib versus naproxen, *Ann. Intern. Med.* **140**(12) (2004), 1059.

[48] R.L. Sah, Y.J. Kim, J.Y. Doong, A.J. Grodzinsky, A.H. Plaas and J.D. Sandy, Biosynthetic response of cartilage explants to dynamic compression, *J. Orthop. Res.* **7** (1989), 619–636.

[49] D.M. Salter and H.S. Lee, Mechanotransduction in normal and osteoarthritic human articular chondrocytes, *J. Med. Sci.* **23**(4) (2003), 183–190.

[50] B.V. Shlopov, W.R. Lie, C.L. Mainardi, A.A. Cole, S. Chubinskaya and K.A. Hasty, Osteoarthritic lesions: involvement of three different collagenases, *Arthritis Rheum.* **40**(11) (1997), 2065–2074.

[51] M. Stefanovic-Racic, T.I. Morales, D. Taskiran, L.A. McIntyre and C.H. Evans, The role of nitric oxide in proteoglycan turnover by bovine articular cartilage organ cultures, *J. Immunol.* **156**(3) (1996), 1213–2120.

[52] M. Stefanovic-Racic, M.O. Mollers, L.A. Miller and C.H. Evans, Nitric oxide and proteoglycan turnover in rabbit articular cartilage, *J. Orthop. Res.* **15**(3) (1997), 442–449.

[53] R. Studer, D. Jaffuers, M. Stefanovic-Racic, P.D. Robbins and C.H. Evans, Nitric oxide in osteoarthritis, *Osteoarthritis Cartilage* **7**(4) (1999), 377–379.

[54] S. Tanaka, C. Hamanishi, H. Kikuchi and K. Fukuda, Factors related to degradation of articular cartilage in osteoarthritis: a review, *Semin. Arthritis Rheum.* **27**(6) (1998), 392–399.

[55] R. Tamura, T. Nakanishi and Y. Kimura, Nitric oxide mediates IL-1 induced matrix degradation and bFGF release in cultured rabbit articular chondrocytes: a possible mechanism of pathophysiological neovascularisation in arthritis, *Endocrinology* **137** (1996), 3729–3737.

[56] D. Taskiran, M. Stefanovic-Racic, H. Georgescu and C. Evans, Nitric oxide mediates suppression of cartilage proteoglycan synthesis by interleukin-1, *Biochem. Biophys. Res. Commun.* **200**(1) (1994), 142–148.

[57] J.M. Weinberg, The lessons of Vioxx, *Cutis* **75**(1), 17.

[58] W. Wu, R.C. Billinghurst, I. Pidoux, J. Antoniou, D. Zukor, M. Tanzer and A.R. Poole, Sites of collagenase cleavage and denaturation of type II collagen in aging and osteoarthritic articular cartilage and their relationship to the distribution of matrix metalloproteinase 1 and matrix metalloproteinase 13, *Arthritis Rheum.* **46**(8) (2002), 2087–2094.

[59] Z. Xu, M.J. Buckely, C.H. Evans and S. Agarwal, Cyclic tensile strain acts as an antagonist of IL-1 beta actions in chondrocytes, *J. Immunol.* **165**(1) (2000), 453–460.

# Treatment of human mesenchymal stem cells with pulsed low intensity ultrasound enhances the chondrogenic phenotype *in vitro*

D. Schumann [a], R. Kujat [a], J. Zellner [a], M.K. Angele [b], M. Nerlich [a], E. Mayr [a] and P. Angele [a,*]

[a] *Department of Trauma Surgery, University Hospital of Regensburg, 93053 Regensburg, Germany*
[b] *Department of Surgery, Ludwig Maximilians University, Klinikum Großhadern, 81377 Munich, Germany*

**Abstract.** This study examined the effects of low intensity pulsed ultrasound (LIPUS) on human bone marrow-derived mesenchymal stem cells undergoing chondrogenic differentiation. Aggregates of mesenchymal stem cells and mesenchymal stem cells seeded in three dimensional matrices were cultured in a defined chondrogenic medium and subjected to LIPUS for the first 7 days of culture. At 1, 7, 14 and 21 days, samples were harvested for histology, immunohistochemistry, RT-PCR, and quantitative DNA and matrix macromolecule analysis. Cell aggregates with daily treatment for 20 minutes showed no significant differences for proteoglycan and collagen content during chondrogenic differentiation. However ultrasound application for 40 minutes daily resulted in a statistically significant increase of the proteoglycan and collagen content after 21 days in culture. Aggregates treated for 20 minutes daily showed decreased expression of chondrogenic genes at all time points. In contrast, 40 minutes of daily treatment of aggregates resulted in a significant increase of chondrogenic marker genes after an initial decrease at day 7 with time in culture. Ultrasound treated cell-scaffold constructs showed a significant increase of chondrogenic marker gene expression and extracellular matrix deposition. This study indicates that LIPUS can be used to enhance the chondrogenesis of mesenchymal stem cells in cell aggregates and cell-scaffold constructs. We have found a dependency on the specific treatment parameters. We hypothesize that LIPUS can be used for an improved *in vitro* preparation of optimized tissue engineering implants for cartilage repair. Furthermore this non-invasive method could also be of potential use *in vivo* for regenerative therapy of cartilage in the future.

Keywords: Chondrogenesis, stem cell, bone marrow, mechanobiology, ultrasound

## 1. Introduction

Articular cartilage possesses little capacity for endogenous repair after having been damaged by disease or trauma. Various surgical procedures depending on ingrowth of mesenchymal stem cells into the defects showed repair with fibrocartilage which is of minor quality and less resistant to physical forces [33,41,43,44]. New treatment options using tissue engineering strategies for cartilage repair showed intriguing results [5,39]. Mesenchymal stem cells were identified as possible repair cells of

*Address for correspondence: Peter Angele, MD, Department of Trauma Surgery, University Hospital of Regensburg, Franz Josef Strauss Allee 11, 93053 Regensburg, Germany. Tel.: +49 941 944 6805; Fax: +499 41 944 6806; E-mail: angelepeter @aol.com.

cartilaginous defects, because under appropriate conditions, they have the ability to differentiate into chondrocytic cells producing cartilaginous extracellular matrix *in vitro* and *in vivo* [1,22,23,40,52,53].

In contrast to chondrocytes and cartilage explants [6,7,16,21,27,31,45,46] there are few studies that examine the influence of biomechanical forces on mesenchymal stem cells undergoing chondrogenesis. It has been found that repeated, cyclic hydrostatic pressure enhanced chondrogenesis of mesenchymal stem cells in an *in vitro* aggregate culture system detected by an increased extracellular matrix deposition of collagen and aggrecan [3]. Similar results could be detected, when mesenchymal stem cells were loaded in three-dimensional matrices and subjected to cyclic, compressive load [2].

Both animal experiments and clinical trials have established the stimulatory effects of low-intensity pulsed ultrasound for fracture repair, treatment of pseudarthrosis, bone fusion and callus distraction [8, 11,17,25,28–30,35]. For cartilage repair, low-intensity pulsed ultrasound has been shown to have a stimulatory effect for the repair of articular cartilage defects in rabbits [9] and for cartilage restoration in experimental osteoarthritis [19,20]. Effects of ultrasound treatment on chondrocytes are mainly based on chondrocytic cells undergoing endochondral ossification [24,34,37,38,51]. Low intensity ultrasound resulted in an increased aggrecan gene expression [38,50,51], with increased cartilaginous matrix production [24,37,38] possibly mediated by an increased calcium incorporation [34,37]. Low intensity ultrasound treatment of chondrocytes from hyaline cartilage, obtained from chick sternum and then loaded in alginate gels, showed an influence on chondrocytic proliferation and differentiation in an intensity-dependent manner [54]. Recently, ultrasound has also been shown to be effective for mesenchymal stem cells undergoing chondrocyte differentiation [13].

The purpose of this *in vitro* study was to evaluate the effects of the applied duration of low-intensity pulsed ultrasound on chondrogenic differentiation of postnatal, human mesenchymal stem cells. Furthermore the effect of ultrasound treatment on chondrogenic differentiation of mesenchymal stem cells in tissue engineering scaffolds was assessed.

## 2. Methods

### 2.1. Preparation of aggregates and cell/scaffold constructs

Human bone marrow was obtained from the iliac crests of patients undergoing spine fusion using a protocol approved by the Ethical Committee of the University Hospital of Regensburg. After Percoll gradient fractionation, Dulbecco's modified Eagle's Medium with 10% fetal bovine serum was added to the aspirate and $10 \times 10^6$ nucleated cells/100 mm dish were plated and grown at 37°C with 5% $CO_2$ until the cells were 80% confluent. Adherent cell colonies were trypsinized, counted, and then introduced into an *in vitro* aggregate culture system [23,52]. In further experiments, mesenchymal stem cells were seeded in cylindrical biodegradable composite scaffolds, containing 70% estered hyaluronan (obtained from Jaloskin, Fidia Advanced Polymers) and 30% gelatin (Sigma) ($2 \times 10^6$/scaffold). The composite scaffolds were manufactured with defined, highly interconnective pores with an average pore size of 350 $\mu$m [1,2,4].

The aggregates and the cell/scaffold constructs were cultured at 37°C/5% $CO_2$ in a defined, serum-free medium, previously shown to induce chondrogenic differentiation of mesenchymal stem cells [1–4, 23,52].

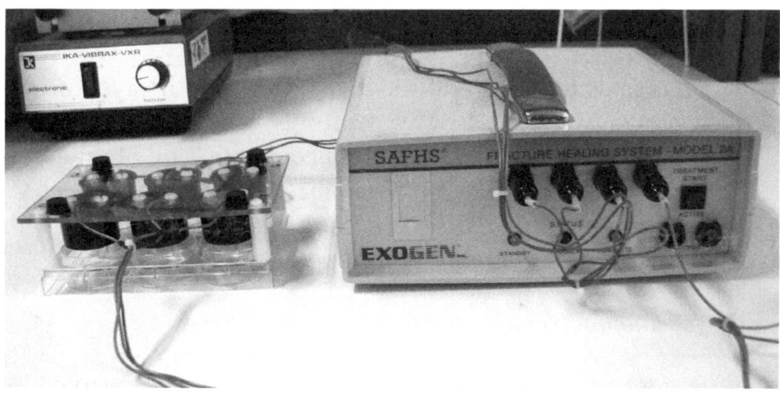

Fig. 1. Low-intensity pulsed ultrasound application unit: The application unit consists of an sonic accelerated fracture healing system (SAFHS®) device and transducers. A 6-well tissue culture plate can be connected using ultrasound contact gel.

## 2.2. Ultrasound treatment

Pulsed low intensity ultrasound (30 mW/cm$^2$) was applied for 0, 20 and 40 minutes per day, for the first 7 consecutive days of culture. The ultrasound application unit consisted of a sonic accelerated fracture healing system (SAFHS®) device (model 2A; Exogen) and transducers connected (using ultrasound coupling gel) to a 6-well tissue culture plate filled with 2.5 ml of tissue culture medium (Fig. 1). The SAFHS® device provides pulsed ultrasound with a carrier frequency of 1.5 MHz and with a 200 $\mu$s tone burst repeating at 1.0 kHz. This setting was used in earlier experiments by Exogen Inc. and has proven to provide ultrasonic waves in the medium [34]. The distance between transducer and samples was less than 2 mm. All experiments were carried out at 37°C.

After each daily loading period, the aggregates or cell/scaffold constructs were transferred back into separate 15 ml polypropylene conical tubes and cultured in 1 ml of fresh standard culture medium for a total of either 7, 14 or 21 days before harvesting for analysis. All experiments were performed on at least three sets of samples.

## 2.3. Real-time PCR analysis

For RT-PCR a protocol was used as described previously [2]. RNA was extracted using RNA-Bee (Biozol, Eching, Germany) according to the single-step acid-phenol guanidinium method. cDNA synthesis was performed by using Superscript RNase H⁻ reverse transcriptase (Invitrogen, Karlsruhe, Germany) in the presence of oligo-dt primers. RT-PCR analysis was performed and monitored using Light Cycler Kits (Roche, Mannheim, Germany). The level of each target gene was normalized to the reference gene glyceraldehyde phosphate dehydrogenase (GAPDH). The primers were as follows:

**GAPDH** *Forward* 5′ - gAA ggT gAA ggT Cgg AgT C *Reverse* 5′ - gAA gAT ggT gAT ggg ATT TC
**Aggrecan** *Forward* 5′ - ACT TCC gCT ggT CAg ATg gA *Reverse* 5′ - TCT CgT gCC AgA TCA TCA CC **Col 1** *Forward* 5′ - Agg gCC AAg ACg AAg ACA TC *Reverse*
5′ - AgA TCA CgT CAT CgC ACA ACA **Col 2** *Forward* 5′ - TTC AgC TAT ggA gAT gAC AAT C *Reverse* 5′ - AgA gTC CTA gAg TgA CTg Ag.

## 2.4. DNA measurement

DNA contents of the aggregates and cell/scaffold constructs were assayed using a two-step fluorometric assay as described previously [2,3]. Briefly, scaffolds were prepared by digestion with papain solution overnight. Aliquots of the digest (100 $\mu$l) were mixed with 1 ml of diluted Hoechst 33258 Dye Solution (1 $\mu$g/ml). The fluorescence emission of the samples was measured with a Cytofluor at $A_{360}$ nm excitation, $A_{460}$ nm emission with a gain of 70. A standard curve was produced with calf thymus DNA (Sigma) and used to determine the DNA content of the experimental samples.

## 2.5. Determination of glycosaminoglycan content

The glycosaminoglycan content of the aggregates and cell/scaffold constructs was quantified with a spectrophotometric assay as described elsewhere [2,3]. Scaffolds were digested in papain solution overnight. After centrifugation the supernatant was discarded and the pellets or cell-scaffold constructs resuspended in distilled water. Samples were allowed to precipitate in Safranin O reagent on nitrocellulose filter (BioRad, USA), which covered wells in a dot-blot apparatus (BioRad, USA). The precipitates were collected on the nitrocellulose filter and then dissolved in 10% cetylpyridinium chloride. The absorbance of the liquid was spectrophotometrically read at 536 nm. For the experimental groups and the unloaded controls (only when matrices were involved) the mean glycosaminoglycan content of cell-free scaffolds was substracted to get a minimal estimation of the produced extracellular proteoglycan even when a complete degradation of the scaffolds would influence the measurement.

## 2.6. Determination of hydroxyproline content

The hydroxyproline content of the aggregates and cell/scaffold constructs was measured as described elsewhere [2,3]. Scaffolds were hydrolyzed with 6 M HCl at 108°C for 24 hours. The neutralized samples were resuspended in hydroxyproline buffer. Chloramine T solution, perchloric acid and P-dimethylaminobenzaldehyde solution were sequentially added. The absorbance was spectrophotometrically read at 560 nm. Cis-4-hydroxy-d-proline was used to construct a standard curve. For the experimental groups and the unloaded controls (only when matrices were involved) the mean collagen content of cell-free scaffolds was substracted to get a minimal estimation of the produced extracellular collagen even when a complete degradation of the scaffolds would influence the measurement.

## 2.7. Statistic analysis

The expression of molecular markers, DNA-, proteoglycan- and collagen-contents with and without application of ultrasound were compared between the groups at each time point using two-way ANOVA with Bonferroni post hoc tests or Wilcoxon Signed Rank tests for multiple comparisons. The tests were performed using Sigma Stat$^{TM}$ Software for Windows Version 11.5, SPSS Inc. Significance was accepted at a level of $p < 0.05$.

## 3. Results

Chondrogenic differentiation of mesenchymal stem cells was achieved by using an *in vitro* aggregate culture system. The increase in expression of collagen type II in unloaded aggregates on day 21 compared to day 1 of culture reflected the occurrence of chondrogenic differentiation of the mesenchymal

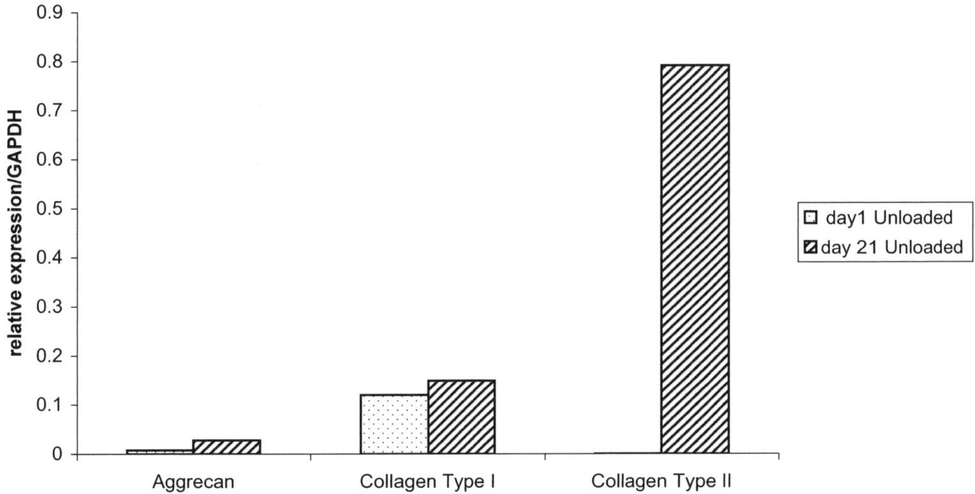

Fig. 2. Relative expression of aggrecan, collagen type I and II of untreated aggregates during chondrogenesis (day 1, day 21).

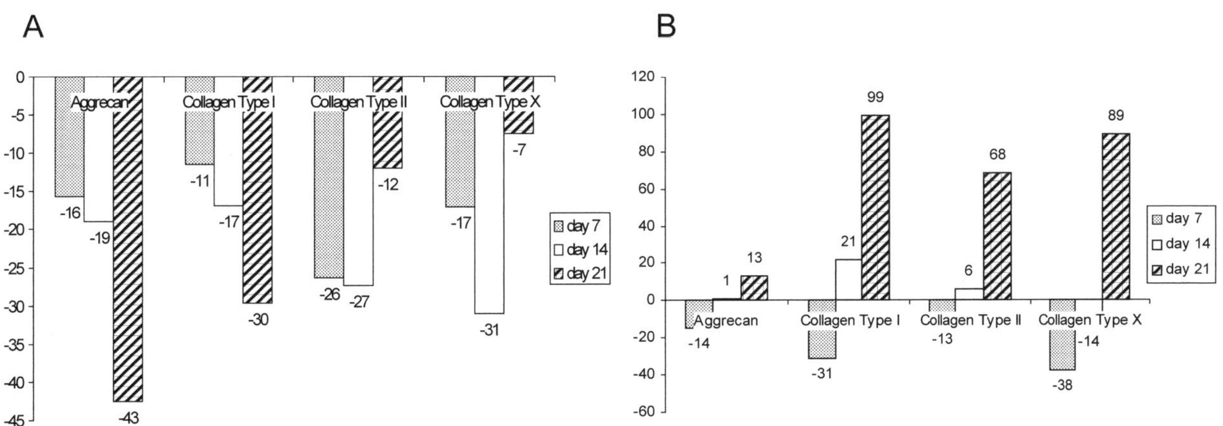

Fig. 3. Relative expression (%) of aggrecan, collagen type I, II, X of ultrasound-treated aggregates versus unloaded controls during chondrogenesis. Samples were exposed to ultrasound for 20 minutes (A) or 40 minutes (B) for the first 7 days of culture.

stem cells (Fig. 2). Aggregates were subjected to low intensity pulsed ultrasound for 20 or 40 minutes for the first 7 days of chondrogenic differentiation, non-treated samples served as controls (Fig. 3). At the mRNA level, 20 minutes of ultrasound treatment resulted in a decrease in chondrogenic marker gene expression at all time points during chondrogenic differentiation compared to unloaded controls (Fig. 3). In contrast, 40 minutes of ultrasound treatment produced an initial decrease, directly after the last ultrasound application (day 7), but an increase in expression levels of chondrogenic marker genes with further time in culture (day 14, 21) (Fig. 3). No significant differences in DNA content during chondrogenesis could be detected between samples treated with ultrasound for 20 minutes per day (ANOVA, $p > 0.45$) or 40 minutes per day (ANOVA, $p > 0.45$), and untreated controls (Fig. 4).

Ultrasound treatment applied for 20 minutes per day, resulted in no significant differences for proteoglycan and collagen content between ultrasound treated samples and untreated controls (Fig. 4). No significant differences between ultrasound application for 40 minutes per day and the untreated controls

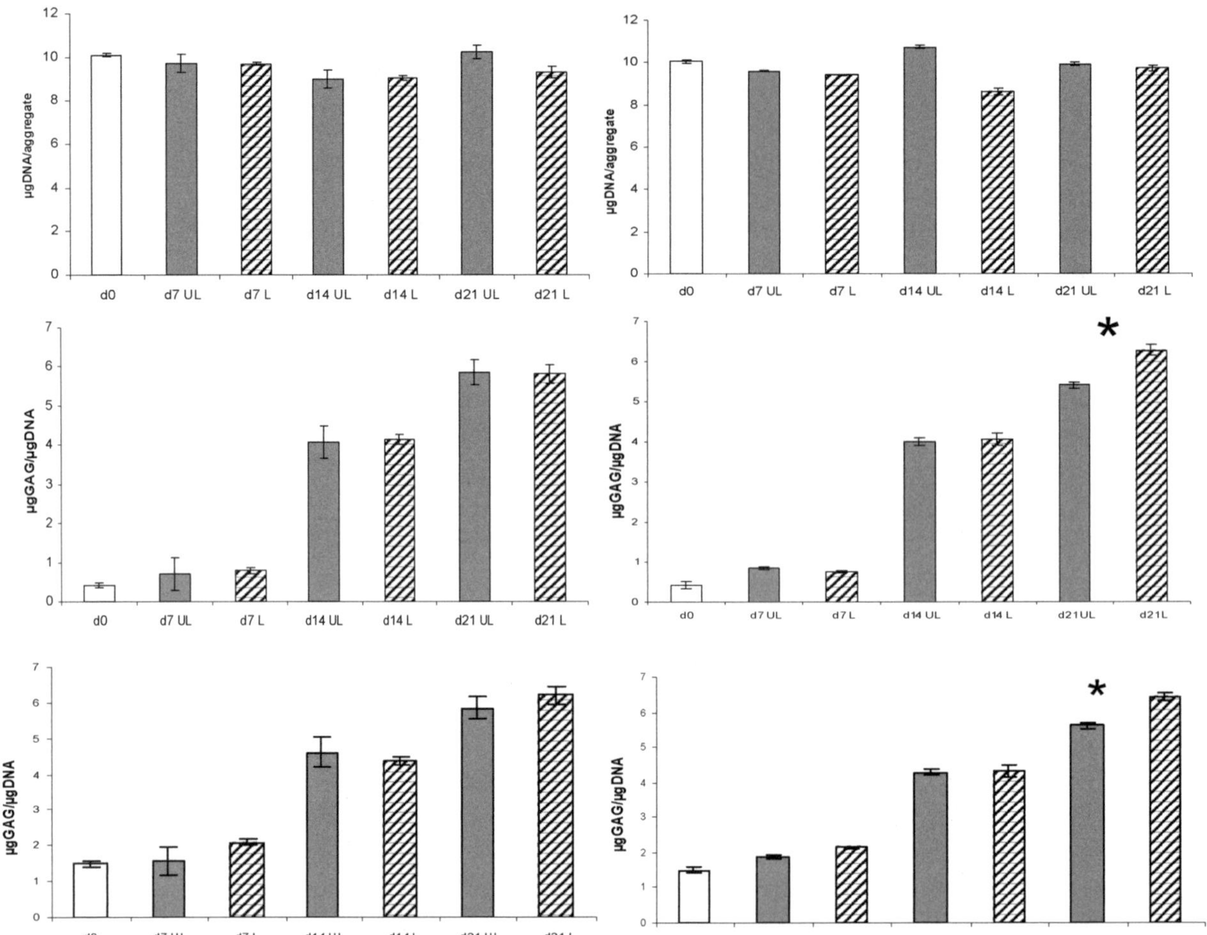

Fig. 4. DNA content (upper panels), proteoglycan content (middle panels), collagen content (lower panels) of aggregates (mean ± standard deviation) treated with (L) or without (UL) ultrasound treatment for 20 (left panels) or 40 minutes per day (right panels) for the first 7 days of culture. (*) Statistically significant difference between control and ultrasound treated aggregates ($p < 0.05$).

could be detected after 7 and 14 days in culture. However, ultrasound application for 40 minutes per day resulted in a significant increase of proteoglycan and collagen content ($p < 0.05$) after 21 days in culture (14 days after the last ultrasound application).

Histological analysis revealed chondrogenic differentiation of mesenchymal stem cells in aggregates after 21 days in culture independently of ultrasound treatment (0, 20, 40 minutes). When the ultrasound was applied for 40 minutes, a trend to an increased aggregate size and chondrogenic extracellular matrix deposition in the aggregates (metachromatic staining) could be detected compared to controls or samples with 20 minutes per day of ultrasound application (not shown).

In order to examine the effects of ultrasound treatment on chondrogenic differentiation of mesenchymal stem cells in a clinically applicable tissue engineering setup, cell loaded polymer scaffolds were subjected to ultrasound treatment. In the untreated cell loaded polymer scaffolds, a significant increase in expression levels of chondrogenic markers (aggrecan, collagen type II) could be detected after 21 days in culture in comparison to unloaded controls on day 1, indicating chondrogenic differentiation of mesenchymal stem cells in the tissue engineering scaffolds. On day 21 in culture compared to day 1 collagen type II expression was 32971 fold increased and the aggrecan expression showed an increase of 278%. The expression of collagen type I revealed only an increase of 40% on day 21 in culture compared to day 1 (Fig. 5).

Low intensity pulsed ultrasound treated cell/scaffold constructs were compared with unloaded samples during chondrogenesis with respect to expression levels of chondrogenic marker genes (Fig. 6). After low intensity ultrasound treatment, aggrecan gene expression was not significantly altered on day 7 or 21 in comparison to unloaded controls. For the collagen type I gene expression no relevant increase after ultrasound treatment could be detected at any time point. The expression of collagen type II was significantly increased in the ultrasound treated samples compared to untreated controls (ANOVA, $p <$ 0.001, Fig. 6).

No statistical significant differences in DNA content were found between untreated and ultrasound treated cell loaded polymer scaffolds during chondrogenesis ($p = 0.09$) (Fig. 7). Independent of the ultrasound treatment, the proteoglycan content was significantly higher after 21 days in culture compared

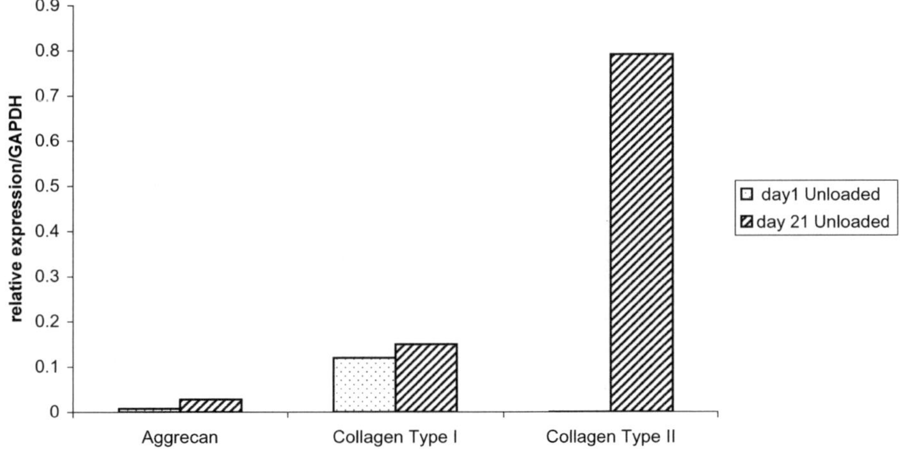

Fig. 5. Relative expression of aggrecan, collagen type I, II of untreated cell/scaffold-constructs during chondrogenesis (day 1, day 21).

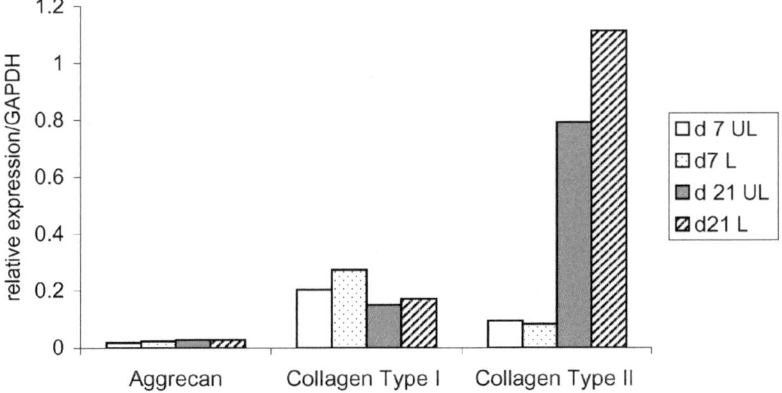

Fig. 6. Relative expression of aggrecan and collagen type I, II of cell/scaffold constructs during chondrogenesis (day 7, 21) after ultrasound-treatment (for the first 7 days; L) and untreated controls (UL).

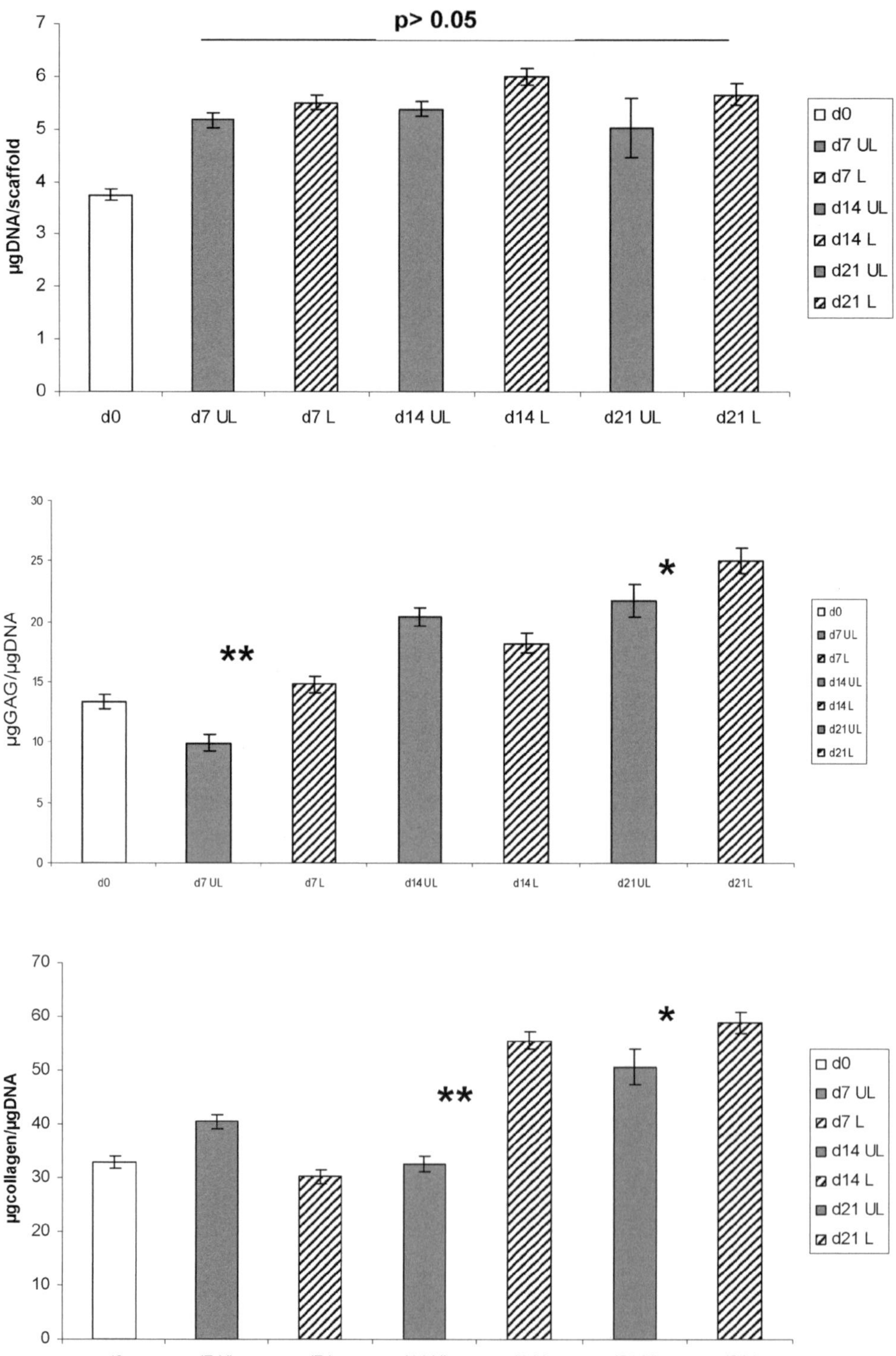

Fig. 7. DNA content, proteoglycan content, collagen content of cell/scaffold constructs (mean $\pm$ standard deviation) treated with (L) or without (UL) ultrasound treatment for the first 7 days of culture. Note statistically significant differences (Wilcoxon signed rank test) between ultrasound treated samples and controls: ($^*$) for $p < 0.05$; ($^{**}$) for $p < 0.001$.

to the corresponding samples on day 0 ($p < 0.0001$) (Fig. 7). At day 7 ($p < 0.001$) and day 21 ($p < 0.05$) a significant increase in proteoglycan content could be detected for ultrasound treated samples in comparison to untreated controls. After 14 days in culture no significant increase for ultrasound treated samples could be seen.

Independently of the ultrasound treatment, the collagen content was significantly higher after 21 days in culture compared to the corresponding samples on day 0 ($p < 0.0001$) (Fig. 7). Between untreated and ultrasound treated samples no significant difference in collagen content could be detected on day 7 ($p = 0.26$), however after 14 and 21 days in culture a significant increase for ultrasound treated samples was detected (day 14: $p < 0.001$; day 21: $p < 0.05$) (Fig. 7).

The histological analysis confirmed chondrogenic differentiation after 21 days by extracellular matrix deposition (Picrosirius red stained sections). The ultrasound treated scaffolds had a qualitative increase in extracellular matrix deposition throughout the scaffolds, especially in the center of the scaffolds, in comparison with untreated controls after 21 days in culture.

## 4. Discussion

To our knowledge this is the first time that ultrasound treatment showed enhancement of chondrogenic differentiation of adult human mesenchymal stem cells in scaffolds appropriate for tissue engineering. The precise application modes seemed to be crucial for successful ultrasound treatment. For treatment of bone defects ultrasound application modes have been established in multiple experimental and clinical trials [8,10,17,25,26,28–30,35,42]. On an experimental level, changes in ultrasound modes resulted in marked differences in chondrocytic cell behaviour [14,54,55]. In a study by Zhang et al. ultrasound treatment of chondrocytic cells, applied for 20 min with an intensity of 2 mW/cm$^2$ resulted in increased cell proliferation and increased extracellular chondrogenic matrix deposition, whereas 30 mW/cm$^2$ showed no significant change in cell proliferation and cell differentiation in comparison to untreated control samples [54]. Recently, the optimal intensity of ultrasound treatment for human mesenchymal stem cells undergoing chondrogenic differentiation was established to be 30 mW/cm$^2$ [13]. In our study we used this optimized intensity mode.

According to the literature, a daily application of 20 minutes duration is most effective for ultrasound treatment of bony tissue [8,15,24,25,32,34,35] and for chondrocytic cells [9,12,19,54,55]. To our knowledge, there are no studies available that have examined the effects of changing the daily ultrasound duration on mesenchymal stem cells undergoing chondrogenic differentiation. We observed that 40 minutes, in contrast to 20 minutes, of daily treatment showed a marked increase in chondrogenic marker expression and chondrogenic extracellular matrix deposition. This is in contrast to a recent study which showed enhancing effects on chondrogenic differentiation with 20 minutes of daily ultrasound application [13]. In comparison to our study they applied ultrasound not exclusively on mesenchymal stem cells but also on chondrocytic cells after their chondrogenic differentiation. As demonstrated by Takahashi et al. this can result in significant differences, because undifferentiated mesenchymal cells and differentiated chondrocytes differ significantly in terms of their responses to biomechanical forces [47]. Comparable results to our study were achieved by application of other mechanical forces to mesenchymal cells, showing increased extracellular matrix production after longer daily loading durations [14].

For the first 7 days no extracellular deposition of cartilage-like matrix was detected in the aggregates and the cell-scaffold constructs [1,2,23,52]. Therefore this timeframe was used in this study to selectively treat mesenchymal stem cells and not chondrocytic cells. Other biomechanical treatments with this timeframe led to enhanced chondrogenesis of mesenchymal stem cells in the aggregate culture system [3] as well as in tissue engineering scaffolds [2]. Recently it has been shown that cyclic mechanical compression could induce the chondrogenic differentiation of mesenchymal stem cells without addition of inductive factors to the medium [18]. In the present study, both the ultrasound treated cell scaffold constructs and the non-treated controls were cultured in conditions that promote the chondrogenic differentiation of mesenchymal stem cells. Therefore, the influence of ultrasound on the initiation of chondrogenic differentiation was not directly explored. Nevertheless, the results of this study point to the importance of load during skeletal tissue regeneration.

In the present study no significant differences could be seen for DNA-content between loaded samples and unloaded controls, consistent with the literature for chondrocyte [48,49] and mesenchymal stem cell [2,3,13] loading, which postulated that mechanically stressed cells may direct their anabolic efforts towards production and maintenance of matrix rather than cell proliferation.

Ultrasound treatment can enhance chondrogenic differentiation of mesenchymal stem cells. We found treatment conditions with consecutive enhanced chondrogenic marker expression that increased extracellular matrix deposition in aggregates and tissue engineering scaffolds. In aggregates, a significant increase of proteoglycan and collagen contents after ultrasound treatment compared to untreated controls could only be seen on day 21 of culture, 14 days after the last ultrasound treatment. One possible explanation for the delay could be the initial decrease in expression of chondrocytic markers after ultrasound treatment in the aggregate culture model. This decrease was followed by an increase in chondrogenic marker expression and extracellular matrix deposition in a rebound effect. The increase in protein deposition after ultrasound treatment was only approximately 20% over untreated controls for both proteoglycan and collagen contents, in contrast to the literature, where ultrasound treatment resulted in a 2 fold increase of aggrecan deposition [13]. Apart from methodological differences, one possible explanation for these differences could be the timing of the application of ultrasound, i.e. not exclusively to mesenchymal stem cells but also to chondrocytic cells after their chondrogenic differentiation.

The tissue engineering scaffolds used in this study had been previously shown to serve as an appropriate environment for chondrogenic differentiation of mesenchymal stem cells *in vitro* [1,2] and *in vivo* [1, 4,36]. Under biomechanical compression this biomaterial allowed further improvement of chondrogenic differentiation of mesenchymal stem cells [2]. In the present study, the tissue engineering scaffolds did not have the initial decrease of chondrogenic markers after ultrasound treatment that was seen in the aggregates. Therefore, compared to untreated controls, the increase of extracellular, chondrogenic matrix deposition after ultrasound treatment occurred earlier in tissue engineering scaffolds than in the aggregate culture system. This was especially true for the expression of collagen type II after ultrasound treatment.

Huang et al. showed that cyclic compressive loading can promote the chondrogenesis of rabbit bone marrow derived mesenchymal stem cells by inducing the synthesis of TGF-$\beta$1, which can stimulate the mesenchymal cells to differentiate into chondrocytes [18]. We did not examine whether ultrasound enhanced transforming growth factor beta expression during chondrocyte differentiation of human mesenchymal stem cells as the enhancement of chondrogensis was done with TGF-$\beta$1 in the medium. It is possible that ultrasound may influence the synthesis of TGF-$\beta$1, as described previously [13], but one can also speculate that ultrasound treatment might lead to an increased access of exogenous added TGF-$\beta$1 and other supplements to the center of the cell–matrix constructs. This could explain the increase in extracellular matrix deposition, especially in the center of the scaffolds.

The mechanism by which ultrasound is sensed by the differentiating cells is presently unknown, as is the mechanism by which the load is transduced into an increase in chondrogenic marker expression and extracellular matrix deposition. This system can be used to elicit mechanoreceptors and signal transduction pathways influenced by mechanical forces during chondrogenesis of mesenchymal cells in tissue engineering scaffolds. In the end this could lead to an improved *in vitro* preparation of optimized tissue engineered implants for cartilage repair. The beneficial effects of ultrasound treatment on chondrogenic differentiation *in vivo* are still under debate [9,12], but with the appropriate application modes ultrasound may have possible applications in the non-surgical *in vivo* cartilage repair in the future.

## Acknowledgements

The authors thank Tanja Weinfurtner and Daniela Drenkard for their excellent technical assistance. This study was supported by intramural funding of the University Hospital of Regensburg.

## References

[1] P. Angele, R. Kujat, M. Nerlich, J. Yoo, V. Goldberg and B. Johnstone, Engineering of osteochondral tissue with bone marrow mesenchymal progenitor cells in a derivatized hyaluronan – gelatin composite sponge, *Tissue Eng.* **5** (1999), 545–554.

[2] P. Angele, D. Schumann, M. Angele, B. Kinner, C. Englert, R. Hente, B. Fuchtmeier, M. Nerlich, C. Neumann and R. Kujat, Cyclic, mechanical compression enhances chondrogenesis of mesenchymal progenitor cells in tissue engineering scaffolds, *Biorheology* **41** (2004), 335–346.

[3] P. Angele, J.U. Yoo, C. Smith, J. Mansour, K.J. Jepsen, M. Nerlich and B. Johnstone, Cyclic hydrostatic pressure enhances the chondrogenic phenotype of human mesenchymal progenitor cells differentiated *in vitro*, *J. Orthop. Res.* **21** (2003), 451–457.

[4] P. Angele, R. Kujat, M. Nerlich, V. Goldberg, J. Yoo and B. Johnstone, Meniscus repair with mesenchymal progenitor cells in a biodegradable composite matrix, *Trans. Ortho. Res. Soc.* **605** (2000).

[5] M. Brittberg, T. Tallheden, B. Sjogren-Jansson, A. Lindahl and L. Peterson, Autologous chondrocytes used for articular cartilage repair: an update, *Clin. Orthop.* (2001), S337–348.

[6] M.D. Buschmann, Y.A. Gluzband, A.J. Grodzinsky and E.B. Hunziker, Mechanical compression modulates matrix biosynthesis in chondrocyte/agarose culture, *J. Cell Sci.* **108**(Pt 4) (1995), 1497–1508.

[7] S.E. Carver and C.A. Heath, Influence of intermittent pressure, fluid flow, and mixing on the regenerative properties of articular chondrocytes, *Biotechnol. Bioeng.* **65** (1999), 274–281.

[8] S.D. Cook, S.L. Salkeld, L.P. Patron, J.P. Ryaby and T.S. Whitecloud, Low-intensity pulsed ultrasound improves spinal fusion, *Spine J.* **1** (2001), 246–254.

[9] S.D. Cook, S.L. Salkeld, L.S. Popich-Patron, J.P. Ryaby, D.G. Jones and R.L. Barrack, Improved cartilage repair after treatment with low-intensity pulsed ultrasound, *Clin. Orthop.* (2001), S231–243.

[10] N. Doan, P. Reher, S. Meghji and M. Harris, In vitro effects of therapeutic ultrasound on cell proliferation, protein synthesis, and cytokine production by human fibroblasts, osteoblasts, and monocytes, *J. Oral Maxillofac. Surg.* **57** (1999), 409–419; discussion 420.

[11] L.R. Duarte, The stimulation of bone growth by ultrasound, *Arch. Orthop. Trauma Surg.* **101** (1983), 153–159.

[12] G.N. Duda, A. Kliche, R. Kleemann, J.E. Hoffmann, M. Sittinger and A. Haisch, Does low-intensity pulsed ultrasound stimulate maturation of tissue-engineered cartilage?, *J. Biomed. Mater. Res.* **68B** (2004), 21–28.

[13] K. Ebisawa, K. Hata, K. Okada, K. Kimata, M. Ueda, S. Torii and H. Watanabe, Ultrasound enhances transforming growth factor beta-mediated chondrocyte differentiation of human mesenchymal stem cells, *Tissue Eng.* **10** (2004), 921–929.

[14] S.H. Elder, S.A. Goldstein, J.H. Kimura, L.J. Soslowsky and D.M. Spengler, Chondrocyte differentiation is modulated by frequency and duration of cyclic compressive loading, *Ann. Biomed. Eng.* **29** (2001), 476–482.

[15] H. Fujioka, J. Tanaka, S. Yoshiya, M. Tsunoda, K. Fujita, N. Matsui, T. Makino and M. Kurosaka, Ultrasound treatment of nonunion of the hook of the hamate in sports activities, *Knee Surg. Sports Traumatol. Arthrosc.* **12** (2004), 162–164.

[16] A.C. Hall, J.P. Urban and K.A. Gehl, The effects of hydrostatic pressure on matrix synthesis in articular cartilage, *J. Orthop. Res.* **9** (1991), 1–10.

[17] J.D. Heckman, J.P. Ryaby, J. McCabe, J.J. Frey and R.F. Kilcoyne, Acceleration of tibial fracture-healing by non-invasive, low-intensity pulsed ultrasound, *J. Bone Joint Surg. Am.* **76** (1994), 26–34.

[18] C.Y. Huang, K.L. Hagar, L.E. Frost, Y. Sun and H.S. Cheung, Effects of cyclic compressive loading on chondrogenesis of rabbit bone-marrow derived mesenchymal stem cells, *Stem Cells* **22** (2004), 313–323.

[19] M.H. Huang, H.J. Ding, C.Y. Chai, Y.F. Huang and R.C. Yang, Effects of sonication on articular cartilage in experimental osteoarthritis, *J. Rheumatol.* **24** (1997), 1978–1984.

[20] M.H. Huang, R.C. Yang, H.J. Ding and C.Y. Chai, Ultrasound effect on level of stress proteins and arthritic histology in experimental arthritis, *Arch. Phys. Med. Rehabil.* **80** (1999), 551–556.

[21] T. Ikenoue, M.C. Trindade, M.S. Lee, E.Y. Lin, D.J. Schurman, S.B. Goodman and R.L. Smith, Mechanoregulation of human articular chondrocyte aggrecan and type II collagen expression by intermittent hydrostatic pressure *in vitro*, *J. Orthop. Res.* **21** (2003), 110–116.

[22] Y. Jiang, B.N. Jahagirdar, R.L. Reinhardt, R.E. Schwartz, C.D. Keene, X.R. Ortiz-Gonzalez, M. Reyes, T. Lenvik, T. Lund, M. Blackstad, J. Du, S. Aldrich, A. Lisberg, W.C. Low, D.A. Largaespada and C.M. Verfaillie, Pluripotency of mesenchymal stem cells derived from adult marrow, *Nature* **418** (2002), 41–49.

[23] B. Johnstone, T.M. Hering, A.I. Caplan, V.M. Goldberg and J.U. Yoo, *In vitro* chondrogenesis of bone marrow-derived mesenchymal progenitor cells, *Exp. Cell Res.* **238** (1998), 265–272.

[24] C.M. Korstjens, P.A. Nolte, E.H. Burger, G.H. Albers, C.M. Semeins, I.H. Aartman, S.W. Goei and J. Klein-Nulend, Stimulation of bone cell differentiation by low-intensity ultrasound – a histomorphometric in vitro study, *J. Orthop. Res.* **22** (2004), 495–500.

[25] T.K. Kristiansen, J.P. Ryaby, J. McCabe, J.J. Frey and L.R. Roe, Accelerated healing of distal radial fractures with the use of specific, low-intensity ultrasound. A multicenter, prospective, randomized, double-blind, placebo-controlled study, *J. Bone Joint Surg. Am.* **79** (1997), 961–973.

[26] A. Lerner, H. Stein and M. Soudry, Compound high-energy limb fractures with delayed union: our experience with adjuvant ultrasound stimulation (exogen), *Ultrasonics* **42** (2004), 915–917.

[27] Z.Y. Ma, [Diagnosis and treatment of discoid meniscus of the knee (author's transl.)], *Zhonghua Wai Ke Za Zhi* **19** (1981), 622–623.

[28] E. Mayr, A. Laule, G. Suger, A. Ruter and L. Claes, Radiographic results of callus distraction aided by pulsed low-intensity ultrasound, *J. Orthop. Trauma* **15** (2001), 407–414.

[29] E. Mayr, C. Mockl, A. Lenich, M. Ecker and A. Ruter, [Is low intensity ultrasound effective in treatment of disorders of fracture healing?], *Unfallchirurg* **105** (2002), 108–115.

[30] E. Mayr, M.M. Rudzki, M. Rudzki, B. Borchardt, H. Hausser and A. Ruter, [Does low intensity, pulsed ultrasound speed healing of scaphoid fractures?], *Handchir. Mikrochir. Plast. Chir.* **32** (2000), 115–122.

[31] S. Mizuno, T. Tateishi, T. Ushida and J. Glowacki, Hydrostatic fluid pressure enhances matrix synthesis and accumulation by bovine chondrocytes in three-dimensional culture, *J. Cell Physiol.* **193** (2002), 319–327.

[32] K. Naruse, A. Miyauchi, M. Itoman and Y. Mikuni-Takagaki, Distinct anabolic response of osteoblast to low-intensity pulsed ultrasound, *J. Bone Miner. Res.* **18** (2003), 360–369.

[33] S. Nehrer, M. Spector and T. Minas, Histologic analysis of tissue after failed cartilage repair procedures, *Clin. Orthop.* (1999), 149–162.

[34] P.A. Nolte, J. Klein-Nulend, G.H. Albers, R.K. Marti, C.M. Semeins, S.W. Goei and E.H. Burger, Low-intensity ultrasound stimulates endochondral ossification *in vitro*, *J. Orthop. Res.* **19** (2001), 301–307.

[35] P.A. Nolte, A. van der Krans, P. Patka, I.M. Janssen, J.P. Ryaby and G.H. Albers, Low-intensity pulsed ultrasound in the treatment of nonunions, *J. Trauma* **51** (2001), 693–702; discussion 702–703.

[36] P. Angele, R. Kujat, H. Faltermeier, D. Schumann, R. Mueller and M. Nerlich, Biodegradable Hyaluronsäureester/Gelatine Kompositmatrix zur osteochondralen Differenzierung mesenchymaler Vorläuferzellen, *Biomaterialien* **4** (2003), 11–18.

[37] J. Parvizi, V. Parpura, J.F. Greenleaf and M.E. Bolander, Calcium signaling is required for ultrasound-stimulated aggrecan synthesis by rat chondrocytes, *J. Orthop. Res.* **20** (2002), 51–57.

[38] J. Parvizi, C.C. Wu, D.G. Lewallen, J.F. Greenleaf and M.E. Bolander, Low-intensity ultrasound stimulates proteoglycan synthesis in rat chondrocytes by increasing aggrecan gene expression, *J. Orthop. Res.* **17** (1999), 488–494.

[39] L. Peterson, T. Minas, M. Brittberg, A. Nilsson, E. Sjogren-Jansson and A. Lindahl, Two- to 9-year outcome after autologous chondrocyte transplantation of the knee, *Clin. Orthop.* (2000), 212–234.

[40] M. Pittenger, A. Mackay, S. Beck, R. Jaiswal, R. Douglas, J. Mosca, M. Moorman, D. Simonetti, S. Craig and D. Marshak, Multilineage potential of adult human mesenchymal stem cells, *Science* **284** (1999), 143–148.

[41] K.H. Pridie, A method of resurfacing osteoarthritic knee joints, *J. Bone Joint Surg.* **41** (1959), 618–619.

[42] J. Schortinghuis, J.L. Ruben, G.M. Raghoebar and B. Stegenga, Therapeutic ultrasound to stimulate osteoconduction; A placebo controlled single blind study using e-PTFE membranes in rats, *Arch. Oral Biol.* **49** (2004), 413–420.

[43] J.R. Steadman, Long-term clinical results with microcrakture (MCFR) and debridement for treatment of full-thickness chondral defects, in: *Annual Meeting of the Amercan Academy of Orthopaedic Surgeons*, 1998.

[44] J.R. Steadman, K.K. Briggs, J.J. Rodrigo, M.S. Kocher, T.J. Gill and W.G. Rodkey, Outcomes of microfracture for traumatic chondral defects of the knee: Average 11-year follow-up, *Arthroscopy* **19** (2003), 477–484.

[45] J.K. Suh, Dynamic unconfined compression of articular cartilage under a cyclic compressive load, *Biorheology* **33** (1996), 289–304.

[46] J.K. Suh, G.H. Baek, A. Aroen, C.M. Malin, C. Niyibizi, C.H. Evans and A. Westerhausen-Larson, Intermittent subambient interstitial hydrostatic pressure as a potential mechanical stimulator for chondrocyte metabolism, *Osteoarthritis Cartilage* **7** (1999), 71–80.

[47] I. Takahashi, G.H. Nuckolls, K. Takahashi, O. Tanaka, I. Semba, R. Dashner, L. Shum and H.C. Slavkin, Compressive force promotes sox9, type II collagen and aggrecan and inhibits IL-1beta expression resulting in chondrogenesis in mouse embryonic limb bud mesenchymal cells, *J. Cell Sci.* **111** (1998), 2067–2076.

[48] P.A. Torzilli, R. Grigiene, C. Huang, S.M. Friedman, S.B. Doty, A.L. Boskey and G. Lust, Characterization of cartilage metabolic response to static and dynamic stress using a mechanical explant test system, *J. Biomech.* **30** (1997), 1–9.

[49] T. Toyoda, B.B. Seedhom, J. Kirkham and W.A. Bonass, Upregulation of aggrecan and type II collagen mRNA expression in bovine chondrocytes by the application of hydrostatic pressure, *Biorheology* **40** (2003), 79–85.

[50] C.W. Wu, Exposure to low-intensity ultrasound stimulates aggrecan gene expression by cultured chondrocytes, *Trans. Orthop. Res. Soc.* (1996), 622.

[51] K.H. Yang, J. Parvizi, S.J. Wang, D.G. Lewallen, R.R. Kinnick, J.F. Greenleaf and M.E. Bolander, Exposure to low-intensity ultrasound increases aggrecan gene expression in a rat femur fracture model, *J. Orthop. Res.* **14** (1996), 802–809.

[52] J.U. Yoo, T.S. Barthel, K. Nishimura, L. Solchaga, A.I. Caplan, V.M. Goldberg and B. Johnstone, The chondrogenic potential of human bone-marrow-derived mesenchymal progenitor cells, *J. Bone Joint Surg. Am.* **80** (1998), 1745–1757.

[53] J.U. Yoo and B. Johnstone, The role of osteochondral progenitor cells in fracture repair, *Clin. Orthop.* (1998), S73–S81.

[54] Z.J. Zhang, J. Huckle, C.A. Francomano and R.G. Spencer, The effects of pulsed low-intensity ultrasound on chondrocyte viability, proliferation, gene expression and matrix production, *Ultrasound Med. Biol.* **29** (2003), 1645–1651.

[55] Z.J. Zhang, J. Huckle, C.A. Francomano and R.G. Spencer, The influence of pulsed low-intensity ultrasound on matrix production of chondrocytes at different stages of differentiation: an explant study, *Ultrasound Med. Biol.* **28** (2002), 1547–1553.

# Part IV: Cartilage engineering

Biorheology 43 (2006) 447–454
IOS Press



# Mesenchymal progenitor cells in adult human articular cartilage

Koji Hiraoka, Shawn Grogan, Tsaiwei Olee and Martin Lotz *
*Division of Arthritis Research, The Scripps Research Institute, La Jolla, CA 92037, USA*

**Abstract.** The transmembrane receptor Notch-1 regulates cell fate and differentiation and was suggested to identify a cell type with progenitor characteristics in newborn bovine articular cartilage. We show that Notch-1 is expressed on >70% of BM-MSC in early passage monolayer culture. We also demonstrate that normal articular cartilage contains Notch-1+ cells and that the frequency is increased in OA. Most Notch-1+ cells in OA cartilage are located in the clusters of proliferating cells. These findings indicate that multipotential mesenchymal progenitor cells are present in articular cartilage from adult humans and that their frequency is increased in OA. This observation has implications for understanding the intrinsic repair capacity of articular cartilage and raises the possibility that these progenitor cells might be involved in the pathogenesis of arthritis.

## 1. Introduction

Articular cartilage contains only cells of mesenchymal lineage that are responsible for production and maintenance of the tissue [21]. Adult articular cartilage has limited capacity for repair and this may in part account for the fact that cartilage is among the tissues with the highest prevalence of aging-associated pathologies [8,23].

When cultured under appropriate conditions, cartilage cells can be induced to form cartilage-like tissue. Monolayer-expanded chondrocytes isolated from human articular cartilage are able to form hyaline-like tissue when implanted into cartilage defects *in vivo* [7]. These observations indicate that cells from adult human cartilage do have some potential for tissue regeneration and repair. It is currently unknown whether this potential is a feature of all cartilage cells or only of certain subpopulations. In addition, cells from adult human articular cartilage that were expanded in monolayer culture were not only able to express a cartilage-like gene and protein expression pattern but also had osteogenic and adipogenic potential when cultured under appropriate conditions [3,13,14,29]. This plasticity is reminiscent of mesenchymal stem cells and subpopulations of cartilage cells express surface markers and functional properties of mesenchymal stem cells [1,17].

Mesenchymal stem cells (MSC) are present in various tissues and most of our current information is based on studies on cells from bone marrow and adipose tissues. MSC can be characterized by the expression of certain cell surface antigens such as Stro-1, CD29, CD44, CD71, CD90, CD105, and CD166. MSC are negative for hematopoietic lineage markers CD14, CD4 and CD45 [5,15,19,25,26]. Despite the identification of these candidate markers there is at present no consensus on a single marker for MSC [24].

*Address for correspondence: Martin Lotz, MD, Division of Arthritis Research, The Scripps Research Institute, 10550 North Torrey Pines Road, La Jolla, CA 92037, USA. Tel.: +1 858 784 8960; Fax: +1 858 784 2744; E-mail: mlotz@scripps.edu.

A recent study demonstrated that the superficial zone of immature bovine articular cartilage contains MSC and that these cells can be identified by the expression of Notch-1 [17]. Notch receptors are transmembrane proteins that are activated by neighboring cells expressing the transmembrane ligands Delta, Jagged and Serrate [2,4]. Notch activation can either promote or inhibit cell differentiation. Expression patterns of Notch and receptor and ligands and results from overexpression or deletion of these genes suggest a role of this signalling pathway in skeletal development. Misexpression of Delta-1 in chick embryos negatively regulated the transition from prehypertrophic to hypertrophic chondrocytes [12]. Notch activation suppressed chondrocytic differentiation and cell proliferation in the prechondrocytic cell line ATDC5 [30].

The objectives of the present study were to examine the expression of Notch-1 in human bone marrow-derived MSC and in articular cartilage.

## 2. Methods

### 2.1. MSC isolation and culture

Mesenchymal stem cells were isolated from the iliac crest bone marrow of normal adult donors. Bone marrow aspirates were mixed with 50 $\mu$l RosetteSep Human Mesenchymal Stem Cell Cocktail (StemCell Technologies) per ml of sample and incubated for 20 minutes at room temperature. The mixture was diluted with twice the volume of PBS containing 2% FBS and 1 mM EDTA. The diluted sample was gently layered on the top of the Ficoll-Plaque and centrifuged for 25 minutes at $300 \times g$ at room temperature. Cells at the interface were collected, and washed 2 times with PBS containing 2% FBS and 1 mM EDTA. After washing, cells were resuspended in MesenCult basal medium for human mesenchymal stem cells with stimulatory supplements (StemCell Technologies), counted, and cultured at a density of $2 \times 10^5$ cells/cm$^2$. Culture medium was replaced at 24 h and 72 h and every 3–4 days thereafter. MSC were subcultured after 10 to 14 days by treatment with trypsin–EDTA (0.05% and 0.53 mM respectively) for 5 min, subsequently washed in medium supplemented and seeded into fresh flasks.

### 2.2. Isolation and culture of cells from cartilage

Cartilage from the femoral condyles and tibial plateaus of the knee joints was obtained at autopsy from donors without known history of joint disease or from healthy organ donors. Tissue was also obtained at the time of total joint replacement surgery from patients with osteoarthritis. All samples were graded according to a modified Mankin scale [20].

Cartilage slices were cut into 2–3 mm$^3$ pieces, washed with McCoy's 5A modified medium (Life Technologies, Rockville, MD), treated with trypsin (10% vol/vol) for 15 minutes in a 37°C shaking waterbath, and washed 3 times with Hank's Balanced Salt Solution (HBSS). The tissues were transferred to McCoy's medium, supplemented with 10% fetal bovine serum (FBS) (Hyclone, Logan, UT), 1 mM Na-pyruvate, 100 Units/ml penicillin, 100 $\mu$g/ml streptomycin, Hepes buffer (25 mM), 2 mg/ml clostridial collagenase type IV (Sigma, St. Louis, MO) and 0.15 mg/ml hyaluronidase (Sigma) and digested overnight on a gyratory shaker. The cells were washed three times with HBSS and cultured at high cell density in McCoy's medium supplemented as above. Cells were used in primary (passage zero) or passaged in monolayer culture as indicated for each set of experiments.

## 2.3. Flow cytometry

After harvesting the cells and preparing a single cell suspension ($10^5$ cells per 50 $\mu$l PBS supplemented with 1% BSA and 0.1% Na-azide), the cells were fixed with 0.2 ml of 2% PFA in PBS at 4°C for 30 minutes and washed twice with PBS + 1% BSA and 0.1% Na-azide. The cells were incubated in 0.2 ml of 70% ETOH for at least 30 minutes at $-20$°C for permeabilization. After washing twice, the cells were stained at 4°C for 60 min with FITC-conjugated anti-CD166 (Ancell), anti-Notch-1 (Santa Cruz), mouse or goat isotype (negative control). Anti-Notch-1 antibody required 2nd antibody (anti-goat IgG FITC conjugate). The cells were then washed 3× in wash buffer (PBS, 1% BSA, 0.1% Na-azide). The cells were subjected to FACS analysis using a Becton Dickinson FACScan and Cell Quest software (Becton Dickinson, San Jose, CA, USA).

## 2.4. Immunohistochemistry

After blocking with 10% normal serum or BSA for 30 minutes, the sections were incubated with primary antibody against the intracellular part of Notch-1 at 2 $\mu$g/ml for 1 to 2 hours at room temperature, or overnight at 4°C. After washing the sections 3 times for 5 minutes in PBS a second blocking was performed for 10 minutes. The sections were incubated for 30 minutes with diluted biotinylated secondary antibody. The slides were washed 3 times in PBS, and incubated for 30 minutes with Vectastain ABC-AP reagent (Vector), or the peroxidase-based Elite ABC system. The slides were washed and the sections incubated for 4 to 20 minutes with alkaline phosphatase substrate solution, or 3,3-diaminobenzidine (DAB) substrate. Slides were rinsed with water, counterstained with diluted Hematoxylin or Methyl Green, rehydrated in graded ethanol or 3 changes of 1-butanol, Hemo-de and mounted with Refrax mounting medium (Anatech).

## 3. Results

### 3.1. Notch-1 expression on human BM-MSC

MSC were isolated from iliac crest-derived bone marrow, expanded in monolayer culture and analyzed for Notch-1 expression by flow cytometry. The majority of cells express Notch-1 in early passage (Fig. 1). There was no evidence of hematopoietic cells in the cultures as the cells were negative for CD45 (not shown). The percentage of Notch-1-positive cells progressively declined during serial passaging (Fig. 1) with a reduction to less than 20% Notch-1 positive cells by passage 7.

These results indicate that Notch-1 is expressed on human BM-MSC and that the levels of Notch-1 expression gradually decrease during MSC expansion in high density culture.

### 3.2. Notch-1 expression on cells from normal and osteoarthritic human cartilage

Normal articular cartilage was collected from knee joints and cells in first passage monolayer culture were analyzed by flow cytometry for Notch-1 expression (Fig. 2). A small percentage of cells were Notch-1 positive and this was higher 3.94 ± 2.7%) in the young individuals (age 22 ± 7) as compared to older normal donors (age 56 ± 9). The percentage of Notch-1 positive cells was higher in OA samples as compared to age-matched controls (Table 1).

Immunohistochemistry showed that Notch-1 positive cells are distributed throughout the different zones of normal cartilage. In osteoarthritic cartilage, many of the cells in the chondrocyte clusters are Notch-1 positive (Fig. 3).

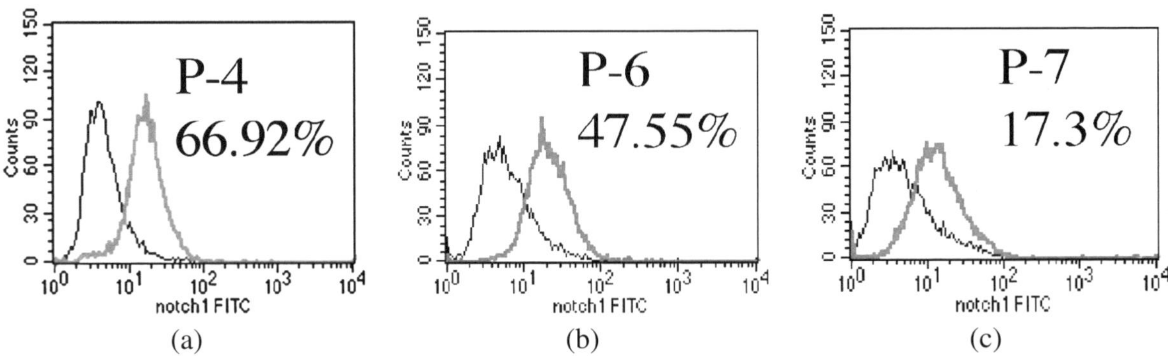

Fig. 1. Notch-1 expression in MSC monolayer culture. MSC were isolated from iliac crest-derived bone marrow and expanded in monolayer culture and analyzed by flow cytometry. The figure shows the same cell preparation at passage 4 (left panel), passage 6 (center panel) and passage 7 (right panel). Additional MSC preparations ($n = 5$) were analyzed between passage 2 and passage 7 and showed a similar decrease in Notch-1 expression.

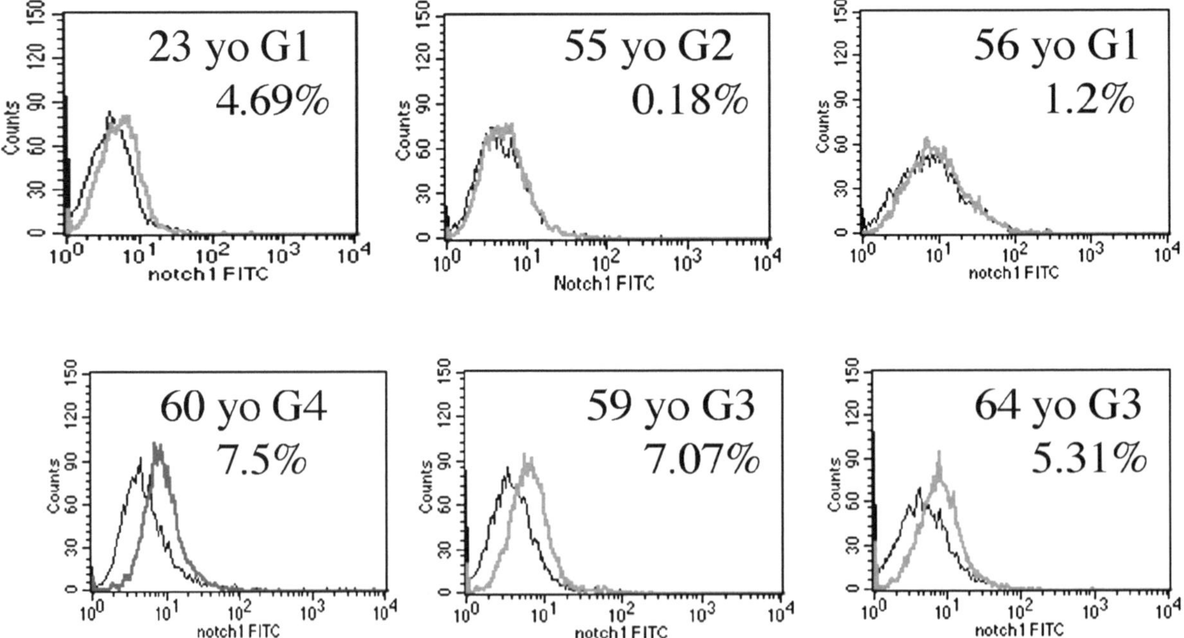

Fig. 2. Notch-1 expression on human articular cartilage cells. Flow cytometry of human bone marrow-derived MSC (passage 4) and cartilage cells (passage 1 from normal human knee cartilage) for Notch-1. (B) Histograms show double staining for CD166 and Notch-1.

Table 1

Effect of cell density on Notch-1 expression in MSC monolayer culture. Human bone marrow-derived MSC in passage 6 were plated at the indicated cell densities in 24 well plates and cultured for 48 h. Notch-1 expression was analyzed by flow cytometry

| Grade | $n$ | Age | % |
| --- | --- | --- | --- |
| 1–2 | 12 | $22 \pm 7$ | $3.94 \pm 2.7$ |
| 1–2 | 7 | $56 \pm 9$ | $1.52 \pm 1.9$ |
| 3–4 | 10 | $70 \pm 13$ | $4.57 \pm 3.1$ |

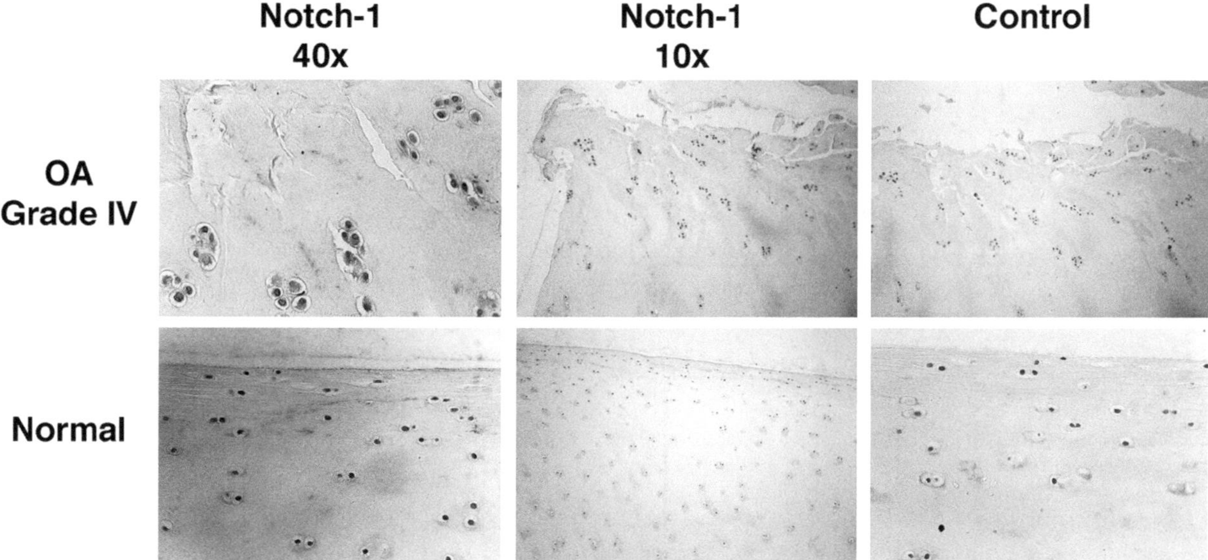


Fig. 3. Notch-1 expression in normal and osteoarthritic human cartilage. Sections of normal and osteoarthritc human knee cartilage were stained with Notch-1 antibody or isotype control. Note the increased staining of cells in the chondrocyte clusters in OA cartilage.

### 3.3. Co-expression of Notch-1 and CD166

CD166 is one of several previously identified MSC cell surface markers [19]. Analysis of double stained BM-MSC and cartilage cells showed that the same cells co-express CD166 and Notch-1 (Fig. 4).

## 4. Discussion

The notion that mature articular cartilage has limited intrinsic repair potential dates back to the work of Hunter in the 18th century and has been maintained with support from experimental models and clinical studies. However, evidence exists that under certain conditions new tissue is formed although it does not completely recapitulate the structural organization and functional properties of articular cartilage [22,23]. Studies on cultured cartilage cells demonstrated that they can be expanded and in response to appropriate stimulation produce cartilage extracellular matrix components. The differentiation status of cartilage cells can be altered by spatial organization cells and by biochemical stimuli. When cells are expanded in monolayer, a gradual loss of cartilage specific gene and protein expression patterns is observed but this is reversed when cells are transferred to a three-dimensional culture [6]. Certain growth factors such as TGF or BMPs enhance the cartilage like signature while inflammatory cytokines or retinoic acid suppress this. Adding further support for the plasticity of cartilage cells are recent findings that monolayer cultured cells are not only able to reexpress a chondrocyte specific differentiation pattern but can also undergo osteogenic and adipogenic differentiation [13,28]. This degree of plasticity and multilineage differentiation potential is the key feature of mesenchymal progenitor or stem cells [9]. Following the discovery that stem cells are present in adult tissues we determined if human articular cartilage contains mesenchymal stem cells [1]. The initial study used the cell surface markers CD166 and CD105 which had been suggested to be co-expressed by bone marrow-MSC [19]. We showed that adult human

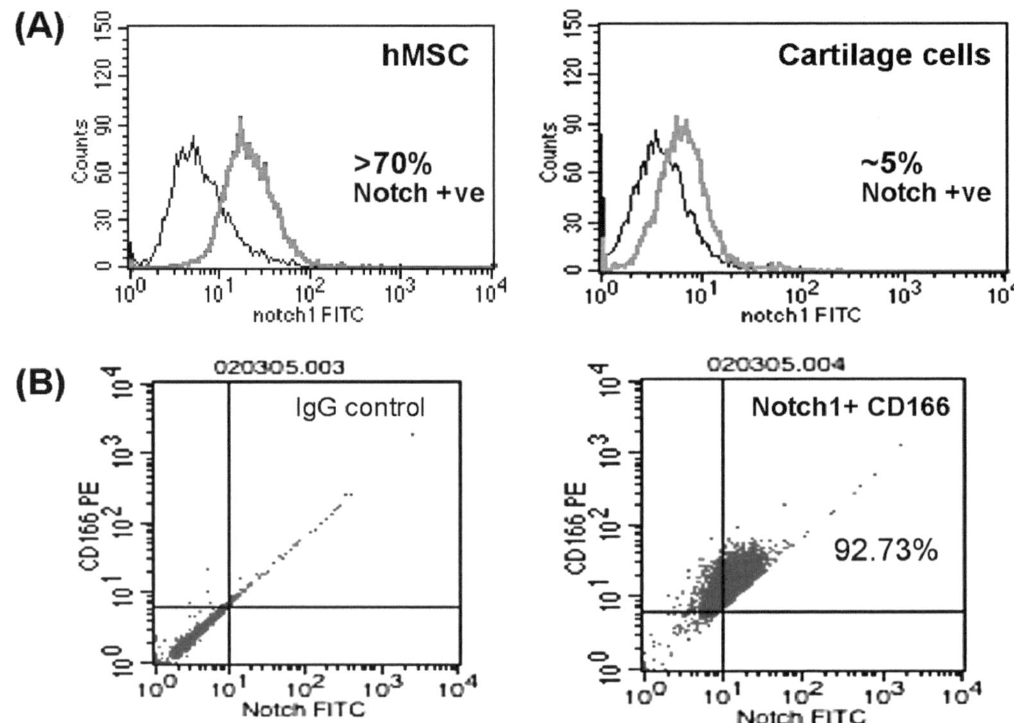



Fig. 4. Co-expression of CD166 and Notch-1 on human BM-MSC. (A) Flow cytometric analysis of human bone marrow-derived MSC (passage 4) and cartilage cells (passage 1 from normal human knee cartilage) for Notch-1. (B) Histograms show double staining for CD166 and Notch-1.

articular cartilage contains a small subpopulation of CD105/CD166 positive cells and that they can be differentiated into chondrocytes, osteoblasts and adipocytes. Immature bovine cartilage also contains a cell population with functional features of MSC. These cells expressed the cell surface receptor Notch-1 and were primarily located in the superficial zone of cartilage [17].

The present results extend these findings and first show that Notch-1 is expressed on human BM-MSC. In early passage the majority of cells is Notch-1 positive and this declines with increasing passage number. Notch-1 activation is a candidate mechanism as it is known in other cell lineages to regulate maintenance of progenitor cells [11]. This is also consistent with suppressive effects of Notch on chondrocytic and osteoblastic differentiation of mesenchymal progenitor cells [16,30].

The results from analysis of human articular cartilage show that normal tissue contains a small number of cells that are Notch-1 positive. We did not observe a specific localization of these cells as has been reported for immature bovine cartilage [17]. However, the increased frequency of Notch-1 positive cells in osteoarthritis cartilage was primarily due to the large number of positive cells in the chondrocyte clusters which are thought to represent cells that have proliferated in response to tissue injury. Results from an experimental model of mechanical cartilage injury also showed that the proliferating cells are Notch-1 positive [18].

In conclusion, Notch-1 is expressed on BM-MSC and in a subpopulation of cells in normal mature human articular cartilage. The increased expression of Notch-1 in the chondrocyte clusters in osteoarthritic cartilage suggests that mesenchymal progenitor cells are activated in this disease and may contribute to the aberrant cartilage remodelling process.

## Acknowledgements

This work was supported by NIH grants AG07996 and AG20923.

## References

[1]  S. Alsalameh, R. Amin, T. Gemba and M. Lotz, Identification of mesenchymal progenitor cells in normal and osteoarthritic human articular cartilage, *Arthritis Rheum.* **50** (2004), 1522–1532.

[2]  S. Artavanis-Tsakonas, M.D. Rand and R.J. Lake, Notch signaling: cell fate control and signal integration in development, *Science* **284** (1999), 770–776.

[3]  A. Barbero, S. Ploegert, M. Heberer and I. Martin, Plasticity of clonal populations of dedifferentiated adult human articular chondrocytes, *Arthritis Rheum.* **48** (2003), 1315–1325.

[4]  M. Baron, An overview of the Notch signalling pathway, *Semin. Cell Dev. Biol.* **14** (2003), 113–119.

[5]  F.P. Barry, R.E. Boynton, S. Haynesworth, J.M. Murphy and J. Zaia, The monoclonal antibody SH-2, raised against human mesenchymal stem cells, recognizes an epitope on endoglin (CD105), *Biochem. Biophys. Res. Commun.* **265** (1999), 134–139.

[6]  P.D. Benya and J.D. Shaffer, Dedifferentiated chondrocytes reexpress the differentiated collagen phenotype when cultured in agarose gels, *Cell* **30** (1982), 215–224.

[7]  M. Brittberg, A. Lindahl, A. Nilsson, C. Ohlsson, O. Isaksson and L. Peterson, Treatment of deep cartilage defects in the knee with autologous chondrocyte transplantation, *N. Engl. J. Med.* **331** (1994), 889–895.

[8]  J.A. Buckwalter and T.D. Brown, Joint injury, repair, and remodeling: roles in post-traumatic osteoarthritis, *Clin. Orthop.* (2004), 7–16.

[9]  A.I. Caplan, Review: mesenchymal stem cells: cell-based reconstructive therapy in orthopedics, *Tissue Eng.* **11** (2005), 1198–1211.

[10]  D.C. Colter, R. Class, C.M. DiGirolamo and D.J. Prockop, Rapid expansion of recycling stem cells in cultures of plastic-adherent cells from human bone marrow, *Proc. Natl. Acad. Sci. USA* **97** (2000), 3213–3218.

[11]  I.M. Conboy, M.J. Conboy, G.M. Smythe and T.A. Rando, Notch-mediated restoration of regenerative potential to aged muscle, *Science* **302** (2003), 1575–1577.

[12]  R. Crowe, J. Zikherman and L. Niswander, Delta-1 negatively regulates the transition from prehypertrophic to hypertrophic chondrocytes during cartilage formation, *Development* **126** (1999), 987–998.

[13]  R. de la Fuente, J.L. Abad, J. Garcia-Castro, G. Fernandez-Miguel, J. Petriz, D. Rubio, C. Vicario-Abejon, P. Guillen, M.A. Gonzalez and A. Bernad, Dedifferentiated adult articular chondrocytes: a population of human multipotent primitive cells, *Exp. Cell Res.* **297** (2004), 313–328.

[14]  F. Dell'Accio, J. Vanlauwe, J. Bellemans, J. Neys, C. De Bari and F.P. Luyten, Expanded phenotypically stable chondrocytes persist in the repair tissue and contribute to cartilage matrix formation and structural integration in a goat model of autologous chondrocyte implantation, *J. Orthop. Res.* **21** (2003), 123–131.

[15]  J.E. Dennis, J.P. Carbillet, A.I. Caplan and P. Charbord, The STRO-1+ marrow cell population is multipotential, *Cells Tissues Organs* **170** (2002), 73–82.

[16]  V. Deregowski, E. Gazzerro, L. Priest, S. Rydziel and E. Canalis, Notch1 overexpression inhibits osteoblastogenesis by suppressing WNT/beta-catenin but not bone morphogenetic protein signaling, *J. Biol. Chem.* (2006).

[17]  G.P. Dowthwaite, J.C. Bishop, S.N. Redman, I.M. Khan, P. Rooney, D.J. Evans, L. Haughton, Z. Bayram, S. Boyer, B. Thomson, M.S. Wolfe and C.W. Archer, The surface of articular cartilage contains a progenitor cell population, *J. Cell Sci.* **117** (2004), 889–897.

[18]  F.M. Henson, E.A. Bowe and M.E. Davies, Promotion of the intrinsic damage-repair response in articular cartilage by fibroblastic growth factor-2, *Osteoarthritis Cartilage* **13** (2005), 537–544.

[19]  M.K. Majumdar, M.A. Thiede, J.D. Mosca, M. Moorman and S.L. Gerson, Phenotypic and functional comparison of cultures of marrow-derived mesenchymal stem cells (MSCs) and stromal cells, *J. Cell. Physiol.* **176** (1998), 57–66.

[20]  H.J. Mankin, H. Dorfman, L. Lippiello and A. Zarins, Biochemical and metabolic abnormalities in articular cartilage from osteo-arthritic human hips. II. Correlation of morphology with biochemical and metabolic data, *J. Bone Joint Surg. Am.* **53** (1971), 523–537.

[21]  H. Muir, The chondrocyte, architect of cartilage. Biomechanics, structure, function and molecular biology of cartilage matrix macromolecules, *Bioessays* **17** (1995), 1039–1048.

[22]  S. Nehrer, M. Spector and T. Minas, Histologic analysis of tissue after failed cartilage repair procedures, *Clin. Orthop. Relat. Res.* (1999), 149–162.

[23]  S. O'Driscoll, Update on cartilage repair, *Orthopedics* **25** (2002), 1342, 1388.

[24]  M.F. Pittenger and B.J. Martin, Mesenchymal stem cells and their potential as cardiac therapeutics, *Circ. Res.* **95** (2004), 9–20.

[25] A. Saalbach, U.F. Haustein and U. Anderegg, A ligand of human thy-1 is localized on polymorphonuclear leukocytes and monocytes and mediates the binding to activated thy-1-positive microvascular endothelial cells and fibroblasts, *J. Invest. Dermatol.* **115** (2000), 882–888.

[26] P.J. Simmons and B. Torok-Storb, Identification of stromal cell precursors in human bone marrow by a novel monoclonal antibody, STRO-1, *Blood* **78** (1991), 55–62.

[27] L.A. Solchaga, K. Penick, J.D. Porter, V.M. Goldberg, A.I. Caplan and J.F. Welter, FGF-2 enhances the mitotic and chondrogenic potentials of human adult bone marrow-derived mesenchymal stem cells, *J. Cell. Physiol.* **203** (2005), 398–409.

[28] T. Tallheden, J.E. Dennis, D.P. Lennon, E. Sjogren-Jansson, A.I. Caplan and A. Lindahl, Phenotypic plasticity of human articular chondrocytes, *J. Bone Joint Surg. Am.* **85-A**(Suppl. 2) (2003), 93–100.

[29] M. Thornemo, T. Tallheden, E. Sjogren Jansson, A. Larsson, K. Lovstedt, U. Nannmark, M. Brittberg and A. Lindahl, Clonal populations of chondrocytes with progenitor properties identified within human articular cartilage, *Cells Tissues Organs* **180** (2005), 141–150.

[30] N. Watanabe, Y. Tezuka, K. Matsuno, S. Miyatani, N. Morimura, M. Yasuda, R. Fujimaki, K. Kuroda, Y. Hiraki, N. Hozumi and K. Tezuka, Suppression of differentiation and proliferation of early chondrogenic cells by Notch, *J. Bone Miner. Metab.* **21** (2003), 344–352.

Biorheology 43 (2006) 455–470
IOS Press



# Dynamic compressive strain influences chondrogenic gene expression in human mesenchymal stem cells

Jonathan J. Campbell, David A. Lee and Dan L. Bader *
*Medical Engineering Division and IRC in Biomedical Materials, Department of Engineering,
Queen Mary University of London, London, UK*

**Abstract.** This study tests the hypothesis that dynamic compressive strain selectively enhances chondrogenic differentiation by human mesenchymal stem cells (MSCs). Primary MSCs were isolated and expended in monolayer culture. The cells were seeded in alginate constructs or in pellet culture. The time course of chondrogenic differentiation was assessed by real-time QPCR of mRNA expression analysis for cartilage specific markers. Collagen types II and X mRNA, not present in undifferentiated MSCs, were detectable by 2–4 days of chondrogenic induction and continued to rise significantly throughout the culture period of 10 days ($p < 0.001$). Basal levels of gene expression for Sox-9 and aggrecan were evident in undifferentiated MSCs, although chondrogenic induction for a period of 8 days resulted in an increased trend in the gene expression levels. The alginate system was also used in mechanical conditioning studies. Dynamic compression was applied, in an intermittent regimen, at a strain amplitude of 15% and frequency of 1 Hz in the presence and absence of 10 ng/ml TGF$\beta_3$, for a period of 8 days. Results indicated significant changes in the levels of mRNA expression for the chondrogenic markers. For example, by day 8, the application of the strain regimen alone caused an up-regulation in all the chondrogenic markers compared to the control samples (no TGF$\beta$, no compression). However, the combined effects of strain and TGF$\beta$ on these markers were more complex than purely additive.

## 1. Introduction

Current cell-based procedures for the repair of articular cartilage typically involve the implantation of autologous chondrocytes into defect sites with or without an associated 3D scaffold. However, there are many problems associated with the use of differentiated cells including cell source, expansion potential, de-differentiation and donor site morbidity. The use of multipotential cells of bone marrow origin is an attractive alternative, provided controlled differentiation into mature chondrocytes can be achieved. Mesenchymal stem cells hold promise in therapeutic applications due to their ease of isolation from sites distant to tissue damage, expansion potential *in vitro*, and ability to differentiate to multiple connective tissue lineages [8,31]. Cartilage has been proposed as a potential target tissue for these cells with tissue engineering repair strategies, thus far utilizing autologous chondrocytes to produce improved cartilage repair [6,7]. In order to use MSCs successfully for this purpose a thorough exploration and optimisation of the governing factors controlling MSC chondrogenesis is required.

---

*Address for correspondence: Dan Bader, PhD, DSc, Department of Engineering, Queen Mary University of London, Mile End Road, London, E1 4NS, UK. Tel.: +44 20 7882 5274; Fax: +44 20 8983 3052; E-mail: D.L.Bader@qmul.ac.uk.

To date, limited studies have explored the influence of mechanical forces on mesenchymal stem cells. There are numerous factors that can be varied including the cell origin and stage of maturity, mechanical loading parameters and culture conditions. Differentiation-specific studies have been performed with various mechanical stimuli namely, hydrostatic pressure [2], equaxial and uniaxial strain [20] and oscillatory fluid flow [27]. The predominant loading mode for articular cartilage is, however, dynamic compression. A recent study investigated the effect of cyclic compressive loading on chondrogenic differentiation of rabbit MSCs [19]. They suggested that compressive loading, in the absence of chondrogenic-induction factor, TGF-$\beta$, could induce chondrogenic differentiation. To the knowledge of the authors, there are no comparable published studies, which have reported on human MSCs.

This study tests the hypothesis that dynamic compressive strain may be used to selectively enhance chondrogenic differentiation and tissue formation by human MSCs. This required establishing an appropriate environment for mechanical stimulation. Accordingly the temporal response of MSCs cultured in either traditional pellet forms or 3D alginate constructs will be compared in the presence or absence of TGF$\beta$.

## 2. Materials and methods

### 2.1. Mesenchymal stem cell culture

Primary mesenchymal stem cells, isolated from healthy bone marrow donors, and characterized for positive expression of CD-105 and 44 and negative for CD34 and 45 were obtained commercially (Cambrex Bio Sciences Ltd, Wokingham, UK). The cells were expanded in monolayer through 5 passages in media consisting of high glucose DMEM supplemented with L-glutamine and sodium pyruvate, 100 U/ml penicillin, 100 $\mu$g/ml streptomycin and 10% FCS (all Sigma-Aldrich, Poole, UK). MSCs were recovered from monolayer culture by enzymatic treatment with trypsin/EDTA (Sigma-Aldrich, Poole, UK) at 95% confluence. The cell number and viability was monitored by use of a haemocytometer in conjunction with Trypan Blue (Sigma-Aldrich, Poole, UK) staining. During passage the cells were plated at a density of 5000 cells/cm$^2$.

### 2.2. MSC chondrogenic pellet culture

Passage 5 MSCs were recovered from monolayer culture, washed and re-suspended in PBS. Aliquots containing $2.5 \times 10^5$ cells were added to 15 ml conical tubes and the cells were then centrifuged at $500 \times g$ to form cell pellets. The PBS was removed and pellets were bathed in 250 $\mu$l of a defined chondrogenic media (Cambrex Bioscience), consisting of high glucose DMEM, 1 $\mu$M dexamethasone, 1 mM sodium pyruvate, 0.17 mM ascorbic acid-2-phosphate, 0.35 mM proline and ITS + 1 premix and 10 ng/ml TGF-$\beta_3$ (all Sigma Poole, UK). Control samples were cultured in chondrogenic media without TGF-$\beta_3$ supplementation. At appropriate time points, representative pellets were removed from culture and frozen in liquid nitrogen for subsequent gene expression analysis.

### 2.3. Cell alginate construct preparation and chondrogenic culture

A 6% (w/v) sodium alginate GMB (Manugel®, ISP Alginates Ltd, Surrey) solution was prepared in Earle's balanced salt solutions, EBSS, (Sigma, Poole, UK) and sterilised by autoclaving. The alginate was cooled to 37°C and an equal volume of a MSC suspension containing $2 \times 10^7$ cells/ml was added to the alginate, to yield a final concentration of $1 \times 10^7$ cells/ml in 3% alginate. To prepare constructs suitable for mechanical loading experiments, the cell/alginate suspension was added by Pasteur pipette to the cylindrical wells of a stainless steel casting chamber of 6 mm diameter and 4 mm depth. This casting chamber was bound at the upper and lower surfaces by dialysis membranes (2.4 nm pores, BDH, Poole, UK) secured in place by wire mesh. Cell/alginate suspensions were cross-linked by submersing the casting chamber in filter-sterilised high glucose DMEM (Sigma, Poole, UK) supplemented with 100 mM $CaCl_2$ and antibiotics for 1½ hours at 37°C. Following incubation, the casting chamber was removed and disassembled. The individual gelled constructs, with a volume of 113.1 $\mu l$, were washed with 2 changes of EBSS and placed in culture vessels containing defined media or DMEM + 10% FCS for further experimentation. The cell-seeded alginate constructs at a concentration of $10^7$ cells/ml were distributed, one per well, in a 24 well plate and bathed overnight in 1 ml of control chondrogenic media (defined chondrogenic media without TGF$\beta$ supplementation). The following day, designated time 0, the constructs were bathed in either 1 ml of chondrogenic media with or without 10 ng/ml of TGF$\beta$ and incubated for periods of up to 10 days with media replaced by Pasteur pipette every 2 days. Samples were taken ever 2 days for analysis of cell viability.

In order to release their cellular content, cell/alginate constructs were incubated for 10 minutes at 37°C in a modified alginate disruption buffer, consisting of 150 mM sodium chloride and 55 mM sodium citrate solution in PBS or RNA*later*. This acted as a calcium chelator, disrupting cross-linking between polysaccharide chains. The released cells were spun to a pellet by centrifugation at 500$g$ for 5 minutes. From this stage RNA isolation, purification and analysis followed.

### 2.4. Intermittent dynamic mechanical stimulation of cell-seeded alginate constructs

The cell-seeded alginate constructs with a cell density at $1 \times 10^7$ cells/ml were incubated overnight in defined chondrogenic media in the absence of TGF-$\beta_3$. Individual cylindrical constructs were arranged, in the centre of each well in a 24-well-plate, as detailed in a previous study [25], and then positioned into the cell-straining apparatus (Zwick-Roel, Redditch, UK). This arrangement was positioned within the custom made cell-straining rig and individual flat-ended indenters at the base of stainless steel pins platens were lowered onto the top surface of the cell-seeded constructs to provide a tare load of 0.028 N. A syringe fitted with an angled gauge 22 needle was used to deliver 1 ml of defined chondrogenic media, in the presence or absence of 10 ng/ml TGF-$\beta_3$ supplementation, into each of the wells. The experiment incorporated 12 constructs (6 controls + 6 dynamically loaded) which were bathed in control chondrogenic media without TGF-$\beta$ supplementation, and a further 12 constructs bathed in chondrogenic media with TGF-$\beta$ supplementation. In this manner, there were 4 separate conditions per plate. Experimental time course conditions were selected as 4 and 8 days culture, with and without dynamic compressive strain.

The assembled straining apparatus with cell-seeded constructs was transferred to an adapted incubator where the loading frame was secured to the loading actuator. Dynamic unconfined compression with an amplitude of 15% strain and frequency of 1 Hz was applied to the experimental constructs with repeated cycles of loading and recovery of 1½ hours and 4½ hours, respectively. In this way, 4 complete loading/unloading cycles of 6 hours per 24 hour period. The medium was changed every 2 days.

### 2.5. Cell viability encapsulated in 3% alginate constructs

3% alginate GMB/MSC constructs at a seeding density of $1 \times 10^7$ cells/ml were prepared under sterile conditions. These were individually distributed into wells of a 24 well plate (Corning, NY, USA) and bathed in defined chondrogenic media (Cambrex Bioscience). The constructs were incubated in a humidified environment at 37°C/5% $CO_2$ for 12 days with media replacement every 2 days. At specified time points, representative cell/alginate constructs were removed culture, sectioned vertically and incubated for 1 hr at 37°C with 5 $\mu$M calcein-AM and 40 $\mu$M EthD-1 (Molecular Probes, OR, USA) in 1 ml defined chondrogenic media. The constructs were laid cut-side down on a cover slip and placed on the stage of an inverted microscope fitted with fluorescent optics (Nikon, Kingston-upon-Thames, UK). Fluorescently stained cells were visualised under UV light at 200× magnification with filter attachments allowing the separation of green (live) and red (dead) cells. A systematic sampling procedure was over the depth of each construct, as described in a recent study [18]. The live-dead cell count was recorded and the mean viable percentage was determined for the whole construct and peripheral and core regions.

## 2.6. Determination of mRNA levels

RNA was isolated and quantified and cDNA was generated according to the following methods. Quantitative gene expression analysis for chondrocytes specific genes was carried out using the relative measurement technique utilising pre-designed Taqman® gene expression assays optimised for 100% efficiency (Applied Biosystems, Foster City, CA, USA). Taqman® probe details are given in Table 1.

Cellular material was isolated from both cell pellets and cell/$Ca^{2+}$ alginate constructs using a modified procedure described previously [35], by disruption in RNA*later* (Sigma, Poole, UK) containing 150 mM sodium chloride and 100 mM sodium citrate. Total RNA was extracted using the RNeasy mini kit (Qiagen, Crawley, UK) and cleared of contaminating genomic DNA with DNA-*free*™ (Ambion, Oxford, UK). Reverse transcription was carried out with 1 $\mu$g of total RNA and Multiscribe™ reverse transcriptase at 2.5 U/$\mu$l with random primers according to the manufacturers protocol (cDNA archive kit, Applied Biosystems, Foster City, CA, USA).

Single-plex real-time PCR reactions were carried out in 30 $\mu$l volumes of TaqMan® Fast Universal Master Mix containing 5 $\mu$M Taqman® primer/probes (FAM dyes, Applied Biosystems, Foster City, CA, USA) and 1 $\mu$l of reverse-transcribed cDNA. Reactions were accomplished using real-time PCR

Table 1

Description of proprietary Taqman™ probes (Applied Biosystems, Foster City, US)

| Gene name | Cat. number (Applied Biosystems) | Accession number | Symbol | Chromosome location | |
|---|---|---|---|---|---|
| Collagen type II | Hs 00156568 | NM 001844 | COL2A1 | Chr. 46,679,969 46,684,528 | 12 – |
| Collagen type X | Hs00166657 | NM 000493 | COL10A1 | Chr. 116,546,814 116,553,363 | 6 – |
| SRY (sex determining region-y)-box 9 | Hs00165814 | NM000346 | SOX9 | Chr. 67,630,455 67,634,156 | 17 – |
| Aggrecan | Hs00202971 | NM013227 | AGC1 | Chr. 87,204,561 87,215,775 | 15 – |

apparatus (Mx3000P, Stratagene, La Jolla, CA, USA) with an amplification profile as follows: initial denaturation/enzyme activation at 95°C for 10 minutes, followed by 40 cycles of denaturation at 95°C for 30 secs and annealing/amplification at 60°C for 1 min.

A measurement of quantitative expression of target sequences, relative to the abundance of the endogenous reference gene (GAPDH) and the day 0 calibrator sample was carried using the $\Delta\Delta C_T$ (threshold cycle) method, as previously described [30]. No template and no-amplification controls were run to check for evidence of contaminating external of internal sources, respectively.

## 2.7. Calculation of relative gene expression

To minimise false-negative and spurious data, only those samples demonstrating significant expression of GAPDH (threshold cycles (Ct) <30) were included for gene expression analysis. Threshold cycle (Ct) values were determined with the equipment software (Mx3000P, Stratagene, La Jolla, CA, USA), using the minimum Ct spread algorithm. This algorithm reports a Ct value for a rise in fluorescence above the standard deviation in background fluorescence for each replicate in a sample and where the spread of these standard deviations is minimal. Individual fluorescence readings were normalised to an internal reference dye to minimise well-to-well variations.

Non-parametric statistical methods were carried out according to published statistical methodology. Non-parametric statistical testing between time and all four treatment conditions was accomplished by the use of the Kruskal–Wallis test for independent measures and multiple treatment conditions:

$$H = \frac{12}{N(N-1)} \times \sum \frac{T^2}{n} - 3(N+1). \tag{1}$$

$N$ = total number of samples,
$n$ = number of samples per treatment group,
$T$ = total of the ranks in a treatment group.

The calculated value of $H$ must exceed the table value of $\chi^2$ with $k-1$ degrees of freedom at the 5% significance level. In the event of a significance difference, post-hoc analysis of statistical relationships was carried out using Dunn's test where statistical significance was apparent where:

$$\frac{T_i}{n_i} - \frac{T_j}{n_j} > Q \times SE. \tag{2}$$

$i/j$ = two conditions under test,
$n$ = number of samples per treatment group,
$T$ = total of the ranks in a treatment group,
$Q$ = statistic for differences in mean ranks at the 5% significance level.

$$SE = \sqrt{\frac{N(N+1)}{12}\left(\frac{1}{n_i} + \frac{1}{n_j}\right)}.$$

## 3. Results

### 3.1. Cell viability in 3% alginate culture

MSCs viability profiles were analysed for a seeding density of $10^7$ cells/ml and bathed in 1 ml of defined chondrogenic media. The results for three sections of the construct are illustrated in Fig. 1. An initial viability of 75% after 24 hours of alginate culture dropped to 30% by day 12. No statistically significant differences were observed between the whole, core or peripheral data sets.

### 3.2. Temporal profiles of gene expression in pellet cultures and free swelling alginate constructs

The temporal profiles for Sox-9 expression by MSCs in pellet culture in the presence or absence of TGF$\beta$, are illustrated in Fig. 2a. Similar profiles were apparent in both groups with an increase up to a peak relative expression at day 8, followed by subsequent decline at day 10. The corresponding profiles for MSCs seeded in alginate constructs are illustrated in Fig. 2b. Sox-9 demonstrated peak expression values at day 8 and 6 for the groups incubated in the absence and presence of TGF$\beta$, respectively. In the presence of TGF$\beta$, MSCs in pellet culture exhibited a monotonic increase in the expression of collagen type II from days 2 to 12, as illustrated in Fig. 3a. A similar trend was observed with MSC-seeded alginate cultures at $10^7$ cells/ml, as shown in Fig. 3b. Type X collagen expression in the absence of TGF$\beta$ was only apparent at day 2 for pellet cultures (Fig. 4a). In the presence of TGF$\beta$, the expression of type X collagen generally increased with time in culture for both pellet and alginate samples (Fig. 4). With respect to aggrecan expression, there was a general increase in expression with culture time by MSCs in pellet cultures and alginate constructs in both the presence and absence of TGF$\beta$, as illustrated in Fig. 5.

### 3.3. The influence of dynamic stimulation on the gene expression in alginate constructs

All constructs remained intact throughout the *in vitro* dynamic mechanical stimulation procedure. Total RNA was successfully isolated from cells embedded in alginate constructs and analysed for purity by ratio measurement of absorbance for nucleic acid ($A_{260}$) against protein ($A_{280}$), and quantification

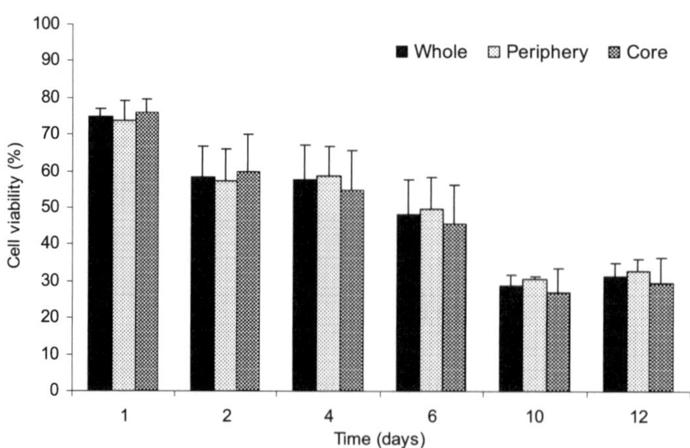


Fig. 1. The temporal and spatial profile of mesenchymal stem cell viability assessed in both central and peripheral regions and across the whole alginate constructs. Each value represents mean + SD.

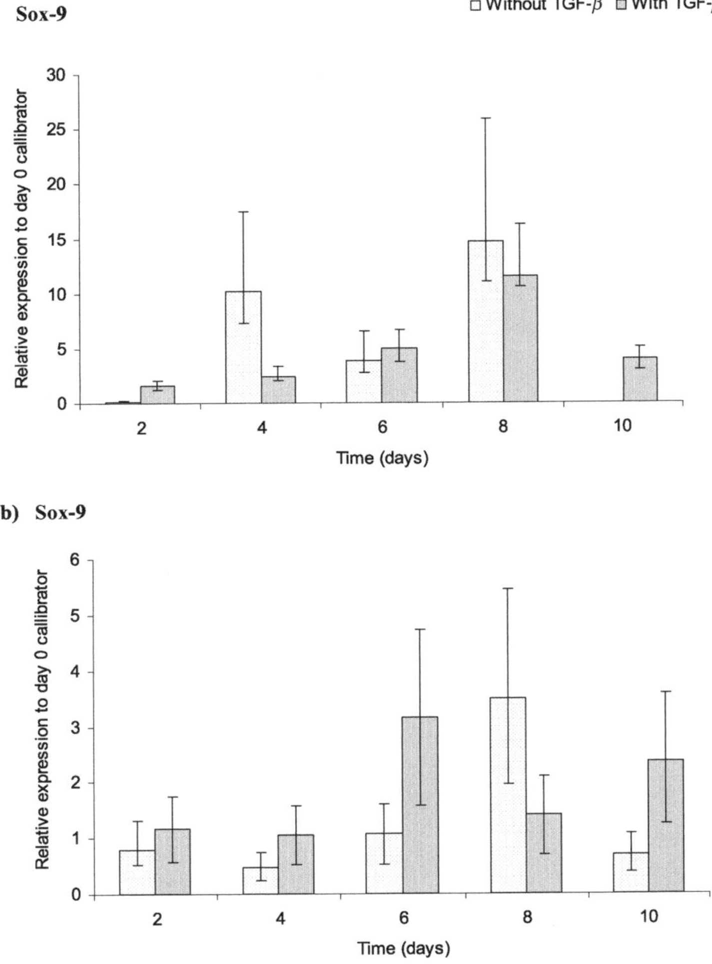


Fig. 2. The relative gene expression of Sox-9 over a 10 day culture period in the presence or absence of 10 ng/ml TGF$\beta$ in (a) pellet culture and (b) alginate constructs. Each value represents the median $\pm$ inter-quartile range.

in $\mu$g/ml. It was found that appropriate mean Ct values for GAPDH at days 4 and 8 were in evidence. Six samples were included for each condition at each time point, but samples failing to show an RNA concentration were not included in the analysis stage. Figures 6–9 provide the quantification of the expression of Sox-9, collagen type II and X, and aggrecan normalised to individual sample GAPDH values and relative to the day 0 control sample (calibrator sample) at days 4 and 8 of culture.

A down-regulation in Sox-9 expression was apparent in all treatment conditions compared to the day 0 control specimens, as evident by values below unity (Fig. 6). No significant differences between strained and unstrained Sox-9 expression was evident, although there was a tendency for a higher expression with the application of dynamic compression on day 4.

There was no evidence of collagen type II expression at day 4 in unstrained samples without TGF-$\beta$ (Fig. 7). The other day 4 specimens yielded collagen type II expression, which was particularly evident with TGF$\beta$ supplementation. However by day 8, all samples demonstrated an up-regulation of collagen type II, indicating that chondrogenesis was in evidence. There was a tendency for stronger expression in

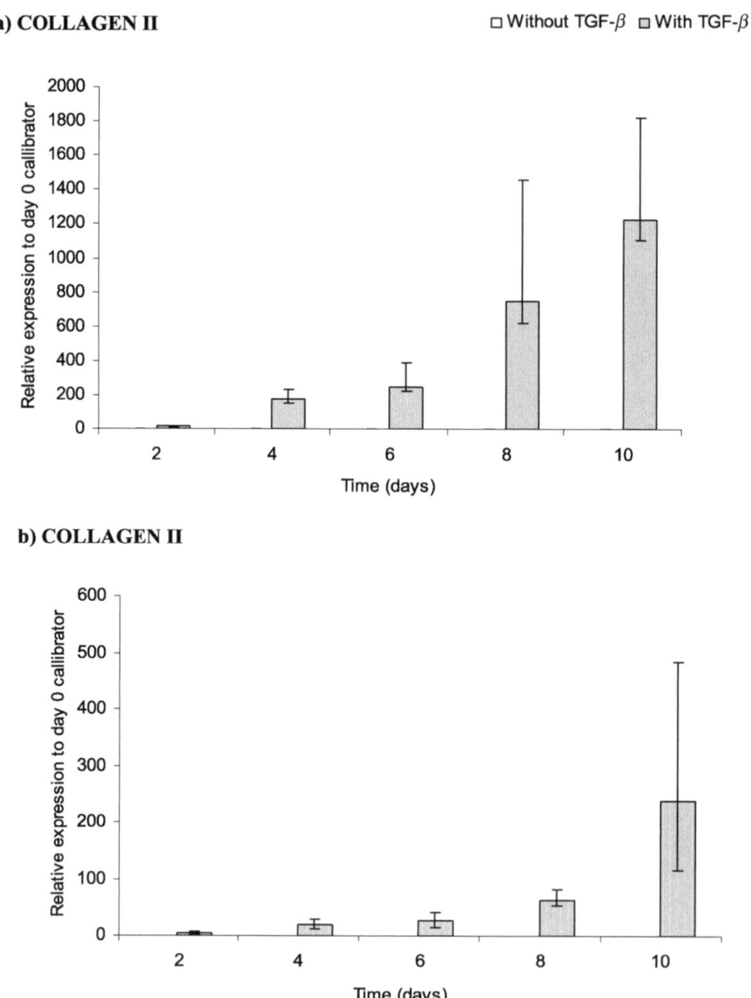



Fig. 3. The relative gene expression of collagen type II over a 10 day culture period in the presence or absence of 10 ng/ml TGF$\beta$ in (a) pellet culture and (b) alginate constructs. Each value represents the median $\pm$ inter-quartile range.

the TGF-$\beta$ treated samples with the highest expression evident when both types of stimuli were applied, although this effect was not statistically significant ($p > 0.05$).

The results for collagen type X (Fig. 8) demonstrate considerable gene expression for all four test conditions. By day 8, dynamic compression in the absence of TGF-$\beta_3$ produced a significant increase in collagen type X expression ($580\times$ day 0 expression) compared to all other treatment conditions ($p < 0.05$). Notwithstanding this effect with dynamic compression alone, the other day 8 samples expressed reduced collagen type X levels compared to corresponding day 4 samples.

Similar to the trends in Sox-9 data, relative values for aggrecan expression were below unity for all treatment conditions (Fig. 9). However, aggrecan expression levels were consistently higher at day 8 compared to day 4 under all treatment conditions ($p < 0.001$). Dynamic compression alone at day 8 produced the highest value of aggrecan expression, which was significantly greater than the combination of compression and TGF-$\beta_3$ on both day 4 ($p < 0.05$) and day 8 ($p < 0.001$).

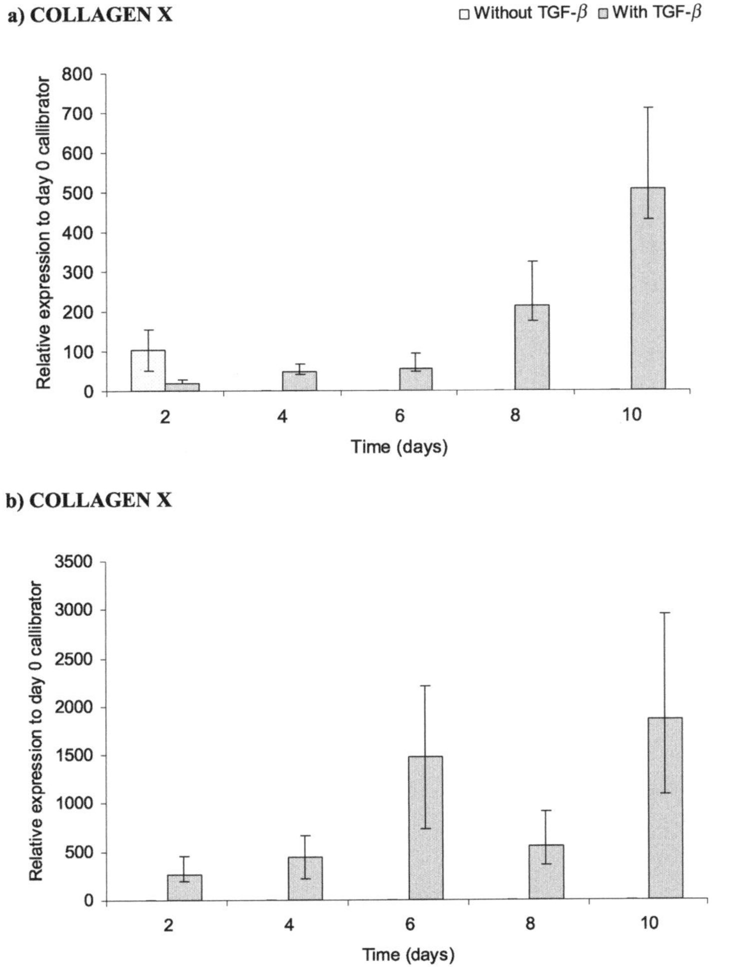



Fig. 4. The relative gene expression of collagen type X over a 10 day culture period in the presence or absence of 10 ng/ml TGFβ in (a) pellet culture and (b) alginate constructs. Each value represents the median ± inter-quartile range.

Table 2 presents a summary of the trends in gene expression with the application of intermittent dynamic compressive strain in the presence and absence of TGF-β. All the chondrogenic genes investigated were up-regulated with the application of strain in the absence of the growth factor. By contrast, a more complex pattern of gene regulation was apparent with the application of dynamic strain in the presence of the growth factor. All the chondrogenic genes were down-regulated at either 4 days or 8 days under these conditions. Aggrecan was down-regulated at both days 4 and 8, collagen type II and Sox-9 at day 8 and collagen type X at day 4.

## 4. Discussion

The present work examines the modulation in gene expression in MSCs in response to mechanical loading. Quantitative real-time gene analysis offered the ability to assess the molecular response of MSCs to intermittent dynamic mechanical loading over an extended (8 day) culture period. The study

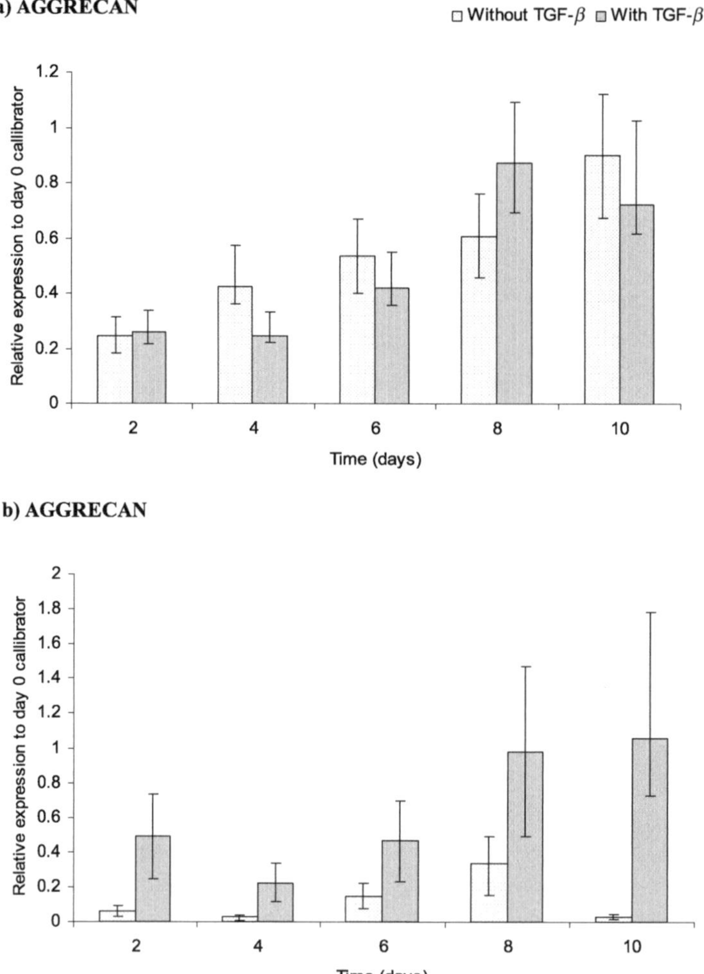



Fig. 5. The relative gene expression of aggrecan over a 10 day culture period in the presence or absence of 10 ng/ml TGFβ in (a) pellet culture and (b) alginate constructs. Each value represents the median ± inter-quartile range.

has used pre-optimised commercially available probes (Taqman®), so that the relative efficiencies of amplicon–primer binding between target and endogenous control genes were equal. Accordingly, the equation for calculating relative expression (Eqs (1), (2)) was appropriate. Non-parametric statistical testing was applied due to the heterogeneous nature of the MSC sub-populations [5,14,32].

A decrease in cell viability over the 12 day culture period was clearly evident in alginate constructs (Fig. 1). However, the spatial distribution suggested that there was no preferential loss across the alginate constructs. These findings suggest that the viability loss may not be attributed to inadequate mass transfer of key nutrients to cells at the centre of the construct, in contrast to a recent study using chondrocyte-seeded alginate constructs of identical dimensions [18].

In the first phase of the work, chondrogenic differentiation was initiated in both pellet and alginate cultures. The pellet culture system, with induction through TGF-β and dexamethasone supplementation, has been previously described [21]. With the commencement of chondrogenesis, the expression of collagens type II and X and aggrecan all increased over the 10 day culture period (Figs 3–5). However, Sox-9

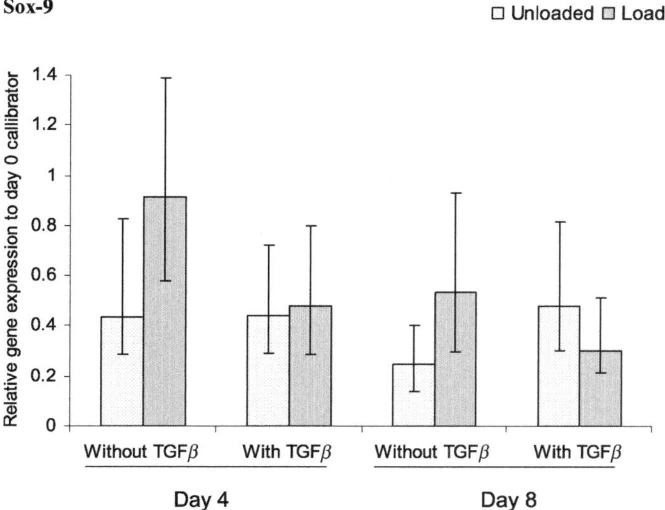



Fig. 6. The relative gene expression of Sox-9 by mesenchymal stem cells seeded in alginate constructs subjected to 15% intermittent dynamic compression at 1 Hz for 4 and 8 days, in medium supplemented with 0 or 10 ng/ml of TGF$\beta$. Each value represents the median $\pm$ inter-quartile range.

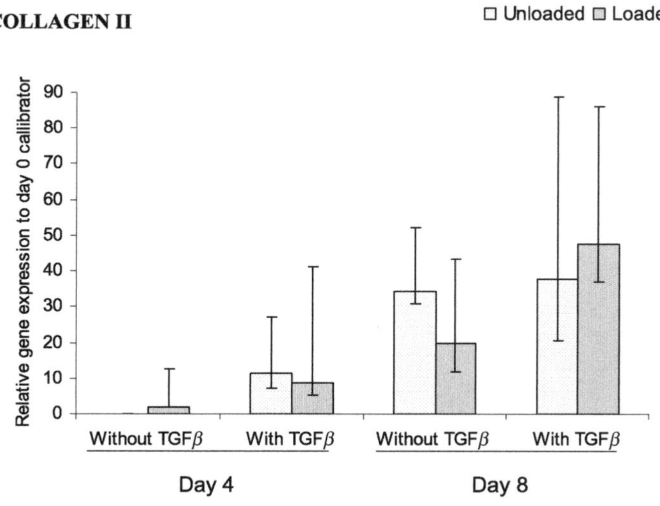



Fig. 7. The relative gene expression of collagen type II by mesenchymal stem cells seeded in alginate constructs subjected to 15% intermittent dynamic compression at 1 Hz for 4 and 8 days, in medium supplemented with 0 or 10 ng/ml of TGF$\beta$. Each value represents the median $\pm$ inter-quartile range.

expression exhibited a rise to peak expression levels by day 8 followed by a decline (Fig. 2). Similar patterns of expression were evident between cell pellet and cell/alginate cultures.

The occurrence and accumulation of collagen type X would indicate differentiation towards the chondrocyte hypertrophy, matrix calcification and endochondral ossification [34]. However, a growing number of studies report the expression of this gene at the early stages of chondrogenesis, particularly under high cell density conditions [9,39]. This has been confirmed by the present study, where expression was reported throughout the early stages of chondrogenesis in both pellet and cell/alginate cultures (Fig. 4).

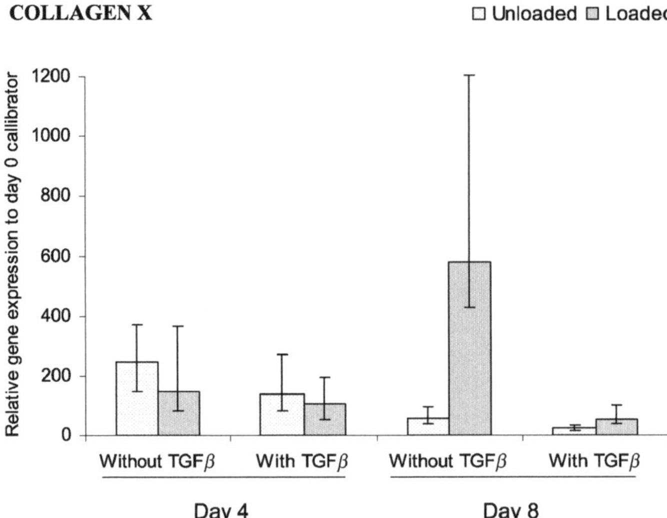


Fig. 8. The relative gene expression of collagen type X by mesenchymal stem cells seeded in alginate constructs subjected to 15% intermittent dynamic compression at 1 Hz for 4 and 8 days, in medium supplemented with 0 or 10 ng/ml of TGFβ. Each value represents the median ± inter-quartile range.

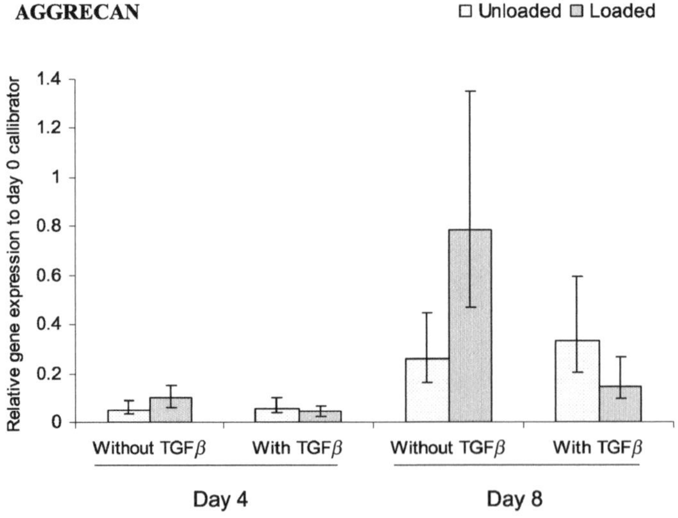


Fig. 9. The relative gene expression of aggrecan by mesenchymal stem cells seeded in alginate constructs subjected to 15% intermittent dynamic compression at 1 Hz for 4 and 8 days, in medium supplemented with 0 or 10 ng/ml of TGFβ. Each value represents the median ± inter-quartile range.

This significant expression of collagen type X (Fig. 4) may have been due to the high cell densities used ($10^7$ cells/ml), as previously demonstrated [9]. Previous studies have reported phenotypic modulation caused by an enforced spherical morphology within hydrogels. For example, collagen type II and aggrecan gene expression was less evident in marrow stromal cells cultured in collagen gels where cells could assume a more stellate morphology, as opposed to alginate and agarose hydrogels [15]. In contrast to this, the rounding of MSCs by numerous scaffold materials including alginate cultures was reported to have no effect on the chondrogenic differentiation [3].

Table 2

Trends in gene regulation with the addition of dynamic compression in the presence and absence of transforming growth factor. Day (4) *Day (8)*

| Gene of interest | Intermittent dynamic compression (1 Hz) | | | |
| --- | --- | --- | --- | --- |
| | Absence of TGF-$\beta$ | | Presence of TGF-$\beta$ | |
| | Day 4 | *Day 8* | Day 4 | *Day 8* |
| Collagen II | Up | *Down* | Down | *Up* |
| Collagen X | Down | *Up*$^{*}$ | Down | *Up*$^{*}$ |
| Sox-9 | Up | *Up* | Up | *Down* |
| Aggrecan | Up | *Up* | Down | *Down*$^{*}$ |

$^{*}$Indicates statistical significance $p \leqslant 0.05$.

It is clear that temporal gene expression for aggrecan and Sox-9 follow a more complex profile than the continuous up-regulation, demonstrated by both collagen II and X (Figs 2–5). The transient expression of Sox-9 with the onset of chondrogenesis has been shown by other studies where peak expression occurred before large increases in collagen type II expression in embryonic stem cells [26,36]. Interestingly, the transient expression of Sox-9 was more pronounced in micromass (peak day 6) than monolayer cultures which demonstrated a more gradual peak (day 7–9) [26]. Moreover, peaks in expression occurred earlier, at day 3, in collagen gel-encapsulated cells [36]. These experiments indicate that Sox-9 expression may be sensitive to factors, such as cell density fluctuation, the extracellular matrix environment and gel encapsulation causing modifications in cell morphology and nutrient delivery.

The transient nature of gene expression must be considered in association with the duration of its action. It is clear from experimental evidence that a reduction in gene expression would be detectable during periods of reduced gene transcription since mRNA degrades spontaneously [17]. Aggrecan and collagen type II mRNA have been shown to have half-lives ranging from 8.5 to 9.7 hours and 7.4 hours, respectively [23,28]. In contrast, Sox-9 mRNA has been shown to have a much shorter half-life, ranging from 1.2 to 2.5 hours [23,33]. Thus, it is evident that halting the transcriptional mechanism may cause reduction to basal values within the sampling frame adopted in the current study. Sox-9 would be particularly sensitive to this process, which may account for the peak values evident at 8 days of chondrogenesis.

Sox-9 expression has been linked to the regulation of other chondrogenic genes investigated in this study. The relationship between Sox-9 and the regulation of collagen type II has been previously investigated [29]. Sox-9 is required to prevent conversion to hypertrophic chondrocytes [1]. This finding is consistent with the present data demonstrating Sox-9 down regulation by day 10 concurrent with the continued up-regulation in collagen type X. Sox-9 has also been shown to up-regulate aggrecan expression, through the generation of an aggrecan promoter sequence [33], potentially explaining the observed pattern for aggrecan expression (Fig. 5).

As discussed previously, Sox-9 and aggrecan, but not collagen type II, were elevated in both pellet and alginate control samples in the absence of TGF$\beta$. Although few studies run adjacent TGF$\beta$-free control specimens, chondrogenesis by expression of aggrecan and collagen type II was reported to occur in alginate-embedded adult bone marrow stromal cells with supplementation of 10% FCS media alone [15]. Aggrecan gene expression has been shown to occur in Poly-L-lactic acid/alginate cell scaffold without the addition of TGF-$\beta$1 [10]. Detailed analyses of Sox-9 function reveal that the regulation of collagen type II gene expression occurs in conjunction with other agents such as perlecan [16] and PPAR-$\gamma$2 [22].

It is possible, therefore, that induction of these contributing factors was absent from specimens cultured without TGF$\beta$.

The intermittent dynamic compression profile adopted, of 1.5 hr loading/4.5 hr unloading, on alginate cultures did not produce measurable lift-off (data not show), a problem identified by the authors associated with continuous dynamic compression [13]. The sensitivity of chondrogenic gene expression to dynamic mechanical stimulation has been recently reported using mesenchymal stem cells [19]. MSCs seeded within agarose hydrogels and exposed to dynamic uniaxial compression for five days exhibited collagen type II and aggrecan expression levels significantly greater than in control samples and comparable to those attained with TGF$\beta_1$ treatment and a combination of stimuli (TGF-$\beta$ + compression) [19]. These results vary from the present findings, particularly with respect to the effects of TGF-$\beta$. The present work used real-time PCR, which measures the abundance of mRNA transcripts more accurately than standard PCR methods incorporating gel electrophoresis, as used in the previous study [19].

Collagen type X has been shown previously to be susceptible to mechanical regulation. Dynamic intermittent cyclic tension and hydrostatic pressure over a 3 day period up-regulates its expression in primary chondrocytes seeded in alginate scaffolds compared to unloaded controls [37]. A similar up-regulation was seen with intermittent dynamic tensile regimes applied to chondrocyte-seeded collagen sponges over an 8 day period [38]. It is thus apparent from these and the present findings that collagen X may be up-regulated with mechanical stimulation, irrespective of the mode in which it is delivered.

The highest expression of collagen type II and Sox-9 occurred in the TGF-$\beta_3$ treated specimens without the application of dynamic strain. This finding might be predicted, since Sox-9 is a transcription factor for collagen type II [4], and would be transiently up-regulated prior to the accelerated expression of the collagen type II gene. It is relevant to note that Sox-9 was expressed in the day 4 un-stimulated control samples, although collagen type II was absent. By day 8, Sox-9 expression was reduced while collagen type II expression had become evident. Although the trends for the combined effects of TGF-$\beta$ and dynamic compression were similar for both culture times, the precise nature of the response was not predicted, given past studies [19]. Thus with the exception of collagen type II expression (Fig. 7), the expression levels of Sox-9, collagen X and aggrecan were all reduced with the provision of TGF-$\beta$3 in addition to dynamic compression when compared to the action of the latter alone.

It is possible that non-significant fluctuations in Sox-9 expression with mechanical loading may have been due to the presence of transient peaks occurring outside the time points chosen for this experiment. For example, the elevated expression for dynamic compression alone at day 4 may have reflected the extremity of a rise in expression that peaked at day 6 and reduced thereafter. Temporal studies with greater resolution may have been able to contribute a more comprehensive analysis of expression patterns, although the half-life values reported for the genes of interest (1.2–9.7 hours) may be too short to contribute consistent results. Indeed, large fluctuations have been a reported problem in temporal gene expression studies using relative gene expression in a mesenchymal progenitor cell line [35]. For these reasons it may be more appropriate to consider a shortened experimental time frame, with gene expression analysis occurring over a 24–48 hour period following chondrogenic induction. This approach has proved valuable for aggrecan expression analysis, where transient up-regulation was shown over a 30 hour period [11] and in the developing chick limb-bud where Sox-9 was upregulated by 1 hour, and collagen type II by 10 hours following chondrogenic induction by TGF-$\beta_1$ implantation [12].

In conclusion this study has demonstrated a similar profile of gene expression between pellet and alginate cultures associated with the induction of chondrogenesis. In addition, both TGF$\beta$ and dynamic compression influence the expression of chondrogenic genes, although the interactions between the two stimuli are complex in nature and requires further investigation.

## Acknowledgement

This work was supported by the Engineering and Physical Sciences Research Council, UK.

## References


[1] H. Akiyama, M.C. Chaboissier, J.F. Martin, A. Schedl and B. de Crombrugghe, The transcription factor Sox9 has essential roles in successive steps of the chondrocyte differentiation pathway and is required for expression of Sox5 and Sox6, *Genes and Development* **16** (2002), 2813–2828.

[2] P. Angele, J.U. Yoo, C. Smith, J. Mansour, K.J. Jepsen, M. Nerlich and B. Johnstone, Cyclic hydrostatic pressure enhances the chondrogenic phentotype of human mesenchymal progenitor cells differentiated in vitro, *J. Orthop. Res.* **21** (2003), 451–457.

[3] H.A. Awad, M.Q. Wickham, H.A. Leddy, J.M. Gimble and F. Guilak, Chondrogenic differentiation of adipose-derived adult stem cells in agarose, alginate and gelatin scaffolds, *Biomaterials* **25** (2004), 3211–3222.

[4] D.M. Bell, K.K. Leung, S.C. Wheatley, L.J. Ng, S. Zhou, K.W. Ling, M.H. Sham, P. Koopman, P.P. Tam and K.S. Cheah, SOX9 directly regulates the type-II collagen gene, *Nature Genetics* **16** (1997), 174–178.

[5] D. Benayahu and J. Sela, Histomorphometric analysis of heterotopic bone formed by stromal- osteogenic subpopulations, *Calcified Tissue International* **59** (1996), 254–258.

[6] G. Bentley, L.C. Biant, R.W. Carrington, M. Akmal, A. Goldberg, A.M. Willaims, J.A. Skinner and J. Pringle, A prospective, randomised comparison of autologous chondrocyte implantation versus mosaicplasty for osteochondral defects of the knee, *J. Bone Jt. Surg.* **85-A** (2003), 223–230.

[7] M. Brittberg, A. Lindahl, A. Nilsson, C. Ohlsson, O. Isakssonand and L. Peterson, Treatment of deep cartilage defects in the knee with autologous chondrocyte transplantation, *N. Eng. J. Med.* **331** (1994), 889–895.

[8] A.I. Caplan, Mesenchymal stem cells, *J. Orthop. Res.* **9** (1991), 641–650.

[9] A.L. Carlberg, B. Pucci, R. Rallapalli, R.S. Tuan and D.J. Hall, Efficient chondrogenic differentiation of mesenchymal cells in micromass culture by retroviral gene transfer of BMP-2, *Differentiation* **67** (2005), 128–138.

[10] E.J. Caterson, L.J. Nesti, W.J. Li, K.G. Danielson, T.J. Albert, A.R. Vaccaro and R.S. Tuan, Three-dimensional cartilage formation by bone marrow-derived cells seeded in polylactide/alginate amalgam, *Journal of Biomedical Materials Research* **57** (2001), 394–403.

[11] J. Chen, W. Yan and L.A. Setton, Static compression induces zonal-specific changes in gene expression for extracellular matrix and cytoskeletal proteins in intervertebral disc cells in vitro, *Matrix Biology* **22** (2004), 573–583.

[12] J. Chimal-Monroy, J. Rodriguez-Leon, J.A. Montero, Y. Ganan, D. Macias, R. Merino and J.M. Hurle, Analysis of the molecular cascade responsible for mesodermal limb chondrogenesis: Sox genes and BMP signalling, *Developmental Biology* **257** (2003), 292–301.

[13] T.T. Chowdhury, D.L. Bader, J.C. Shelton and D.A. Lee, Temporal regulation of chondrocyte metabolism in agarose constructs subjected to dynamic compression, *Archives Biochem. Biophys.* **417** (2003), 105–111.

[14] R.J. Deans and A.B. Moseley, Mesenchymal stem cells: Biology and potential clinical uses, *Experimental Hematology* **28** (2000), 875–884.

[15] D.R. Diduch, L.C.M Jordan, C.M. Mierisch and G. Balian, Marrow stromal cells embedded in alginate for repair of osteochondral defects, *Arthroscopy: The Journal of Arthroscopic & Related Surgery* **16** (2000), 571–577.

[16] M.M. French, S.E. Smith, K. Akanbi, T. Sanford, J. Hecht, M.C. Farach-Carson and D.D. Carson, Expression of the heparan sulfate proteoglycan, perlecan, during mouse embryogenesis and perlecan chondrogenic activity *in vitro*, *Journal of Cell Biology* **145** (1999), 1103–1115.

[17] M. Grunberg-Manago, Messenger RNA stability and its role in control of gene expression in bacteria and phages, *Annual Review of Genetics* **33** (1999), 193–227.

[18] H.K. Heywood, D.A. Lee and D.L. Bader, Cellular utilisation determines viability and matrix distribution profiles in chondrocyte-seeded agarose constructs, *Tissue Engineering* **10** (2004), 1467–1479.

[19] C.Y. Huang, K.L. Hagar, L.E. Frost, Y. Sun and H.S. Cheung, Effects of cyclic compressive loading on chondrogenesis of rabbit bone-marrow derived mesenchymal stem cells, *Stem Cells* **22** (2004), 313–323.

[20] M. Jagodzinski, M. Drescher, J. Zeichen, S. Hankemeier, C. Krettek, U. Bosch and M. van Griensven, Effects of cyclic longitudinal mechancal strain and dexamethasone on osteogenic differentiation of human bone marrow stromal cells, *Eur. Cells and Materials* **7** (2004), 35–41.

[21] B. Johnstone, T.M. Hering, A.I. Caplan, V.M. Goldberg and J.U. Yoo, In vitro chondrogenesis of bone marrow-derived mesenchymal progenitor cells, *Exp. Cell Res.* **238** (1998), 265–272.

[22] Y. Kawakami, M. Tsuda, S. Takahashi, N. Taniguchi, C.R. Esteban, M. Zemmyo, T. Furumatsu, M. Lotz, J.C. Belmonte and H. Asahara, Transcriptional coactivator PGC-1alpha regulates chondrogenesis via association with Sox9, *Proceedings of the National Academy of Sciences of the United States of America* **102** (2005), 2414–2419.

[23] M.D. Kinkel and W.E. Horton, Coordinate down-regulation of cartilage matrix gene expression in Bcl-2 deficient chondrocytes is associated with decreased SOX9 expression and decreased mRNA stability, *J. Cell Biochem.* **88** (2003), 941–953.

[24] R. Lane Smith, S.F. Rusk, B.E. Ellison, P. Wessells, K. Tsuchiya, D.R. Carter, W.E. Caler, L.J. Sandell and D.J. Schurman, In vitro stimulation of articular chondrocyte mRNA and extracellular matrix synthesis by hydrostatic pressure, *J. Orthop. Res.* **14** (1996), 53–60.

[25] D.A. Lee and D.L. Bader, Compressive strains at physiological frequencies influence the metabolism of chondrocytes seeded in agarose, *J. Orthop. Res.* **15** (1997), 181–188.

[26] C.L. Lengner, C. Lepper, A.J. van Wijnen, J.L. Stein, G.S. Stein and J.B. Lian, Primary mouse embryonic fibroblasts: A model of mesenchymal cartilage formation, *J. Cell Physiol.* **200** (2004), 327–333.

[27] Y.J. Li, N.N. Batra, L. You, S.C. Meier, I.A. Coe, C.E. Yellowley and C.R. Jacobs, Oscillatory fluid flow affects human marrow stromal cell proliferation and differentiation, *Journal of Orthopaedic Research* **22** (2004), 1283–1289.

[28] D.J. McQuillan, C.J. Handley and H.C. Robinson, Control of proteoglycan biosynthesis. Further studies on the effect of serum on cultured bovine articular cartilage, *Biochem. J.* **237** (1986), 741–747.

[29] L.J. Ng, S. Wheatley, G.E. Muscat, J. Conway-Campbell, J. Bowles, E. Wright, D.M. Bell, P.P. Tam, K.S. Cheah and P. Koopman, SOX9 binds DNA, activates transcription, and coexpresses with type II collagen during chondrogenesis in the mouse, *Developmental Biology* **183** (1997), 108–121.

[30] M.W. Pfaffl, Quantification strategies in Real-Time PCR, in: *A–Z of Quantitative PCR*, S.A. Bustin, ed., La Jolla, 2004, pp. 87–120.

[31] M.F. Pittenger, A.M. Mackay, S.C. Beck, R.K. Jaiswal, R. Douglas, J.D. Mosca, M.A. Moorman, D.W. Simonetti, S. Craig and D.R. Marshak, Multilineage potential of adult human mesenchymal stem cells, *Science* **284** (1999), 143–147.

[32] N.P. Rhodes, J.K. Srivastava, R.F. Smith and C. Longinotti, Heterogeneity in proliferative potential of ovine mesenchymal stem cell colonies, *J. Mat. Sci. Mole. Med.* **15** (2004), 397–402.

[33] I. Sekiya, D.C. Colter and D.J. Prockop, BMP-6 enhances chondrogenesis in a subpopulation of human marrow stromal cells, *Biochem. Biophys. Res. Commun.* **284** (2001), 411–418.

[34] G. Shen, The role of type X collagen in facilitating and regulating endochondral ossification of articular cartilage, *Orthodontics and Craniofacial Research* **8** (2005), 11–17.

[35] A. Steinart, M. Weber, A. Dimmler, C. Julius, N. Schutze, U. Noth, H. Cramer, J. Eulert, U. Zimmermann and C. Hendrich, Chondrogenic differentiation of mesnchymal progenitor cells encapsulated in ultrahigh-viscosity alginate, *Journal of Orthopaedic Research* **21** (2003), 1090–1097.

[36] I. Takahashi, G.H. Nuckolls, K. Takahashi, O. Tanaka, I. Semba, R. Dashner, L. Shum and H.C. Slavkin, Compressive force promotes sox9, type II collagen and aggrecan and inhibits IL-1beta expression resulting in chondrogenesis in mouse embryonic limb bud mesenchymal cells, *Journal of Cell Science* **111** (1998), 2067–2076.

[37] M. Wong, M. Siegrist and K. Goodwin, Cyclic tensile strain and cyclic hydrostatic pressure differentially regulate expression of hypertrophic markers in primary chondrocytes, *Bone* **33** (2003), 685–693.

[38] Q.Q. Wu and Q. Chen, Mechanoregulation of chondrocyte proliferation, maturation, and hypertrophy: ion-channel dependent transduction of matrix deformation signals, *Exp. Cell. Res.* **256** (2000), 383–391.

[39] J.U. Yoo and B. Johnstone, The role of osteochondral progenitor cells in fracture repair, *Clin. Orthop.* **355** (1998), S73–S81.

Biorheology 43 (2006) 471–480
IOS Press



# Human chondrocytes and mesenchymal stem cells grown onto engineered scaffold

Andrea Facchini [a,b], Gina Lisignoli [a], Sandra Cristino [a], Livia Roseti [a],
Luciana De Franceschi [a], Emanuele Marconi [a] and Brunella Grigolo [a,*]

[a] *Laboratorio di Immunologia e Genetica, Istituto di Ricerca Codivilla Putti, Istituti Ortopedici Rizzoli,
Via di Barbiano 1/10, 40136 Bologna, Italy*
[b] *Dipartimento di Medicina Interna e Gastroenterologia, Università di Bologna, Via Massarenti 9,
40138 Bologna, Italy*

**Abstract.** The development of improved methods for treatment of chondral defects using autologous cells in combination with biomaterials leads to a new generation of implantable devices. Their association gives rise to a hybrid construct combining biological and material components that can be specifically committed. The comprehension of cellular and molecular mechanisms of cartilage repair and the use of biomaterials in combination with chondrocytes or mesenchymal stem cells in the treatment of cartilage defects has opened a new era of therapeutical strategies. Recently, their applicability in the treatment of early lesions in osteoarthritis is under investigation. To obtain new information on the behaviour of chondrocytes and mesenchymal stem cells grown on a hyaluronan derivative scaffold (Hyaff®-11) already used in cartilage repair, we analysed a series of molecules expressed by these cells by Real-Time RT-PCR and immunohistochemical analyses. The data obtained with this work showed that this biomaterial is able to reduce the expression of some catabolic molecules by human chondrocytes and provide a good environment to support the differentiation of mesenchymal stem cells in chondrogenic sense. These observations confirm Hyaff®-11 as a suitable scaffold both for chondrocytes and mesenchymal stem cells for the treatment of articular cartilage defects.

Keywords: Human chondrocytes, mesenchymal stem cells, hyaluronan-based scaffold, chondrogenic differentiation

## 1. Introduction

An important problem in orthopaedics is represented by cartilage tissue lesions as fractures or defects created after injury or by cartilage lesions that occur during osteoarthritic processes [4,12,17,22]. Cartilage repair require a multistep controlled activation and silencing of genes producing extracellular matrix, growth factors, adhesion molecules involved in tissue formation [25].

The recent advances in biology and materials have pushed tissue engineering to the forefront of new treatments also for cartilage repair [18,32]. This technique combines isolated cells with scaffold/cell carriers in order to improve autologous cell transplantation procedure for cartilage formation and repair [8]. A vast amount of research has been focused on the design of new scaffolds for the success of such therapeutical strategy [7,33]. Moreover, the use of mature chondrocytes isolated from articular cartilage or mesenchymal stem cells (MSCs) isolated from bone marrow has been evaluated by different authors [10,16,19,28].

*Address for correspondence: Brunella Grigolo, PhD, Laboratorio di Immunologia e Genetica, Istituto di Ricerca Codivilla Putti, I.O.R., Via di Barbiano 1/10, 40136 Bologna, Italy. Tel.: +39 051 6366803/6366802; Fax: +39 051 6366807; E-mail: grigolob@alma.unibo.it.

*In vitro* and *in vivo* data indicate that the differentiation of isolated chondrocytes grown onto a scaffold allows the progressive increase for known constituents of cartilage matrix and the repair of the damaged tissue [9,10]. MSCs seem to be a good alternative to chondrocytes for cartilage regeneration [31]. Chondrogenic potential of MSCs to produce cartilage-like tissues has been widely demonstrated in animal and human models [2,11,20,29]. The MSCs population contains chondroprogenitors capable of *in vitro* proliferation and differentiation in the presence of specific factors (e.g. TGF$\beta$ and BMP-7) that significantly increase the frequency of tissue formation [14,23]. MSCs commitment to chondrogenesis happens when the proliferation and phenotypic expression of cells are affected and different genes involved in these processes are modulated [5]. Hyaluronic-acid (HA) based biodegradable polymers have been shown to provide successful cell scaffolds for tissue-engineered repair of cartilage [10,26]. HA, a naturally-occurring molecule present in all soft tissues, plays an essential role in the maintenance of the normal extracellular matrix structure and it has been shown to provide good results when used *in vivo* animal models, since it favoured the formation of new cartilage tissue [10].

Aim of our study was to verify if a hyaluronan derivative could be not only a suitable scaffold for chondrocytes allowing the re-expression of differentiated phenotype [9], but also able to reduce the expression of some catabolic factors involved in cartilage degeneration as observed in osteoarthritis [13].

Moreover, we wanted also to verify if such a biomaterial could provide a good environment to support the differentiation of MSCs in chondrogenic sense.

A better understanding of the repair mechanisms induced by the cultured chondrocytes and the regulatory mechanisms controlling tissue degeneration combined with identification and culture of stem cells with chondrogenic potential will be the key to new cartilage treatments.

## 2. Materials and methods

### 2.1. Test material

The scaffold used in this study was made of Hyaff®-11, a polymer derived from the total esterification of sodium hyaluronate (80–200 kDa) with the benzyl alcohol on the free carboxyl groups of glucuronic acid along the polymeric chain. The configuration used was a non-woven mesh that is a pad composed of a random array of polymer fibers having a diameter of 40 $\mu$m and kindly provided by F.A.B. S.r.l. (FIDIA Advanced Biopolymers, Abano Terme, Italy).

### 2.2. Human chondrocytes isolation and seeding on the biomaterial

Human articular cartilage specimens were obtained from the knees of 4 patients (mean age 22 years) undergoing joint replacement surgery. Fragments of the excised cartilaginous tissues were put in Dulbecco's Modified Eagle's Medium (DMEM) (Gibco BRL, Grand Island, NY, USA) and chondrocytes were isolated by sequential enzymatic digestions: 30 min with 0.1% hyaluronidase (Sigma, St. Louis, MO, USA), 1 hour with 0.5% pronase (Sigma) and 1 hour with 0.2% collagenase (Sigma) at 37°C in DMEM with 25 mM HEPES (Sigma), 100 units/ml penicillin (Biological Industries, Kibbutz, Israel), 100 $\mu$g/ml streptomycin (Biological Industries), 50 $\mu$g/ml gentamicin (Flow Laboratoires, Biaggio, Switzerland), 2.5 $\mu$g/ml amphotericin B (Biological Industries). The isolated chondrocytes were filtered by 100 $\mu$m and 70 $\mu$m nylon meshes, washed, and centrifuged. The cell number and viability were assessed by the Tripan Blue dye exclusion method and the cells were cultured under conventional monolayer culture conditions. Once sufficient cells were available, usually after the 3rd to 4th passage, chondrocytes were seeded onto Hyaff®-11 non-woven meshes at a density of $1 \times 10^6$ cells/cm² in 35 mm Petri dishes. The time points evaluated were: 1 h, 24 hrs, 7 days, and 14 and 21 days for molecular biology analyses and 1 day and 30 days for immunohistochemical evaluations.

## 2.3. Human MSCs isolation

Human bone marrow-derived mesenchymal stromal cells were obtained from 5 ml of marrow aspirate during hip surgery of 5 post-traumatic patient donors (mean age 49 years). Bone marrow was collected in 15 ml tubes containing DMEM (low glucose) (Gibco) supplemented with 15% FBS (Sigma), gentamicin sulphate (Biological Industries), 50 mg/ml and antibiotics (pen-streptomycin, 100 U/ml–100 $\mu$g/ml) (Gibco) plus heparin (50 U/ml) (Sigma). Briefly, as previously described [15], MSCs were isolated using Ficoll Hypaque (Pharmacia, Biotech, Uppsala, Sweden) density gradient ($d = 1.077$ gr/ml). Cells were washed twice, resuspended in DMEM low glucose with 15% FBS, counted and plated at a concentration of $2 \times 10^6$ cells/T150 flask. After 1 week non-adherent cells were removed and the adherent MSCs expanded *in vitro*. The expression of haematopoietic and non-haematopoietic markers was analysed using monoclonal antibodies anti CD34, CD14, CD45, CD105, CD44 to test the purity of the isolated cells.

After the expansion period, MSCs were seeded on non-woven Hyaff®-11 ($2.5 \times 10^5$ cells/0.5 cm³) and incubated in 24-well non-adherent cell culture Petri for 4 hours at 37°C, and then analyzed under the inverted microscope to check cells adhesion to the polymer surface. 3 ml of chondrogenic medium (DMEM high glucose supplemented with ITS (Biosciences, Bedford, MA, USA) + Premix: 6.25 $\mu$g/ml insulin (Sigma), 6.25 $\mu$g/ml transferrin (Sigma), 5.33 $\mu$g/ml linoleic acid (Sigma), and 1.25 mg/ml bovine serum albumin (Sigma), $10^{-7}$ dexamethasone (Sigma), 37.5 $\mu$g/ml ascorbate-2 phosphate (Wako, Collingwood, Australia), 1 mM sodium pyruvate (Sigma), and pen-streptomycin 100 U/ml–100 $\mu$g/ml (Gibco) was then added to the samples. 24 hours later (Day 0), the medium was substituted with complete chondrogenic medium with or without TGF$\beta$-1 (R&D Systems, Minneapolis, MN, USA) at the concentration of 20 ng/ml. Cell culture medium was changed twice a week. The molecular biology analyses were performed at 0, 7, 14, 21 and 28 days; the immunohistological analyses were performed on day 14 and 28.

## 2.4. Real Time PCR analyses

Hyaff®-11 scaffolds seeded with the cells (chondrocytes or MSCs) were collected at the different experimental time points, placed in microcon 100 filtration devices and centrifuged at $1500 \times g$ for 5 min in order to remove the liquid medium. Cells were lysed directly in the culture scaffold by the addition of 0.5 ml of RNAWIZ™ (Ambion, Austin, TX, USA) reagent and total RNA was subsequently isolated using the single-step guanidinium thiocyanate–phenol–chloroform method. Complementary DNA was synthesized from 1 $\mu$g of total RNA per sample with 45 min incubation at 42°C, using Moloney murine leukemia virus reverse transcriptase (Perkin Elmer, Norwalk, CT, USA) and oligo-(dT) priming. PCR primers for the housekeeping gene glyceraldehyde-3-phosphate dehydrogenase (GAPDH) used as an internal control were obtained from published references [1]. PCR primers for: collagen type I, collagen type II, aggrecan I, MMP-1 and MMP-13, were designed using LightCycler Probe design Software (Roche Molecular Biochemicals, Mannheim, Germany) and reported in Table 1. Real-time PCR was run

Table 1

RT-PCR primers description

| RNA template | Primer sequences | Amplicon size (base pairs) | Annealing temperature ($^\circ$C) |
|---|---|---|---|
| Aggrecan | 5′-TCG AGG ACA GCG AGG CC<br>3′-TCG AGG GTG TAG CGT GTA GAG A | 85 | 60 |
| Type I collagen | 5′-AGG TGC TGA TGG CTC TCC T<br>3′-GGA CCA CTT TCA CCC TTG T | 105 | 58 |
| Type II collagen | 5′-GAC AAT CTG GCT CCC AAC<br>3′-ACA GTC TTG CCC CAC TTA C | 257 | 58 |
| GAPDH | 5′-TGG TAT CGT GGA AGG ACT CAT GAC<br>3′-ATG CCA GTG AGC TTC CCG TTC AGC | 190 | 60 |
| MMP-1 | 5′-TGG ACC TGG AGG AAA TCT TG<br>3′-CCG CAA CAC GAT GTA AGT TG | 125 | 56 |
| MMP-13 | 5′-TCA CGA TGG CAT TGC T<br>3′-GCC GGT GTA GGT GTA GA | 277 | 56 |

in a LightCycler Instrument (Roche) using the QuantiTech™ SYBR® Green PCR Kit (Qiagen, GmbH, Germany) with the following protocol: initial activation of HotStarTaq™ DNA Polymerase at 94°C for 15 min, 45 cycles of 94°C for 15 s, 56–60°C for 20 s and 72°C for 10 s. The increase in PCR product was monitored for each amplification cycle by measuring the increase in fluorescence caused by the binding of SYBR Green I dye to dsDNA. For each sample at the different experimental times the threshold cycle ($C_T$) values (i.e. the cycle number at which the detected fluorescence exceeds background levels) were determined. Specificity of the amplicons was confirmed by melting curve analysis and agarose gel electrophoresis.

For each target gene mRNA levels were normalized to GAPDH and expressed as relative gene expression.

## 2.5. Immunohistochemistry

Immunohistochemical staining for collagen type I and type II was performed in engineered tissues. Specimens were embedded in OCT, snap-frozen in liquid nitrogen ad stored at −80°C. Engineered tissues were sectioned into 5 $\mu$m cryostat sections, air dried and stored at −80°C. Sections were transferred at room temperature, air dried for 15 minutes and fixed in acetone at 4°C for 10 minutes. Air dried fixed samples were rehydrated and incubated at room temperature for 30 minutes with monoclonal antibodies (MoAb) anti-human collagen type I, anti-human collagen type II (Chemicon International, Temecula, CA, USA), that were diluted both 1 : 10, in 0.04 M pH 7.6 TBS containing 2% bovine serum albumin (BSA) (Sigma). An enzymatic pre-treatment with hyaluronidase 0.1% (Sigma) at 37°C for 5 minutes was performed only for the immunostaining of human collagen type II. Slides were then rinsed and incubated for 30 minutes at room temperature with secondary goat anti-mouse/rabbit antibody (Dako, Glostrup, Denmark), rinsed again and treated with new-fuchsin substrate detection Kit (Dako). Finally, samples were counterstained with hematoxylin. Negative control sections were obtained by omitting the primary antibody.

# 3. Results

## 3.1. Real-Time PCR analysis: Chondrocytes

Fold changes in gene expression calculated on the basis of $\Delta C_T$ values and evaluated at the different experimental times are shown in Fig. 1. Type collagen I expression increased up to day 7 and then

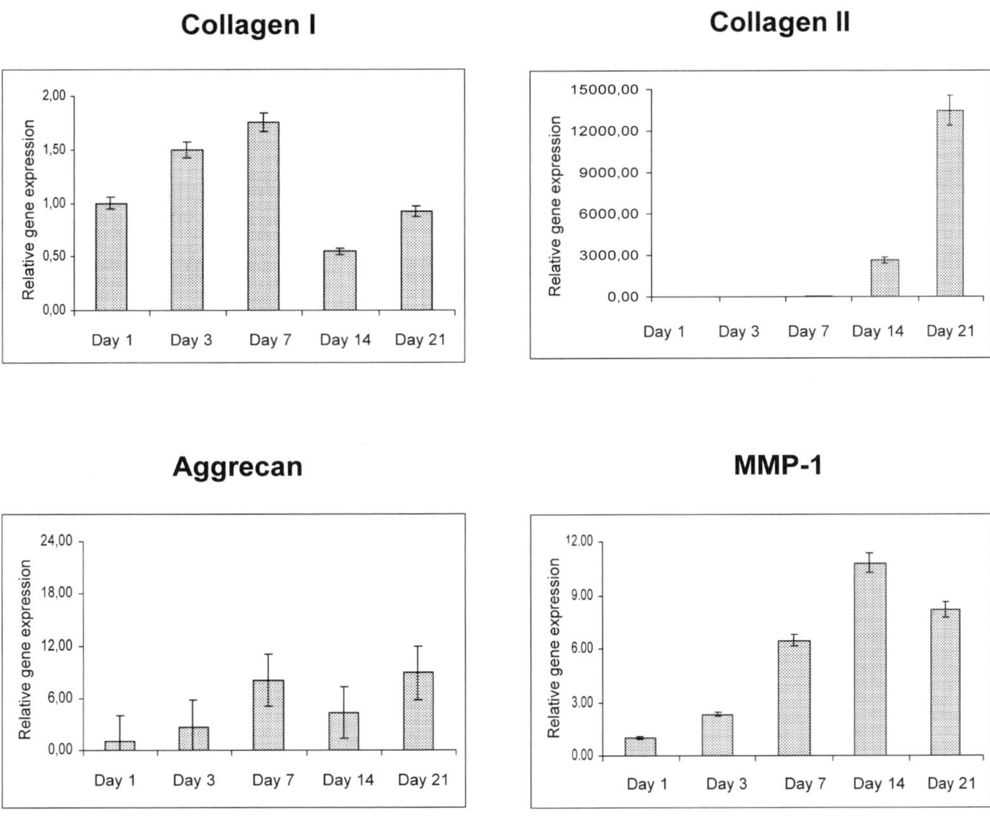

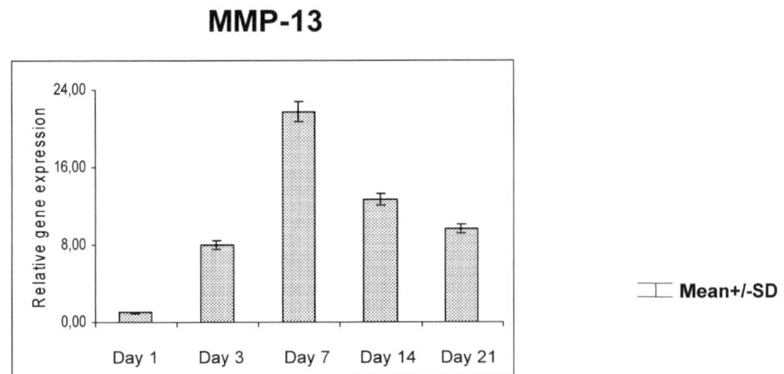

Fig. 1. Kinetics for collagen I, collagen II, aggrecan, MMP-1 and MMP-13 mRNAs expression in chondrocytes grown onto Hyaff®-11 scaffolds at different experimental times. For each target gene mRNA levels were normalized to GAPDH and expressed as relative gene expression. Data are expressed as mean ± SD.

decreased up to day 21; type II collagen mRNA was expressed at very low level until day 7 and increased only from day 14 and by day 21. Aggrecan expression did not show a regular trend, however its levels increased compared to day 1. MMP-1 and MMP-13 expressions increased during the first culture period (up to day 7), but progressively decreased up to day 21.

## 3.2. Real-Time PCR analysis: MSCs

As shown in Fig. 2, there is a biphasic variation in collagen type I gene expression level during time culture. Firstly, there is an approximately ten-fold increase in collagen type I mRNA expression from day 0 to day 7, then, expression starts to be downregulated until day 28. Conversely to collagen type I, collagen type II and aggrecan gene expression show a significant trend which increase during time culture. For both mRNAs, expression is gradually increasing from day 0 to day 21 and then very importantly in the last week of culture. Collagen type II was detectable from day 7 whereas aggrecan was already expressed on day 0.

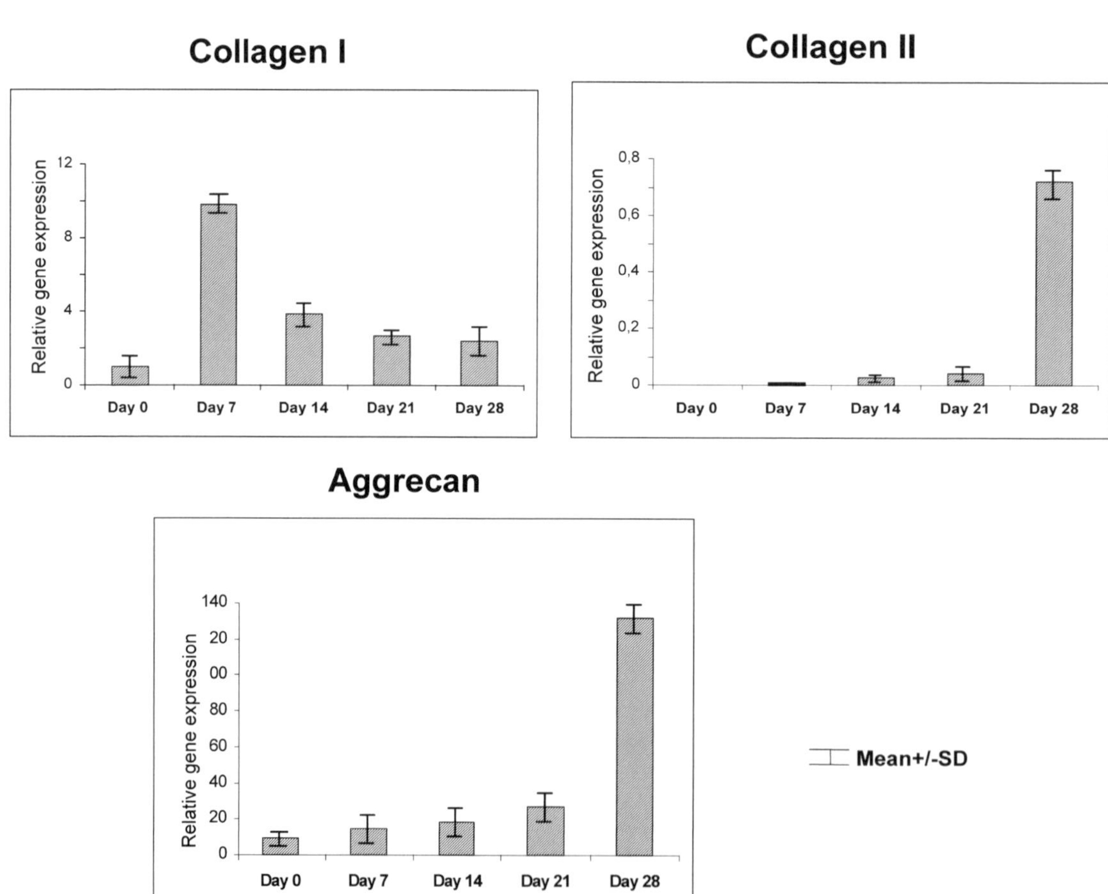


Fig. 2. Kinetics for type I collagen, type II collagen and aggrecan mRNAs expression in MSCs grown onto Hyaff®-11 scaffolds at different experimental times. For each target gene mRNA levels were normalized to GAPDH and expressed as relative gene expression. Data are expressed as mean ± SD.

### 3.3. Immuhistochemistry: Chondrocytes

Collagen type II, which is the main marker of chondrocyte phenotype was re-expressed by the cells that had been grown first in monolayer cultures and then seeded and grown onto the biomaterial. On the contrary, collagen type I, which is the marker of fibroblastic phenotype became completely undetectable at the end of the culture period (Fig. 3).

### 3.4. Immuhistochemistry: MSCs

Two different observers analysed the slides both outside and inside the scaffolds. Results were expressed as the score of positive cells. As shown in Table 2, collagen type I was positive at day 7, inside and outside the scaffold, while at day 28 this marker was negative on cells. Analysis of collagen type II showed that h-MSCs were negative at day 7 and became positive to collagen type II only starting from day 21, both outside and inside the scaffold.

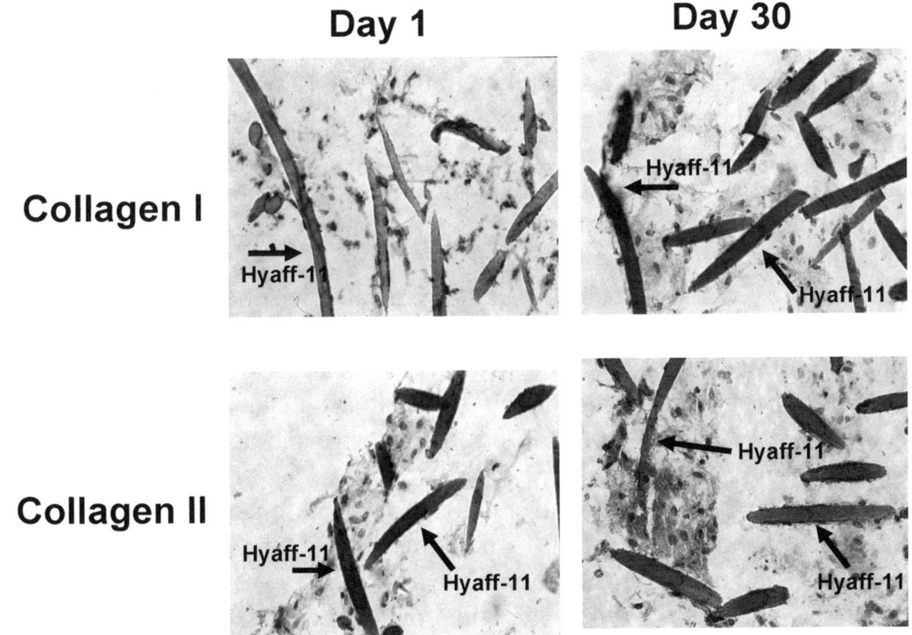


Fig. 3. Collagen I and collagen II immunostaining of chondrocytes grown on Hyaff®-11 at day 1 and 30 from the seeding. The images are from one representative patient. Collagen I and collagen II were developed using new fucsin (red color is positive stain). Hyaff®-11 fibers are indicated with arrows.

Table 2

Immunohistochemical analysis of human mesenchymal stem cells (HuMSCs) grown onto Hyaff®-11 scaffold

| Markers | Day 7 | Day 28 |
| --- | --- | --- |
| Collagen type I | ++ | Neg. |
| Collagen type II | Neg. | +++ |

Results were expressed as the score of positive cells.
Neg. = not expressed, + = low number of positive cells, ++ = intermediate number of positive cells, +++ = high number of positive cells.

## 4. Discussion

The repair of damaged cartilage by means of tissue engineering approach consists of the use of a suitable scaffold, which provides a biodegradable three-dimensional structure for tissue development, and its association with autologous cells [7,33]. The goal is to regenerate a cartilage with the structural and mechanical properties of the original one, result that can not be obtained with the traditional therapeutical approaches used before [3].

The methodology involves the *in vitro* cultivation of chondrocytes that can be isolated from healthy cartilage, expanded *in vitro* and grown onto the biomaterial or, recently, the use of MSCs that can be selected from bone marrow, expanded in culture and addressed to chondrogenic phenotype by the use of specific growth factors [2,11].

The use of autologous chondrocyte transplantation is limited to the repair of cartilage damaged after traumatic lesions or in patients with osteochondritis dissecans [27] but this approach is now addressing also to the treatment of degenerative lesions in osteoarthritic patients which are usually treated with or without drugs [24].

Hyaff-11 scaffold seeded with autologous differentiated chondrocytes is now utilized by surgeons for the reconstruction of cartilage defects in traumatic patients [26]. Clinical and histological findings showed an improvement of the patient condition and the formation of a new cartilage with many features of hyaline tissue [6].

The data obtained with this study indicate that human articular chondrocytes grown onto Hyaff-11 scaffold are able not only to re-differentiate expressing the characteristic molecules of their mature phenotype, but also to down-regulate the expression of some molecules involved in the catabolic pathways during early osteoarthritic processes, such as metalloproteinases which are able to degrade the extracellular matrix.

The combination of the hyaluronan scaffold with MSCs may be a valid alternative to mature cells. Compared to the use of chondrocytes which requires the harvest of a healthy cartilage sample, the isolation and growth of cells in culture, that of MSCs is less invasive and do not require an expansion phase *in vitro*.

Our data indicated that human MSCs grown onto the same hyaluronan scaffold used for chondrocytes, are able to differentiate in chondrogenic sense, and this is also favoured by the presence of TGF$\beta$ in a way similar to that occurring during *in vivo* chondrogenesis.

The use of hyaluronan with both chondrocytes and progenitor cells seems to be effective to regenerate articular cartilage lesions after traumatic events or in the early degenerative process.

This is not really a surprise, since hyaluronan is a major constituent of the extracellular matrix of articular hyaline cartilage, and is highly abundant in the mesenchyme of the developing embryo [21]. Hence, the biomaterial made of hyaluronan, allows the recapitulation of the events that occur during cartilage maturation providing not only structural but also biological support to reparative processes [30].

## Acknowledgements

We thank Patrizia Rappini and Graziella Salmi for their assistance in the preparation of the manuscript and Luciano Pizzi for his technical assistance. We thank Fidia Advanced Biopolymers for kindly providing the scaffolds.

# References


[1]  F.J. Blanco, Y. Geng and M. Lotz, Differentiation-dependent effects of IL-1 and TGF-$\beta$ on human articular chondrocyte proliferation are related to inducible nitric oxide synthase expression, *J. Immunol.* **154** (1995), 4018–4026.

[2]  D. Bosnakovski, M. Mizuno, G. Kim, T. Ishiguro, M. Okumura, T. Iwanaga, T. Kadosawa and T. Fujinaga, Chondrogenic differentiation of bovine bone marrow mesenchymal stem cells in pellet culture system, *Exp. Hematol.* **32** (2004), 502–509.

[3]  M. Brittberg, A. Lindahl and A. Nilsson, Treatment of deep cartilage defects in the knee with autologous chondrocyte transplantation, *N. Engl. J. Med.* **331** (1994), 889–895.

[4]  J.A. Buckwalter, L.C. Rosenberg and E.B. Hunziker, Articular cartilage: composition, structure, response to injury and methods of repair, in: *Articular Cartilage and Knee Joint Function: Basic Science and Arthroscopy.* J.W. Ewing, ed., Raven Press, New York, 1990, pp. 19–56.

[5]  J. Chimal-Monroy, J. Rodriguez-Leon, J.A. Montero, Y. Ganan, D. Macias, R. Merino and J.M. Hurle, Analysis of the molecular cascade responsible for mesodermal limb chondrogenesis: Sox genes and BMP signaling, *Dev. Biol.* **257** (2003), 292–301.

[6]  S.C. Dickinson, T.J. Sims, L. Pittarello, C. Soranzo, A. Pavesio and A.P. Hollander, Quantitative outcome measures of cartilage repair in patients treated by tissue engineering, *Tissue Eng.* **11** (2005), 277–287.

[7]  S.R. Frenkel and P.E. Di Cesare, Scaffolds for articular cartilage repair, *Ann. Biomed. Eng.* **32** (2004), 26–34.

[8]  L. Galois, A.M. Freyria, L. Grossin, P. Hubert, D. Mainard, D. Herbage, J.F. Stoltz, P. Netter, E. Dellacherie and E. Payan, Cartilage repair: surgical techniques and tissue engineering using polysaccharide- and collagen-based biomaterials, *Biorheology* **41** (2004), 433–443.

[9]  B. Grigolo, G. Lisignoli, A. Piacentini, M. Fiorini, P. Gobbi, G. Mazzotti, M. Duca, A. Pavesio and A. Facchini, Evidence for redifferentiation of human chondrocytes grown on a hyaluronan-based biomaterial (Hyaff$^{\circledR}$-11): Molecular, immunohistochemical and ultrastructural analysis, *Biomaterials* **23** (2002), 1187–1195.

[10]  B. Grigolo, L. Roseti, M. Fiorini, M. Fini, G. Giavaresi, N. Nicoli Aldini, R. Giardino and A. Facchini, Transplantation of chondrocytes seeded on a hyaluronan derivative (Hyaff$^{\circledR}$-11) into cartilage defects in rabbits, *Biomaterials* **22** (2001), 2417–2424.

[11]  B.C. Heng, T. Cao and E.H. Lee, Directing stem cell differentiation into the chondrogenic lineage in vitro, *Stem Cells* **22** (2004), 1152–1167.

[12]  E.B. Hunziker, Articular cartilage repair: basic science and clinical progress. A review of the current status and prospects, *Osteoarthritis Cart.* **10** (2002), 432–463.

[13]  F. Iannone and G. Lapadula, The pathophysiology of osteoarthritis, *Aging Clin. Exp. Res.* **15** (2003), 364–372.

[14]  N. Indrawattana, G. Chen, M. Tadakoro, L.H. Shann, H. Ohgushi, T. Tateishi, J. Tanaka and A. Bunyaratvej, Growth factor combination for chondrogenic induction from human mesenchymal stem cell, *Biochem. Biophys. Res. Commun.* **320** (2004), 914–919.

[15]  T. Kaisho, K. Oritani, J. Ishikawa, M. Tanabe, O. Muraoka, T. Ochi and T. Hirano, Human bone marrow stromal cell lines from mieloma and rheumatoid arthritis that can support murine pre-B cell growth, *J. Immunol.* **149** (1992), 4088–4095.

[16]  K.W. Kavalkovich, R.E. Boynton, J.M. Murphy and F. Barry, Chondrogenic differentiation of human mesenchymal stem cells within an alginate layer culture system, *In Vitro Cell Dev. Biol. Anim.* **38** (2002), 457–466.

[17]  K.E. Kuettner, Cartilage integrity and homeostasis, in: *Rheumatology,* J. Klippel and P.A. Dieppe, eds, Vol. 2, Mosby International, London, 1998, pp. 6.1–6.16.

[18]  W.J. Landis, R. Jacquet, J. Hillyer, J. Zhang, L. Siperko, S. Chubinskaya, S. Asamura and N. Isogai, The potential of tissue engineering in orthopedics, *Orthop. Clin. North. Am.* **36** (2005), 97–104.

[19]  W.J. Li, R. Tuli, W. Okafor, A. Derfoul, K.G. Danielson, D.J. Hall and R.S. Tuan, A three-dimensional nanofibrous scaffold for cartilage tissue engineering using human mesenchymal stem cells, *Biomaterials* **26** (2005), 599–609.

[20]  M.K. Majumdar, V. Banks, D.P. Peluso and E.A. Morris, Isolation, characterization, and chondrogenic potential of human bone marrow-derived multipotential stromal cells, *J. Cell Physiol.* **185** (2000), 98–106.

[21]  M.P. Maleski and C.B. Knudson, Hyaluronan-mediated aggregation of limb bud mesenchyme and mesenchymal condensation during chondrogenesis, *Exp. Cell Res.* **2251** (1996), 55–66.

[22]  J.A. Martin and J.A. Buckwalter, Aging, articular cartilage chondrocyte senescence and osteoarthritis, *Biogerontology* **3** (2004), 257–264.

[23]  Y. Miura, J. Parvizi, J.S. Fitzsimmonsn and S.W. O'Driscoll, Brief exposure to high-dose transforming growth factor-$\beta$1 enhances periosteal chondrogenesis in vitro. A preliminary report, *J. Bone J. Surg. (Am.)* **84** (2002), 793–799.

[24]  R.W. Moskowitz, M.A. Kelly, D.G. Lewallen and C.T. Vangsness, Nonsurgical approaches to pain management for osteoarthritis of the knee, *Am. J. Orthop.* **33** (2004), 10–14.

[25]  K. Okazaki and L.J. Sandell, Extracellular matrix gene regulation, *Clin. Orthop. Relat. Res.* **427** (2004), S123–128.

[26] Pavesio, G. Abatangelo, A. Borrione, D. Brocchetta, A.P. Hollander, E. Kon, F. Torasso, S. Zanasi and M. Marcacci, Hyaluronan-based scaffolds (Hyalograft C) in the treatment of knee cartilage defects: preliminary clinical findings, *Novartis Found Symp.* **249** (2003), 203–217.

[27] L. Peterson, T. Minas, M. Brittberg and A. Lindahl, Treatment of osteochondritis dissecans of the knee with autologous chondrocyte transplantation: results at two to ten years, *J. Bone Joint Surg. (Am.)* **85** (2003), 17–24.

[28] M.S. Ponticiello, R.M. Schinagl, S. Kadiyala and F.P. Barry, Gelatin-based resorbable sponge as a carrier matrix for human mesenchymal stem cells in cartilage regeneration therapy, *J. Biomed. Mater. Res.* **52** (2000), 246–255.

[29] B. Schmitt, J. Ringe, T. Haupl, M. Notter, R. Manz, G.R. Burmester, M. Sittinger and C. Kaps, BMP2 initiates chondrogenic lineage development of adult human mesenchymal stem cells in high-density culture, *Differentiation* **71** (2003), 567–577.

[30] L.A. Solchaga, J.S. Temenoff, J. Gao, A.G. Mikos, A.I. Caplan and W.M. Goldberg, Repair of osteochondral defects with hyaluronan- and polyester-based scaffolds, *Osteoarthritis Cart.* **13** (2005), 297–309.

[31] R.S. Tuan, G. Boland and R. Tuli, Adult mesenchymal stem cells and cell-based tissue engineering, *Arthritis Res. Ther.* **5** (2003), 32–45.

[32] C.A. Vacanti and J.P. Vacanti, The science of tissue engineering, *Orthop. Clin. North. Am.* **31** (2000), 351–356.

[33] X. Wang, S.P. Grogan, F. Rieser, V. Winkelmann, V. Maquet, M.L. Berge and P. Mainil-Varlet, Tissue engineering of biphasic cartilage constructs using various biodegradable scaffolds: an in vitro study, *Biomaterials* **25** (2004), 3681–3688.

Biorheology 43 (2006) 481–488
IOS Press



# Uniform tissues engineered by seeding and culturing cells in 3D scaffolds under perfusion at defined oxygen tensions

D. Wendt [a,*,1], S. Stroebel [a,1], M. Jakob [a], G.T. John [b] and I. Martin [a]

[a] *Departments of Surgery and of Research, University Hospital, Basel, Switzerland*
[b] *PreSens GmbH, Regensburg, Germany*

**Abstract.** In this work, we assessed whether culture of uniformly seeded chondrocytes under direct perfusion, which supplies the cells with normoxic oxygen levels, can maintain a uniform distribution of viable cells throughout porous scaffolds several milimeters in thickness, and support the development of uniform tissue grafts. An integrated bioreactor system was first developed to streamline the steps of perfusion cell seeding of porous scaffolds and perfusion culture of the cell-seeded scaffolds. Oxygen tensions in perfused constructs were monitored by in-line oxygen sensors incorporated at the construct inlet and outlet. Adult human articular chondrocytes were perfusion-seeded into 4.5 mm thick foam scaffolds at a rate of 1 mm/s. Cell-seeded foams were then either cultured statically in dishes or further cultured under perfusion at a rate of 100 $\mu$m/s for 2 weeks. Following perfusion seeding, viable cells were uniformly distributed throughout the foams. Constructs subsequently cultured statically were highly heterogeneous, with cells and matrix concentrated at the construct periphery. In contrast, constructs cultured under perfusion were highly homogeneous, with uniform distributions of cells and matrix. Oxygen tensions of the perfused medium were maintained near normoxic levels (inlet $\cong$ 20%, outlet > 15%) at all times of culture. We have demonstrated that perfusion culture of cells seeded uniformly within porous scaffolds, at a flow rate maintaining a homogeneous oxygen supply, supports the development of uniform engineering tissue grafts of clinically relevant thicknesses.

Keywords: Bioreactor, fluid flow, chondrocyte, mass transport, functional tissue engineering

## 1. Introduction

Tissues engineered *in vitro* by seeding and culturing cells into 3D porous scaffolds are frequently reported to have an inhomogeneous structure, consisting of a dense layer of cells and extracellular matrix (ECM) concentrated along the periphery, and a necrotic interior region. Such an inhomogeneous structure may limit the initial mechanical functionality and subsequent *in vivo* development of grafts of clinically relevant size.

In order to generate homogeneous tissue grafts, cells may have to be initially seeded into porous 3D scaffolds with a uniform distribution [7,8], thereby establishing a template for spatially uniform ECM deposition. The cells seeded within the interior regions must then be supplied with sufficient nutrients during prolonged 3D culture in order to maintain viability and support the production of ECM; possibly

---

*Address for correspondence: David Wendt, Institute for Surgical Research & Hospital Management, University Hospital Basel, Hebelstrasse 20, ZLF, Room 405, 4031 Basel, Switzerland. Tel.: +41 61 265 2379; Fax: +41 61 265 3990; E-mail: dwendt@uhbs.ch.
[1] Equally contributing authors.

requiring the application of convective fluid flow. Since oxygen has a low solubility in culture medium, and is likely to be supplied to cells within the scaffold can help to determine whether the flow rate applied is sufficient.

We previously described a bioreactor for the perfusion of cell suspensions through the pores of 3D scaffolds in alternate directions, resulting in high efficiency and high spatial uniformity of cell seeding [8]. In this work, we first developed an integrated bioreactor system to allow prolonged perfusion of culture medium through 3D constructs following cell seeding by perfusion, within a single device. To continuously monitor the range of oxygen levels in the perfused constructs, in-line oxygen sensors were incorporated within the culturing pathway near the inlet and outlet of the constructs. Using the developed bioreactor system, we then tested the hypothesis that cells seeded uniformly throughout a 3D scaffold, when cultured under direct perfusion that supplies a normoxic range of oxygen, will generate a homogeneous graft with a uniform distribution of cells and ECM. The hypothesis was tested using a *human* chondrocyte/foam scaffold model system.

## 2. Methods

### 2.1. Integrated perfusion bioreactor system

The bioreactor system is designed to perfuse first a cell suspension (Fig. 1a), and subsequently culture media (Fig. 1b), through the pores of a 3D scaffold within a single closed system. To avoid any risk of mechanically induced cell damage from a pump, we aimed to design a seeding flow path which did not recirculate the cell suspension through a peristaltic pumphead. Therefore, based on our previous seeding bioreactor, the seeding pathway is designed to instead pump the headspace above the cell suspension back and forth from one Teflon column to the other, thereby generating an alternating flow of the cell suspension through the scaffold. Cell settling and cell attachment to bioreactor components are minimized by its vertical orientation, component material properties, and by minimizing the surface area of horizontal surfaces where cells tend to accumulate. Scaffolds are lightly press-fit and clamped within a scaffold chamber (Fig. 1c), such that fluid flow cannot deviate around the scaffold but must flow through its pores. The chamber is manufactured from polycarbonate and polished until translucent, thus permitting the detection of possible air bubbles. Teflon FEP tubes (6 mm i.d.; Cole Parmer) are connected to disposable three-way stopcocks (Hi-Flow™; Medex GmbH) via polypropylene luer adaptors (EM-Technik GmbH), and stopcocks are then connected to the scaffold chamber via its luer connections.

Following the cell seeding phase, stopcocks can simply be rotated to divert flow through a separate perfusion loop for prolonged culture. Although the seeding path minimizes cell attachment to the bioreactor components, the separate pathway for prolonged culture eliminates the inadvertent culture of any attached cells which would nonetheless proliferate on the bioreactor components and consume vital nutrients. Gas exchange for medium oxygenation and pH buffering is achieved through platinum-cured silicon tubing (1/32″ i.d., 1/16″ o.d., 2 m length; Cole Parmer). Media is exchanged through Interlink® injection sites (Becton Dickinson) connected to the reservoir bottle and is performed without removing the system from the incubator.

To ensure accurate, controlled, and reproducible perfusion through each construct (i.e., avoiding either preferential channelling or negligible flow through one particular construct), each construct can be seeded and cultured independently from the others using independent flow pathways and a multi-channel peristaltic pump. Up to eight integrated bioreactor units can be placed into a standard-sized incubator.

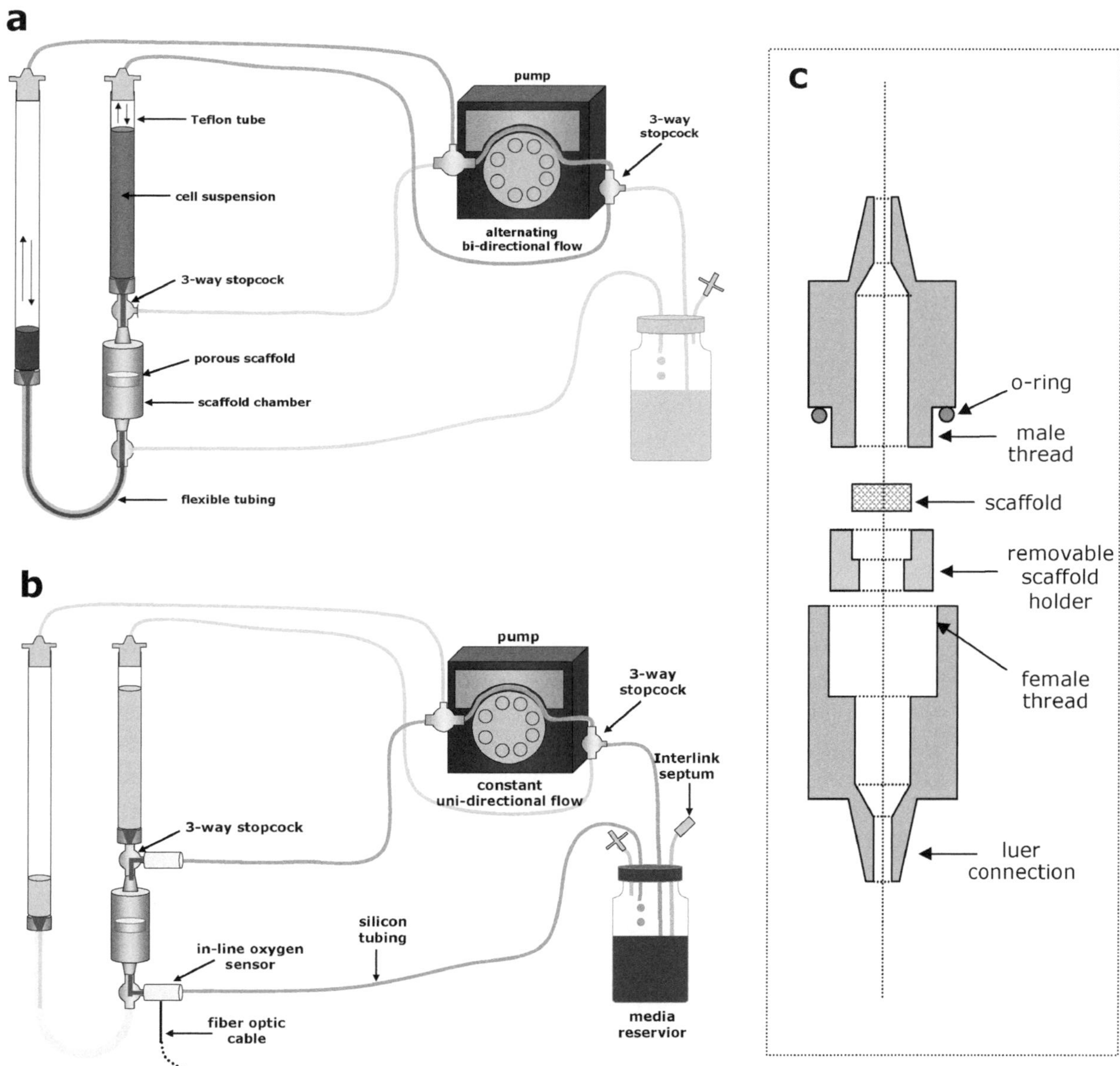


Fig. 1. Integrated perfusion bioreactor system. The bioreactor system streamlines the phases of cell seeding and prolonged culture of the cell-seeded scaffolds within a single device. (a) Cell seeding pathway. Alternating bi-directional flow of the cell suspension through the scaffold is generated by pumping the headspace above the cell suspension back and forth from one Teflon column to the other. In this way, cells are not recirculated through the pump, eliminating potential mechanically induced cell damage from the pumphead. (b) Prolonged culturing pathway. Following the cell seeding phase, the bioreactor system remains within the incubator, cell-seeded scaffolds are not handled or transferred to a separate bioreactor, but instead, stopcocks simply divert flow to a separate perfusion loop for prolonged culture. (c) Scaffold chamber. The scaffold is inserted into a removable holder, which simplifies sterile assembly and construct harvesting. The top of the chamber clamps the scaffold in place by its outer 1 mm periphery. The flow path contains a gradual expansion and contraction to reduce flow separation, and has a straight cylindrical region to fully develop the flow before reaching the construct.

Excluding the scaffold chamber, all components were selected for disposable use, eliminating risks of contamination and endotoxin build up associated with reuse. All bioreactor components were assessed for cytotoxicity in accordance with ISO10993-5.

## 2.2. Oxygen sensors

In order to eliminate artifacts (i.e., inadvertent media re-oxygenation) associated with sampling and off-line analysis, *in-line* oxygen sensors were required. However, a perfusion system for the current tissue engineering application imposes unique requirements for the sensors (e.g., small size, accurate at low flow rates, long-term stability), necessitating non-traditional sensor technologies. Therefore, chemo-optic flow-through micro-oxygen sensors (FTC-PSt-3; PreSens GmbH, Germany), based on the quenching of luminescence by oxygen, which do not consume oxygen, are independent of flow rate, and maintain long-term stability, are incorporated into the culturing perfusion loop (Fig. 1b). A fiber optic cable transmits the optical signals between the sensor and oxygen meter (Fibox-3; PreSens GmbH).

Sensors are connected directly to the stopcocks at the inlet and outlet of the scaffold chamber, in order to prevent oxygen ingress into the system through tubing and other components, as seen in preliminarily experiments if sensors were placed further upstream/downstream. Nitrogen-sparged water (i.e., 0% oxygen) was perfused through the system to demonstrate that negligible oxygen from the incubator was diffusing into the bioreactor between the inlet and outlet sensor.

## 2.3. Experimental model system

In two independent experiments, culture expanded adult human articular chondrocytes (AHAC) (1.3E+07 AHAC suspended in 8 ml media) were perfusion seeded onto foam disks made of poly(ethylene glycol terephthalate)/poly(butylene terephthalate) (PolyActive, 300/55/45 composition, compression molded, 8 mm diameter, 4.5 mm thick; IsoTis OrthoBiologics, The Netherlands) ($n = 12$ per experiment) at a superficial velocity of 1 mm/s overnight using an oscillating bi-directional flow regime (0.008 Hz). Following perfusion seeding, four constructs were examined by confocal microscopy to determine the cell viability (staining with LIVE/DEAD® Viability/Cytotoxicity Kit; Molecular Probes) as previously described [8], and were assessed histologically for the uniformity of the cell distribution (hematoxylin and eosin (H&E) stained cross-sections). Four seeded foams were transferred to agarose-coated dishes for static culture, and four remained within the bioreactor system for perfusion culture with continuous uni-directional flow at a superficial velocity of 100 $\mu$m/s. Constructs were cultured in 25 ml of DMEM (4.5 g/l glucose with nonessential amino acids) supplemented with 10% FBS, 10 mM HEPES, 2 mM glutamine, 1 mM sodium pyruvate, 100 U/ml penicillin, 100 $\mu$g/ml streptomycin, 0.1 mM ascorbic acid 2-phosphate, 1 U/ml insulin, and 10 ng/ml TGF$\beta$-3 for two weeks in a 20% oxygen/5% $CO_2$ incubator, with media exchanged three times per week. Cultured constructs were examined with confocal microscopy for the distribution of viable cells and histologically for the uniformity of the cell and matrix distribution (H&E staining).

## 3. Results

### 3.1. Cell seeding into porous scaffolds

Following cell seeding, essentially all AHAC detected in the foams were viable, likely due to efficient nutrient transport and the flushing of dead cells out from the scaffold pores resulting from perfused

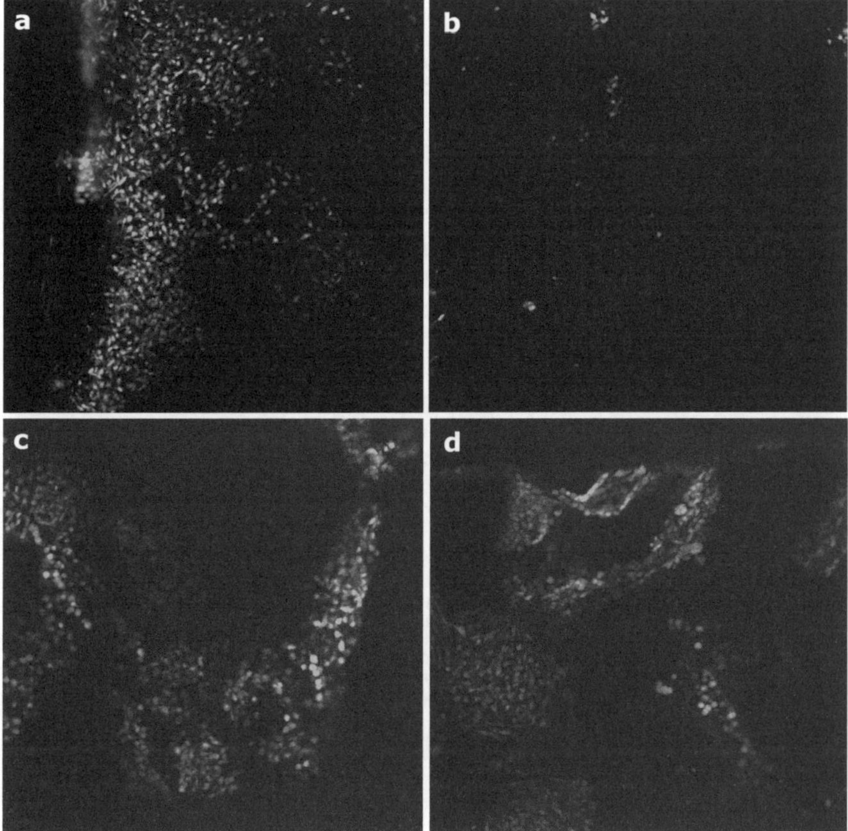

Fig. 2. Cell viability after two weeks of culture. Presented confocal microscopy images show only the channel for viable cell staining. Statically cultured constructs contained (a) a high density of viable cells concentrated at the periphery, and (b) an essentially non-viable internal region. In contrast, perfusion cultured constructs contained high densities of cells both (c) at their periphery, and (d) throughout their internal region.

flow. Furthermore, cells were seeded with a highly uniform distribution throughout the cross-sections, consistent with results obtained using a previously described perfusion bioreactor [8].

## 3.2. Static culture of cell-scaffold constructs

Perfusion seeded foams cultured statically for 2 weeks contained a dense layer of viable cells along the outer 0.5–1.0 mm periphery of the construct which encapsulated an internal region with predominantly non-viable cells (Fig. 2a). Histologically, constructs cultured statically were highly heterogeneous, containing cells and ECM concentrated only along the outer 1 mm periphery (Fig. 3a).

## 3.3. Perfusion culture of cell-scaffold constructs

At all times throughout the culture period, oxygen tensions measured at the inlet remained near saturation levels (i.e., 20%), and those measured at the outlet remained above 15% (Fig. 4). These data confirmed that at the flow rate used, the bioreactor maintained an efficient and rather homogeneous oxygen supply to the cells within the perfused constructs, close to normoxic levels.

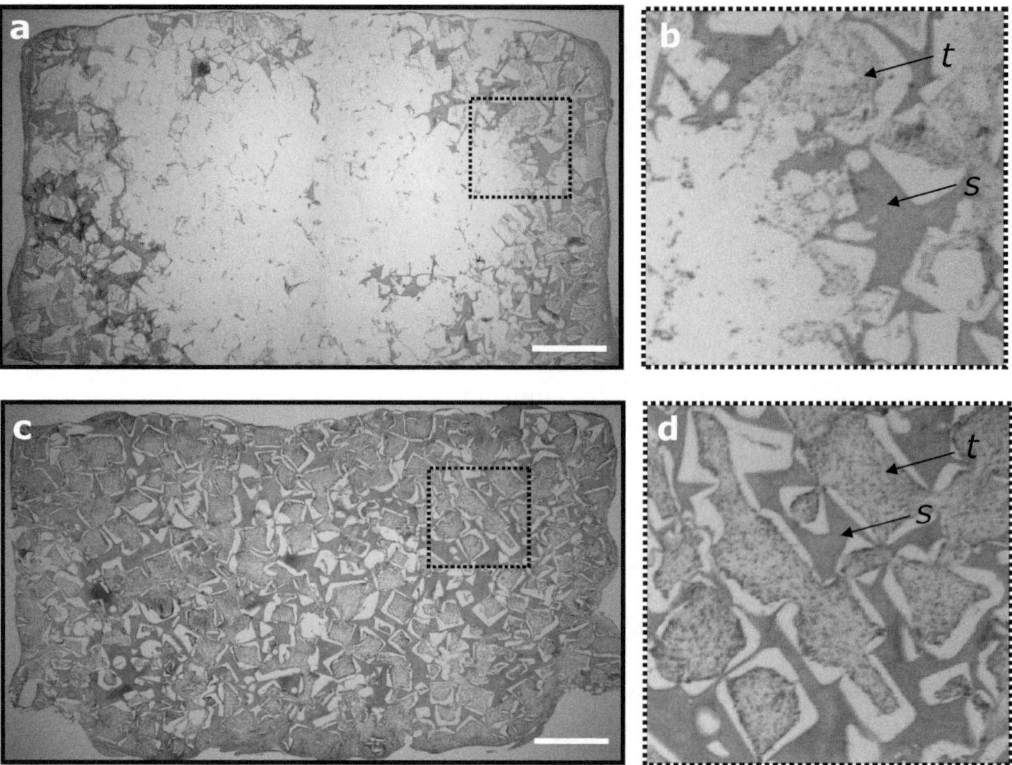



Fig. 3. Representative H&E stained cross-sections following two weeks of culture. (a,b) Statically cultured constructs; (c,d) perfusion cultured constructs; (a,c) low magnification images show the tissue distribution throughout the entire cross-section (scale bar = 1 mm); (b,d) higher magnification images identify the tissue "*t*" and scaffold "*s*" within the cross-sections. Statically cultured constructs contained cells and matrix only at the construct surface, reaching a depth of approximately 1 mm into the scaffold. In contrast, perfusion cultured constructs were highly homogeneous, containing a uniform distribution of cells and matrix throughout the cross-section.

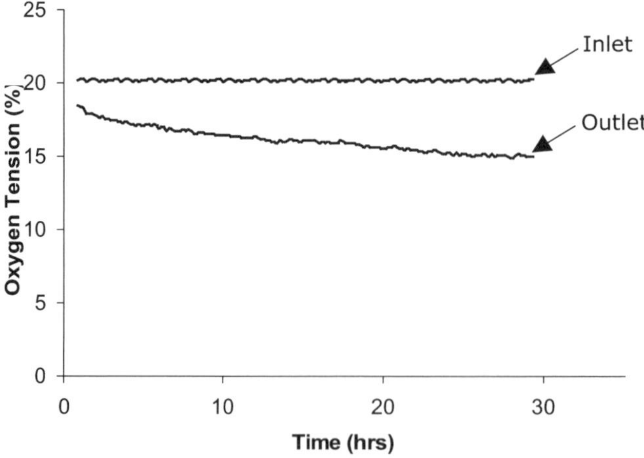



Fig. 4. Profiles of the oxygen tensions measured at the inlet and outlet of the scaffold chamber. Inlet oxygen concentrations remained near saturation levels throughout the two week culture period. Oxygen levels measured at the outlet were less than 5% lower than the inlet, indicating that cells within the construct were exposed to relatively homogeneous oxygen tensions.

Perfusion seeded foams cultured under perfusion contained viable AHAC distributed throughout both the exterior and interior regions of the foams, with few non-viable cells detected (Fig. 2b). In contrast to static cultures, constructs cultured under perfusion were remarkably homogeneous, containing a uniform distribution of both cells and ECM throughout the entire cross-section (Fig. 3b). However, despite the remarkable tissue uniformity in perfused constructs, only in small localized areas did AHAC have a rounded cell morphology, characteristic of differentiated chondrocytes.

## 4. Discussion

In this paper, we described the design of an integrated bioreactor system, which streamlines within a single device the phases of perfusion cell seeding and prolonged perfusion culture of the cell-seeded scaffolds. Using the developed bioreactor we then demonstrated that culture of uniformly seeded AHAC under direct perfusion, maintaining cells in a normoxic range of oxygen levels, can maintain a uniform distribution of viable cells throughout thick porous scaffolds, and support the development of uniform tissue grafts is necessary to generate a homogeneous tissue with a uniform distribution of both cells and ECM.

Previous studies have used perfusion systems to culture chondrocytes into porous scaffolds, but either did not address the uniformity of the resulting tissues [1,5,6] or reported the formation of heterogeneous tissues, despite using scaffolds that were rather thin [2]. The discrepancy between the latter study and our results could be explained by the fact that in our experiments perfusion culture was introduced following the uniform seeding of cells by perfusion, and/or by the higher flow rate (approximately 10-fold higher) used in our system, which may have provided a more efficient and homogeneous oxygen supply to the chondrocytes within the 3D constructs.

Despite initially having a uniform distribution of cells, statically cultured foams contained only a thin layer of cells and matrix concentrated at the construct periphery. Although oxygen tensions within statically cultured constructs were not assessed in this study, the heterogeneity could be explained by the steep oxygen gradients (from 20% at the surface to 2% at a depth of 1 mm from the surface) previously predicted [3] and measured [4] using the same polyactive foams cultured in the absence of perfusion. The dramatically improved tissue uniformity generated under perfusion as compared to static culture may be attributed to the maintenance of a normoxic range of oxygen levels across the constructs, as well as to the increase of mass transport of other nutrients (e.g., glucose) and metabolic waste products. In this context, studies are ongoing, using the developed bioreactor system, to understand the specific influence of oxygen on chondrocyte metabolism and function within a 3D construct.

In conclusion, we have developed an integrated perfusion bioreactor system for the engineering of uniform 3D grafts based on AHAC. The described device could be used in conjunction with other cell types, not only for generating uniform tissue grafts, but also as a controlled model system to investigate fundamental mechanisms of cell function in a 3D environment.

## Acknowledgements

This work was supported by the Swiss Federal Office for Education and Science (B.B.W.) under the Fifth European Framework Growth Program (project SCAFCART, grant No. G5RD-CT-1999-00050). We are grateful to Jeroen Pieper from IsoTis OrthoBiologics for kindly providing the Polyactive foams.

# References

[1] T. Davisson, R.L. Sah and A. Ratcliffe, Perfusion increases cell content and matrix synthesis in chondrocyte three-dimensional cultures, *Tissue Eng.* **8** (2002), 807–816.

[2] N.S. Dunkelman, M.P. Zimber, R.G. LeBaron, R. Pavelec, M. Kwan and A.F. Purchio, Cartilage production by rabbit articular chondrocytes on polyglycolic acid scaffolds in a closed bioreactor system, *Biotechnol. Bioeng.* **46** (1995), 299–305.

[3] M.C. Lewis, B.D. Macarthur, J. Malda, G. Pettet and C.P. Please, Heterogeneous proliferation within engineered cartilaginous tissue: the role of oxygen tension, *Biotechnol. Bioeng.* **91** (2005), 607–615.

[4] J. Malda, J. Rouwkema, D.E. Martens, E.P. Le Comte, F.K. Kooy, J. Tramper, C.A. van Blitterswijk and J. Riesle, Oxygen gradients in tissue-engineered PEGT/PBT cartilaginous constructs: measurement and modeling, *Biotechnol. Bioeng.* **86** (2004), 9–18.

[5] D. Pazzano, K.A. Mercier, J.M. Moran, S.S. Fong, D.D. DiBiasio, J.X. Rulfs, S.S. Kohles and L.J. Bonassar, Comparison of chondrogenesis in static and perfused bioreactor culture, *Biotechnol. Prog.* **16** (2000), 893–896.

[6] M.T. Raimondi, F. Boschetti, L. Falcone, F. Migliavacca, A. Remuzzi and G. Dubini, The effect of media perfusion on three-dimensional cultures of human chondrocytes: integration of experimental and computational approaches, *Biorheology* **41** (2004), 401–410.

[7] G. Vunjak-Novakovic, B. Obradovic, I. Martin, P.M. Bursac, R. Langer and L.E. Freed, Dynamic cell seeding of polymer scaffolds for cartilage tissue engineering, *Biotechnol. Prog.* **14** (1998), 193–202.

[8] D. Wendt, A. Marsano, M. Jakob, M. Heberer and I. Martin, Oscillating perfusion of cell suspensions through three-dimensional scaffolds enhances cell seeding efficiency and uniformity, *Biotechnol. Bioeng.* **84** (2003), 205–214.

Biorheology 43 (2006) 489–496
IOS Press



# Evaluation of human MSCs cell cycle, viability and differentiation in micromass culture

Jing-Wei Yang [a,b,1], Natalia de Isla [a,1], Céline Huselstein [a], Marie-Nathalie Sarda-Kolopp [c], Na Li [a], Yin-Ping Li [a,b], Ou-Yang Jing-Ping [b], Jean-François Stoltz [a] and Assia Eljaafari [a,*]

[a] *Research Group on Cell and Tissue Engineering and Mechanics, LEMTA UMR CNRS 7563, Faculty of Medicine, UHP Nancy 1, Vandoeuvre-lès-Nancy, France*
[b] *Department of Pathophysiology, Medical College of Wuhan University, Wuhan 430071, China*
[c] *Immunology Laboratory, Faculty of Medicine, UHP Nancy 1, Vandoeuvre-lès-Nancy, France*

**Abstract.** Mesenchymal stem cells (MSCs) have the potential to differentiate into distinct mesenchymal tissue cells. They are easy to expand while maintaining their undifferentiated state, which suggests that these cells could be an attractive cell source for tissue engineering of cartilage. *In vitro* high density micromass culture has been widely used for chondrogenesis induction. Our objective was to investigate human MSCs cell cycle, viability and differentiation in these conditions. Therefore, to induce human MSCs chondrogenesis, micromasses were cultured in the presence of transforming growth factor-$\beta 1$ in serum free medium for 21 days. Cell cycle, cell viability and cell phenotype were analyzed by flow cytometry. From day 0 to 7, the G0/G1 phase increased, whereas the S phase decreased gradually, but cell cycle phases (S, G0/G1 and G2/M) did not significantly change after day 7. Less than 10% of cells were apoptotic, but no necrosis was observed, even at day 21. We observed a decrease in CD90 and CD105 expression, from day 0 to 21. In conclusion, our results demonstrate a good viability of human MSCs in micromass culture during the whole period of culture. Moreover, micromass culture allowed human MSCs to be synchronized at the G0/G1 phase, while their phenotype suggested some degree of differentiation.

Keywords: Mesenchymal stem cell, TGF-$\beta 1$, viability, cell cycle, differentiation

## 1. Introduction

Mesenchymal stem cells (MSCs) have the potential to differentiate into distinct mesenchymal tissue cells, such as osteoblasts, adipocytes, chondrocytes, myoblasts and early neural progenitor cells. They are easy to expand in culture while maintaining their undifferentiated state, which suggests that these cells could be an attractive cell source for cartilage tissue engineering approaches. A number of studies has shown that bone marrow (BM) stroma-drived mesenchymal stem cells (MSCs) represent a promising candidate cell type for cartilage tissue engineering, instead of primary chondrocytes which are difficult to culture and expand [1,7].

---

[1] Both authors contributed equally to this work.

[*] Address for correspondence: Dr. Assia Eljaafari, Research Group on Cell and Tissue Engineering and Mechanics, LEMTA UMR CNRS 7563, Faculty of Medicine, UHP Nancy 1, Vandoeuvre-lès-Nancy, France. Tel.: +33 6 25 84 40 92; E-mail: Eljaafari@aol.com.

Human MSCs can be characterized by the expression of several cell surface antigens. However, their cell surface phenotype borrows features of endothelial, epithelial, and muscle cells and do not express any specific antigen. Therefore, the combination of several monoclonal antibodies is classically used in an effort to characterize and isolate human MSCs. For instance, antibodies specific to Stro-1, SH2, SH3, Thy-1, VCAM and LFA-3 can be used to positively label MSC, whereas antibodies specific to hematopoietic stem cells, such as, CD45 and CD34, can be used as negative markers [5]. A recent study has shown that when human articular chondrocytes (HACs) dedifferentiate in monolayer expansion, they express CD105 and CD90 at similar levels than undifferentiated MSCs, and acquire the ability to differentiate towards various mesenchymal lineages. This suggests that modification in the expression of CD105 and CD90 may be relevant to evaluate differentiation of MSCs into chondrocytes [3].

Precartilage condensation is induced in *vivo* by a process that involves upregulation of fibronectin. MSCs accumulate in regions of increased cell–matrix interactions and initiate chondrogenesis in *vivo*. *In vitro* high density micromass culture is an assay system which mimics the first step of MSC condensation, moreover the changes in extracellular matrix composition which occur in this culture system parallel those seen in *vivo*, making this a good model to study the early steps of chondrogenesis. Therefore, micromass culture has been widely used for chondrogenesis induction and also for investigating factors and signaling events involved in chondrogenesis [8,12]. In contrast to monolayer culture and other three-dimensional culture systems in biomaterial, micromasses are very compact with very little interspace among cells. Thus, it is possible that these conditions have some effects on cell viability and activity. But few studies have investigated this possibility. Moreover, the effect of micromass culture treatment on human MSCs cell cycle has not been resolved.

Most studies on chondrogenic differentiation of human MSCs have described synthesis of aggrecan, and other components of cartilage extra-cellular matrix, such as type II collagen, link protein, fibromodulin, cartilage oligomeric matrix protein, decorin, and chondroadherin [6,10]. In our study, we were interested in monitoring viability and cell cycle of human MSCs in micromass culture, during differentiation of human MSCs into chondrocytes. Moreover, we also studied the changes in CD105 and CD90 surface markers in micromass culture during chondrogenesis, in the presence of TGF-$\beta$1.


## 2. Methods

### 2.1. Isolation and cultures of human MSCs

To isolate human MSCs, bone marrow was harvested from normal adult donors after informed consent. Human MSCs were cultured in complete culture medium (alpha minimal essential medium, ($\alpha$-MEM, Gibco); 10% fetal bovine serum (FBS, Gibco); 2 mM L-glutamine, 100 U/ml penicillin/streptomycin, 1 $\mu$g/ml amphotericin B (Gibco)) in 25-cm$^2$ flask at 37°C with 5% $CO_2$. Culture medium was changed every 3 to 4 days. Adherent cells were further expanded in 75 cm$^2$ flasks. When cells grew to 80% confluence, they were harvested with 0.25% trypsin and 1 mM EDTA for 5 minutes at 37°C, and subcultured in complete culture medium. After the fourth subculture, chondrogenesis was induced in micromass cultures.

### 2.2. Human MSCs identification

Cells in culture were detached using 0.25% trypsin-EDTA, stained with fluorescein isothiocyanate (FITC) or phycoerythrin (PE)-conjugated antibodies against CD90, CD73, CD105, CD45 and CD34 (monoclonal antibodies, BD), washed and analyzed using flow cytometry.

## 2.3. Micromass culture and chondrogenesis induction

Human MSCs were trypsinized, counted, tested for viability by trypan blue exclusion, and resuspended in culture medium at a density of $2.0 \times 10^7$ viable cells/ml. Micromasses were obtained by pipetting 15 $\mu$l droplets of cell suspension in the center of each well in a 24-well plastic cell culture plate [2]. The plates were placed in a humidified incubator at 37°C with 5% $CO_2$. Cells were allowed to attach to plastic for 3 hours, and then they were flooded carefully with 1 ml of medium. Micromasses were cultured in a serum free medium (DMEM/F12, Gibco) containing 10 ng/ml transforming growth factor-$\beta$1, 0.1 mM L-ascorbic acid, 1 mM sodium pyruvate, $10^{-7}$ M dexamethasone, 0.5 mg/ml BSA, 4.7 $\mu$g/ml linoleic acid, 40 $\mu$g/ml L-proline and 1% ITS (Sigma). Medium changes were carried out twice a week up to 21 days. When appropriate, cells in micromass were isolated using 2% collagenase B (Roche) at 37°C for 30 minutes and then were used for the experiments.

## 2.4. Cell viability assay

Cells issued from micromass cultures were stained with the Vybrant™ Apoptosis Assay Kit #2 (Molecular Probes). Briefly, $2 \times 10^5$ cells were resuspended in 100 $\mu$l of 1× annexin-binding buffer. 5 $\mu$l of Alexa Fluor 488-annexin V, 1 $\mu$l of the 100 $\mu$g/ml PI were added to each cell suspension and incubated at room temperature for 15 minutes. After the incubation period, 400 $\mu$l of 1× annexin-binding buffer was added. Samples were mixed gently and kept on ice, and as soon as possible the stained cells were analyzed by flow cytometry (EPICS XL Coulter; Beckman-Coulter).

## 2.5. Cell cycle assay

Cells issued from micromass cultures were stained with the Coulter DNA-Prep Reagents Kit (Beckman Coulter). Cells were resuspended in 50 $\mu$l of DMEM W/O phenol red (Gibco), and 50 $\mu$l DNA-Prep LPR was added. Samples were agitated for 30–60 seconds. Then cells were stained with DNA-Prep stain and agitated for 30–60 seconds. Cells were kept at 4°C until analyzed stained cells by flow cytometry (EPICS XL Coulter; Beckman-Coulter). Data analysis was performed with Dean and Jett model Multi Cycle AV (Phoenix Flow System, San Diego, USA).

## 2.6. Phenotypic analysis

After harvesting the cells and preparing a single cell suspension ($2 \times 10^5$ cells per 200 $\mu$l DMEM W/O phenol red (Gibco) supplemented with 0.4% bovine serum albumin (BSA; Sigma)), cells were stained at 4°C for 45 minutes with CD105 (monoclonal antibody; Ancell), and CD90 (monoclonal antibody, R&D). Cells were then washed 3 times in DMEM W/O phenol red and cell pellets were resuspended in DMEM W/O phenol red supplemented with 0.4% BSA, cells were stained at room temperature for 45 minutes with Alexa-488 (Molecular Probes) in the dark. Cells were then washed 3 times in DMEM W/O phenol red and cell pellets were resuspended in 4% PBS-paraformaldehyde. Cells were kept at 4°C before analysis by flow cytometry (EPICS XL Coulter; Beckman-Coulter).

For each surface marker, percentage of positive cells and the level of marker expression were calculated. The level of marker expression was calculated as the difference between geometric mean fluorescence intensity (MFI) of sample cells and that of isotype control. Quantitative expression of number of CD105 and CD90 per cell was daily calculated by using Qifikit® beads (Becton Dickinson).

## 3.  Results

### 3.1.  Cell viability

High density human MSCs were pipetted into micromass and then differentiated in serum-free medium in the presence of TGF-$\beta$1. Cell viability was detected by Vybrant™ Apoptosis Assay Kit #2. With this assay, apoptotic cells show green fluorescence, dead cells show red and green fluorescence, and live cells show little or no fluorescence. The cell viability assay indicated that among the cells obtained from micromasses, less than 10% cells were apoptotic, throughout the whole period of cultures. No significant necrosis was observed, even at day 21 (Fig. 1).

### 3.2.  Cell cycle

The cell cycle transition from the G0/G1 phase to the S phase indicated cell proliferation. Flow cytometry analysis of propidium iodide-stained cells in micromass demonstrated that the percentage of cells in the G0/G1 phase increased from 84.1% at day 1 to 91.8% at day 7. In contrast, the percentage of cells in the S phase decreased from 14.3% at day 1 to 6.3% at day 7 (Fig. 2). However, cell cycle phases (S, G0/G1 and G2/M) did not significantly change after day 7, until day 21 (Fig. 2). These results suggest that proliferation of cells in micromasses decreased during the first week, and then was maintained until day 21.

### 3.3.  Phenotypic analysis of cell surface markers

For phenotypic characterization of human MSCs, cell surface phenotypic analysis was performed and cells were analyzed using different cell surface markers. Phenotypic characterization of human MSCs showed, as expected, that human MSCs were positive for CD90, CD73 and CD105, but negative for CD45 and CD34 (Fig. 3).

In order to assess the differentiation status of human MSCs in micromass culture with TGF-$\beta$1, the levels of expression of CD105 and CD90 were measured by flow cytometry at different time points. Our results showed that the percentage of CD105 positive cells significantly decreased from day 1 to day 7, and did not change significantly after day 7. Starting from day 1, the percentage of CD90 positive cells continuously decreased until day 21. Moreover the quantitative expression of CD105 and CD90

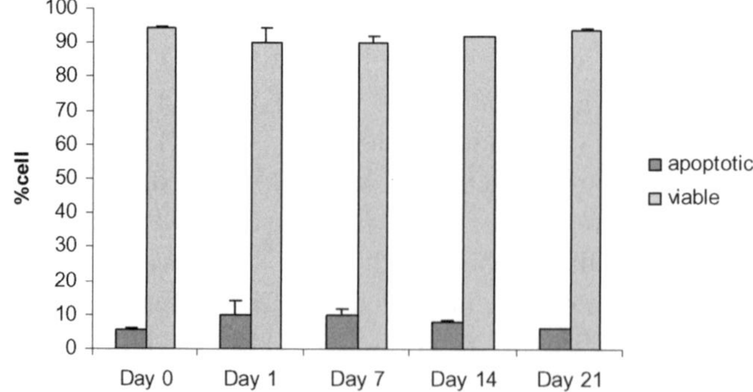


Fig. 1. Cell viability of human MSCs was analyzed by flow cytometry during the whole period of micromass culture. Among the cells, and throughout cultures, less than 10% of cells were apoptotic. The remaining cells were viable, and no necrosis was observed, even at day 21. Results are means ± SD of 3 different experiments.

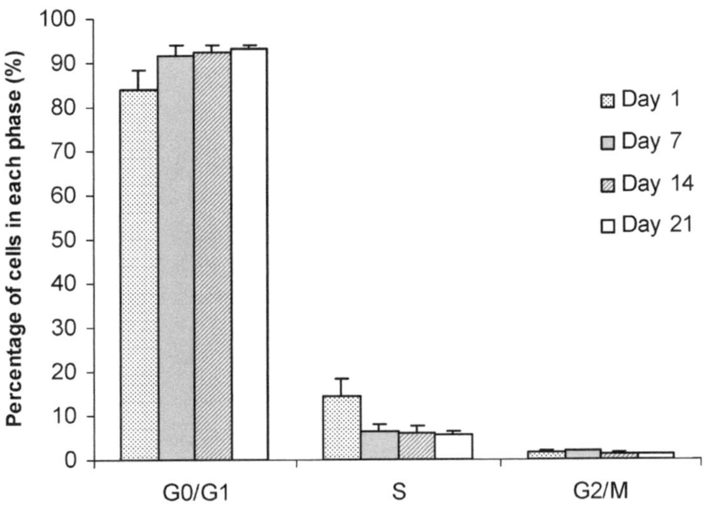



Fig. 2. Cell cycle analyse of human MSCs in micromass culture. The percentage of cells in each phase of the cell cycle was analyzed by flow cytometry. Results are means ± SD of 3 different experiments. The percentage of cells in the G0/G1 phase increased from day 1 to day 7. In contrast, the percentage of cells in the S phase decreased. However, cell cycle phases (S, G0/G1 and G2/M) didn't change significantly after day 7 till day 21.

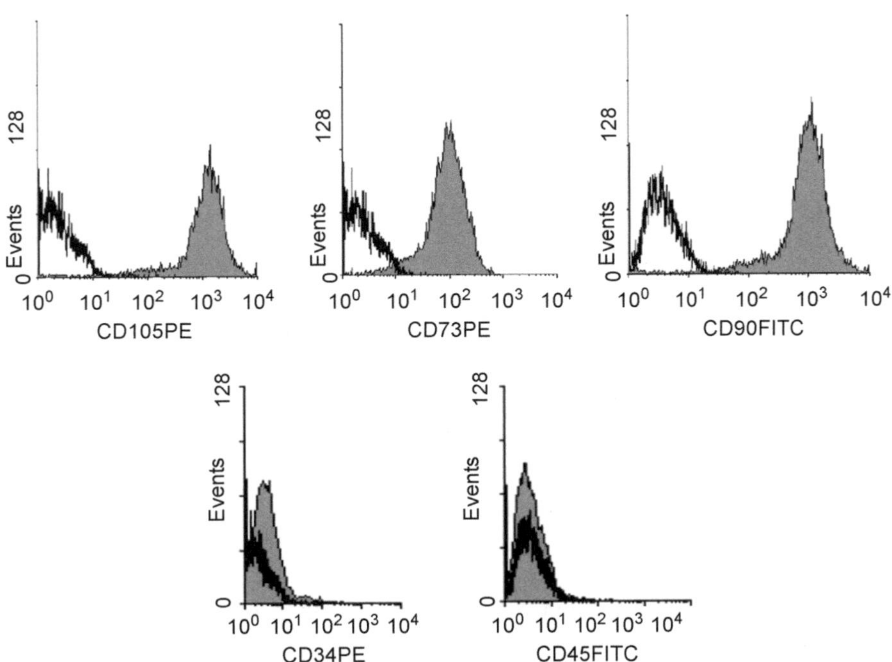



Fig. 3. Flow cytometric analyses of adhesion molecule surface expression of human MSCs. Cells were labelled with monoclonal antibodies specific for molecules indicated in each flow cytometric histogram. FACSort using Cell-Quest software analyzed labeled cells. Open and filles areas in each histogram represent cells labeled with isotype-matched Ig and specific monoclonal antibodies, respectively. The profiles are representative of human MSCs labeled from at least 10 donors.

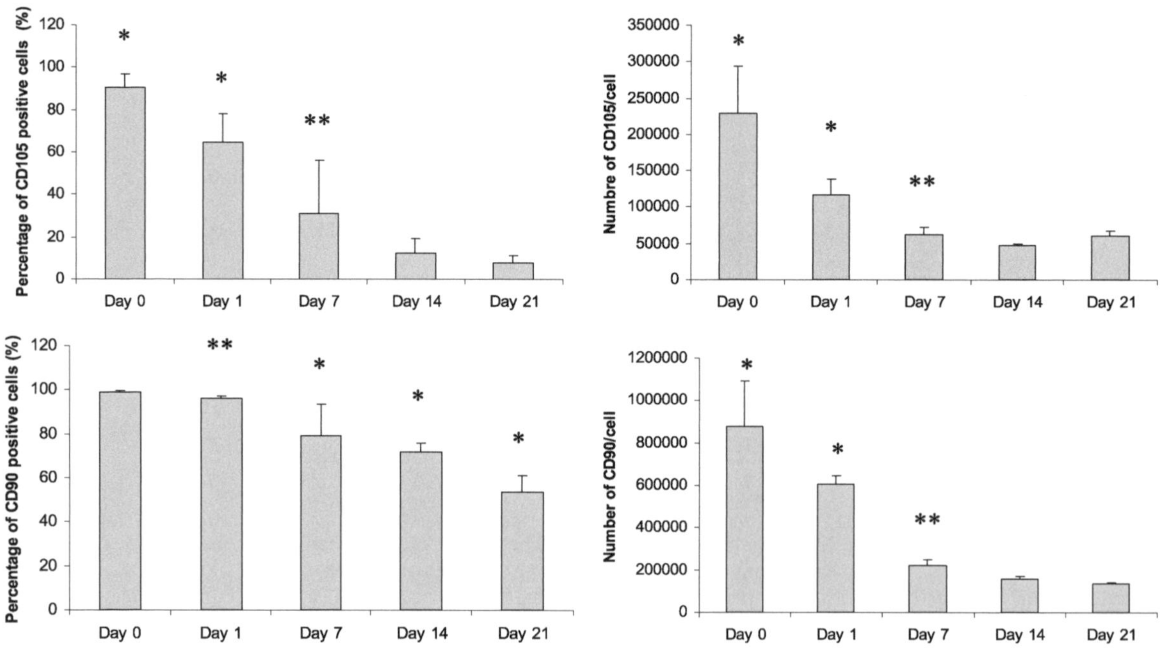

Fig. 4. Expression of CD105 and CD90 of human MSCs in micromass culture. The percentage of positive cells and the quantitative expression of number of CD105 and CD90 per cell at each time of culture were analyzed by flow cytometry. Results are presented as the mean ± SD of 3 different experiments. The groups marked with * were compared with the other groups, $p < 0.05$; The groups marked with ** were compared with the groups marked *, $p < 0.05$.

decreased significantly from day 1 to day 7, but did not change significantly after day 7. Thus these results strongly suggest a certain degree of human MSCs differentiation (Fig. 4).

## 4. Discussion

MSCs are multi-potential cells that can differentiate into bone, cartilage, fat, muscle, bone marrow stroma, and other tissue types when induced by the appropriate biological approach *in vitro*. Chondrogenesis of embryonic mesenchymal stem cells follows several steps beginning with a mesenchymal cell condensation phase during which mesenchymal cells aggregate at high density in the pre-cartilage core. Cell–cell interaction mediated by N-cadherin is an important regulator of this critical cellular condensation step, which precedes the appearance of chondrocytes that produce cartilage-specific extracellular matrix proteins, such as type II collagen and sulfated proteoglycan. In our study to induce chondrogenic differentiation, MSCs were cultured at high density in a three-dimensional environment, in order to enhance cell–cell interactions favorable for chondrogenesis, similar to embryonic mesenchymal cells that undergo chondrogenesis.

The G0 phase is a specific stem-cell phase that allows them to retain their ability to engraft when they reenter the cell cycle secondary of a stimulus [13]. The transition of cells from G0 phase into G1 phase is related to cell fate, such as differentiation, proliferation or apoptosis. Following stimulation with mitotic signal, stem cells move from the G0 to the G1 stage, otherwise, cells undergo differentiation or apoptosis. In our study, cell cycle results showed that the percentage of G0/G1 cell increased from Day 1 to Day 7, and was maintained at 90% until day 21. Analysis of the cell cycle showed that cell proliferation was very low during the whole period of culture. There are two possible explanations for this low cell proliferation. First, high density culture may have a deleterious effect on the growth of human MSCs, and induce apoptosis or necrosis. Second, a large percentage of human MSCs in the G0 phase is in a quiescent state, in which cells demonstrate weak metabolism. Our results showed that viability of mesenchymal stem cells in micromass culture was maintained during the whole period of culture, and less than 10% of cells were apoptotic with no appearance of necrosis. Therefore, these results suggest that human MSCs in micromass culture were in a quiescent state with low level of proliferation in the presence of TGF-$\beta$1.

Endoglin/CD105, an accessory receptor of the transforming growth factor-$\beta$ (TGF-$\beta$) receptor complex, has been recently found on chondrocytes. Members of the TGF-$\beta$ superfamily are mediators of cell proliferation and differentiation and play regulatory roles in cartilage and bone formation [9,4]. CD90/Thy-1 is a glycosylphosphatidylinositol (GPI)-anchored glycoprotein previously reported on a minority of chondrocytes in normal articular cartilage [11]. However, the biological role of CD90 remains enigmatic. A recent study showed that the expression of CD105 and CD90 is upregulated in chondrocyte monolayer culture, and that this upregulation is correlated with the dedifferentiation of chondrocyte towards various mesenchymal lineages. In contrast, endoglin/CD105 and CD90/Thy-1 are lost when MSC differentiation is activated. In our study we analyzed the coexpression of CD105 and CD90 to evaluate human MSCs differentiation status. Results showed that the decrease in CD90 and CD105 expression started as soon as day 1, and gradually decreased up to day 21, assessing thus the differentiation of human MSCs. In conclusion, our results demonstrate a good viability of human MSCs in micromass culture during the whole period of culture. Moreover, human MSCs remained in a resting state while their phenotype suggested to us some degree of differentiation.

# References


[1] F.P. Barry and J.M. Murphy, Mesenchymal stem cells: clinical applications and biological characterization, *Review, Int. J. Biochemistry Cell. Biology* **36** (2004), 568–584.

[2] C. De Bari, F. Dell'Accio and F.P. Luyten, Failure of *in vitro*-differentiated mesenchymal stem cells from the synovial membrane to form ectopic stable cartilage *in vivo*, *Arthritis & Rheumatism* **1** (2004), 142–150.

[3] J. Diaz-Romero, J.P. Gaillard, S.P. Grogan, D. Nesic, T. Trub and P. Mainil-Varlet, Immunophenotypic analysis of human articular chondrocytes: Changes in surface markers associated with cell expansion in monolayer culture, *J. Cell Physiology* **202** (2005), 731–742

[4] M. Jakob, O. Demarteau, D. Schafer, B. Hintermann, W. Dick, M. Heberer and I. Martin, Specific growth factors during the expansion and redifferentiation of adult human articular chondrocytes enhance chondrogenesis and cartilaginous tissue formation *in vitro*, *J. Cell. Biochem.* **81**(2) (2001), 368–377.

[5] E.A. Jones, S.E. Kinsey, A English, R.A. Jones, L. Straszynski, D.M. Meredith, A.F. Markham, J. Andrew, P. Emery and D. McGonagle, Isolation and characterization of bone marrow multipotential mesenchymal progenitor cells, *Arthritis & Rheumatism* **12** (2002), 3349–3360.

[6] K.W. Kavalkovich, R.E. Boynton, J.M. Murphy and F. Barry, Chondrogenic differentiation of human mesenchymal stem cells within an alginate layer culture system, *In Vitro Cell. Dev. Biol. Anim.* **9** (2002), 457–466.

[7] J.J. Minguell, A. Erices and P. Conget, Mesenchymal stem cells, *Exp. Biol. Med.* **6** (2001), 507–520.

[8] H. Nakahara, J.E. Dennis, S.P. Bruder, S.E. Haynesworth, D.P. Lennon and A.I. Caplan, *In vitro* differentiation of bone and hypertrophic cartilage from periosteal-derived cells, *Exp. Cell Res.* **8** (1991), 492–503.

[9] W.L. Parker, M.B. Goldring and A. Philip, Endoglin is expressed on human chondrocytes and forms a heteromeric complex with betaglycan in a ligand and type II TGFbeta receptor independent manner, *J. Bone Miner. Res.* **2** (2003), 289–302.

[10] I. Sekiya, J.T. Vuoristo, B.L. Larson and D.J. Prockop, *In vitro* cartilage formation by human adult stem cells from bone marrow stroma defines the sequence of cellular and molecular events during chondrogenesis, *PNAS* **99** (2002), 4397–4402.

[11] K.L. Summers, J.L. O'Donnell, M.S. Hoy, M. Peart, J. Dekker, A. Rothwell and D.N. Hart, Monocyte-macrophage antigen expression on chondrocytes, *J. Rheumatol.* **7** (1995), 1326–1334.

[12] A.D. Weston, V. Rosen, R.A. Chandraratna and T.M. Underhill, Regulation of skeletal progenitor differentiation by the BMP and retinoid signaling pathways, *J. Cell Biol.* **148** (2000), 679–690.

[13] I. Wilmut, A.E. Schnieke, J. McWhir, A.J. Kind and K.H.S. Campbell, Viable offspring derived from fetal and adult mammalian cells, *Nature* **385** (1997), 810–813.

Biorheology 43 (2006) 497–507
IOS Press



# Dynamic deformational loading results in selective application of mechanical stimulation in a layered, tissue-engineered cartilage construct

Kenneth W. Ng [a], Robert L. Mauck [a,*], Lauren Y. Statman [a], Evan Y. Lin [a],
Gerard A. Ateshian [b] and Clark T. Hung [a,**]

[a] *Cellular Engineering Laboratory, Department of Biomedical Engineering, Columbia University,
New York, NY, USA*
[b] *Musculoskeletal Biomechanics Laboratory, Departments of Biomedical and Mechanical Engineering,
Columbia University, New York, NY, USA*

**Abstract.** The application of dynamic physiologic loading to a bilayered chondrocyte-seeded agarose construct with a 2% (wt/vol) top layer and 3% (wt/vol) bottom layer was hypothesized to (1) improve overall construct properties and (2) result in a tissue that mimics the mechanical inhomogeneity of native cartilage. Dynamic loading over the 28 day culture period was found to significantly increase bulk mechanical and biochemical properties versus free-swelling culture. The initial depth-distribution of the compressive Young's modulus ($E_Y$) reflected the intrinsic properties of the gel in each layer and a similar trend to the native tissue, with a softer 2% gel layer and a much stiffer 3% gel layer. After 28 days in culture, free-swelling conditions maintained this general trend while loaded constructs possessed a reverse profile, with significant increases in $E_Y$ observed only in the 2% gel. Histological analysis revealed preferential matrix formation in the 2% agarose layer, with matrix localized more pericellularly in the 3% agarose layer. Finite element modeling revealed that, prior to significant matrix elaboration, the 2% layer experiences increased mechanical stimuli (fluid flow and compressive strain) during loading that may enhance chondrocyte stimulation and nutrient transport in that layer, consistent with experimental observations. From these results, we conclude that due to the limitations in 3% agarose, the use of this type of bilayered construct to construct depth-dependent inhomogeneity similar to the native tissue is not likely to be successful under long-term culture conditions. Our study underscores the importance of other physical properties of the scaffold that may have a greater influence on interconnected tissue formation than intrinsic scaffold stiffness.

Keywords: Cartilage, tissue engineering, inhomogeneity, digital image correlation, agarose, scaffold design

## 1. Introduction

Articular cartilage is the lubricated, load-bearing surface in diarthrodial joints with marked differences in cellular, biochemical, and mechanical properties through the depth [2–4,13,19,29,32,33]. There has

---

*Current address: McKay Orthopaedic Research Laboratory, Department of Orthopaedic Surgery, University of Pennsylvania, Philadelphia, PA, USA.

**Address for correspondence: Clark T. Hung, PhD, 351 Engineering Terrace, MC8904, 1210 Amsterdam Avenue, Department of Biomedical Engineering, Columbia University, New York, NY, 10027, USA. Tel.: +1 212 854 6542; Fax: +1 212 854 8725; E-mail: cth6@columbia.edu.

been increasing focus on studying and replicating aspects of the organization of articular cartilage for tissue engineering purposes [15,17,27]. Constructs replicating the native arrangement of chondrocyte populations have shown success in preserving the phenotype of superficial, middle, and deep zone cells in engineered cartilage [15,17]. In addition, constructs fabricated with inhomogeneous depth-dependent material properties have indicated that differences in scaffold properties (e.g., Young's modulus, permeability, porosity) may be utilized to help control tissue development [27]. As chondrocytes respond to a multitude of mechanical stimuli [6–9,20], recreating both the depth-dependent cellular and mechanical characteristics of native cartilage may be an important factor in developing a fully functional replacement tissue.

Agarose has been used as a model scaffold for its well documented maintenance of chondrocyte phenotype, support of cartilage tissue development, and ability to transmit mechanical stimuli [5,6,18,24]. Studies by our laboratory using chondrocyte-seeded agarose hydrogels have found that in engineered constructs that are initially homogeneous, the resulting tissue develops a depth-dependent, mechanical inhomogeneity over time in culture [14,27]. The degree of inhomogeneity is affected by the application of deformational loading and may be due to a combination of mechanical stimuli and nutrient transport factors [14]. In our recent study, a bilayered tissue construct was created with a top layer of 2% agarose and a bottom layer of 3% agarose [27]. It was found that due to the disparate mechanical properties of the agarose layers, this construct initially possessed distinct material properties for each layer. After time in culture and tissue elaboration, in the absence of applied mechanical loading, the construct developed a smoother distribution of inhomogeneous mechanical properties that appeared to reflect the degree of tissue development in each layer, as well as the initial scaffold layer properties. This finding indicated that the mechanical inhomogeneity of the engineered tissue can be influenced by the initial scaffold design. In the current study, it is hypothesized that the application of dynamic loading to this type of bilayered construct can (1) improve overall construct properties and (2) further accentuate the development of mechanical inhomogeneity of the resulting tissue to mimic that of native cartilage. For the latter hypothesis, in addition to experimental studies, an FEM model was constructed based on biphasic theory to gain better understanding of the spatiotemporal physical environment during loading of the bilayered construct.

## 2. Materials and methods

### 2.1. Biphasic Finite Element Model (FEM)

To gain an appreciation for the physical environment that chondrocytes in each gel layer would experience during applied loading prior to significant matrix elaboration, a custom finite element mesh was defined with a commercial software package (I-DEAS, SDRC, Plano, TX) to model a bilayered construct with a 2% agarose top layer and a 3% agarose bottom layer ($\varnothing$ 5.0 mm $\times$ 2 mm, 1 mm thickness per layer). Meshes were defined as axisymmetric and contained 600 elements per mesh and eight nodes per element, with the distribution of elements biased towards the free edge of the construct. Each gel layer was assumed homogeneous, with a linear isotropic elastic solid matrix, with 2% gel parameters: Young's modulus ($E_Y$) = 10 kPa, $v = 0.3$, $k = 1 \times 10^{-13}$ m$^4$/N s and 3% gel region parameters: $E_Y = 20$ kPa, $v = 0.3$, $k = 1 \times 10^{-13}$ m$^4$/N s. These parameters were chosen to approximate experimental values found in the literature [1,11,14,27]. For comparison, a homogeneous 2% agarose disk of the same size and 2% gel parameters above was modeled under the same conditions. A custom finite element modeling (FEM) program [28] incorporating biphasic theory [25] was used to solve the problem of unconfined deformational loading of constructs with an applied sinusoidal deformation of 10% of the construct thickness at a frequency of 1 Hz. Results from this analysis were output at the point of maximal axial deformation ($t = 0.5$ s) for dynamic deformational loading.

## 2.2. Creation and culture of agarose constructs

Articular bovine chondrocytes were isolated via enzymatic digestion as described previously [22]. Briefly, chondrocytes were isolated from the carpometacarpal joint of 2–3 month old calves via serial digestion of full thickness cartilage slices in 0.25% pronase (Calbiochem, San Diego, CA) and 0.05% collagenase (Sigma Chemicals, St. Louis, MO). Cells were resuspended and mixed in equal parts with agarose (type VII, Sigma) in phosphate buffered saline (PBS) at 40°C to yield 2% and 3% (wt/vol) hydrogel suspensions with $60 \times 10^6$ cells/ml. Bilayered constructs were created as previously described [27] by adding 50 $\mu$l of 3% agarose cell suspension into a custom template ($\varnothing$ 6.35 × 2.3 mm) immediately followed by 50 $\mu$l of 2% agarose suspension, with the two layers permitted to gel together at room temperature for 20 minutes. Disks ($\varnothing$ 4.76 × 2.3 mm) were cored with a sterile dermal punch from the center region of the template and cultured in 100 mm Petri dishes ($\sim$15–20 per dish) with 35 ml of fully supplemented (essential and non-essential amino acids, sodium bicarbonate, HEPES, TES, BES, penicillin–streptomycin, 20% fetal bovine serum [FBS]) high glucose Dulbecco's Modified Eagle's Medium (DMEM, Sigma) with fresh (daily) 50 $\mu$g/ml ascorbate (Sigma). All constructs were maintained at 37°C and 5% $CO_2$ with media changed daily. With each media change, construct orientation was changed (i.e., disks were flipped over) in order to reduce any bias due to orientation in culture. Constructs were cultured for two days before the start of mechanical testing (designated as day 0) to allow the cells to acclimate to their *in vitro* environment.

## 2.3. Daily dynamic loading

Dynamic loading (DL) of bilayered disks was carried out on a custom deformational loading bioreactor in 5 ml of fresh culture media as above with a loading regime of $\sim$10% strain applied at 1 Hz for 3 hrs/day, 5 days/wk [14]. Free-swelling (FS) controls were maintained adjacent to the bioreactor during the loading period in similar media conditions. After this loading period, all disks were returned to 35 ml of fresh medium for overnight culture.

## 2.4. Bulk mechanical testing

Bulk mechanical testing was performed in unconfined compression between two impermeable platens in a custom material testing device as previously described [22]. Constructs ($n = 4$–5 per group, per time point) were first equilibrated under a creep tare load of $\sim$0.02 N followed by a stress relaxation test with a ramp displacement of 1 $\mu$m/sec to 10% strain (based on the measured post-creep thickness). After equilibrium was reached (2000 sec), a sinusoidal displacement of 40 $\mu$m amplitude was applied at 1 Hz. Bulk compressive $E_Y$ was determined from the equilibrium response of the stress relaxation test by dividing the equilibrium stress (minus the tare stress) by the applied static strain. Bulk dynamic modulus ($G^*$) at 1 Hz was calculated from the ratio of the measured stress amplitude and the applied strain amplitude of the dynamic testing. Following bulk mechanical testing, samples were allowed to recover in culture media described above for 30 minutes prior to local mechanical testing (see below).

## 2.5. Depth-dependent mechanical testing

Local compressive $E_Y$ measurements of bilayered constructs were carried out on a microscopy system for mechanical testing and image correlation, as described previously [32]. Each disk was cut in half diametrically and one half was loaded onto a custom unconfined compression device mounted on the motorized stage of an inverted microscope. The initial uncompressed thickness ($h_0$) of the specimen was measured optically using a calibrated $4\times$ objective (1.66 $\mu$m/pixel). A tare strain of 5% of the initial sample thickness was applied at 1 $\mu$m/s and the sample was allowed to equilibrate for 20 minutes, where multiple overlapping images of the sample cross-section were then taken. An additional compression of 5% of $h_0$ was then applied and a second set of images were acquired after allowing the sample to equilibrate again for 20 minutes. Images of the sample cross-section were stitched using Panavue Image Assembler (Panavue, PQ, Canada). Image analyses were performed using an optimized digital image correlation technique producing accurate axial displacement and strain fields ($\varepsilon_{zz}(z)$) [32], where $E_Y(z) = \sigma_{zz}/\varepsilon_{zz}(z)$ and $\sigma_{zz}$ is the normal stress measured on the specimen surface.

## 2.6. Biochemical composition and histology

Following mechanical testing, one half of each sample was fixed, dehydrated, and embedded for histological analysis of GAG and collagen via Safranin O and Picrosirius Red, respectively [22]. The two layers in the remaining half were sharply dissected at the interface and then digested with papain for biochemical analysis of each layer as described previously [22]. GAG content was assessed using the 1,9-dimethylmethylene blue dye-binding assay [10] scaled for microplates. Collagen content was assessed by measuring orthohydroxyproline (OHP) via the dimethylaminobenzaldehyde and chloramine T assay [31], with a collagen : OHP ratio of 10 : 1 used as a conversion factor [12].

## 2.7. Statistical analysis

Statistics were performed using the Statistica (Statsoft, Inc., Tulsa, OK) software package. At least 4–5 samples in each group were analyzed for each data point, with data reported as the mean and standard deviation. For mechanical and biochemical data, groups were examined using multivariate analysis of variance with $E_Y$, $G^*$, GAG, or collagen as the dependent variables, and culture time, loading, and axial position (for local modulus data only) as the independent variables. Fisher's least-significant difference (LSD) post hoc tests were carried out with statistical significance set at $\alpha = 0.05$.

# 3. Results

## 3.1. FEM results – Dynamic loading of bilayered constructs

Finite element models of bilayered agarose constructs under applied dynamic deformation showed quantitative and qualitative differences in the resulting magnitude and distribution of mechanical stimuli compared to homogeneous gel constructs (Fig. 1). Homogeneous 2% agarose constructs exhibited greatest radial strain and fluid pressurization in the central region ($z = 0$–0.0025) with maximal fluid flow along the edges directed outward (Fig. 1A,B). Predicted compressive strain in the homogeneous construct during loading was uniform throughout the construct (not shown). These trends for the homogeneous constructs showed no differences with depth. Bilayered constructs, however, displayed a more

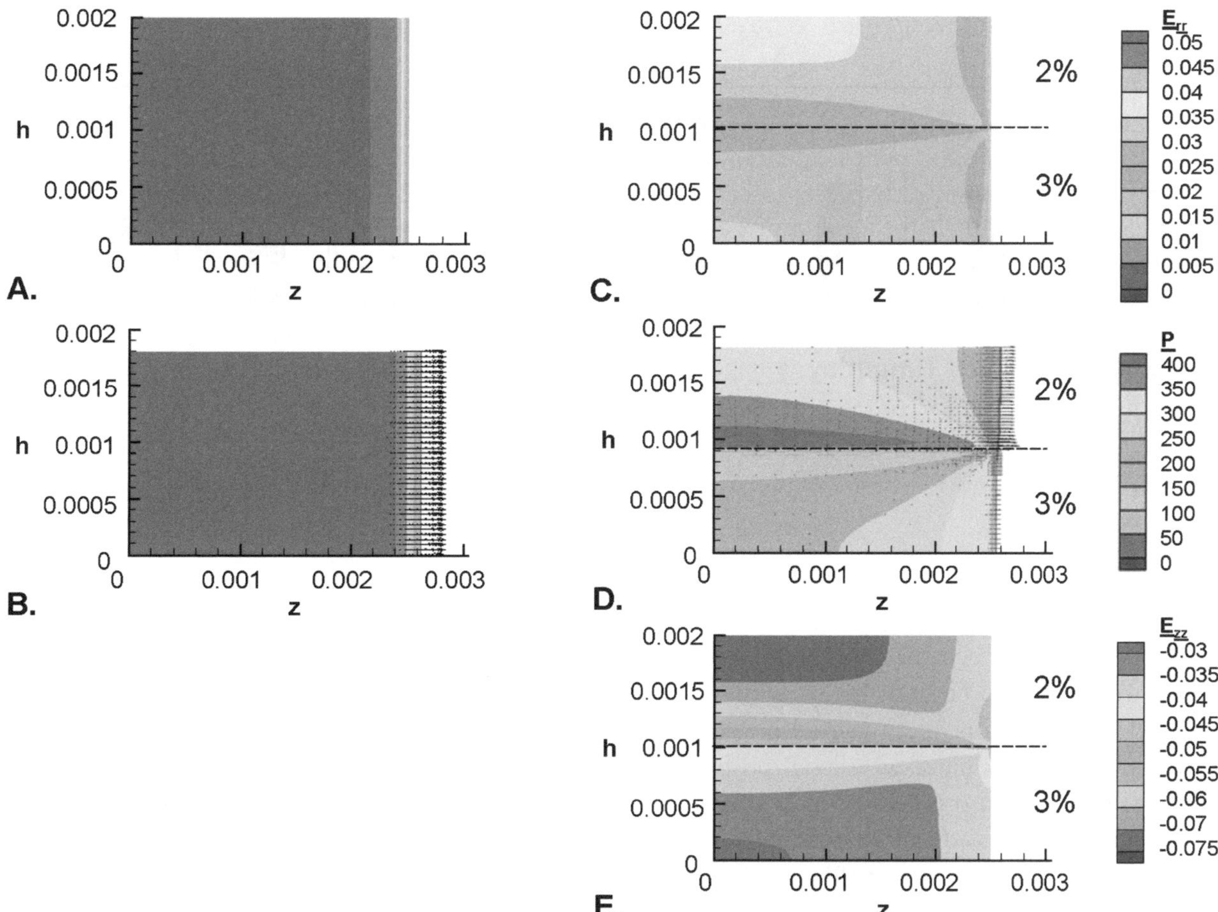

Fig. 1. Finite element predictions of radial strain (A, C; contour mapping, $E_{rr}$), hydrostatic pressure (B, D; contour mapping, $P$: Pa), and fluid flows (B, D; arrows) of a homogeneous 2% agarose construct and a bilayered construct during unconfined compressive loading. The bilayered construct shows a more inhomogeneous distribution of generated mechanical signals compared to the uniform gel construct. In addition, greater compressive strain was found in the 2% agarose layer than the 3% agarose layer (E; contour mapping, $E_{zz}$). Predicted compressive strain for the homogeneous construct was uniform throughout (not shown). Dotted line (C, D, E) indicates interface between layers, $z$ = radial position (m), $h$ = axial position (m).

inhomogeneous distribution of mechanical stimuli in both the depth and radial directions, with greater radial strain, compressive strain, fluid flow, and pressure in the upper 2% agarose layer than the lower 3% agarose layer (Fig. 1C,D,E). A small amount of fluid flow was also observed to be directed downward through the upper layer into the lower one.

## 3.2. Bulk mechanical properties of bilayered constructs

After 28 days in culture, dynamically loaded constructs possessed a significantly higher $E_Y$ than free-swelling constructs ($31.17 \pm 2.48$ kPa vs. $26.08 \pm 2.65$ kPa, $p < 0.05$), with both construct groups stiffer than on day 0 ($E_Y$: $21.64 \pm 1.62$ kPa, $p < 0.05$) (Fig. 2A). This difference was accentuated in $G^*$, with significant increases resulting only in the dynamically loaded group (DL: $0.18 \pm 0.03$ MPa, FS: $0.12 \pm 0.03$ MPa, Day 0: $0.10 \pm 0.02$ MPa, $p < 0.05$) (Fig. 2B). No constructs were found to delaminate under either free-swelling or dynamic loading conditions.

    *K.W. Ng et al. / Dynamic loading of a bilayered engineered cartilage construct*

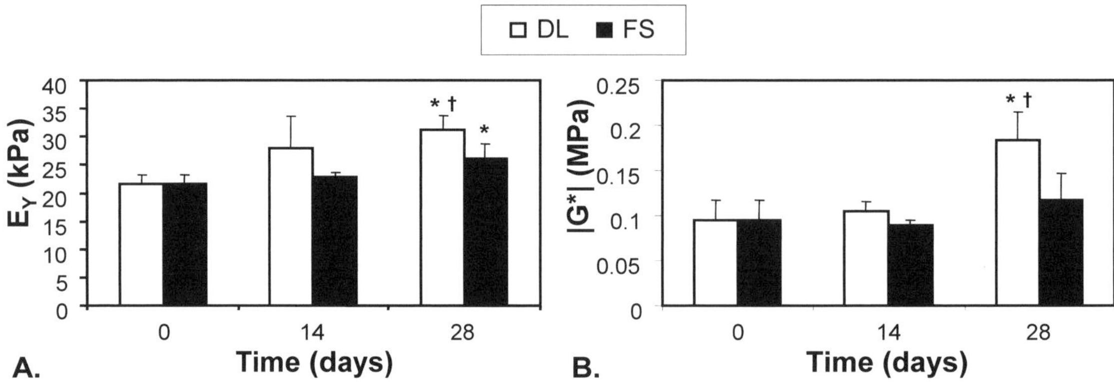


Fig. 2. Constructs showed significant increases in bulk compressive Young's modulus ($E_Y$, A) after 28 days in culture. At this time point, dynamic loading resulted in significant increases in both bulk $E_Y$ (A) and dynamic modulus ($|G^*|$, B) compared to day 0 and free-swelling values. $n = 4$–5; $^*p < 0.05$ vs. respective day 0; $^\dagger p < 0.05$ vs. respective FS.

### 3.3. Depth-dependent material properties

Construct bulk mechanical properties obtained from both the microscopy and bulk testing device above were found to be statistically similar for all constructs tested (data not shown, $p > 0.05$). Evaluation of bilayered construct material properties using optical microscopy and digital image correlation revealed an initially piece-wise axial profile in $E_Y$, with a significantly softer 2% agarose layer compared to the 3% agarose layer (Fig. 3 top, $p < 0.05$). With time in culture, this profile became much smoother, though significant increases in $E_Y$ were only observed in the 2% layer in both groups (Fig. 3 bottom, $p < 0.05$). In free-swelling culture, it was noted that the 2% layer still remained softer at the outer edge (Fig. 3 bottom, $z/h = 0$, $p < 0.05$). The trend observed in the free-swelling culture at all time points were consistent with previously reported results [27]. This trend was partially reversed with the application of dynamic loading, resulting in significant increases in local $E_Y$ at positions $z/h = 0$ and 0.2 (Fig. 3, middle, bottom) that by day 28 were significantly stiffer than its entire 3% layer and also stiffer than the FS 2% agarose layer at position $z/h = 0$ (Fig. 3 bottom, $p < 0.05$).

### 3.4. Biochemistry and histology

Construct GAG and collagen content (normalized to wet weight) was found to increase over time in culture (Table 1). Dynamic loading significantly improved the GAG and collagen of each layer over time in culture versus free-swelling controls (Table 1, $p < 0.05$). In each culture condition, GAG and collagen content were similar in either layer over time in culture.

Histological staining revealed a more diffuse GAG and collagen distribution in the 2% layer in both dynamically loaded and free-swelling constructs. Matrix, especially GAG, was found to be more localized pericellularly in the 3% layer of both construct groups. A highly cellular and matrix-rich layer of tissue approximately 100 $\mu$m thick was found present on the surface of the 2% layer only in free-swelling constructs on day 28 (Fig. 4) that was not present at earlier time points (not shown). This layer was noticeably absent in dynamically loaded constructs.

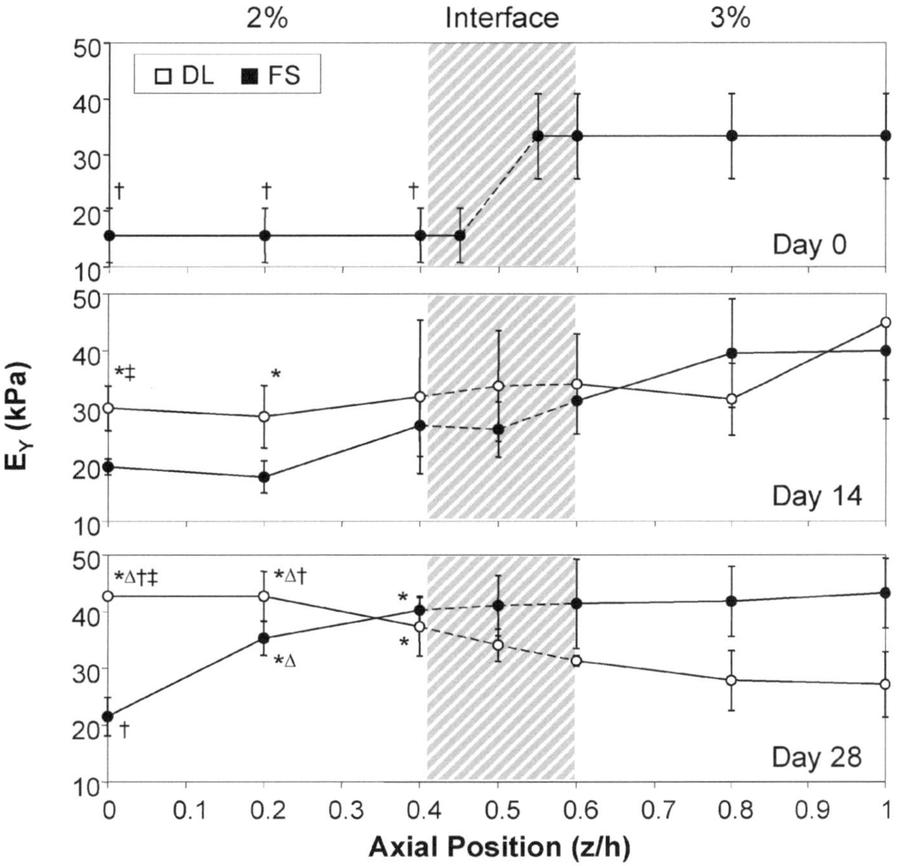



Fig. 3. Time in culture and dynamic loading were found to modulate the distribution of local compressive $E_Y$ through the depth of the bilayered construct. In both free-swelling and in dynamically loaded constructs, the 2% gel layer was the only layer of the bilayered construct to exhibit any significant increases in local $E_Y$ over time in culture. $n = 4$–5; $^*p < 0.05$ vs. day 0; $\Delta = p < 0.05$ vs. day 14; $^\dagger p < 0.05$ vs. 3% layer; $^\ddagger p < 0.05$ vs. respective FS.

Table 1

Biochemical content (normalized to wet weight) of engineered cartilage constructs increased over the 28 day culture period, with no significant differences between layers in either culture condition. The application of dynamic loading was found to significantly increase both GAG and collagen content in each layer over time in culture. $n = 4$–5

| | GAG (%ww) | | Collagen (%ww) | |
|---|---|---|---|---|
| DL | 2% | 3% | 2% | 3% |
| Day 0 | $0.06 \pm 0.02$ | $0.06 \pm 0.02$ | $0.00 \pm 0.00$ | $0.05 \pm 0.06$ |
| Day 14 | $0.35 \pm 0.10^*$ | $0.39 \pm 0.08^{*\dagger}$ | $0.42 \pm 0.15^{*\dagger}$ | $0.38 \pm 0.09^{*\dagger}$ |
| Day 28 | $0.95 \pm 0.20^{*\dagger}$ | $0.90 \pm 0.18^{*\dagger}$ | $1.23 \pm 0.39^{*\dagger}$ | $1.20 \pm 0.35^{*\dagger}$ |
| FS | 2% | 3% | 2% | 3% |
| Day 0 | $0.06 \pm 0.02$ | $0.06 \pm 0.02$ | $0.00 \pm 0.00$ | $0.05 \pm 0.06$ |
| Day 14 | $0.13 \pm 0.13$ | $0.06 \pm 0.03$ | $0.14 \pm 0.12$ | $0.05 \pm 0.07$ |
| Day 28 | $0.32 \pm 0.11^*$ | $0.21 \pm 0.18$ | $0.22 \pm 0.09^*$ | $0.09 \pm 0.16$ |

$^*p < 0.05$ vs. previous time points, $^\dagger p < 0.05$ vs. respective FS of same time point.

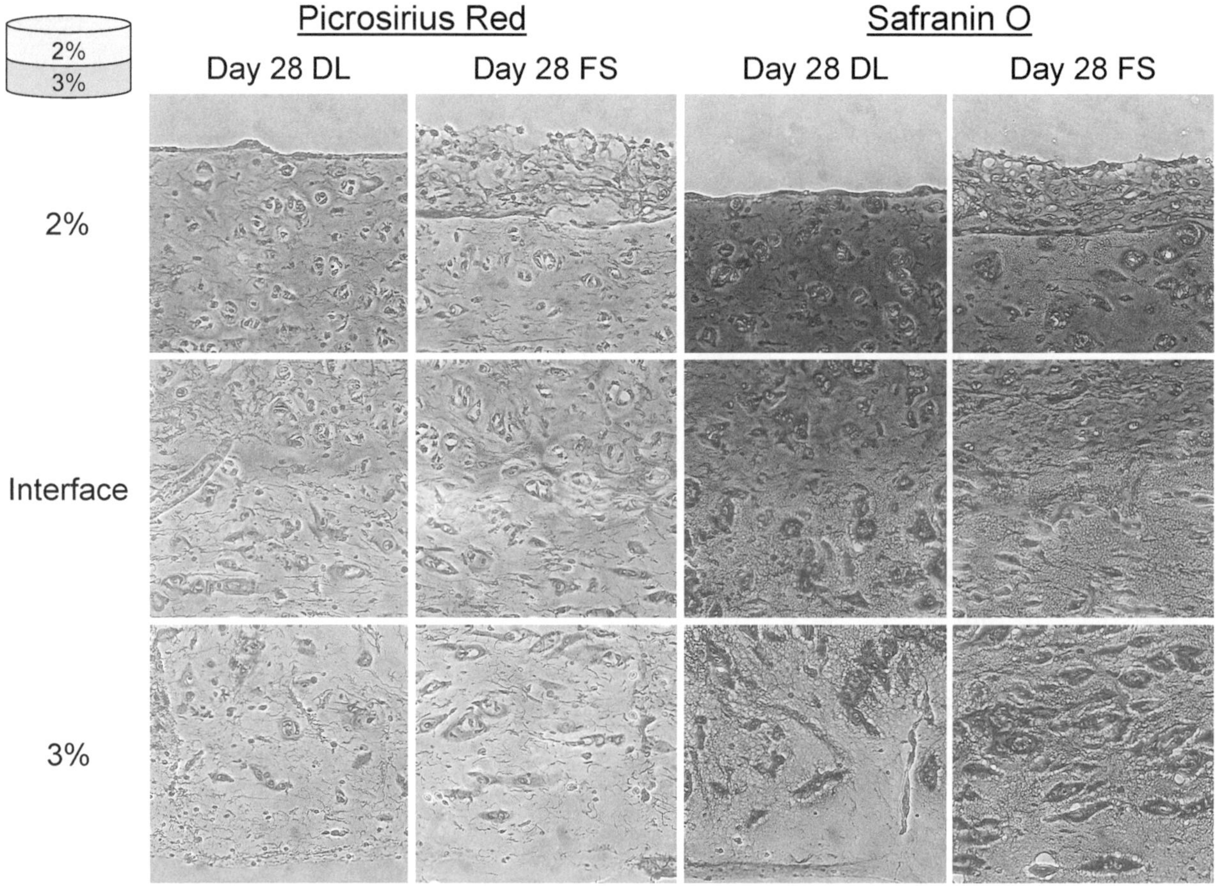


Fig. 4. Representative sections of bilayered constructs using Picrosirius Red (for collagen) and Safranin O (for GAG) dyes. Histological staining revealed diffuse matrix distribution in the 2% layer and a more pericellular matrix distribution in the 3% layer in both culture conditions. A highly cellular, matrix-rich layer of tissue was found present on the surfaces of the 2% layer of free-swelling constructs. Scale bar = 100 $\mu$m.

## 4. Discussion

It has been previously shown that the mechanical inhomogeneity of an engineered cartilage construct is dependent on a combination of tissue elaboration and the initial scaffold material properties [27]. In the present study, the application of dynamic deformational loading to a bilayered, engineered cartilage construct increased bulk construct mechanical properties and biochemical content and was found to further modulate the development of the local tissue mechanical properties. Specifically, it was found that under loading, the developing tissue in the 2% agarose layer became much stiffer than that in the 3% agarose layer over time in culture, reversing the trend observed in free-swelling. The FEM model of the bilayered construct using the initially prescribed gel mechanical properties under 10% deformational loading showed that the amount of compressive strain and fluid flow was considerably greater in the 2% compared to the 3% agarose layer. These observations suggest that cells seeded in each layer experience differing levels of mechanical stimuli during deformational loading, and that signaling mechanisms and solute transport would be more significant in the 2% layer, consistent with the greater tissue development observed in that layer. This model, however, can only predict the resulting mechanical forces in the initial construct before tissue development. As more information is collected on the tissue development over time in culture, the FEM model can be expanded to study the gradients of mechanical stimuli through the entire culture period. In addition, this FEM model does not address cellular deformation, another source of stimulation for the chondrocytes. In our preliminary cell deformation measurements, it was found that under an applied static strain of 20%, chondrocytes seeded in the 2% scaffold at day 0 experienced an equilibrium cell deformation of $\sim$20% whereas chondrocytes in the 3% layer saw little or no cell deformation [26]. As chondrocytes have been shown to respond to cell deformation [18], this is another source of stimulation that needs further investigation in regards to the results of the presented study.

Dynamic loading was found to elevate the mechanical properties and biochemical content of the construct as a whole over free-swelling conditions. These bulk results may be directly explained by the improved nutrient and waste transport through the construct under dynamic deformational loading, as predicted by a previous theoretical study [21]. Some experimental confirmation of this hypothesis is given in a past study examining cell density and nutrient supply [22]. In that study, dynamic loading in the presence of increased FBS concentrations did not significantly increase the mechanical and biochemical properties of construct with a low initial seeding density (10 million cells/ml). However, in constructs with high seeding density (60 million cells/ml), it was found that maximal increases in bulk construct properties were obtained in the presence of both dynamic loading and increased FBS concentration. The results of that study would indicate that sufficient access to nutrients and stimulatory molecules is one of key factors in maximizing cartilaginous tissue development.

While dynamic loading of constructs is likely to enhance convective transport of solutes, results show that the 2% layer was the only region to exhibit significant increases in mechanical stiffness even under free-swelling conditions. This finding suggests that the differences in pore size and structure between 2% and 3% agarose also affect diffusive transport mechanisms, as hypothesized in our earlier free-swelling study [27]. In our earlier study, measurements of the diffusion of 70 kDa dextran in agarose demonstrated significantly lower concentrations in 3% agarose compared to 2% agarose at 1 h and 6 h [27] and it is expected that this effect would be exacerbated with larger molecular weight matrix products synthesized by the chondrocytes. Further support for this hypothesis is provided by the study of Sengers et al., who used a finite element model to simulate a 2% agarose hydrogel seeded with 10 million cells/ml developing over time in culture [30]. They found that with reduced diffusion of matrix molecules, predicted matrix accumulation was mostly pericellular and that such constructs had a significantly lower predicted aggregate modulus compared to cases of increased molecular diffusion. In the current study, matrix formation in the 3% agarose layer in both loaded and free-swelling constructs was found to mimic the numerical predictions of Sengers et al., with matrix formation localized pericellularly, and little increase in the local material properties of the 3% layer.

These theoretical and experimental results taken together imply that the scaffold permeability and porosity may be more important factors than its initial mechanical stiffness. They also suggest that there is a critical concentration of agarose above which interferes with interconnected matrix formation, probably due to hindered transport, and that this critical concentration falls between 2% and 3%. Based on these results, we conclude that a strategy for engineering cartilage constructs that exhibit depth-dependent inhomogeneity using layered agarose gels of varying concentrations is not likely to be successful. These results suggest the hypothesis that there is an optimal concentration of agarose which sufficiently facilitates transport of matrix products away from cells and into the inter-territorial matrix, without causing excessive loss into the culture media.

Histological examination of the constructs revealed a layer of cell outgrowth found on the surfaces of the free-swelling constructs that was only apparent on the surfaces of the 2% agarose layer (Fig. 4).

The compression of this layer, which is without any scaffold support, may explain the softer measured 2% agarose properties at $z = 0$ that are inconsistent with previous comparisons of chondrocyte-seeded 2% and 3% agarose hydrogels [27]. The layer of outgrowing cells was found to occur in free-swelling agarose hydrogel systems [14,16,22] and in cartilage explant systems [23]. Dynamically loaded bilayered constructs did not possess this tissue outgrowth and this observation, along with the measured changes in mechanical properties of the loaded 2% layer, is consistent with previous findings [22]. The lack of the tissue layer may be due to an inhibition of proliferation resulting from physical contact with the loading platen or micro-motion at the platen-construct contact surface sloughing off any outgrowth. In a previous study on agarose construct inhomogeneity [14], a coarser level of local property measurements was used (5 regions through the depth with properties averaged in each region) that may have masked the effects of the outgrowth layer on the edge measurements versus the finer increments used in this paper.

This study demonstrates that dynamic loading of a chondrocyte-seeded layered agarose construct produces an inhomogeneous compressive modulus over time in culture which does not mimic the inhomogeneity of the scaffold at day 0 or over time in free-swelling culture. Matrix elaboration occurred preferentially in the layer with the lower agarose concentration, suggesting that agarose transport properties play a much more important role than initial stiffness in regulating tissue production. Though the concept that intelligent scaffold design in combination with other factors such as applied stimuli or selective cell populations can help steer tissue development remains promising, the experimental results do not support the hypothesis that the application of dynamic loading to this type of bilayered construct can accentuate the development of mechanical inhomogeneity of the resulting tissue to mimic that of native cartilage.

## Acknowledgements

This study was supported by grants from the National Institutes of Health (AR46532, AR46568) and a pre-doctoral fellowship from the Whitaker Foundation.

## References

[1] N.A. Andarawis, S.L. Seyhan, R.L. Mauck, M.A. Soltz, G.A. Ateshian and C.T. Hung, A novel device for direct permeation measurements of hydrogels and soft hydrated tissues, in: *Advances in Bioengineering, BED 51*, B.B. Lieber, ed., Vol., ASME, New York, 2001, pp. 299–300.

[2] M.B. Aydelotte, R.R. Greenhill and K.E. Kuettner, Differences between sub-populations of cultured bovine articular chondrocytes. II. Proteoglycan metabolism, *Connect. Tissue Res.* **18**(3) (1988), 223–234.

[3] M.B. Aydelotte and K.E. Kuettner, Differences between sub-populations of cultured bovine articular chondrocytes. I. Morphology and cartilage matrix production, *Connect. Tissue Res.* **18**(3) (1988), 205–222.

[4] J.A. Buckwalter and H.J. Mankin, Articular cartilage: tissue design and chondrocyte–matrix interactions, *Instr. Course Lect.* **47** (1998), 477–486.

[5] M.D. Buschmann, Y.A. Gluzband, A.J. Grodzinsky, J.H. Kimura and E.B. Hunziker, Chondrocytes in agarose culture synthesize a mechanically functional extracellular matrix, *J. Orthop. Res.* **10**(6) (1992), 745–758.

[6] M.D. Buschmann, Y.A. Gluzband, A.J. Grodzinsky and E.B. Hunziker, Mechanical compression modulates matrix biosynthesis in chondrocyte/agarose culture, *J. Cell Sci.* **108**(Pt 4) (1995), 1497–1508.

[7] S.E. Carver and C.A. Heath, Increasing extracellular matrix production in regenerating cartilage with intermittent physiological pressure, *Biotechnol. Bioeng.* **62**(2) (1999), 166–174.

[8] P.G. Chao, Z. Tang, E. Angelini, A.C. West, K.D. Costa, and C.T. Hung, Dynamic osmotic loading of chondrocytes using a novel microfluidic device, *J. Biomech.* **38**(6) (2005), 1273–1281.

[9] T.T. Chowdhury, D.L. Bader, J.C. Shelton and D.A. Lee, Temporal regulation of chondrocyte metabolism in agarose constructs subjected to dynamic compression, *Arch. Biochem. Biophys.* **417**(1) (2003), 105–111.

[10] R.W. Farndale, C.A. Sayers and A.J. Barrett, A direct spectrophotometric microassay for sulfated glycosaminoglycans in cartilage cultures, *Connect. Tissue Res.* **9**(4) (1982), 247–248.

[11] W.Y. Gu, H. Yao, C.Y. Huang and H.S. Cheung, New insight into deformation-dependent hydraulic permeability of gels and cartilage, and dynamic behavior of agarose gels in confined compression, *J. Biomech.* **36**(4) (2003), 593–598.

[12] A.P. Hollander, T.F. Heathfield, C. Webber, Y. Iwata, R. Bourne, C. Rorabeck and A.R. Poole, Increased damage to type II collagen in osteoarthritic articular cartilage detected by a new immunoassay, *J. Clin. Invest.* **93**(4) (1994), 1722–1732.

[13] E.B. Hunziker, T.M. Quinn and H.J. Hauselmann, Quantitative structural organization of normal adult human articular cartilage, *Osteoarthritis Cartilage* **10**(7) (2002), 564–572.

[14] T.A. Kelly, K.W. Ng, C.C. Wang, G.A. Ateshian and C.T. Hung, Spatial and temporal development of chondrocyte-seeded agarose constructs in free-swelling and dynamically loaded cultures, *J. Biomech.* **39**(8) (2006), 1489–1497.

[15] T.K. Kim, B. Sharma, C.G. Williams, M.A. Ruffner, A. Malik, E.G. McFarland and J.H. Elisseeff, Experimental model for cartilage tissue engineering to regenerate the zonal organization of articular cartilage, *Osteoarthritis Cartilage* **11**(9) (2003), 653–664.

[16] J.D. Kisiday, B. Kurz, M.A. DiMicco and A.J. Grodzinsky, Evaluation of medium supplemented with insulin–transferrin–selenium for culture of primary bovine calf chondrocytes in three-dimensional hydrogel scaffolds, *Tissue Eng.* **11**(1–2) (2005), 141–151.

[17] T.J. Klein, B.L. Schumacher, T.A. Schmidt, K.W. Li, M.S. Voegtline, K. Masuda, E.J. Thonar and R.L. Sah, Tissue engineering of stratified articular cartilage from chondrocyte subpopulations, *Osteoarthritis Cartilage* **11**(8) (2003), 595–602.

[18] D.A. Lee and M.M. Knight, Mechanical loading of chondrocytes embedded in 3D constructs: in vitro methods for assessment of morphological and metabolic response to compressive strain, *Methods Mol. Med.* **100** (2004), 307–324.

[19] H.J. Mankin, V.C. Mow, J.A. Buckwalter, J.P. Iannotti and A. Ratcliffe, Articular cartilage structure, composition, and function, in: *Orthopaedic Basic Science. Biology and Biomechanics of the Musculoskeletal System*, J.A. Buckwalter, T.A. Einhorn and S.R. Simon, eds, Vol., American Academy of Orthopaedic Surgeons, Rosemont, 2000, pp. 443–470.

[20] R.L. Mauck, M.A. Soltz, C.C. Wang, D.D. Wong, P.H. Chao, W.B. Valhmu, C.T. Hung and G.A. Ateshian, Functional tissue engineering of articular cartilage through dynamic loading of chondrocyte-seeded agarose gels, *J. Biomech. Eng.* **122**(3) (2000), 252–260.

[21] R.L. Mauck, C.T. Hung and G.A. Ateshian, Modeling of neutral solute transport in a dynamically loaded porous permeable gel: implications for articular cartilage biosynthesis and tissue engineering, *J. Biomech. Eng.* **125** (5)(2003), 602–614.

[22] R.L. Mauck, C.C. Wang, E.S. Oswald, G.A. Ateshian and C.T. Hung, The role of cell seeding density and nutrient supply for articular cartilage tissue engineering with deformational loading, *Osteoarthritis Cartilage* **11**(12) (2003), 879–890.

[23] M. Moretti, D. Wendt, D. Schaefer, M. Jakob, E.B. Hunziker, M. Heberer and I. Martin, Structural characterization and reliable biomechanical assessment of integrative cartilage repair, *J. Biomech.* **38**(9) (2005), 1846–1854.

[24] J.K. Mouw, N.D. Case, R.E. Guldberg, A.H. Plaas and M.E. Levenston, Variations in matrix composition and GAG fine structure among scaffolds for cartilage tissue engineering, *Osteoarthritis Cartilage* **13**(9) (2005), 828–836.

[25] V.C. Mow, S.C. Kuei, W.M. Lai and C.G. Armstrong, Biphasic creep and stress relaxation of articular cartilage in compression? Theory and experiments, *J. Biomech. Eng.* **102**(1) (1980), 73–84.

[26] K.W. Ng, C.C. Wang, X.E. Guo, G.A. Ateshian and C.T. Hung, Characterization of inhomogeneous bi-layered chondrocyte-seeded agarose constructs of differing agarose concentrations, *Trans. ORS* **28** (2003), 960.

[27] K.W. Ng, C.C. Wang, R.L. Mauck, T.A. Kelly, N.O. Chahine, K.D. Costa, G.A. Ateshian and C.T. Hung, A layered agarose approach to fabricate depth-dependent inhomogeneity in chondrocyte-seeded constructs, *J. Orthop. Res.* **23**(1) (2005), 134–141.

[28] S. Park, C.T. Hung and G.A. Ateshian, Mechanical response of bovine articular cartilage under dynamic unconfined compression loading at physiological stress levels, *Osteoarthritis Cartilage* **12**(1) (2004), 65–73.

[29] R.M. Schinagl, D. Gurskis, A.C. Chen and R.L. Sah, Depth-dependent confined compression modulus of full-thickness bovine articular cartilage, *J. Orthop. Res.* **15**(4) (1997), 499–506.

[30] B.G. Sengers, C.C. Van Donkelaar, C.W. Oomens and F.P. Baaijens, The local matrix distribution and the functional development of tissue engineered cartilage, a finite element study, *Ann. Biomed. Eng.* **32**(12) (2004), 1718–1727.

[31] H. Stegemann and K. Stalder, Determination of hydroxyproline, *Clin. Chim. Acta* **18**(2) (1967), 267–273.

[32] C.C. Wang, J.M. Deng, G.A. Ateshian and C.T. Hung, An automated approach for direct measurement of two-dimensional strain distributions within articular cartilage under unconfined compression, *J. Biomech. Eng.* **124**(5) (2002), 557–567.

[33] C.C. Wang, N.O. Chahine, C.T. Hung and G.A. Ateshian, Optical determination of anisotropic material properties of bovine articular cartilage in compression, *J. Biomech.* **36**(3) (2003), 339–353.

Biorheology 43 (2006) 509–513
IOS Press



# Three-dimensional tissue constructs built by bioprinting

Karoly Jakab [a], Brook Damon [a], Adrian Neagu [b], Anatolij Kachurin [c] and Gabor Forgacs [a,d,*]

[a] *Department of Physics, University of Missouri, Columbia, MO 65211, USA*

[b] *Department of Biophysics and Medical Informatics, Victor Babes University of Medicine and Pharmacy Timisoara, 1900 Timisoara, Romania*

[c] *Sciperio Inc., 2721 Discovery Dr. Suite 400, Orlando, FL 32826, USA*

[d] *Department of Biology, University of Missouri, Columbia, MO 65211, USA*

**Abstract.** Bioprinting is an evolving tissue engineering technology. It utilizes computer controlled three-dimensional printers for rapid and high-precision construction of three-dimensional biological structures. We employed discrete and continuous bioprinting to build three-dimensional tissue constructs. In the former case bioink particles – spherical cell aggregates composed of many thousands of cells – are delivered one by one into biocompatible scaffolds, the biopaper. Structure formation takes place by the subsequent fusion of the bioink particles due to their liquid-like and self-assembly properties. In the latter case a mixture of cells and scaffold material is extruded from the biocartridge akin to toothpaste to arrive at the desired construct. Specifically, we built rectangular tissue blocks of several hundred microns in thickness as well as tubular structures of several millimeters in height. The physical basis of structure formation was studied by computer simulations.

Keywords: Tissue engineering, bioprinter, spherical cell aggregate, tissue liquidity, self-assembly

## 1. Introduction

It has recently been suggested and demonstrated that three-dimensional living structures of specific geometry can be constructed using the technology of bioprinting [2,3,8,9]. One approach utilizes inkjet printers to deliver individual cells into the biocompatible scaffold [8]. Another technique is based on mechanical printers that extrude discrete bioink particles, which typically are cellular spheroids of varying size, composed of a single cell type (i.e. "single-color bioink") or of several cell types, as well as extracellular matrix ("multi-color bioink") [3,6]. In yet another technology, also utilizing mechanical extruders, the cell-scaffold mixture is printed in a continuous manner [9]. The advantage of resorting to printing of cells or their aggregates, as opposed to seeding them into scaffolds as in the more traditional tissue engineering, is the possibility of employing rapid prototyping for building biological structures of complex shape. The ultimate goal in this technology is to mitigate the chronic shortage of replacement organs, by "printing organs" or organ modules [4].

Here we demonstrate both discrete and continuous bioprinting. First, we build tissue blocks by preparing the initial cellular arrangement by printing individual spherical cell aggregates into sheets. As described earlier [1,2,6] cellular spheroids composed of adhesive and motile cells possess liquid-like

---

*Address for correspondence: Gabor Forgacs, Department of Physics, University of Missouri, Columbia, MO 65211, USA. Tel.: +1 573 882 3036; Fax: +1 573 882 4195; E-mail: forgacsg@missouri.edu.

properties. The molecular basis of such behavior is provided by the Differential Adhesion Hypothesis (DAH) [7], which stipulates that tissues possess measurable surface or interfacial tension, whose origin is in the adhesive properties of their constituent cells. Indeed, it has been shown that postprinting structure formation proceeds via the fusion of the spherical bioink particles and the final outcome of the process strongly depends on the properties of the embedding gel-biopaper [2]. Furthermore, computer simulations revealed that the single parameter that controls structure formation, consistent with DAH, is the gel-aggregate interfacial tension [2,6]. Finally, we print tubular structures by continuously extruding cell–gel paste.

## 2. Discrete bioprinting: Thick cellular sheets

Tissues with layered structures abound in living organisms. Perhaps the most common such tissue is skin. Replacement of skin in burn patients, depending on the extent of injury, presents numerous challenges [5]. Skin grafts with thickness of several millimeters maybe necessary. Typically, these are harvested (and often reharvested) from donor sites by removing skin from healthy regions of the body (if such still exist), leaving behind exposed areas with risks of infection. Furthermore, transplanted grafts, even if they heel, often do so with considerable scars.

Printing cellular sheets, in principle, may be ideally suited for replacing damaged skin. True, a certain number of healthy cells are needed, but these can be harvested from several locations, thus creating little disturbance in healthy skin. After expanding these cells, aggregates are prepared with diameter comparable to the thickness of the desired planar graft. Printing and postprinting structure formation of centimeter-size patches takes 2–3 days.

For the purpose of demonstration of such a process, we prepared uniform size (475 micron diameter) spherical aggregates of Chinese Hamster Ovary (CHO) cells and packaged them into micropipettes of comparable inner diameter (Fig. 1). These micropipettes serve as cartridges for the bioprinter depicted

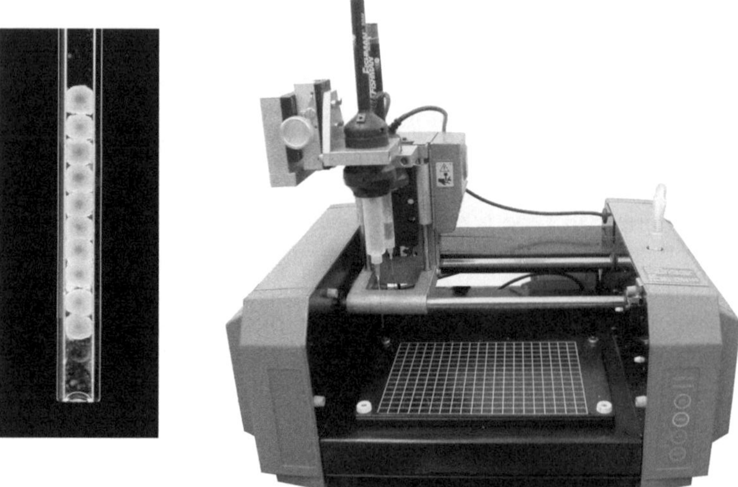

Fig. 1. Right: A bioink cartridge, a micropipette (the one shown has 500 micron inner diameter) with the spherical aggregate-bioink particles. Right: Three-dimensional bioprinter with two mechanically driven extruders (i.e. deposition heads). In discrete printing one of the extruders hosts the bioink cartridge (shown on the left). The other extruder prints the biopaper-hydrogel scaffold. The x–y stage and the z directional motion of the extruders are fully computer controlled.

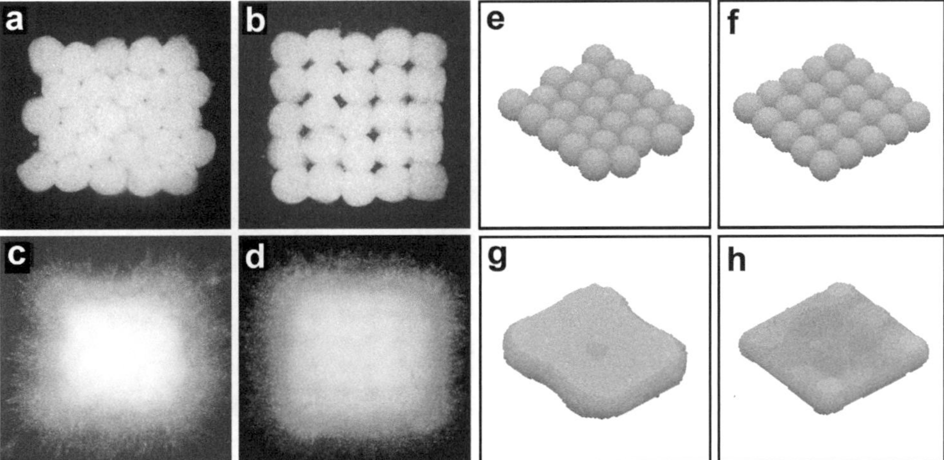

Fig. 2. Sheet formation depends on the initial configuration and the tissue–matrix interfacial tension. Two initial structures of 25 aggregates of CHO cells (500 micron in diameter) were embedded in 1.0 mg/ml collagen type I in a hexagonal (a) and square lattice (b) arrangements. Compact sheets after 144 hours of incubation are shown in panels (c) and (d). Similar initial states, made of model cell aggregates, 925 cells each, (e) and (f) where prepared by computer. Monte Carlo Simulations (based on a simple statistical mechanical model of a system of cell and gel particles, characterized by a single parameter, the gel–cell interfacial tension) lead to the final configurations shown in panels g and h, respectively (arrived at after 250,000 Monte Carlo Steps). For details of the simulations, see references [2,6].

in Fig. 1, which can simultaneously host two cartridges. In the actual printing process one cellular cartridge and another one containing the material of the biopaper-gel are used. (Note, in this process the gel-biopaper itself is printed.) First, a square collagen bed is laid down by the printer, into which subsequently cell aggregates are deposited in a square or hexagonal pattern as shown in Fig. 2. Using 1 mg/ml collagen gel, postprinting structure formation consistently resulted in fused cellular sheets. Fusion was faster (at room temperature) starting from the hexagonal initial state, since in this case coalescence of neighboring aggregates required less cellular motion.

As mentioned, structure formation is controlled by the mutual interfacial properties of the biopaper and the bioink. Similarly to earlier results, a computer simulation (for details see [2,6]) based on DAH and a specific model with a single parameter (the interfacial tension between the biopaper and bioink) reproduced structure formation (Fig. 2).

The stability of the final pattern depends on the embedding scaffold material [2]. Indeed, as can be seen in Figs 2c,d, CHO cells can migrate into the gel. Increasing collagen concentration considerably enhances this tendency. Eventually the final pattern reduces to a single large spheroidal aggregate (consistent with evolution to a lower energy state) and some cells dispersed in the scaffold [2]. The real challenge of the printing technology, or more generally, of any tissue engineering activity to produce functional tissue, is to optimize the properties of the embedding material for the given cell type.

## 3. Continuous bioprinting: Cellular tubes

Similarly to skin, blood vessels represent layered tissues. In addition, they have tubular geometry, considerably more complicated than the simple planar arrangements discussed above. The lumen-facing inner layer of mid-size or large blood vessels is composed of endothelial cells, whereas the outer layer is muscle. Sandwiched between the two is an extracellular layer of varying thickness. Engineering blood

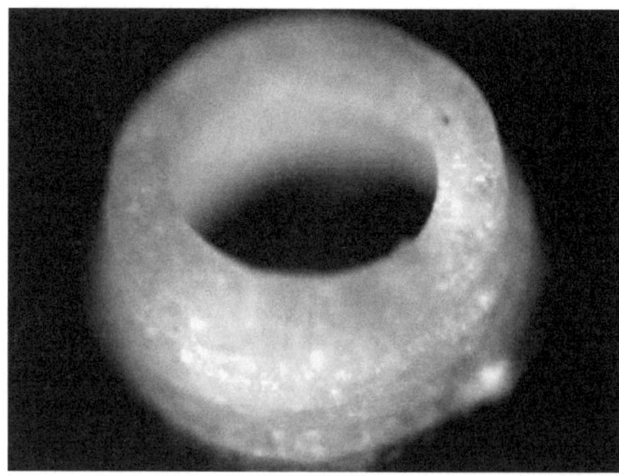

Fig. 3. Continuously printed cellular tube. For details, see text.

vessels is of crucial importance, because most tissues and all organs require vasculature. It is not surprising that success in tissue engineering so far has been achieved predominantly in producing avascular tissues (e.g. cartilage, tendon).

We have reported the (manual) printing of simple tubular structures (using aggregates of a single cell type) using discrete printing [2]. Here we illustrate continuous printing by building a more complicated tubular structure, made of a mixture of cells and scaffold material (Fig. 3). Quail-derived QCE6 endothelial cells were mixed with pluronic F-127 dissolved in cell-free culture medium to arrive at a homogeneous 27% w/w cellular solution. A hypodermic 1 CC syringe pre-cooled on ice was charged with the cellularized pluronic gel and inserted in one of the positive displacement heads of the bioprinter in Fig. 1. The cellularized gel was deposited in a spiral mode onto a polystyrene slide fixed on the XY stage of the printer and kept at room temperature (23°C). The linear speed of the deposition was 200 microns/sec. The deposition nozzle was a 30 GP stainless steel capillary (EFD), 350 micron O.D. The deposition head was gradually rising 100 microns up at the end of each spiral loop. The tubular construct shown in Fig. 3 contains 15 spiral loops total, with dimensions ~2.7 mm O.D., ~2.2 mm I.D., ~1.7 mm total height. The time taken to build the tube was 10 min and 45 sec.

## 4. Conclusions

We have demonstrated in this work the printing of living biological structures with specific geometry. We illustrated the technology by both discrete and continuous deposition. In the former case we used spherical cell aggregates as bioink particles to produce a cellular sheet. In the latter case a cell–gel paste was employed to arrive at a tubular structure. Both the sheet and the tube represent prototype structures for numerous biologically relevant tissue shapes and could be employed as building blocks for more complicated patterns, in particular organs.

## Acknowledgement

This work was partially supported by NASA and NSF.

# References

[1] K. Jakab, A. Neagu, V. Mironov and G. Forgacs, Organ printing: fiction or science, *Biorheology* **41** (2004), 371–375.

[2] K. Jakab, A. Neagu, V. Mironov, R.R. Markwald and G. Forgacs, Engineering biological structures of prescribed shape using self-assembling multicellular systems. *Proc. Natl. Acad. Sci. USA* **101** (2004), 2864–2869.

[3] V. Mironov, T. Boland, T. Trusk, G. Forgacs and R.R. Markwald, Organ printing: Computer-aided jet-based 3D tissue engineering. *Trends in Biotechnology* **21** (2003), 157–161.

[4] V. Mironov, R. Markwald and G. Forgacs, Organ printing: self-assembling cell aggregates as "bioink", *Science and Medicine* **9** (2003), 69–71.

[5] F.A. Navarro, M.L. Stoner, H.B. Lee, C.S. Park, F.M. Wood and D.P. Orgill, Melanocyte repopulation in full-thickness wounds using a cell spray apparatus, *J. Burn Care Rehabil.* **22** (2001), 41–46.

[6] A. Neagu, K. Jakab, R. Jamison and G. Forgacs, The role of physical mechanisms in biological self-organization, *Phys. Rev. Lett.* **95** (2005), 178104-1-4.

[7] M.S. Steinberg, Does differential adhesion govern self-assembly processes in histogenesis? Equilibrium configurations and the emergence of a hierarchy among populations of embryonic cells, *J. Exp. Zool.* **173** (1970), 395–433.

[8] T. Xu, J. Jin, C. Gregory, J.J. Hickman and T. Boland, Inkjet printing of viable mammalian cells, *Biomaterials* **26** (2005), 93–99.

[9] Y. Yan, X. Wang, Z. Xiong, H. Liu, F. Liu, F. Lin, R. Wu, R. Zhang and Q. Du, Direct construction of a three-dimensional structure with cells and hydrogel, *J. Bioact. Compat. Polym.* **20** (2005), 259–269.

# Part V: Clinical applications

Biorheology 43 (2006) 517–521
IOS Press



# Post-traumatic osteoarthritis: The role of stress induced chondrocyte damage

J.A. Martin and J.A. Buckwalter [*]

*University of Iowa Department of Orthopaedics and Rehabilitation, Iowa City, IA, USA*

**Abstract.** Post-traumatic osteoarthritis is the form of osteoarthritis (OA) that develops following joint injury. Although its end-stage is indistinguishable from idiopathic OA, many patients with post-traumatic OA are younger than those with idiopathic OA, and they have a well-defined precipitating insult. Clinical and experimental studies suggest that excessive acute impact energy or chronic mechanical overload cause the degeneration of the articular surface responsible for post-traumatic OA. Yet, the mechanisms by which excessive mechanical force causes OA remain unknown. For these reasons it has not been possible to develop effective methods of preventing or decreasing the risk of post-traumatic OA. We hypothesized that mechanical loading that exceeds the tolerance of the articular surface causes chondrocyte damage due to oxidative stress. Our *in vitro* tests of human articular cartilage samples showed that shear stress causes chondrocyte death and that anti-oxidants decrease the shear stress induced cell death. These observations suggest that specific patterns of loading are particularly damaging to articular surfaces and that improved treatments of joint injuries may include mechanical methods of minimizing shear stresses and biologic methods of minimizing oxidative damage.

Keywords: Osteoarthritis, trauma, post-traumatic osteoarthritis, oxidative damage, chondrocytes, anti-oxidants

## 1. Introduction

Post-traumatic OA occurs after a variety of joint injuries [1,2,7]: most commonly and predictably following injuries that disrupt the articular surface, or injuries that lead to joint instability [11,19]. Post-traumatic OA appears to be directly related to both the energy delivered to the articular surface acutely at the time of injury, and to excessive chronic or repetitive mechanical stresses applied to articular surfaces [1,3,11]. The energy delivered to the articular surface is primarily a function of the intensity of the acute force pulse causing the joint damage. The chronic or repetitive excessive mechanical stress is caused by either residual joint incongruity due to displacement or loss of parts of the articular surface, or by joint instability due to joint incongruity and/or joint ligament, meniscal or capsule injuries. Since the relationships between acute energy delivered to an articular surface, versus chronic increased articular surface contact stress, and post-traumatic OA have not been investigated in rigorous clinical and basic scientific investigations, the mechanisms responsible for post-traumatic OA remain unknown. This lack of knowledge leaves clinicians with little guidance – other than accumulated empirical observations – in their efforts to forestall post-traumatic OA in patients with joint injuries [11].

Preliminary studies in our laboratory suggested that elevated shear stress releases reactive oxygen species (ROS) from chondrocytes [12–16]. Oxidative damage resulting from chronically elevated ROS

---

[*]Address for correspondence: Joseph A. Buckwalter, 01013 Pappajohn Pavilion, Department of Orthopaedics, University of Iowa College of Medicine, Iowa City, IA 52242, USA. Tel.: +1 319 356 2595; Fax: +1 319 356 8999; E-mail: joseph-buckwalter@uiowa.edu.

can accelerate cell senescence and induce apoptosis, suggesting a link between chondrocyte damage and subsequent cartilage degeneration. Based on these findings, we hypothesized that excessive mechanical shear stress causes chondrocyte death *via* an oxidative mechanism. To test this, we cultured human cartilage explants in a mechanically active bioreactor, the Triaxial Compression Vessel (TCV), which is capable of modulating shear stress levels [9]. This device was used to test the effects of shear stress on chondrocyte viability. We also tested the effect of adding an antioxidant, $n$-acetyl cysteine (NAC), to determine the role of oxidants in mediating shear stress effects.

## 2. Methods

Human cartilage explants (4 mm diameter) were harvested from non-osteoarthritic ankle joints from 8 donors. A subset of the explants was pretreated overnight with 2.5 mM NAC, and all explants were placed in the TCV for mechanical stress treatment. The TCV is a novel culture device capable of imposing variable shear stress states at quasi-physiologic levels [9]. A water chamber applies transverse compression, and a piston applies axial compression. The interplay between axial and transverse compression determines shear stress levels. In this study, four different treatments (900 cycles at 1 Hz) were applied: (1) non-stressed control, (2) 5 MPa axial compression only (maximum shear stress), (3) 5 MPa axial plus 5 MPa transverse compression (minimal shear stress), and (4) 5 MPa axial plus 2.5 MPa transverse compression (intermediate shear stress).

Explants were incubated in antioxidants for 2 hours prior to mechanical stress exposure. These included $n$-acetylcysteine (NAC) at a concentration of 2.5 mM, and vitamin E at a concentration of 100 $\mu$M. Additional explants were incubated for 2 hours before TCV treatment with the nitric oxide (NO) synthase inhibitor N-Nitro-L-Arginine Methyl Ester (L-NAME), which blocks NO-induced apoptosis. L-NAME was used at a concentration of 1.0 mM.

Following mechanical stress treatment in the TCV, explants were removed from the device and incubated overnight in calcein AM to stain viable cells. After cryosectioning, slides were mounted with DAPI to stain cell nuclei. Micrographs of the superficial, middle, and deep zones of the cartilage sections were taken under UV light (to image DAPI-stained nuclei) and under 488 nm light (to image calcein-stained cells) using an Olympus BX60 epifluoresence microscope. Total cells (DAPI-stained) and live cells (calcein-stained) were counted using an custom designed automated MATLAB based image analysis program to determine percent viability. The program automatically identifies and counts fluorescent-labeled cells in high-resolution composite images of full thickness cartilage sections (4 mm wide $\times$ 1–3 mm thick). The program also automatically segments images into superficial (top 15%), middle (16–65%), and deep (66–100%) zones.

For some experiments replicate cryosections from explants were stained with an anti-p53 antibody or stained for apoptosis using a commercial *in situ* fluoresence TUNEL assay kit (Roche). These stains were analyzed as described above for viability stains.

Each experiment was performed with at least 6 explants. One-way ANOVA and Tukey's test were used to evaluate the statistical significance of differences between treatment groups.

## 3. Results

Chondrocyte viability in all zones was high (>95% of unstressed controls) when explants were treated with 2 MPa axial compression alone or 2 MPa axial compression with either 1 or 2 MPa transverse

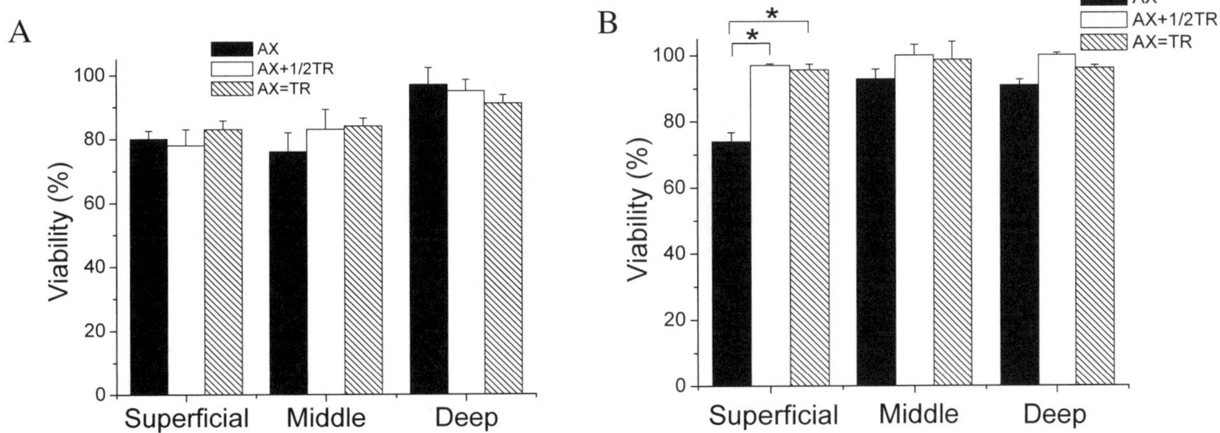

Fig. 1. The effect of variable confinement on chondrocyte viability. (A) Cartilage explants exposed to 900 cycles with 2 MPa axial compression alone (AX), or in combination with 1.0 MPa transverse compression (AX+1/2TR) or 2 MPa transverse compression (AX+TR). (B) Cartilage explants exposed to 900 cycles with 5 MPa axial compression alone (AX), or in combination with 2.5 MPa transverse compression (AX+1/2TR) or 5 MPa transverse compression (AX+TR). Bars and asterisks above the columns indicate a significant difference ($p < 0.05$ by ANOVA).

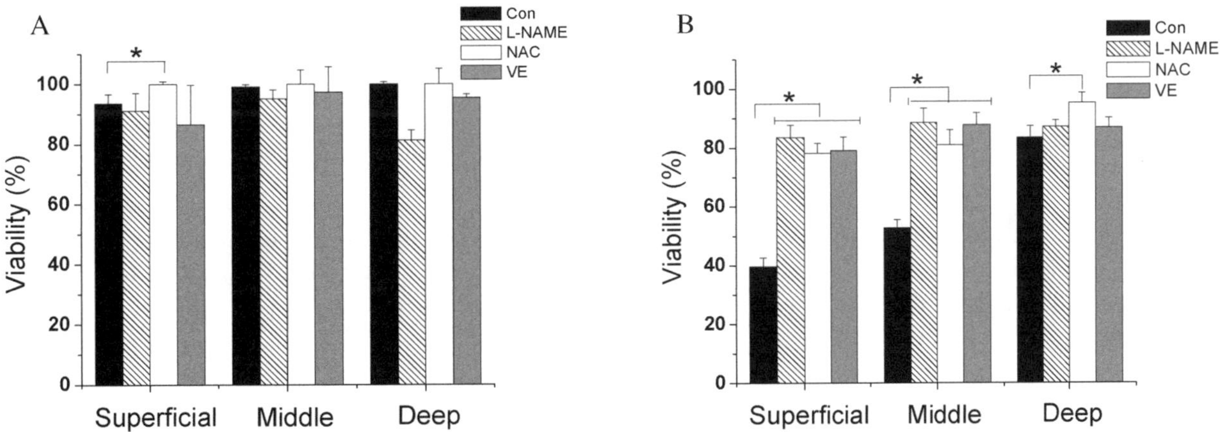

Fig. 2. Effects of NAC, VE, and L-NAME on chondrocyte viability after axial compression. (A) 2 MPa axial compression. (B) 5 MPa axial compression. Controls (Con) were not treated before stress application. Columns represent means and standard errors (error bars) based on analysis of >6 sections. Bars and asterisks indicate significant differences ($p < 0.05$).

compression (Fig. 1A). In contrast, there was marked loss of viability when explants were treated with 5 MPa axial compression alone (high shear stress) (Fig. 1B). The highest loss of chondrocyte viability occurred in the superficial zone, where viability was reduced to less than 75% of unstressed controls. Viability was >90% in groups treated with combinations of 5 MPa axial compression and either 5 MPa or 2.5 MPa transverse compression (low to moderate shear stress).

Pre-incubation of explants with NAC, vitamin E, or L-NAME prior to high shear stress treatment significantly increased viability in the superficial and middle zones of cartilage explants (Fig. 2). All three agents improved viability from 39% to ~80% in the superficial zone and from 52% to >80% in the middle zone. NAC also had a significant effect on deep zone viability, which improved from 83% to 97%. Neither vitamin E, nor L-NAME had significant effects in the deep zone.

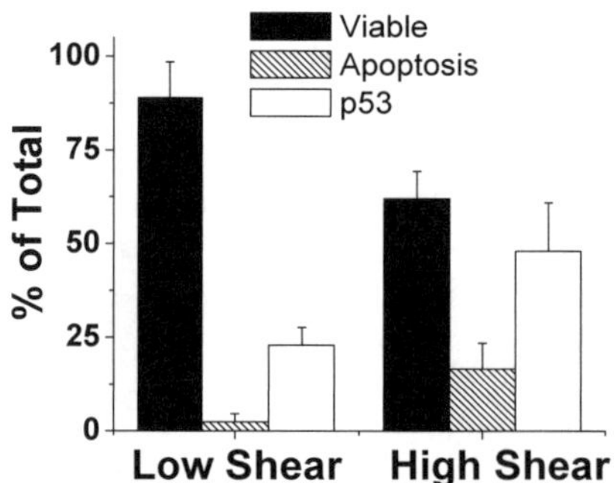


Fig. 3. High shear stress induces p53 expression and apoptosis. Explants were exposed to high shear stress (5 MPa axial compression alone) or low shear stress (5 MPa Axial compression with 2.5 MPa transverse compression). Histologic stains for viability, apoptosis, and p53 were measured by quantitative image analysis. Columns represent mean and standard errors.

Superficial zone chondrocyte apoptosis increased from 2.4% in the low shear stress group (5 MPa axial compression with 5 MPa transverse compression) to 16.7% in the high shear stress group (5 MPa axial compression alone) (Fig. 3). In the same specimens, p53 expression increased from 23% to 48% (Fig. 3). These shear stress-induced changes were paralleled by losses in chondrocyte viability, which declined in the superficial zone from 89% in low shear stress to 62% in high shear stress. Shear stress-related differences in viability, apoptosis and p53 expression were significant ($p < 0.05$). No significant changes were seen in the deep zone (data not shown).

## 4. Discussion

Exposure to high shear stress in the TCV induces chondrocyte death in the superficial zone of cartilage explants. Treating the explants with antioxidants significantly reduces stress-induced cell death, supporting the hypothesis that oxidative stress mediates the damaging effects of shear stress on chondrocytes. These findings suggest that exaggerated shear stresses at sites of articular incongruities or in unstable joints lead to low cell density, a factor that could contribute to cartilage degeneration in post-traumatic OA. Vitamin E, NAC, and L-NAME all showed similar anti-apoptosis activity in explants exposed to high shear stress. These data indicate that oxidative damage to lipid membranes plays a role in shear stress-induced apoptosis. The findings support the hypothesis that vitamin E and other antioxidants may be used to prevent high shear stress-induced apoptosis and potentially to delay the onset of OA in joints where instability or incongruity lead to such abnormal mechanical stresses.

There is a clear need to develop and implement strategies to prevent post-traumatic OA [8]. With better understanding of the pathogenesis of post-traumatic OA, clinical and basic research could be focused on developing new minimally invasive and non-surgical treatments of joint injuries that have a high risk of post-traumatic osteoarthritis, and on treatments that prevent or delay the development of OA in patients at high risk of post-traumatic OA. If further work confirms the damaging effects of high shear stress, efforts should be focussed on developing treatments of joint injuries that minimize repetitive shear stress due to articular surface incongruity, joint instability or joint mal-alignment. In addition, there is increasing evidence that biologic interventions can decrease mechanical stress induced chondrocyte damage [4,6,10]. For example, the work of D'Lima and co-workers shows that caspase inhibition can decrease mechanically induced chondrocyte apoptosis [4–6], and Haut and colleagues have reported that P188 surfactant can limit chondrocyte necrosis following impact loading [17,18]. Our investigations of the effects of anti-oxidants in preventing mechanically induced chondrocyte damage and those of others [10] show that some of these agents have the potential to decrease the deleterious effects of mechanical loads on articular cartilage. Use of agents that limit oxidative damage to chondrocytes may have a role in decreasing the risk of OA following joint injuries. The results of this study suggest that the combination of mechanical treatments of injured joints that minimize repetitive high shear stresses and biologic treatments that minimize oxidative damage could decrease the risk of post-traumatic OA.

## Acknowledgement

The work reported in this manuscript was supported by award P50 AR48939 National Institutes of Health Specialized Center on Research for OA. http://poppy.obrl.uiowa.edu/Specialized Center of Research/SCOR.htm.

## References

[1] J.A. Buckwalter, Mechanical injuries of articular cartilage, in: *Biology and Biomechanics of the Traumatized Synovial Joint*, G. Finerman, ed, American Academy of Orthopaedic Surgeons, Park Ridge, IL, 1992, pp. 83–96.

[2] J.A. Buckwalter, Osteoarthritis and articular cartilage use, disuse and abuse: experimental studies, *J. Rheumatol.* **22**(Suppl. 43) (1995), 13–15.

[3] J.A. Buckwalter and T.D. Brown, Joint injury, repair, and remodeling: roles in post-traumatic osteoarthritis, *Clin. Orthop. Relat. Res.* (2004), 7–16.

[4] D.D. D'Lima, S. Hashimoto, P.C. Chen, C.W. Colwell, Jr. and M.K. Lotz, Human chondrocyte apoptosis in response to mechanical injury, *Osteoarthritis Cartilage* **9** (2001), 712–719.

[5] D.D. D'Lima, S. Hashimoto, P.C. Chen, C.W. Colwell, Jr. and M.K. Lotz, Impact of mechanical trauma on matrix and cells, *Clin. Orthop. Relat. Res.* (2001), S90–99.

[6] D.D. D'Lima, S. Hashimoto, P.C. Chen, M.K. Lotz and C.W. Colwell, Jr., Prevention of chondrocyte apoptosis, *J. Bone Joint Surg. Am.* **83–A** (Suppl. 2), (2001), 25–26.

[7] D.R. Dirschl, J.L. Marsh, J.A. Buckwalter, R. Gelberman, S.A. Olson, T.D. Brown and A. Llinias, Articular fractures. *J. Am. Acad. Orthop. Surg.* **12** (2004), 416–423.

[8] A.C. Gelber, M.C. Hochberg, L.A. Mead, N.Y. Wang, F.M. Wigley and M.J. Klag, Joint injury in young adults and risk for subsequent knee and hip osteoarthritis, *Ann. Intern. Med.* **133** (2000), 321–328.

[9] A.D. Heiner and J.A. Martin, Cartilage responses to a novel triaxial mechanostimulatory culture system, *J. Biomech.* **37** (2004), 689–695.

[10] B. Kurz, A. Lemke, M. Kehn, C. Domm, P. Patwari, E.H. Frank, A.J. Grodzinsky and M. Schunke, Influence of tissue maturation and antioxidants on the apoptotic response of articular cartilage after injurious compression, *Arthritis Rheum.* **50** (2004), 123–130.

[11] J.L. Marsh, J. Buckwalter, R. Gelberman, D. Dirschl, S. Olson, T. Brown and A. Llinias, Articular fractures: does an anatomic reduction really change the result?, *J. Bone Joint Surg. Am.* **84–A** (2002), 1259–1271.

[12] J.A. Martin, T. Brown, A. Heiner and J.A. Buckwalter, Post-traumatic osteoarthritis: the role of accelerated chondrocyte senescence, *Biorheology* **41** (2004), 479–491.

[13] J.A. Martin, T.D. Brown, A.D. Heiner and J.A. Buckwalter, Chondrocyte senescence, joint loading and osteoarthritis, *Clin. Orthop. Relat. Res.* (2004), S96–103.

[14] J.A. Martin and J.A. Buckwalter, The role of chondrocyte senescence in the pathogenesis of osteoarthritis and in limiting cartilage repair, *J. Bone Joint Surg. Am.* **85-A** (Suppl. 2) (2003), 106–110.

[15] J.A. Martin and J.A. Buckwalter, Telomere erosion and senescence in human articular cartilage chondrocytes, *J. Gerontol. A. Biol. Sci. Med. Sci.* **56** (2001), B172–179.

[16] J.A. Martin, A.J. Klingelhutz, F. Moussavi-Harami and J.A. Buckwalter, Effects of oxidative damage and telomerase activity on human articular cartilage chondrocyte senescence, *J. Gerontol. A. Biol. Sci. Med. Sci.* **59** (2004), 324–337.

[17] D.M. Phillips and R.C. Haut, The use of a non-ionic surfactant (P188), to save chondrocytes from necrosis following impact loading of chondral explants, *J. Orthop. Res.* 22 (2004), 1135–1142.

[18] S.A. Rundell, D.C. Baars, D.M. Phillips and R.C. Haut, The limitation of acute necrosis in retro-patellar cartilage after severe blunt impact to the in vivo rabbit patello-femoral joint, *J. Ortho. Res.* 23 (2005) (on line).

[19] V. Wright, Post-traumatic osteoarthritis – a medico-legal minefield, *Brit. J. Rheum.* 29 (1990), 474–478.

Biorheology 43 (2006) 523–535
IOS Press



# Acoustic properties of articular cartilage under mechanical stress

Heikki J. Nieminen [a,b,*], Juha Töyräs [c], Mikko S. Laasanen [b,d] and Jukka S. Jurvelin [a,d]

[a] *Department of Physics, University of Kuopio, POB 1627, 70211 Kuopio, Finland*
[b] *Department of Anatomy, University of Kuopio, POB 1627, 70211 Kuopio, Finland*
[c] *Department of Clinical Neurophysiology, Kuopio University Hospital and University of Kuopio, POB 1777, 70211 Kuopio, Finland*
[d] *Department of Clinical Physiology and Nuclear Medicine, Kuopio University Hospital, POB 1777, 70211 Kuopio, Finland*

**Abstract.** Mechano-acoustic and elastographic techniques may provide quantitative means for the *in vivo* diagnostics of articular cartilage. These techniques assume that sound speed does not change during tissue loading. As articular cartilage shows volumetric changes during compression, acoustic properties of cartilage may change affecting the validity of mechano-acoustic measurements. In this study, we examined the ultrasound propagation through human, bovine and porcine articular cartilage during stress-relaxation in unconfined compression. The time of flight (*TOF*) technique with known cartilage thickness (true sound speed) as well as *in situ* calibration method [Suh, Youn, Fu, *J. Biomech.* **34** (2001), 1347–1353] were used for the determination of sound speed. Ultrasound speed and attenuation decreased in articular cartilage during ramp compression, but returned towards the level of original values during relaxation. Variations in ultrasound speed induced an error in strain and compressive moduli provided that constant ultrasound speed and time-of-flight data was used to determine the tissue thickness. Highest errors in strain ($-11.8 \pm 12.0\%$) and dynamic modulus ($15.4 \pm 17.9\%$) were recorded in bovine cartilage. *TOF* and *in situ* calibration methods yielded different results for changes in sound speed during compression. We speculate that the variations in acoustic properties in loaded cartilage are related to rearrangement of the interstitial matrix, especially to that of collagen fibers. In human cartilage the changes, are, however relatively small and, according to the numerical simulations, mechano-acoustic techniques that assume constant acoustic properties for the cartilage will not be significantly impaired by this phenomenon.

Keywords: Articular cartilage, osteoarthrosis, mechanical indentation, ultrasound speed, ultrasound attenuation

## 1. Introduction

Mechano-acoustic and elastographic techniques have been introduced as means to characterize composition and properties of articular cartilage [10,16,31–35]. These techniques could provide potential means for the *in vivo* diagnostics of cartilage integrity. Laboratory tests have already shown the potential of the mechano-acoustic (ultrasound) indentation methods [16,31]. As the ultrasound indentation technique can provide information on the tissue thickness at the site of indentation it enables calculation of true material properties for the tissue. Thereby, the technique overcomes the uncertainties related to effects of finite, unknown tissue thickness that potentially impair the validity of traditional mechanical indentation measurements [2,18,20].

---

*Address for correspondence: Heikki J. Nieminen, Department of Physics, University of Kuopio, POB 1627, 70211 Kuopio, Finland. Tel.: +358 17 162341; Fax: +358 17 162585; E-mail: heikki.nieminen@uku.fi; Web: www.luotain.uku.fi.

The mechano-acoustic techniques assume that acoustic properties, such as sound speed and attenuation, are constant and do not depend on the stress–strain state of the tissue. Based on these assumptions, a prototype of arthroscopic ultrasound indentation instrument was introduced in our previous study [16]. This instrument is capable of simultaneous measurement of stress, strain and tissue thickness enabling determination of dynamic modulus of articular cartilage. *In situ* calibration technique [31] has been proposed for the simultaneous determination of tissue equilibrium Young's modulus and ultrasound speed. However, in this technique and in the technique presented by Laasanen et al. [16] the sound speed is assumed to be constant during mechanical indentation. Thus, accuracy of the mechano-acoustic techniques depends critically on the invariability of ultrasound speed during compression.

Since cartilage is structurally inhomogeneous, mechanically anisotropic and poroviscoelastic [4,19], acoustic properties of the tissue may vary as a function of imposed stress and strain [34]. However, it is not known if that is the case. If ultrasound speed and, consequently cartilage thickness and deformation, could be measured reliably, sound attenuation in cartilage could be determined, provided that the changes in ultrasound beam geometry during indentation are known. Earlier laboratory measurements have suggested that both the sound speed and attenuation, when accurately measured, are significantly related to tissue structure and composition [1,7,11,22,24,32]. Further, by compensation of the sound attenuation in the overlying cartilage the acoustic properties of subchondral bone could be more accurately evaluated. The parameters such as integrated reflection coefficient (*IRC*) [7,30], i.e. the mean ultrasound reflection over frequency range, and the ultrasound roughness index (*URI*) [30] could provide information on structural and morphological pathology of subchondral bone manifested in early osteoarthrosis (OA) [9,15,26]. Broadband ultrasound backscatter (*BUB*) [7,29], a parameter that is related to structure [29], composition [5] and functional properties [13] of bone, could also be determined for subchondral bone provided that the attenuation in cartilage is known. This could significantly improve arthroscopic diagnostics and follow-up of osteoarthrotic changes in subchondral bone.

In the present study, sound speed and attenuation were measured during unconfined compression of articular cartilage with a mechano-acoustic material testing device. The study was established to reveal whether the acoustic properties of cartilage depend on the mechanical stress (or strain) the tissue is subjected to as well as to evaluate how possible stress-related changes would affect the mechano-acoustic characterization of articular cartilage and subchondral bone. Human, bovine and porcine patellar cartilage samples, known to exhibit significantly different structural, compositional and functional characteristics [28], were used to provide a comprehensive view on the acoustic properties of loaded articular cartilage. Further, different techniques for determination of ultrasound speed, i.e. the time of flight (*TOF*) technique with known cartilage thickness (true sound speed) as well as *in situ* calibration method (*ISCM*) [31] were compared.

## 2. Materials and methods

### 2.1. Sample preparation

Frozen human cadaver knees (age = 20–78 years) were obtained from the Jyväskylä Central Hospital, Jyväskylä, Finland, and thawed before sample preparation. A permission to use the human samples was granted by the national authority (National Authority for Medicolegal Affairs, Helsinki, Finland, permission 1781/32/200/01). In addition, fresh bovine (age = 1–3 years) and porcine (age = approx. 4 months) knees were obtained from a local abattoir (Atria Oyj, Kuopio, Finland) within few hours from slaughtering.

Osteochondral samples (dia. = 16 mm) were drilled from the lateral upper quadrant of human ($n = 6$), bovine ($n = 6$) and porcine ($n = 6$) patellae with intact articular cartilage. During preparation the samples were moistened with phosphate buffered saline (PBS) supplemented with enzyme inhibitors: 5 mM ethylenediaminetetraacetic acid (EDTA) (Riedel-de-Haen, Seelze, Germany) and 5 mM benzamide HCl (Sigma Chemical Co., St. Louis, MO, USA). A full thickness cartilage disk was extracted from the centre of the osteochondral plug using a biopsy punch (dia. = 4 mm) and a razor blade. After the preparation, samples were frozen ($-20°$C) for later mechano-acoustic measurements.

An elastomer sample (dia. = 4 mm) (Teknikum Oy, Vammala, Finland) was used as reference material. For acoustic measurements also vacuum degassed distilled water was used as reference medium.

## 2.2. Biomechanical testing and analyses

The mechanical and acoustic properties of the cartilage and elastomer disks were determined with a custom-made apparatus [16]. In immersion bath (PBS), the samples were placed between the ultrasound transducer and a metallic plate with the cartilage surface facing the transducer (Fig. 1). The metallic plate was lifted upwards with a high-precision actuator (PM1A, Newport, Irvine, CA) (precision: 0.1 $\mu$m) to make contact between the cartilage and the transducer. A load cell, attached to the ultrasound transducer, was used to measure the contact load. After the first contact, 5% pre-strain was applied to ensure uniform contact at the transducer–cartilage and cartilage–metallic plate interfaces. This was followed by three subsequent stress–relaxation steps (strain: 5%, ramp velocity: 1 $\mu$m/s, relaxation criteria: 39 Pa/min). Young's modulus at equilibrium was determined from the equilibrium response of the tissue. Finally, the sample was subjected to dynamic loading (frequency: 1 Hz, strain amplitude: 1%) and the dynamic modulus for the sample was calculated.

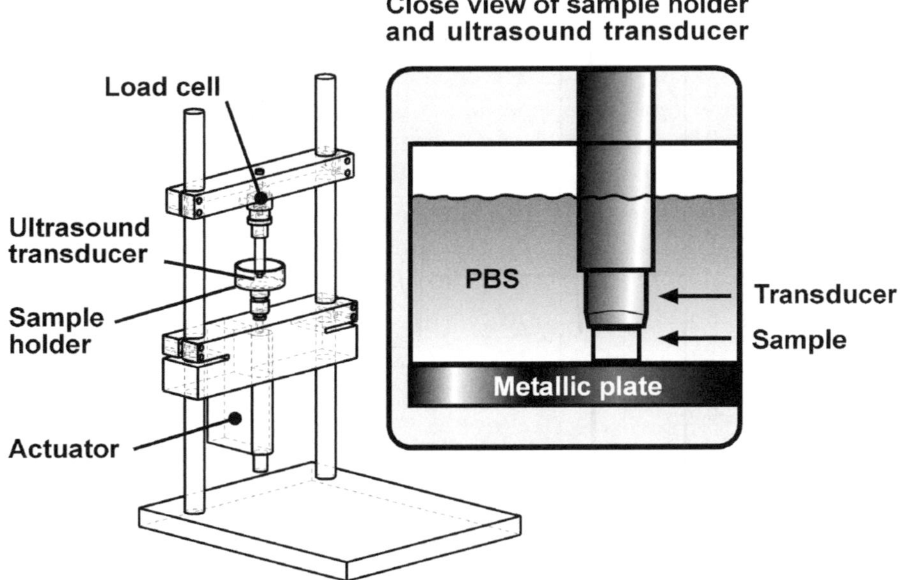


Fig. 1. Mechano-acoustic material testing device and geometry used in the measurements of mechanical and real-time acoustic parameters of articular cartilage. A cartilage sample with no subchondral bone was immersed in PBS bath and placed between the compressive metallic plate and the ultrasound transducer. The actuator movement produced tissue strain and the compressive load was registered with a load cell. Ultrasonic measurements were conducted simultaneously in intervals of 10 s.

### 2.3. Ultrasonic measurements

Ultrasonic measurements were conducted during the stress–relaxation test with 10-second-intervals using the A-mode *pulse-echo* method and a non-focused contact ultrasound transducer (VM-116, Panametrics, Waltham, MA) (center frequency 10.3 MHz, 7.1–14.2 MHz, −3 dB [32]). Acoustic pulses were electrically excited using a 0.5 MHz to 100 MHz pulser-receiver board (PAC-IPR-100, Physical Acoustic Corporation, Princeton, NJ). The ultrasound reflected back from the cartilage–metallic plate interface, as observed by the ultrasound transducer, was received with the pulser-receiver board and digitized at a 500 MHz sampling frequency using an 8-bit A/D-board (PAC-AD-500, Physical Acoustic Corporation). The data was digitally stored for later analysis.

### 2.4. Ultrasound data analyses

Ultrasound attenuation in time domain, as determined from the signal amplitudes, was calculated using Eq. (1):

$$\alpha_{\text{amp}}(t) = \frac{10}{x(t)} \log_{10} \frac{A_1(t)}{A_2(t)}, \tag{1}$$

where $x(t)$ is cartilage thickness at a given time $t$. $A_1(t)$ and $A_2(t)$ are the peak-to-peak amplitudes of the ultrasound pulse that have travelled once and twice back and forth the sample, respectively (Fig. 2).

Attenuation in frequency domain was calculated as follows:

$$\alpha_{\text{int}}(t) = \frac{1}{\Delta f} \int_{\Delta f} \frac{10}{x(t)} \log_{10} \frac{A_1(t, f)}{A_2(t, f)}, \tag{2}$$

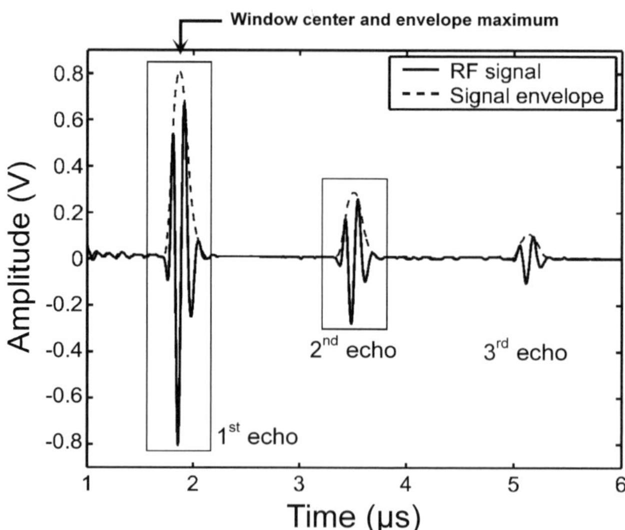


Fig. 2. Typical ultrasound signal measured through the bovine articular cartilage. The first echo was registered after the ultrasound pulse had travelled once back and forth through the sample. The second echo was registered after the pulse had travelled twice back and forth through the sample. A 300-sample-window was positioned by matching the echo envelope maximum with the window centre. Peak-to-peak amplitudes for amplitude attenuation analysis were determined as the difference between the maximum and the minimum signal amplitudes.

where $A_1(t, f)$ and $A_2(t, f)$ are the amplitude spectra of pulses that have travelled once and twice back and forth the sample, respectively. $\Delta f$ indicates the frequency range (here 5–9 MHz), where the attenuation was found to be linearly dependent on frequency [22]. The spectra were calculated for 300-sample-windows using the fast Fourier transform (FFT). The centre of this 0.6 $\mu$s window was positioned at the maximum of the pulse envelope (Fig. 2). The selected pulse was zero-padded to 4096 samples prior to FFT.

Ultrasound speed, $c_T(t)$ ("true ultrasound speed"), was calculated during stress–relaxation measurements using the equation

$$c_T(t) = \frac{2x(t)}{TOF(t)}, \tag{3}$$

where $TOF(t)$ is the time-of-flight of ultrasound pulse as it travels back and forth the sample. $TOF(t)$ was determined using the cross-correlation method [6,22]. In addition, ultrasound speed was determined using the *in situ* calibration method [31], according to the following equation:

$$c_{ISCM}(t) = \frac{2[x(0) - x(t)]}{TOF(0) - TOF(t)}, \tag{4}$$

where $TOF(0)$ is the time-of-flight at the beginning of measurement, i.e. at the first contact of the transducer and sample surface.

Finally, error analyses were conducted to reveal how strain, elastic indentation moduli and attenuation (i.e. amplitude attenuation) are affected by the compression-related variation of ultrasound speed [16], provided that the time-of-flight and constant ultrasound speed are used to define the tissue thickness. These analyses are presented in detail in Appendix.

All data processing was conducted with a custom-made software programmed with LabVIEW (version 6.1, National Instruments, Austin, TX) and Matlab (versions 5.3.1 and 6.5.1, Mathworks Inc., Natick, MA).

## 2.5. Statistical analyses

The non-parametric Friedman *post hoc* test for $n$ related samples was used to study strain dependence of parameters within each species. Kruskal–Wallis *post hoc* test was used for testing differences in parameter values between the species. Wilcoxon signed ranks test for paired samples was used to test differences in errors of strain and Young's modulus between ramp end and mechanical equilibrium. Statistical analyses were conducted using the SPSS software (version 11.5.1, SPSS Inc. Chigaco, IL) and a custom-made Matlab program (version 6.5.1, Mathworks Inc., Natick, MA).

## 3. Results

Ultrasound speed, as determined under the preload (before the ramp compression) using the *TOF*-method or *in situ* calibration method, showed no statistically significant variations ($p > 0.05$) across the species (Fig. 3). However, significant differences ($p < 0.01$) were found in amplitude and integrated attenuation between the bovine and porcine samples.

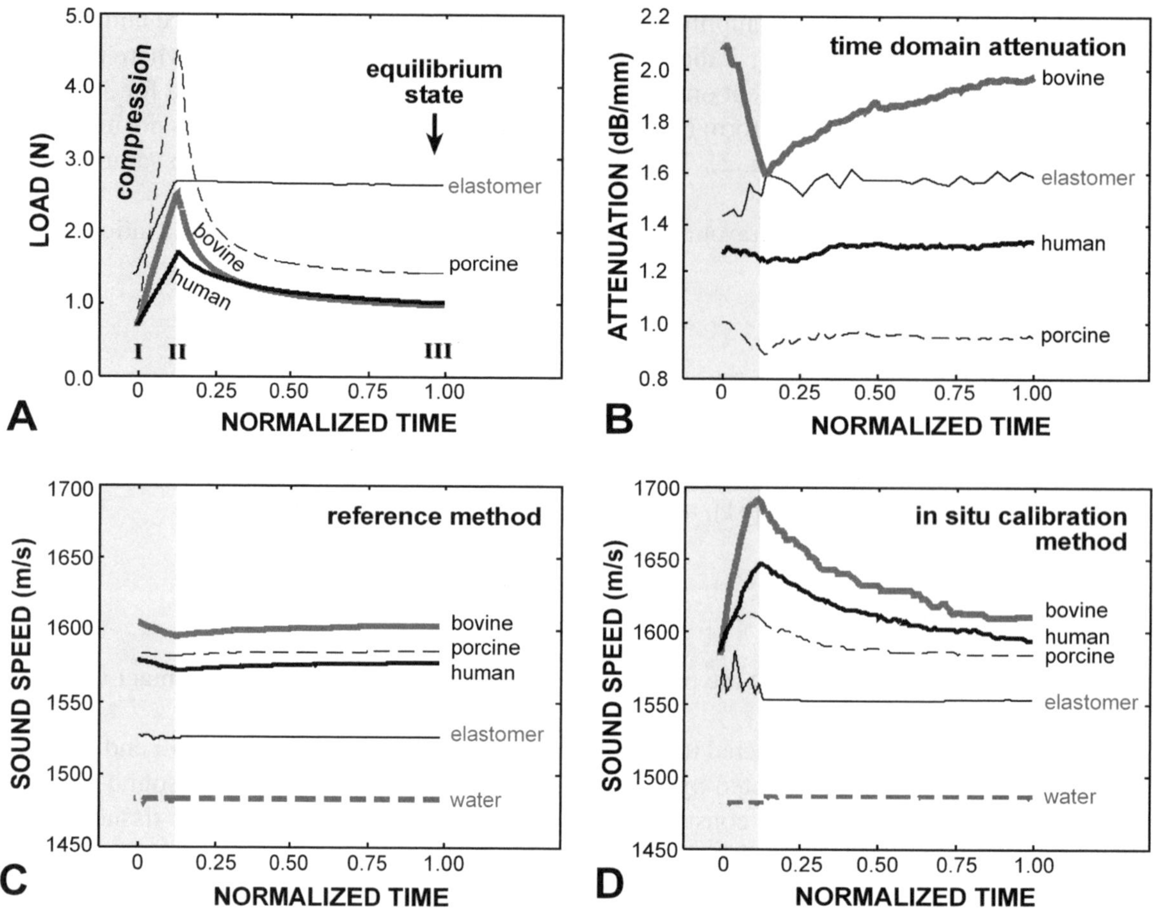

Fig. 3. Mean values as a function of stress–relaxation time in unconfined compression for stress (A), time-domain attenuation (B), the ultrasound speed as determined with the *TOF*-method (C) and the *in situ* calibration method (D). The data was registered during the first stress–relaxation step. Ultrasound attenuation as well as true ultrasound speed decreased during the ramp phase and recovered towards the initial equilibrium value during subsequent stress relaxation.

During the ramp compression amplitude and integrated attenuation of ultrasound decreased in bovine cartilage samples ($-0.44$ dB/mm and $-0.98$ dB/mm, $p < 0.05$) while no statistically significant changes were determined in human or porcine samples (Fig. 3). After ramp, i.e. during the stress–relaxation, values of ultrasound attenuation were restored towards the level prior to ramp compression in all sample groups (Tables 1 and 2). Qualitatively, compression induced changes in attenuation at all frequencies (5–9 MHz) (Fig. 4).

The true ultrasound speed, $c_T(t)$, showed a statistically significant variation in bovine ($p < 0.05$) and human samples ($p < 0.01$) during ramp compression and stress–relaxation, while no change was detected in porcine tissue (Fig. 3). Similarly, significant variation ($p < 0.05$) in ultrasound speed, $c_{ISCM}(t)$, as determined using *in situ* calibration method, was observed during ramp compression and stress–relaxation in bovine and human cartilage, whereas in porcine cartilage no significant changes were detected. Reference measurements of degassed distilled water using the similar loading protocol as for cartilage samples showed no alterations in the ultrasound speed or attenuation values. Similarly, only negligible changes were detected in the acoustic properties of elastomer during the stress–relaxation

Table 1

Changes (%, ±SD) in amplitude attenuation at the end of ramp compression (strain: 5%, ramp velocity: 1 $\mu$m/s, relaxation criteria: 39 Pa/min, 3 steps) and after full stress relaxation at equilibrium, as compared to values before compression

|  | STEP 1 | | STEP 2 | | STEP 3 | |
|---|---|---|---|---|---|---|
|  | Ramp end | Equilibrium | Ramp end | Equilibrium | Ramp end | Equilibrium |
| Human | $-3.9 \pm 18.4$ | $2.5 \pm 10.4$ | $-15.3 \pm 14.5$ | $-2.8 \pm 10.7$ | $-22.1 \pm 8.6$ | $-7.6 \pm 9.0$ |
| Bovine | $-19.4 \pm 11.9$ | $-4.4 \pm 17.8$ | $-14.1 \pm 17.8$ | $12.2 \pm 14.4$ | $2.8 \pm 17.3$ | $3.6 \pm 13.6$ |
| Porcine | $-8.9 \pm 9.5$ | $-4.4 \pm 8.7$ | $-10.9 \pm 6.9$ | $-5.4 \pm 6.1$ | $-10.6 \pm 9.8$ | $-3.6 \pm 7.1$ |

Table 2

Changes (%, ±SD) in integrated attenuation at the end of ramp compression (strain: 5%, ramp velocity: 1 $\mu$m/s, relaxation criteria: 39 Pa/min, 3 steps) and after full stress relaxation at equilibrium, as compared to values before compression

|  | STEP 1 | | STEP 2 | | STEP 3 | |
|---|---|---|---|---|---|---|
|  | Ramp end | Equilibrium | Ramp end | Equilibrium | Ramp end | Equilibrium |
| Human | $-0.7 \pm 39.9$ | $1.1 \pm 19.0$ | $-19.3 \pm 15.9$ | $-3.6 \pm 15.6$ | $-27.4 \pm 12.1$ | $-10.6 \pm 14.7$ |
| Bovine | $-31.2 \pm 20.2$ | $-16.5 \pm 21.6$ | $-17.5 \pm 19.0$ | $-16.6 \pm 15.1$ | $0.4 \pm 19.5$ | $0.6 \pm 15.1$ |
| Porcine | $-10.0 \pm 10.4$ | $-7.3 \pm 10.9$ | $-12.1 \pm 7.0$ | $-6.2 \pm 8.9$ | $-8.1 \pm 8.4$ | $-3.6 \pm 4.8$ |

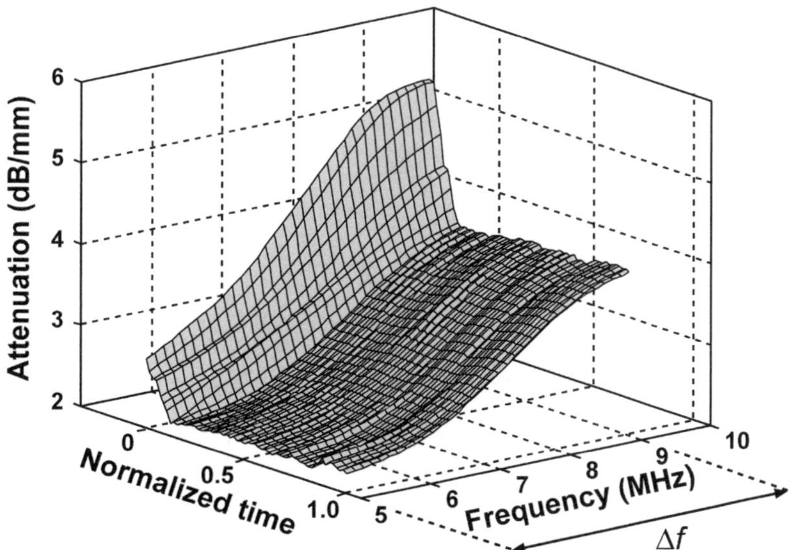


Fig. 4. Typical ultrasound attenuation spectrum measured for bovine cartilage as a function of time during stress relaxation test in unconfined compression. Tissue compression affected ultrasound attenuation at all frequencies.

test. Use of constant ultrasound speed induced only minute error ($-0.6 \pm 0.6\%$ to $-0.2 \pm 0.2\%$ across all species) on attenuation values determined at the ramp end (Table 3).

Dynamic and equilibrium moduli showed no statistically significant correlation with the ultrasound speed or attenuation values. At mechanical equilibrium, after long term loading, speed and attenuation values were not significantly related to the induced strain (Fig. 5). In contrast, provided that a constant ultrasound speed was assumed, errors in determined values of strain or elastic moduli were induced as true alterations in ultrasound speed took place during ramp compression (Table 3). The largest errors were seen in bovine cartilage at the ramp end, i.e. $-11.8 \pm 12.0\%$ for the strain and $15.4 \pm 17.9\%$ for the

Table 3

Errors (%, ±SD) in measured strain, related to use of constant sound speed, modulus and amplitude attenuation at the end of ramp compression end and mechanical equilibrium

| | Error in measured strain | | Error in measured modulus | | Error in measured attenuation | |
| --- | --- | --- | --- | --- | --- | --- |
| | Ramp end | Equilibrium | Ramp end | Equilibrium | Ramp end | Equilibrium |
| Human | $-8.3 \pm 3.2$ | $-1.5 \pm 3.5$ | $9.2 \pm 3.9$ | $1.6 \pm 3.8$ | $-0.4 \pm 0.2$ | $-0.1 \pm 0.2$ |
| Bovine | $-11.8 \pm 12.0$ | $-2.5 \pm 7.3$ | $15.4 \pm 17.9$ | $3.0 \pm 7.1$ | $-0.6 \pm 0.6$ | $-0.1 \pm 0.4$ |
| Porcine | $-3.2 \pm 4.5$ | $0.1 \pm 2.6$ | $3.4 \pm 4.7$ | $0.0 \pm 2.6$ | $-0.2 \pm 0.2$ | $0.0 \pm 0.1$ |

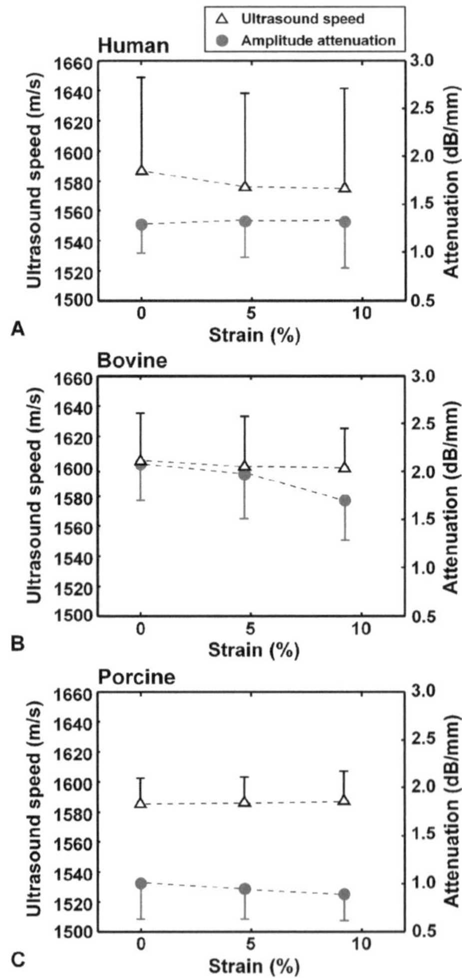


Fig. 5. Mean values (±SD) of ultrasound attenuation and speed for each species as a function of compressive strain at equilibrium. No statistically significant strain dependent changes were revealed in attenuation or speed values within any species.

dynamic modulus. As the tissue equilibrated these errors in bovine cartilage decreased to $-2.5 \pm 7.3\%$ and $3.0 \pm 7.1\%$, respectively. In all species errors in strain and moduli were significantly different ($p < 0.05$) at ramp end as compared to equilibrium. Errors in strains did not vary significantly across the species (Table 3).

## 4. Discussion

In the present study, unconfined compression tests of human, bovine and porcine patellar cartilage samples with simultaneous 10.3 MHz pulse-echo ultrasound measurements were conducted using a custom made mechano-acoustic material testing device. The device was validated by reference measurements in degassed distilled water bath (without sample) and with an elastomer sample. In contrast to single phase elastic material (elastomer), the experimental results indicated that the dynamic loading affected the acoustic properties of articular cartilage. These alterations in acoustic parameters may not be interpreted by simple mixture laws.

Under a high-rate unconfined compression test, cartilage efficiently maintains its total volume (behaves like incompressible, Poisson's ratio $\nu = 0.5$, elastic material) and both axial and lateral strains are generated [3]. Lateral expansion of the sample re-orientates collagen fibers to the direction more parallel to articular surface, i.e. in the direction perpendicular to ultrasound beam. As the stress relaxation takes place under constant strain, the recoil of lateral expansion [3] obviously realigns collagen fibers towards the original orientation. In the present study, this could be qualitatively visualized by a cartilage deformation pattern under unconfined compression. The deformation pattern was reconstructed from the polarized microscopic images that reveal collagen network architecture in tissue (Fig. 6). At the same time the insterstitial water is squeezed out from the tissue and cartilage water content decreases. While the ultrasound speed in the solid matrix, i.e. collagen fibers, is higher than in water, the apparent solid matrix content increases and, according to simple mixture laws, sound speed should increase in the loaded cartilage. This was not confirmed in the present study but the true sound speed decreased during ramp compression and then, to some degree, increased during stress–relaxation phase. We suggest that the changes in sound speed, instead of changes in solid matrix content, were related to rearrangement of collagen matrix. Our results may be explained with the earlier findings demonstrating that the transversally measured ultrasound speed is higher in the superficial zone with transversally oriented collagen fibers than in the deeper zones where the collagen fibers are longitudinally oriented [1]. Goss & O'Brien [12] also showed that the sound speed in mouse tail, with the collagen strongly oriented along the tail, is higher along the tail than across it.

The ultrasound attenuation in bovine articular cartilage was more sensitively affected by mechanical loading, as compared to human or porcine tissue. As the acoustic wave is a moving mechanical disturbance, it is likely that the mechanical and structural anisotropy produces acoustic anisotropy as well. Our results indicate that, physically, either dissipation and/or scattering of sound decreased during ramp compression, especially in bovine cartilage. The re-arrangement of collagen fibers, known to act as significant ultrasound scatterers [7,21,23,25], may decrease ultrasound scattering lowering the attenuation in cartilage.

The reference measurements with distilled degassed water and elastomer suggested that the ultrasound speed values of loaded cartilage, determined by using the *TOF*-method (true ultrasound speed) were accurate. However, the changes in ultrasound speed during compression, as determined using the *in situ* calibration method (*ISCM*), were opposite to those recorded by the *TOF*-method. Systematically, the compression increased values of *ISCM* sound speed in all species, whereas the true sound speed was found to decrease during compression. The values determined using *ISCM* are sensitive to small changes in $TOF(t)$. Even though the ultrasound speed values differ between *TOF* and *ICSM* methods during stress–relaxation, they are similar at the mechanical equilibrium. Clinical *in vivo* mechano-acoustic diagnostics of cartilage must be rapid and thus be based on short-term indentations, not long-term static compression. The demand for the instantaneous measurement may invalidate the clinical use of *ISCM*.

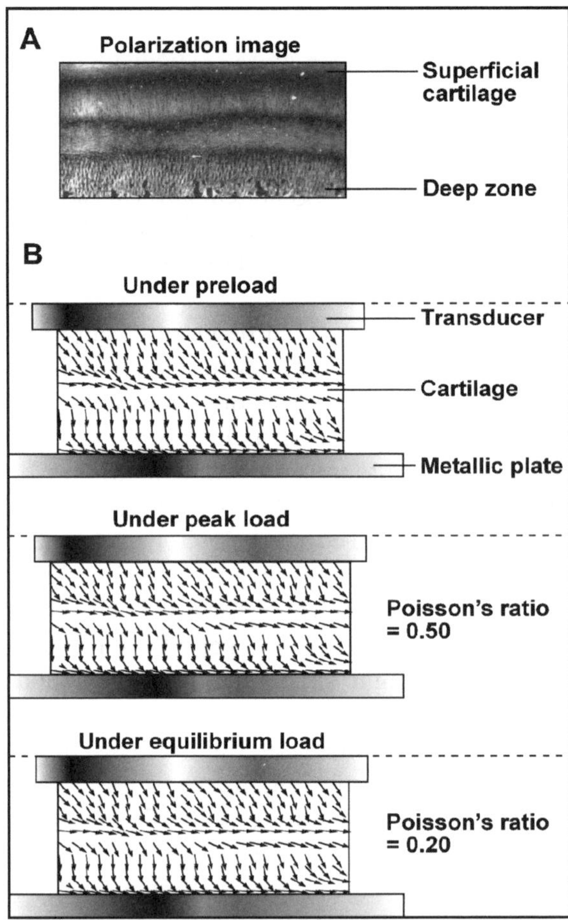


Fig. 6. Graphical simulation of interstitial deformation pattern of bovine articular cartilage under unconfined compression. (A) Typical polarization microscopic image[1] of bovine articular cartilage used for deriving collagen fibre orientation in the tissue (B, under preload). Collagen orientation (range: 0–90°) is presented as a vector field (B). Vector fields under peak load (at the end of ramp compression) and under equilibrium load (at mechanical equilibrium) were derived using the following assumptions: cartilage–transducer and cartilage–metallic plate interfaces are frictionless and the deformation is homogeneous. The values of Poisson's ratio (lateral-to-axial strain ratio) were assumed to be 0.50 under peak load indicating incompressibility and 0.20 at equilibrium load which is a typical value for bovine articular cartilage [17].

[1]The methodology for polarized microscopic imaging, image processing and calculation of the local collagen orientation has been presented earlier in detail [8,27].

As constant ultrasound speed is assumed changes in ultrasound speed during short-term compression induces noticeable errors in the determined strain and elastic modulus. However, the errors were minor at the mechanical equilibrium. The errors in strain and elastic moduli values were highest in bovine cartilage, while in human and porcine cartilage changes were relatively small. In human cartilage maximal errors in strain and elastic modulus were $-8.3 \pm 3.2\%$ and $9.2 \pm 3.9\%$, respectively. Thereby, the use of constant ultrasound speed in calculations would induce a negative and positive bias to strain and elastic modulus, respectively. In contrast, errors in attenuation were small in all species when constant ultrasound speed was assumed.

It is notable that the variations in ultrasound speed may be specific to the measurement geometry and protocol applied in the present study. It should be further studied whether the changes in ultrasound speed depend on loading geometry, e.g. unconfined compression and indentation or the loading rate (ramp velocity). In clinical *in vivo* applications accurate methods for determining true ultrasound speed would eliminate traditional errors in the determination of mechanical parameters. Since such methods are currently not available, further development of methodology for ultrasound speed measurement in *in vivo* articular cartilage should be conducted.

The results of the present study suggest that the ultrasound speed and attenuation in articular cartilage vary during ramp compression and stress–relaxation. If the acoustic properties of loaded cartilage are known, the mechano-acoustic measurements could provide more information on the tissue integrity. They would also enable more accurate information on cartilage thickness, thereby improving the accuracy of calculated compressive moduli. By assuming a constant sound speed the errors in mechanical parameters can be significant. However, further studies are needed to highlight the cartilage acoustic behaviour under true instantaneous loading. Moreover, small uncertainties may be acceptable as the natural (biological) variation of mechanical parameters in intact and especially in osteoarthrotic cartilage is large.

## Acknowledgements

Financial support from the Techonology Development Center (TEKES), Kuopio University Hospital, Kuopio, Finland (EVO 5103), the Graduate School for Musculoskeletal Diseases, Finland and The Finnish Cultural Foundation, Helsinki, Finland, is acknowledged. Ilkka Kiviranta, Jyväskylä Central Hospital, Jyväskylä, Finland and Atria Oyj, Kuopio, Finland are acknowledged for the help in providing the tissue. Jarno Rieppo, University of Kuopio, Kuopio, Finland, is acknowledged for assistance in polarized light microscopy.

## Appendix

By assuming a constant speed for ultrasound during cartilage compression potential errors to mechano-acoustic determination of strain, elastic modulus and attenuation are introduced. The following analyses were applied to examine to errors during short term compression, i.e. at the end of ramp compression as well as after long term compression, i.e. at mechanical equilibrium.

When constant ultrasound speed is assumed, the measured strain in tissue is expressed by the following equation

$$\varepsilon_{\text{meas}} = \frac{c_0(TOF_0 - TOF_1)}{c_0 TOF_0} \tag{A1}$$

whereas for determination of true strain the variation of ultrasound speed must be considered. Hence, the true strain is

$$\varepsilon_{\text{true}} = \frac{c_0 TOF_0 - c_1 TOF_1}{c_0 TOF_0}, \tag{A2}$$

where $c_0$ is the ultrasound speed and $TOF_0$ time-of-flight before compression. Respectively, $c_1$ is ultrasound speed and $TOF_1$ time-of-flight after compression. The relative error can now be expressed as

$$\Delta\varepsilon = \frac{\varepsilon_{\text{meas}} - \varepsilon_{\text{true}}}{\varepsilon_{\text{true}}} = \frac{TOF_1(c_1 - c_0)}{c_0 TOF_0 - c_1 TOF_1}. \tag{A3}$$

Similarly, elastic modulus of the tissue, as determined using ultrasound indentation technique [16] can be derived if constant ultrasound speed is assumed:

$$E_{\text{meas}} = \frac{P(1 - \nu^2)}{a\kappa(c_0 TOF_0 - c_0 TOF_1)}, \tag{A4}$$

where $P$ is load, $\nu$ the Poisson's ratio, $a$ the indenter diameter and $\kappa$ is the scale factor due to finite and variable tissue thickness [14]. The true modulus can be presented as

$$E_{\text{true}} = \frac{P(1 - \nu^2)}{a\kappa(c_0 TOF_0 - c_1 TOF_1)}. \tag{A5}$$

As $P$, $\nu$, $a$ and $\kappa$ are constants, the following formulation can be derived for the error of elastic modulus:

$$\Delta E = \frac{E_{\text{meas}} - E_{\text{true}}}{E_{\text{true}}} = \frac{c_0 TOF_0 - c_1 TOF_1}{c_0 TOF_0 - c_0 TOF_1} - 1. \tag{A6}$$

Amplitude attenuation was calculated with constant ultrasound speed as follows:

$$\alpha_{\text{meas}}(t) = \frac{20}{c_0 TOF_1} \log_{10} \frac{A_1(t)}{A_2(t)}. \tag{A7}$$

The true amplitude attenuation was determined with Eq. (1) presented in Materials and methods. If $x(t)$ in Eq. (1) is expressed as $c_1 TOF_1/2$ the following formula for error in attenuation can be derived:

$$\Delta\alpha = \frac{\alpha_{\text{meas}} - \alpha_{\text{true}}}{\alpha_{\text{true}}} = \frac{c_1 - c_0}{c_0}. \tag{A8}$$

## References


[1] D.H. Agemura, W.D. O'Brien, Jr., J.E. Olerud, L.E. Chun and D.E. Eyre, Ultrasonic propagation properties of articular cartilage at 100 MHz, *J. Acoust. Soc. Am.* **87** (1990), 1786–1791.

[2] R.C. Appleyard, M.V. Swain, S. Khanna and G.A. Murrell, The accuracy and reliability of a novel handheld dynamic indentation probe for analysing articular cartilage, *Phys. Med. Biol.* **46** (2001), 541–550.

[3] C.G. Armstrong, W.M. Lai and V.C. Mow, An analysis of the unconfined compression of articular cartilage, *J. Biomech. Eng.* **106** (1984), 165–173.

[4] J.A. Buckwalter, L.C. Rosenberg and E.B. Hunziker, in: *Articular Cartilage: Composition, Structure, Response to Injury, and Methods of Facilitating Repair*, J.W. Ewing, ed., Raven Press Ltd., New York, 1990, pp. 19–56.

[5] S. Chaffaî, F. Peyrin, S. Nuzzo, R. Porcher, G. Berger and P. Laugier, Ultrasonic characterization of human cancellous bone using transmission and backscatter measurements: relationships to density and microstructure, *Bone* **30** (2002), 229–237.

[6] R.E. Challis and R.I. Kitney, Biomedical signal processing (in four parts). Part 1. Time-domain methods, *Med. Biol. Eng. Comput.* **28** (1990), 509–524.

[7] E. Chérin, A. Saïed, P. Laugier, P. Netter and G. Berger, Evaluation of acoustical parameter sensitivity to age-related and osteoarthritic changes in articular cartilage using 50-MHz ultrasound, *Ultrasound Med. Biol.* **24** (1998), 341–354.

[8] E. Collett, *Polarized Light: Fundamentals and Applications*, Marcel Dekker Inc., New York, 1993, 581 pp.

[9] T. Farkas, R.D. Boyd, M.B. Schaffler, E.L. Radin and D.B. Burr, Early vascular changes in rabbit subchondral bone after repetitive impulsive loading, *Clin. Orthop.* (1987), 259–267.

[10] M. Fortin, M.D. Buschmann, M.J. Bertrand, F.S. Foster and J. Ophir, Dynamic measurement of internal solid displacement in articular cartilage using ultrasound backscatter, *J. Biomech.* **36** (2003), 443–447.

[11] S.A. Goss, R.L. Johnston and F. Dunn, Comprehensive compilation of empirical ultrasonic properties of mammalian tissues, *J. Acoust. Soc. Am.* **64** (1978), 423–457.

[12] S.A. Goss and W.D. O'Brien, Jr., Direct ultrasonic velocity measurement of mammalian collagen threads, *J. Acoust. Soc. Am.* **85** (1979), 507–511.

[13] M.A. Hakulinen, J. Töyräs, S. Saarakkala, J. Hirvonen, H. Kröger and J.S. Jurvelin, Ability of ultrasound backscattering to predict mechanical properties of bovine trabecular bone, *Ultrasound Med. Biol.* **30** (2004), 919–927.

[14] W.C. Hayes, L.M. Keer, G. Herrmann and L.F. Mockros, A mathematical analysis for indentation tests of articular cartilage, *J. Biomech.* **5** (1972), 541–551.

[15] A. Hulth, Does osteoarthrosis depend on growth of the mineralized layer of cartilage?, *Clin. Orthop.* (1993), 19–24.

[16] M.S. Laasanen, J. Töyräs, J. Hirvonen, S. Saarakkala, R.K. Korhonen, M.T. Nieminen, I. Kiviranta and J.S. Jurvelin, Novel mechano-acoustic technique and instrument for diagnosis of cartilage degeneration, *Physiol. Meas.* **23** (2002), 491–503.

[17] M.S. Laasanen, J. Toyras, R.K. Korhonen, J. Rieppo, S. Saarakkala, M.T. Nieminen, J. Hirvonen and J.S. Jurvelin, Biomechanical properties of knee articular cartilage, *Biorheology* **40** (2003), 133–140.

[18] T. Lyyra, J. Jurvelin, P. Pitkänen, U. Väätäinen and I. Kiviranta, Indentation instrument for the measurement of cartilage stiffness under arthroscopic control, *Med. Eng. Phys.* **17** (1995), 395–399.

[19] A.F. Mak, The apparent viscoelastic behavior of articular cartilage – the contributions from the intrinsic matrix viscoelasticity and interstitial fluid flows, *Transactions of the ASME. J. Biomech. Eng.* **108** (1986), 123–130.

[20] M.Q. Niederauer, S. Cristante, G.M. Niederauer, R.P. Wilkes, S.M. Singh, D.F. Messina, M.A. Walter, B.D. Boyan, J.C. Delee and G. Niederauer, A novel instrument for quantitatively measuring the stiffness of articular cartilage, *Transact. Orthop. Res. Soc.* **23** (1998), 905.

[21] H.J. Nieminen, J. Töyräs, J. Rieppo, M.T. Nieminen, J. Hirvonen, R. Korhonen and J.S. Jurvelin, Real-time ultrasound analysis of articular cartilage degradation in vitro, *Ultrasound Med. Biol.* **28** (2002), 519–525.

[22] H.J. Nieminen, S. Saarakkala, M.S. Laasanen, J. Hirvonen, J.S. Jurvelin and J. Töyräs, Ultrasound attenuation in normal and spontaneously degenerated articular cartilage, *Ultrasound Med. Biol.* **30** (2004), 493–500.

[23] M. O'Donnell, J.W. Mimbs and J.G. Miller, Relationship between collagen and ultrasonic backscatter in myocardial tissue, *J. Acoust. Soc. Am.* **69** (1981), 580–588.

[24] B. Pellaumail, V. Dewailly, A. Watrin, D. Loeuille, P. Netter, G. Berger and A. Saïed, Attenuation coefficient and speed of sound in immature and mature rat cartilage: a study in the 30–70 MHz frequency range, *IEEE Ultrasonics Symposium* (1999), 1361–1365.

[25] J. Pohlhammer and W.D. O'brien, Jr., Dependence of the ultrasonic scatter coefficient on collagen concentration in mammalian tissues, *J. Acoust. Soc. Am.* **69** (1981), 283–285.

[26] E.L. Radin and R.M. Rose, Role of subchondral bone in the initiation and progression of cartilage damage, *Clin. Orthop.* (1986), 34–40.

[27] J. Rieppo, J. Hallikainen, J.S. Jurvelin, H.J. Helminen and M.M. Hyttinen, Novel quantitative polarization microscopic assesment of cartilage and bone collagen birefringence, orientation and anisotropy, *Transact. Orthop. Res. Soc.* **28** (2003), 0570.

[28] J. Rieppo, E.P. Halmesmäki, U. Siitonen, M.S. Laasanen, J. Töyräs, I. Kiviranta, M.M. Hyttinen, J.S. Jurvelin and H.J. Helminen, Histological differences of human, bovine and porcine cartilage, *Transact. Orthop. Res. Soc.* **28** (2003), 0589.

[29] C. Roux, V. Roberjot, R. Porcher, S. Kolta, M. Dougados and P. Laugier, Ultrasonic backscatter and transmission parameters at the os calcis in postmenopausal osteoporosis, *J. Bone Miner. Res.* **16** (2001), 1353–1362.

[30] S. Saarakkala, J. Töyräs, J. Hirvonen, M.S. Laasanen, R. Lappalainen and J.S. Jurvelin, Ultrasound indentation of normal and spontaneously degenerated bovine articular cartilage, *Ultrasound Med. Biol.* **30** (2004), 783–792.

[31] J.K. Suh, I. Youn and F.H. Fu, An in situ calibration of an ultrasound transducer: a potential application for an ultrasonic indentation test of articular cartilage, *J. Biomech.* **34** (2001), 1347–1353.

[32] J. Töyräs, M.S. Laasanen, S. Saarakkala, M. Lammi, J. Rieppo, J. Kurkijärvi, R. Lappalainen and J.S. Jurvelin, Speed of sound in normal and degenerated bovine articular cartilage, *Ultrasound Med. Biol.* **29** (2003), 447–454.

[33] Y.P. Zheng and A.F. Mak, An ultrasound indentation system for biomechanical properties assessment of soft tissues in-vivo, *IEEE Trans. Biomed. Eng.* **43** (1996), 912–918.

[34] Y.P. Zheng, A.F.T. Mak, K.P. Lau and L. Qin, An ultrasonic measurement for in vitro depth-dependent equilibrium strains of articular cartilage in compression, *Phys. Med. Biol.* **47** (2002), 3165–3180.

[35] Y.P. Zheng, S.L. Bridal, J. Shi, A. Saied, M.H. Lu, B. Jaffre, A.F.T. Mak and P. Laugier, High resolution ultrasound elastomicroscopy imaging of soft tissues: system development and feasibility, *Phys. Med. Biol.* **49** (2004), 3925–3938.

Biorheology 43 (2006) 537–545
IOS Press



# The mechanical environment of chondrocytes in articular cartilage

Michael A. Adams[*]

*University of Bristol, Department of Anatomy, UK*

**Abstract.** There is a growing literature concerning chondrocyte responses to mechanical loading, but relatively little is known about the mechanical environment these cells experience in a living joint. Calculations indicate that high forces are applied to limb joints whenever the joints are flexed, because flexion can cause body weight to act on long lever arms compared to the joint centre, whereas the muscles which extend the joint act on much shorter lever arms. As a result, joint reaction forces (which compress the cartilage) can rise to 3–6 times body weight during activities such as stair climbing. Articular cartilage tends to spread this load evenly over the joint surface, but is too thin to do this well, and compressive stresses can rise to 10–20 MPa. Within cartilage, matrix stresses vary locally, possibly as a result of variation in composition or undulations in the subchondral bone, and further modifications of stress occur within each chondron. Articular cartilage is a fibrous solid and cells within it are deformed by mechanical loading rather than subjected to a hydrostatic pressure. The mechanical environment of chondrocytes can best be reproduced *in vitro* by direct compression of the articular surface of cartilage which is supported naturally by adjacent cartilage and subchondral bone.

Keywords: Cartilage, mechanics, mechanobiology, matrix, stress

## 1. Introduction

There is a large and growing literature concerning chondrocyte responses to mechanical loading. However, comparatively little is known about the mechanical environment that these cells experience in a living joint, and fundamental questions have only recently been tackled. For example, what forces are applied to living joints, and how are they distributed across the articular surface? How do stresses vary within cartilage, and within the chondron, and what stresses and strains do chondrocytes actually experience? Do they experience a hydrostatic pressure? How is their mechanical environment affected by tissue disruption (such as fibrillation) or by water loss?

The purpose of this short review is to attempt to answer these and related questions, and then apply the information to the problem of how best to reproduce the mechanical environment of chondrocytes in laboratory experiments.

## 2. Structure and function of articular cartilage

Articular cartilage has a complex structure based on a three-dimensional collagen architecture (Fig. 1) and consequently its material properties vary markedly with location and with direction [30]. The surface

---

[*]Address for correspondence: Dr. Michael A. Adams, Department of Anatomy, University of Bristol, Southwell Street, Bristol BS2 8EJ, UK. Tel.: +44 117 9288363; Fax: +44 117 9254794; E-mail: M.A.Adams@bris.ac.uk.

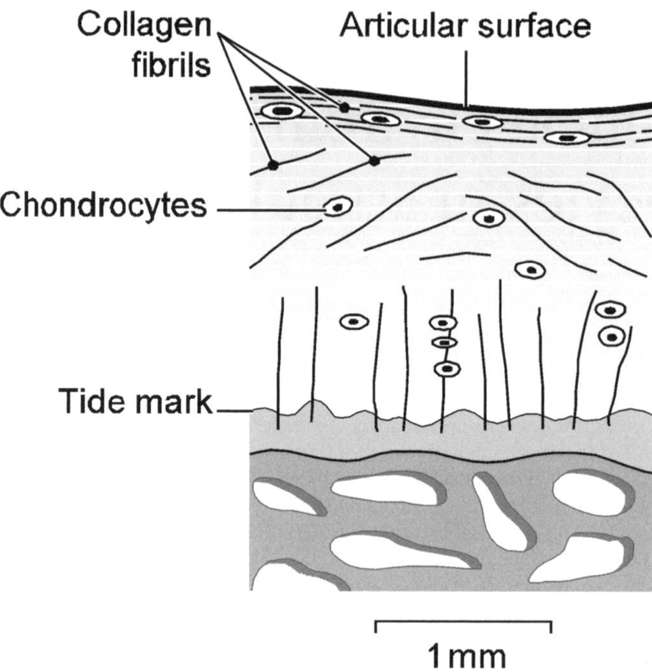


Fig. 1. Cartilage structure. From [1] with permission.

zone has a high tensile strength parallel to the surface and so acts to "keep the lid on" the softer material underneath [29] and prevent excessive water loss from it [38]. The main mechanical functions of articular cartilage are (a) to form a low-friction surface which allows easy joint movement with a minimum of heat and loss of material ("wear"), and (b) to deform sufficiently to spread out applied loading over the articular surface, reducing contact stress (force per unit area) on the underlying bone.

## 3. Forces acting on synovial joints

The low friction of cartilage ensures that joint reaction forces act on it more or less perpendicular to its surface. The joint reaction force can greatly exceed body weight, because it arises primarily from muscle tension, which can be considerable during vigorous activity. Muscles generally have their insertions close to the centre of rotation of joints, so that muscle contraction can produce a rapid angular movement of the joint. The disadvantage of this, however, is that the muscle must act on a shorter lever arm than superincumbent body-weight. In order to balance moments, therefore, the muscle force must exceed superincumbent body weight by a factor of 3–6, with the highest muscle forces being required when the joint is most flexed [12].

Muscle forces are further increased if they are acting to accelerate the body, perhaps in a sprint or jump, or to cause the body to change direction. According to Newton's 3rd Law, "Force = mass × acceleration" so muscle forces increase in direct proportion to the acceleration, which is the change in velocity per unit time. Similarly, any deceleration of body weight, such as occurs in a fall on the buttocks or collision with an external object, will create high inertial forces, proportional to the rate at which the body is slowed down. In the case of a fall, this depends on the length of a person's legs (which largely determines the body's velocity at impact), and the softness of the landing (which determines how quickly the body is brought to rest).

A third situation that leads to high muscle forces on a joint is the need to stabilise a joint by simultaneous contraction of antagonist muscles. A watchmaker, for example, needs antagonistic contraction to carefully control small movements of his fingers, and the resulting forces on the finger joints can then be considerable. Antagonistic muscle activity has been shown to increase the compressive force acting on the spine by up to 45% [17].

Under extreme circumstances, muscle forces on lower limb joints can probably rise to $10\times$ body weight, although precise experimental verification of this would be difficult to obtain. The fact that muscle forces generally contribute much more to joint loading than does superincumbent body weight makes the distinction between "weight-bearing" and "non-weight-bearing" somewhat arbitrary. However, it should be realised that muscle forces tend to be proportional to body weight, and this probably explains why obesity is so closely linked to osteoarthritis in certain joints.

## 4. Distribution of forces across the articular surface

To a certain extent, the resultant force acting on a joint can be distributed by the shape of the joint surfaces. If the joint is congruent (both surface have the same curvature) the load will generally be spread more evenly, because contact will take place over a larger area. However, it is sometimes better for the joints to be slightly incongruous, as in the hip joint where the acetabulum has a tendency to a "gothic arch" shape compared to the spherical femoral head [5]. This can be an advantage when the resultant force usually acts in one direction (in this case, more or less vertically) because it ensures that peak forces are not concentrated in the central (approximately horizontal) region of the acetabulum which lies perpendicular to the (approximately vertical) applied force. The gothic arch ensures that this central area does not make contact at low loads and that, at high loads, the force is spread more evenly over the entire acetabulum [6,15,43].

The relative softness of cartilage compared to bone enables it to deform and increase the area of contact between articulating surfaces, thereby reducing contact stress. This is especially important in incongruent joints such as the knee, and probably explains why knee cartilage is so thick. However, if cartilage were very soft, it would distribute stress evenly at low loads, but not at high loads, and it is obviously important that it does the latter. Also, the requirement for it to act as an efficient low-friction bearing material also limits its softness and thickness. Therefore cartilage tends to be fairly stiff and thin, and rather than equalise stress at moderate load levels it merely reduces the size of stress concentrations acting on the articular surfaces. This has been demonstrated by inserting pressure-sensitive paper between the articular surfaces of cadaveric hip joints while loading the joints to simulate phases of the walking cycle: marked localised stress concentrations rising to 6–10 MPa were found [6,43] as shown in Fig. 2. Measurements from an instrumented hip prosthesis *in vivo* have shown that peak stresses reach 18 MPa when rising from a chair [21].

The highest stresses acting on articular cartilage during vigorous activities can be gauged from the stresses required to damage the tissue (assuming that no skeletal tissue would be so "over-engineered" that its strength greatly exceeded the highest stresses normally applied to it). When human knee joint cartilage is compressed to failure, damage is first detected within the tissue at an average stress of 15 MPa [34], and bovine knee cartilage fails at an average 36 MPa [3,27]. Thin (healthy) cartilage is stiffer than thick (healthy) cartilage [40], and may also be stronger. Compressive failure involves rupture of the cartilage surface, apparently in tension [3,16,26,46] as shown in Fig. 3.

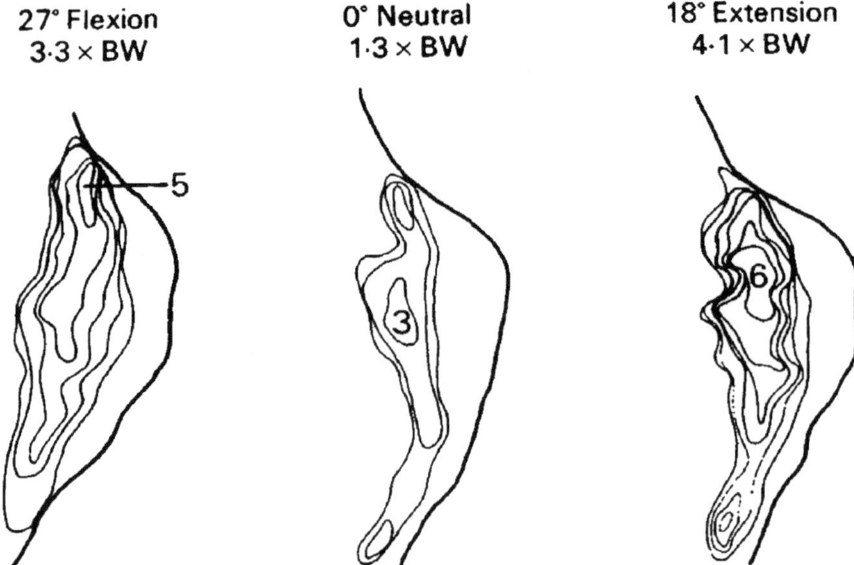


Fig. 2. Distribution of compressive stress acting on the acetabulum of a cadaveric hip joint. The joint was loaded with various multiples of bodyweight (BW) while positioned to simulate three stages in the walking cycle [6]. Stress concentrations of up to 6 MPa indicate that cartilage is unable to equalise stress across an entire joint surface.

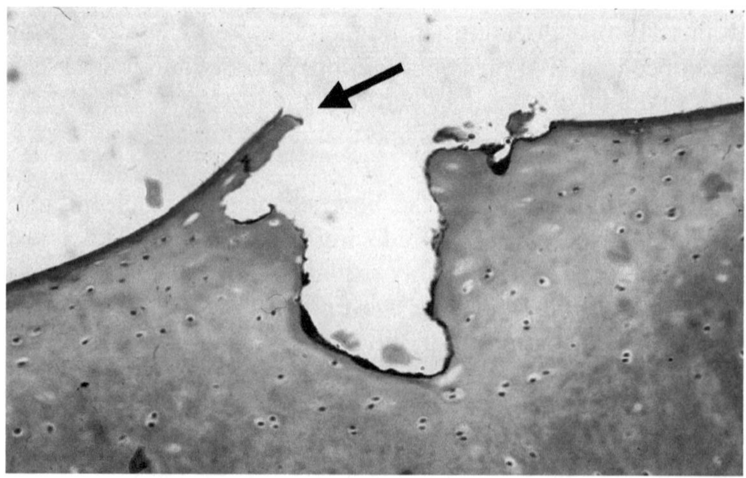

Fig. 3. Rupture of the surface of articular cartilage following compressive overload. It appears that the surface has failed in tension (arrow). Histological section of bovine knee cartilage, compressed using a flat metal indenter *in vitro* [26].

## 5. Stress concentrations within cartilage

If a miniature pressure transducer is inserted into cartilage and pulled parallel to its surface, it can detect localised concentrations of compressive "stress" within the tissue. Validation tests show that the transducer output is linearly related to applied load [2] and that, in fibrocartilage, it is approximately equal to the average compressive stress acting perpendicular to its membrane [31]. Stress concentrations probably correspond to irregularities in subchondral bone, or to local variations in cartilage composition.

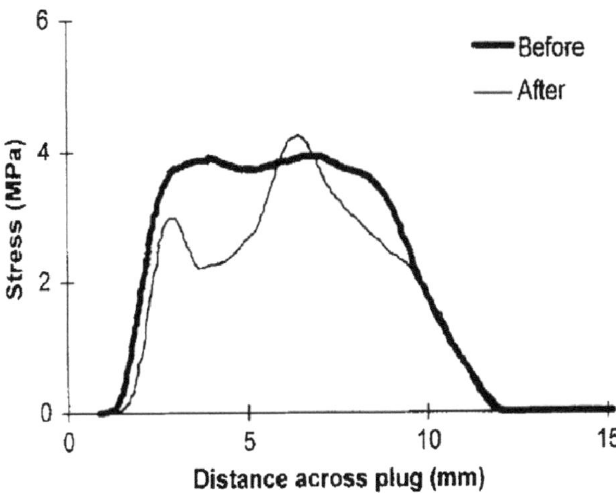


Fig. 4. Local stress concentrations can be observed in articular cartilage by pulling a pressure-sensitive transducer through the tissue, parallel to the articular surface [2]. Note that stress concentrations are increased following a period of compressive creep loading which drives water from the tissue.

They increase markedly following sustained "creep" loading [2] as shown in Fig. 4. Creep loading at 2 MPa *in vitro* reduces cartilage thickness by 45% in 30 minutes [3] and just 20 seconds of squatting *in vivo* reduces patella cartilage volume and thickness by 5% [14].

Finite element models suggest that even more localised variations in stress occur in the immediate environment of the chondrocyte [20] and they appear to influence cell deformation and phenotype [37]. Stresses within the chondron – the collagen "basket" which surrounds chondrocytes [22] – may be further influenced by altered composition and collagen organisation in the pericellular and territorial matrix.

## 6. Chondrocytes do not experience hydrostatic pressure

Load-carriage in articular cartilage depends on fluid pressurisation in microscopic pores, as well as by deformation of the solid matrix [35], and "biphasic" theories are able to explain slow deformations of cartilage in terms of fluid phase flowing relative to the solid phase [32]. Unfortunately, some biologists have assumed that this implies that chondrocytes in articular cartilage experience a hydrostatic pressure, in which the intensity of loading is equal in all directions. Such loading would not deform chondrocytes, which contain no gaseous voids and behave like viscoelastic solids [18]. The highly hydrated nucleus pulposus and inner annulus regions of intervertebral discs do exhibit a true hydrostatic pressure [4] but articular cartilage does not: it is a fibrous solid with a relatively rigid three-dimensional collagen network which prohibits fluid behaviour. Fluid can flow *within* the matrix, in notional "pores" with dimensions of the order of nanometres, from regions of high stress to low stress, and this fluid flow could conceivably be detected by cells. However, under normal loading conditions, the fluid pressure varies from one location to another (i.e. it is not hydro*static*) and stresses in the solid phase also vary with location and direction, so cells embedded in the matrix will be pressed on by different amounts from different sides. Confocal microscopy studies confirm that chondrocytes are indeed deformed by mechanical loading, being squashed perpendicular to the principal direction of loading, which is usually perpendicular to the cartilage surface [18,19].

## 7. Matrix and cell deformations vary with depth

The composition and mechanical properties of articular cartilage vary with depth from the surface, with the surface zone being particularly stiff and strong when stretched parallel to that surface [25]. Cells are very much softer than their surrounding matrix [23], and deform according to the deformation of the hole in the matrix which they occupy. It would be expected therefore that cell deformations will vary with depth in cartilage. In addition, compressive stresses generally decrease with distance form the surface, because the compressive force effectively spreads out as it is transmitted from the articular surface to bone, to an extent that depends on the ratio of cartilage thickness to the diameter of the compressed area. This effect will exaggerate the differences in cell deformation and loading with depth.

It is not surprising therefore that mechanical indentation of cartilage *in vitro* kills more cells near the surface compared to the deep zone [10] and that more apoptotic cells are found close to the surface in osteoarthritis [39].

The mechanical properties of the cartilage matrix also change with age [25], and following non-enzymatic glycation [42], and following sustained loading which expels water from the tissues [3]. All of these effects would therefore influence cell responses to mechanical loading.

## 8. Reproducing the mechanical environment of chondrocytes *in vitro*

Reproducing the physiological mechanical environment of cartilage cells in the laboratory is not an easy task. Because cell deformation under load depends on matrix deformation, the latter should be kept as realistic as possible. The main component of applied loading should be applied perpendicular to the cartilage surface, as in life, and loaded cartilage should be supported naturally by adjacent cartilage and by subchondral bone. A lack of lateral support would allow unnatural "barrel" deformations of the loaded tissue, which probably facilitates full-depth vertical splitting under load [45]. Removing cartilage from bone would also allow radial (lateral) expansion and disturb the integrity of the collagen network. The extremes of "confined" and "unconfined" compression are both artificial, although they can be useful for providing experimental measurements that can easily be compared with the predictions of mathematical models. Cartilage hydration should be preserved *in vitro* because it affects both fluid flow within, and deformation of, the matrix.

In life, cartilage is loaded by cartilage, but this is not easy to achieve in the laboratory while at the same time controlling the area of contact (and hence the compressive stress). Opposing plugs of cartilage can be used [45] but it is more convenient to use a flat indenter, whose area can be carefully controlled [27, 34]. As discussed previously [34], an indenter should be hard, smooth, have low friction, and be impermeable in order to minimise friction and unnatural fluid flow. A bevel around the edge of the indenter is useful in reducing artifactual stress concentrations at its edge [9,26], but at the cost of introducing some uncertainty in the size of the area of contact, and hence the contact stress.

The force under an indenter spreads outwards with increasing cartilage depth, as discussed above, so care should be taken not to choose "control" cartilage from the region of tissue just outside the loaded area. Experiments have shown that this apparently unloaded tissue can exhibit increased collagen denaturation [9] and increased cell death [10] although this latter finding is not due entirely to mechanical causes [11].

Finally, loading magnitude will be considered. The normal physiological range of compressive stress for human hip joint cartilage during walking is probably 1–10 MPa [6]. The principles of adaptive

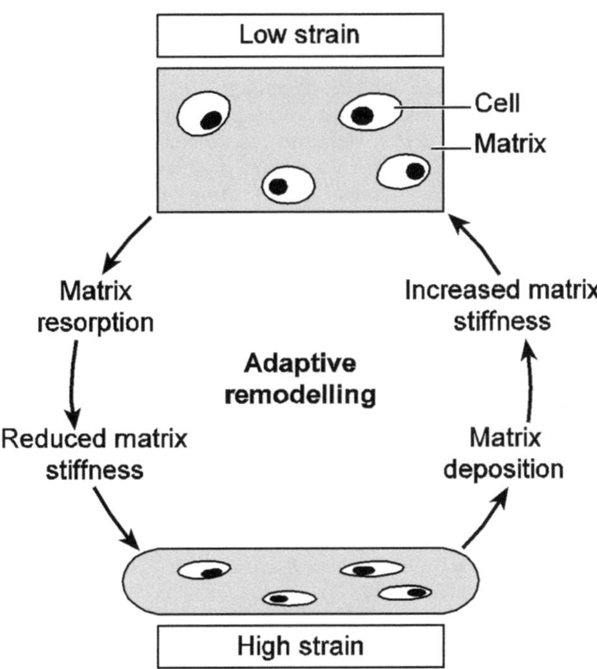


Fig. 5. In adaptive remodelling, connective tissue cells adapt the stiffness of their extracellular matrix to suit the mechanical demands placed upon it. From [1] with permission.

remodelling (Fig. 5) have been shown to apply to articular cartilage [8,28,36,41,44] – even though excessive loading can be detrimental [7,13,24,33] – and they suggest that loading below 1 MPa should induce a catabolic stimulus, not an anabolic one. Several published experiments have shown evidence for anabolic activity at stresses below 1 MPa, suggesting that, although the applied stress was low, the applied strain (deformation) was not. This could indicate inadequate support for the loaded tissue. Two wrongs do not make a right, and interpretation of mechanobiology experiments would be simpler if both the applied stress and cartilage strain were kept within physiological limits.

## Acknowledgement

The work of the author is funded by the BBSRC (UK).

## References

[1] M.A. Adams, N. Bogduk, K. Burton and P. Dolan, *The Biomechanics of Back Pain*, Churchill Livingstone, Edinburgh, 2002.

[2] M.A. Adams, A.J. Kerin, L.S. Bhatia, G. Chakrabarty and P. Dolan, Experimental determination of stress distributions in articular cartilage before and after sustained loading, *Clin. Biomech.* **14** (1999), 88–96.

[3] M.A. Adams, A.J. Kerin and M.R. Wisnom, Sustained loading increases the compressive strength of articular cartilage, *Connect. Tissue Res.* **39** (1998), 245–256.

[4] M.A. Adams, D.S. McNally and P. Dolan, 'Stress' distributions inside intervertebral discs. The effects of age and degeneration, *J. Bone Joint Surg. Br.* **78** (1996), 965–972.

[5] N.Y. Afoke, P.D. Byers and W.C. Hutton, The incongruous hip joint. A casting study, *J. Bone Joint Surg. [Br.]* **62-B** (1980), 511–514.

[6] N.Y. Afoke, P.D. Byers and W.C. Hutton, Contact pressures in the human hip joint, *J. Bone Joint Surg. [Br.]* **69** (1987), 536–541.

[7] J. Arokoski, I. Kiviranta, J. Jurvelin, M. Tammi and H.J. Helminen, Long-distance running causes site-dependent decrease of cartilage glycosaminoglycan content in the knee joints of beagle dogs, *Arthritis Rheum.* **36** (1993), 1451–1459.

[8] J.P. Arokoski, J.S. Jurvelin, U. Vaatainen and H.J. Helminen, Normal and pathological adaptations of articular cartilage to joint loading [In Process Citation], *Scand. J. Med. Sci. Sports* **10** (2000), 186–198.

[9] K. Clements, A. Hollander, M. Sharif and M. Adams, Cyclic loading can denature type II collagen in articular cartilage, *Connect. Tissue Res.* **45** (2004), 174–180.

[10] K.M. Clements, Z.C. Bee, G.V. Crossingham, M.A. Adams and M. Sharif, How severe must repetitive loading be to kill chondrocytes in articular cartilage?, *Osteoarthritis Cartilage* **9** (2001), 499–507.

[11] K.M. Clements, N. Burton-Wurster and G. Lust, The spread of cell death from impact damaged cartilage: lack of evidence for the role of nitric oxide and caspases, *Osteoarthritis Cartilage* **12** (2004), 577–585.

[12] P.A. Costigan, K.J. Deluzio and U.P. Wyss, Knee and hip kinetics during normal stair climbing, *Gait Posture* **16** (2002), 31–37.

[13] F. Eckstein, S. Faber, R. Muhlbauer, J. Hohe, K.H. Englmeier, M. Reiser and R. Putz, Functional adaptation of human joints to mechanical stimuli, *Osteoarthritis Cartilage* **10** (2002), 44–50.

[14] F. Eckstein, B. Lemberger, T. Stammberger, K.H. Englmeier and M. Reiser, Patellar cartilage deformation in vivo after static versus dynamic loading, *J. Biomech.* **33** (2000), 819–825.

[15] F. Eckstein, B. Merz, P. Schmid and R. Putz, The influence of geometry on the stress distribution in joints – a finite element analysis, *Anat. Embryol. (Berl.)* **189** (1994), 545–552.

[16] R. Flachsmann, N.D. Broom and A.E. Hardy, Deformation and rupture of the articular surface under dynamic and static compression, *J. Orthop. Res.* **19** (2001), 1131–1139.

[17] K.P. Granata and W.S. Marras, The influence of trunk muscle coactivity on dynamic spinal loads, *Spine* **20** (1995), 913–919.

[18] F. Guilak, Compression-induced changes in the shape and volume of the chondrocyte nucleus, *J. Biomech.* **28** (1995), 1529–1541.

[19] F. Guilak, The deformation behavior and viscoelastic properties of chondrocytes in articular cartilage, *Biorheology* **37** (2000), 27–44.

[20] F. Guilak and V.C. Mow, The mechanical environment of the chondrocyte: a biphasic finite element model of cell-matrix interactions in articular cartilage, *J. Biomech.* **33** (2000), 1663–1673.

[21] W.A. Hodge, K.L. Carlson, R.S. Fijan, R.G. Burgess, P.O. Riley, W.H. Harris and R.W. Mann, Contact pressures from an instrumented hip endoprosthesis, *J. Bone Joint Surg. [Am.]* **71** (1989), 1378–1386.

[22] A.K. Jeffery, G.W. Blunn, C.W. Archer and G. Bentley, Three-dimensional collagen architecture in bovine articular cartilage, *J. Bone Joint Surg. [Br.]* **73** (1991), 795–801.

[23] W.R. Jones, H.P. Ting-Beall, G.M. Lee, S.S. Kelley, R.M. Hochmuth and F. Guilak, Alterations in the Young's modulus and volumetric properties of chondrocytes isolated from normal and osteoarthritic human cartilage, *J. Biomech.* **32** (1999), 119–127.

[24] J. Jurvelin, I. Kiviranta, A.M. Saamanen, M. Tammi and H.J. Helminen, Indentation stiffness of young canine knee articular cartilage – influence of strenuous joint loading, *J. Biomech.* **23** (1990), 1239–1246.

[25] G.E. Kempson, Age-related changes in the tensile properties of human articular cartilage: a comparative study between the femoral head of the hip joint and the talus of the ankle joint, *Biochim. Biophys. Acta* **1075** (1991), 223–230.

[26] A.J. Kerin, A. Coleman, M.R. Wisnom and M.A. Adams, Propagation of surface fissures in articular cartilage in response to cyclic loading in vitro, *Clin. Biomech.* **18** (2003), 960–968.

[27] A.J. Kerin, M.R. Wisnom and M.A. Adams, The compressive strength of articular cartilage, *Proc. Inst. Mech. Eng. [H]* **212** (1998), 273–280.

[28] I. Kiviranta, J. Jurvelin, M. Tammi, A.M. Saamanen and H.J. Helminen, Weight bearing controls glycosaminoglycan concentration and articular cartilage thickness in the knee joints of young beagle dogs, *Arthritis Rheum.* **30** (1987), 801–809.

[29] R.K. Korhonen, M. Wong, J. Arokoski, R. Lindgren, H.J. Helminen, E.B. Hunziker and J.S. Jurvelin, Importance of the superficial tissue layer for the indentation stiffness of articular cartilage, *Med. Eng. Phys.* **24** (2002), 99–108.

[30] M.S. Laasanen, J. Toyras, R.K. Korhonen, J. Rieppo, S. Saarakkala, M.T. Nieminen, J. Hirvonen and J.S. Jurvelin, Biomechanical properties of knee articular cartilage, *Biorheology* **40** (2003), 133–140.

[31] D.W. McMillan, D.S. McNally, G. Garbutt and M.A. Adams, Stress distributions inside intervertebral discs: the validity of experimental "stress profilometry", *Proc. Inst. Mech. Eng. [H]* **210** (1996), 81–87.

[32] V.C. Mow, M.C. Gibbs, W.M. Lai, W.B. Zhu and K.A. Athanasiou, Biphasic indentation of articular cartilage – II. A numerical algorithm and an experimental study, *J. Biomech.* **22** (1989), 853–861.

[33] R.C. Murray, C.F. Zhu, A.E. Goodship, K.H. Lakhani, C.M. Agrawal and K.A. Athanasiou, Exercise affects the mechanical properties and histological appearance of equine articular cartilage, *J. Orthop. Res.* **17** (1999), 725–731.

[34] E.M. Obeid, M.A. Adams and J.H. Newman, Mechanical properties of articular cartilage in knees with unicompartmental osteoarthritis, *J. Bone Joint Surg. Br.* **76** (1994), 315–319.

[35] A. Oloyede and N. Broom, The biomechanics of cartilage load-carriage, *Connect. Tissue Res.* **34** (1996), 119–143.

[36] I.G. Otterness, J.D. Eskra, M.L. Bliven, A.K. Shay, J.P. Pelletier and A.J. Milici, Exercise protects against articular cartilage degeneration in the hamster, *Arthritis Rheum.* **41** (1998), 2068–2076.

[37] L.A. Setton and J. Chen, Cell mechanics and mechanobiology in the intervertebral disc, *Spine* **29** (2004), 2710–2723.

[38] L.A. Setton, W. Zhu and V.C. Mow, The biphasic poroviscoelastic behavior of articular cartilage: role of the surface zone in governing the compressive behavior [see comments], *J. Biomech.* **26** (1993), 581–592.

[39] M. Sharif, A. Whitehouse, P. Sharman, M. Perry and M. Adams, Increased apoptosis in human osteoarthritic cartilage corresponds to reduced cell density and expression of caspase-3, *Arthritis Rheum.* **50** (2004), 507–515.

[40] D.E. Shepherd and B.B. Seedhom, Thickness of human articular cartilage in joints of the lower limb, *Ann. Rheum. Dis.* **58** (1999), 27–34.

[41] B. Vanwanseele, F. Eckstein, H. Knecht, E. Stussi and A. Spaepen, Knee cartilage of spinal cord-injured patients displays progressive thinning in the absence of normal joint loading and movement, *Arthritis Rheum.* **46** (2002), 2073–2078.

[42] N. Verzijl, J. DeGroot, Z.C. Ben, O. Brau-Benjamin, A. Maroudas, R.A. Bank, J. Mizrahi, C.G. Schalkwijk, S.R. Thorpe, J.W. Baynes, J.W. Bijlsma, F.P. Lafeber and J.M. TeKoppele, Crosslinking by advanced glycation end products increases the stiffness of the collagen network in human articular cartilage: a possible mechanism through which age is a risk factor for osteoarthritis, *Arthritis Rheum.* **46** (2002), 114–123.

[43] R. von Eisenhart, C. Adam, M. Steinlechner, M. Muller-Gerbl and F. Eckstein, Quantitative determination of joint incongruity and pressure distribution during simulated gait and cartilage thickness in the human hip joint [see comments], *J. Orthop. Res.* **17** (1999), 532–539.

[44] J.Q. Yao and B.B. Seedhom, Mechanical conditioning of articular cartilage to prevalent stresses, *Br. J. Rheumatol.* **32** (1993), 956–965.

[45] N.B. Zimmerman, D.G. Smith, L.A. Pottenger and D.R. Cooperman, Mechanical disruption of human patellar cartilage by repetitive loading in vitro, *Clin. Orthop.* (1988), 302–307.

[46] V. Morel, C. Berutto and T.M. Quinn, Effects of damage in the articular surface on the cartilage response to injurious compression *in vitro*, *J. Biomech.* **39** (2006), 924–930.

Biorheology 43 (2006) 547–551
IOS Press



# Age-related quantitative MRI changes in healthy cartilage: Preliminary results

Jean Christophe Goebel [a], Astrid Watrin-Pinzano [a], Isabelle Bettembourg-Brault [b],
Freddy Odille [c], Jacques Felblinger [c], Isabelle Chary-Valckenaere [a,b], Patrick Netter [a],
Alain Blum [a,d], Pierre Gillet [a,b,*] and Damien Loeuille [a,b]

[a] *Department of Pharmacology, Faculté de Médecine, UMR 7561 CNRS University Nancy I,
Avenue de la Forêt de Haye, 54505 Vandoeuvre lès Nancy, France*
[b] *Clinique Rhumatologique (Pr. Jacques Pourel), Chu Nancy brabois, 54511 Vandoeuvre cedex, France*
[c] *Imagerie Adaptative Diagnostique et Interventionnelle (Pr. Jacques Felblinger), Chu de Nancy
Brabois, Tour Drouet 4ème étage, 54511 Vandoeuvre lés Nancy cedex, France*
[d] *Service de Radiologie Guilloz (Pr. Alain Blum), Chu de Nancy, Hôpital Central, CO 34,
54000 Nancy cedex, France*

**Abstract.** *Objectives*: As the early form of OA is characterized by elevated water content in the cartilage tissue, the purpose of this study was to verify *in vivo* if age-related changes in patellar cartilage in healthy volunteers can be detected using quantitative MRI with T2 mapping and volume measurement MRI methods. *Design*: Thirty healthy volunteers of various classes of age (18 to 65 years old) were enrolled in this study. MR images of the patellar cartilage were acquired at 1.5T. Patellar cartilage volume and T2 maps were determined. *Results*: Despite non-significance, there was a trend in reducing cartilage volume with ageing ($r$: $-0.25$). In contrast global T2 slightly increased with ageing ($r$: 0.46). BMI ($r$: 0.51) and bone volume ($r$: 0.69) are well correlated to cartilage volume. *Conclusion*. Age-related physiologic changes in the water content of patellar cartilage can be detected using MRI. The proposed T2-mapping method, coupled with other non-invasive MR cartilage imaging techniques, could aid in the early diagnosis of OA.

Keywords: Cartilage, MRI, T2 map, volume, patella, age-related changes

## 1. Introduction

Magnetic Resonance Imaging (MRI) of articular cartilage has attracted intense interest and been the subject of numerous research studies over the past 10 years [13]. There are several reasons: the essential role articular cartilage plays in the function of the diarthrodial joints of the body, the high prevalence of degeneration and traumatic injury of articular cartilage, and the recent development of new surgical procedures that hold the promise of forming repair tissue that is hyaline or hyaline-like cartilage. In fact, MRI is the optimal diagnostic method for non-invasive evaluation of chondral lesions.

Because MRI can directly visualize articular cartilage, it is likely to be a useful modality in the study of cartilage ageing and osteoarthritis (OA). Current clinical MRI techniques demonstrate joint anatomy, and can be used to determine morphologic parameters such as cartilage volume, thickness, and presence

---

*Address for correspondence: Pierre Gillet. E-mail: Pierre.Gillet@medecine.uhp-nancy.fr.

of focal, superficial cartilage lesions. More recently techniques have been described for generating spatially localized quantitative maps of MRI relaxation times of cartilage. Such MRI parametric mapping techniques have the potential to identify and localize specific biochemical and structural changes within the extracellular cartilage matrix [7].

Some new quantitative MR imaging techniques [4], such as cartilage T2 mapping [8] and volume assessment [3] demonstrate sensitivity to age- and OA-related biochemical and structural changes in the extracellular cartilage matrix and thus have the potential to serve as image markers of OA versus physiologic senescence [11,15]. The purpose of the study is thus to determine if an age-dependent variation in patellar cartilage T2 and patellar cartilage volume occurs in asymptomatic volunteers.

## 2. Patients and methods

### 2.1. Recruitment of volunteers

Thirty healthy volunteers (11 females and 19 males) were screened to exclude those with a known contraindication for MRI: pacemaker, cerebral aneurysm clip, cochlear implant, presence of metal/shrapnel in strategic locations, such as in the eye, claustrophobia, and inability to cooperate with study requirements and give informed consent. Healthy adults over the age of 18 years were included in the study if (1) they had no history of pain or traumatic injury of the studied knee and (2) if the studied knee has a normal pattern on a preliminary MRI (Spin Echo T2 weighted sequence (T2wSE)). The studied population was stratified by age into 5 cohorts: (A) 18–25 years ($n = 6$), (B) 26–35 years ($n = 7$), (C) 36–45 years ($n = 7$), (D) 46–55 years ($n = 5$) and 56–65 years ($n = 5$). Additional demographic data collected at time of the MRI study included height and weight. Body mass index (BMI) was calculated by dividing the subject's weight in kilograms by the square of the height in meters. The study design was approved by the Institutional Review Board of University Hospital Center of Nancy.

### 2.2. MR protocol

The patella was examined in the axial plane with a 1.5 T whole body magnetic resonance unit (Signa Advantage HiSpeed GE Medical Systems, Milwaukee, WI) using a commercial coil dedicated to the knee. The following sequences and parameters were used:

*T2 map.* A four echo T2w SE imaging with TR (repetition time) constant at 3500 msec with four different TE (echo time, 16, 32, 48 and 64 msec); field of view $16 \times 16$ cm; acquisition matrix $256 \times 192$ pixels; resolution $0.47 \times 0.47$ mm; slice thickness 5 mm; spacing between slices 0.5 mm; 48 adjoining slices, number of excitation: 1; acquisition time: 25 minutes. T2 values were calculated from the four echoes by using a non-linear least squares curve fitting on a pixel by pixel basis assuming a monoexponential T2 decay. Obtained T2 values (global T2) were the averaged T2 data over the entire patellar cartilage thickness.

*Volume assessment.* A T1 weighted fat suppressed 3D gradient recall acquisition was performed with the following parameters: flip angle 15 degrees; TR 17.7 msec; TE 6.5 ms; FOV $12 \times 12$ cm; matrix acquisition $320 \times 320$ matrix; acquisition time 10 min; number of excitation: 1.5. Axial images were obtained at a partition thickness of 2 mm (60 adjoining slices). Patellar cartilage and bone volumes were determined by image processing after semi-automatical segmentation on an independent workstation using the software program Osirix [14].

Global and regional T2 values as well as volume were then studied, the whole patellar cartilage being digitally divided in 8 Regions Of Interest (ROIs): medial and lateral facets of the patella being equally subdivided into 4 parts in the axial direction.

Table 1
Summary of the parameters studied, by age group cohort. Results are expressed as mean $\pm$ s.e.m.

| Group | BMI | Bone (ml) | Cartilage (ml) | Global T2 cartilage (msec) |
|---|---|---|---|---|
| A (18–25 years) | $24.4 \pm 1.23$ | $20.6 \pm 1.70$ | $3.7 \pm 0.34$ | $27.9 \pm 0.66$ |
| B (26–35 years) | $25.1 \pm 2.16$ | $16.8 \pm 1.52$ | $3.4 \pm 0.38$ | $30.2 \pm 0.80$ |
| C (36–45 years) | $22.7 \pm 0.98$ | $15.3 \pm 0.71$ | $3.0 \pm 0.26$ | $27.2 \pm 0.98$ |
| D (46–55 years) | $23.2 \pm 1.23$ | $17.7 \pm 3.16$ | $3.2 \pm 0.40$ | $30.5 \pm 1.81$ |
| E (56–65 years) | $25.9 \pm 1.79$ | $17.4 \pm 0.88$ | $2.9 \pm 0.13$ | $31.9 \pm 0.92$ |

## 3. Results

No clear difference was observed between groups in terms of BMI and bone volume (Table 1). Despite non-significance, there was a trend in reducing cartilage volume with ageing ($r$: $-0.25$). In contrast, global T2 slightly increased with ageing ($r$: 0.46). BMI ($r$: 0.51) and bone volume ($r$: 0.69) are well correlated to cartilage volume.

ROIs study confirmed that cartilage volume was more abundant in the center of the patella especially in the lateral part. In this particular area, age-related T2 variation was more pronounced. In contrast it remained only a non-significant trend in age-related volume decrease.

## 4. Discussion

Our study reveals age-dependant changes in T2 relaxation time of patellar cartilage in healthy asymptomatic volunteers, like previously demonstrated by Mosher et al. [9,10]. In contrast, no concomitant significant variation in cartilage volume was depicted. The T2 relaxation time of articular cartilage is a function of the water content of the tissue [6]. To measure the T2 relaxation time with a high degree of accuracy, care must be taken with the MR technique. Typically, a multi-echo, spin-echo acquisition is used and signal levels are fitted to one degree or more decaying exponential, depending upon whether it is felt that there is more than one distribution of T2 within the sample.

The T2 relaxation time of articular cartilage characterizes the interactions of cartilage fluid (protons) with the solid matrix (collagen and proteoglycan [16]). In this tissue, there are several distinct "pools" of water molecules: the molecules associated with the collagen fibrils, the molecules hydrogen-bonded to proteoglycans by means of electrostatic attraction, and free water molecules. Only this last pool is responsible for the cartilage signal intensity.

During MRI T2-mapping, an image of the T2 relaxation times is then generated (Fig. 1), either with a color map or gray scale representing the relaxation times (see extensive review in [8]). Several investigators have measured in clinics the spatial distribution of T2 relaxation times within cartilage. In humans, ageing appears to be associated with a symptomatic increase in T2 relaxation times in the transitional zone of patellar cartilage [9]. Additionally, focal increases in T2 relaxation times within cartilage have been associated with matrix damage, particularly loss of collagen matrix [5]. Concerning our study, we can postulate that the age-related changes in T2 are related to "physiological" senescence denaturation process of the extracellular matrix and not to OA, even asymptomatic, because a normal MRI of the knee (T2wSE) was required to be included in the study.

*Finally*, in healthy asymptomatic volunteers, age-related T2 increase seems more sensible and site-dependant than cartilage volume decrease. It probably reflects asymptomatic early degeneration in articular cartilage, e.g. changes in water, proteoglycan and collagen contents before a "volumic" impact (i.e.

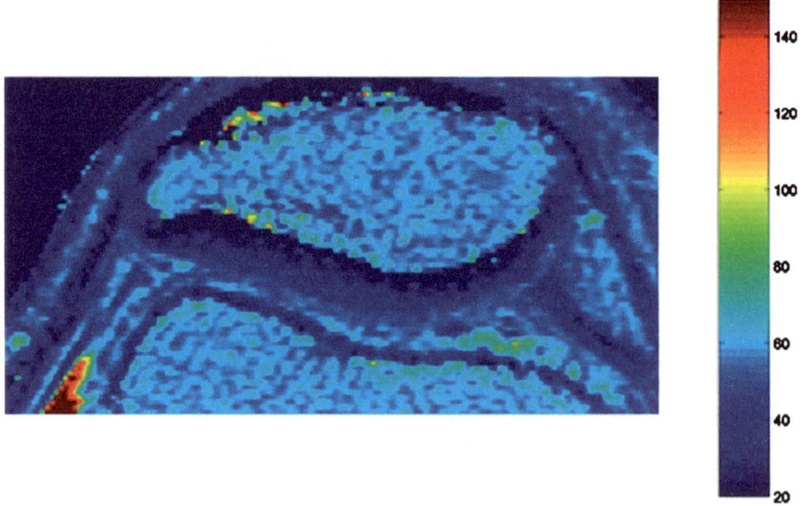

Fig. 1. Typical T2 map of a normal patella (23 years) at 1.5 T.

cartilage loss). In addition to cartilage volume assessment [1–3,12], *in vivo* cartilage T2 mapping can improve detection of cartilage ageing [10] and evaluation of new structure modifying pharmaceuticals and cartilage engineering.

## Acknowledgements

We thank Docteur Anne Christine RAT for her assistance with the statistical analysis. This study was supported by grants from Région Lorraine and CPRC CHU Nancy.

## References

[1] S. Amin, M.P. LaValley, A. Guermazi, M. Grigoryan, D.J. Hunter, M. Clancy, J. Niu, D.R. Gale and D.T. Felson, The relationship between cartilage loss on magnetic resonance imaging and radiographic progression in men and women with knee osteoarthritis, *Arthritis Rheum.* **52** (2005), 3152–3159.

[2] F.M. Cicuttini, A.J. Teichtahl, A.E. Wluka, S. Davis, B.J. Strauss and P.R. Ebeling, The relationship between body composition and knee cartilage volume in healthy, middle-aged subjects, *Arthritis Rheum.* **52** (2005), 461–467.

[3] F. Eckstein and C. Glaser, Measuring cartilage morphology with quantitative magnetic resonance imaging, *Semin. Musculoskelet. Radiol.* **8** (2004), 329–353.

[4] G.E. Gold and C.F. Beaulieu, Future of MR imaging of articular cartilage, *Semin. Musculoskelet. Radiol.* **5** (2001), 313–327.

[5] D.W. Goodwin, Y.Z. Wadghiri and J.F. Dunn, Micro-imaging of articular cartilage: T2, proton density, and the magic angle effect, *Acad. Radiol.* **5** (1998), 790–798.

[6] C. Liess, S. Lusse, N. Karger, M. Heller and C.C. Gluer, Detection of changes in cartilage water content using MRI T2-mapping in vivo, *Osteoarthritis Cartilage* **10** (2002), 907–913.

[7] D. Loeuille, P. Olivier, A. Watrin, L. Grossin, P. Gonord, G. Guillot, S. Etienne, A. Blum, P. Netter and P. Gillet, The biochemical content of articular cartilage: an original MRI approach, *Biorheology* **39** (2002), 269–276.

[8] T.J. Mosher and B.J. Dardzinski, Cartilage MRI T2 relaxation time mapping: overview and applications, *Semin. Musculoskelet. Radiol.* **8** (2004), 355–368.

[9] T.J. Mosher, B.J. Dardzinski and M.B. Smith, Human articular cartilage: influence of aging and early symptomatic degeneration on the spatial variation of T2 – preliminary findings at 3 T, *Radiology* **214** (2000), 259–266.

[10] T.J. Mosher, Y. Liu, Q.X. Yang, J. Yao, R. Smith, B.J. Dardzinski and M.B. Smith, Age dependency of cartilage magnetic resonance imaging T2 relaxation times in asymptomatic women, *Arthritis Rheum.* **50** (2004), 2820–2828.

[11] P. Olivier, D. Loeuille, A. Watrin, F. Walter, S. Etienne, P. Netter, P. Gillet and A. Blum, Structural evaluation of articular cartilage: potential contribution of magnetic resonance techniques used in clinical practice, *Arthritis Rheum.* **44** (2001), 2285–2295.

[12] J.P. Raynauld, J. Martel-Pelletier, M.J. Berthiaume, F. Labonte, G. Beaudoin, J.A. de Guise, D.A. Bloch, D. Choquette, B. Haraoui, R.D. Altman, M.C. Hochberg, J.M. Meyer, G.A. Cline and J.P. Pelletier, Quantitative magnetic resonance imaging evaluation of knee osteoarthritis progression over two years and correlation with clinical symptoms and radiologic changes, *Arthritis Rheum.* **50** (2004), 476–487.

[13] M.P. Recht, D.W. Goodwin, C.S. Winalski and L.M. White, MRI of articular cartilage: revisiting current status and future directions, *AJR Am. J. Roentgenol.* **185** (2005), 899–914.

[14] A. Rosset, L. Spadola and O. Ratib, OsiriX: an open-source software for navigating in multidimensional DICOM images, *J. Digit Imaging* **17** (2004), 205–216.

[15] A. Watrin, J.P. Ruaud, P.T. Olivier, N.C. Guingamp, P.D. Gonord, P.A. Netter, A.G. Blum, G.M. Guillot, P.M. Gillet and D.H. Loeuille, T2 mapping of rat patellar cartilage, *Radiology* **219** (2001), 395–402.

[16] A. Watrin-Pinzano, J.P. Ruaud, P. Olivier, L. Grossin, P. Gonord, A. Blum, P. Netter, G. Guillot, P. Gillet and D. Loeuille, Effect of proteoglycan depletion on T2 mapping in rat patellar cartilage, *Radiology* **234** (2005), 162–170.

Biorheology 43 (2006) 553–560
IOS Press



# Bi-zonal cartilaginous tissues engineered in a rotary cell culture system

A. Marsano[a], D. Wendt[a], T.M. Quinn[b], T.J. Sims[c], J. Farhadi[a], M. Jakob[a], M. Heberer[a] and I. Martin[a,*]

[a] *Departments of Surgery and of Research, University Hospital Basel, Switzerland*
[b] *Cartilage Biomechanics Group, Ecole Polytechnique Fédérale de Lausanne (EPFL), Switzerland*
[c] *University of Bristol, Academic Rheumatology, Southmead Hospital, Bristol, UK*

**Abstract.** In this study, we aimed at validating a rotary cell culture system (RCCS) bioreactor with medium recirculation and external oxygenation, for cartilage tissue engineering. Primary bovine and human culture-expanded chondrocytes were seeded into non-woven meshes of esterified hyaluronan (HYAFF®-11), and the resulting constructs were cultured statically or in the RCCS, in the presence of insulin and TGF$\beta$3, for up to 4 weeks. Culture in the RCCS did not induce significant differences in the contents of glycosaminoglycans (GAG) and collagen deposited, but markedly affected their distribution. In contrast to statically grown tissues, engineered cartilage cultured in the RCCS had a bi-zonal structure, consisting of an outgrowing fibrous capsule deficient in GAG and rich in collagen, and an inner region more positively stained for GAG. Structurally, trends were similar using primary bovine or expanded human chondrocytes, although the human cells deposited inferior amounts of matrix. The use of the presented RCCS, in conjunction with the described medium composition, has the potential to generate bi-zonal tissues with features qualitatively resembling the native meniscus.

Keywords: Tissue engineering, bioreactor, hydrodynamic flow, articular chondrocytes, rotating wall vessel, meniscus

## 1. Introduction

Hydrodynamic flow conditions generated using different culture systems have been demonstrated to modulate the development of engineered cartilage [5]. In particular, the use of Rotary Cell Culture Systems (RCCS), often referred to as rotating wall vessels, has been reported to promote the development of cartilaginous tissues with increased amounts of glycosaminoglycans (GAG) and collagen, and more uniform spatial distributions of extracellular matrix as compared to statically cultured constructs [13–16]. These findings were attributed to the dynamic laminar flow in the RCCS, which enhanced mass transport around the constructs, while generating minimal turbulent eddies and shear stresses.

The version of the RCCS typically used for scaffold-based tissue engineering approaches (i.e., the Slow Turning Lateral Vessel, STLV) has recently been extended beyond the basic *batch mode* models towards more integrated configurations in which culture medium is continuously recirculated through the culture vessel and an external flow loop (Fig. 1). These perfused models allow for medium sampling and exchange without interrupting bioreactor rotation, facilitate inline monitoring of the culture media components [6,17], and provide a means for external gas exchange. Since medium is recirculated through an external oxygenator, the role of the inner co-axial cylinder as an oxygenator is obsolete in perfused

---

*Address for correspondence: Ivan Martin, Institute for Surgical Research and Hospital Management, University Hospital Basel, Hebelstrasse 20, 4031 Basel, Switzerland. Tel.: +41 61 265 2384; Fax: +41 61 265 3990; E-mail: imartin@uhbs.ch.

 *A. Marsano et al. / Engineering of bi-zonal cartilage tissues*

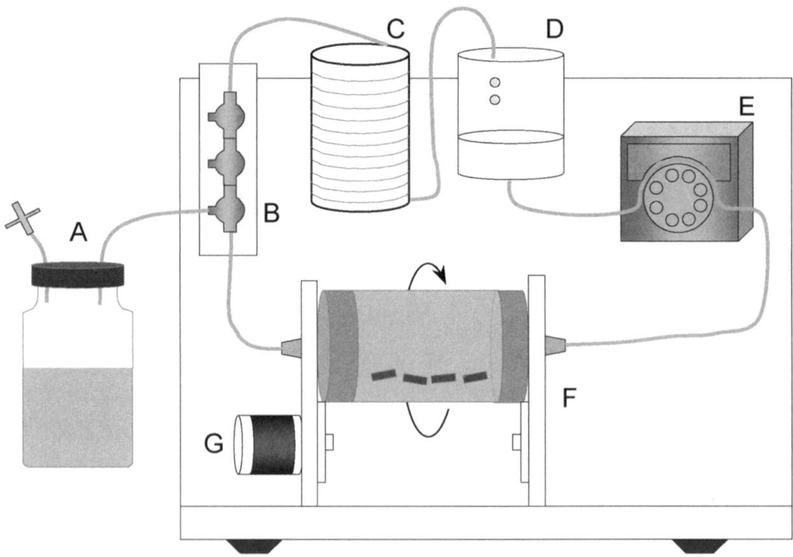

Fig. 1. Schematic diagram of the RCCS with medium re-circulating system. A: Bottle for waste medium; B: Valve manifold; C: Silicon tubing oxygenator; D: Bubble trap; E: Peristaltic pump; F: Cylindrical culture vessel, without inner core; G: Stepper motor.

systems. Removal of this cylinder would not only reduce the frequency of damaging collisions between constructs and vessel walls, but would allow for more flexible bioreactor designs. Larger constructs could be cultured without increasing the radial dimensions of the outer vessel, which would otherwise lead to a significant increase in the volume of required medium. This is particularly relevant in terms of operating costs if expensive medium supplements (e.g., growth factors) are to be used.

In this paper we aimed at validating a perfused STLV RCCS configuration without the concentric inner cylinder, designed with a relatively small (40 ml) culture chamber, for the engineering of cartilaginous tissues. The system was first tested with primary bovine chondrocytes, typically used in studies performed with other RCCS designs, and then with human culture-expanded chondrocytes, representing a clinically relevant cell source. Cells were seeded in porous scaffolds made of esterified hyaluronan (HYAFF®-11) and the resulting constructs were cultured in the RCCS in the presence of soluble factors (i.e., insulin and TGF$\beta$) known to enhance GAG and collagen deposition [2,10].

## 2. Materials and methods

### 2.1. Bioreactor configuration

The RCCS used in this study, manufactured by Synthecon Inc. (TX, USA), consisted of a culture vessel and a medium recirculation system (Fig. 1). The cylindrical vessel was 3.5 cm in length and 3.8 cm in diameter, with a culturing volume of 40 ml. The medium recirculation system was composed of a peristaltic pump (314D pumphead, 4 roller; Watson Marlow), a valve manifold, a bubble trap, and an oxygenator. The valve manifold was used for the initial filling of the bioreactor system and for the periodic changes of the culture medium. Recirculating medium entered the bubble trap (maintained approximately one-third full with medium) through a top port and exited through a port at the base to eliminate small bubbles entrapped in the culture medium supply lines. The oxygenator consisted of a 2 m length coil of silicon tubing (Tygon 3350; 3/32″ i.d., 5/32″ o.d.). A total volume of 75 ml culture medium was used to fill the culture vessel and recirculation system. The RCCS was placed into a conventional humidified 5% $CO_2$ incubator.

## 2.2. Cell isolation and expansion

Bovine articular cartilage was collected from the femoropatellar grooves of two 6 months-old cows. Human articular cartilage was collected from the femoral condyle of two cadavers (20 and 32 years of age), with no known clinical history of joint disorders, within 24 hours after death, after informed consent of the relatives and approval by the local ethical committee. Cartilage tissues were finely minced and digested by incubation for 22 hours at 37°C in 0.15% type II collagenase (Worthington Biochemical Corporation, Lakewood, NJ) (1 ml solution per 100 mg tissue). Isolated cells were resuspended in "complete medium", composed of Dulbecco's Modified Eagle Medium (DMEM; 4.5 g/l glucose with nonessential amino acids), 10% fetal bovine serum, 0.1 mM nonessential amino acids, 1 mM sodium pyruvate, 100 mM HEPES buffer, 100 U/ml penicillin, 100 $\mu$g/ml streptomycin and 0.29 mg/ml L-glutamine. Human articular chondrocytes were expanded for two passages in complete medium further supplemented with 1 ng/ml Transforming growth factor-$\beta$1 (TGF-$\beta$1), 5 ng/ml Fibroblast growth factor-2 (FGF-2) and 10 ng/ml Platelet-derived growth factor-bb (PDGF-bb), previously shown to increase human chondrocyte proliferation rate and post-expansion chondrogenic capacity [1,9].

## 2.3. Cell seeding and culture in porous scaffolds

Expanded human or freshly isolated bovine articular chondrocytes were seeded into non-woven meshes (5.5 mm diameter, 2 mm thick disks) made of esterified hyaluronan (HYAFF®-11, Fidia Advanced Biopolymers, Abano Terme, Italy), at a density of 4E + 06 cells/scaffold (7E + 07 cells/cm$^3$). Briefly, 4E + 06 cells were resuspended in 28 $\mu$l of complete medium and slowly dispersed over the top surface of the dry meshes with a micropipette. The seeded scaffolds were cultured for 2 days in 12 well-plates before being transferred either to the described RCCS or to 6 well-plates. Cell-scaffold constructs were then cultured for 4 weeks in complete medium further supplemented with 10 ng/ml of TGF$\beta$3, 0.1 mM ascorbic acid 2-phosphate and 10 $\mu$g/ml of human insulin, in order to enhance chondrogenesis [2,10], with medium changes twice a week.

During culture in the RCCS, culture medium was continuously recirculated through the system at a flow rate of 0.6 ml/min. The angular velocity of the vessel was increased from 16 rpm at the beginning of the culture to 50 rpm after 4 weeks in order to maintain constructs in a continual free-fall condition [11]. For each experiment, at least three specimens per experimental group were assessed histologically and biochemically as described below.

## 2.4. Histological assessment

Engineered tissues were rinsed in phosphate buffered saline, fixed in 4% buffered formalin for 24 h at 4°C, embedded in paraffin, and cross-sectioned (7 $\mu$m thick). Sections were stained with Safranin-O for GAG and with elastica-van gieson for collagen.

## 2.5. Biochemical assays

Engineered tissues were digested with 1 ml protease K solution (1 mg/ml protease K in 50 mM Tris with 1 mM EDTA, 1 mM iodoacetamide, and 10 mg/ml pepstatin-A) for 15 h at 56°C [8]. DNA was quantified with the CyQUANT® Cell Proliferation Assay Kit (Molecular Probes, Eugene, OR), with calf thymus DNA as a standard. GAG was quantified with the dimethylmethylene blue colorimetric assay, with chondroitin sulfate as a standard [4]. Total collagen was quantified as follows: digested samples were hydrolyzed for 24 hours at 110°C in constant boiling hydrochloric acid and excess acid removed by lyophilization. Dried hydrolyzates were analyzed using a Biochrom 20 Plus amino acid analyzer equipped with post column Ninhydrin detection to measure the hydroxyproline amount. Collagen content was calculated using a hydroxyproline to collagen ratio of 1 : 10 [8].

## 2.6. Statistical analysis

Differences between the experimental groups were evaluated by non-parametric Mann Whitney $U$ tests and considered to be statistically significant with $P < 0.05$.

## 3. Results

### 3.1. Primary bovine articular chondrocytes

Cultivation of chondrocyte-scaffold constructs in the RCCS generated bi-zonal tissues, with distinct patterns of spatial distribution of GAG and collagen (Fig. 2). In particular, the central region was intensely stained for GAG, whereas the external region was negatively stained for GAG and contained a higher density of elongated cells. The external capsule was outgrowing from the original scaffold, as indicated by the lack of scaffold fibers, and was more intensely stained for collagen than the central core. In statically grown constructs, the intensities of GAG and collagen stain were relatively uniform throughout the cross-sections and generally lower than those observed in the central region of the RCCS-cultured tissues. The dry weight fractions of GAG and total collagen were similar in constructs cultured statically and in the RCCS for 2 or 4 weeks, with no significant increase with time (Fig. 3).

### 3.2. Expanded human articular chondrocytes

As compared to primary bovine chondrocytes, expanded human chondrocytes generated tissues which were less intensely stained for GAG, but with a similar pattern in the GAG and collagen distribution (Fig. 4). In particular, culture in the RCCS resulted in the formation of a fibro-cartilaginous tissue outgrowing from the original scaffold area, although more irregular in shape than using bovine chondrocytes. The outgrowing tissue was negatively stained for GAG but was more intensely stained for collagen than the construct internal region. The central construct region, corresponding to the original scaffold area, was more intensely stained for GAG following RCCS than static culture. Statically grown constructs did not display a preferential pattern of GAG and collagen distribution.

Similar to constructs generated from bovine cells, the dry weight fractions of GAG and total collagen were comparable in constructs cultured statically and in the RCCS (Fig. 5). However, the fractions of GAG and collagen increased between 2 and 4 weeks of culture, likely due to the progressive re-differentiation of the expanded cells.

## 4. Discussion

In this paper we describe the use of a perfused RCCS configuration for the engineering of cartilage tissues. While culture in the RCCS did not significantly affect the overall fractions of GAG and collagen

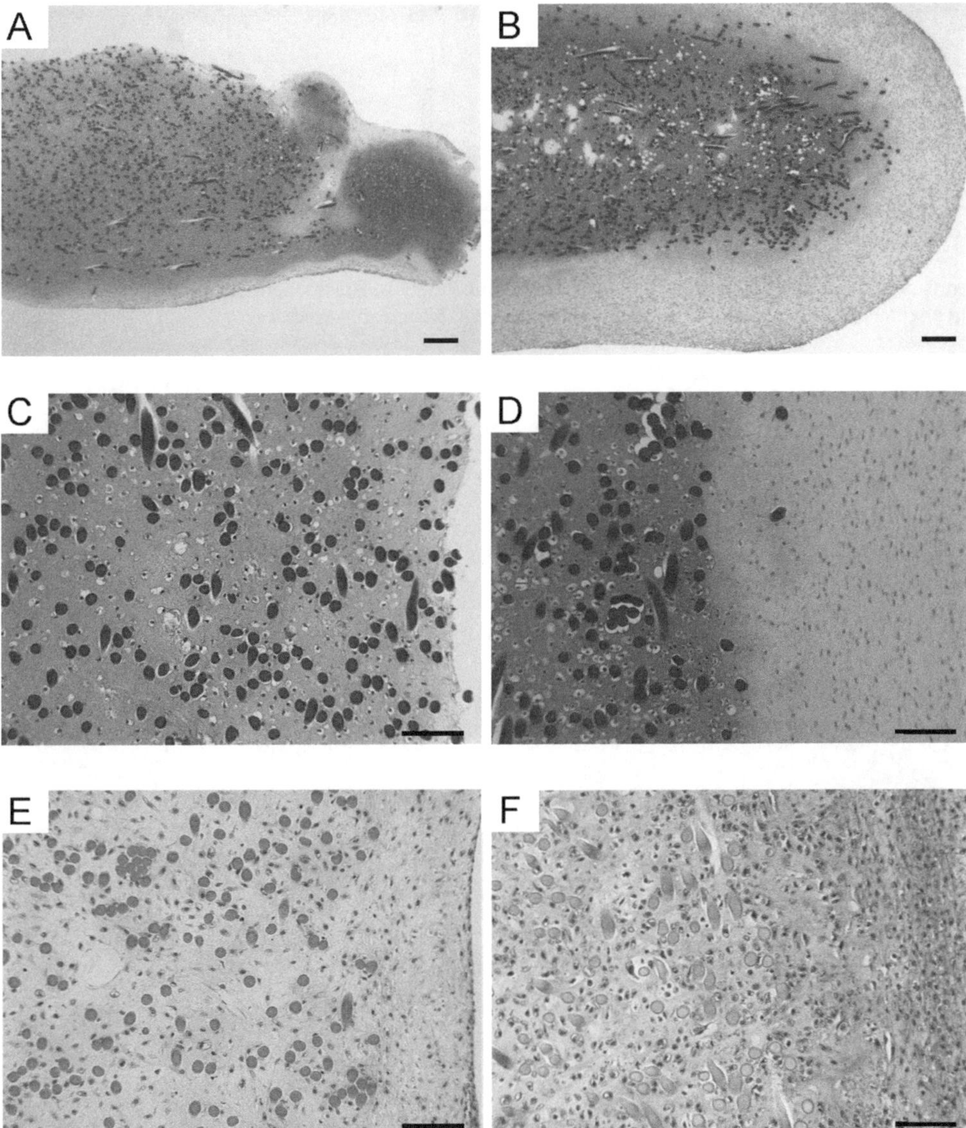

Fig. 2. Histology of primary bovine chondrocyte-based engineered tissues. Representative sections of engineered tissues, generated after 4 weeks of culture statically (A, C, E) or in the RCCS (B, D, F). Sections were stained for GAG with Safranin O (A, B, C, D) or for collagen with elastica-van gieson (E, F). Scale bar = 200 $\mu$m (A, B) or 100 $\mu$m (C, D, E, F). The darker spots correspond to undegraded polymer fibers.

in the engineered tissue, their distribution was strongly influenced, resulting in the formation of bi-zonal tissues, with an inner region stained predominantly for GAG and an outer capsule stained predominantly for collagen.

Previous studies using a standard tissue culture RCCS vessel (i.e., STLV unit with concentric cylinder arrangement and without medium recirculation) reported the formation of relatively uniform cartilaginous tissues [13–16]. Considering that hydrodynamics can have a dramatic impact on the development of engineered cartilage, we speculate that the fluid dynamics generated in the two configurations of the

 *A. Marsano et al. / Engineering of bi-zonal cartilage tissues*

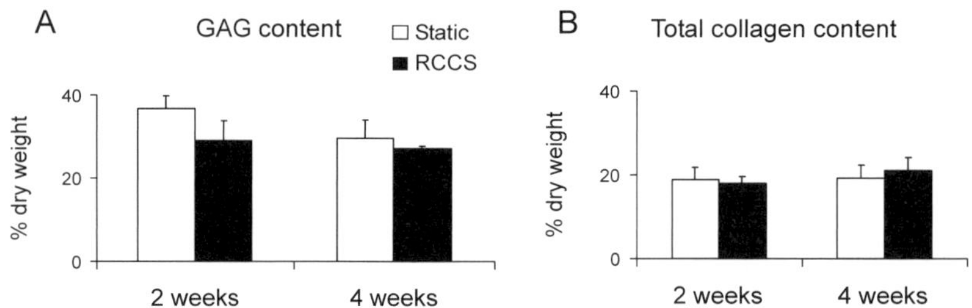

Fig. 3. Biochemistry of primary bovine chondrocyte-based engineered tissues. Biochemical quantification of GAG (A) and total collagen (B) in engineered tissues cultured statically or in the RCCS for 2 or 4 weeks.

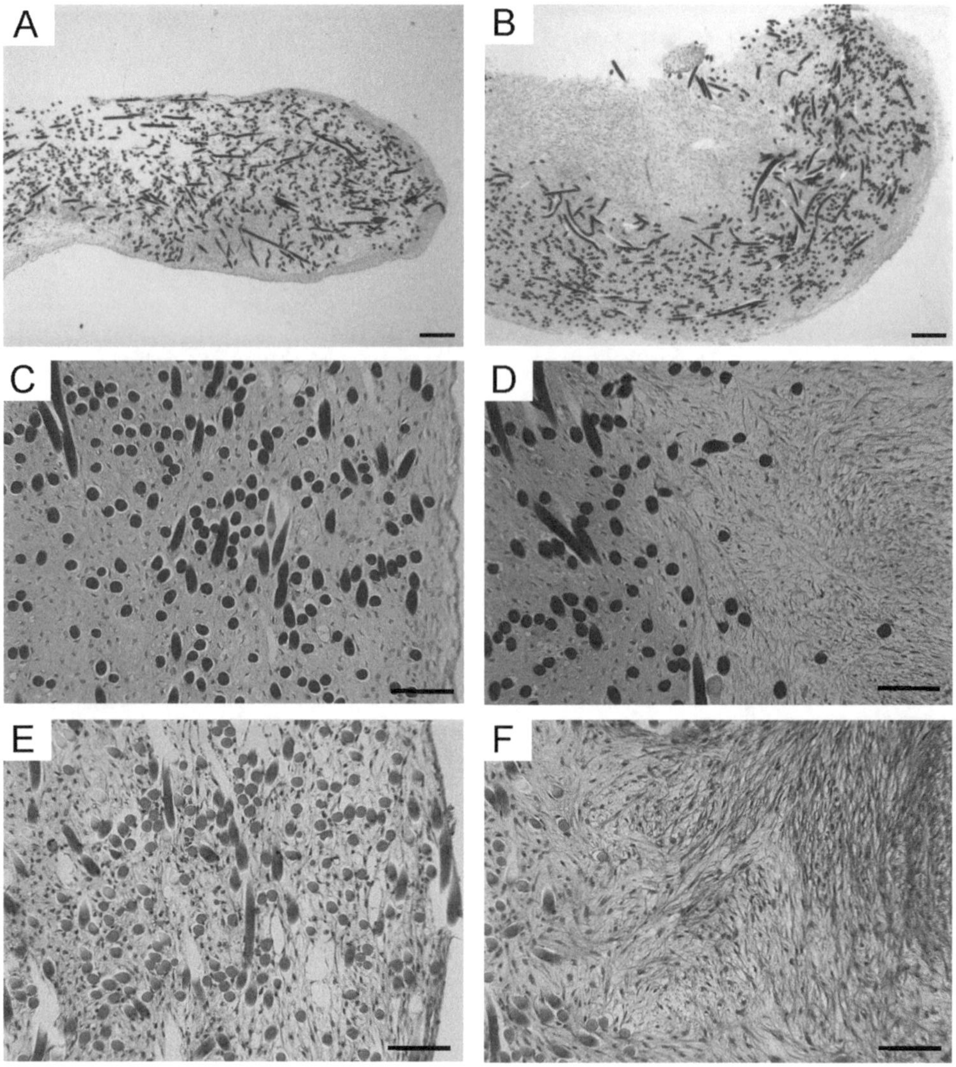

Fig. 4. Histology of expanded human chondrocyte-based engineered tissues. Representative sections of engineered tissues, generated after 4 weeks of culture statically (A, C, E) or in the RCCS (B, D, F). Sections were stained for GAG with Safranin O (A, B, C, D) or for collagen with elastica-van gieson (E, F). Scale bar = 200 $\mu$m (A, B) or 100 $\mu$m (C, D, E, F). The darker spots correspond to undegraded polymer fibers.

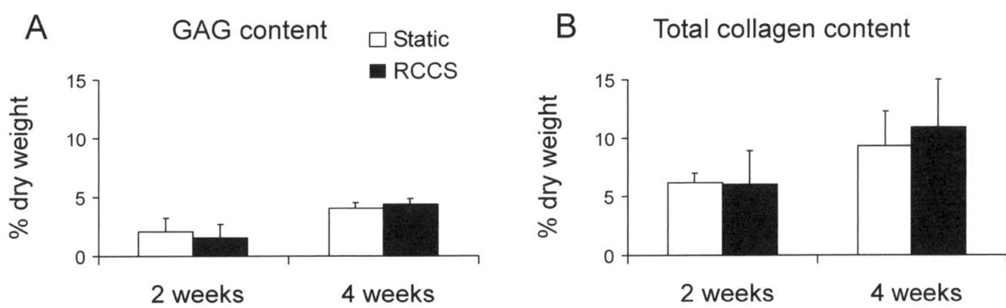



Fig. 5. Biochemistry of expanded human chondrocyte-based engineered tissues. Biochemical quantification of GAG (A) and total collagen (B) in engineered tissues cultured statically or in the RCCS for 2 or 4 weeks.

RCCS, with and without the central cylinder, may account for the apparent inconsistency. In the absence of 3D constructs, rotation of both vessel configurations would generate solid body rotation of the culture medium. However, with the introduction of 3D constructs into the flow (dynamically freefalling), it is likely that the mixing patterns and dissipation of turbulent eddies would be altered without the inner concentric cylinder. Interestingly, in these same studies, constructs engineered in mixed flask bioreactors were surrounded by fibrous external capsules, similar to those reported here. Perhaps the turbulent eddies and shear stresses in our RCCS are more similar to those in a mixed flask than in the standard RCCS design. Although the dynamic fluctuations of the freefalling construct represent a considerable challenge for a computational fluid dynamics model, flow visualization techniques could be used to better characterize the local velocity profiles and shear stresses around constructs in both vessel configurations.

Alternatively, the differing results could be explained by the medium supplementation with soluble factors not used in the previous studies. The presence of TGF$\beta$3, in conjunction with fluid flow, could have supported chondrocyte proliferation [12], which is known to be associated with a de-differentiated and fibroblastic phenotype [3,7]. The specific scaffold used is not likely to have determined the result, since uniform tissues were obtained when non-woven meshes of esterified hyaluronan were used in the standard RCCS design [14]. The outgrowth of a fibrous tissue in the RCCS occurred when either freshly isolated bovine or expanded de-differentiated human chondrocytes were used, and thus appeared to be generated by a mechanism common to various chondrocytic cells. Further experimental studies, in conjunction with flow visualization techniques in the new configuration of the RCCS vessel, are necessary to elucidate the mechanisms underlying our findings.

The bi-zonal tissues generated in this study, with an outer fibrocartilage-like capsule and a central region containing larger amounts of GAG, in some respects resemble the structure of the meniscus. In this context, studies are ongoing to more precisely characterize the molecular composition and mechanical function of the engineered tissues and to compare them with those of native meniscus. Efforts in this direction could lead to the controlled use of hydrodynamic culture conditions and soluble factors for the engineering of meniscus substitutes starting from articular chondrocytes.

## Acknowledgements

This work was supported by the Swiss Federal Office for Education and Science (B.B.W.) under the Fifth European Framework Growth Program (Meniscus Regeneration, Project No. GRD1-2001-40401). We are grateful to C. Seemayer and L. Tornillo from the Institute of Pathology at the University of Basel for the elastica-van gieson stains, to Enrico Tognana (Fidia Advanced Biopolymers, I) for the generous supply of HYAFF®-11 meshes, and to Jan Steels (Cellon, B) for the cooperation in the custom-production of the RCCS unit.

# References


[1] A. Barbero, S. Ploegert, M. Heberer and I. Martin, Plasticity of clonal populations of dedifferentiated adult human articular chondrocytes, *Arthritis Rheum.* **48** (2003), 1315–1325.

[2] T. Blunk, A.L. Sieminski, K.J. Gooch, D.L. Courter, A.P. Hollander, A.M. Nahir, R. Langer, G. Vunjak-Novakovic and L.E. Freed, Differential effects of growth factors on tissue-engineered cartilage, *Tissue Eng.* **8** (2002), 73–84.

[3] K.R. Brodkin, A.J. Garcia and M.E. Levenston, Chondrocyte phenotypes on different extracellular matrix monolayers, *Biomaterials* **25** (2004), 5929–5938.

[4] R.W. Farndale, D.J. Buttle and A.J. Barrett, Improved quantitation and discrimination of sulphated glycosaminoglycans by use of dimethylmethylene blue, *Biochim. Biophys. Acta* **883** (1986), 173–177.

[5] L.E. Freed, I. Martin and G. Vunjak-Novakovic, Frontiers in tissue engineering. In vitro modulation of chondrogenesis, *Clin. Orthop. Relat. Res.* **367** (1999), S46–S58.

[6] F.G. Gao, A.S. Jeevarajan and M.M. Anderson, Long-term continuous monitoring of dissolved oxygen in cell culture medium for perfused bioreactors using optical oxygen sensors, *Biotechnol. Bioeng.* **86** (2004), 425–433.

[7] J. Glowacki, E. Trepman and J. Folkman, Cell shape and phenotypic expression in chondrocytes, *Proc. Soc. Exp. Biol. Med.* **172** (1983), 93–98.

[8] A.P. Hollander, T.F. Heathfield, C. Webber, Y. Iwata, R. Bourne, C. Rorabeck and A.R. Poole, Increased damage to type II collagen in osteoarthritic articular cartilage detected by a new immunoassay, *J. Clin. Invest.* **93** (1994), 1722–1732.

[9] M. Jakob, O. Demarteau, D. Schafer, B. Hintermann, W. Dick, M. Heberer and I. Martin, Specific growth factors during the expansion and redifferentiation of adult human articular chondrocytes enhance chondrogenesis and cartilaginous tissue formation in vitro, *J. Cell Biochem.* **81** (2001), 368–377.

[10] K. Kellner, M.B. Schulz, A. Gopferich and T. Blunk, Insulin in tissue engineering of cartilage: a potential model system for growth factor application, *J. Drug Target* **9** (2001), 439–448.

[11] M. Lappa, Organic tissues in rotating bioreactors: fluid-mechanical aspects, dynamic growth models, and morphological evolution, *Biotechnol. Bioeng.* **84** (2003), 518–532.

[12] P. Malaviya and R.M. Nerem, Fluid-induced shear stress stimulates chondrocyte proliferation partially mediated via TGF-beta1, *Tissue Eng.* **8** (2002), 581–590.

[13] I. Martin, B. Obradovic, L.E. Freed and G. Vunjak-Novakovic, Method for quantitative analysis of glycosaminoglycan distribution in cultured natural and engineered cartilage, *Ann. Biomed. Eng.* **27** (1999), 656–662.

[14] M. Pei, L.A. Solchaga, J. Seidel, L. Zeng, G. Vunjak-Novakovic, A.I. Caplan and L.E. Freed, Bioreactors mediate the effectiveness of tissue engineering scaffolds, *FASEB J.* **16** (2002), 1691–1694.

[15] G. Vunjak-Novakovic, I. Martin, B. Obradovic, S. Treppo, A.J. Grodzinsky, R. Langer and L.E. Freed, Bioreactor cultivation conditions modulate the composition and mechanical properties of tissue-engineered cartilage, *J. Orthop. Res.* **17** (1999), 130–138.

[16] G. Vunjak-Novakovic, B. Obradovic, I. Martin and L.E. Freed, Bioreactor studies of native and tissue engineered cartilage, *Biorheology* **39** (2002), 259–268.

[17] Y. Xu, J. Sun, G. Mathew, A.S. Jeevarajan and M.M. Anderson, Continuous glucose monitoring and control in a rotating wall perfused bioreactor, *Biotechnol. Bioeng.* **87** (2004), 473–477.

Biorheology 43 (2006) 561–575
IOS Press



# Effect of Peroxisome Proliferator Activated Receptor (PPAR)$\gamma$ agonists on prostaglandins cascade in joint cells

David Moulin, Paul-Emile Poleni, Mélanie Kirchmeyer, Sylvie Sebillaud, Meriem Koufany, Patrick Netter, Bernard Terlain, Arnaud Bianchi and Jean-Yves Jouzeau [*]
*Laboratoire de Physiopathologie et Pharmacologie Articulaires, UMR 7561 CNRS-UHP, Nancy, France*

**Abstract.** In response to inflammatory cytokines, chondrocytes and synovial fibroblasts produce high amounts of prostaglandins (PG) which self-perpetuate locally the inflammatory reaction. Prostaglandins act primarily through membrane receptors coupled to G proteins but also bind to nuclear Peroxisome Proliferator-Activated Receptors (PPARs). Amongst fatty acids, the cyclopentenone metabolite of $PGD_2$, 15-deoxy-$\Delta^{12,14}PGJ_2$ (15d-$PGJ_2$), was shown to be a potent ligand of the PPAR$\gamma$ isotype prone to inhibit the production of inflammatory mediators. As the stimulated synthesis of $PGE_2$ originates from the preferential coupling of inducible enzymes, cyclooxygenase-2 (COX-2) and membrane PGE synthase-1 (mPGES-1), we investigated the potency of 15d-$PGJ_2$ to regulate prostaglandins synthesis in rat chondrocytes stimulated with interleukin-1$\beta$ (IL-1$\beta$). We demonstrated that 15d-$PGJ_2$, but not the high-affinity PPAR$\gamma$ ligand rosiglitazone, decreased almost completely $PGE_2$ synthesis and mPGES-1 expression. The inhibitory potency of 15d-$PGJ_2$ was unaffected by changes in PPAR$\gamma$ expression and resulted from inhibition of NF-$\kappa$B nuclear binding and I$\kappa$B$\alpha$ sparing, secondary to reduced phosphorylation of IKK$\beta$. Consistently with 15d-$PGJ_2$ being a putative endogenous regulator of the inflammatory reaction if synthesized in sufficient amounts, the present data confirm the variable PPAR$\gamma$-dependency of its effects in joint cells while underlining possible species and cell types specificities.

Keywords: PPAR$\gamma$, eicosanoids, 15-deoxy-$\Delta^{12,14}$-prostaglandin $J_2$, thiazolidinediones, chondrocyte, synoviocyte

## 1. Introduction

Prostaglandins (PG) are well-known lipid mediators that reproduce the cardinal signs of inflammation [76] and are secreted by cyclooxygenases (COX) in excessive amounts within inflammatory joints [50]. They have pathophysiological relevance to various diseases [32] and inhibition of their biosynthesis is thought to account for most of the therapeutical properties of NSAIDs [68]. In arthritis, the pathological contribution of PG is supported mainly by $PGE_2$, which is the major mediator produced by macrophages [7], synovial fibroblasts and chondrocytes [41] in response to an inflammatory stimulus. The effects of $PGE_2$ on joint tissues vary with the differentiation status of chondrocytes (growth plate or hyaline cartilage) and are often estimated indirectly through the consequences of PG inhibition by NSAIDs [15,27].

[*]Address for correspondence: Jean-Yves Jouzeau, PharmD, PhD, Laboratoire de Physiopathologie et Pharmacologie Articulaires, UMR 7561 CNRS-UHP, Avenue de la forêt de Haye, BP 184, 54505 Vandœuvre-Lès-Nancy, France. Tel.: +33 3 83 68 39 50; Fax: +33 3 83 68 39 59; E-mail: jean-yves.jouzeau@pharma.uhp-nancy.fr.

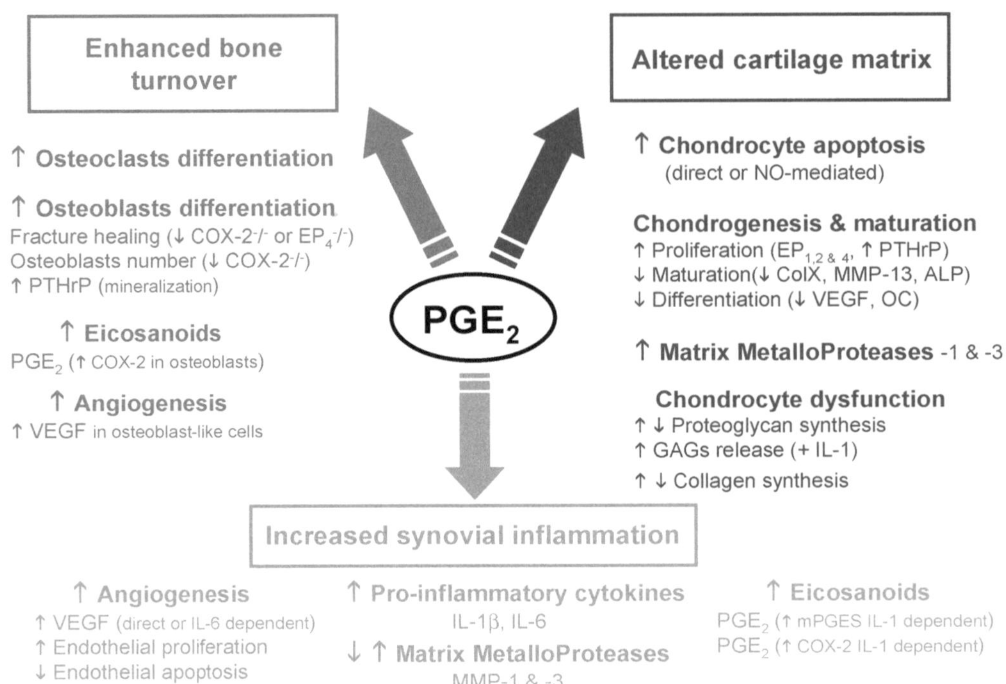


Fig. 1. Multiple effects of PGE$_2$ on articular tissues.

This is one of the main limitation when interpreting their potency on cartilage matrix since beside inhibition of COX enzymes in chondrocytes [3], NSAIDs have chondroprotective properties independent from arachidonic acid metabolism [19,34]. Indeed, the modulation of proteoglycan turnover by NSAIDs is affected variably by the synthetic PGE$_1$ analog misoprostol [18] whereas the reversal of their effects on matrix components by addition of exogenous PGE$_2$ [31] is found scarcely in the literature.

Despite the limits mentioned above, it is generally accepted that PGE$_2$ can have pathophysiological relevance to joint diseases by disrupting homeostasis of either cartilage, synovium or bone (Fig. 1). Thus, there is evidence that PGE$_2$ contributes to the formation of an altered cartilage matrix by: (i) favouring chondrocyte apoptosis, with a variable dependency on nitric oxide [59]; (ii) promoting chondrogenesis by inducing the proliferation of stem cells through EP receptors [13]; (iii) reducing maturation and terminal differentiation of growth plate chondrocytes [13]; (iv) activating matrix metalloproteases (MMP) -1 and -3 [63]; (v) modifying collagen synthesis in growth plate and cartilage chondrocytes [29,50]; (vi) stimulating glycosaminoglycans (GAG) release in interleukin-1 (IL-1) stimulated chondrocytes [31]; (vii) affecting variably proteoglycan synthesis [25,47]. There is also evidence that PGE$_2$ contributes to the maintenance of synovial inflammation by: (i) promoting angiogenesis, either by a direct or IL-6 dependent induction of vascular endothelium growth factor (VEGF), or by stimulating endothelial cells proliferation while reducing apoptosis [50]; (ii) stimulating the release of pro-inflammatory cytokines [35,80]; (iii) affecting variably the production or activity of MMP-1 and -3 [33,60]; (iv) self perpetuating its production by triggering IL-1 dependent induction of inducible enzymes of the arachidonic acid cascade [24,44]. Finally, there is evidence that PGE$_2$ contributes to an enhanced turnover of bone by: (i) promoting osteoclastogenesis [42,46]; (ii) favouring osteoblast formation and differentiation [79,81]; (iii) stimulating expression of VEGF in osteoblasts [38]; (iv) promoting its own release by osteoblast-like cells [64]. As a consequence, there is a great interest in studying the potency of drugs which are able to reduce $PGE_2$ synthesis in articular cells and do not belong to the pharmacological class of NSAIDs.

In these two last decades, the dual discovery of an inducible COX isoenzyme (COX-2) playing a major role in inflammation [20] and of a preferential enzymatic coupling between constitutive and inducible isoforms of phospholipases $A_2(PLA_2)$, COXs and terminal PG synthases [74] has open insight to a new pharmacological modulation of the arachidonic acid cascade. Of primary importance, Prostaglandin E synthase-1 (PGES-1), the enzyme converting the COX-derived prostaglandin $H_2$ into $PGE_2$, has been found in multiple forms having distinct enzymatic properties, modes of expression, subcellular localizations and intracellular functions [54]. One of its isoforms, mPGES-1, is a perinuclear membrane-associated protein belonging to the microsomal glutathione S-transferase (GST) family, which expression is induced by pro-inflammatory cytokines, growth factors, bacterial endotoxins and phorbol esters while being down-regulated by anti-inflammatory corticosteroids [54]. As indicated before, mPGES-1 is preferentially linked with inducible COX-2 and contributes to the synthesis of PG in inflammatory conditions [74], whereas little is known about the pathophysiological role of the newly cloned mPGES-2 isoform [73].

Although each inducible enzyme of the arachidonic acid cascade is rate limiting by controlling the bioavailability of substrate to downstream effectors, inhibition of this new inducible step in PG synthesis has therapeutical relevance for at least two reasons: (i) mPGES-1 should be the main trigger of the stimulated synthesis of $PGE_2$ [43]; (ii) inhibition of mPGES-1 could favour the biotransformation of $PGH_2$ into prostaglandins other than $PGE_2$, depending on the cellular content in specific isomerases [20]. If the cell type co-expresses PGD synthase, as reported for human chondrocytes [66], inhibition of mPGES-1 could favour indirectly the synthesis of PGD2 which, in the presence of proteins, would transform into its cyclopentenone by-product 15-deoxy-$\Delta^{12,14}$prostaglandin $J_2$ [67]. Very importantly, 15d-$PGJ_2$ was shown to be synthesized in the late phase of acute experimental inflammation [28] and to have anti-inflammatory properties in various experimental models [17,39]. This fatty acid, thought to be an endogenous regulator of inflammation [77], is a natural agonist of peroxisome proliferator-activated receptor $\gamma$ (PPAR$\gamma$), a ligand-activated nuclear transcription factor belonging to the nuclear hormone receptor superfamily [61]. Once activated, PPAR$\gamma$ regulates, as a heterodimer with retinoid X receptor (RXR), the expression of numerous target genes having a peroxisome proliferator response element (PPRE) in their promoters [4]. In addition to differentiation of adipocytes and glucose homeostasis [11], PPAR$\gamma$ agonists were shown recently to control inflammation [4], as well in joint cells exposed to inflammatory stimulus [6,69] as in experimental arthritis [16,39]. We hypothesized therefore that 15d-$PGJ_2$ could regulate the inducible arachidonic acid cascade by activating PPAR$\gamma$ in chondrocytes, thus providing a possible feed-back loop as in rheumatoid synoviocytes [75].

In the present study, we investigated whether the natural PPAR$\gamma$ agonist 15d-$PGJ_2$ and the high-affinity PPAR$\gamma$ agonist rosiglitazone could affect PG synthesis induced by IL-1$\beta$ in rat chondrocytes. As inhibition of COX-2 by PPAR$\gamma$ ligands was shown previously to be moderate in articular cells [6,69], special care was given to the contribution of mPGES-1 to the kinetics of PG production.

The present work demonstrated an early moderate induction of COX-2 and a delayed strong induction of mPGES-1 by IL-1$\beta$ in rat chondrocytes. The stimulated synthesis of $PGE_2$ fitted well the kinetics and extent of mPGES-1 expression whereas the production of 6-keto-$PGF_{1\alpha}$ (the stable metabolite of $PGI_2$) remained lower and below the extent of COX-2 induction. In our experimental conditions, 15d-$PGJ_2$ reduced 6-keto-$PGF_{1\alpha}$ level and COX-2 expression but suppressed almost completely $PGE_2$ level and mPGES-1 expression, supporting that mPGES-1 was the most rate limiting step in $PGE_2$ synthesis. The inhibitory potency of 15d-$PGJ_2$ was not reproduced by rosiglitazone and was neither reduced by the blockade of PPARγ with antagonist GW-9662 nor enhanced by the overexpression of PPARγ. Consistent with a PPARγ-independent mechanism, we demonstrated finally that 15d-PGJ$_2$ (but not rosiglitazone) reduced IL-1β-induced translocation/binding of NF-κB and IκBα degradation by maintaining IKKβ in a nonphosphorylated state, both of which being necessary for induction of mPGES-1 and stimulation of PGE$_2$ synthesis by IL-1β in rat chondrocytes.

## 2. Material and methods

### 2.1. Isolation and culture of rat chondrocytes

Chondrocytes were isolated from femoral heads of young healthy Wistar male rats (Charles River, Saint-Aubin-les-Elbeuf, France) killed according to national animal care guidelines. Cells were obtained by sequential digestion with pronase and collagenase [45], then washed two times in phosphate-buffered saline (PBS) and cultured to confluence in 75-cm$^2$ flasks at 37°C in a humidified atmosphere containing 5% CO$_2$. The culture medium was DMEM/Ham's F-12 supplemented with L-glutamine (2 mM), penicillin (100 U/ml), streptomycin (100 μg/ml) and either 10% heat-inactivated fetal calf serum (FCS, Life Technologies) during subcultures or 1% FCS during experiments. Chondrocytes were used between passages 1 and 3 to prevent dedifferentiation.

### 2.2. Experimental design

Chondrocytes maintained in low (1%) FCS medium were stimulated with 10 ng/ml interleukin-1β (IL-1β, Sigma, St Quentin Fallavier, France) in the presence or absence of PPAR agonists added 4 h before IL-1. In a preliminary kinetic study, mRNA levels of COX-2 and mPGES-1 were determined from 6 h to 48 h after IL-1β challenge whereas levels of 6-keto-PGF$_{1α}$ and PGE$_2$ were assayed from 6 h to 36 h in culture medium. In subsequent experiments, COX-2 mRNA level was checked 12 h after IL-1β exposure and levels of mPGES-1 mRNA, 6-keto-PGF$_{1α}$ and PGE$_2$ at 24 h. PPARγ agonists rosiglitazone (Cayman, Ann Arbor, USA) or 15d-PGJ$_2$ (Calbiochem, Meudon, France) were used in the range of 0.1 to 10 μM whereas PPARγ antagonist GW-9662 (Cayman) was used at 10 μM.

### 2.3. RNA extraction and real-time PCR analysis

Chondrocytes total RNA was isolated with Trizol® (Invitrogen, Cergy-Pontoise, France) before reverse transcription of 2 μg for 90 min at 37°C with 200 units Moloney Murine Leukemia Virus reverse transcriptase (Invitrogen) and hexamer random primers. Levels of COX-2 and mPGES-1 mRNAs were quantified by real time polymerase chain reaction (RT-PCR) with the Lightcycler® (Roche) technology. After amplification, PCR products were characterized by their melting temperatures and their sizes in a 2% agarose gel stained with 0.5 μg/ml of ethidium bromide. Each run included standard dilutions, positive and negative reaction controls and the mRNA level of each gene of interest was determined comparatively to the ribosomal protein S29, chosen as housekeeping gene, to provide normalized ratio of mRNA levels. The gene-specific primer pairs used were: mPGES-1, sense 5′-TCGCCTGGATACATTTCCTC-3′, antisense 5′-GTCCCCCATTGTGGTATCTG-3′; COX-2, sense 5′-TACAAGCAGTGGCAAAGGCC-3′, antisense 5′-CAGTATTGAGGAGAACAGATGGG-3′; S29, sense 5′-AAGATGGGTCACCAGCAGCTCTACG-3′, antisense 5′-AGACGCGGCAAGAGCGAGAA-3′.

## 2.4. Transient transfection

Chondrocytes were seeded at $5 \times 10^5$ cells/well in 6-well plates, then transfected with 500 ng of a PPARγ expression vector (pcDNA3.1 PPARγ, a generous gift from Dr H. Fahmi [Centre Hospitalier de l'Université de Montréal, Montréal, Canada]) for 2 h using 10 $\mu$l of polyethyleneimine reagent (Euromedex, Souffelweyersheim, France) in 1 ml of culture medium. Experiments with IL-1β and PPARγ agonists were performed 24 h after transfection.

## 2.5. NF-κB transactivation analysis

Nuclear proteins were prepared with the TransAM® nuclear extract kit according to the manufacturer's protocol (Active Motif Europe, Rixensart, Belgium). Protein concentration was determined by a Bradford-based assay (Biorad Laboratories, Marne-la-Coquette, France) and NF-κB activation using the TransAM® ELISA kit (Active Motif Europe). Briefly, 5 $\mu$g of each nuclear extract was added to a well coated with an oligonucleotide sharing a NF-κB consensus binding site. After a short incubation under smooth agitation and successive washings, specific nuclear proteins were detected by a p65 antibody, whose fixation was revealed, after an additional washing, by a diluted horseradish peroxidase (HRP)-conjugated antibody. Conjugate fixation was quantified by adding 100 $\mu$l of 3,3′,5,5′-tetramethylbenzidine substrate solution, then stopping the reaction by addition of 100 $\mu$l 0.5 M $H_2SO_4$ and reading final absorbance at 450 nm on a microplate reader Multiskan® (Labsystems, Montigny-le-Bretonneux, France).

## 2.6. Assays for $PGE_2$ and 6-keto-$PGF_{1\alpha}$

Levels of $PGE_2$ and 6-keto-$PGF_{1\alpha}$ were determined in culture supernatants using Assay Design® ELISA kits (Oxford Biomedical Research, Ann Arbor, USA) according to manufacturer's instructions. Limit of detection was 10 pg/ml and 1.4 pg/ml for $PGE_2$ and 6-keto-$PGF_{1\alpha}$ respectively, and cross-reactivity was negligible with $PGE_1$ and $PGF_{2\alpha}$ respectively (manufacturer's data).

## 2.7. Western blot analysis

Cells were scrapped off the flask in $1\times$ Laemmli Blue, then disrupted by sonication and centrifuged at 3000 rpm for 10 minutes. Total protein extracts were analysed by SDS/PAGE (10% of acrylamide) and electroblotted onto PVDF membrane. After 1 h in blocking buffer, membranes were blotted overnight at 4°C with rabbit antibodies against IκBα (Ozyme, St Quentin en Yvelines, France), phosphorylated IKKα/β (Ozyme) or β-actin (Sigma). After successive washings with Tris Buffered Saline (TBS)–Tween, blots were incubated for 1 h at room temperature with horseradish peroxidase (HRP)-conjugated antibody (Cell Signaling, Beverly, USA). After additional washings, protein bands were detected by chemiluminescence with the Phototope Detection system according to manufacturer's recommendations (Cell Signaling, Beverly, USA).

## 2.8. Statistical analysis

Results are expressed as the mean value $\pm$ SD of at least three assays. Comparisons were made by ANOVA, followed by the Fisher PLSD *post-hoc* test using the Statview™ 5.0 software (SAS Institute Inc). A value of $p$ less than 0.05 was considered as significant ([*,†]$p < 0.05$ *vs* control or [#]$p < 0.05$ *vs* IL-1β stimulation).

## 3. Results

### 3.1. Kinetics of prostaglandins cascade in IL-1$\beta$ stimulated rat chondrocytes

The spontaneous release of $PGE_2$ and 6-keto-$PGF_{1\alpha}$ and the level of COX-2 and mPGES-1 mRNAs were very low in resting cells (Table 1). In IL-1$\beta$ stimulated cells, $PGE_2$ levels increased earlier than 6-keto-$PGF_{1\alpha}$ levels, although both were maximal at 24 h with a higher range of variation for $PGE_2$ (70-fold) than for 6-keto-$PGF_{1\alpha}$ (11-fold) (Table 1). Expression of inducible genes was detected from 6 h, with a maximal 37-fold induction at 12 h for COX-2 and 68-fold induction at 24 h for mPGES-1 (Table 1).

### 3.2. Effect of PPAR$\gamma$ agonists on IL-1$\beta$-induced prostaglandins cascade

In IL-1$\beta$ stimulated chondrocytes, 15d-$PGJ_2$ (10 $\mu$M) reduced the release of prostaglandins by approximately 90% for $PGE_2$ and 65% for 6-keto-$PGF_{1\alpha}$ whereas rosiglitazone was ineffective (Table 2). The level of COX-2 and mPGES-1 mRNAs was reduced by around 40% and 90% respectively by 10 $\mu$M

Table 1

Time course of prostaglandins cascade in IL-1$\beta$ stimulated chondrocytes

| | Controls | $T = 6$ h | $T = 12$ h | $T = 24$ h |
|---|---|---|---|---|
| **Mediators (pg/ml)** | | | | |
| $PGE_2$ | $50 \pm 1$ | $250 \pm 15^*$ | $1002 \pm 52^*$ | $3500 \pm 50^*$ |
| 6 keto-$PGF_{1\alpha}$ | $82 \pm 2$ | $89 \pm 2$ | $175 \pm 5^\dagger$ | $898 \pm 76^\dagger$ |
| **mRNA (gene of interest/S29)** | | | | |
| COX-2 | $0.01 \pm 0.01$ | $0.08 \pm 0.06^*$ | $0.37 \pm 0.09^*$ | $0.18 \pm 0.02^*$ |
| mPGES-1 | $0.01 \pm 0.01$ | $0.05 \pm 0.01^\dagger$ | $0.09 \pm 0.05^\dagger$ | $0.68 \pm 0.11^\dagger$ |

Rat cells were exposed to 10 ng/ml IL-1$\beta$ for 6, 12, 24, 36 or 48 h before total RNA extraction and collection of culture supernatant. Prostaglandins levels [$PGE_2$, 6-keto $PGF_{1\alpha}$] were assayed by ELISA in culture supernatant; relative abundances of COX-2 and mPGES-1 mRNAs were analysed by Real time PCR and normalized to S29 mRNA. Prostaglandins levels and PCR COX-2/S29 or mPGES-1/S29 mRNAs ratios presented in board are expressed as mean $\pm$ SD from four independent experiments. Statistically significant differences from controls ($p < 0.05$) are indicated as $^*$ for $PGE_2$ or COX-2 and $^\dagger$ for 6-keto $PGF_{1\alpha}$ or mPGES-1.

Table 2

Effect of PPAR$\gamma$ agonists on IL-1$\beta$-induced prostaglandins cascade in chondrocytes

| | Controls | IL-1$\beta$ (10 ng/ml) | IL-1$\beta$ + 15d (10 $\mu$M) | IL-1$\beta$ + Rosi (10 $\mu$M) |
|---|---|---|---|---|
| **Mediators (pg/ml)** | | | | |
| $PGE_2$ | $200 \pm 40$ | $5100 \pm 50^*$ | $505 \pm 70^\#$ | $4500 \pm 85$ |
| 6 keto-$PGF_{1\alpha}$ | $93 \pm 4$ | $1257 \pm 45^*$ | $424 \pm 8^\#$ | $1380 \pm 48$ |
| **mRNA (gene of interest/S29)** | | | | |
| COX-2 | $0.02 \pm 0.01$ | $0.78 \pm 0.08^*$ | $0.47 \pm 0.08^\#$ | $1.02 \pm 0.2^\#$ |
| mPGES-1 | $0.01 \pm 0.01$ | $0.6 \pm 0.1^*$ | $0.05 \pm 0.07^\#$ | $0.54 \pm 0.013$ |

After a 4 hours pre-treatment with 10 $\mu$M of 15d-$PGJ_2$ or rosiglitazone, chondrocytes were incubated with 10 ng/ml of IL-1$\beta$ for 12 or 24 h. $PGE_2$ and 6-keto-$PGF_{1\alpha}$ levels were assayed by ELISA in culture supernatant; relative abundances of COX-2 and mPGES-1 mRNAs were analysed by Real time PCR and normalized to S29 mRNA. Prostaglandins levels and PCR COX-2/S29 or mPGES-1/S29 mRNAs ratios presented in board are expressed as mean $\pm$ SD from four independent experiments. Statistically significant differences ($p < 0.05$) are indicated as $^*$ for comparison with non-stimulated controls and $^\#$ for comparison with IL-1$\beta$ stimulated cells.

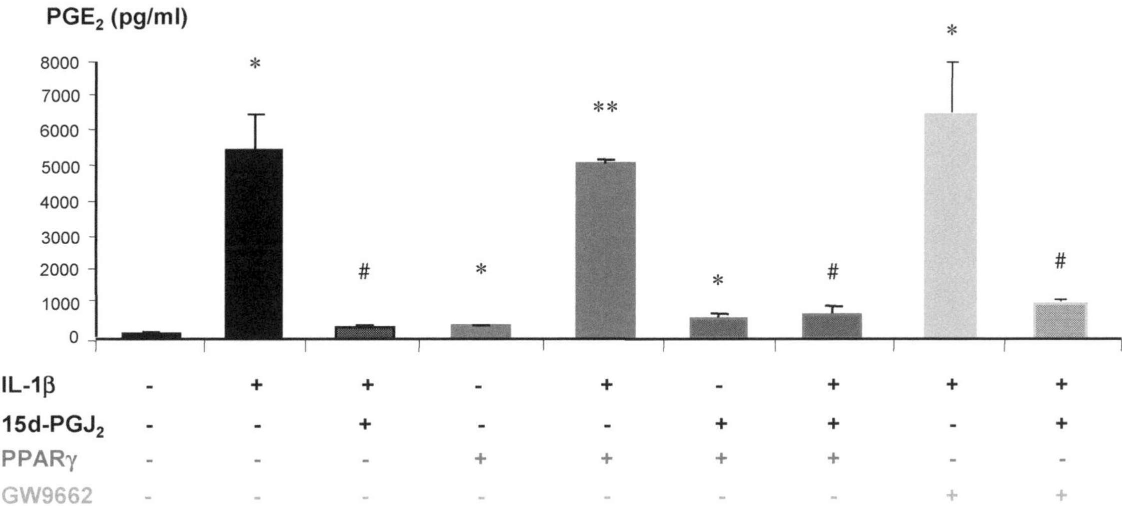


Fig. 2. Effect of PPARγ blockade or overexpression on the inhibition of IL-1β induced responses by 15d-PGJ₂. Chondrocytes in 6-well plates were transfected with pcDNA3.1 PPARγ construct (500 ng) for 36 hrs. Thereafter, cells were pre-treated for 4 hours with 10 $\mu$M of 15d-PGJ₂, and then stimulated with 10 ng/ml IL-1β for 24 h before collection of culture supernatant. In an another set of experiments, cells were co-treated with 15d-PGJ₂ in the presence or absence of 10 $\mu$M of GW-9662 (a specific PPARγ antagonist), then stimulated with 10 ng/ml IL-1β for 24 h before collection of culture supernatant. PGE₂ levels were assayed by ELISA in culture supernatant. Prostaglandins levels presented in histograms are expressed as mean $\pm$ SD from four independent experiments. Statistically significant differences ($p < 0.05$) are indicated as * for comparison with non-stimulated controls, # for comparison with IL-1β stimulated cells.

of 15d-PGJ₂, whereas rosiglitazone increased COX-2 mRNAs by 35% while reducing mPGES-1 mR-NAs only marginally (Table 2). The basal levels of prostaglandins and mRNAs were unaffected by both PPARγ agonists (data not shown).

### 3.3. Contribution of PPARγ to the inhibitory potency of 15d-PGJ₂ on inducible prostaglandins cascade

When 15d-PGJ₂ (10 $\mu$M) was tested in combination with an equivalent concentration of GW9662, an irreversible PPARγ antagonist, its inhibitory effect on IL-1β-induced PGE₂ and mPGES-1 mRNA levels was not reduced (Fig. 2). Resting chondrocytes overexpressing PPARγ showed a limited increase in PGE₂ level and mPGES-1 expression while responding normally to IL-1β challenge (Fig. 2). However, the inhibitory effect of 15d-PGJ₂ (10 $\mu$M) on IL-1β-induced PGE₂ and mPGES-1 mRNA levels was not improved in PPARγ overexpressing cells (Fig. 2). Control experiments confirmed that adiponectin mRNA level increased when chondrocytes were exposed to 10 $\mu$M of 15d-PGJ₂ alone but not in combination with GW9662 and that this increase was triggered in cells transfected with PPARγ expression vector (data not shown).

### 3.4. Contribution of NF-κB pathway to the inhibitory potency of 15d-PGJ₂ on inducible prostaglandins cascade

As shown in Fig. 3A, IL-1β stimulated the translocation of NF-κB complex to the nucleus and this activation was reduced strongly by 15d-PGJ₂ (10 $\mu$M) but not by Rosiglitazone (10 $\mu$M). Western-blot analysis revealed that IL-1β provoked a rapid disappearance of IκBα from the cytosol, which was prevented partly by 15d-PGJ₂ but not by Rosiglitazone (Fig. 3B). In addition, IL-1β induced the phospho-

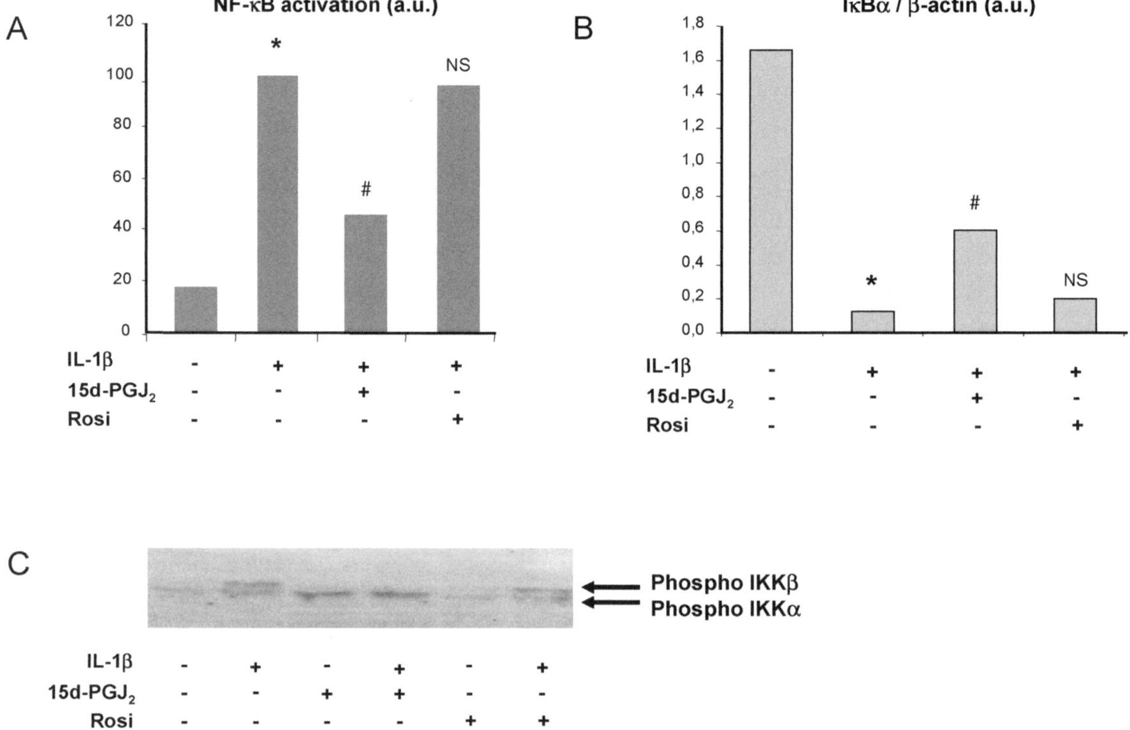


Fig. 3. Contribution of NF-κB pathway to IL-1β induced responses and 15d-PGJ$_2$ inhibitory effects. In one set of experiments (A), chondrocytes cultured in 6-well plates were exposed to 10 ng/ml IL-1β for 15 min in the presence or absence of 10 μM of 15d-PGJ$_2$ or Rosiglitazone before extraction of nuclear proteins. Activation of NF-κB was determined by ELISA using the TransAm[®] technology; results are expressed as relative arbitrary units setting a value of 100 for IL-1β treatment and are representative of three different experiments. Statistically significant differences ($p < 0.05$) are indicated as * for comparison with non-stimulated controls and # for comparison with IL-1β stimulated cells. (B, C), Cells were incubated four hours in the presence of PPARγ ligands, then with 10 ng/ml IL-1β for 5 min. Total proteins were extracted and immunoblot analysis was performed with specific antibodies. Total lysates were analysed with antibodies against non-phosphorylated IκBα (B), phosporylated IKKα/β (C) and β-actin (B, C). The results shown are representative of three independent experiments.

rylation of the IκBKinase complex and the phosphorylation of the β subunit was inhibited by 15d-PGJ$_2$ but not by Rosiglitazone (Fig. 3C).

## 4. Discussion

In the present work, we confirmed that rat chondrocytes produced high amounts of prostaglandins in response to IL-1β stimulation [56], with a kinetics similar to human osteoarthritic chondrocytes [6,43]. In basal conditions, chondrocytes produced both PGE$_2$ and 6keto-PGF$_{1\alpha}$ but the extent of variation was maximal for PGE$_2$ after IL-1β challenge [41,56]. Although PGE$_2$ synthesis was associated with induction of COX-2 in IL-1β-stimulated joint cells [6,56], mPGES-1 was demonstrated recently to be coordinately up-regulated but with a different time-course [44,70]. Our kinetics study confirmed an early induction of COX-2 and a delayed induction of mPGES-1 in IL-1β-stimulated chondrocytes [43], thereby mimicking the time course reported in inflamed rat tissues [30]. The increase in PGE$_2$ level fitted well with the extent of mPGES-1 gene expression but not with that of COX-2, whereas changes in 6keto-PGF$_{1\alpha}$ level were much lower than the extent of COX-2 induction. Of course, each inducible enzyme of the arachidonic acid cascade is rate limiting by controlling the bioavailability of substrate to downstream effectors [20,74]. However, our results support strongly that mPGES-1 expression was the most limiting step in PGE$_2$ synthesis, consistent with previous experiments with MK-886 [43], a five-lipoxygenase activating protein (FLAP) inhibitor having *in vitro* inhibitory potency on mPGES-1 [48]. As the stimulated synthesis of 6keto-PGF$_{1\alpha}$ requires a successive metabolization by COX-2 and prostacyclin synthase (PGIS), the lower than expected increase could reflect a limited induction of PGIS in rat chondrocyte. Thus, induction of PGIS by IL-1$\beta$ was less than 2-fold in rat non-articular cells [36] despite its selective up-regulation by COX-2 induction in human endothelial cells [9]. A decrease in PGIS expression, contrasting with an increase in mPGES-1 expression, was also reported in inflamed tissues of rat with adjuvant polyarthritis [14]. Alternatively, other metabolic pathways may have been favoured as the conversion of cyclic endoperoxides into other prostaglandins, depending on the substrate concentration dependencies of the terminal synthases [9,20], or of arachidonic acid into hydroxylated non-prostaglandins metabolites [40], depending on the balance between COX and lipoxygenases pathways [49]. Nonetheless, IL-1$\beta$ stimulated all inducible steps of the arachidonic acid cascade to produce PGE$_2$ maximally in rat chondrocytes.

The study of COX-2 or mPGES-1 expression and prostaglandins release in activated chondrocytes showed that 15d-PGJ$_2$ was inhibitory whereas the PPARγ comparator rosiglitazone was inactive in the same concentration range. These results were irrespective of the binding affinity of agonists to PPARγ [78], suggesting that this isotype was not necessarily required for such pharmacological modulation [57,58]. As expected, when we tried to reduce the inhibitory potency of 15d-PGJ$_2$ by preventing its binding to PPARγ with the irreversible antagonist GW9662, we failed to observe any changes in COX-2 and mPGES-1 mRNAs as well as PGE$_2$ levels. Complementary, the overexpression of PPARγ didn't enhance the potency of 15d-PGJ$_2$ in our experimental system. Finally, despite existence of a PPRE consensus site in the promoter of human COX-2 [52] and evidence that PPARγ agonists stimulated COX-2 gene expression in synovial fibroblasts [37], 15d-PGJ$_2$ failed to stimulate the basal production of PGE$_2$ in rat chondrocytes, as was the case in human osteoarthritic chondrocytes [23]. As we observed that 15d-PGJ$_2$ and rosiglitazone were able to activate a specific PPARγ target gene in our culture system (data not shown), these results support strongly that 15d-PGJ$_2$ was acting independently of PPARγ. Very few data are available in the rat species, but a PPARγ-dependent inhibition of inducible arachidonic acid cascade was reported in cardiac myocytes stimulated with IL-1$\beta$ [53]. As the inhibitory potency of 15d-PGJ$_2$ was closely similar in both studies, we suggest that this discrepancy was supported by cell type specificities. Indeed, inhibition of prostacyclin metabolites was very different for the same level of COX-2 inhibition whereas the synthetic PPARγ agonist troglitazone was inhibitory in cardiac myocytes while being ineffective in chondrocytes [6]. When considering the inhibitory potency of 15d-PGJ$_2$ in human chondrocytes, a decrease of PGE$_2$ levels comparable to that of rat cells was reported [23], but this was supported by a stronger inhibition of COX-2. For the control of other inflammatory mediators [22] and apoptosis [66], the dose-dependent effect of 15d-PGJ$_2$ was thought to be mainly supported by activation of PPARγ in human chondrocytes. The biological responses to PPAR agonists are well known to differ between species [4], as for example the carcinogenicity of fibrates in rodents. However, our results suggest that the potency of PPARγ agonists on joint cells is influenced by both cell type and species differences. Consistently, 15d-PGJ$_2$ and troglitazone were shown to inhibit PGE$_2$ production and mPGES-1 expression in IL-1$\beta$-stimulated human synovial fibroblasts [12] whereas troglitazone was totally ineffective on LPS-induced COX-2 expression in rat cells [69]. Finally, one may underline that the contribution of PPARγ may also depend on 15d-PGJ$_2$ concentration, since inhibition of $PGE_2$ production was reported to be PPAR$\gamma$-dependent in the nanomolar range while becoming PPAR$\gamma$-dependent in the micromolar range [2]. Despite a variable contribution of the PPAR$\gamma$ isotype to its pharmacological potency, the present work confirms that 15d-$PGJ_2$ down-regulates inducible steps of the arachidonic acid cascade in joint cells, thereby contributing likely to its anti-arthritic properties [39].

We next investigated whether the dose-dependent inhibitory effect of 15d-$PGJ_2$ could be supported by an interaction with the NF-$\kappa$B pathway, which is known to be one of its major targets in many cell types [5,65]. Previous study of the mouse mPGES-1 promoter indicated that it lacked binding sites for NF-$\kappa$B, CRE, and E-box that have been implicated in COX-2 induction, implying that the mechanisms for inducible expression of COX-2 and mPGES-1 were distinct in this species [55]. In human synovial fibroblasts, the transcriptional regulation of mPGES-1 gene by IL-1$\beta$ was shown to be closely dependent on the transcription factor early growth response factor-1 (Egr-1) [12], although AP-1 and SP-1 binding sites were also found [21]. In human chondrocytes, IL-1$\beta$ was demonstrated to use overlapping, but distinct, signalling pathways to induce COX-2 and mPGES-1 with a major role of ERK1/2 and p38$\beta$ MAPK for controlling the later [51]. In a non-articular human cell type, a substantial role for NF-$\kappa$B was demonstrated recently in the co-ordinate induction of COX-2 and mPGES-1 by IL-1$\beta$ [8]. As indicated before, some of these signalling pathways can be inhibited in a PPAR$\gamma$-dependent manner, possibly secondary to the squelching of transcription factors as CBP/p300 by protein–protein interaction with PPAR$\gamma$ [72]. Consequently, such mechanism is unlikely to explain the PPAR$\gamma$-independent inhibitory potency of 15d-$PGJ_2$ in our system. Although the promoter of rat mPGES-1 has not been explored to date, our data with mutated I$\kappa$B$\alpha$ support a major role of NF-$\kappa$B in the control of its transcriptional activity. We showed further that 15d-$PGJ_2$ inhibited IL-1$\beta$-induced NF-$\kappa$B nuclear binding (EMSA) and transactivation (TransAM$^\circledR$ assay). This inhibitory effect was consistent with the ability of 15d-$PGJ_2$ to inhibit I$\kappa$B kinase (IKK), by limiting the phosphorylation of its catalytic subunit IKK$\beta$ [5], and to prevent IkB$\alpha$ degradation by the proteasome. We cannot exclude that inhibition of COX-2 by 15d-$PGJ_2$ may participate to its inhibitory potency on mPGES-1, since the rate of $PGE_2$ itself could enhance induction of mPGES-1 by IL-1$\beta$ [44], but we suggest that 15d-$PGJ_2$ was interfering directly with NF-$\kappa$B. Indeed, its reactive cyclopentenone ring renders 15d-$PGJ_2$ highly reactive with substances containing nucleophilic groups such as cysteinyl thiol group of proteins, a chemical reaction known as Michael's addition [26]. Therefore, 15d-$PGJ_2$ has been shown to inhibit IKK function by binding covalently to it [62] or to inhibit the NF-$\kappa$B binding to DNA in a direct manner, via alkylation of a conserved cysteine residue located in the p65 subunit DNA-binding domain [62,71]. A direct chemical interaction with NF-$\kappa$B components is further sustained by the ability of 15d-$PGJ_2$ to suppress induction of COX-2 in PPAR$\gamma$ deficient macrophages [6,10]. Although occurring by a PPAR$\gamma$-independent mechanism, the present data support that 15d-$PGJ_2$ is a multi-step inhibitor of the NF-$\kappa$B pathway (Fig. 4).

One intriguing question, beyond the scope of this study, is the possible therapeutical relevance of our results. Although we did not check for 15d-$PGJ_2$ levels in IL-1$\beta$-stimulated chondrocytes, several lines of evidence must be considered. Firstly, it is well established that 15d-$PGJ_2$ is synthesized *in vivo* [28,67] and that it could contribute to the resolution of experimental acute inflammation [28]. Secondly, the successive chemical dehydratation of $PGD_2$ into 15d-$PGJ_2$ is favoured by albumin [26] but can even occur in the absence of proteins [67]. Thirdly, 15d-$PGJ_2$ is found in the picomolar to the nanomolar range in biological fluids [1,66], whereas most of its anti-inflammatory effects occur above one micromolar. Fourthly, PGD synthase is expressed in cytokines-stimulated human OA chondrocytes [66]. Fifthly, 15d-$PGJ_2$ was found to be located mainly intracellularly by immunohistochemical analysis [67]. So, it is not

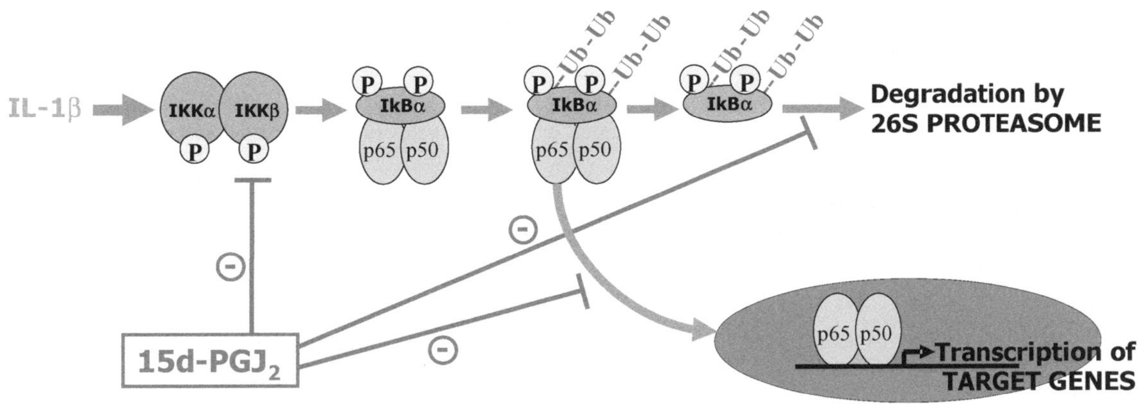


Fig. 4. Multistep inhibition of NF-κB pathway by 15d-PGJ₂.

possible to conclude if 15d-PGJ₂ would exert a negative feed-back loop in pathological situation, but our data suggest that 15d-PGJ₂ could behave as an endogenous regulator of inflammation, if synthesized locally in sufficient amounts.

# References


[1] L.C. Bell-Parikh, T. Ide, J.A. Lawson, P. McNamara, M. Reilly and G.A. FitzGerald, Biosynthesis of 15-deoxy-delta12,14-PGJ2 and the ligation of PPARgamma, *J. Clin. Invest.* **112** (2003), 945–955.

[2] E.B. Berry, J.A. Keelan, R.J. Helliwell, R.S. Gilmour and M.D. Mitchell, Nanomolar and micromolar effects of 15-deoxy-delta12,14-prostaglandin J2 on amnion-derived WISH epithelial cells: differential roles of peroxisome proliferator-activated receptors gamma and delta and nuclear factor kappaB, *Mol. Pharmacol.* **68** (2005), 169–178.

[3] F.J. Blanco, R. Guitian, J. Moreno, F.J. de Toro and F. Galdo, Effect of antiinflammatory drugs on COX-1 and COX-2 activity in human articular chondrocytes, *J. Rheumatol.* **26** (1999), 1366–1373.

[4] C. Blanquart, O. Barbier, J.C. Fruchart, B. Staels and C. Glineur, Peroxisome proliferator-activated receptors: regulation of transcriptional activities and roles in inflammation, *J. Steroid Biochem. Mol. Biol.* **85** (2003), 267–273.

[5] S. Boyault, A. Bianchi, D. Moulin, S. Morin, M. Francois, P. Netter, B. Terlain and K. Bordji, 15-Deoxy-delta(12,14)-prostaglandin J(2) inhibits IL-1beta-induced IKK enzymatic activity and IkappaBalpha degradation in rat chondrocytes through a PPARgamma-independent pathway, *FEBS Lett.* **572** (2004), 33–40.

[6] S. Boyault, M.A. Simonin, A. Bianchi, E. Compe, B. Liagre, D. Mainard, P. Becuwe, M. Dauca, P. Netter, B. Terlain and K. Bordji, 15-Deoxy-delta12,14-PGJ2, but not troglitazone, modulates IL-1beta effects in human chondrocytes by inhibiting NF-kappaB and AP-1 activation pathways, *FEBS Lett.* **501** (2001), 24–30.

[7] T.G. Brock, R.W. McNish and M. Peters-Golden, Arachidonic acid is preferentially metabolized by cyclooxygenase-2 to prostacyclin and prostaglandin E2, *J. Biol. Chem.* **274** (1999), 11660–11666.

[8] M.C. Catley, J.E. Chivers, L.M. Cambridge, N. Holden, D.M. Slater, K.J. Staples, M.W. Bergmann, P. Loser, P.J. Barnes and R. Newton, IL-1beta-dependent activation of NF-kappaB mediates PGE2 release via the expression of cyclooxygenase-2 and microsomal prostaglandin E synthase, *FEBS Lett.* **547** (2003), 75–79.

[9] G.E. Caughey, L.G. Cleland, P.S. Penglis, J.R. Gamble and M.J. James, Roles of cyclooxygenase (COX)-1 and COX-2 in prostanoid production by human endothelial cells: selective up-regulation of prostacyclin synthesis by COX-2, *J. Immunol.* **167** (2001), 2831–2838.

[10] A. Chawla, Y. Barak, L. Nagy, D. Liao, P. Tontonoz and R.M. Evans, PPAR-gamma dependent and independent effects on macrophage-gene expression in lipid metabolism and inflammation, *Nat. Med.* **7** (2001), 48–52.

[11] A. Chawla, E.J. Schwarz, D.D. Dimaculangan and M.A. Lazar, Peroxisome proliferator-activated receptor (PPAR) gamma: adipose-predominant expression and induction early in adipocyte differentiation, *Endocrinology* **135** (1994), 798–800.

[12] S. Cheng, H. Afif, J. Martel-Pelletier, J.P. Pelletier, X. Li, K. Farrajota, M. Lavigne and H. Fahmi, Activation of peroxisome proliferator-activated receptor gamma inhibits interleukin-1beta-induced membrane-associated prostaglandin E2 synthase-1 expression in human synovial fibroblasts by interfering with Egr-1, *J. Biol. Chem.* **279** (2004), 22057–22065.

[13] C.A. Clark, E.M. Schwarz, X. Zhang, N.M. Ziran, H. Drissi, R.J. O'Keefe and M.J. Zuscik, Differential regulation of EP receptor isoforms during chondrogenesis and chondrocyte maturation, *Biochem. Biophys. Res. Commun.* **328** (2005), 764–776.

[14] D. Claveau, M. Sirinyan, J. Guay, R. Gordon, C.C. Chan, Y. Bureau, D. Riendeau and J.A. Mancini, Microsomal prostaglandin E synthase-1 is a major terminal synthase that is selectively up-regulated during cyclooxygenase-2-dependent prostaglandin E2 production in the rat adjuvant-induced arthritis model, *J. Immunol.* **170** (2003), 4738–4744.

[15] P.R. Colville-Nash and D.A. Willoughby, COX-1, COX-2 and articular joint disease: a role for chondroprotective agents, *Biorheology* **39** (2002), 171–179.

[16] S. Cuzzocrea, E. Mazzon, L. Dugo, N.S. Patel, I. Serraino, R. Di Paola, T. Genovese, D. Britti, M. De Maio, A.P. Caputi and C. Thiemermann, Reduction in the evolution of murine type II collagen-induced arthritis by treatment with rosiglitazone, a ligand of the peroxisome proliferator-activated receptor gamma, *Arthritis Rheum.* **48** (2003), 3544–3556.

[17] S. Cuzzocrea, N.S. Wayman, E. Mazzon, L. Dugo, R. Di Paola, I. Serraino, D. Britti, P.K. Chatterjee, A.P. Caputi and C. Thiemermann, The cyclopentenone prostaglandin 15-deoxy-Delta(12,14)-prostaglandin J(2) attenuates the development of acute and chronic inflammation, *Mol. Pharmacol.* **61** (2002), 997–1007.

[18] J.T. Dingle, Prostaglandins in human cartilage metabolism, *J. Lipid Mediat.* **6** (1993), 303–312.

[19] J.T. Dingle, The effect of nonsteroidal antiinflammatory drugs on human articular cartilage glycosaminoglycan synthesis, *Osteoarthritis Cartilage* **7** (1999), 313–314.

[20] R.N. Dubois, S.B. Abramson, L. Crofford, R.A. Gupta, L.S. Simon, L.B. Van De Putte and P.E. Lipsky, Cyclooxygenase in biology and disease, *FASEB J.* **12** (1998), 1063–1073.

[21] L. Ekstrom, L. Lyrenas, P.J. Jakobsson, R. Morgenstern and M.J. Kelner, Basal expression of the human MAPEG members microsomal glutathione transferase 1 and prostaglandin E synthase genes is mediated by Sp1 and Sp3, *Biochim. Biophys. Acta* **1627** (2003), 79–84.

[22] H. Fahmi, J.A. Di Battista, J.P. Pelletier, F. Mineau, P. Ranger and J. Martel-Pelletier, Peroxisome proliferator–activated receptor gamma activators inhibit interleukin-1beta-induced nitric oxide and matrix metalloproteinase 13 production in human chondrocytes, *Arthritis Rheum.* **44** (2001), 595–607.

[23] H. Fahmi, J.P. Pelletier, F. Mineau and J. Martel-Pelletier, 15d-PGJ(2) is acting as a 'dual agent' on the regulation of COX-2 expression in human osteoarthritic chondrocytes, *Osteoarthritis Cartilage* **10** (2002), 845–848.

[24] W.H. Faour, Y. He, Q.W. He, M. de Ladurantaye, M. Quintero, A. Mancini and J.A. Di Battista, Prostaglandin E(2) regulates the level and stability of cyclooxygenase-2 mRNA through activation of p38 mitogen-activated protein kinase in interleukin-1 beta-treated human synovial fibroblasts, *J. Biol. Chem.* **276** (2001), 31720–31731.

[25] K. Fukuda, K. Ohtani, H. Dan and S. Tanaka, Interleukin-1 inhibits keratan sulfate production by rabbit chondrocytes: possible role of prostaglandin E2, *Inflamm. Res.* **44** (1995), 178–181.

[26] M. Fukushima, Biological activities and mechanisms of action of PGJ2 and related compounds: an update, *Prostaglandins Leukot. Essent. Fatty Acids* **47** (1992), 1–12.

[27] P. Ghosh, Nonsteroidal anti-inflammatory drugs and chondroprotection. A review of the evidence, *Drugs* **46** (1993), 834–846.

[28] D.W. Gilroy, P.R. Colville-Nash, S. McMaster, D.A. Sawatzky, D.A. Willoughby and T. Lawrence, Inducible cyclooxygenase-derived 15-deoxy(Delta)12-14PGJ2 brings about acute inflammatory resolution in rat pleurisy by inducing neutrophil and macrophage apoptosis, *FASEB J.* **17** (2003), 2269–2271.

[29] M.B. Goldring, L.F. Suen, R. Yamin and W.F. Lai, Regulation of collagen gene expression by prostaglandins and interleukin-1beta in cultured chondrocytes and fibroblasts, *Am. J. Ther.* **3** (1996), 9–16.

[30] J. Guay, K. Bateman, R. Gordon, J. Mancini and D. Riendeau, Carrageenan-induced paw edema in rat elicits a predominant prostaglandin E2 (PGE2) response in the central nervous system associated with the induction of microsomal PGE2 synthase-1, *J. Biol. Chem.* **279** (2004), 24866–24872.

[31] M.M. Hardy, K. Seibert, P.T. Manning, M.G. Currie, B.M. Woerner, D. Edwards, A. Koki and C.S. Tripp, Cyclooxygenase 2-dependent prostaglandin E2 modulates cartilage proteoglycan degradation in human osteoarthritis explants, *Arthritis Rheum.* **46** (2002), 1789–1803.

[32] A.N. Hata and R.M. Breyer, Pharmacology and signaling of prostaglandin receptors: multiple roles in inflammation and immune modulation, *Pharmacol. Ther.* **103** (2004), 147–166.

[33] W. He, J.P. Pelletier, J. Martel-Pelletier, S. Laufer and J.A. Di Battista, Synthesis of interleukin 1beta, tumor necrosis factor-alpha, and interstitial collagenase (MMP-1) is eicosanoid dependent in human osteoarthritis synovial membrane explants: interactions with antiinflammatory cytokines, *J. Rheumatol.* **29** (2002), 546–553.

[34] E.V. Hess and J.H. Herman, Cartilage metabolism and anti-inflammatory drugs in osteoarthritis, *Am. J. Med.* **81** (1986), 36–43.

[35] H. Inoue, M. Takamori, Y. Shimoyama, H. Ishibashi, S. Yamamoto and Y. Koshihara, Regulation by PGE2 of the production of interleukin-6, macrophage colony stimulating factor, and vascular endothelial growth factor in human synovial fibroblasts, *Br. J. Pharmacol.* **136** (2002), 287–295.

[36] A. Itoh, J. Nishihira, H. Makita, K. Miyamoto, E. Yamaguchi and M. Nishimura, Effects of IL-1beta, TNF-alpha, and

macrophage migration inhibitory factor on prostacyclin synthesis in rat pulmonary artery smooth muscle cells, *Respirology* **8** (2003), 467–472.

[37] T. Kalajdzic, W.H. Faour, Q.W. He, H. Fahmi, J. Martel-Pelletier, J.P. Pelletier and J.A. Di Battista, Nimesulide, a preferential cyclooxygenase 2 inhibitor, suppresses peroxisome proliferator-activated receptor induction of cyclooxygenase 2 gene expression in human synovial fibroblasts: evidence for receptor antagonism, *Arthritis Rheum.* **46** (2002), 494–506.

[38] Y. Kanno, H. Tokuda, K. Nakajima, A. Ishisaki, T. Shibata, O. Numata and O. Kozawa, Involvement of SAPK/JNK in prostaglandin E(1)-induced VEGF synthesis in osteoblast-like cells, *Mol. Cell Endocrinol.* **220** (2004), 89–95.

[39] Y. Kawahito, M. Kondo, Y. Tsubouchi, A. Hashiramoto, D. Bishop-Bailey, K. Inoue, M. Kohno, R. Yamada, T. Hla and H. Sano, 15-deoxy-delta(12,14)-PGJ(2) induces synoviocyte apoptosis and suppresses adjuvant-induced arthritis in rats, *J. Clin. Invest.* **106** (2000), 189–197.

[40] J.S. Kerr, T.M. Stevens, G.L. Davis, J.A. McLaughlin and R.R. Harris, Effects of recombinant interleukin-1 beta on phospholipase A2 activity, phospholipase A2 mRNA levels, and eicosanoid formation in rabbit chondrocytes, *Biochem. Biophys. Res. Commun.* **165** (1989), 1079–1084.

[41] I. Knott, M. Dieu, M. Burton, A. Houbion, J. Remacle and M. Raes, Induction of cyclooxygenase by interleukin 1: comparative study between human synovial cells and chondrocytes, *J. Rheumatol.* **21** (1994), 462–466.

[42] Y. Kobayashi, T. Mizoguchi, I. Take, S. Kurihara, N. Udagawa and N. Takahashi, Prostaglandin E2 enhances osteoclastic differentiation of precursor cells through protein kinase A-dependent phosphorylation of TAK1, *J. Biol. Chem.* **280** (2005), 11395–11403.

[43] F. Kojima, H. Naraba, S. Miyamoto, M. Beppu, H. Aoki and S. Kawai, Membrane-associated prostaglandin E synthase-1 is upregulated by proinflammatory cytokines in chondrocytes from patients with osteoarthritis, *Arthritis Res. Ther.* **6** (2004), R355–365.

[44] F. Kojima, H. Naraba, Y. Sasaki, M. Beppu, H. Aoki and S. Kawai, Prostaglandin E2 is an enhancer of interleukin-1beta-induced expression of membrane-associated prostaglandin E synthase in rheumatoid synovial fibroblasts, *Arthritis Rheum.* **48** (2003), 2819–2828.

[45] K.E. Kuettner, B.U. Pauli, G. Gall, V.A. Memoli and R.K. Schenk, Synthesis of cartilage matrix by mammalian chondrocytes in vitro. I. Isolation, culture characteristics, and morphology, *J. Cell Biol.* **93** (1982), 743–750.

[46] X.H. Liu, A. Kirschenbaum, S. Yao and A.C. Levine, Cross-talk between the interleukin-6 and prostaglandin E(2) signaling systems results in enhancement of osteoclastogenesis through effects on the osteoprotegerin/receptor activator of nuclear factor-{kappa}B (RANK) ligand/RANK system, *Endocrinology* **146** (2005), 1991–1998.

[47] G.N. Lowe, Y.H. Fu, S. McDougall, R. Polendo, A. Williams, P.D. Benya and T.J. Hahn, Effects of prostaglandins on deoxyribonucleic acid and aggrecan synthesis in the RCJ 3.1C5.18 chondrocyte cell line: role of second messengers, *Endocrinology* **137** (1996), 2208–2216.

[48] J.A. Mancini, K. Blood, J. Guay, R. Gordon, D. Claveau, C.C. Chan and D. Riendeau, Cloning, expression, and up-regulation of inducible rat prostaglandin e synthase during lipopolysaccharide-induced pyresis and adjuvant-induced arthritis, *J. Biol. Chem.* **276** (2001), 4469–4475.

[49] J. Martel-Pelletier, F. Mineau, H. Fahmi, S. Laufer, P. Reboul, C. Boileau, M. Lavigne and J.P. Pelletier, Regulation of the expression of 5-lipoxygenase-activating protein/5-lipoxygenase and the synthesis of leukotriene B(4) in osteoarthritic chondrocytes: role of transforming growth factor beta and eicosanoids, *Arthritis Rheum.* **50** (2004), 3925–3933.

[50] J. Martel-Pelletier, J.P. Pelletier and H. Fahmi, Cyclooxygenase-2 and prostaglandins in articular tissues, *Semin. Arthritis Rheum.* **33** (2003), 155–167.

[51] K. Masuko-Hongo, F. Berenbaum, L. Humbert, C. Salvat, M.B. Goldring and S. Thirion, Up-regulation of microsomal prostaglandin E synthase 1 in osteoarthritic human cartilage: critical roles of the ERK-1/2 and p38 signaling pathways, *Arthritis Rheum.* **50** (2004), 2829–2838.

[52] E.A. Meade, T.M. McIntyre, G.A. Zimmerman and S.M. Prescott, Peroxisome proliferators enhance cyclooxygenase-2 expression in epithelial cells, *J. Biol. Chem.* **274** (1999), 8328–8334.

[53] M. Mendez and M.C. LaPointe, PPARgamma inhibition of cyclooxygenase-2, PGE2 synthase, and inducible nitric oxide synthase in cardiac myocytes, *Hypertension* **42** (2003), 844–850.

[54] M. Murakami, Y. Nakatani, T. Tanioka and I. Kudo, Prostaglandin E synthase, *Prostaglandins other Lipid Mediat.* **68–69** (2002), 383–399.

[55] H. Naraba, C. Yokoyama, N. Tago, M. Murakami, I. Kudo, M. Fueki, S. Oh-Ishi and T. Tanabe, Transcriptional regulation of the membrane-associated prostaglandin E2 synthase gene. Essential role of the transcription factor Egr-1, *J. Biol. Chem.* **277** (2002), 28601–28608.

[56] E. Nedelec, A. Abid, C. Cipolletta, N. Presle, B. Terlain, P. Netter and J. Jouzeau, Stimulation of cyclooxygenase-2-activity by nitric oxide-derived species in rat chondrocyte: lack of contribution to loss of cartilage anabolism, *Biochem. Pharmacol.* **61** (2001), 965–978.

[57] M. Okada, S.F. Yan and D.J. Pinsky, Peroxisome proliferator-activated receptor-gamma (PPAR-gamma) activation suppresses ischemic induction of Egr-1 and its inflammatory gene targets, *FASEB J.* **16** (2002), 1861–1868.

[58] E.J. Park, S.Y. Park, E.H. Joe and I. Jou, 15d-PGJ2 and rosiglitazone suppress Janus kinase-STAT inflammatory signaling

through induction of suppressor of cytokine signaling 1 (SOCS1) and SOCS3 in glia, *J. Biol. Chem.* **278** (2003), 14747–14752.

[59] J.P. Pelletier, J.C. Fernandes, D.V. Jovanovic, P. Reboul and J. Martel-Pelletier, Chondrocyte death in experimental osteoarthritis is mediated by MEK 1/2 and p38 pathways: role of cyclooxygenase-2 and inducible nitric oxide synthase, *J. Rheumatol.* **28** (2001), 2509–2519.

[60] M.H. Pillinger, P.B. Rosenthal, S.N. Tolani, B. Apsel, V. Dinsell, J. Greenberg, E.S. Chan, P.F. Gomez and S.B. Abramson, Cyclooxygenase-2-derived E prostaglandins down-regulate matrix metalloproteinase-1 expression in fibroblast-like synoviocytes via inhibition of extracellular signal-regulated kinase activation, *J. Immunol.* **171** (2003), 6080–6098.

[61] W.S. Powell, 15-Deoxy-delta12,14-PGJ2: endogenous PPARgamma ligand or minor eicosanoid degradation product?, *J. Clin. Invest.* **112** (2003), 828–830.

[62] A. Rossi, P. Kapahi, G. Natoli, T. Takahashi, Y. Chen, M. Karin and M.G. Santoro, Anti-inflammatory cyclopentenone prostaglandins are direct inhibitors of IkappaB kinase, *Nature* **403** (2000), 103–108.

[63] T. Sadowski and J. Steinmeyer, Effects of non-steroidal antiinflammatory drugs and dexamethasone on the activity and expression of matrix metalloproteinase-1, matrix metalloproteinase-3 and tissue inhibitor of metalloproteinases-1 by bovine articular chondrocytes, *Osteoarthritis Cartilage* **9** (2001), 407–415.

[64] Y. Sakuma, Z. Li, C.C. Pilbeam, C.B. Alander, D. Chikazu, H. Kawaguchi and L.G. Raisz, Stimulation of cAMP production and cyclooxygenase-2 by prostaglandin E(2) and selective prostaglandin receptor agonists in murine osteoblastic cells, *Bone* **34** (2004), 827–834.

[65] J.U. Scher and M.H. Pillinger, 15d-PGJ2: the anti-inflammatory prostaglandin?, *Clin. Immunol.* **114** (2005), 100–109.

[66] Z.Z. Shan, K. Masuko-Hongo, S.M. Dai, H. Nakamura, T. Kato and K. Nishioka, A potential role of 15-deoxy-delta(12,14)-prostaglandin J2 for induction of human articular chondrocyte apoptosis in arthritis, *J. Biol. Chem.* **279** (2004), 37939–37950.

[67] T. Shibata, M. Kondo, T. Osawa, N. Shibata, M. Kobayashi and K. Uchida, 15-deoxy-delta 12,14-prostaglandin J2. A prostaglandin D2 metabolite generated during inflammatory processes, *J. Biol. Chem.* **277** (2002), 10459–10466.

[68] D.L. Simmons, R.M. Botting and T. Hla, Cyclooxygenase isozymes: the biology of prostaglandin synthesis and inhibition, *Pharmacol. Rev.* **56** (2004), 387–437.

[69] M.A. Simonin, K. Bordji, S. Boyault, A. Bianchi, E. Gouze, P. Becuwe, M. Dauca, P. Netter and B. Terlain, PPAR-gamma ligands modulate effects of LPS in stimulated rat synovial fibroblasts, *Am. J. Physiol. Cell Physiol.* **282** (2002), C125–133.

[70] D.O. Stichtenoth, S. Thoren, H. Bian, M. Peters-Golden, P.J. Jakobsson and L.J. Crofford, Microsomal prostaglandin E synthase is regulated by proinflammatory cytokines and glucocorticoids in primary rheumatoid synovial cells, *J. Immunol.* **167** (2001), 469–474.

[71] D.S. Straus, G. Pascual, M. Li, J.S. Welch, M. Ricote, C.H. Hsiang, L.L. Sengchanthalangsy, G. Ghosh and C.K. Glass, 15-deoxy-delta 12,14-prostaglandin J2 inhibits multiple steps in the NF-kappa B signaling pathway, *Proc. Natl. Acad. Sci. USA* **97** (2000), 4844–4849.

[72] K. Subbaramaiah, D.T. Lin, J.C. Hart and A.J. Dannenberg, Peroxisome proliferator-activated receptor gamma ligands suppress the transcriptional activation of cyclooxygenase-2. Evidence for involvement of activator protein-1 and CREB-binding protein/p300, *J. Biol. Chem.* **276** (2001), 12440–12448.

[73] N. Tanikawa, Y. Ohmiya, H. Ohkubo, K. Hashimoto, K. Kangawa, M. Kojima, S. Ito and K. Watanabe, Identification and characterization of a novel type of membrane-associated prostaglandin E synthase, *Biochem. Biophys. Res. Commun.* **291** (2002), 884–889.

[74] T. Tanioka, Y. Nakatani, N. Semmyo, M. Murakami and I. Kudo, Molecular identification of cytosolic prostaglandin E2 synthase that is functionally coupled with cyclooxygenase-1 in immediate prostaglandin E2 biosynthesis, *J. Biol. Chem.* **275** (2000), 32775–32782.

[75] Y. Tsubouchi, Y. Kawahito, M. Kohno, K. Inoue, T. Hla and H. Sano, Feedback control of the arachidonate cascade in rheumatoid synoviocytes by 15-deoxy-Delta(12,14)-prostaglandin J2, *Biochem. Biophys. Res. Commun.* **283** (2001), 750–755.

[76] D.A. Willoughby, P.R. Colville-Nash and M.P. Seed, Inflammation, prostaglandins, and loss of function, *J. Lipid Mediat.* **6** (1993), 287–293.

[77] D.A. Willoughby, A.R. Moore, P.R. Colville-Nash and D. Gilroy, Resolution of inflammation, *Int. J. Immunopharmacol.* **22** (2000), 1131–1135.

[78] T.M. Willson, J.E. Cobb, D.J. Cowan, R.W. Wiethe, I.D. Correa, S.R. Prakash, K.D. Beck, L.B. Moore, S.A. Kliewer and J.M. Lehmann, The structure-activity relationship between peroxisome proliferator-activated receptor gamma agonism and the antihyperglycemic activity of thiazolidinediones, *J. Med. Chem.* **39** (1996), 665–668.

[79] K. Yoshida, H. Oida, T. Kobayashi, T. Maruyama, M. Tanaka, T. Katayama, K. Yamaguchi, E. Segi, T. Tsuboyama, M. Matsushita, K. Ito, Y. Ito, Y. Sugimoto, F. Ushikubi, S. Ohuchida, K. Kondo, T. Nakamura and S. Narumiya, Stimulation of bone formation and prevention of bone loss by prostaglandin E EP4 receptor activation, *Proc. Natl. Acad. Sci. USA* **99** (2002), 4580–4585.

[80] T. Yoshida, H. Sakamoto, T. Horiuchi, S. Yamamoto, A. Suematsu, H. Oda and Y. Koshihara, Involvement of prostaglandin

E(2) in interleukin-1alpha-induced parathyroid hormone-related peptide production in synovial fibroblasts of patients with rheumatoid arthritis, *J. Clin. Endocrinol. Metab.* **86** (2001), 3272–3278.

[81] X. Zhang, E.M. Schwarz, D.A. Young, J.E. Puzas, R.N. Rosier and R.J. O'Keefe, Cyclooxygenase-2 regulates mesenchymal cell differentiation into the osteoblast lineage and is critically involved in bone repair, *J. Clin. Invest.* **109** (2002), 1405–1415.

Biorheology 43 (2006) 577–587
IOS Press



# Inhibition of interleukin-1$\beta$-induced activation of MEK/ERK pathway and DNA binding of NF-$\kappa$B and AP-1: Potential mechanism for Diacerein effects in osteoarthritis

F. Domagala [b,*], G. Martin [a], P. Bogdanowicz [a], H. Ficheux [b] and J.-P. Pujol [a]

[a] *Laboratory of Connective Tissue Biochemistry, Faculty of Medicine, Caen Cedex, France*
[b] *Laboratoire Negma-Lerads, Toussus-le-Noble, Magny-les-Hameaux Cedex, France*

**Abstract.** In the present report we have shown that bovine articular chondrocytes cultured in low oxygen tension, i.e. in conditions mimicking their hypoxic *in vivo* environment, respond to IL-1$\beta$ (10 ng/ml) by an increased DNA binding activity of NF-$\kappa$B and AP-1 transcription factors. Incubation of the cells with $10^{-5}$ M Rhein, the active metabolite of Diacerhein, for 24 h was found to reduce this activity particularly in the case of AP-1. Mitogen activated kinases (ERK-1 and ERK-2) were activated by exposure of the chondrocytes to a 1 h treatment with IL-1$\beta$. This effect was greater in hypoxia (3% $O_2$) than in normoxia (21% $O_2$). Rhein was capable of reducing the IL-1$\beta$-stimulated ERK1/ERK2 pathway whatever the tension of oxygen present in the environment. The mRNA steady-state levels of collagen type II (COL2A1) and aggrecan core protein were found to be significantly increased by a 24-h treatment with $10^{-5}$ M Rhein. This stimulating effect was also observed in the presence of IL-1$\beta$, suggesting that the drug could prevent or reduce the IL-1$\beta$-induced inhibition of extra cellular matrix synthesis. IL-1-induced collagenase (MMP1) expression was significantly decreased by Rhein under the same conditions. In conclusion, Rhein can effectively inhibit the IL-1-activated MAPK pathway and the binding of NF-$\kappa$B and AP-1 transcription factors, two key factors involved in the expression of several pro-inflammatory genes by chondrocytes. In addition, the drug can reduce the procatabolic effect of the cytokine, by reducing the MMP1 synthesis, and enhance the synthesis of matrix components, such as type II collagen and aggrecan. These results may explain the anti-osteoarthritic properties of Rhein and its disease-modifying effects on OA cartilage, in spite of the absence of inhibition at prostaglandin level.

## 1. Introduction

Osteoarthritis (OA) is a common disorder with enormous social and economic consequences, in which ageing, genetic, hormonal and mechanical factors are major contributors to its progression [6,14]. OA is commonly described as a non-inflammatory disease; however, inflammation is increasingly recognized as contributing to the symptoms and progression of this pathology [40]. Histological studies have shown that there is an ongoing inflammation which contributes to damaging the osteoarthritic joint through

---

*Address for correspondence: F. Domagala, Preclinical Department, Laboratoire Negma-Lerads, Toussus- le-Noble, Magny-les-Hameaux Cedex, France. Tel.: +33 1 39258018; Fax: +33 1 39258025; E-mail: f.domagala@negma-lerads.fr.

cartilage degradation. Moreover, synovial inflammation may be detected with the presence of mild or severe cartilage changes in OA [15].

In normal joint, smooth linings of articular cartilage cover the ends of the bones. The joint space is also filled with fluid that helps lubricate the joint. Cartilage consists of cells called chondrocytes which synthesize an extra cellular matrix, rich in collagen and proteoglycans. The healthy cartilage is in a state of balance between degradation and synthesis processes. Extra cellular matrix synthesis is tightly regulated by cytokines such as interleukin 1 (IL-1$\beta$) and tumour necrosis factor alpha (TNF$\alpha$) [17] which inhibit matrix synthesis and by growth factors such as transforming growth factor, insulin like growth factor and bone morphogenetic proteins (BMPs) which stimulate the synthesis.

However, in some pathologies such as OA, a disruption of the balance leads to an increase in cytokine-induced catabolic processes and to cartilage erosion. OA is characterized by quantitative and qualitative changes in the architecture and composition of all the joint structures. In that pathology, homeostasis is lost and the synovial fluid becomes less viscous, leaving the cartilage and synovium exposed to mechanical and inflammatory damage.

Local inflammatory activity is a well-recognized component of OA in which IL-1$\beta$ holds a crucial role [13,20] and it has been shown that levels of cytokines are increased in OA synovial fluid [21,29, 43]. But IL-1$\beta$ can also increase the expression of other mediators of inflammation, all contributing to cartilage injury [32,38,41]. It is now clear that IL-1$\beta$ induces a cascade of events that leads to cartilage damage. First, the binding of IL-1$\beta$ to cell membrane receptors (Toll-like/IL-1) stimulates intracellular signal transduction molecules such as the IL-1 Receptor-Associated Kinase (IRAK) and TNF Receptor-Associated Factor (TRAF), which activates I$\kappa$B kinase. This enzyme induces the phosphorylation of I$\kappa$K and then its degradation, allowing the translocation of the active nuclear factor-kappa B (NF-$\kappa$B) dimmers to the nucleus where it interacts with the promoter regions of inflammatory genes.

IL-1$\beta$ triggers several signalling cascades in chondrocytes, including the activation of mitogen activated protein kinases (MAPKs) which induce the DNA binding activity of NF-$\kappa$B and activator protein-1 (AP-1) transcription factors, implicated in the transcription of many pro inflammatory genes [7,22]. In particular, a high NF-$\kappa$B and AP-1 binding activity in the synovium of patients with OA has been observed [1,19], leading to inflammatory process and degradation. Then, IL-1 induces the expression of cyclooxygenase (COX-2) [12] and inducible nitric oxide synthase (iNOS) [41], which synthesize prostaglandins and nitric oxide (NO), promoting inflammation. Matrix metalloprotease (MMP) production in chondrocytes and synoviocytes is also up-regulated [4,36], while its natural inhibitor (TIMP) is decreased [8], inducing the degradation of types I & II collagen.

Diacerein, a disease-modifying drug with anthraquinonic structure has been shown to inhibit the *in vitro* and *in vivo* production and activity of IL-1$\beta$ [30] and ameliorates the course of osteoarthritis by preventing or reducing the lesions of the joint tissue [3]. Diacerein, which is completely metabolized by humans into Rhein, an active metabolite, is currently used in the treatment of OA and showed significant structure modifying effects in hip OA in the 3-year ECHODIAH trial [10]. The cellular mechanisms underlying these beneficial effects are not completely understood, but Diacerein has multiple mechanisms of action in catabolic and anabolic pathways. *In vitro*, Diacerein inhibits the IL-1$\beta$-induced stimulation of collagenase expression by cultured rabbit chondrocytes [2], down-regulates the expression of IL-1$\beta$ receptor in articular chondrocytes [24,44], enhances the expression of TGF-$\beta$ [11] and increases collagen and aggrecan [2,39].

The objective of this work was to investigate the effect of Diacerein and its active metabolite Rhein on the IL-1$\beta$ signalling pathways, particularly on the activation of the transcription factors, but also on the expression of MMP-1 (collagenase), type II collagen and aggrecan in bovine articular chondrocytes in low oxygen tension by evaluating the steady state levels of corresponding mRNAs.

## 2. Materials and methods

### 2.1. Reagents

Mouse anti-phospho MAP kinase (specific for threonine and tyrosine phosphorylated residues of ERK1/ERK2) and rabbit anti-MAP kinase 1/2 monoclonal antibodies (ERK1/2-CT) were purchased from Upstate Biotechnology Inc. (Lake Placid, NY, USA). Rabbit polyclonal anti-IkB$\alpha$ and rabbit polyclonal anti-c-jun were obtained from Santa Cruz Biotechnology Inc. (Santa Cruz, CA, USA). Primary antibodies were revealed with anti-rabbit or anti-mouse horseradish peroxidase-labeled secondary antibodies (Santa Cruz), using an ECL + Plus Western Blot detection kit (Amersham Pharmacia Biotech Europe. Orsay, France). The consensus oligonucleotide probes: AP1 (5′-5′CGCTTGATGAGTCAGCCC GGAA-3′) and NF-kB (5′-AGTTGAGGGGACTTTCCCAGGC-3′) were supplied by Life Technologies, Inc. (Cergy Pontoise, France). All other chemicals were of the highest purity available and ordered from Sigma-Aldrich Co. (St Quentin-Fallavier, France). Human recombinant IL-1$\beta$ was a generous gift from Dr Soichiro Sato, Shizuoka, Japan.

### 2.2. Culture and treatment of articular chondrocytes

Chondrocytes were isolated from the knee joints of freshly slaughtered calves by enzymatic treatment. Briefly, slices of cartilage were dissected out and kept in Earle's balanced salt solution. Chondrocytes were released by digestion with type XIV protease (4 mg/ml) for 1.5 h and type I collagenase (1 mg/ml) overnight in DMEM at 37°C. The cells were centrifuged and seeded in DMEM (Life Technologies, Inc.) supplemented with 10% heat-inactivated fetal calf serum (FCS) and antibiotics: penicillin (100 IU/ml), streptomycin (100 $\mu$g/ml), and fungizone (0.25 $\mu$g/ml). The cells were allowed to recover for 48 h at 37°C in a humidified atmosphere supplemented with 5% $CO_2$.

For experiments in hypoxic environment, cartilage samples were dissected in a special plastic chamber previously equilibrated at 3% oxygen and the subsequent procedure was performed under the same conditions. The $O_2$ level of the culture medium was reduced from 21% to 3% $O_2$, by bubbling a gas mixture without oxygen for 1.5 hour. Because of its antioxidant properties, phenol red, generally used as pH indicator, was not added to the medium.

### 2.3. Preparation of cytoplasmic and nuclear extracts

At the end of the experiments, the chondrocytes were rinsed twice with ice-cold phosphate-buffered saline (PBS) and scraped in a hypotonic buffer. The cell suspension was centrifuged at $10,000 \times g$ for 10 min and the resulting supernatants used as cytoplasmic extracts. The pellet was incubated in hypertonic buffer for 4 h at 4°C and centrifuged at $10,000 \times g$ for 30 min to provide nuclear extracts. The hypotonic buffer contained: 10 mM Hepes (pH 7.9), 0.1% NP-40, 1.5 mM $MgCl_2$, 10 mM KCl, 1 mM EGTA, 1 mM EDTA, 5 mM NaF, 0.5 mM DDT, 0.5 mM PMSF, leupeptin, pepsatin A and aprotinin at 10 $\mu$g/ml, 1 mM $Na_3VO_4$. The hypertonic buffer was composed of: 20 mM Hepes (pH 7.6), 25% glycerol, 1 mM $MgCI_2$, 420 mM NaCI, 0.2 mM EDTA, 0.25 mM DTT, 0.5 mM PMSF, leupeptin, pepstacin A and aprotinin at 10 $\mu$g/ml, 1 mM $Na_3VO_4$. The protein amount was determined by Bradford's colorimetric procedure (Bio-Rad S.A., lvry sur Seine, France).

## 2.4. Western blot analysis

Cellular extracts (15 $\mu$g protein) were subjected to sodium dodecylsulfate–polyacrylamide gel electrophoresis (SDS-PAGE) under reducing conditions and electrophoretically transferred to polyvinylidene difluoride transfer membranes (PVDF, NEN Life Sciences Products, Zawentem, Belgium). The membranes were blocked for 1 h at room temperature in Tris-buffered saline (pH 7.6) with 0.1% Tween-20 (TBS-T) and 10% non-fat dry milk. They were then rinsed twice in TBS-T and incubated overnight at 4°C with the primary antibody. After washing with TBS-T, they were incubated for 1 h with the appropriate secondary antibody. The membranes were developed with the ECL + Plus chemo-luminescence detection kit. To check for the presence of equal amounts of protein or to analyse the expression of several proteins, blots were stripped (100 mM 2-mercaptoethanol, 2% SDS, 62.5 mM Tris-HCL (pH 6.7) for 30 min at 50°C under stirring and used again.

## 2.5. Electrophoretic mobility shift assays (EMSA)

Nuclear extracts (10 $\mu$g protein) were incubated in binding buffer for 30 min at 25°C with the cDNA probes, previously labelled with $^{32}$P-$\gamma$-ATP (25 fmoles) using T4 polynucleotide kinase (Life Technologies, Inc.). The final binding reaction for NF-kB was performed, in: 20 mM Hepes (pH 7.5), 50 mM KCl, 4 mM $MgCl_2$, 0.2 mM EDTA, 0.5 mM DTT, 0.05% NP40. 20% glycerol, 1 mg/ml BSA, 0.025 mM poly dI-dC. The binding reaction for AP-1 was composed of Hepes 20 mM, Tris-HCl pH 8.2 mM, DTT 1 mM, NaCI 80 mM, glycerol 10% (v/v), PMSF 0.2 mM, EDTA 0.4 mM, EGTA 0.2 mM. The samples were then submitted to 8% PAGE in 0.5 × TBE (45 mM Tris (pH 7.8), 45 mM boric acid and 1 mM EDTA) and visualized by autoradiography.

## 2.6. RNA extraction and Northern blot

Total RNA from the chondrocyte cultures were extracted by the guanidium isothiocyanate-phenol-chloroform procedure. RNA was fractionated by electrophoresis on 1% agarose/MOPS/formaldehyde gel and transferred to a nylon membrane (Biodyne B, NEN, Boston, USA). RNA samples were fixed on the membrane by UV exposure. Type II collagen, aggrecan core and GAPDH probes were generated by reverse transcription-polymerase chain reaction, using the following primers: type II collagen, forward primer: GACCCCATGCAGTACATG, reverse primer: GACCGTCTTGCCCCACTT: aggrecan core protein: f.p.: CCCTGGACTTTGACAGGGC, r.p.: AGGAAACTCGTCCTTGTCTCC; GAPDH: f.p.: TGGTATCGTGGAAGGACTCATGAC, r.p.: ATGCCAGTGAGCTTCCCGTTCAGC. The probes were $^{32}$P-labeled, using a random priming kit (Invitrogen, Inc.). Prehybridization (1 h at 42°C) and hybridization (18 h at 48 to 60°C) were performed in 2/3 $Na_2HPO_4$ pH 7.2–7.4, 1/3 SDS 20% (v/v). Blots were washed several times in 2 × SSC (standard saline citrate) and 0.1% SDS at 42°C. Final washes were in 0.1 × SSC plus 0.1% SDS at 42°C. The $^{32}$P-labeled cDNA-mRNA hybrids were visualized by autoradiography. The relative density of the detected signals was measured by densitometric scans of X-ray film (Kodak, X-Omat, NY, USA) and quantified by the Image Quant program (Molecular Dynamics, Sunnyvale, USA).

# 3. Results

## 3.1. Effect of IL-1 treatment on NF-κB and AP-1 DNA binding activity in normoxia and hypoxia

The treatment of chondrocytes with IL-1$\beta$ for 1 and 24 h induced the binding of NF-κB complexes to the labelled oligonucleotide containing the consensus sequence corresponding to this transcription factor. Two protein complexes were observed corresponding to the subunits p65 and p50 of the heterodimer NF-κB. This binding was greater in low-oxygen environment, especially after 1 h of IL-1$\beta$ incubation.

IL-1$\beta$ was also able to enhance the binding of proteins to the consensus sequence of AP-1. This effect of IL-1$\beta$ was also superior in low-oxygen environment but was higher after 24 h suggesting that the activation of AP-1 requires more time than that of NF-κB (Fig. 1).

## 3.2. Effect of Rhein on IL-1β-induced NF-kB and AP-1 activation

Considering the higher effect of IL-1$\beta$ in hypoxic cultures (3% $O_2$), a situation close to the environment of chondrocytes within the cartilage, we chose this condition to determine the effect of Rhein.

The pre-treatment of chondrocytes for 18 h with $10^{-5}$ M of Rhein, slightly decreased IL-1$\beta$-induced NF-κB binding to the corresponding consensus probe, whereas it significantly reduced that of AP-1 (Fig. 2).

## 3.3. Effect of Rhein on IκB activation and ERK 1/2 activation

As expected, IL-1$\beta$ induced a decrease in the amount of IκB-$\alpha$ protein detected by Western blotting in cytoplasmic extracts of hypoxic cultures.

Rhein alone increased the amount of IκB-$\alpha$ and reduced the IL-1$\beta$ effect when added in combination with the cytokine. This effect suggests that the drug could prevent or delay the degradation of IκB-$\alpha$ in hypoxic chondrocytes cultures (Fig. 3).

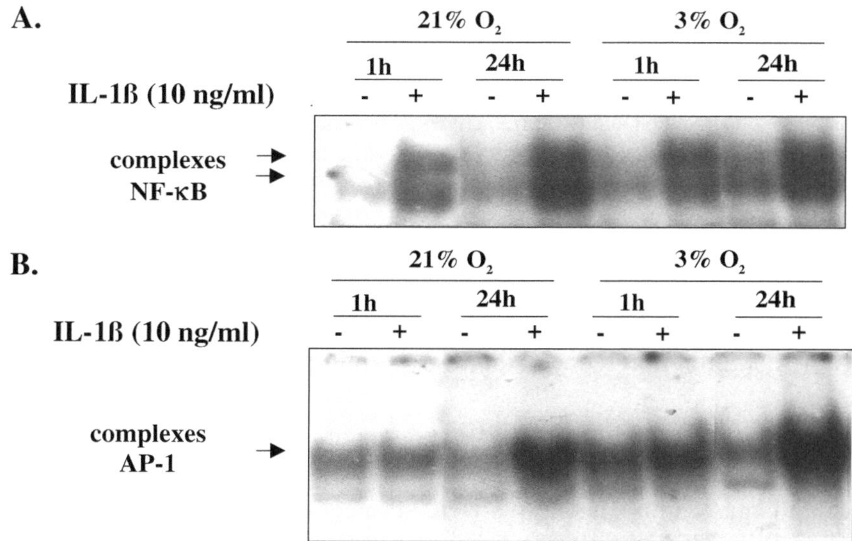


Fig. 1. Effect of IL-1$\beta$ treatment on NF-κB (A) and AP-1 (B) DNA binding activity. Chondrocytes cultures were incubated for 1 h or 24 h in either normoxia (21% $O_2$) or hypoxia (3% $O_2$). Representation of three independent experiments.

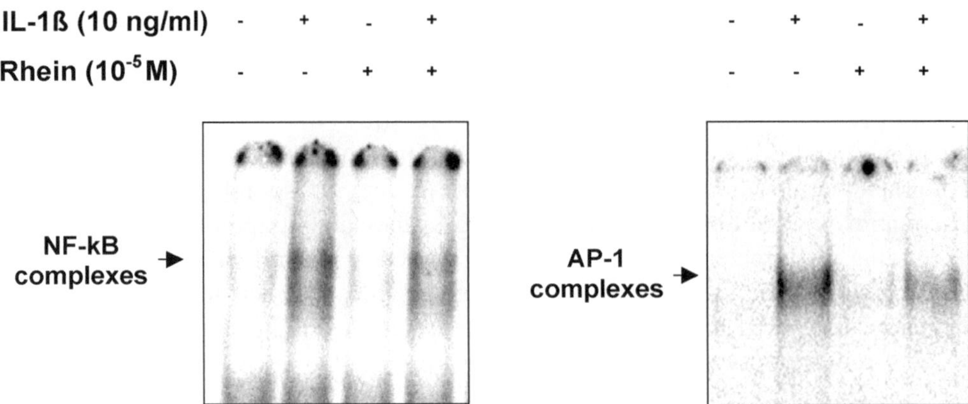


Fig. 2. Effect of Rhein ($10^{-5}$ M) on IL-1-induced NF-$\kappa$B and AP-1 DNA binding activity. Chondrocytes were cultured in hypoxia and incubated with Rhein for 18 h in 1% FCS-containing medium. Then, cells were incubated for further 24 h in the presence of IL-1$\beta$ (10 ng/ml) and the drug. Representation of three independent experiments.

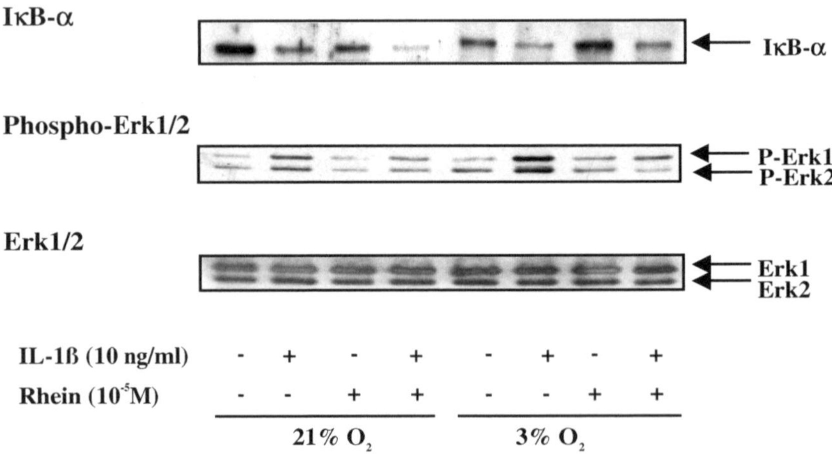


Fig. 3. Effect of Rhein on I$\kappa$B degradation and ERK 1/2 activation. Chondrocytes were cultured in either normoxia (21% $O_2$) or hypoxia (3% $O_2$) and incubated with Rhein $10^{-5}$ M for 18 h. Then, cells were incubated for further 1 h in the presence of IL-1$\beta$ (10 ng/ml) and the drug. Representation of three independent experiments.

Then, we determined the effects of IL-1$\beta$ incubation on the activity of MAP kinases. We found that the treatment of cells with IL-1$\beta$ for 1 h caused an increased phosphorylation of ERK-1 and -2, reflecting the activation of these kinases which is more effective in hypoxia than in normoxia. The incubation of cells with $10^{-5}$ M of Rhein for 18 h reduced the IL-1$\beta$-stimulated MAP kinase activity (Fig. 3).

## 3.4. Effect of Rhein on the type II collagen and aggrecan core protein mRNA steady-state levels

Type II collagen and aggrecan mRNA levels were strongly increased in cells previously treated with $10^{-5}$ M of Rhein for 18 h. This stimulating effect was still visible in chondrocytes further exposed to IL-1$\beta$ for 24 h (Fig. 4), indicating that Rhein is capable of preventing the inhibitory effect of IL-1$\beta$ on type II collagen and aggrecan expression.

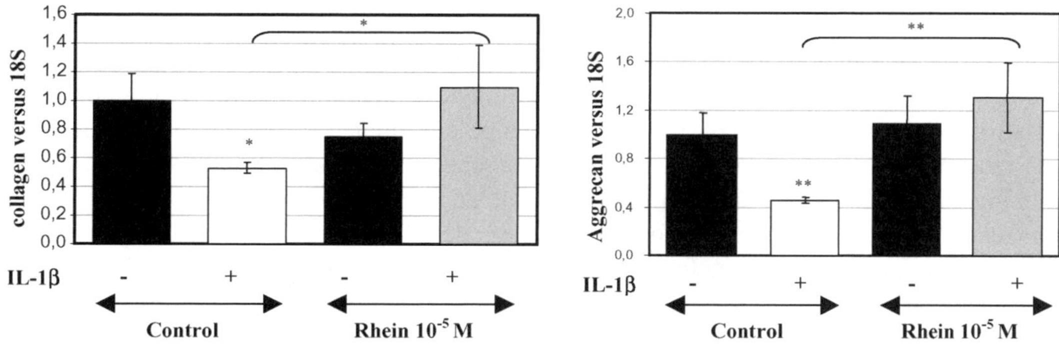

Fig. 4. Effect of Rhein on the type II collagen and aggrecan core protein mRNA levels. Chondrocytes were treated for 18 h with or without $10^{-5}$ M Rhein. Then, they were incubated for 24 h with or without IL-1$\beta$ (50 ng/ml). Total RNA was extracted and levels were determined using 18S as an internal control. Representation of three independent experiments. Significance of the differences between treated and control was estimated by the Student's test ($^{*}p < 0.05$, $^{**}p < 0.01$).

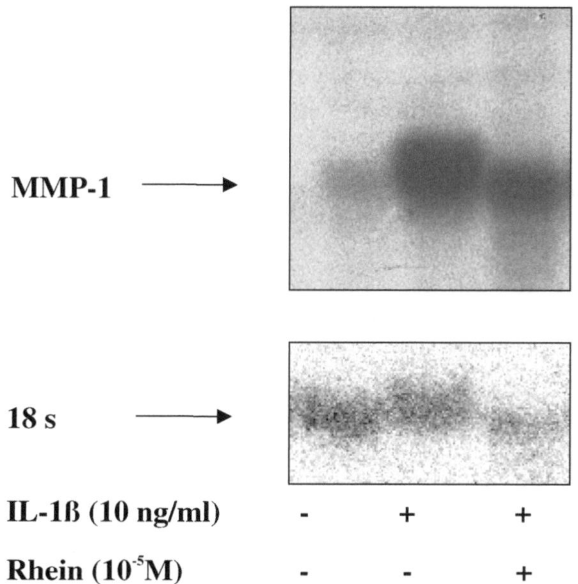


Fig. 5. Effect of Rhein on the collagenase (MMP-1) mRNA levels.

## 3.5. Effect of Rhein on the collagenase (MMP-1) mRNA levels

Since IL-1-induced activation of MAP kinases and AP-1 may up-regulate the expression of metalloproteases, including MMP-1, it was of interest to determine the effect of Rhein. Northern blots performed on RNA extracted from chondrocytes showed that Rhein significantly reduced the strong IL-1$\beta$-induced collagenase (MMP-1) expression (Fig. 5).

## 4. Discussion

In chondrocytes two qualitatively distinct functional programs can be distinguished. The catabolic program induced by pro-inflammatory stimuli and characterized by the secretion of proteases, the suppression of matrix synthesis, and the inhibition of chondrocytes proliferation. The anabolic program is associated with the production of extra cellular matrix, protease inhibitors, and cell replication. IL-1$\beta$ is the prototypic inducer of catabolic responses in chondrocytes and stimulates the expression of proteases including stromelysin and collagenase. IL-1$\beta$ is also a potent inducer of PG and NO synthesis, and this is related to the induction of cyclooxygenase 2 (COX-2) and inducible nitric oxide synthase (iNOS) gene expression via the activation of transcription factors such as NF$\kappa$B and AP-1.

Cellular responses to IL-1$\beta$ stimulation trigger a cascade of protein kinases that transmit signals that ultimately regulate gene expression. In mammalian cells, three subgroups of the MAPKs have been detected: ERK1/2, JNKs, and P38. The mammalian ERK1/2 is activated by multiple stimuli, such as growth factors, neurotrophic factors, neurotransmitters, and cytokines [23] and it has also been shown that the MAPKs are activated by IL-1$\beta$ [18]. Cumulated results suggest that activation of ERK is able to induce AP-1 activity [16].

The first part of the study examined the effect of IL-1$\beta$ treatment on NF-$\kappa$B and AP-1 DNA binding activity in bovine articular chondrocytes cultures in low oxygen conditions (3% $O_2$), which reflect the *in vivo* hypoxic situation within cartilage, in comparison with normoxia (21% $O_2$). Indeed, it has been shown that gene expression and the response of articular chondrocytes to cytokines such as IL-1$\beta$ were highly dependent on the environmental oxygen level of chondrocytes [26]. It has been particularly shown that reduced oxygen tension significantly up-regulated collagen II and aggrecan core protein expression [31]. In this study, we have confirmed that IL-1$\beta$-induced the DNA binding of NF-$\kappa$B and AP-1 and that these effects were significantly higher in hypoxic cultures than in normoxia. Then oxygen level and re-oxygenation stress significantly modulate gene expression and the response of articular chondrocytes to cytokines such as IL-1$\beta$.

In our study, we also found that the IL-1$\beta$ induced extra cellular signal-regulated kinase (ERK) 1/2 phosphorylation, which is in accordance with previous data [9], and that this effect was higher in hypoxic conditions.

Then we used chondrocyte cultures in low oxygen tension to reproduce *in vitro* the life conditions of chondrocytes for further experimentations.

The second part examined the role of Diacerein and its active metabolite Rhein on IL-1$\beta$ signalling pathways, particularly on the activation of NF-$\kappa$B and AP-1 transcription factors, but also on the expression of MMP-1 (collagenase), type II collagen and aggrecan by evaluating the steady state levels of corresponding mRNAs.

The results demonstrated that the activation of MAP kinases is responsible for the subsequent stimulation of AP-1 and NF-$\kappa$B DNA binding leading to MMP induction, iNOS synthesis and down regulation of type II collagen and were in agreement with previous studies which showed the activation of the MAPKs with IL-1$\beta$ treatment in hypoxia [25].

We have shown that Rhein and Diacerein inhibit the activation of ERK kinase signal transduction, NF-$\kappa$B and AP-1 transcription factors induced by IL-1$\beta$ in bovine articular chondrocytes. These data could explain the inhibitory effect exerted by Rhein on the IL-1$\beta$ induced down-regulation of aggrecan expression and on the IL-1$\beta$ induced MMP-1 expression observed in our study.

Matrix metalloproteinases belong to a family of enzymes that degrades the extra cellular matrix components and play an important role in tissue repair, tumour invasion, and metastasis. It has been previously shown that the AP-1 site is required for the basal level expression of the human stromelysin gene [34] and the human collagenase-3 gene [5]. Moreover, it has been shown that drugs inhibited MMP gene expression by blocking signals leading to IKK activation, which in turn inhibits I$\kappa$B$\alpha$ phosphorylation and NF-$\kappa$B activation [35]. NF-kB translocation can also enhance the transcription of the collagenase-3 gene in IL-1-stimulated synoviocytes [27]. Therefore the signal transduction cascade after cytokine stimulation of synoviocytes results in the activation of parallel kinase cascades regulating AP-1 and NF-$\kappa$B [42].

However, another mechanism contributing to the beneficial effect of both drugs could be the inhibition of NO production as it has previously been shown that Diacerein and Rhein significantly decrease NO production [28,37]. These results could also explain the inhibition of iNOS expression previously shown with Rhein [33], by preventing NF-$\kappa$B activation. Then, Diacerhein reduces OA cartilage damage through multiple mechanisms of action in catabolic and anabolic pathways which explains its anti-inflammatory and anti-osteoarthritic properties.

In conclusion, we have shown that Diacerein and its active metabolite Rhein act at an early step of the IL-1 signaling cascade, leading from MAP kinase activation to modulation of the transcription factor NF-$\kappa$B and AP-1 activities and subsequent regulation of transcription of genes, including iNOS, MMPs, type II collagen and aggrecan. By its ability to inhibit the IL-1$\beta$-induced NF-$\kappa$B and AP-1 activation, Rhein could contribute to reduce the deleterious effects exerted by the cytokine on the cartilage matrix. Taken that NF-$\kappa$B- and AP-1- dependent gene expression is also playing a great role in inflammatory diseases, Diacerein and Rhein, could be of some interest in the global treatment of chronic inflammatory diseases and their modifying effects on cartilage, such as OA.

# References


[1] H. Asahara, K. Fujisawa, T. Kobata, T. Hasunuma, T. Maeda, M. Asanuma, N. Ogawa, H. Inoue, T. Sumida and K. Nishioka, Direct evidence of high DNA binding activity of transcription factor AP-1 in rheumatoid arthritis synovium, *Arthritis Rheum.* **40**(5) (1997), 912–918.

[2] M. Boittin, F. Redini, G. Loyau and J.P. Pujol, Effect of Diacerhein (ART 50) on the matrix synthesis and collagenase secretion by cultured joint chondrocytes in rabbits, *Rev. Rhum. Ed. Fr.* **60**(6 Pt 2) (1993), 68S–76S.

[3] K.D. Brandt, G. Smith, S.Y. Kang, S. Myers, B. O'Connor and M. Albrecht, Effects of Diacerhein in an accelerated canine model of osteoarthritis, *Osteoarthritis Cartilage* **5**(6) (1997), 438–449.

[4] C.E. Brinckerhoff, Regulation of metalloproteinase gene expression: implications for osteoarthritis, *Crit. Rev. Eukaryot. Gene Expr.* **2** (1992), 145–164.

[5] G. Buttice, S. Quinones and M. Kurkinen, The AP-1 site is required for basal expression but is not necessary for TPA-response of the human stromelysin gene, *Nucleic Acids Res.* **19**(13) (1991), 3723–3731.

[6] F.M. Cicuttini and T.D. Spector, Osteoarthritis in the aged. Epidemiological issues and optimal management, *Drugs Aging* **6** (1995), 409–420.

[7] W. Conca, P.B. Kaplan and S.M. Krane, Increases in levels of procollagenase messenger RNA in cultured fibroblasts induced by human recombinant interleukin 1 beta or serum follow c-jun expression and are dependent on new protein synthesis, *J. Clin. Invest.* **83**(5) (1989), 1753–1757.

[8] J.A. DiBattista, J.P. Pelletier, M. Zafarullah, N. Fujimoto, K. Obata and J. Martel-Pelletier, Coordinate regulation of matrix metalloproteases and tissue inhibitor of metalloproteinase expression in human synovial fibroblasts, *J. Rheumatol. Suppl.* **43** (1995), 123–128.

[9] Dong Qian, Hai-Yan Lin, Hong-Mei Wang, Xuan Zhang, Dong-Lin Liu, Qing-Lei Li and Cheng Zhu, Normoxic induction of the hypoxic-inducible factor-1$\alpha$ by interleukin-1$\beta$ involves the extracellular signal-regulated kinase 1/2 pathway in normal human cytotrophoblast cells, *Biol. Reprod.* **70**(6) (2004), 1822–1827.

[10] M. Dougados, M. Nguyen, L. Berdah, B. Mazieres, E. Vignon and M. Lequesne, ECHODIAH Investigators Study Group Evaluation of the structure-modifying effects of diacerein in hip osteoarthritis: ECHODIAH, a three-year, placebo-controlled trial. Evaluation of the chondromodulating effect of Diacerein in OA of the hip, *Arthritis Rheum.* **44**(11) (2001), 2539–2547.

[11] N. Felisaz, K. Boumediene, C. Ghayor, J.F. Herrouin, P. Bogdanowicz, P. Galerra and J.P. Pujol, Stimulating effect of diacerein on TGF-beta1 and beta2 expression in articular chondrocytes cultured with and without interleukin-1, *Osteoarthritis Cartilage* **7**(3) (1999), 255–264.

[12] Y. Geng, F.J. Blanco, M. Cornelisson and M. Lotz, Regulation of cyclooxygenase-2 expression in normal human articular chondrocytes, *J. Immunol.* **155** (1995), 796–801.

[13] M.B. Goldring, The role of cytokines as inflammatory mediators in osteoarthritis: lessons from animal models, *Connect. Tissue. Res.* **40** (1999), 1–11.

[14] D. Hamerman, Aging and osteoarthritis: basic mechanisms, *J. Am. Geriatr. Soc.* **41** (1993), 760–770.

[15] L. Haywood, D.F. McWilliams, C.I. Pearson et al., Inflammation and angiogenesis in osteoarthritis, *Arthritis Rheum.* **48** (2003), 2173–2177.

[16] C. Huang, W.Y. Ma, M.R. Young, N. Colburn and Z. Dong, Shortage of mitogen-activated protein kinase is responsible for resistance to AP-1 transactivation and transformation in mouse JB6 cells, *Proc. Natl. Acad. Sci. USA* **95**(1) (1998), 156–161.

[17] B.L. Kidd and L.A. Urban, Mechanisms of inflammatory pain, *Br. J. Anaesth.* **87** (2001), 3–11.

[18] J.M. Kyriakis and J. Avruch, Protein kinase cascades activated by stress and inflammatory cytokines, *Bioessays* **18** (1996), 567–577.

[19] T. Lehmann, L.Q. Nguyen and M.L. Handel, Synovial fluid induced nuclear factor-kappaB DNA binding in a monocytic cell line, *J. Rheumatol.* **27**(12) (2000), 2769–2776.

[20] M. Lotz, F.J. Blanco, J. von Kempis, J. Dudler, R. Maier, P.M. Villiger and Y. Geng, Cytokine regulation of chondrocyte functions, *J. Rheumatol. Suppl.* **43** (1995), 104–108.

[21] G. Loyau and J.P. Pujol, The role of cytokines in the development of osteoarthritis, *Scand. J. Rheumatol. Suppl.* **81** (1990), 8–12.

[22] W.J. Lukiw and N.G. Bazan, Strong nuclear factor-kappaB-DNA binding parallels cyclooxygenase-2 gene transcription in aging and in sporadic Alzheimer's disease superior temporal lobe neocortex, *J. Neurosci. Res.* **53**(5) (1998), 583–592.

[23] C.J. Marshall, Specificity of receptor tyrosine kinase signaling: transient versus sustained extracellular signal-regulated kinase activation, *Cell* **80** (1995), 179–185.

[24] J. Martel-Pelletier, F. Mineau, F.C. Jolicoeur, J.M. Cloutier and J.P. Pelletier, In vitro effects of Diacerhein and Rhein on interleukin 1 and tumor necrosis factor-alpha systems in human osteoarthritic synovium and chondrocytes, *J. Rheumatol.* **25**(4) (1998), 753–762.

[25] G. Martin, R. Andriamanalijaona, S. Grassel, R. Dreier, M. Mathy-Hartert, P. Bogdanowicz, K. Boumediene, Y. Henrotin, P. Bruckner and J.P. Pujol, Effect of hypoxia and reoxygenation on gene expression and response to interleukin-1 in cultured articular chondrocytes, *Arthritis Rheum.* **50**(11) (2004), 3549–3560.

[26] G. Martin, R. Andriamanalijaona, S. Grassel, R. Dreier, M. Mathy-Hartert, P. Bogdanowicz, K. Boumediene, Y. Henrotin, P. Bruckner and J.P. Pujol, Effect of hypoxia and reoxygenation on gene expression and response to interleukin-1 in cultured articular chondrocytes, *Arthritis Rheum.* **50**(11) (2004), 3549–3560.

[27] J.A. Mengshol, M.P. Vincenti, C.I. Coon, A. Barchowsky and C.E. Brinckerhoff, Interleukin-1 induction of collagenase 3 (matrix metalloproteinase 13) gene expression in chondrocytes requires p38, c-Jun N-terminal kinase, and nuclear factor kappaB: differential regulation of collagenase 1 and collagenase 3, *Arthritis Rheum.* **43** (2000), 801–811.

[28] C.J. Menkes, D. Borderie, A. Hernvann and O. Ekindjian, Inhibition by Rhein of nitrosothiol production of human chondrocytes from osteoarthritis in culture, *Bull. Acad. Natl. Med.* **183**(4) (1999), 785–795; discussion 795–796.

[29] J. Middleton, A. Manthey and J. Tyler, Insulin-like growth factor (IGF) receptor, IGF-I, interleukin-1 beta (IL-1 beta), and IL-6 mRNA expression in osteoarthritic and normal human cartilage, *J. Histochem. Cytochem.* **44** (1996), 133–141.

[30] A.R. Moore, K.J. Greenslade, C.A. Alam and D.A. Willoughby, Effects of Diacerhein on granuloma induced cartilage breakdown in the mouse, *Osteoarthritis Cartilage* **6**(1) (1998), 19–23.

[31] C.L. Murphy and A. Sambanis, Effect of oxygen tension and alginate encapsulation on restoration of the differentiated phenotype of passaged chondrocytes, *Tissue Eng.* **7**(6) (2001), 791–803.

[32] R.M. Palmer, M.S. Hickery, I.G. Charles, S. Moncada and M.T. Bayliss, Induction of nitric oxide synthase in human chondrocytes, *Biochem. Biophys. Res. Commun.* **193**(1) (1993), 398–405.

[33] J.P. Pelletier, F. Mineau, J.C. Fernandes, N. Duval and J. Martel-Pelletier, Diacerhein and Rhein reduce the interleukin 1beta stimulated inducible nitric oxide synthesis level and activity while stimulating cyclooxygenase-2 synthesis in human osteoarthritic chondrocytes, *J. Rheumatol.* **25**(12) (1998), 2417–2424.

[34] A.M. Pendas, M. Balbin, E. Llano, M.G. Jimenez and C. Lopez-Otin, Structural analysis and promoter characterization of the human collagenase-3 gene (MMP13), *Genomics* **40**(2) (1997), 222–233.

[35] S. Philip, A. Bulbule and G.C. Kundu, Matrix metalloproteinase-2: mechanism and regulation of NF-kappaB-mediated activation and its role in cell motility and ECM-invasion, *Glycoconj. J.* **21**(8–9) (2004), 429–441.

[36] P. Reboul, J.P. Pelletier, G. Tardif, J.M. Cloutier and J. Martel-Pelletier, The new collagenase, collagenase-3, is expressed and synthesized by human chondrocytes but not by synoviocytes: a role in osteoarthritis, *J. Clin. Invest.* **97** (1996), 2011–2019.

[37] C. Sanchez, M. Mathy-Hartert, M.A. Deberg, H. Ficheux, J.Y. Reginster and Y.E. Henrotin, Effects of Rhein on human articular chondrocytes in alginate beads, *Biochem. Pharmacol.* **65**(3) (2003), 377–388.

[38] J.D. Sandy, Proteolytic degradation of normal and osteoarthritic cartilage matrix, in: K.D. Brandt, M. Doherty, L.S. Lohmander, eds, *Osteoarthritis*, Oxford University Press, New York, 2003, pp. 82–91.

[39] G.N. Smith, Jr., S.L. Myers, K.D. Brandt, E.A. Mickler and M.E. Albrecht, Diacerhein treatment reduces the severity of osteoarthritis in the canine cruciate-deficiency model of osteoarthritis, *Arthritis Rheum.* **42**(3) (1999), 545–554.

[40] T.D. Spector, D.J. Hart, D. Nandra et al., Low-level increases in serum C-reactive protein are present in early osteoarthritis of the knee and predict progressive disease, *Arthritis Rheum.* **40** (1997), 723–727.

[41] J. Stadler, M. Stefanovic-Racic, T.R. Billiar, R.D. Curran, L.A. McIntyre, H.I. Georgescu, R.L. Simmons and C.H. Evans, Articular chondrocytes synthesize nitric oxide in response to cytokines and lipopolysaccharide, *J. Immunol.* **147** (1991), 3915–3920.

[42] P.P. Tak and S.G. Firestein, NF-$\kappa$B: a key role in inflammatory diseases, *J. Clin Invest.* **107**(1) (2001), 7–11.

[43] C.I. Westacott, J.T. Whicher, I.C. Barnes, D. Thompson, A.J. Swan and P.A. Dieppe, Synovial fluid concentration of five different cytokines in rheumatic diseases, *Ann. Rheum. Dis.* **49**(9) (1990), 676–681.

[44] M. Yaron, I. Shirazi and I. Yaron, Anti-interleukin-1 effects of diacerein and Rhein in human osteoarthritic synovial tissue and cartilage cultures, *Osteoarthritis Cartilage* **7**(3) (1999), 272–280.

Biorheology 43 (2006) 589–594
IOS Press



# Studies in animal models of osteoarthritis as predictors of a structure-modifying effect of diacerhein in humans with osteoarthritis

Kenneth D. Brandt

*Indiana University School of Medicine, Division of Rheumatology, 1110 West Michigan Street,
Room LO 545, Indianapolis, IN 46202, USA
Tel.: +1 317 274 0672; Fax: +1 317 278 8882; E-mail: kbrandt@iupui.edu*

In addressing the use of animal models of osteoarthritis (OA) for predicting a structure-modifying effect of the anthraquinone derivative, diacerhein (DAR) (Fig. 1), in humans with OA, it is worth considering at the outset the context for the use of animal models, in general, as predictors of drug effects in humans with OA. In this vein, the excellent chapter by Doherty et al. [6] on the limitations of animal models in OA research provides excellent background.

Paradigms of drug development based on animal models rely chiefly on demonstration of therapeutic activity in an animal model of the human disease to justify moving on to evaluation of the compound in man. In contrast, mechanism-based paradigms involve identification of a specific target (e.g., cytokine, enzyme, receptor) that is presumed to be important in the human disease and the selection of compounds that have appropriate activities against the target *in vitro* in assays utilizing appropriate human reagents, ensuring that the compound possesses the necessary pharmacodynamic properties. Promising compounds may then be evaluated in humans without a requirement for demonstration of therapeutic efficacy in an animal model of the disease.

Drug development schemes that rely on animal models can be confounded by false positive and false negative results, partly because species differences with respect to the relative contribution of various mediators, receptors or enzymes to the pathology are common. For example, antihistamines are efficacious in treating asthma in guinea pigs but not in humans; prostaglandin inhibitors (e.g., indomethacin)

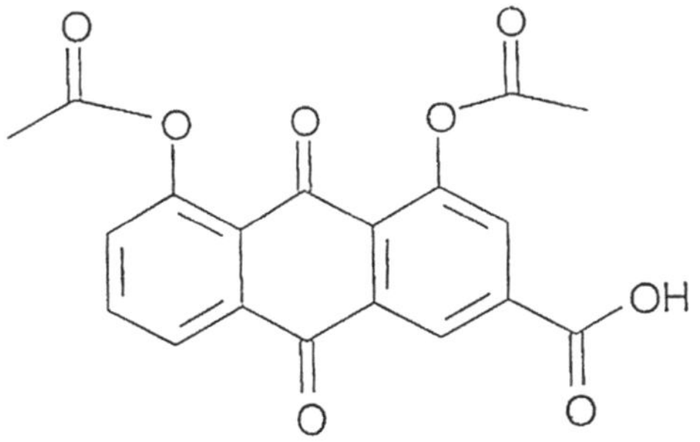

Fig. 1. Structure of DAR.

are effective in preventing joint damage in animal models of inflammatory arthritis, but not in humans.

Even though results in animal models of OA may be misleading, models play an important role in the evaluation of disease-modifying osteoarthritis drugs (DMOADs). Elucidation of the mechanisms underlying development of joint pathology in an animal model can suggest the *possibility* that a particular enzyme, cytokine or receptor plays a role in human OA. Evidence that pharmacologic disruption of a specific disease mechanism has therapeutic benefit in an animal model provides powerful support for proceeding to a randomized clinical trial in humans.

*In vitro* evidence that DAR stimulated production of collagen in chondrocyte cultures, increased hyaluronic acid synthesis by human synovial fibroblasts, reduced the effect of interleukin 1-$\beta$ (IL1-$\beta$) on glycosaminoglycan synthesis, and inhibited production of IL-1 by mouse peritoneal microphages and human OA synovial membrane led to a study that demonstrated the efficacy of DAR in New Zealand rabbits in which OA was produced by closed contusion of the patella caused by the impact of a 1-kg weight dropped from a height of 1 m [10]. The score of morphologic severity of OA, based on gross and microscopic changes, was significantly lower in the control group than in the contusion group ($p < 0.03$) and in the DAR group than in untreated contused rabbits ($p < 0.05$). Average scores for gross pathology were, in fact, lower in the DAR-treated animals than in the controls [10].

Similarly, Bendele et al. [1] showed a beneficial effect of DAR on the development of OA in guinea pigs. In that study, the development of joint pathology in Hartley guinea pigs, a strain that is genetically predisposed to spontaneous OA, was accelerated by surgical transection of the quadriceps tendon and removal of the right patella of 2-month old male animals, which led to increased loading of the contralateral (left) hind limb. Guinea pigs treated with DAR exhibited a decrease of approximately 50% in the severity of structural changes of OA, in comparison with untreated animals.

We have established that transection of the anterior cruciate ligament in the dog is not merely a model of cartilage injury and repair but a model of progressive OA that, over a period of about 4 years, leads to full-thickness ulceration of the articular cartilage in the unstable joint [2]. Furthermore, not only the cartilage but other tissues in the joint (e.g., synovium, subchondral bone) are involved [4,12].

Placebo-controlled trials in the canine cruciate-deficiency model of OA have also shown DAR to be efficacious [15]. Although some 4 years are needed for development of full-thickness cartilage loss in the OA knee of the neurologically intact cruciate-deficient dog, if the sensory input from the ipsilateral hind limb is interrupted, e.g., by articular nerve transection or L4–S1 dorsal root ganglionectomy [13,18], the process can be greatly accelerated and features of end-stage OA becomes apparent as early as 6–8 weeks after transection of the cruciate ligament.

Notably, whereas DAR was effective in the neurologically intact cruciate-deficient dog Fig. 2, it was ineffective in dogs which sensory input from the ipsilateral hind limb had been interrupted prior to transection of the cruciate ligament, resulting in much more severe and rapidly progressive joint damage [5].

In an ovine bilateral meniscectomy model of OA, treatment with DAR reduced the thickening of the subchondral bone but did not affect osteophyte formation [9]. The scores for severity of the OA pathology in DAR-treated and untreated meniscectomized sheep were similar. In DAR-treated animals, the thickness of articular cartilage and of subchondral bone in the lesion zone of the lateral tibial plateau was significantly lower than that in the analogous area of untreated meniscectomized animals at 3 months; however, at 9 months the subchondral bone at this site in the DAR-treated animals was not different from that in non-operated controls. The density of both subchondral and trabecular bone in the lesion

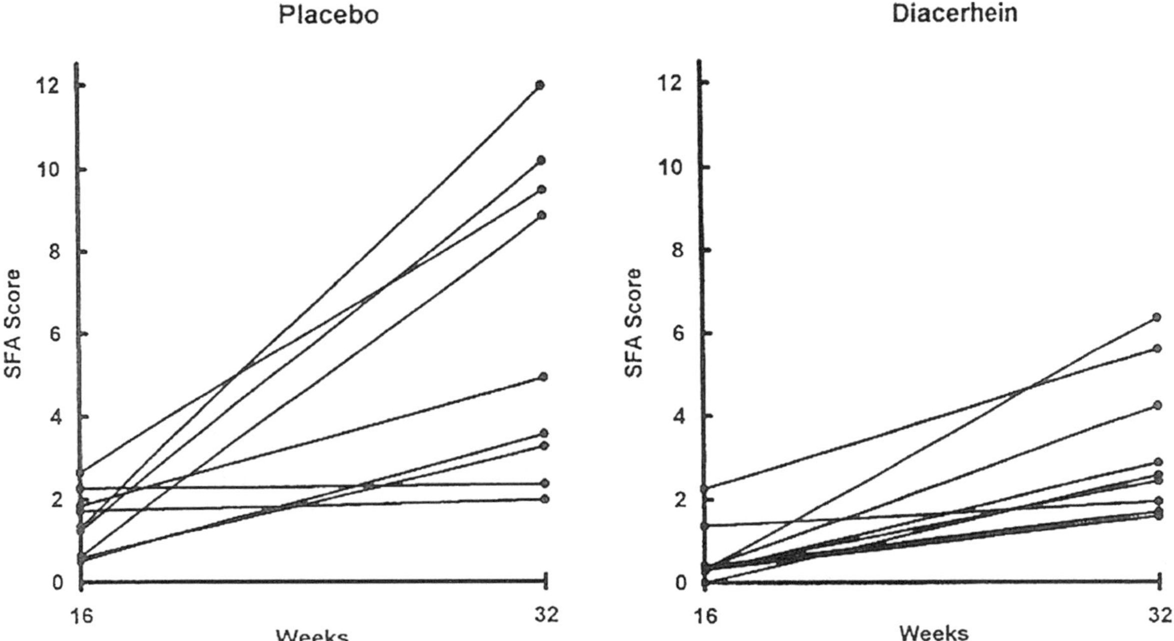


Fig. 2. Progression of osteoarthritis from week 16 to week 32 after anterior cruciate ligament transaction. Each dog was examined arthroscopically at 16 weeks, at which time an SFA (Société Française d'Arthroscopie) score was assigned. At 32 weeks, a score was similarly determined for the gross pathologic changes visible on direct examination. In every dog, the score at 32 weeks was higher than that at 16 weeks. The increase between week 16 and week 32 was less marked in the DAR-treated dogs than in placebo-treated dogs. From [15].

zone increased significantly in the DAR treatment group, relative to that in the non-operated controls. In contrast, in the outer zone of the lateral tibial plateau and lateral femoral condyle (i.e., sites of osteophyte formation) the thickness of subchondral bone in the meniscectomized group, in both DAR-treated and untreated sheep, increased, indicating that DAR does not interfere with endochondral ossification. This is consistent with our observation that treatment with bisphosphonates, which reduce turnover of subchondral bone, had no effect on osteophyte formation [11]. The results are not surprising: in subchondral bone, formation is linked to resorption. Osteophyte formation, in contrast, depends on endochondral ossification.

Other studies in the canine cruciate-deficiency model showed that addition of DAR or of its active metabolite, rhein, to explant cultures of OA cartilage reduced the level of fragmentation of chondrocyte DNA and cell death, and levels of caspace-3 in the tissue [14] (Fig. 3). The authors speculated that the decrease in caspace-3 was related to a reduction in the level of inducible nitric oxide (NO) synthase and, consequently, of NO. Work in rats with adjuvant-induced arthritis showed that administration of DAR, starting at the time of adjuvant injection, significantly suppressed both the development of joint pathology and the associated increase in plasma NO concentration [17]. *In vitro*, rhein was shown to significantly inhibit NO production by chondrocytes that were stimulated with interleukin-1$\alpha$. Taken together, these data suggest that the inhibitory effect of DAR in adjuvant-induced arthritis is related, at least in part, to reduction of NO synthesis.

In a placebo-controlled 3-year clinical trial of DAR in humans with hip OA (the ECHODIAH study), the rate of joint space narrowing (a surrogate for loss of articular cartilage) among completers in the active treatment group was significantly lower than that in the placebo group and a significantly

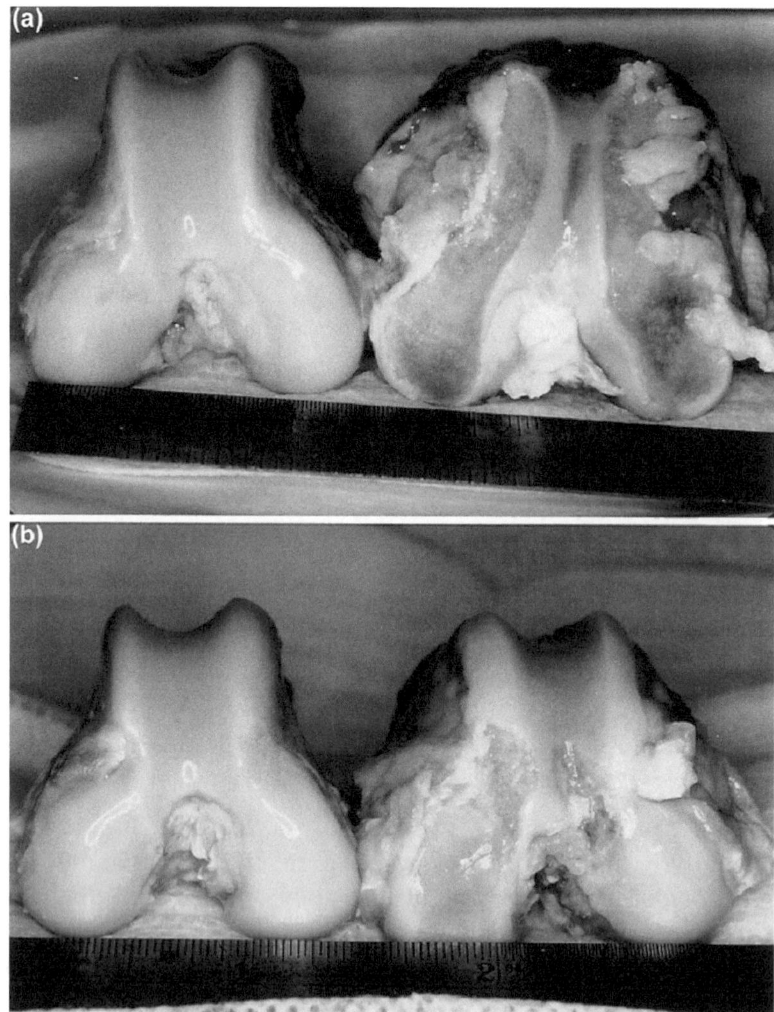

Fig. 3. (a) Specimens from a dog in the untreated control group. Note the full-thickness ulceration of articular cartilage involving virtually all of the medial and lateral trochlear ridge and both femoral condyles of the OA knee. The contralateral knee is grossly normal. (b) Specimens from a dog in the DAR-treatment group, showing moderate ulceration of the medial trochlear ridge and medial femoral condyle of the OA knee. The contralateral knee is grossly normal. Although cartilage ulceration tended to be less severe in the DAR group than in the untreated controls, the difference was not statistically significant. From [5].

higher proportion of patients in the DAR treatment arm showed no radiologic progression (joint space loss < 0.5 mm) [7]. However, a high proportion of all subjects who were randomized to treatment had rapidly progressive hip OA and about 25% underwent total hip arthroplasty during the 3-year trial. DAR appeared to have a DMOAD effect only in subjects who had more slowly progressive hip disease, and not in the rapid progressors. This mirrors the results of DAR treatment in the two canine cruciate-deficiency models of OA described above [5,15]. In retrospect, the DMOAD effect of DAR in humans could have been anticipated on the basis of the positive results obtained in the dog and guinea pig models of OA cited above. Notably, both of these models have also demonstrated the efficacy of doxycycline as a DMOAD [8,16], and those observations were recently confirmed in a randomized placebo-controlled trial of doxycycline in humans with knee OA [3].

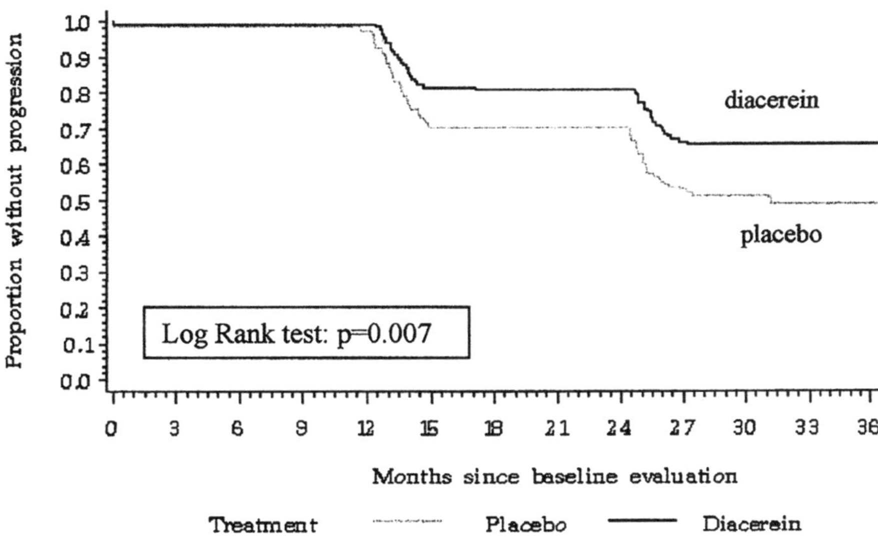


Fig. 4. Proportion of patients in the completer population without radiologic progression (i.e., a change in minimal joint space width ⩾0.5 mm) during the study. The 2 time-to-event curves (obtained according to the method of Kaplan and Meier) were compared using the log rank test. From [7].

## Acknowledgement

Kathie Lane provided excellent clerical support during the preparation of this manuscript.

## References


[1]  A.-M. Bendele, R.A. Bendele, J.F. Hulman and B.P. Swann, Effets bénéfiques d'un traitement par la diacerhéine chez des cobayes atteints d'arthrose, *Rev. Praticien (Paris)* **46** (1996), S35–S39.

[2]  K.D. Brandt, E.M. Braunstein, D.M. Visco, B. O'Connor, D. Heck, B. Katz and M. Albrecht, Anterior (cranial) cruciate ligament transection in the dog: A bona fide model of osteoarthritis, not merely of cartilage injury and repair, *J. Rheumatol.* **18** (1991), 436–446.

[3]  K.D. Brandt, S.A. Mazzuca, B.P. Katz, K.A. Lane, K.A. Buckwalter, D.E. Yocum, F. Wolfe, T.J. Schnitzer, L.W. Moreland, S. Manzi, J.D. Bradley, L.J. Sharma, C.V. Oddis, S.T. Hugenberg and L.W. Heck, Effect of doxycycline on progression of osteoarthritis, *Arthritis Rheum.* **52** (2005), 2015–2025.

[4]  K.D. Brandt, S.L. Myers, D. Burr and M. Albrecht, Osteoarthritic changes in canine articular cartilage, subchondral bone, and synovium fifty-four months after transection of the anterior cruciate ligament, *Arthritis Rheum.* **34** (1991), 1560–1570.

[5]  K.D. Brandt, G. Smith, S.Y. Kang, S. Myers, B. O'Connor and M. Albrecht, Effects of diacerhein in an accelerated canine model of osteoarthritis, *Osteoarthritis Cartilage* **5** (1997), 438–449.

[6]  N.S. Doherty, R.J. Griffiths and E.R. Pettipher, The role of animal models in the discovery of novel disease-modifying osteoarthritis drugs (DMOADs), in: *Osteoarthritis*, K.D. Brandt, M. Doherty and L.S. Lohmander, eds, 1st edn, Oxford University Press, Oxford, UK, 1998, pp. 439–449.

[7]  M. Dougados, M. Nguyen, L. Berdah, B. Maziéres, E. Vignon and M. Lequesne, For the ECHODIAH Investigators Study Group. Evaluation of the structure-modifying effects of diacerein in hip osteoarthritis. ECHODIAH, a three-year, placebo-controlled trial, *Arthritis Rheum.* **44** (2001), 2539–2547.

[8]  R.A. Greenwald, Treatment of destructive arthritis disorders with MMP inhibitors, *Ann. NY Acad. Sci.* **732** (1994), 181–198.

[9]  S.-Y. Hwa, D. Burkhardt, C. Little and P. Ghosh, The effects of orally administered diacerein on cartilage and subchondral bone in an ovine model of osteoarthritis, *J. Rheumatol.* **28** (2001), 825–834.

[10]  B. Maziéres, M. Berdail, M. Trechart and G. Viggier, Etude de la diacetylrheine sur un modele post-contusif d'arthrose expérimentale chez le lapin, *Rev. Rhum* [Ed Fr] 1993 **60** (1993), 77S–81S.

[11]  S.L. Myers, K.D. Brandt, D.B. Burr, B.L. O'Connor and M. Albrecht, Effects of a bisphosphonate on bone histomorphometry and dynamics in the canine cruciate-deficiency model of osteoarthritis, *J. Rheumatol.* **26** (1999), 2645–2653.

[12] S.L. Myers, K.D. Brandt, B.L. O'Connor, D.M. Visco and M.E. Albrecht, Synovitis and osteoarthritic changes in canine articular cartilage after cruciate ligament transection. Effect of surgical hemostasis, *Arthritis Rheum.* **33** (1990), 1406–1415.

[13] B.L. O'Connor, D.M. Visco, K.D. Brandt, S.L. Myers and L. Kalasinski, Neurogenic acceleration of osteoarthritis: The effects of prior articular nerve neurectomy on the development of osteoarthritis after anterior cruciate ligament transection in the dog, *J. Bone Joint Surg.* **74A** (1992), 367–376.

[14] J.-P. Pelletier, F. Mineau, C. Boileau and J. Martel-Pelletier, Diacerein reduces the level of cartilage chondrocyte DNA fragmentation and death in experimental dog osteoarthritic cartilage at the same time that it inhibits caspase-3 and inducible nitric oxide synthase, *Clin. Exp. Rheum.* **21** (2003), 171–177.

[15] G.N. Smith, Jr, S.L. Myers, K.D. Brandt, E.A. Mickler and M.E. Albrecht, Diacerhein treatment reduces the severity of osteoarthritis in the canine cruciate-deficiency model of osteoarthritis, *Arthritis Rheum.* **42** (1999), 545–554.

[16] G.N. Smith, Jr, S.L. Myers, K.D. Brandt, E.A. Mickler and M.E. Albrecht, Diacerhein treatment reduces the severity of osteoarthritis in the canine cruciate-deficiency model of osteoarthritis, *Arthritis Rheum.* **42** (1999), 545–554.

[17] T. Tamura and K. Ohmori, Diacerein suppresses the increase in plasma nitric oxide in rat adjuvant-induced arthritis, *Eur. J. Pharmacol.* **419** (2001), 269–274.

[18] J.A. Vilensky, B.L. O'Connor, K.D. Brandt, E.A. Campbell and P.I. Rogers, Serial kinematic analysis of the canine knee after L4-S1 dorsal root ganglionectomy: implications for the cruciate-deficiency model of osteoarthritis, *J. Rheumatol.* **21** (1994), 2113–2117.

Biorheology 43 (2006) 595–601
IOS Press



# IL-1$\beta$ synthesis by chondrocyte analyzed by 3D microscopy and flow cytometry: Effect of Rhein

N.G. de Isla [a,1], J.W. Yang [a,c,1], C. Huselstein [a], S. Muller [a] and J.F. Stoltz [a,b,*]

[a] *Mécanique et Ingénierie Cellulaire et Tissulaire, UMR CNRS 7563, Lemta, Faculté de Médecine, UHP Nancy 1, Vandoeuvre-lès-Nancy, France*
[b] *Unité Therapie Cellulaire et Tissues, CHU, 54500, Vandoeuvre-lès-Nancy, France*
[c] *Department of Pathophysiology, Medical College of Wuhan University, Wuhan, 430071, China*

**Abstract.** Several factors are known to be involved in the destruction of the articular cartilage. Interleukin-1 (IL-1) plays an important role in the pathogenesis of osteoarthritis (OA) either directly or through the stimulation of catabolic factors. The action of IL-1 on articular cartilage is multifaceted and it most likely plays an important role in the mechanism of cartilage destruction. IL-1 suppresses the synthesis of the cartilage matrix components and promotes the degradation of cartilage matrix macromolecules. Diacerein is an anthraquinone molecule that has been shown to reduce the severity of OA, both in man and in animal models. The present study was designed to evaluate *in vitro* effects of diacerein on IL-1$\beta$ expression in LPS or IL-1$\alpha$ stimulated chondrocytes. Intracellular IL-1$\beta$ production was analysed in articular chondrocytes cultured in monolayer or in alginate 3D-biosystems in the presence of lipopolysaccharide (LPS) or IL-1$\alpha$, with or without diacerein. The results show that LPS and IL-1$\alpha$ increase intracellular IL-1$\beta$ and Diacerein inhibited LPS-induced and IL-1$\alpha$ induced IL-1$\beta$ production by articular chondrocytes. Moreover, the effect of mechanical stimulation was analysed. An inhibitory effect of DAR at therapeutic concentrations on IL-1$\beta$ production in articular chondrocytes is suggested.

Keywords: Articular chondrocyte, interleukin-1$\beta$, diacerein, mechanical stress, osteoarthritis

## 1. Introduction

Several factors are known to be involved in the destruction of the articular cartilage. Interleukin-1 (IL-1) plays an important role in the pathogenesis of osteoarthritis (OA) by stimulating inducible NO synthase (iNOS), cyclo-oxygenase II (COX-II) and proteases [1,2]. IL-1 also suppresses the synthesis of the cartilage matrix components, mainly collagen type II [3] and aggrecan [4].

Drugs which interfere with factors known to initiate and/or contribute to the breakdown of the articular cartilage may provide therapeutic benefit in the treatment of OA. Diacerein is a drug belonging to the anthraquinone chemical class and employed in OA treatment. This drug is a low molecular weight heterocyclic compound designated as 4,5-bis(acetyloxy)-9,10-dioxo-2 anthracene) carboxylic acid. Its mechanism of action appears to be different from that described for a classical nonsteroidal

---

[1]Both authors contributed equally to this work.

*Address for correspondence: Professor Jean-François Stoltz, Mécanique et Ingénierie Cellulaire et Tissulaire, UMR CNRS 7563, Lemta and IFR 111, Faculté de Médecine, UHP Nancy 1, BP 184, Vandoeuvre-lès-Nancy, France. Tel.: +33 383 683457; Fax: +33 383 683459; E-mail: jf.stoltz@chu-nancy.fr.

anti-inflammatory drug, in which diacerein and its active metabolite rhein inhibit IL-1 production and activity [5]. In animal models, diacerein has demonstrated protective effects on cartilage matrix degradation [6]. In clinical trials, its oral administration was associated with symptomatic improvement in the majority of patients with OA [7,8].

During normal activity, articular cartilage is subject to dynamic loading applied perpendicular to the articular surface [9]. Compression of cartilage causes deformation of cells and of extracellular matrix (ECM), gradients in hydrostatic pressure and intratissue fluid flow. These mechanical changes can alter chondrocyte behaviour and ECM homeostasis.

The present study was designed to evaluate *in vitro* effects of diacerein on IL-1$\beta$ production by LPS and IL-1$\alpha$ stimulated articular chondrocytes. Cells in monolayer and 3D culture were analysed. Moreover, the effects of mechanical stimulation was studied.

## 2. Materials and methods

### 2.1. Cells and cell culture

Articular chondrocytes were isolated enzymatically from femoral heads of male Wistar rats (10 weeks old) as described previously [10]. Cells were cultured in complete medium, DMEM F-12 supplemented with glutamine (2 mM), fungizone (2.5 $\mu$g/ml), penicillin (100 U/ml), streptomycin (100 $\mu$g/ml) and fetal bovine serum (10%), all purchased from Life Technologies. After a subculture, cells were cultured in monolayer or in alginate beads (3D).

#### 2.1.1. Monolayer culture
Cells were cultured in LabTek chamber slides (Nunc, USA) until confluence. Then, cells were incubated with LPS (Sigma, France) at 2 $\mu$g/ml and/or DAR (Negma-Lerads, France) at 20 $\mu$g/ml during 48 h.

#### 2.1.2. 3D-culture
Cells were suspended at $3 \times 10^6$ cells/ml in a sodium alginate solution (medium viscosity, Sigma, 20 g/l in 0.9% NaCl). The cell suspension was slowly passed through a 18 gauges needle in a dropwise fashion into a 100 mM $CaCl_2$. After instantaneous gelation, the beads were allowed to polymerize further for 10 minutes in this solution. Thereafter, they were washed two times with a saline solution and cultured for 21 days in a humidified atmosphere of 5% $CO_2$ at 37°C. After this period, beads were incubated with recombinant rat IL-1$\alpha$ (R&D, USA) at 2 ng/ml and/or DAR (20 $\mu$g/ml) during 48 h. For each condition, half of the beads were subject to mechanical stimulation. When appropriate, beads were treated with Sodium citrate 55 (mM) for 15 minutes to obtain cells.

### 2.2. Mechanical stimulation

An agitator 10°/tridimensional Polymax 1040 (STR 9 STUART, Est LAB, France) was used to stimulate the cells embedded in alginate beads [11]. 10 beads were deposed in sterile 15 ml conical tubes containing 10 ml of medium. Tubes were aligned parallel to the agitation plate. The agitator was set at 30 cycles per minute at 37°C, during 48 hours.

## 2.3. Fluorescence staining

Cells were fixed (PAF 1%, 10 minutes), permeabilized (Triton 0.1%, 3 minutes), labelled with first antibody (IgG1 monoclonal, mouse anti-rat IL-1β, R&D) for 45 min and then labelled with secondary antibody (anti-mouse IgG coupled to Alexa 488, Molecular Probes) for 45 minutes.

## 2.4. Flow cytometry

Determinations were performed using an EPICS XL-MCL: Coulter cytomètre. Samples were gated on the basis of their Forward Scatter and Side Scatter signals and the "Analysis Region" was set before each experience with control samples. Fluorescence profiles (F) were collected only on those cells appearing in the analyse region. For each sample 10,000 events were collected at 530 nm in logarithmic mode. Nonspecific binding of antibodies and autofluorescence was controlled by labelling cells with mouse isotype control and conjugate goat anti-mouse antibody. Photomultiplier tube (PMT) voltage for the fluorescence detector was adjusted prior to running samples to obtain a negative fluorescence signal for control samples (negative and isotype). This source power value was fixed for all measurements. Fluorescence intensity of each sample is automatically acquired by the device.

## 2.5. Confocal fluorescence microscopy

Samples were observed in a confocal microscope LEICA SP2.

## 2.6. Statistical analysis

Results are expressed as mean ± standard deviation (SD) of at least 3 experiments. One-way analysis of variance followed by Student's comparison test was used for statistical analyses. A $p$ value of $<0.05$ was considered as significant.

# 3. Results

## 3.1. Intracellular IL-1β production by LPS stimulated chondrocytes cultured in monolayer

Intracellular IL-1β distribution and production was analysed in monolayer rat articular chondrocytes stimulated with LPS for 48 h. In the absence of LPS, a weak intracellular IL-1β expression was detected but there was a significant increase when LPS was added to the culture medium (Figs 1 and 2). At the same time, DAR decreases intracellular IL-1β production by LPS-stimulated chondrocytes cultured in monolayer.

## 3.2. Intracellular IL-1β production by IL-1α stimulated chondrocytes cultured in alginate beads

In monolayers, chondrocytes undergo a gradual differentiation characterized by a change in their spherical shape into a fibroblastic appearance. At the same time they replace the synthesis of cartilage-specific collagens type II, IX and XI with molecules that are normally expressed by fibroblasts or pre-chondrocytes like collagen type I, II and V [12]. When chondrocytes are culture in alginate bead, they maintain a rounded shape, underwent active cell division and retain a stable chondrogenic phenotype

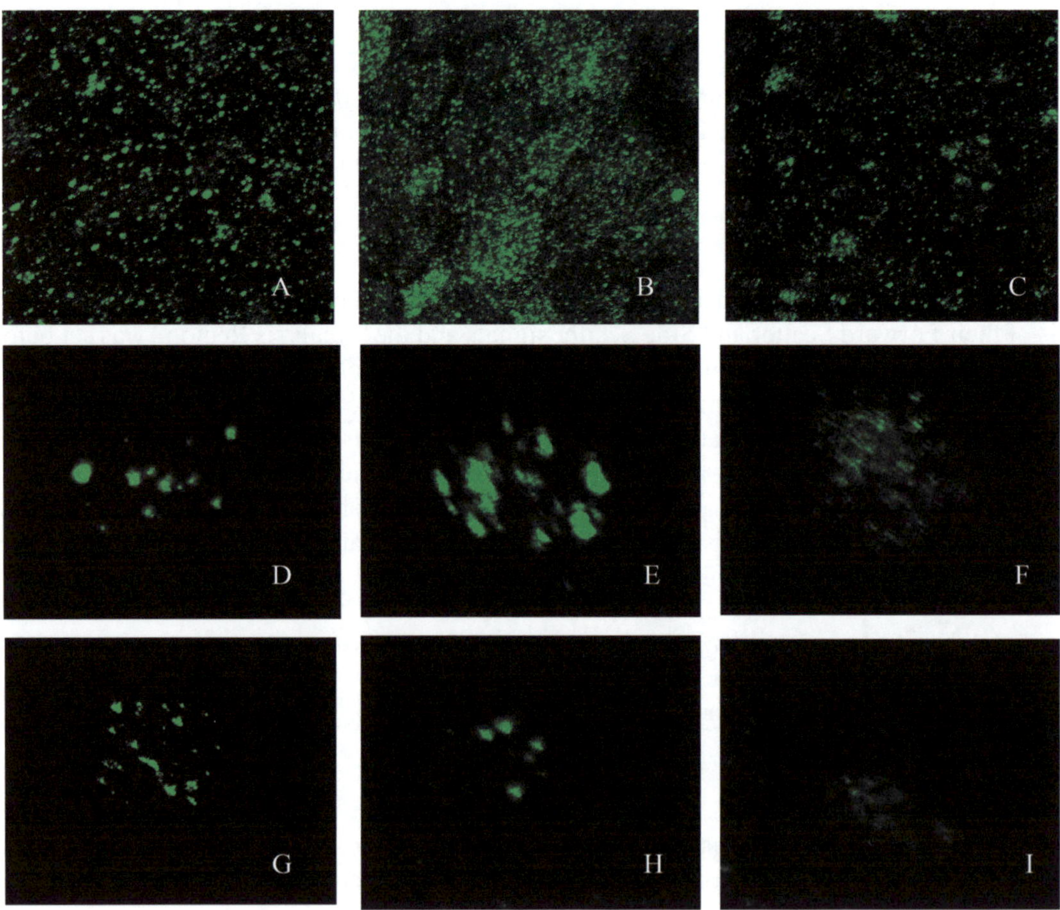

Fig. 1. IL-1$\beta$ production by 48 hours LPS stimulated rat articular chondrocytes cultured in monolayer (A–C) or by 48 hours IL-1$\alpha$-stimulate chondrocytes cultured in alginate beads (D–I). Observations by fluorescence confocal microscopy (Leica SP2, Obj. 40×, NA 0.8). A: control, B: LPS for 48 hs, C: Diacerein + LPS for 48 hs, D: control, E: IL-1$\alpha$ for 48 hs, F: IL-1$\alpha$ + DAR for 48 hs, G: control knocking, H: IL-1$\alpha$ for 48 hs under knocking, I: IL-1$\alpha$ + DAR for 48 hs under knocking.

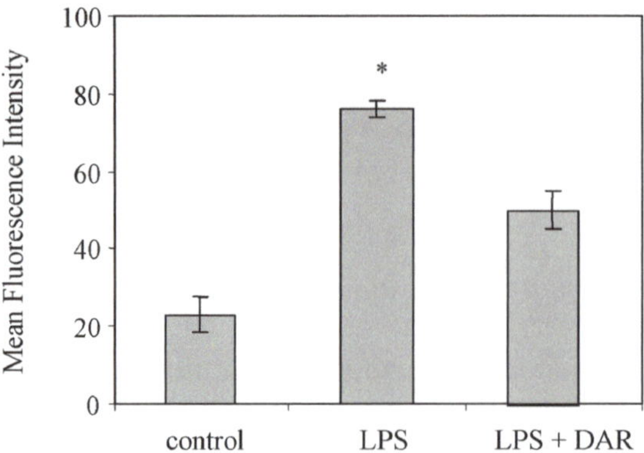

Fig. 2. Analyse by flow cytometry of intracellular IL-1$\beta$ production by rat articular chondrocyte cultured in monolayer. Cells were stimulated for 48 hours with LPS and/or Diacerein (DAR). $^*p < 0.05$ (LPS versus the others).

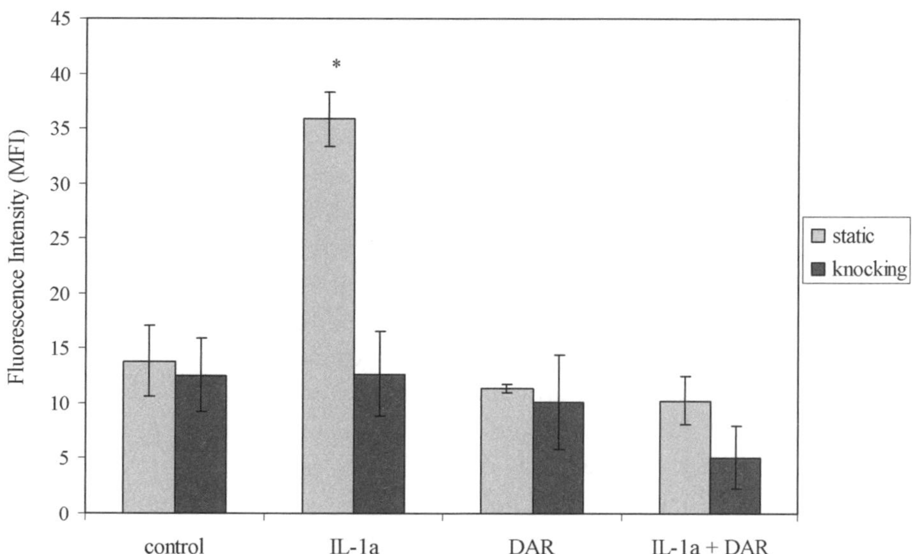


Fig. 3. Analyse by flow cytometry of intracellular IL-1β production by rat articular chondrocyte cultured in alginate beads. After 21 days of culture, cells were stimulated for 48 hours with IL-1α and/or knocking. The effect of Diacerein (DAR) was analyzed. $^*p < 0.05$ (static versus knocking).

[13]. In this work, chondrocyte were cultured for 21 days in alginate beads in order to keep stable their phenotype and allow the synthesis of the extracellular matrix [11]. Intracellular IL-1β distribution was observed with confocal microscopy in 3D-cultured rat articular chondrocytes stimulated with IL-1α during 48 h. In the absence of IL-1 α, a weak IL-1β expression was detected but there was an increase (intensity and number of spots) when IL-1α was added (Fig. 1). DAR decreases intracellular IL-1β production by IL-1α-stimulated chondrocytes cultured in alginate beads. Similar results when obtained when semi-quantification of intracellular IL-1β produced by chondrocytes in alginate beads was achieved by flow cytometry (Fig. 3).

### 3.3. Effect of mechanical stimulation on intracellular IL-1β production by IL-1α stimulated chondrocytes cultured in alginate beads

To test the hypothesis that chondrocytes are more responsive to DAR under mechanical stimulation, a simple technique based on bead shock developed by Gigant-Huselstein [11] was used in our study. Confocal microscopy images (Fig. 1) show that mechanical stimulation decreases IL-1α-induced IL-1β production by articular chondrocyte. Moreover, IL-1β production was almost undetectable in IL-1α stimulated cells treated with DAR and subject to mechanical stimulation. Similar results were obtained by flow cytometry studies (Fig. 3).

## 4. Discussion and conclusion

IL-1β is a major catabolic pro-inflammatory cytokine involved in cartilage destruction during OA. In this work, the production of intracellular IL-1β production by LPS- or IL-1α-stimulated rat articular chondrocytes was visualised by confocal fluorescence microscopy and quantified by flow cytometry. Furthermore, the effect of DAR and mechanical stimulation on IL-1β induced production was analysed.

When chondrocytes cultured in monolayer were stimulated with LPS, an increase in IL-1$\beta$ production was observed. DAR added to culture medium decreases LPS induced intracellular IL-1$\beta$ production. These results are in concordance with previous studies achieved by other techniques [5].

Culture of chondrocyte in alginate beads is useful not only for maintaining a differentiated phenotype, but also for promoting restoration of dedifferentiated chondrocytes to a normal phenotype [12]. Moreover, as the effect of cytokines on cartilage depends on the differentiation state of chondrocytes [13], culture in alginate beads provides a relevant model for the study of chondrocyte biology in presence of mediators of cartilage destruction in OA and pharmacological agents [14,15]. So, to analyse the effect of DAR on IL-1$\alpha$-stimulated chondrocytes, cells were cultured for 21 days in alginate beads in order to keep stable their phenotype and allow the synthesis of the extracellular matrix. Similarly to LPS, IL-1$\alpha$ increase intracellular IL-1$\beta$ production by articular chondrocytes and DAR abolish this effect.

Articular cartilage is subject in permanence to dynamic loading applied perpendicular to the articular surface which can alter chondrocyte behaviour and ECM homeostasis. In this work chondrocytes were exposed to mechanical stimulation by knocking the biosystems containing cells in a fluid flow. The mechanical stress achieved by knocking biosystems together stimulates traumatisms resulting from accidental shocks or from intensive physical exercise. Our results show that mechanical stimulation decrease IL-1$\alpha$-induced IL-1$\beta$ production in articular chondrocyte. Moreover, IL-1$\beta$ production was almost undetectable in IL-1$\alpha$ stimulated cells treated with DAR and subject to mechanical stimulation.

In conclusion, our results suggest an inhibitory effect of DAR at therapeutic concentrations on IL-1$\beta$ production in articular chondrocytes. Moreover, mechanical stimulation could increase this response.

## References


[1] A.M. Badger, M.N. Cook, M.W. Lark, T.M. Newman-Tarr, B.A. Swift, A.H. Nelson, F.C. Barone and S. Kumar, SB 203580 inhibits p38 mitogen-activated protein kinase, nitric oxide production, and inducible nitric oxide synthase in bovine cartilage-derived chondrocytes, *J. Immunol.* **161** (1998), 467–473.

[2] S.G. Hwang, S.S. Yu, H. Poo and J.S. Chun, c-Jun/activator protein-1 mediates interleukin-1$\beta$-induced dedifferentiation but not cyclooxygenase-2 expression in articular chondrocytes, *J. Biol. Chem.* **280** (2005), 29780–29787.

[3] C. Chadjichristos, C. Ghayor, M. Kypriotou, G. Martin, E. Renard, L. Ala-Kokko, G. Suske, B. de Crombrugghe, J.P. Pujol and P. Galera, Sp1 and Sp3 transcription factors mediate interleukin-1 beta down-regulation of human type II collagen gene expression in articular chondrocytes, *J. Biol. Chem.* **278** (2003), 39762–39772.

[4] D. Pfander, N. Heinz, P. Rothe, H.D. Carl and B. Swoboda, Tenascin and aggrecan expression by articular chondrocytes is influenced by interleukin 1beta: a possible explanation for the changes in matrix synthesis during osteoarthritis, *Ann. Rheum. Dis.* **63** (20043), 240–244.

[5] M. Yaron, I. Shirazi and I. Yaron, Anti-interleukin-1 effects of diacerein and rhein in human osteoarthritic synovial tissue and cartilage cultures, *Osteoarthr. Cartilage* **7** (1999), 272–280.

[6] G.N. Smith, Jr., S.L. Myers, K.D. Brandt, E.A. Mickler and M.E. Albrecht, Diacerhein treatment reduces the severity of osteoarthritis in the canine cruciate-deficiency model of osteoarthritis, *Arth. Rheum.* **42** (1999), 545–554.

[7] J.P. Pelletier, M. Yaron, B. Haroui, P. Cohen, M.A. Nahir, D. Choquette, I. Wigler, I.A. Rosner and A.D. Beaulieu, Efficacy and safety of diacerein in osteoarthritis of the knee. A Double-Blind, Placebo-Controlled Trial, *Arth. Rheum.* **43** (2000), 2339–2348.

[8] M. Dougados, M. Nguyen, L. Berdah, B. Maziéres, E. Vignon and M. Lequesne, Evaluation of the structure-modifying effects of diacerein in hip osteoarthritis. ECHODIAH, a Three-Year, Placebo-Controlled Trial, *Arth. Rheum.* **44** (2001), 2539–2547.

[9] D.A. Lee, M.M. Knight, J.F. Bolton, B.D. Idowu, M.V. Kayser and D.L. Bader, Chondrocyte deformation within compressed agarose construct at the cellular and sub-cellular levels, *J. Biomech.* **33** (2000), 81–95.

[10] G. Miralles, R. Baudoin, D. Dumas, D. Baptiste, P. Hubert, J.F. Stoltz, E. Dellacherie, D. Mainard, P. Netter and E. Payan, Sodium alginate sponges with or without sodium hyaluronate: in vitro engineering of cartilage, *J. Biomed. Mater. Res.* **57** (2001), 268–278.

[11] C. Gigant-Huselstein, D. Dumas, P. Hubert, D. Baptiste, E. Dellacherie, D. Mainard, P. Netter, E. Payan and J.F. Stoltz, Influence of mechanical stress on extracellular matrixes synthesized by chondrocytes seeded onto alginate and hyaluronate-based 3D biosystems, *J. Mech. Med. Biol.* **3** (2003), 59–70.

[12] P.D. Benya and J.D. Shaffer, Dedifferentiated chondrocytes reexpress the differentiated collagen phenotype when cultured in agarose gels, *Cell* **30** (1982), 215–224.

[13] H. Liu, Y.W. Lee and M.F. Dean, Re-expression of differentiated proteoglycan phenotype by dedifferentiated human chondrocytes during culture in alginate beads, *BBA* **1425** (1998), 505–515.

[14] F. Lamre, N. Steimberg, C. Le Griel, S. Demignot and M. Adolphe, Dedifferentiated chondrocytes cultured in alginate beads: restoration of the differentiated phenotype and of the metabolic responses to interleukin-1β, *J. Cell Physiol.* **176** (1998), 303–313.

[15] D. Dumas, C. Gigant, N. Presle, C. Cipolletta, G. Miralles, E. Payan, J.Y. Jouzeau, D. Mainard, B. Terlain, P. Netter and J.F. Stoltz, The role of 3D-microscopy in the study of chondrocyte-matrix interaction (alginate bead or sponge, rat femoral head cap, human osteoarthritic cartilage) and pharmacological application, *Biorheology* **37** (2000), 165–176.

[16] C. Sanchez, M. Mathy-Hartert, M.A. Deberg, H. Ficheux, J.-Y.L. Reginster and Y.E. Herontin, Effects of rhein on human articular chondrocytes in alginate beads, *Biochem. Pharmacol.* **65** (2003), 377–388.

Biorheology 43 (2006) 603–609
IOS Press



# Perspectives on chondrocyte mechanobiology and osteoarthritis

Joseph A. Buckwalter*, James A. Martin and Thomas D. Brown
*University of Iowa Department of Orthopaedics and Rehabilitation, Iowa City, IA, USA*

**Abstract.** Osteoarthritis, the clinical syndrome of joint pain and dysfunction due to joint degeneration, is among the most frequent and symptomatic medical problems for middle aged and older people, and it is the most common cause of long term disability in most populations of people over 65. Currently there are no effective methods of preventing or curing osteoarthritis. Post-traumatic OA, the joint degeneration, pain and dysfunction that develop following joint injury, is the form of OA that is most directly related to elevated articular surface contact stress. However, mechanical stress that exceeds the tolerance of the articular surface can cause or accelerate the progression of joint degeneration in all individuals and in all synovial joints. In some patients, decreasing mechanical forces on degenerated joint surfaces stimulates formation of a new biologic articular surface. The advances in understanding of the effects of mechanical forces on chondrocytes and cartilage presented and discussed at the 4th Symposium on Mechanobiology: Cartilage and Chondrocyte will help in the efforts to develop new methods of preventing and treating osteoarthritis.

## 1. Introduction

Osteoarthritis (OA), the clinical syndrome of joint pain and dysfunction caused by joint degeneration, has multiple risk factors, including joint dysplasia, genetic and developmental abnormalities, age, and joint injuries. The etiology, or etiologies, of OA have not been well-defined, but mechanical stress that exceeds the tolerance of the articular surface has an important role in the development and progression of joint degeneration in all forms of OA. However, mechanistic relationships have yet to be defined between human OA and mechanical stress on articular surfaces. Understanding of these relationships will rapidly improve approaches to the prevention and treatment of OA.

## 2. Epidemiologic and clinical observations

Normal synovial joints can withstand repetitive loading during normal activities for a lifetime without developing OA [18,20]. However, considerable clinical, epidemiological, and experimental evidence supports the concept that mechanical demand that exceeds the tolerance of the articular surface has a major – if not the major – role in the development and progression of OA. Surveys of individuals with physically demanding occupations, including farmers, construction workers, metal workers, miners, and pneumatic drill operators, suggest that repetitive intense joint loading is associated with early onset

---

*Address for correspondence: Joseph A. Buckwalter, 01013 Pappajohn Pavilion, Department of Orthopaedics, University of Iowa College of Medicine, Iowa City, IA 52242, USA. Tel.: +1 319 356 2595; Fax: +1 319 356 8999; E-mail: joseph-buckwalter@uiowa.edu.

of joint degeneration [3,8,26,28,37,44,48,51,52,55,70,82,84]. Investigations of the relationship between participation in sports and OA indicate that those sports which subject joints to intense impact loading increase the risk of OA [18,20,50]. Other work has shown an association between obesity and knee OA, and suggests that increased mechanical demand resulting from obesity is the principal reason for the association between obesity and knee OA [3,43]. Furthermore, weight loss appears to significantly reduce the risk of developing symptomatic knee OA [36]. Currently clinicians recommend weight reduction, shoe modifications, osteotomies and joint distraction treatments for OA based on the concept that reducing mechanical stress will decrease symptoms, slow the progression of joint degeneration, and possibly stimulate restoration of some form of articular surface [2,16,21,22,27,47,57,58,68,72,74,88].

## 3. Articular surface contact stresses and joint degeneration

Experimental studies show that excessive mechanical stress can directly damage the articular cartilage and subchondral bone, and that it can adversely alter chondrocyte function including the balance between synthetic and degradative activity [5,17,18,22,34]. While direct measurements of *in vivo* cartilage-on-cartilage contact stresses in human joints have not been made, considerable experimental and computational evidence indicates the ranges of probable physiological contact stresses. These peak and spatial mean contact stresses appear approximately similar in various weight-bearing joints under habitual peak loading. *In vitro*, peak contact stresses are in the range of 6–9 MPa [1,13,14,46,63,64]. The spatial mean contact stresses are considerably less, in the range of 2–4 MPa [14,65]. Repo and Finlay [76] found that cartilage could not survive more than 25 MPa of impulsive contact stress. However, since peak physiological contact stresses are typically 3–4 times less than this threshold for permanent damage from impact loading, there apparently is a built-in "safety factor". The available *in vivo* data come from an instrumented hip endoprosthesis, which recorded contact stresses for cartilage-on-metal; [45] peak local stresses were reported as ranging from 1.4 MPa during bed rest, to as high as 18 MPa during rising from a chair. (These values probably are not reflective of *in vivo* cartilage-on-cartilage contact stresses.) The *in vivo* tolerance of human articular cartilage to contact stresses (or some function thereof), over time, is not known. However, Hadley et al. [41] demonstrated an association between hip joint degeneration and habitually high mechanical demand, and identified a quantitative threshold (25 MPa-years), below which degeneration usually did not occur.

## 4. Effects of impact loading on articular surfaces

Work by Haut, Borrelli, Tozilli and others, dealing primarily with joints subjected to blunt impact trauma at sub-fracture levels, has highlighted important considerations directly pertinent to the effects of impact loading on articular surfaces [4–7,9–12,25,35,53,54,56,60–62,66,67,78,86]. First, the nuances of joint apposition/engagement during an impact event play a key role in the distribution and severity of damage [4], making it difficult if not impossible to predict impact sequelae simply from the global dynamics of an impact event. Second, they have shown that impact severity far below the level needed to produce intra-articular fracture suffices to cause substantial acute micro-damage (subchondral bone and trabecular micro-fractures, cartilage fissures, chondrocyte death, proteoglycan release), often subsequently progressing to detectable compromise of the mechanical integrity of cartilage [35,67]. Third, from a more technical biomechanics perspective, their systematic laboratory investigations of blunt impact have highlighted the importance of considering loading rate effects [35], solid–fluid interactions in the cartilage matrix, shear-stress-related failure mechanisms [6,7], energy- and/or stress-based damage thresholds [35,66], and spatial distribution of damage [35].

## 5. Mechanical forces and post-traumatic osteoarthritis

The joint degeneration, pain and dysfunction that often develops following acute joint injury [19,26,33,38,69,83,89], is the form of OA that is most clearly and directly related to excessive acute or repetitive mechanical demand. The mechanisms responsible for the development of OA in humans following acute joint injury are not well understood. However, observational studies confirm that both pulses of acute energy and chronic mechanical overload damage articular surfaces, and cause or accelerate the degeneration of articular cartilage [22,23,32,34,35,41,73,75,76,81,85,90].

Orthopaedic surgeons routinely perform extensive surgical procedures, some having substantial complication rates, in an effort to restore the alignment and congruity of articular surfaces following intra-articular fractures [59]. The purpose of these procedures is to decrease residual joint incongruity and thereby to decrease focal elevations of contact stress presumed to be responsible for post-traumatic OA. These widely accepted practices are based largely on the assumption that joints are less likely to develop OA if the peak stresses on focal areas of the articular surface are reduced [59]. However, there is little evidence to guide surgeons in determining how much stress the articular surface can tolerate, either in the form of acute impact or chronically increased stress, and the potential for human joints to repair and remodel the articular surface after injury is poorly understood [15,19,59]. While experimental studies of articular incongruities have focused on load magnitudes, other experimental evidence has shown that bone and cartilage metabolism is particularly sensitive to load rate [24,39,40,42,77,79,80,87]. Traditional mechanical testing of articular incongruity using static testing methods ignores the potential importance of time-dependent loads. Cartilage loading rates and transient peak stresses, both on the articular surface and within the substance of the cartilage, have not been ascertainable with traditional testing methods.

With better understanding of the role of mechanical forces in the pathogenesis of OA, clinical and basic research could be focused on developing new minimally invasive and non-surgical treatments of joint injuries that have a high risk of post-traumatic osteoarthritis, and on treatments that prevent or delay the development of OA in patients at high risk of post-traumatic OA. There is increasing evidence that biologic interventions can decrease mechanical stress induced chondrocyte damage [29,31,49]. For example, the work of D'Lima and co-workers shows that caspase inhibition can decrease mechanically induced chondrocyte apoptosis [29–31], and Haut and colleagues have reported that P188 surfactant can limit chondrocyte necrosis following impact loading [71,78]. Our investigations of the effects of anti-oxidants in preventing mechanically induced chondrocyte damage (reported in this volume) and those of others [49] show that some of these agents have the potential to decrease the deleterious effects of mechanical loads on articular cartilage. New treatments might include (1) innovative methods of decreasing the loading of injured joint surfaces while maintaining joint motion; (2) early biologic joint resurfacing following injury; (3) inhibition of enzymes that degrade the articular cartilage matrix or stimulation of matrix synthesis (4) reduction of oxidative damage in injured joints and (5) novel biologic adjuncts targeted at newly identified mechano-response pathways.

## 6. Conclusions

Taken together, the epidemiologic, clinical and experimental evidence shows that excessive acute impact energy or chronic mechanical overload cause the degeneration of the articular surface that leads to the clinical syndrome of OA. Treatment of osteoarthritic joints by procedures that decrease articular surface loading (osteotomies, muscle releases and joint distraction) stimulates restoration of biologic articular surfaces in some patients. Despite the clear importance of mechanical forces in causing and the possibility altering mechanical forces can decrease joint pain and improve function in patients with OA, the effects of mechanical forces on chondrocytes and articular cartilage remain poorly understood. For these reasons it has not been possible to develop predictably effective biological and mechanical methods of preventing or decreasing the risk of post-traumatic OA. The fourth International Sympmosium organized by Professor J.F. Stoltz dedicated to the mechanobiology of chondrocytes and cartilage have focussed attention on this critically important subject and promoted productive exchanges of ideas and collaborative work. There is little doubt that these symposia will make important contributions to advances in the prevention and treatment of osteoarthritis.

## Acknowledgement

The work reported in this manuscript was supported by award P50 AR48939 National Institutes of Health Specialized Center on Research for OA. http://poppy.obrl.uiowa.edu/Specialized Center of Research/SCOR.htm.

## References

[1] D. Adams and S.A.V. Swanson, Direct measurement of local pressures in the cadaveric human hip joint during simulated level walking, *Ann. Rheum. Dis.* **44** (1985), 658–666.

[2] R. Aldegheri, G. Trivella and M. Saleh, Articulated distraction of the hip, *Clin. Orthop. Rel. Res.* **301** (1994), 94–101.

[3] J.J. Anderson and D.T. Felson, Factors associated with osteoarthritis of the knee in the first national Health and Nutrition Examination Survey (HANES I): Evidence for an association with overweight, race and physical demands of work, *Am. J. Epidemiol.* **127** (1988), 179–189.

[4] P.J. Atkinson and R.C. Haut, Injuries produced by blunt trauma to the human patellofemoral joint vary with flexion angle of the knee, *J. Orthop. Res.* **19** (2001), 827–833.

[5] P.J. Atkinson and R.C. Haut, Subfracture insult to the human cadaver patellofemoral joint produces occult injury, *J. Orthop. Res.* **13** (1995), 936–944.

[6] T.S. Atkinson, R.C. Haut and N.J. Altiero, Impact-induced fissuring of articular cartilage: an investigation of failure criteria, *J. Biomech. Eng.* **120** (1998), 181–187.

[7] T.S. Atkinson, R.C. Haut and N.J. Altiero, An investigation of biphasic failure criteria for impact-induced fissuring of articular cartilage, *J. Biomech. Eng.* **120** (1998), 536–537.

[8] B. Axmacher and H. Lindberg, Coxarthrosis in farmers, *Acta Orthop. Scand.* **64**(Suppl. 253) (1993), 59.

[9] J. Borrelli, Jr. and W.M. Ricci, Acute effects of cartilage impact, *Clin. Orthop. Relat. Res.* (2004), 33–39.

[10] J. Borrelli, Jr., K. Tinsley, W.M. Ricci, M. Burns, I.E. Karl and R. Hotchkiss, Induction of chondrocyte apoptosis following impact load, *J. Orthop. Trauma.* **17** (2003), 635–641.

[11] J. Borrelli, Jr., P.A. Torzilli, R. Grigiene and D.L. Helfet, Effect of impact load on articular cartilage: development of an intra-articular fracture model, *J. Orthop. Trauma.* **11** (1997), 319–326.

[12] J. Borrelli, Jr., Y. Zhu, M. Burns, L. Sandell and M.J. Silva, Cartilage tolerates single impact loads of as much as half the joint fracture threshold, *Clin. Orthop. Relat. Res.* (2004), 266–273.

[13] T.D. Brown, D.D. Anderson, J.V. Nepola, R.J.S. RJ, D.R. Pedersen and R.A. Brand, Contact stress aberrations following imprecise reduction of simple tibial plateau fractures, *J. Orthop. Res.* **6** (1988), 851–862.

[14] T.D. Brown and D.T. Shaw, In vitro contact stress distributions in the natural human hip, *J. Biomech.* **16** (1983), 373–384.

[15] J.A. Buckwalter, Articular cartilage injuries, *Clin. Orthop. Relat. Res.* (2002), 21–37.

[16] J.A. Buckwalter, Joint distraction for osteoarthritis, *Lancet* **347** (1996), 279–280.

[17] J.A. Buckwalter, Mechanical injuries of articular cartilage, in: *Biology and Biomechanics of the Traumatized Synovial Joint*, G. Finerman, ed., American Academy of Orthopaedic Surgeons, Park Ridge, IL, 1992, pp. 83–96.

[18] J.A. Buckwalter, Osteoarthritis and articular cartilage use, disuse, and abuse: experimental studies, *J. Rheumatol. Suppl.* **43** (1995), 13–15.

[19] J.A. Buckwalter and T.D. Brown, Joint injury, repair, and remodeling: roles in post-traumatic osteoarthritis, *Clin. Orthop. Relat. Res.* (2004), 7–16.

[20] J.A. Buckwalter, N.E. Lane and S.L. Gordon, Exercise as a cause of osteoarthritis, in: *Osteoarthritic Disorders*, K.E. Kuettner and V.M. Goldberg, eds, American Academy of Orthopaedic Surgeons, Rosemont, IL, 1995, pp. 405–417.

[21] J.A. Buckwalter and S. Lohmander, Operative treatment of osteoarthrosis: Current practice and future development, *J. Bone Joint Surg.* **76A** (1994), 1405–1418.

[22] J.A. Buckwalter, H.J. Mankin, Articular cartilage I. Tissue design and chondrocyte–matrix interactions, *J. Bone Joint Surg.* **79A** (1997), 600–611.

[23] J.A. Buckwalter and V.C. Mow, Cartilage repair in osteoarthritis, in: *Osteoarthritis: Diagnosis and Management*, 2nd edn, R.W. Moskowitz, D.S. Howell, V.M. Goldberg and H.J. Mankin, eds, Saunders, Philadephia, 1992, pp. 71–107.

[24] M.D. Buschmann, Y.A. Gluzband, A.J. Grodzinsky and E.B. Hunziker, Mechanical compression modulates matrix biosynthesis in chondrocyte/agarose culture, *J. Cell Sci.* **108**(Pt 4) (1995), 1497–1508.

[25] C.T. Chen, M. Bhargava, P.M. Lin and P.A. Torzilli, Time, stress, and location dependent chondrocyte death and collagen damage in cyclically loaded articular cartilage, *J. Orthop. Res.* **21** (2003), 888–898.

[26] C. Cooper, P. Croft, D. Coggon, C. Wickham and M. Cruddas, Farming and osteoarthritis of the hip, *Acta Orthop. Scand.* **64**(Suppl. 253) (1993), 58–59.

[27] M.B. Coventry, D.M. Ilstrup and S.L. Wallrichs, Proximal tibial osteotomy. A critical long-term study of eighty-seven cases, *J. Bone Joint Surg.* **75A** (1993), 196–201.

[28] P. Croft, C. Cooper, C. Wickham and D. Coggon, Osteoarthritis of the hip and occupational activity, *Scand. J. Work Envrion. Health* **18** (1992), 59–63.

[29] D.D. D'Lima, S. Hashimoto, P.C. Chen, C.W. Colwell, Jr. and M.K. Lotz, Human chondrocyte apoptosis in response to mechanical injury, *Osteoarthritis Cartilage* **9** (2001), 712–719.

[30] D.D. D'Lima, S. Hashimoto, P.C. Chen, C.W. Colwell, Jr. and M.K. Lotz, Impact of mechanical trauma on matrix and cells, *Clin. Orthop. Relat. Res.* (2001) S90–99.

[31] D.D. D'Lima, S. Hashimoto, P.C. Chen, M.K. Lotz and C.W. Colwell, Jr., Prevention of chondrocyte apoptosis, *J. Bone Joint Surg. Am.* **83-A**(Suppl. 2) (2001), 25–26.

[32] P. Dieppe, The classification and diagnosis of osteoarthritis, in: *Osteoarthritic Disorders*, K.E. Kuettner and V.M. Goldberg, eds, American Academy of Orthopaedic Surgeons, Rosemont, IL, 1995, pp. 5–12.

[33] D.R. Dirschl, J.L. Marsh, J.A. Buckwalter, R. Gelberman, S.A. Olson, T.D. Brown and A. Llinias, Articular fractures, *J. Am. Acad. Orthop. Surg.* **12** (2004), 416–423.

[34] J.M. Donohue, D. Buss, T.R. Oegema and R.C. Thompson, The effects of indirect blunt trauma on adult canine articular cartilage, *J. Bone Joint Surg.* **65-A** (1983), 948–956.

[35] B.J. Ewers, D. Dvoracek-Driksna, M.W. Orth and R.C. Haut, The extent of matrix damage and chondrocyte death in mechanically traumatized articular cartilage explants depends on rate of loading, *J. Orthop. Res.* **19** (2001), 779–784.

[36] D.T. Felson, Y. Zhang, J.M. Anthony, A. Naimark and J.J. Anderson, Weight loss reduces the risk for symptomatic knee osteoarthritis in women, *Ann. Int. Med.* **116** (1992), 535–539.

[37] K. Forsberg and B.E. Nilsson, Coxarthrosis on the island of Gotland: Increased prevalence in a rural population, *Acta Orthop. Scand.* **63** (1992), 1–3.

[38] A.C. Gelber, M.C. Hochberg, L.A. Mead, N.Y. Wang, F.M. Wigley and M.J. Klag, Joint injury in young adults and risk for subsequent knee and hip osteoarthritis, *Ann. Intern. Med.* **133** (2000), 321–328.

[39] S.A. Goldstein, L.S. Matthews, J.L. Kuhn and S.J. Hollister, Trabecular bone remodeling: an experimental model, *J. Biomech.* **24**(Suppl. 1) (1991), 135–150.

[40] M.L. Gray, A.M. Pizzanelli, A.J. Grodzinsky and R.C. Lee, Mechanical and physiochemical determinants of the chondrocyte biosynthetic response, *J. Orthop. Res.* **6** (1988), 777–792.

[41] N.A. Hadley, T.D. Brown and S.L. Weinstein, The effects of contact pressure elevations and aseptic necrosis on the long-term outcome of congenital hip dislocation, *J. Orthop. Res.* **8** (1990), 504–513.

[42] A.C. Hall, J.P.G. Urban and K.A. Gehl, The effects of hydrostatic pressure on matrix synthesis in articular cartilage, *J. Orthop. Res.* **9** (1991), 1–10.

[43] A.J. Hartz, M.E. Fisher, G. Bril, S. Kelber, D. Rupley, B. Oken and A.A. Rimm, The association of obesity with joint pain and osteoarthritis in the HANES data, *J. Chronic. Dis.* **39** (1986), 311–319.

[44] M. Heliovaara, M. Makela, O. Impivaara, P. Knekt, A. Aroma and K. Sievers, Association of overweight, trauma and workload with coxarthrosis: a health survey of 7,217 persons, *Acta Orthop. Scand.* **64** (1993), 513–518.

[45] W.A. Hodge, K.L. Carlson, R.S. Fijhan, R.G. Burgess, P.O. Riley, W.H. Harris and R.A. Mann, Contact pressures from an instrumented hip endoprothesis, *J. Bone Joint Surg.* **71A** (1989), 1378–1386.

[46] A. Iglic, V.K. Iglic, V. Antolic, F. Srakar and U. Stanic, Effect of periacetabular osteotomy on the stress on the human hip joint articular surface, *IEEE Trans. Rehab. Engrn.* **1** (1993), 107–212.

[47] M. Itoman, M. Yamamoto, K. Yonemoto, M. Sekiguchi and H. Kai, Histological examination of surface repair tissue after successful osteotomy for osteoarthritis of the hip joint, *Int. Orthop.* **16** (1992), 118–121.

[48] J.H. Kellergren and J.S. Lawrence, Osteoarthritis and disk degeneration in an urban population, *Ann. Rheum. Dis.* **12** (1958), 5.

[49] B. Kurz, A. Lemke, M. Kehn, C. Domm, P. Patwari, E.H. Frank, A.J. Grodzinsky and M. Schunke, Influence of tissue maturation and antioxidants on the apoptotic response of articular cartilage after injurious compression, *Arthritis Rheum.* **50** (2004), 123–130.

[50] N.E. Lane and J.A. Buckwalter, Exercise: A cause of osteoarthritis?, in: *Rheumatic Disease Clinics of North America*, R. Moskowitz, ed., Saunders, Philadelphia, 1993, pp. 617–633.

[51] J.S. Lawrence, Rheumatism in coal miners. III. Occupational factors, *Brit. J. Ind. Med.* **12** (1955), 249–251.

[52] J.S. Lawrence, Rheumatism in cotton operatives, *Brit. J. Ind. Med.* **18** (1961), 270.

[53] A.S. Levin, C.T. Chen and P.A. Torzilli, Effect of tissue maturity on cell viability in load-injured articular cartilage explants, *Osteoarthritis Cartilage* **13** (2005), 488–496.

[54] P.M. Lin, C.T. Chen and P.A. Torzilli, Increased stromelysin-1 (MMP-3), proteoglycan degradation (3B3- and 7D4) and collagen damage in cyclically load-injured articular cartilage, *Osteoarthritis Cartilage* **12** (2004), 485–496.

[55] H. Lindberg and F. Montgomery, Heavy labor and the occurrence of gonarthrosis, *Clin. Orthop. Rel. Res.* **214** (1987), 235–236.

[56] E. Lucchinetti, C.S. Adams, W.E. Horton, Jr. and P.A. Torzilli, Cartilage viability after repetitive loading: a preliminary report, *Osteoarthritis Cartilage* **10** (2002), 71–81.

[57] A.C. Marijnissen, P.M. van Roermund, J. van Melkebeek and F.P. Lafeber, Clinical benefit of joint distraction in the treatment of ankle osteoarthritis, *Foot Ankle Clin.* **8** (2003), 335–346.

[58] A.C. Marijnissen, P.M. Van Roermund, J. Van Melkebeek, W. Schenk, A.J. Verbout, J.W. Bijlsma and F.P. Lafeber, Clinical benefit of joint distraction in the treatment of severe osteoarthritis of the ankle: proof of concept in an open prospective study and in a randomized controlled study, *Arthritis Rheum.* **46** (2002), 2893–2902.

[59] J.L. Marsh, J. Buckwalter, R. Gelberman, D. Dirschl, S. Olson, T. Brown and A. Llinias, Articular fractures: does an anatomic reduction really change the result?, *J. Bone Joint Surg. Am.* **84-A** (2002), 1259–1271.

[60] D. Milentijevic, D.L. Helfet and P.A. Torzilli, Influence of stress magnitude on water loss and chondrocyte viability in impacted articular cartilage, *J. Biomech. Eng.* **125** (2003), 594–601.

[61] D. Milentijevic, I.F. Rubel, A.S. Liew, D.L. Helfet and P.A. Torzilli, An in vivo rabbit model for cartilage trauma: a preliminary study of the influence of impact stress magnitude on chondrocyte death and matrix damage, *J. Orthop. Trauma.* **19** (2005), 466–473.

[62] D. Milentijevic and P.A. Torzilli, Influence of stress rate on water loss, matrix deformation and chondrocyte viability in impacted articular cartilage, *J. Biomech.* **38** (2005), 493–502.

[63] Y. Miyanaga, T. Fukubayashi and H. Kurosawa, Contact study of the hip joint: Load-deformation pattern, contact area and contact pressure, *Arch. Orthop. Trauma Surg.* **103** (1984), 13–17.

[64] J. Mizrahi, L. Solomon, B. Kaufman and T. O'Duggan, An experimental method for investigating load distribution in the cadaveric human hip, *J. Bone Joint Surg.* **63B** (1981), 610–613.

[65] B.H. Nelson, D.D. Anderson, R.A. Brand and T.D. Brown, Effect of osteochondral defects on articular cartilage: Contact pressures studied in dog knees, *Acta Orthop. Scand.* **59** (1988), 574–579.

[66] W.N. Newberry, J.J. Garcia, C.D. Mackenzie, C.E. Decamp and R.C. Haut, Analysis of acute mechanical insult in an animal model of post-traumatic osteoarthrosis, *J. Biomech. Eng.* **120** (1998), 704–709.

[67] W.N. Newberry, C.D. Mackenzie and R.C. Haut, Blunt impact causes changes in bone and cartilage in a regularly exercised animal model, *J. Orthop. Res.* **16** (1998), 348–354.

[68] S. Odenbring, N. Egund, A. Lindstand, L.S. Lohmander and H. Wilen, Cartilage regeneration after proximal tibial osteotomy for medial gonarthrosis, *Clin. Orthop. Relat. Res.* **277** (1992), 210–216.

[69] S.A. Olson and J.L. Marsh, Posttraumatic osteoarthritis, *Clin. Orthop. Relat. Res.* (2004), 2.

[70] R.E.H. Partridge and J.J.R. Duthie, Rheumatism in dockers and civil servants: a comparison of heavy manual and sedentary workers, *Ann. Rheum. Dis.* **27** (1968), 559–568.

[71] D.M. Phillips and R.C. Haut, The use of a non-ionic surfactant (P188) to save chondrocytes from necrosis following impact loading of chondral explants, *J. Orthop. Res.* **22** (2004), 1135–1142.

[72] J.J. Ploegmakers, P.M. van Roermund, J. van Melkebeek, J. Lammens, J.W. Bijlsma, F.P. Lafeber and A.C. Marijnissen, Prolonged clinical benefit from joint distraction in the treatment of ankle osteoarthritis, *Osteoarthritis Cartilage* **13** (2005), 582–588.

[73] E.L. Radin, M.G. Ehrlich, R. Chernack et al., Effect of repetitive impulsive loading on the knee joints of rabbits, *Clin. Orthop. Relat. Res.* **131** (1978), 288–293.

[74] E.L. Radin, P. Maquet and H. Park, Rationale and indications for the "hanging hip" procedure. A clinical and experimental study, *Clin. Orthop.* **112** (1975), 221–230.

[75] E.L. Radin, R.B. Martin, D.B. Burr et al., Effects of mechanical loading on the tissues of the rabbit knee, *J. Orthop. Res.* **2** (1984), 221–234.

[76] R.U. Repo and J.B. Finlay, Survival of articular cartilage after controlled impact, *J. Bone Joint Surg.* **59-A** (1977), 1068–1075.

[77] C.T. Rubin and L.E. Lanyon, Regulation of bone formation by applied dynamic loads, *J. Bone Joint Surg. Am.* **66** (1984), 397–402.

[78] S.A. Rundell, D.C. Baars, D.M. Phillips and R.C. Haut, The limitation of acute necrosis in retro-patellar cartilage after severe blunt impact to the in vivo rabbit patello-femoral joint, *Joint Orthop. Res.* **23** (2005) (on line).

[79] R.L. Sah, Y.J. Kim, J.Y. Doong, A.J. Grodzinsky, A.H. Plaas and J.D. Sandy, Biosynthetic response of cartilage explants to dynamic compression, *J. Orthop. Res.* **7** (1989), 619–636.

[80] R. Schneiderman, D. Keret and A. Maroudas, Effects of mechanical and osmotic pressure on the rate of glycosaminoglycan synthesis in the human adult femoral head cartilage: an in vitro study, *J. Orthop. Res.* **4** (1986), 393–408.

[81] Q. Shi, H. Hasizume, H. Inoue, T. Miyake and N. Nagayama, Finite element analysis of the pathogenesis of osteoarthritis in the first carpometacarpal joint, *Acta Medica Okayama* **49** (1995), 43–51.

[82] H.H.G. Templaar and J.V. Breeman, Rheumatism and occupation, *Acta Rheumatol.* **4** (1932), 36.

[83] S. Tepper and M.C. Hochberg, Factors associated with hip osteoarthritis: data from the First National Health and Nutrition Examination Survey (NHANES-I), *Am. J. Epidemiol.* **137** (1993), 1081–1088.

[84] A. Thelin, Hip joint arthrosis: an occupational disorder among farmers, *Am. J. Ind. Med.* **18** (1990), 339–343.

[85] R.C. Thompson, T.R. Oegema, J.L. Lewis and L. Wallace, Osteoarthritic changes after acute transarticular load: an animal model, *J. Bone Joint Surg.* **73A** (1991), 990–1001.

[86] P.A. Torzilli, R. Grigiene, J. Borrelli, Jr. and D.L. Helfet, Effect of impact load on articular cartilage: cell metabolism and viability, and matrix water content, *J. Biomech. Eng.* **121** (1999), 433–441.

[87] J.P. Urban, A.C. Hall and K.A. Gehl, Regulation of matrix synthesis rates by the ionic and osmotic environment of articular chondrocytes, *J. Cell Physiol.* **154** (1993), 262–270.

[88] A.A. van Valburg, P.M. van Roermund, A.C. Marijnissen, J. van Melkebeek, J. Lammens, A.J. Verbout, F.P. Lafeber and J.W. Bijlsma, Joint distraction in treatment of osteoarthritis: a two-year follow-up of the ankle, *Osteoarthritis Cartilage* **7** (1999), 474–479.

[89] V. Wright, Post-traumatic osteoarthritis – a medico-legal minefield, *Brit. J. Rheum.* **29** (1990), 474–478.

[90] N.B. Zimmerman, D.G. Smith, L.A. Pottenger and D.R. Cooperman, Mechanical disruption of human patellar cartilage by repetitive loading in vitro, *Clin. Orthop. Relat. Res.* **229** (1988), 302–307.

# Author Index